Biopolymers

Edited by A. Steinbüchel

Volume 5
Polysaccharides I
Polysaccharides from Prokaryotes

Edited by E. J. Vandamme, S. De Baets and A. Steinbüchel

Biopolymers

Edited by A. Steinbüchel

Volume 5
Polysaccharides I
Polysaccharides from Prokaryotes

Edited by E. J. Vandamme, S. De Baets and A. Steinbüchel

Editors:

Prof. Dr. Alexander Steinbüchel
Institut für Mikrobiologie
Westfälische Wilhelms-Universität
Corrensstrasse 3
D-48149 Münster
Germany

Prof. Dr. Erick J. Vandamme
Ir. Sophie De Baets
Department of Biochemical and Microbial Technology
Gent University
Coupure links 653
B-9000 Gent
Belgium

Library of Congress Card No: applied for

British Library Cataloguing-in-Publication Data: A catalogue record for this book is available from the British Library

Die Deutsche Bibliothek – CIP-Cataloguing-in-Publication Data
A catalogue record for this publication is available from Die Deutsche Bibliothek

Printed in the Federal Republic of Germany
Printed on acid-free paper.

Composition, Printing, and Bookbinding:
Konrad Triltsch
Print und digitale Medien GmbH,
Ochsenfurt-Hohestadt

ISBN 3-527-30226-3

Preface

Biopolymers and their derivatives are diverse, abundant, important for life, they exhibit fascinating properties and are of increasing importance for various applications. Living matter is able to synthesize an overwhelming variety of polymers, which can be divided into eight major classes according to their chemical structure: (1) nucleic acids such as ribonucleic acids and deoxyribonucleic acids, (2) polyamides such as proteins and poly(amino acids), (3) polysaccharides such as cellulose, starch and xanthan, (4) organic polyoxoesters such as poly(hydroxyalkanoic acids), poly(malic acid) and cutin, (5) polythioesters, which were reported only recently, (6) inorganic polyesters with polyphosphate as the only example, (7) polyisoprenoids such as natural rubber or Gutta Percha and (8) polyphenols such as lignin or humic acids.

Biopolymers occur in any organism, and in most organisms they contribute to the by far major fraction of the cellular dry matter. Biopolymers possess a wide range of different essential or beneficial functions for the organisms: conservation and expression of genetic information, catalysis of reactions, storage of carbon, nitrogen, phosphorus and other nutrients and of energy, defense and protection against the attack of other cells or hazardous environmental or intrinsic factors, sensors of biotic and abiotic factors, communication with the environment and other organisms, mediators of adhesion to surfaces of other organisms or of non-living matter and many more. In addition, many biopolymers are structural components of cells, tissues, and whole organisms.

To fulfil all these different functions, biopolymers must exhibit rather diverse properties. They must very specifically interact with a large variety of different substances, components and materials, and often they must have extraordinarily high affinities to them. Finally, many of them must have a high strength. Some of these properties are utilized directly or indirectly for various applications. This and the possibility to produce them from renewable resources, as living matter mostly does, make biopolymers interesting candidates to industry.

Basic and applied research have already revealed much knowledge on the enzyme systems catalyzing biosynthesis, degradation and modification of biopolymers as well as on the properties of biopolymers. This has also resulted in an increased interest in biopolymers for various applications in industry, medicine, pharmacy, agriculture, electronics and various other areas. However, considering the developments during the last two decades and reviewing the literature shows that our knowledge is still scarce. The genes for the biosynthesis pathways of many biopolymers are still not available or were identified only recently, many new biopolymers have just been described, and from only a minor fraction of biopolymers the biological, chemical, physical and material properties have been investigated. Often promising biopolymers are not available in sufficient amounts. Nevertheless, polymer chemists, engineers and material scientists in academia and industry have discovered biopolymers as chemicals and materials for many new applications, or they consider biopolymers as models to design novel synthetic polymers.

The first edition of this multivolume handbook comprehensively reviews and compiles information on biopolymers in 10 volumes covering (a) occurrence, synthesis, isolation and production, (b) properties and applications, (c) biodegradation and modification not only of natural but also of synthetic polymers, and (e) the relevant analysis methods to reveal the structures and properties. Volumes 1-8 are structured according to the chemical classes of biopolymers, whereas Volume 9 focusses on aspects of the biodegradation of synthetic polymers and Volume 10 deals with general aspects related to biopolymers.

This book series will hopefully be helpful to many scientists, physicians, pharmaceutics, engineers and other experts in a wide variety of different disciplines, in academia and in industry. It may not only support research and development but may be also suitable for teaching.

Publishing of this book series was achieved by chosing volume editors and authors of the individual volumes and chapters for their recognized expertise and for their excellent contributions to the various fields of research. I am very grateful to these scientists for their willingness to contribute to this reference work and for their engagement. Without them and without their comitment and enthusiasm it would have not been possible to compile such a book series.

I am also very grateful to the publisher WILEY-VCH for recognizing the demand for such a book series, for taking the risk to start such a big new project and for realizing the publication of *Biopolymers* in excellent quality. Special thanks are due to Karin Dembowsky and many of her WILEY-VCH colleagues, especially from production and marketing, for their constant effort, their helpful suggestions, constructive criticism, and wonderful ideas.

Last but not least I would like to thank my family for their patience, and I have to excuse for the many hours the preparation of this book series kept me away from them.

Münster, February 2001 Alexander Steinbüchel

Introduction

Polysaccharides comprise a distinct class of biopolymers, produced universally among living organisms. They exhibit a large variety of unique and in most cases rather complex chemical structures, different physiological functions and a wide range of (potential) applications. This volume deals specifically with polysaccharides of prokaryotes, covering both Eubacteria and Archaea, whereas the accompanying Volume 6 deals with polysaccharides of eukaryotes.

After an introductory overview, the bacterial storage polysaccharide glycogen is subject of the second chapter. Subsequently, a wide range of bacterial extracellular polysaccharides with real or potential industrial interest is covered, treating in depth both, those produced by Gram-negative and those produced by Gram-positive bacteria. Included are bacterial alginates, alternan, bacterial cellulose, curdlan, dextran, exopolysaccharides of lactic acid bacteria, glycolipids, hyaluronan, poly-(1,4)-β-D-glucuronan, levan, the sphingan group of exopolysaccharides, surface active polysaccharides and xanthan. The volume concludes with chapters dealing with important, but different classes of bacterial and archaeal cell wall polysaccharides such as murein, pseudomurein and other cell wall polysaccharides from Archaea as well as teichoic acids and teichuronic acids.

In compiling this volume, it has been our intention to provide the scientific and industrial community with a comprehensive view of the current state of knowledge on polysaccharides of prokaryotes. In this respect, this volume attempts to review what is currently known about these fascinating bacterial (exo)polysaccharides, with respect to their discovery, occurrence among bacteria, chemical and physical properties, analysis, biosynthesis, molecular genetics, physiological role, fermentative production, isolation, purification and application. Every possible attempt has been made to collect the most recently published scientific data up to late 2001. A few bacterial polysaccharides could not be described in detail in this volume because we did not receive the manuscripts from the authors in due time. These missing chapters (e.g., on lipopolysaccharides and acetan) will be included in a supplementary volume to be published soon.

From the onset of this endeavor, it was obvious to us that the range of topics we wished to include virtually necessitated our drawing on the expertise of a considerable number of scientists. We were aware that by choosing a multi-author format we entailed risks that might have negative impact on the uniformity of the book and on the timetable for its completion. On the other hand, viewing this book in its totality, we believe that the overall quality and usefulness of its contents vindicates our decision to ask colleagues from all over the world to contribute to this volume. In this respect, their willingness to impart their knowledge to a broad scientific public is gratefully acknowledged. The expertise, enthusiasm and the costly time, which they devoted to their chapters, although all of them have many other obligations and duties, is highly appreciated. The positive interaction with all the contributing authors during the handling of their manuscripts has been invaluable in assembling this volume within a surprisingly short time frame.

Last but not least, we would like to thank WILEY-VCH for publishing *Biopolymers* with their customary professionality and excellence and for their excellent help throughout the gestation and birth of this volume. Special thanks are due to Karin Dembowsky and her colleagues; without their constant effort the book could not have been published.

Ghent, Belgium and	E.J. Vandamme
Münster, Germany	S. De Baets
January 2002	A. Steinbüchel

Contents

1
Polysaccharides from Microorganisms, Plants and Animals

Prof. Ian W. Sutherland
Institute of Cell and Molecular Biology, University of Edinburgh,
Edinburgh EH9 3JH, UK; Tel.: +44-131-650-5331; Fax: +44-131-650-5392;
E-mail: i.w.sutherland@ed.ac.uk

EPS extracellular polysaccharide (exopolysaccharide)
LBG locust bean gum
LAB lactic acid bacteria
M_r molecular weight.

1 Introduction

Polysaccharides are renewable resources which offer a wide variety of potentially useful products to man. In chemical terms, they offer a very wide range of glycosidically linked structures based on about 40 different monosaccharides. Most polysaccharides are formed from a relatively limited range of hexoses and pentoses, but especially among prokaryotes, a number of rare and unusual sugars are found. This provides an extensive group of different architectures resulting in sheets, spirals and single, double and triple helices. Polysaccharides form major structural components of the walls of marine crustaceans, plants, algae and microorganisms. They may also provide carbon and energy reserves for many types of cell or they may be excreted as plant exudates, or as microbial exopolysaccharides. A number of plant polysaccharides such as starch have been very widely used in food and other applications for a very long time indeed. Cellulose and its derivatives have also found many uses because of their easy availability and cheapness. Such polysaccharides and their derivatives can be used as bulk chemicals in place of more expensive products. More recently, other plant polysaccharides such as gluco- and galactomannans have found use in the food industry and elsewhere. Similarly, marine algae have yielded such useful products as agar, alginate, and carrageenans. While cellulose, chitin and some other polysaccharides are insoluble, many of these polymers are water-soluble and are capable of significantly altering the rheology of aqueous based solutes and find a wide range of applications for this reason.

Within the past 40 years, microorganisms have been seen as a further, renewable source of utilizable polysaccharides, and the fermentation industry has provided a small number of novel products with distinctive properties for food and nonfood use as well as materials with specific biological properties for the pharmaceutical industry. All these products taken together represent a minute fraction of those potentially available or those which have been investigated. Further interest in microbial polysaccharides has derived from their significance in many pathogenic species and their employment as essential components of protective vaccines against such pathogens as *Haemophilus influenzae* type B, *Streptococcus pneumoniae* and *Neisseria meningitidis*. They also play a major role in the complex microbial community structures of flocs and biofilms (Sutherland, 1999b, 2001a, b, c) and are thus significant elements in water and sewage purification. Other bacterial polysaccharides are important natural ingredients of traditional fermented foods, in which they play a vital role in the development of texture, gelation, or viscosity. However, the current situation sees a small number of microbial exopolysaccharides as well-established products of biotechnology either with unique market niches or competing with the established plant, algal, or synthetic products. A few more such products may be developed in the future, but the costs of development and production are likely to limit these to a very small number. Many academic papers make exaggerated claims for the potential value of new polysaccharides, but few contain realistic appraisals! Most of our requirements are likely to continue to be met from the traditional plant or algal sources, while new properties can be obtained through chemical modification as exemplified by the widespread applications of carboxymethyl or hydroxyethyl cellulose and carboxymethylguar.

Nevertheless, the chemical composition and structure of an increasing number of microbial and plant polysaccharides is regularly reported. As a result, novel sugars and substituents are being found and are expanding our knowledge of polysaccharides from all sources. Additionally, new and improved methods are providing more accurate structures as well as giving more information on the relationships between chemical structure and physical properties.

2 Polysaccharide Composition and Structure

Polysaccharides may be homopolysaccharides composed of a single monosaccharide unit. Many of these are glucans, containing only D-glucopyranose residues. Alternatively, they may be heteropolysaccharides containing two or more sugars. Additionally, various substituents such as acyl groups, amino acids, or inorganic residues may be attached. The polymers themselves may be linear or branched, with some having very complex branching patterns such as those that are found in glycogen. Even the linear macromolecules may vary between those in which there is a single linkage and monosaccharide type, to other homopolysaccharides with mixed linkages as exemplified by elsinan and by barley and other glucans. In some heteropolysaccharides, there are relatively complex repeat units, perhaps with side-chains of varying length. Alginates present a different feature again, that of "block copolymers" in which the two monosaccharides, D-mannuronic acid and L-guluronic acid, are present in contiguous sequences of varying length. Given that even a disaccharide may be linked in any of eight different ways, the possible complexities of polysaccharide structures are almost infinite.

2.1 Microbial Polysaccharide Composition and Structure

The microbial cell generally contains various polysaccharide structures contributing to its shape and rigidity. In prokaryotes, these include peptidoglycan (composed of repeat units of *N*-acetyl-D-glucosamine and uniquely *N*-acetyl-D-muramic acid, found in almost all Eubacteria), lipopolysaccharides (in Gram-negative bacteria) and teichoic and teichuronic acids (in Gram-positive bacteria). Within the Archaea, a more diverse range of polymers is found – some contain a polymer analogous to peptidoglycan which has been termed pseudomurein, while others possess a variety of polysaccharides and glycoproteins including some with sulfate substituents. One Archaeal species, *Halobacterium halobium* is unique in that it carries protein-linked polysaccharides in which iduronic acid is a component (Wieland et al., 1986). While this uronic acid is present in chondroitin and other animal polysaccharides it is rarely found in Eubacterial polymers, although it as been reported in an extracellular polysaccharide (EPS) from *Butyrivibrio fibrisolvens* (Stack et al., 1988). Cyanobacteria are frequently surrounded by polysaccharide sheaths or envelopes, although in other respects they are typical Gram-negative cells containing lipopolysaccharides in their outer membranes. Further polysaccharides – EPS (or exopolysaccharides) – are often found outside the outermost structures of the prokaryotic or eukaryotic cell. They may be tightly associated with the microbial cells in the form of discrete capsules or else excreted as slime unattached to the cell surface. A number of bacterial species are now known to be capable of synthesizing several EPS, although usually only one is expressed under any specific physiological conditions. Other polysaccharides may be associated with resting cells such as microcysts in *Azotobacter* spp. or with fruiting body formation, as is observed in the Myxobacteria. Within the bacterial cell, glycogen is the sole polysaccharide which may be present as an intracellular carbon and energy storage product. Starch is absent, but some oligosaccharides including trehalose and sophorodextrins may be found. Production of such compounds is usually dependent on the physiological conditions under which the microorganisms are grown.

Bacterial extracellular polysaccharides contain a very wide range of monosaccharides, although a small number are most commonly found. These include D-glucose, D-galactose, and D-mannose. Bacterial polymers are mainly found to contain hexoses or methylpentoses (usually L-fucose or L-rhamnose) together with uronic acids, of which D-glucuronic and D-galacturonic acids are the most common. Pentoses are much less common in bacteria, except in sheaths and other EPS associated with the Cyanobacteria. A few unusual sugars such as aminouronic acids are found. While certain genera yield EPS which contain aminosugars almost always as the acetylaminohexosamine derivatives, others contain *O*-methyl pentoses. A wider range of monosaccharides is found in the lipopolysaccharides of Gram-negative cell walls. These include ketodeoxyoctonic acid, heptoses and the unusual 3,6-dideoxyhexoses – colitose, tyvelose, and abequose – as well as a number of rare aminosugars. The teichoic acids in Gram- positive bacteria are unusual in that many contain glycerol or ribitol phosphate as well as carrying D-alanyl substituents; some teichuronic acids are composed solely of aminouronic acids.

The majority of bacterial exopolysaccharides are either homopolysaccharides composed of a single monosaccharide unit, or heteropolysaccharides in which regular repeat units are formed from two to eight monosaccharides. They thus differ from the highly branched types of polysaccharides associated with many plants, although some of the bacterial polysaccharides do carry regular branches ranging from single monosaccharides to tetrasaccharides and larger. Acyl groups or inorganic substituents such as phosphate or sulfate may also be present (Table 1); a feature absent from most eukaryotic exopolysaccharides. The polysaccharides may be either neutral or polyanionic; one unusual cationic polymer from *Staphylococcus epidermidis* has also been identified (Mack et al., 1996). The anionic polymers owe their charge either to uronic acids, inorganic substituents or to pyruvate ketals or succinyl half-esters. Although sulfate is quite commonly found in wall polymers from marine algae and is also present in some cyanobacterial polysaccharides, it is very rare in other bacterial polymers except some EPS from the Archaea (Parolis et al., 1996a). It is also in a polysaccharide linked to peptidoglycan in an *Arthrobacter* sp. (Yamazaki et al., 1998) and may yet be found in other prokaryotes as more polymers are characterized. Some of the homopolysaccharides, with the exception of dextrans and levans, also possess regular repeating structures. Bacterial alginates, like their marine algal counterparts, are formed from irregular sequences of D-mannuronic acid and L-guluronic acid residues, but differ from the algal material in being heavily acetylated on many of the D-mannuronosyl residues (Sutherland, 1990).

Much more complex repeat unit types of polymer are found among some Cyanobacteria. These can be highly branched, carrying various side-chains of differing length, as demonstrated by Wolk (2000). They also differ from other prokaryotic EPS in frequently containing pentoses such as arabinose and xylose. Thus, the EPS from

Tab. 1 Non-carbohydrate substituents of microbial and other polysaccharides

Acetate	Widely found in bacterial polysaccharides
Glycerate	In a few bacterial polysaccharides
Phosphate	Mainly in polysaccharides of Gram positive bacteria
Pyruvate	Bacterial and algal polysaccharides
Succinate	Bacterial polysaccharides
Sulfate	Polysaccharides of algae, Archaea, Cyanobacteria (Also animal polymers such as heparin, chondroitin)

Cyanospira capsulata is composed of octasaccharide repeat units in which three branches exist from the main chain and in which there are seven different monosaccharides including L-arabinose and 4-*O*-(1-carboxyethyl)-mannose (Garozzo et al., 1998). Similarly, the viscous EPS from *Nostoc commune* proved to be composed of a 1,4-linked xylogalactoglucan backbone to which D-ribofuranose and 3-*O*-[(R)-1-carboxethyl]-D-glucuronic acid groups were attached (Helm et al., 2000).

Most microorganisms are found in close association with other microbial cells, commonly in the form of biofilms or flocs. EPS of the contributing microbial flora provide much of the dry matter of biofilms, flocs, and related structures. They also play major roles in determining the structures and physical properties of any of these microbial aggregates (Sutherland, 2001b, c). Studies of biofilms, flocs, and related structures reveals the enormous variety of environments, the great variety of microorganisms which may be components of these, and the range of polysaccharides which they may synthesize. Some exopolysaccharides, despite their chemical simplicity can nevertheless display a very wide range of linkage types and acylation patterns. As a result, considerable differences in physical properties are also seen within polymers with the same gross composition.

Relatively few polysaccharides are common to prokaryotes and eukaryotes. These include the storage polymer glycogen, which is found in many bacteria, in yeasts, and in animal cells. There are also a small number of bacterial EPS which are either identical to or closely resemble polysaccharides found in eukaryotic macroalgae or in animals. While cellulose is found as an EPS in *Gluconacetobacter xylinum* and some other bacteria and has no structural role, it is of course a major structural component in plant cell walls. Another EPS found in a limited number of bacterial species is hyaluronic acid; this is essentially similar to the material found in the synovial fluid of joints or the vitreous humor of the eye. Although bacterial alginates are at first apparently similar to the products obtained from marine algae, they are heavily acetylated and of higher mass, while the products from *Pseudomonas* spp. contain L-guluronic acid but lack any polyguluronic acid sequences. Some bacterial strains also secrete EPS which are effectively desulfatoheparin derivatives or analogs of the heparin precursor *N*-acetylheparosan (Vann et al., 1981) and have been chemically modified in the preparation of heparin-like compounds (Casu et al., 1994). Some other bacterial polysaccharides bear some structural similarities to chondroitin.

2.2 Composition and Structure of Yeast and Other Fungal Polysaccharides

In yeast walls, there are glucans and mannans or mannoproteins along with small amounts of chitin. In most fungi, and to a limited extent in some yeasts, chitin is present in greater amounts. Fungal and yeast polysaccharides represent a wide range of chemical structures, but as yet have not received such intensive study as have prokaryotes. Fungal walls mostly contain some chitin, but several possess cellulose (e.g. *Oomycetes*), while many yield β-D-glucans either as components of the cell walls or as extracellular polymers, with interesting antitumor and immunological effects (Kulicke et al., 1997). Although these glucans are absent from the walls of Zygomycetes, they are often the major cell-wall component of other fungi, and the linkages found in these polymers have been summarized by Seviour et al. (1992). *Saccharomyces cerevisiae* contains only small amounts of chitin in the bud scars, but the walls also have mannoproteins and glucans. In the walls of other yeast species, including *Hansenula*, mannans or phosphomannans may be present. The α-D-glucan pseudonigeran is found in the walls of *Aspergillus niger, Tremella mesenterica* and *Schizophyllum commune*, while *Aureobasidium pullulans* secretes an extracellular α-D-glucan (pullulan) with possible biotechnological potential. It is one of the few fungal extracellular polymers which has been exhaustively studied. Another α-D-glucan is nigeran, in which linkages are alternately 1,3 and 1,4. This polysaccharide can be extracted from the walls of *Aspergillus* and *Penicillium* spp. Under appropriate growth conditions, yeasts may contain intracellular glycogen as carbon and energy storage molecule, while trehalose is also frequently encountered.

Fungal β-D-glucans may also be found as extracellular products in media with high carbon:nitrogen ratios (Stasinopoulos and Seviour, 1989). One of these, from *Sclerotium rolfsii*, resembles the bacterial EPS curdlan in its 1,3-linked backbone, but also carries single glucosyl side-chains which thus yield a soluble, highly viscous polymer in aqueous solution, although it also adopts a triple helical conformation (Bluhm et al., 1982). However, xylose is found as a component of EPS in some other yeasts, including *Cryptococcus neoformans* (Turner and Cherniak, 1991). This polymer consisted of a linear $(1 \rightarrow 3)$-α-D-Man backbone to which glucuronosyl and xylose residues were linked. In marked contrast to this, the EPS from *Pichia holstii* proved to be a phosphorylated mannan (Parolis et al., 1996b). However in general, relatively few of the EPS structures from yeasts have been accurately elucidated.

2.3 Composition and Structure of Plant and Algal Polysaccharides

The properties of all the polysaccharides which find applications in food and in other industries derive from their chemical composition and structures. They vary greatly in their solubility and their ability to form gels. This is all dependent on the primary chemical structure, and on the conformation which is adopted in both the aqueous and solid states. Traditionally, many polysaccharides from plants and animals have been used in food and nonfood applications. Our modern technological society makes increasing use of such materials, as well as developing new products from microorganisms or through chemical synthesis. Starch is available from a wide range of plant sources in which it is present as a storage polymer in the form of water-insoluble granules of characteristic shape. The sources include potato tubers and commercial species of the Graminae, including maize, barley and wheat, in all of which it is a storage polymer within the seeds. Alternative sources include rice and cassava. Typically, starch comprises about 20–25% amylose and up to 75% amylopectin. Fructans are also present as storage polysaccharides in higher plants and may replace starch in this role. Plant seed gums including gluco- and galactomannans are obtained from various temperate or tropical leguminous plants in which they represent nonstarch storage products in the endosperm. Their production involves milling and careful separation from protein and fiber, a difficult process which leads to products that vary widely in quality. These products include konjac mannan, locust bean gum (LBG) and guar gum. All possess linear mannan chains to which glucose or galactose side-chains are attached with varying frequency. Thus, many of the plant polysaccharides which find extensive usage are either homopolysaccharides comprised of glucose, or neutral heteropolysaccharides in which there are perhaps two sugars. Wide differences in physical properties can be seen in the different glucans–the 1,4 β-linked cellulose and the 1,4;1,3 β-linked-glucans from wheat or barley are highly crystalline, insoluble polymers. The highly branched α-glucans show some solubility in water, although some such as amylose can be used to form rigid gels, even at fairly low concentrations. Plant exudate gums are complex natural products obtained by manual harvesting and are typified by gum arabic and gum tragacanth derived from *Acacia* trees and *Astragalus* spp., respectively. Both polysaccharides possess complex side-chains attached to backbones that are composed mainly of galactose.

Other plant polysaccharides are wall components, including cellulose, pectins, xylans, and arabinogalactans. These vary therefore from the rigid linear homopolymer cellulose to much more complex, highly branched heteropolysaccharides. Pectins fall into this latter group and are composed of backbones of D-galacturonic acid and its methyl ester derivative, rhamnose and side-chains which may also include ferulic acid (4-hydroxy-3-methoxycinnamic acid). Pectins form the major cell-wall components of many fruits. As an increasing number of species are investigated, other types of polysaccharide are being found. Thus, a recent report describes a galactoglucomannan from the primary cell walls of the fruit of *Actinidia deliciosa* (Schröder et al., 2001).

Although many algae possess cellulosic wall materials, in others–including siphonous green algae–microfibrillar xylans and mannans have been found (Kaihou et al., 1993). The insoluble mannans of *Codium latum* are linear (1 → 4)-β-linked macromolecules ranging in M_r from 28,000 to 84,000 daltons. Polysaccharides from the red and brown marine macroalgae have been used in traditional oriental foods, but are now much more widely employed in numerous foods as well as in various technological applications (Renn, 1997). They provide several well-established commercial products obtained through extraction of marine algal biomass. Alginates are found in many species of marine brown algae and are extracted from *Laminaria* and *Macrocystis* species in various parts of the temperate world. Alginates can be used for various functions or can be chemically modified. Like their bacterial equivalents, algal alginates are composed of linear molecules in which the varying sequences of D-mannuronic acid and L-guluronic acid residues confer a range of physical properties. Carrageenans are sulfated polysaccharides obtained from Rhodophyceaea; their monosaccharide constituents are galactose and anhydrogalactose. Their capacity to form gels is utilized in the food and related industries. Ulvan is a sulfated polysaccharide from green marine seaweeds that is found in *Ulva* spp. and contains both D-glucuronic acid and L-iduronic acid, together with xylose and rhamnose (Lahaye et al., 1997). Although ulvan is capable of forming soft gels, it has not yet found significant applications. Fucans are polysaccharides found within the Pheophyceaea, and are sulfated polymers composed of fucose, xylose, and glucuronic acid; they have some similarities to heparin.

3 Polysaccharide Synthesis

3.1 Bacterial Synthesis of Polysaccharides

The examination of many bacteria has indicated that, under the conditions which have been carefully studied, most strains are most likely capable of synthesizing only one polymer. However, an increasing number of bacteria are now known to be capable of forming several chemotypes of exopolysaccharide; indeed, it is probably less common to find simultaneous expression of more than one polysaccharide. In *Agrobacterium tumefaciens* or *Rhizobium meliloti*, it has long been recognized that pure cultures of many strains can produce either the heteropolysaccharide succinoglycan or the homopolysaccharide (1→3)-α-D-glucan, curdlan. Glucksmann et al. (1993a, b) noted that 19 *exo* genes were required for the regulation and production of the polysaccharide succinoglycan. Other strains possess the cryptic ability for the formation of a new polymer; they may synthesize either a new chemotype or an altered repeat unit structure if the original type has been eliminated by mutagenesis. Some of the activated sugar nucleotide precursors may be common requirements for synthesis of the original and new polysaccharides. Thus, the galactoglucan produced by several strains of *Rhizobium meliloti* (e.g. Glazebrook and Walker, 1989; Levery et al., 1991), requires the same carbohydrate precursors, UDP-D-glucose and UDP-D-galactose, as does succinoglycan. However, the repeat unit of the galactoglucan is formed from disaccharides instead of octasaccharides. As both products contain acetyl and pyruvate adornments, synthesis of both polymers therefore additionally uses phosphoenolpyruvate and acetyl CoA; succinyl half-esters are lacking from the simpler structure. The second mutant polysaccharide from a *Rhizobium meliloti* strain (M5N1 CS) was an acetylated (1→4)-β-linked homopolymer of D-glucuronic acid; in some of its physical properties, this resembles alginate (Courtois et al., 1994). Christensen et al. (1985) studied a marine *Pseudomonas* sp. that synthesized one polysaccharide containing glucuronic and galacturonic acids in the exponential phase of batch culture, but yielded a totally different polymer in the stationary phase. The second polymer contained ketodeoxyoctonic acid, a component normally associated with the lipopolysaccharides of Gram-negative bacterial walls. It was not clear whether the synthesis of different polymers also occurred at different stages of biofilm development. In another example of multiple polysaccharide synthesis, a *Hyphomonas* strain produced a polysaccharide adhesin for a limited period, while the second polymer was produced throughout the bacterial growth cycle (Langille and Weiner, 1998).

Production of the exopolysaccharide colanic acid is widespread in the Enterobacteriaceae under appropriate physiological conditions (Grant et al., 1969), yet few isolates form this material under the growth conditions normally employed in the laboratory. Stimulation of expression of the polysaccharide requires various environmental stresses such as increased osmotic pressure, reduced incubation temperature, or other alterations to the environment. Thus, in bacterial alginate synthesis in *Pseudomonas aeruginosa*, the sigma factor RpoS (σ^s) functions as a general stress response regulator (Suh et al., 1999). Some *Escherichia coli* K12 derivative strains, as well as natural isolates of *Enterobacter cloacae*, can synthesize large amounts of colanic acid under all conditions tested. Other species such as *Salmonella enterica* also produced colanic acid, albeit in smaller amounts (Grant et al., 1969). Alginate production among Pseudomonads is similar; a few species or isolates yield relatively large amounts of alginate under standard laboratory culture conditions, yet the cryptic capacity for production of this polymer is widespread in strains of *P. aeruginosa* and related *Pseudomonas* spp. Environmental stress yields alginate-forming clones of varying stability synthesizing very large amounts of polymer. Davies et al. (1993) suggested that the increase in alginate production observed on solid media in comparison with planktonic growth of *P. aeruginosa* was due to activation of the *algRI* gene product by adhesins which triggered alginate synthesis. In the case of both colanic acid and alginate complex, regulatory systems are involved in the control of expression and production of the respective polysaccharides (e.g. Brill et al., 1988; Gottesman and Stout, 1991; Goldberg and Dahnke, 1992; Chitnis and Ohman, 1993). The biosynthetic gene for alginate *algD* is controlled through three regulatory genes–*algR1*, *algR2*, and *algR3*.

One unusual feature found in prokaryotic polysaccharide synthesis, is the association of genes for polysaccharide degradation in close association to those for biosynthesis. Romeo et al. (1988) observed this in studies on bacterial glycogen synthesis and suggested that such localization of biosynthetic and degradative genes might facilitate regulation. Although absent from most Enterobacterial strains, quite a number of EPS-synthesizing bacteria are now known to possess a gene for a polysaccharase or polysaccharide lyase in the gene cassette required for EPS biosynthesis. The bacteria include *Rhizobium* spp. producing succinoglycan, cellulose-synthesizing *Gluconacetobacter xylinum*, and hyaluronic acid-forming Streptococci (Sutherland, 1999a).

Dextran or levan production differs from that of other microbial polysaccharides in that the fermentation process is microaerophilic. As extracellular enzymes are involved, high substrate concentrations can be used and addition of 10% sucrose can yield 42% conversion to polysaccharide. No sugar nucleotides are involved, but synthesis occurs only in the presence of sucrose. The products are polymers of very high molecular mass formed by sequential additions of glucose or fructose respectively from the substrate. In dextran synthesis, the fructose released from sucrose by the enzyme dextransucrase is also a useful byproduct.

3.2 Synthesis of Plant and Animal Polysaccharides

The biosynthesis of most plant and animal polysaccharides requires activated nucleotide diphosphate sugars in a manner analogous to that seen for prokaryotes. Starch is found as granules that vary greatly in their size, shape, and structure. It is synthesized in the amyloplast in the form of two distinct components: (1) amylose, which consists of chains of $\alpha(1\rightarrow4)$-linked D-glucose residues; and (2) amylopectin, in which the linear chains are shorter and carry $\alpha(1\rightarrow6)$-branches. The synthesis of these macromolecules is a relatively simple process. It involves synthesis of the glycosyl donor ADP-D-glucose by ADP-glucose pyrophosphorylase, synthesis and elongation of the linear $\alpha(1\rightarrow4)$-linked glucan by starch synthase and formation by the branching enzyme (Q enzyme) of the side-chains present in amylopectin and starch synthetase. The ADP-glucose pyrophosphorylase is important not only in the synthesis of the key glucosyl precursor for starch, but also in the overall regulation of starch biosynthesis (Kacser, 1992). Glycogen in animals and microorganisms is in many respects similar, also being synthesized from an activated glucosyl donor by a simple enzymatic system of a synthase and a branching enzyme. A major difference is seen however in prokaryotes, where the activated donor for glycogen synthesis is ADP-glucose (in contrast to UDP-glucose in eukaryotes). The genetic systems required for synthesis of these homopolysaccharides are thus much simpler than those needed for the more complex heteropolysaccharides of plants and microorganisms. However, knowledge of cellulose synthesis and its regulation has mainly derived from studies on *Acetobacter xylinum* rather than eukaryotic systems.

Karnezis et al. (2000) have recently reviewed the process of β-glycan synthesis. They showed that within the glycosyl transferase family 2, the formation of a new glycosidic linkage involved inversion of the α-anomeric configuration of the glycosyl residue in the sugar nucleotide donor. This family of enzymes included bacterial, fungal, and animal proteins, including those involved in synthesis of cell wall components of fungi and plants. As might be expected, the amino acid sequences reveal common structural features. In plants, several genes with apparently similar function may be found, whereas in bacteria usually a single synthase gene is present. Thus in *Arabidopsis thaliana*, at least 12 cellulase synthase genes are detectable in addition to other cellulose-synthase-like genes, the products of which show some similarity to those of cellulase synthase (Cutler and Somerville, 1997).

4
The Physical Properties and Function of Polysaccharides

Relationships exist between the glycosidic linkage geometry of polysaccharides and their conformation. Cellulose, mannans and chitin all possess diequatorially linked sequences from which flat ribbons are formed. These then undergo hydrogen bonding to form sheets, which in turn may stack together in different ways. This means that the physical properties are dependent not just on the primary structure or even on the properties of single polysaccharide chains, but on the way in which these chains associate together either with similar molecules or with other types of polysaccharides.

In most industrial applications, polysaccharides are used to control the rheology of aqueous-phase materials, one such role being as viscosifiers. The range of polymers used for this includes dextrans, xanthan, sodium alginate, carboxymethylcellulose and λ-carrageenan. Polysaccharides are also very widely employed as gelling agents. Agar, calcium alginate, gellan, and carrageenan are found in this role. The ability of alginates to form thermally irreversible gels with Ca^{2+} provides a mild procedure for cell and enzyme immobilization which is used in a number of biotechnological applications. Polysaccharides vary greatly in mass; even those from prokaryotic cells may range from ~10 kDa to 1–2 mDa. The polymer produced by a single microbial culture may vary depending on the physiological conditions, or it may be partially degraded by enzymes present in the culture fluid. Plant and animal polysaccharides also vary considerably, but this may in part be due to extraction and purification procedures. Because of their macromolecular nature, many polysaccharides are readily soluble and form viscous aqueous solutions; others may form gels either *per se* or in the presence of multivalent cations. A small number yield gels when mixed with other polysaccharides. Gelation is a common feature of water-soluble polysaccharides, especially at higher concentrations. This frequently involves adoption of a coil to helix transition (Chandrasekaran, 1999). The helix may be a duplex, or more rarely – as with the fungal polymer schizophyllan–a triple helix is formed (Fuchs et al., 1997). The physical properties of the polysaccharide will depend primarily on the polymer composition and structure. However, they will also depend on the shape adopted in the helical conformation and on interactions with other macromolecules, with water and with ions. Many microbial polysaccharides undergo transition from an ordered state at lower temperatures and in the presence of ions, to a disordered state at elevated temperature on under low ionic environments (e.g. Nisbet et al., 1984). In the case of some polymers, this represents change from a gel to a sol state. Slight changes may cause considerable differences in physical properties. The side-chains found on many linear polysaccharides promote conformational disorder and inhibit any ordered assembly. This results in solubility in aqueous solutions. Thus, xanthan – which possesses a cellulose backbone with trisaccharide side-chains on alternate glucose residues – has been described as a natural water-soluble cellulose derivative (Christensen et al., 1995).

Even for relatively simple structures such as bacterial galactoglucans, charged residues may be found mainly on the exterior of the extended molecules, and may thus readily promote interaction with ions and macromolecules (Chandrasekaran et al., 1994). This may account for the extensive binding of heavy metal cations ascribed to *Zoogloea* flocs and polysaccharides (Friedman and Dugan, 1967; Kuhn and Pfister, 1989). Normally, the amount of cation bound is relatively small, the binding is non-selective, and the models of galactoglucans fail to indicate how selectivity might be achieved. However, some polysaccharides have been found to bind ions preferentially from a mixture. Bacterial alginates resemble the algal analogs in binding calcium in preference to magnesium ions, with binding and selectivity being greatly enhanced by the removal of acetyl groups (Geddie and Sutherland, 1994). Alteration to physical properties may also occur within the complex; thus, if heterologous bacterial species produce enzymes modifying the polysaccharide structure, changes may be observed (Sutherland, 1999b). Acetyl groups play a major role in determining some of the physical properties of bacterial exopolysaccharides (Sutherland, 1997, 1999c). Some bacteria are capable of producing esterases which show wide specificity, removing acyl groups from several bacterial polymers as well as from other esters (Cui et al., 1999). Release of such enzymes or of polysaccharases either during growth or following cell lysis in aging biofilm or floc structures could therefore alter the physical properties of the structure, either locally or to a greater extent if extensive degradation occurs. Deacylation of the bacterial polysaccharide succinoglycan was shown to improve pseudoplasticity in aqueous solution as well as increasing the cooperativity of the order–disorder transition (Ridout et al., 1997). On the other hand, deacylation of some polysaccharides may lead to loss of any ordered conformation (Villain-Simonnet et al., 2000).

Gels may be formed through intra- or intermolecular associations, and the structural entities of the polysaccharides which cause such gelation may vary greatly. Hydroxyl or methyl groups may be involved, as may carboxyl, sulfate, or phosphate groups in anionic polymers. Inter- or intramolecular hydrogen bonding may lead to gelation, possibly also acting through water molecules at the exterior of polymeric duplexes (Tako, 2000). As many of the bacterial polysaccharides possess side-chains of varying structure and length, these can also be expected to participate in intramolecular and interchain reactions, causing differences in the physical properties of polymers with extensive structural similarities (Talashek and Brant, 1987).

In the formation of cell aggregates and initiation of flocculation and similar processes, exopolysaccharides play a fundamental role. The levan from *Bacillus polymyxa* is thus essential for soil aggregation in the plant rhizosphere (Bezzate et al., 2000). Biopolymers also play a major role in wastewater treatment (Houghton and Quarmby, 1999). This may involve divalent cations, and the hydrophobicity of some of the molecules almost certainly plays a significant role, perhaps acting as a focus for flocculent particle formation and propagation. The nature of the cell surfaces and their exposed macromolecules also affect the outcome. Thus, studies on the fine structure of different Gram-negative bacteria revealed that the capsules were comprised of densely packed fine fibers with the capsules themselves varying in thickness from 10 to 160 nm (Amako et al., 1988). These types of structure, with their associated cellular material, could well stimulate flocculation. In oral biofilms, the presence of polysaccharide synthesized by one species may provide the focus for aggregation of other species through lectin binding or similar mechanisms. Analogous interactions may also occur in wastewater treatment. Very similar interactions between polysaccharides, other macromolecules and lipids and ions occur in natural and processed foods.

5 Commercialization of Polysaccharides

As they are so widely used, some polysaccharides are produced in very large quantities indeed. Clearly the cheapest and most abundant are those from plant sources. Starch and cellulose are available in almost all parts of the world from a variety of local sources and are cheap, bulk chemicals. Even within the European community, in excess of 6 million tonnes of starch is produced annually from maize, potatoes, and wheat. Although most cellulose is used in the form of wood, some 4–5% of wood pulp is converted to cellulose derivatives and regenerated cellulose. Gum guar is also produced in large quantities in the Indian subcontinent (70,000–80,000 tonnes per annum), and is also relatively cheap. When purified, it yields a white powder giving clear, transparent solutions. Gum guar is used in a wide range of food and nonfood applications, and considerable quantities are also used in the oil industry for drilling processes. In these roles, the low and relatively stable cost of guar gum provides a price advantage over the more expensive LBG. Exudate gums such as gum arabic have long been collected manually from *Acacia* growing in arid regions such as the edge of the Sahel. Indeed, it is probably the oldest commercially exploited gum. In the middle of the price range are the products from marine algae–alginates and carrageenan. All are well-established in their product areas (Table 2), and any new polysaccharide has to compete against them in physical properties, quality, cost, and availability. Although a multitude of microbial polysaccharides are available, the number which have been commercialized is very small indeed (see Table 2). Nor is it likely to increase greatly in the foreseeable future. Of these newer products, only xanthan is produced on a large scale by a number of

Tab. 2 Major commercial microbial, plant and algal polysaccharides and their sources

Alginate	Marine algae (Phaeophyceae)
Carrageenan	Marine algae (Rhodophyceae)
Cellulose	Bacteria (*Gluconacetobacter*) and plants
Cellulose and derivatives	Plants
Curdlan	Bacteria (*Agrobacterium*)
Dextran and derivatives	Bacteria (*Leuconostoc*)
Gellan	Bacteria (*Sphingomonas*)
Gum arabic	Plants (*Acacia* spp.)
Gum guar	Plants (*Cyamopsis tetragonolobus*)
Hyaluronic acid	Bacteria (*Streptococcus* spp.) and animals
Konjac mannan	Plants (*Amorphophallus konjac*)
Laminaran	Marine algae (*Laminaria* spp.)
Locust bean gum	Plants (*Ceretonia siliqua*)
Pectin	Plants (citrus fruits and other sources)
Pullulan	Fungi (*Aureobasidium pullulans*)
Starch	Plants
Xanthan	Bacteria (*Xanthomonas campestris*)

companies, and its worldwide supply has been estimated at 20,000 tonnes per annum. Compared with many plant polymers, the microbial products are relatively expensive and, with the exception of xanthan, yields may be low even when produced under optimal conditions. A further barrier to their extensive use in the food industry is the very great cost of obtaining regulatory approval. Currently, only xanthan and gellan are approved food additives in North America and in the European Community. Although dextran was earlier approved for food use, the lack of suitable applications has seen it discarded from this role. The attitude in Japan is rather different and there, microbial polysaccharides are regarded as natural products that can be used in food. An alternative approach now receiving much attention is to develop polysaccharides from bacterial species such as lactic acid bacteria (LAB), which are already permissible for food production. The current drawbacks to this approach are the low yields and difficulties in production.

A few microbial polysaccharides are used for specific purposes, or as substitutes for eukaryotic materials. Thus, bacterial cellulose has a small but specialized market niche and is used for acoustic membranes and wound dressings. Hyaluronic acid from Gram-positive Streptococci is used to replace material obtained with difficulty from animal sources, and finds applications in ophthalmic surgery as wells as in joint replacements and repair of the damage to cartilage caused by osteoarthritic conditions. Both polysaccharides are very high-value products of high purity, and low yields are therefore less important. The production of each presents distinct problems, however. Dextran finds several different applications, but for all of these must first be either chemically modified (as in the case of cross-linked Sephadex adsorbents for separation science) or partially hydrolyzed with acid to yield the correct mass for use as a plasma substitute (D70). Dextran also finds uses in photography and cosmetic preparations.

In some applications, there is an overlap between plant or algal and microbial products. Thus, κ-carrageenan, xanthan, and carboxymethylcellulose all find applications in products such as toothpaste, where flexible, elastic suspending agents are required. It should also not be forgotten that, because of the synergistic interactions between them, polysaccharides are frequently used in mixtures. Xanthan and LBG provide one such example, yielding mixed gels at low total polysaccharide concentrations when neither forms a true gel on its own (Ross-Murphy et al., 1996). This property is widely used for processed human and animal foods.

Several microbial polysaccharides have been produced commercially without attracting adequate market niches. Thus, succinoglycan was manufactured for a time for the oil industry, but is no longer in production. Significant claims for emulsan as a biodegradable detergent were also made, but it too has failed to establish itself, possibly because of its relatively high cost. Some of these macromolecules are effectively lipopolysaccharide-like, amphipathic molecules capable of stabilizing oil-in-water emulsions (Rosenberg and Ron, 1997).

6 New Products

Research into the properties of novel plant and microbial polysaccharides continues. The major obstacle to commercialization of any of these lies in finding novel or superior properties to products already on the market. It is very unlikely that many will survive the rigors of extensive evaluation and market research to reach production. During the past 50 years, a number of bacterial polysaccharides have been introduced, yet few of these remain commercially available as major products. Xanthan, and more recently gellan, prove this point; their unique properties, safety and moderate cost have proved acceptable. Various attempts have been made to genetically engineer xanthan and acetan, and these have yielded products with shortened side-chains and altered physical properties. Such work is exemplified by the acetan mutant with an acetylated pentasaccharide repeat unit obtained by Colquoun et al. (2001), this polymer still undergoing a helix–coil transition at 70 °C. Earlier studies on xanthan yielded mutant products with greater or lesser viscosity than the wild-type polysaccharide, but they suffered from lower yields (Vanderslice et al., 1989). Novel environments such as deep-sea thermal vents, have yielded new bacterial strains and polysaccharides. A few of these are claimed to have interesting properties, but none is yet marketed. They may perhaps find specialized applications in nonfood sectors and in cosmetics. Further synergistic interactions may also find applications for mixtures of microbial and plant polysaccharides, but these will most probably be based on the use of current products rather than novel ones. Evaluation of the antitumor and similar properties of various plant and fungal products may certainly yield some interesting and potentially valuable products, but these will generally not require large-scale production. Although derivatives such as carboxymethylcellulose and cross-linked dextran and agarose have found many applications, especially in biochemistry and biotechnology, few other polysaccharide derivatives are widely used. As well as being the starting material for many microbially derived products, starch has the potential for inclusion in biodegradable thermoplastics. This and other modified forms of polysaccharides offer a possible area for future development, but the cost of development and of production may again be a limiting factor. Another possible approach is the enzymatic upgrading of existing polysaccharide products. Thus, use of an α-Dgalactosidase to remove some of the side-chains from guar introduced enhanced interaction with xanthan (McLeary et al., 1984). This provides a possible replacement of LBG which is of higher cost and in increasingly short supply, with the cheaper and more readily available guar. Another example of potential enzymatic upgrading lies in alginate. The cloning of seven distinct poly-D-mannuronate epimerases from *Azotobacter vinelandii* provided a source of enzymes capable of introducing guluronic acid residues into alginates with low guluronosyl content (Ramstad et al., 1999, 2001). This altered the gelling properties of the polysaccharide, and again may represent a potential upgrading of the product which could originate from either algal or bacterial sources.

The most promising developments are possibly in devising novel applications for many of the existing commercialized polysaccharides from plants and microorganisms, and thus expanding the market for these materials. As more becomes known about the structure–function relationships within these compounds, the choice may increasingly be based on scientifically proven properties rather than empirical decisions. Ultimately, the cosmetics and pharmaceutical industries may provide new markets for existing or novel products. It should also be remembered that an increasing number of polysaccharides have been shown to possess biological activity. Although polysaccharide immunomodulators were first reported over 40 years ago, some polysaccharides are now known to have marked effects on the regulation of immune responses to infectious disease. Through interaction with T cells, macrophages and polymorphonuclear lymphocytes, the polysaccharides can affect both innate and cell-mediated immunity (Tzianabos, 2000). Further screening may well extend the range of these, especially in the case of polymers with antitumor or immune system-promoting or wound-healing activities.

An alternative approach to the direct development of new polysaccharide products can be seen in the current interest in polysaccharides from lactic acid bacteria. The bacterial strains producing these polymers are widely used in the preparation of fermented milk products such as cheeses, yogurt, and kefir. In these products, the polymers not only contribute to flavor, texture, and other sensory attributes, but are also important in determining the stability of the product and its shelf life. Numerous structures for these polysaccharides have been reported in recent years (e.g. De Vuyst & Degeest, 1999; Faber et al., 2001; Marshall et al., 2001), and they seem to comprise a rather limited range of monosaccharides in which D-glucose, D-galactose, and L-rhamnose predominate. Considerable effort has gone into determining the genomic information needed for the synthesis of this group of polysaccharides (e.g. Stingele et al., 1999), and the bacteria have the presumed advantage of being acceptable for food use. As polysaccharide yields are generally low, it is unlikely that the polymers will be produced as separate products; rather, the bacteria will be used to produce polysaccharides *in situ* as part of the complete fermentation product. Some of these polysaccharides may also have significant probiotic activity, potentially enhancing the health of the consumer!

7 References

Amako, K., Meno, Y., Takade, A. (1988) Fine structures of the capsules of *K. pneumoniae* and *Escherichia coli* K1, *J. Bacteriol.* **170**, 4960–4962.

Bezzate, S., Aymerich, S., Chambert, R., Czarnes, S., Berge, O., Heulin, T. (2000) Disruption of the *Paenobacillus polymyxa* levansucrase gene impairs its ability to aggregate soil in the wheat rhizosphere, *Environ. Microbiol.* **2**, 333–342.

Bluhm, T. L., Deslandes, Y., Marchessault, R. H., Perez, S., Rinaudo, M. (1982) Solid state and solution conformation of scleroglucan, *Carbohydr. Res.* **100**,117–130.

Brill, J. A., Quinlan-Walshe, C., Gottesman, S. (1988) Fine structure mapping and identification of two regulators of capsule synthesis in *Escherichia coli* K12, *J. Bacteriol.* **170**, 2599–2611.

Casu, B., Grazioli, G., Razi, N., Guerrini, M., Naggi, A., Torri, G., Oreste, P., Tursi, F., Zoppetti, G., Lindahl, U. (1994) Heparin-like compounds prepared by chemical modification of capsular polysaccharide from *Escherichia coli* K5, *Carbohydr. Res.* **263**, 271–284.

Chandrasekaran, R. (1999) X-ray and molecular modeling studies on the structure–function correlations of polysaccharides, *Macromol. Symp.* **140**, 17–29.

Chandrasekaran, R., Lee, E. J., Thailambal, V. G., Zevenhuizen, L. P. T. M. (1994) Molecular architecture of a galactoglucan from *Rhizobium meliloti*, *Carbohydr. Res.* **261**, 279–295.

Chitnis, C. E., Ohman, D. E. (1993) Genetic analysis of the alginate biosynthetic gene cluster of *Pseudomonas aeruginosa* shows evidence of an operonic structure, *Mol. Microbiol.* **8**, 583–590.

Christensen, B. E., Kjosbakken, J., Smidsrød, O. (1985) Partial chemical and physical characterization of two extracellular polysaccharides produced by marine periphytic *Pseudomonas* sp. strain NCMB 2021, *Appl. Environ. Microbiol.* **50**, 837–845.

Christensen, B. J., Stokke, B. T., Smidsrød, O. (1995) Xanthan – the natural water soluble cellulose derivative, in: *Cellulose and Cellulose Derivatives: Physico-chemical Aspects and Industrial Applications* (Kennedy, J. F., Phillips, G. O., Williams, P. A., Eds.), Chester, England: Woodhead Publishing, pp. 265–278.

Colquhoun, I. J., Jay, A. J., Eagles, J., Morris, V. J., Edwards, K. J., Griffin, A. M., Gasson, M. J. (2001) Structure and conformation of a novel genetically engineered polysaccharide P2, *Carbohydr. Res.* **330**, 325–333.

Courtois, J., Seguin, J.-P., Roblot, C., Heyraud, A., Gey, C., Dantas, L., Barbotin, J.-N., Courtois, B. (1994) Exopolysaccharide production by the *Rhizobium meliloti* M5N1CS strain. Location and quantitation of the sites of *O*-acetylation, *Carbohydr. Polym.* **25**, 7–12.

Cui, W., Winter, W. T., Tanenbaum, S. W., Nakas, J. P. (1999) Purification and characterization of an intracellular carboxylesterase from *Arthrobacter viscosus* NRRL B-1973, *Enzyme Microb. Technol.* **24**, 200–208.

Cutler, S., Somerville, C. (1997) Cellulose synthesis: cloning *in silico*, *Curr. Biol.* **7**, R108–R111.

Davies, D. G., Chakrabarty, A. M., Geesey, G. G. (1993) Exopolysaccharide production in biofilms: substrate activation of alginate genes by *Pseudomonas aeruginosa*, *Appl. Environ. Microbiol.* **59**, 1181–1186.

De Vuyst, L., Degeest, B. (1999) Heteropolysaccharides from lactic acid bacteria, *FEMS Microbiol. Rev.* **23**, 153–177.

Faber, E. J., van den Haak, M. J., Kamerling, J. P., Vliegenhart, J. F. G. (2001) Structure of the exopolysaccharide produced by *Streptococcus thermophilus* S3, *Carbohydr. Res.* **331**, 173–182.

Friedman, B. A., Dugan, P. R. (1967) Concentration and accumulation of metallic ions by the bacterium Zoogloea, *Dev. Ind. Microbiol.* **9**, 381–388.

Fuchs, T., Richtering, W., Burchard, W., Kajiwara, K., Kitamura, S. (1997) Gel point in physical gels: rheology and light scattering from thermoreversible gelling schizophyllan, *Polymer Gel Network* **5**, 541–559.

Garozzo, D., Impallomeni, G., Spina, E., Sturiale, L. (1998) The structure of the exocellular polysaccharide from the cyanobacterium *Cyanospira capsulata*, *Carbohydr. Res.* **307**, 113–124.

Geddie, J. L., Sutherland, I. W. (1994) The effect of acetylation on cation binding by algal and bacterial alginates, *Biotechnol. Appl. Biochem.* **20**, 117–129.

Glazebrook, J., Walker, G. C. (1989) A novel exopolysaccharide can function in place of calcofluor-binding exopolysaccharide in nodulation by *Rhizobium meliloti*, *Cell* **56**, 661–672.

Glucksman, M. A., Reuber, T. L., Walker, G. C. (1993a) Family of glycosyl transferases needed for the synthesis of succinoglycan by *Rhizobium meliloti*, *J. Bacteriol.* **175**, 7033–7044.

Glucksman, M. A., Reuber, T. L., Walker, G. C. (1993b) Genes needed for the modification, polymerization, export and processing of succinoglycan by *Rhizobium meliloti*: a model for succinoglycan biosynthesis, *J. Bacteriol.* **175**, 7045–7055.

Goldberg, J. B., Dahnke, T. (1992) *Pseudomonas aeruginosa* AlgB, which modulates the expression of alginate, is a member of the NtrC subclass of prokaryotic regulators, *Mol. Microbiol.* **6**, 59–66.

Gottesman, S., Stout, V. (1991). Regulation of capsular polysaccharide synthesis in *Escherichia coli* K12, *Mol. Microbiol.* **5**, 1599–1606.

Grant, W. D., Sutherland, I. W., Wilkinson, J. F. (1969) Exopolysaccharide colanic acid and its occurrence in the Enterobacteriaceae, *J. Bacteriol.* **100**, 1187–1193.

Helm, R. F., Huang, Z., Edwards, D., Leeson, H., Peery, W., Potts, M. (2000) Structural characterization of the released polysaccharide of desiccation tolerant *Nostoc commune* DRH1, *J. Bacteriol.* **182**, 974–982.

Houghton, J. I., Quarmby, J. (1999) Biopolymers in wastewater treatment, *Curr. Opin. Biotechnol.* **2**, 259–261.

Kacser, H. (1992) in: *The Biochemistry of Plants* (Davies, D., Ed.). New York: Academic Press, vol. 11, 39–67.

Kaihou, S., Hayashi, T., Otsuru, O., Maeda, M. (1993) Studies on the cell wall mannan of the syphonous green algae, *Codium latum*, *Carbohydr. Res.* **240**, 207–218.

Karnezis, T., Mcintosh, M., Stanisich, V. A., Stone, B. A. (2000) The biosynthesis of β-glycans, *Trends Glycosci. Glycotechnol.* **12**, 211–227.

Kuhn, S. P., Pfister, R. M. (1989) Adsorption of mixed metals and cadmium by calcium alginate immobilised *Zoogloea ramigera*, *Appl. Microbiol.* **31**, 613–618.

Kulicke, W.-M., Lettau, A. I., Thielking, H. (1997) Correlation between immunological activity, molar mass, and molecular structure of different (1→3)-β-glucans, *Carbohydr. Res.* **297**, 135–143.

Lahaye, M., Brunel, M., Bonnin, E. (1997) Fine chemical structure analysis of oligosaccharides produced by an ulvan-lyase degradation of the water-soluble cell-wall polysaccharides from *Ulva* sp. (Ulvales, Chlorophya), *Carbohydr. Res.* **304**, 325–333.

Langille, S. E., Weiner, R. M. (1998) Spatial and temporal deposition of *Hyphomonas* strain VP-6 capsules involved in biofilm formation, *Appl. Env. Microbiol.* **64**, 2906–2913.

Levery, S. B., Zhan, H., Lee, C. C., Leigh, J. A., Hakamori, S. (1991) Structural analysis of a second acidic exopolysaccharide of *Rhizobium meliloti* that can function in alfalfa root nodule invasion, *Carbohydr. Res.* **210**, 339–347.

Mack, D., Fischer, W., Krokotsch, A., Leopold, K., Hartmann, R., Egge, H., Laufs, R. (1996) The intercellular adhesin involved in biofilm accumulation of *Staphylococcus epidermidis* is a linear β-1,6-linked glucosaminoglycan: purification and structural analysis, *J. Bacteriol.* **178**, 175–183.

Marshall, V. M., Dunn, H., Elvin, M., McLay, N., Gu, Y., Laws, A. P. (2001) Structural characterisation of the exopolysaccharide produced by *Streptococcus thermophilus* EU20, *Carbohydr. Res.* **331**, 413–422.

McCleary, B. V., Dea, I. C. M., Windust, J., Cooke, D. (1984) Interaction properties of D-galactose-depleted guar galactomannan samples, *Carbohydr. Polym.* **4**, 253–270.

Nisbet, B. A., Sutherland, I. W., Bradshaw, I. J., Kerr, M., Morris, E. R., Shepperson, W. A. (1984) XM6, a new gel-forming bacterial polysaccharide, *Carbohydr. Polym.* **4**, 377–394.

Parolis, H., Parolis, L. A. S., Boan, I. F., Rodriguez-Valera, F., Widmalm, G., Manca, M. C., Jansson, P.-E., Sutherland, I. W. (1996a) The structure of the exopolysaccharide produced by the halophilic Archaeon *Haloferax mediterranea* strain R4 (ATCC 33500), *Carbohydr. Res.* **295**, 147–156.

Parolis, L. A. S., Duus, J. O., Parolis, H., Meldal, M., Bock, K. (1996b) The extracellular polysaccharide of *Picia (Hansenula) holstii* NRRLY-2448: the structure of the phosphomannan backbone, *Carbohydr. Res.* **293**, 101–117.

Ramstad, M. V., Ellingsen, T. E., Josefsen, K. D., Hoidal, H. K., Valla, S., SkjåkBræk, G., Levine, D. W. (1999) Properties and action pattern of the recombinant mannuronan C-5-epimerase *alge2*, *Enzyme Microb. Technol.* **24**, 636–646.

Ramstad, M. V., Markussen, S., Ellingsen, T. E., SkjåkBræk, G., Levine, D. W. (2001) Influence of environmental conditions on the activity of the recombinant mannuronan C-5 epimerase AlgE2, *Enzyme Microb. Technol.* **28**, 57–69.

Renn, D. (1997) Biotechnology and the red seaweed polysaccharide industry: status, needs and prospects, *Trends Biotechnol.* **15**, 9–14.

Ridout, M. J., Brownsey, G. J., York, G. M., Walker, G. C., Morris, V. J. (1997) Effect of O-acyl substituents on the functional behaviour of *Rhizobium meliloti* succinoglycan, *Int. J. Biol. Macromol.* **20**, 1–7.

Romeo, T., Kumar, A., Preiss, J. (1988) Analysis of the *Escherichia coli* glycogen gene cluster suggests that catabolic enzymes are encoded among the biosynthetic genes, *Gene* **70**, 363–376.

Rosenberg, E., Ron, E. Z. (1997) Bioemulsans – microbial polymeric emulsifiers. *Curr. Opin. Biotechnol.* **8**, 313–316.

Ross-Murphy, S. B., Shatwell, K. P., Sutherland, I. W., Dea, I. C. M. (1996) Influence of acyl substituents on the interaction of xanthans with plant polysaccharides, *Food Hydrocolloids* **10**, 117–122.

Schröder, R., Nicolas, P., Vincent, S. J. F., Fischer, M., Reymond, S., Redgewell, R. J. (2001) Purification and characterisation of a galactoglucomannnan from kiwifruit (*Actinidia deliciosa*), *Carbohydr. Res.* **331**, 291–306.

Seviour, R. J., Stasinopoulos, S. J., Auer, D. P. F., Gibbs, P. A. (1992) Production of pullulan and other exopolysaccharides by filamentous fungi, *Crit. Rev. Biotechnol.* **12**, 279–298.

Stack, R. J., Plattner, R. D., Cote, G. L. (1988) Identification of L-iduronic acid as a constituent of the major extracellular polysaccharide produced by *Butyrivibrio fibrisolvens* strain X6C61. *FEMS Microbiol. Lett.* **51**, 1–6.

Stasinopoulos, S. J., Seviour, R. J. (1989) Exopolysaccharide formation by isolates of Cephalosporium and acremonium, *Mycol. Res.* **92**, 55–60.

Stingele, F., Vincent, S. J. F., Faber, E. J., Newell, J. W., Kamerling, J. P., Neeser, J.-R. (1999) Introduction of the exopolysaccharide gene cluster from *Streptococcus thermophilus* Sfi6 into *Lactococcus lactis* MG1363: production and characterization of an altered polysaccharide, *Mol. Microbiol.* **32**, 1287–1295.

Suh, S., Silo-Suh, L., Woods, D. E., Hassett, D. J., West, S. E. H., Ohman, D. E. (1999) Effect of *rpoS* mutation on the stress response and expression of virulence factors in *Pseudomonas aeruginosa*, *J. Bacteriol.* **181**, 3890–3897.

Sutherland, I. W. (1990) *Biotechnology of Exopolysaccharides*, Cambridge: Cambridge University Press.

Sutherland, I. W. (1997) Microbial exopolysaccharides – structural subtleties and their consequences, *Pure Appl. Chem.* **69**, 1911–1917.

Sutherland, I. W. (1999a) Polysaccharases for microbial polysaccharides. *Carbohydr. Polym.* **38**, 319–328.

Sutherland, I. W. (1999b) Biofilm exopolysaccharides, in: *Microbial Extracellular Polymeric Substances* (Wingender, J., Neu, T. R., Flemming, H.-C., Eds.), Berlin: Springer-Verlag, 73–92.

Sutherland, I. W. (1999c) Polysaccharases in biofilms – sources – action – consequences, in: *Microbial Extracellular Polymeric Substances* (Wingender, J., Neu, T. R., Flemming, H.-C., Eds.), Berlin: Springer-Verlag, 201–216.

Sutherland, I. W. (2001a) Biofilm exopolysaccharides – a strong and sticky framework, *Microbiology* **147**, 3–9.

Sutherland, I. W. (2001b) Exopolysaccharides in biofilms, flocs and related structures, *Water Sci. Technol.* **43**, 77–86.

Sutherland, I. W. (2001c) The Biofilm Matrix – an immobilised but dynamic microbial environment, *Trends Microbiol.* **9**, 222–227.

Tako, M. (2000) Structural principles of polysaccharide gels, *J. Appl. Glycosci.* **47**, 49–53.

Talashek, T. A., Brant, D. A. (1987) The influence of side-chains on the calculated dimensions of three related bacterial polysaccharides, *Carbohydr. Res.* **160**, 303–316.

Turner, S. H., Cherniak, R. (1991) Glucuronoxylomannan of *Cryptococcus neoformans*, *Carbohydr. Res.* **211**, 103–116.

Tzianabos, T. (2000) Polysaccharide immunomodulators as therapeutic agents: structural aspects and biologic function, *Clin. Microb. Rev.* **13**, 522–533.

Vanderslice, R. W., Doherty, D. H., Capage, M., Betlach, M. R., Hassler, R. A., Henderson, N. M., Ryan-Graniero, J., Tecklenberg, M. (1989) Genetic engineering of polysaccharide structure in *Xanthomonas campestris*, in: *Biomedical and Biotechnological Advances in Industrial Polysaccharides*

(Crescenzi, V., Dea, I. C. M., Paoletti, S., Stivala, S. S., Sutherland, I. W., Eds.), New York: Gordon and Breach, 145–156.

Vann, W. F., Schmidt, M. A., Jann, B., Jann, K. (1981) The structure of the capsular polysaccharide (K5 antigen) of urinary tract infective *Escherichia coli* O10:K5:H4, *Eur. J. Biochem.* **116**, 359–364.

Villain-Simonnet, A., Milas, M., Rinaudo, M. (2000) A new bacterial polysaccharide (YAS34). I. Characterization of the conformations and conformational transition, *Int. J. Biol. Macromol.* **27**, 65–75.

Wieland, F., Lechner, J., Sumper, M. (1986) Iduronic acid: constituent of sulphated dolichyl phosphate oligosaccharides in halobacteria. *FEBS Lett.* **195**, 77–81.

Wolk, C. P. (2000). Heterocyst formation in *Anabaena*, in: *Prokaryotic Development* (Brun, Y. V., Shimkets, L. J., Eds.), Washington: ASM, 83–104.

Yamazaki, K., Inukai, K., Suzuki, M., Kuga, H., Korenaga, H. (1998) Structural studies on a sulfated polysaccharide from an *Arthrobacter* sp. by NMR spectroscopy and methylation analysis, *Carbohydr. Res.* **305**, 253–260.

2
Glycogen Synthesis and its Regulation in Bacteria

Prof. Dr. Jack Preiss
Department of Biochemistry and Molecular Biology, Michigan State University, East Lansing, Michigan 48824, USA; Tel.: 01-517-353-3137; Fax: 01-517-353-9334; E-mail: preiss@pilot.msu.edu

ADP-pyridoxal	adenosine diphosphopyridoxal
3PGA	3-phospho-glycerate
8N3ATP	8-azido-ATP
ADPGlc PPase	ADP-glucose pyrophosphorylase
MW	molecular weight

1 Introduction

A number of bacteria accumulate glycogen in stationary phase or under limited growth conditions with excess carbon in the medium. For example, the rate of growth and the quantity of glycogen accumulation in *Escherichia coli* are inversely related when cells are grown in media containing glucose as the carbon source and limited in nitrogen. A limited number of bacteria, however may synthesize glycogen during exponential growth. Glycogen does not seem to be required for growth, as bacterial mutants having defective structural genes for the glycogen biosynthetic enzymes and do not synthesize glycogen, grow as well as their normal parent strains.

Glycogen or other similar α-1,4 glucans have been reported in more than 40 different bacterial species (Preiss, 1989). Glycogen accumulation is not restricted to any class of bacteria, as many Gram-negative and Gram-positive bacteria as well as archaebacteria have been reported to accumulate glycogen.

This chapter reviews evidence for the energy-storage role of glycogen, as well as the reactions involved in their biosynthesis and subsequent utilization. In addition, the current state of knowledge of regulation of glycogen synthesis is discussed. Various aspects of energy storage role of glycogen, the enzymology and regulation of bacterial glycogen synthesis, including genetic regulation, have been previously reviewed (Preiss, 1969, 1973, 1978, 1984, 1989, 1996, 2000; Krebs and Preiss, 1975; Preiss and Walsh, 1981; Preiss et al., 1983; Preiss and Romeo, 1989, 1994).

2 Characterization of Bacterial Glycogen Structure

Microbial glycogen is very similar to mammalian glycogen in that it is composed mainly of α-1,4- glucosyl linkages and is a branched polysaccharide with about 8–12% of the glucosyl linkages being α-1,6. Most bacterial glycogens studied in this manner have chain lengths of ~10–13 glucose units and I_2 spectra with maximum absorption of ~410–480 nm.

3 The Function of Glycogen as an Energy-Storage Compound

Three criteria for classification of compounds having energy-storage function were proposed by Wilkinson (1959). The first criterion is that the compound accumulates intracellularly under conditions where the energy supply for growth of the organism is excessive. The second is that the reserve polymer is used when the components of energy or nutrients in the media are no longer available for sustaining growth and other processes required for maintenance of viability. Wilkinson (1959) suggested various cell functions that would require energy or carbon, for example, maintenance of a functioning semipermeable cytoplasmic membrane, energy for replacement of proteins and nucleic acids during turnover, maintenance of intracellular pH, or for other processes induced for bacterial survival, such as sporulation and encystment.

The last criterion is that the storage compound is used by the cell for energy production to enable it to survive in the nonsupportive environment. The energy requirement for survival is thus known as energy of maintenance. This last criterion is perhaps the most important, to distinguish the reserve storage compound from other substances produced for other functions (Wilkinson, 1959).

Various strains of *E. coli, Enterobacter aerogenes* and *Streptococcus mitis* containing glycogen have been shown to have a more prolonged survival rate than was observed in strains not containing glycogen. It was also shown that, under starvation conditions, the glycogen-less enteric cells degraded their RNA and protein constituents to ammonia. This either did not occur or occurred at a lower rate in cells containing glycogen. Therefore, when glycogen was available to the cell, its use as an energy source minimized the use of RNA and protein degradation for the production of energy.

In various clostridia and *Bacillus*, and also in *Streptomyces viridochromogenes*, a glycogen-like molecule accumulated up to 60% of the cell's dry weight prior to the onset of sporulation. The polysaccharide was then degraded during formation and maturation of the spore. In these cases glycogen served as an endogenous source of carbon and energy for spore formation.

Many experiments thus suggest that glycogen plays a role in the survival of bacteria, but its precise function remains unclear. Further information on the role of bacterial glycogen is needed.

On a bad note it has been shown that synthesis and later degradation of glycogen by oral bacteria may be an important factor in the development of dental caries. Consistent with this postulate is the finding that organisms capable of synthesizing glycogen, produce more acid when exogenous carbohydrate is present, and can produce acid in the absence of exogenous carbohydrate when compared with oral bacteria unable to synthesize glycogen. The acid formed from polysaccharide catabolism may be of significance in production of dental caries. It is formed over a considerable period of time and may consequently be responsible for the lower resting pH observed in plaque from individuals with active caries.

4 Enzymatic Reactions Involved in Glycogen Synthesis

In 1964 it was demonstrated that extracts of several bacteria contained both an ADP-glucose pyrophosphorylase (reaction 1), as well as an ADP-glucose-specific glycogen synthase (reaction 2) (Greenberg and Preiss, 1964):

$$\text{ATP} + \alpha\text{-glucose-1-P} \longleftrightarrow \text{ADP-glucose} + \text{PPi} \qquad \text{(reaction 1)}$$

$$\text{ADP-glucose} + \alpha\text{-glucan} \longleftrightarrow \alpha\text{-1,4-glucosyl-glucan} + \text{ADP} \qquad \text{(reaction 2)}$$

Branching enzyme activity (reaction 3) was also later shown to be present in extracts of many bacteria.

$$\text{Elongated } \alpha\text{-1,4-glucan} \rightarrow \text{Branched } \alpha\text{-1,4, } \alpha\text{-1,6-glucosyl-glucan} \qquad \text{(reaction 3)}$$

5 Properties of the Glycogen Biosynthetic Enzymes

5.1 ADP-Glucose Pyrophosphorylase

5.1.1 Structure

ADP-glucose pyrophosphorylase (ADPGlc PPase) has been purified from a number of microorganisms, and with one exception the enzyme has been found to be homotetrameric in structure with a subunit molecular size of about 50 kDa. The exception, *Bacillus stearothermophilus* ADPGlc PPase (Takata et al., 1997) is also tetrameric but is composed of two different subunits. One subunit is of 387 amino acids, GlgC, 43.3 kDa, and the other, GlgD, is of 343 amino acids, 39 kDa. The *Bacillus* enzyme is thus similar to the higher-plant ADPGlc PPases which have been shown also to be heterotetrameric, $\alpha_2\beta_2$ (Preiss and Sivak, 1998). The plant small subunit is also known as the catalytic subunit, while the large subunit is known as the regulatory subunit. The potato tuber ADPGlc PPase catalytic subunit expressed alone in *E. coli* is highly active, while the large subunit by itself is inactive (Preiss and Sivak, 1998). Expression of both potato tuber ADPGlc PPase subunits results in a heterotetramer having higher affinity for the allosteric activator and lower affinity for the inhibitor. Therefore, the large subunit is regarded as the regulatory subunit. The *B. stearothermophilus* GlgC subunit shows a 42–70% similarity with other ADPGlc PPases, and when expressed in *E. coli* alone has catalytic activity. The *Bacillus* GlgD subunit has no activity and its function is unknown. However, it seems to increase the V_{max} of the GlgC activity and slightly increases the apparent affinity of the enzyme for its substrates.

5.1.2
Allosteric Properties

An important point for both bacterial and plant ADPGlc PPases is that they are allosteric enzymes, and the allosteric function is important for the regulation of the synthesis of bacterial glycogen synthesis and plant starch. Over 50 ADPGlc PPases (mainly bacterial but also from plant tissues) have been studied with respect to their regulatory properties. In almost all cases, glycolytic intermediates activate ADPGlc synthesis while AMP, ADP, or Pi are inhibitors. Glycolytic intermediates in the cell may be considered as indicators of carbon excess and therefore, under conditions of limited growth with excess carbon in the media, accumulation of glycolytic intermediates would be positive effectors for the activation of ADPGlc synthesis. For most of the ADPGlc PPases studied, the activator glycolytic intermediate increases the enzyme's apparent affinity for the substrates, ATP and glucose-1-P and increasing concentrations of activator reverse the inhibition caused by the inhibitors, AMP, ADP or Pi.

The activator specificity of the bacterial and plant ADPGlc PPases can be listed into seven groups (Table 1) on the basis of their specificity of activation by the glycolytic intermediates (Preiss, 1984, 1989). The variation of activator specificity observed correlates with the nature of the dominant carbon assimilation pathway in the bacterium or plant tissue. For example, *E. coli* or *Salmonella typhimurium* obtain their energy mainly through glycolysis. The primary activator for their ADPGlc PPases is fructose 1,6-bisP, while 5′-adenylate is the major inhibitor; their ADPGlc synthetic activity is regulated by the [fructose 1,6-bisP]/[AMP] ratio. In plants and in cyanobacteria, the activator is 3-P-glycerate (3PGA) and inhibitor is Pi. 3PGA is the primary product of the carbon dioxide fixation in these photosynthetic organisms/tissues.

Evidence has accumulated to suggest that the observed *in-vitro* kinetic studies of allosteric activation and the inhibition also occur *in vivo*. A class of mutants of *E. coli* and of *S. typhimurium* LT-2 are affected in their ability to accumulate glycogen. This mutant class has ADPGlc PPases with altered regulatory properties. Mutants with ADPGlc PPases having higher affinity for the activator, fructose 1,6-bisphosphate and/or a lower affinity for the allosteric inhibitor, AMP, accumulate glycogen at a faster rate than the parent wild-type strain. Mutants with enzymes having a lower affinity for the activator accumulate glycogen at a slower rate than the parent strain.

Tab. 1 Classes of ADP-glucose pyrophosphorylase activator specificity

Activator(s)	*Microorganism(s)*	*Predominant carbon assimilation pathway*
3-P-Glycerate	Cyanobacteria, (higher plants)	Photosynthetic Calvin cycle
Pyruvate	*Rhodocyclus purpureus, Rhodospirillum* sp.	Reductive pyruvate cycle
Pyruvate, Fructose-6-P	Some anaerobic photosynthetic bacteria, *Agrobacterium, Arthrobacter*	Reductive pyruvate cycle Entner-Doudoroff pathway
Pyruvate, Fructose-6P	Some *Rhodopseudomonas* sp.	Reductive fructose-1,6-bis-P pyruvate cycle
		Entner-Doudoroff pathway Glycolysis
Fructose-6-P, Fructose-1,6-bis-P	*Rhodopseudomonas viridis*	Glycolysis
	Aeromonads	
	Mycobacterium smegmatis	
Fructose-1,6-bis-P	Enterics	Glycolysis
None	*Serratia* sp., *Enterobacter hafniae, Clostridium pasteuranium*	Glycolysis

The allosteric properties of the mutant ADPGlc PPases that have been studied, together with their maximum level of glycogen accumulation in the stationary phase, are summarized in Table 2. In *E. coli* there is a direct relationship between the affinity of the enzyme for the activator and the ability of the mutant to accumulate glycogen. If the apparent affinity for the activator, fructose-1,6-bis-P, is higher, then maximal glycogen accumulation by the mutant is higher. If the apparent affinity for activator is lower, as seen for mutant *E. coli* SG14 enzyme, then glycogen accumulation is lower in the mutant than the parent strain. The two *S. typhimurium* mutant strains have ADPGlc

Tab. 2 Allosteric kinetic constants of wild-type *E. coli* and *S. typhimurium* LT2 and allosteric mutant ADPGlc PPases, their mutation and their glycogen accumulation rates

Strain	*Glycogen accumulation (mg g^{-1} cells)*[a]	*Fructose-bis-P $A_{0.5}$ (mM)*[b]	*AMP $I_{0.5}$ (mM)*[c]	*Mutation*
E. coli B	20	0.068	0.075	–
Mutant SG5	35	0.022	0.17	P295S
Mutant 618	70	0.015	0.86	G336D
Mutant CL1136	74	0.005	0.68	R67C
Mutant SG14	8.4	0.82	0.50	A44T
S. typhimurium				
LT2	12	0.095	0.11	–
Mutant JP51	20	0.084	0.49	–
Mutant JP23	15	No activation	0.25	–

[a]The bacterial strains were grown in minimal media with 0.75% glucose; data are expressed as maximal mg anhydroglucose units per gram (wet wt.) of cells in stationary phase. [b]$A_{0.5}$ is the fructose 1,6-bis-P concentration causing 50% of maximal activation. [c]$I_{0.5}$ is the AMP concentration required for 50% inhibition.

PPases that are more affected in their affinity for the inhibitor. Both JP23 and JP51 enzymes have lesser affinity for the inhibitor, and these mutants accumulate higher amounts of glycogen than does the parent strain. Of interest is that JP23 ADPGlc PPase is fully active in the absence of activator. Addition of activator fructose-1,6-bis-P causes no further increase in activity.

The above studies, plus another (Dietzler et al., 1974) which showed a direct relationship between the fructose 1,6-bisP concentration in the *E. coli* cell and the rate of glycogen accumulation, clearly demonstrate that *in vivo*, fructose 1,6-bisP is an allosteric activator of ADPGlc PPase and a physiological activator of glycogen synthesis in *E. coli* and in *S. typhimurium*..

5.1.3 Substrate and Allosteric Effector Sites

Chemical modification and site-directed mutagenesis studies of the ADPGlc PPases have provided good evidence for the amino acid residues important for the binding of the activator, the inhibitor and the substrates. Pyridoxal-P has been used as an analogue for either the activator, fructose1,6-bisP or for the substrate, glucose-1-P. The ATP analogue used was the photoaffinity reagent 8-azido-ATP ($8N_3ATP$) that was shown to be a substrate for the *E. coli* enzyme. $8N_3AMP$ was an effective inhibitor analogue.

The amino-acid residue involved in binding the activator was Lys39, and the amino-acid involved in binding the adenine portion of the substrates (ADPGlc and ATP) was Tyr114. Tyr114 was also the major binding site for the adenine ring of the inhibitor, AMP. Lys195 is protected from reductive phosphopyridoxylation by the substrate, ADPGlc or by ATP and glucose 1-P, and thus was considered to be involved in substrate binding. In site-directed mutagenesis experiments it was shown to be the Glc-1-P binding site (Hill et al., 1991).

Similar experiments were also carried out to elucidate the activator, 3PGA, inhibitor, Pi, and substrate, Glc-1-P sites in *Anabaena* PC7120 and in the potato tuber and spinach leaf ADPGlc PPases (Preiss and Sivak, 1998).

Since several of the ADPGlc PPase genes have been cloned and their sequences determined, the identification of the amino-acid sequence about the modified residues enabled one to determine the location of the modified residue in the primary structure of the enzymes. The amino acid sequences of the substrate and allosteric sites identified via chemical modification in the bacteria, *E. coli* and *Anabaena*, as well as in the higher plant enzymes are shown in Table 3. The sequence for the Glc-1-P substrate site is highly conserved for both bacterial and higher plant ADPGlc PPases.

The ADPGlc PPase of *Agrobacterium tumefaciens* has been cloned and expressed in *E. coli* (Utarro and Ugalde, 1994; Utarro et al., 1998). This enzyme has as activators, fructose-6-P and pyruvate an activator specificity different from the *E. coli* ADPGlc PPase. It would be of interest to determine the activator site(s) of this enzyme and compare it with those of the *E. coli* enzyme.

5.1.4 Aspartate Residue 142 is important for Catalysis by *E. coli* ADPGlc PPase

Structural prediction of several bacterial and plant ADP-glucose pyrophosphorylases, as well as of other sugar-nucleotide pyrophosphorylases (Frueauf et al., 2001), was used for comparison with the three-dimensional (3-D) structures of two crystallized pyrophosphorylases (Brown et al., 1999; Blankenfeldt et al., 2000). This comparison led to the discovery of highly conserved residues throughout the super-family of sugar

Tab. 3 Amino acid sequences involved in binding of substrates and effectors of ADPGlc PPase. The amino acids shown by chemical modification and site-directed mutagenesis to be involved in binding are in bold letters

Organism/ tissue	*Activator sites*	*Sequence*
Anabaena	3-P-glycerate	
PC7120	Site 1	412SGIVVVL**K**NAV
	Site 2	375QRRAIID**K**NAR
Potato tuber	3-P-glycerate	
	Site 1	434SGIVTVI**K**DAL
	Site 2	397IKRAIID**K**NAR
E. coli	Fru-1,6-bis-P	38N**K**RAKPAV
	Inhibitor sites	
E. coli	5′-AMP	113W**Y**RGTADAV
Anabaena PC7120	Phosphate	291TRA**R**YLPPTK
	Substrate sites	
E. coli	ATP	113W**Y**RGTADAV
E. coli	Glucose-1-P	190IEFVE**K**PAN
Potato tuber	Glucose-1-P	193IEFAE**K**PQG

nucleotide pyrophosphorylases, despite the low overall homology. One of those residues, Asp142 in the ADP-glucose pyrophosphorylase from *E. coli*, was predicted to be near the substrate site. In order to elucidate the function that Asp142 might play in the *E. coli* ADP-glucose pyrophosphorylase, aspartate was replaced by alanine, asparagine, or glutamate using site-directed mutagenesis. Kinetic analysis in the direction of synthesis or pyrophosphorolysis of the purified mutants showed a decrease in specific activity of up to four orders of magnitude (Frueauf et al., 2001). Comparison of other kinetic parameters, including the apparent affinities for substrates and allosteric effectors, showed no significant changes, thereby excluding this residue from the specific role of ligand binding. Only the D142E mutant exhibited altered K_m values, but none was as pronounced as the decrease in specific activity. These results showed that residue Asp142 is important in the catalysis of the *E. coli* ADP-glucose pyrophosphorylase, and perhaps that the conserved equivalent Asp residue found in all sugar nucleotide pyrophosphorylases is an acid/base catalytic residue for the reaction.

5.1.5 Cloning of ADPGlc PPases with Altered Allosteric Properties from *E. coli* Mutants Affected in Glycogen Synthesis

As indicated above (see Table 2), a class of mutants of *E. coli* and of *S. typhimurium* with altered rates of glycogen accumulation was found to have ADPGlc PPases that were affected in their allosteric properties. To gain insight with respect to amino-acid residues or domains involved in maintaining allosteric function, these allosteric mutant ADPGlc PPases were cloned (Preiss, 1996).

The various amino-acid replacements in the allosteric mutants that have been cloned and analyzed are shown in Table 2. The mutations causing large changes in the allosteric properties of the enzyme occur throughout the ADPGlc PPase sequence. These changes of amino acids in the enzyme affect not only the affinity for the allosteric effectors (Table 2) but also the apparent affinities for the substrate ATP and Mg^{2+}. Therefore, many domains are affected by the mutations.

Site-directed mutagenesis of the ADPGlc PPase gene and analysis of various allosteric mutant genes have provided extensive information in the structure–function relationships of the substrate and catalytic sites. Knowledge of the 3-D structure of the enzyme would provide greater detail of the allosteric and catalytic mechanisms.

5.2 Bacterial Glycogen Synthase

The glycogen synthases in bacteria are specific for ADP-glucose. In systems where the enzyme has been studied, the subunit size is ~50 kDa, and the native form of the enzyme is either dimeric or homotetrameric.

An affinity analog of ADPGlc, adenosine diphosphopyridoxal (ADP-pyridoxal), was used to identify the ADPGlc binding site (Preiss, 1996). Incubation of the enzyme with the analog plus sodium borohydride led to inactivation of activity. About 1 mole of analog per mole of enzyme subunit led to 100% inactivation. A labeled peptide was isolated after tryptic hydrolysis, and the modified lysine residue was identified as Lys15. The sequence, Lys-X-Gly-Gly, where lysine is the amino acid modified by ADP-pyridoxal, has been found to be conserved in the mammalian glycogen synthase as well as in the plant starch synthases.

The structural gene for glycogen synthase, *glg*A, has been cloned from both *E. coli, S. typhimurium, A. tumefaciens* and *B. stearothermophilus* (Uttaro and Ugalde, 1994; Preiss, 1996; Takata et al., 1997) and their nucleotide sequences have been determined. The *E. coli* glycogen synthase sequence consists of 1431 bp specifying a protein of 477 amino acids with a molecular weight (MW) of 52,412 daltons. Replacement of other amino acids for lysine at residue 15 suggested that the lysine residue is mainly involved in binding the phosphate residue adjacent to the glycosidic linkage of the ADPGlc, and not in catalysis (Furukawa et al., 1990, 1993). The major effect on the kinetics of the mutants at residue 15 was the elevation of the K_m of ADPGlc, by about 30- to 50-fold when either Gln or Glu were the substituted amino acids. Substitution of Ala for Gly at residue 17 decreased the catalytic rate constant (K_{cat}) by about three orders of magnitude compared to the wild-type enzyme. Substitution of Ala for Gly18 only decreased the rate constant 3.2-fold. The K_m effect on the substrates, glycogen and ADPGlc, were minimal. It was postulated that the two glycyl residues in the conserved Lys-X-Gly-Gly sequence participated in the catalysis, either by assisting in maintaining the correct conformational change of the active site, or by stabilizing the transition state.

There is still appreciable catalytic activity of the Lys15Gln mutant. The ADP-pyridoxal modification was then repeated with the K15Q mutant, and in this instance about 30-times higher concentration was needed for effective inactivation of the enzyme by ADP-pyridoxal. The enzyme was maximally inhibited by about 80%, and tryptic analysis of the modified enzyme yielded one peptide containing the affinity analogue and having a sequence, Ala-Glu-Asn-modified Lys-Arg. The modified Lys was identified as Lys277 (Furukawa et al., 1994). Site-directed mutagenesis of Lys277 to form a Gln mutant was carried out, and the K_m for ADPGlc was essentially unchanged, though K_{cat} was decreased 140-fold. It was concluded that Lys277 was more involved in the catalytic reaction than in substrate binding.

5.3 Branching Enzyme

The structural genes of various branching enzymes have been cloned from many bacteria (Kiel et al., 1990, 1991, 1992, 1994; Rumbak et al., 1991; Takata et al., 1994, 1996; Preiss, 1996). The nucleotide and deduced amino acid sequences of the *E. coli glg*B gene consisted of 2181 bp specifying a protein of 727 amino acids, and with a MW of 84,231 daltons.

The relationship in amino acid sequences between that of branching enzymes and amylolytic enzymes in the α-amylase family of enzymes has been compared (Jesperson et al., 1993; Svensson, 1994). This family includes enzymes such as isoamylase, pullulanase and cyclodextrin glucanotransferase. There is a marked conservation in the amino acid sequence of the four putative catalytic regions of the amylolytic enzymes in the branching enzymes, whether they are of bacterial, plant or mammalian source. As shown in Table 4, four regions that putatively constitute the catalytic regions of the amylolytic enzymes are conserved in the plant branching isoenzymes and the glycogen branching enzymes of *E. coli*. Analysis of this high conservation in the α-amylase family has been pointed out (Romeo et al., 1988), and subsequently greatly expanded by Jesperson et al. (1993) and by Svensson (1994), not only with respect to sequence homology but also in the prediction the (β/α)8-barrel structural domains with a highly symmetrical fold of eight inner, parallel β-strands, surrounded by eight helices, in the various groups of enzymes in the family. The (β/α)8-barrel structural domain was determined from the crystal structure of some α-amylases and cyclodextrin glucanotransferases.

Conservation of the putative catalytic sites of the α-amylase family in the bacterial and higher-plant branching enzymes is anticipated as the branching enzyme catalyzes two reactions in synthesizing α-1,6-glucosidic linkages by cleavage of an α-1,4-glucosidic linkage in an 1,4-α-D-glucan to form a nonreducing end oligosaccharide chain that is transferred to a C-6 hydroxyl group of the same or other 1,4-α-D-glucan. The eight highly conserved amino acid residues of the α-amylase family are also functional in branching enzyme catalysis (Libessart and Preiss, 1998). Recently, the crystal structure of a truncated form of the *E. coli* branching enzyme has been elucidated (Abad et al., 2001). Analysis of the 3-D structure of the branching enzyme is in process to determine the precise functions and nature of its catalytic residues and mechanism. The C-terminal and N-terminal portions of branching enzymes from various bacteria, as well as higher plants, are dissimilar in sequence and in size. Studies by Kuriki et al. (1997) indicated that these amino acid sequence regions are functional with respect to branching enzyme specificity and to substrate preference (amylose or amylopectin), as well as in the size of oligosaccharide chain transferred and the extent of branching.

Tab. 4 Primary structures of branching enzymes (BE) showing identity with the four conserved regions of the α-amylase family

	Region 1	*Region 2*	*Region 3*	*Region 4*
Maize endosperm BE I	277**D**VVHS**H**	347GF**RFD**GVTS	402TVVA**E**DVS	470CIAYAES**HD**
Maize endosperm BE II	315**D**VVHS**H**	382GF**RFD**GVTS	437VTIG**E**DVS	501CVTYAES**HD**
E. coli glycogen BE	335**D**WVPG**H**	400AL**R**V**D**AVAS	453VTMA**E**EST	517NVFLPLN**HD**
B. subtilis α-amylase	100**D**AVIN**H**	171GF**RFD**AAKH	204FQYG**E**ILQ	261LVTWVES**HD**
P. amylodermosa isoamylase	291**D**VVYN**H**	369DGF**RFD**LAS	412RILR**E**FTV	499SINFIDV**HD**

The sequences have been derived from references indicated in the text. Two examples of enzymes from the amylase family are shown for comparison. Over 40 enzymes ranging from amylases, glucosidases, various α-1,6-debranching enzymes as well as four branching enzymes are compared (Jesperson et al., 1993; Svensson, 1994). The amino acid residues believed to be involved in catalysis are in bold letters.

Indeed, truncation of 113 amino acids of the N-terminal region of the *E. coli* branching enzyme causes it to transfer longer branch-chains than the wild-type enzyme (Binderup et al., 2000, 2002). This finding also indicated that the N terminal region was involved in specifying the size of chain transferred.

6 Genetic Regulation of Glycogen Synthesis in *E. coli*

The enzymes of the glycogen biosynthetic pathway are induced in the stationary phase. The rate of glycogen synthesis is inversely related to growth rate when growth is limited for certain nutrients, e.g., nitrogen. Consistent with this are the findings that the levels of glycogen biosynthetic enzymes in *E. coli* increase as cultures enter the stationary phase (Krebs and Preiss, 1975; Preiss, 1969, 1984, 1996; Preiss and Romeo, 1994).

In cells grown in an enriched medium containing 1% glucose, the specific activities of ADPGlc PPase and glycogen synthase increase 11- to 12-fold, while branching enzyme increases five-fold, as cultures enter the stationary phase (reviewed in Krebs and Preiss, 1975; Preiss and Romeo, 1994; Preiss, 1996). In a defined medium, the ADPGlc PPase and glycogen synthase activities are elevated in the exponential phase, with only about a two-fold increase in specific activity, and branching enzyme is fully induced in the exponential phase. The same phenomena are also seen with the glycogen biosynthetic enzyme levels in *S. typhimurium* (Steiner and Preiss, 1977).

The gene encoding the branching enzyme seems to be regulated differently from the genes for ADPGlc PPase and glycogen synthase.

The structural genes for glycogen biosynthesis are clustered in two adjacent operons, which also contain genes for glycogen catabolism. The structural genes for glycogen synthesis were shown to be located at approximately 75 min on the *E. coli* K-12 chromosome, and the gene order at this location was subsequently established by transduction to be *glg*P-*glg*A-*glg*C-*glg*X-*glg*B-*asd* (Latile-Damotte and Lares, 1977; Romeo et al., 1988). *glg*A encodes glycogen synthase, *glg*C, ADPGlc PPase and *glg*B encodes glycogen branching enzyme. *glg*X encodes an isoamylase and *glg*P encodes glycogen phosphorylase. The genes are close to *asd*, the structural gene for the enzyme, aspartate semialdehyde dehydrogenase (EC 1.2.1.11).

The arrangement of the glycogen biosynthetic genes encoded in plasmid pOP12 has also been determined by deletion-mapping experiments (Okita et al., 1981), and the nucleotide sequence of the entire *glg* gene cluster was determined (for review, see Preiss, 1996). The continuous nucleotide sequence of over 15 kb of this region of the genome has been determined, and it includes the sequences of the flanking genes *asd* and *glp*D (glycerol phosphate dehydrogenase; EC 1.1.99.5; 1.1.1.8) (Austin and Larson, 1991).

The genetic regulation of the glycogen biosynthetic enzymes has been extensively reviewed elsewhere (Romeo and Preiss, 1994; Preiss, 1996) and can be summarized here:

- Glycogen biosynthesis is under the direct control of at least three global regulatory systems, catabolite repression (stimulation *glg*ABC expression by cAMP); stringent response (stimulation *glg*ABC expression by pGpp), and repression of the *glg*ABC expression by the *csr*A gene product, a 61-amino acid polypeptide (Liu et al., 1995, 1997; Romeo and Gong,

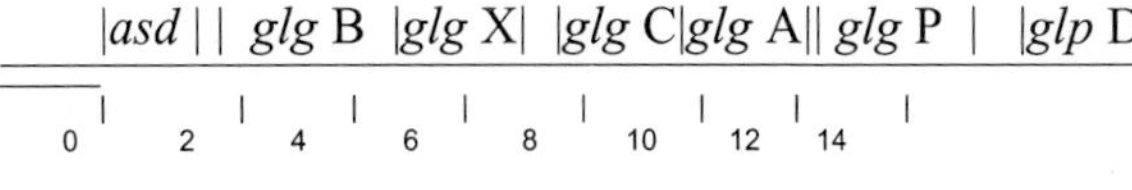

Fig. 1 Structure of the glycogen cluster in *E. coli*. The restriction map is constructed from known contiguous sequences. All of the genes are transcribed from left to right (counterclockwise on the genome except for *glp*D).

1993; Romeo et al., 1993; Yang et al., 1996).

- Although starvation for nitrogen as well as other nutrients leads to enhanced glycogen biosynthesis, the *glg*ABC genes are not part of the nitrogen starvation regulon.
- Transcription of *glg*CA expression appears to involve initiation events at least at four or more separate sites. All of the observed transcripts are in higher relative concentration in stationary phase versus exponential phase. Mutations leading to higher levels of the biosynthetic enzymes also result in enhanced accumulation of specific transcripts.
- The *csr*A gene maps to 58 min on the *E. coli* chromosome (Romeo and Gong, 1993). The *csr*A gene product negatively modulates post-transcriptionally by facilitating decay of *glg*CA mRNAs (Romeo et al., 1993). The *csr*A gene product is a specific mRNA-binding protein binding to *csr*B RNA (Liu et al., 1995, 1997; Gudapaty et al., 2001). Binding of the *csr*A gene product to *csr*B RNA inhibits the repression of the of the *glg*CA operon and the *glg*B genes. Since *csr*B RNA increases in the stationary phase, it is believed that repression of the glycogen biosynthetic genes is relieved by the binding of *csr*A gene product by the *csr*B RNA.

The genes known presently to be involved in glycogen metabolism in *E. coli* are listed in Table 5; the *E. coli* glycogen gene cluster is shown in Figure 1.

Tab. 5 Genes affecting glycogen metabolism in *E. coli*

	Gene product	Map site (min)	Comments
Regulatory gene			
cya	Adenylate cyclase	85	Regulates catabolite repression *crp*
	Cyclic AMP receptor protein	74	Regulates catabolite repression
*rel*A	(p)ppGpp synthase I	60	Mediates stringent response and
*spo*T	(p)ppGpp3′-pyrophosphosphohydrolase	60	response to carbon/energy
*csr*A	6.8 kDa polypeptide (CsrA)	58	Regulation of gluconeogenesis and *glg*CA and *glg*B transcription
csrB	A noncoding RNA molecule that binds to CsrA protein		Binds to CsrA protein
*kat*F	(rpoS) σs	59	Required for expression of *glg*S, not *glg*CA, pleiotropic
*glg*Q	Trans-acting factor (unidentified)	(?)	Transcriptional regulation of *glg*CA and B
*glg*R	Cis-acting site	75	Transcriptional regulation of *glg*CA
Structural gene			
I. Biosynthetic			
*glg*C	ADP-glucose pyrophosphorylase	75	Synthesis of glucosyl donor
*glg*A	Glycogen synthase	75	α-1,4-glucosyl transferase
*glg*B	Glycogen branching enzyme	75	α-1,6 branch synthesis
*glg*S	7.9 kDa polypeptide	67	Function?
II. Degradative			
*glg*P (*glg*Y)	Glycogen phosphorylase	75	Phosphorolysis
*amy*A	α-amylase	43	Hydrolysis
*glg*X	Glucan hydrolase/transferase	75	Isoamylase

7 References

Abad, M.C., Binderup, K., Preiss, J., Geiger, J.H. (2001) Crystallization and preliminary X-ray diffraction studies of *Escherichia coli* branching enzyme, *Acta Crystallogr. D* (in press).

Austin, D., Larson, T.J. (1991) Nucleotide sequence of the glpD gene encoding aerobic *sn* glycerol 3-phosphate dehydrogenase of *Escherichia coli*, *J. Bacteriol.* **173**, 101–107.

Binderup, K., Mikkelsen. R., Preiss, J. (2000) Limited proteolysis of branching enzyme from *Escherichia coli*, *Arch. Biochem. Biophys.* **377**, 366–371.

Binderup, K., Mikkelsen, R., Preiss, J. (2002) Truncation of the amino terminus of branching enzyme changes its chain transfer pattern, *Arch. Biochem. Biophys.* (in press).

Dietzler, D. N., Leckie, M. P., Lais, C. J., Magnani, J. L. (1974) Evidence for the regulation of bacterial glycogen biosynthesis *in vivo*, *Arch. Biochem. Biophys.* **162**, 602–606.

Blankenfeldt, W., Asuncion, M., Lam, J. S., Naismith, J. H. (2000) The structural basis of the catalytic mechanism and regulation of glucose-1-phosphate thymidylyltransferase (RmlA), *EMBO J.* **19**, 6652–6663.

Brown, K., Pompeo, F., Dixon, S., Mengin-Lecreulx, D., Cambillau, C., Bourne, Y. (1999) Crystal structure of the bifunctional *N*-acetylglucosamine 1-phosphate uridyltransferase from *Escherichia coli*: a paradigm for the related pyrophosphorylase superfamily, *EMBO J.* **18**, 4096–4107.

Frueauf, J., Ballicora, M. A., Preiss, J. (2001) Aspartate residue 142 is important for catalysis by ADP-glucose pyrophosphorylase from *Escherichia coli*, *J. Biol. Chem.* (in press).

Furukawa, H., Tagaya, M., Inouye, M., Preiss, J., Fukui, T. (1990) Identification of lysine 15 at the active site in *Escherichia coli* glycogen synthase, *J. Biol. Chem.* **265**, 2086–2090.

Furukawa, H., Tagaya, M., Tanazawa, K., Fukui, T. (1993) Role of the conserved Lys-X Gly-Gly sequence at the ADPglucose binding site in *Escherichia coli* glycogen synthase, *J. Biol. Chem.* **268**, 23837–23842.

Furukawa, H., Tagaya, M., Tanazawa, K., Fukui, T. (1994) Identification of Lys277 at the active site of *Escherichia coli* glycogen synthase. Application of affinity labeling combined with site-directed mutagenesis, *J. Biol. Chem.* **269**, 868–871.

Greenberg, E., Preiss, J. (1964) The occurrence of adenosine diphosphate glucose:glycogen transglucosylase in bacteria, *J. Biol. Chem.* **239**, 4314–4315.

Gudapaty, S., Kazushi, S., Wang, X., Babitzke, P., Romeo, T. (2001) Regulatory interactions of Csr components: the RNA binding protein CsrA activates *csrB* transcription in *Escherichia coli*, *J. Bacteriol.* **183**, 6017–6027.

Hill, M. A., Kaufmann, K., Otero, J., Preiss, J. (1991) Biosynthesis of bacterial glycogen: mutagenesis of a catalytic site residue of ADPglucose pyrophosphorylase from *Escherichia coli*, *J. Biol. Chem.* **266**, 12455–12460.

Jesperson, H. M., Macgregor, E. A., Henrissat, B., Sierks, M. R., Svensson, B. (1993) Starch- and glycogen-debranching and branching enzymes; prediction of structural features of the catalytic (β/α)8-barrel domain and evolutionary relationship to other amylolytic enzymes, *J. Protein Chem.* **12**, 791–805.

Kiel, J. A. K. W., Boels, J. M., Beldman, G., Venema, G. (1990) Nucleotide sequence of the *Synechococcus* sp. PCC7942 branching enzyme gene (*glg* B): expression in *B. subtilis*, *Gene* **89**, 77–84.

Kiel, J. A. K. W., Boels, J. M., Beldman, G., Venema, G. (1991) Molecular cloning and nucleotide sequence of the branching enzyme gene (*glg* B) from *Bacillus stearothermophilus* and expression in *Escherichia coli* and *Bacillus subtilis*, *Mol. Gen. Genet.* **230**, 136–144.

Kiel, J. A. K. W., Boels, J. M., Beldman, G., Venema, G. (1992) The *glg* B gene from the thermophile *Bacillus caldolyticus* encodes a thermolabile branching enzyme, *DNA Seq.* **3**, 221–232.

Kiel, J. A. K. W., Boels, J. M., Beldman, G., Venema, G. (1994) Glycogen in *Bacillus subtilis*: molecular characterization of an operon encoding enzymes involved in glycogen synthesis and degradation, *Mol. Microbiol.* **11**, 203–218.

Krebs, E. G., Preiss, J. (1975) Regulatory mechanisms in glycogen metabolism, in: *Biochemistry of Carbohydrates* (Whelan, W.J., Ed.), MTP International Review of Science, Baltimore, MD: University Park Press, Vol. 5, Chapter 7, 337–389.

Kuriki, T., Stewart, D. C., Preiss, J. (1997) Construction of chimeric enzymes out of maize endosperm branching enzymes I and II: activity and properties, *J. Biol. Chem.* **272**, 28999–29004.

Latil-Damotte, M., Lares, C. (1977) Relative order of glg mutations affecting glycogen biosynthesis in *Escherichia coli* K12, *Mol. Gen. Genet.* **150**, 325–329.

Libessart, N., Preiss, J. (1998) Arginine residue 384 at the catalytic center is important for branching enzyme II from maize endosperm, *Arch. Biochem. Biophys.* **360**, 135–141.

Liu, M. Y., Yang, H., Romeo, T. (1995) The product of the pleiotropic *Escherichia coli* gene *csrA* modulates glycogen biosynthesis via effects on mRNA stability, *J. Bacteriol.* **177**, 2663–2672.

Liu, M. Y., Gui, G., Wei, B., Preston, J. F., III, Qakford, L., Yüksel, U., Giedroc, D. P., Romeo, T. (1997) The RNA molecule Csr B binds to the global regulatory protein CsrA and antagonizes its activity in *Escherichia coli*, *J. Biol. Chem.* **272**, 17502–17510.

Okita, T. W., Rodriguez, R. L., Preiss, J. (1981) Cloning of the glycogen biosynthetic enzyme structural genes of *Escherichia coli*, *J. Biol. Chem.* **256**, 6944–6952.

Preiss, J. (1969) The regulation of the biosynthesis of α-l,4 glucans in bacteria and plants, in: *Current Topics in Cellular Regulation* (Horecker, B. L., Stadtman, E. R., Eds.), New York: Academic Press, Vol. 1, 125–160.

Preiss, J. (1973) ADP-glucose pyrophosphorylase, in: *The Enzymes* (Boyer, P. D., Ed.), 3rd edition. New York: Academic Press, Vol. 8, 73–119.

Preiss, J. (1978) Regulation of ADP-glucose pyrophosphorylase, in: Advances in Enzymology and Related Areas of Molecular Biology (Meister, A., Ed.), New York: John Wiley & Sons, Vol. 46, 317–381.

Preiss, J. (1984) Bacterial glycogen synthesis and its regulation, *Annu. Rev. Microbiol.* **38**, 419–458.

Preiss, J. (1989) Chemistry and metabolism of intracellular reserves, in: *Bacteria in Nature. A treatise on the interactions of bacteria and their habitats* (Leadbetter, E. R., Poindexter, J. S., Eds.), New York: Plenum Press, Vol. 3, 189–258.

Preiss, J. (1996) Regulation of glycogen synthesis, in: *Escherichia coli and Salmonella typhimurium: Cellular and Molecular Biology*, 2nd edition (Neidhardt, F. C., Ed.), Washington, DC: American Society for Microbiology, Vol. 1, 1015–1024.

Preiss, J. (2000) Glycogen biosynthesis, in: *Encyclopedia of Microbiology*, 2nd edition (Lederburg, J., Ed.), San Diego, CA: Academic Press, Vol. 2, 541–555.

Preiss, J., Romeo, T. (1989) Physiology, biochemistry and genetics of bacterial glycogen synthesis, in: *Advances in Bacterial Physiology*, Vol. 30, 184–238.

Preiss, J., Romeo, T. (1994) Molecular biology and regulation of bacterial glycogen biosynthesis, in: *Progress in Nucleic Acids Research and Molecular Biology* (Moldave, K., Cohn, W. E., Eds.), Vol. 47, 299–329.

Preiss, J., Sivak, M. N. (1998) Biochemistry, molecular biology and regulation of starch synthesis, in: *Genetic Engineering, Principles and Methods* (Setlow, J. K., Ed.), New York: Plenum Press, Vol. 20, 177–223.

Preiss, J., Walsh, D.A. (1981) The comparative biochemistry of glycogen and starch metabolism and their regulation, in: *Biology of Complex Carbohydrates* (Ginsburg, V., Ed.), New York: J. Wiley & Sons, Vol. I, Chapter 5, 199–314.

Preiss, J., Yung, S. G., Baecker, P. A. (1983) Regulation of bacterial glycogen synthesis, *J. Mol. Cell. Biochem.* **57**, 61–80.

Romeo, T., Gong. M. (1993) Genetic and physical mapping of the regulatory gene *csrA* on the *Escherichia coli* K-12 chromosome. *J. Bacteriol.* **175**, 5740–5741.

Romeo, T., Preiss, J. (1989) Genetic regulation of glycogen biosynthesis in *Escherichia coli*: *in vitro* effects of cyclic AMP and guanosine 5′-diphosphate and analysis of *in vivo* transcripts, *J. Bacteriol.* **171**, 2773–2782.

Romeo, T., Kumar, A., Preiss, J. (1988) Analysis of the *Escherichia coli* glycogen gene cluster suggests that catabolic enzymes are encoded among the biosynthetic genes, *Gene* **70**, 363–376.

Romeo, T., Gong, M., Liu, M. Y., Brun-Zinkernagel, A.-M. (1993) Identification and molecular characterization of *csrA*, a pleiotropic gene from *Escherichia coli* that affects glycogen biosynthesis, gluconeogenesis, cell size, and surface properties, *J. Bacteriol.* **175**, 4744–4755.

Rumbak, E., Rawlings, D. E., Lindsay, G. G., Woods, D. R. (1991) Characterization of the *Butyrivibrio fibrisolvens* glgB gene which encodes a glycogen-branching enzyme with starch clearing activity, *J. Bacteriol.* **173**, 6732–6741.

Steiner, K. E., Preiss, J. (l977) Biosynthesis of bacterial glycogen. Genetic and allosteric regulation of glycogen biosynthesis in *Salmonella typhimurium LT.2*, *J. Bacteriol.* **129**, 246–253.

Svensson, B. (1994) Protein engineering in the α-amylase family: catalytic mechanism, substrate specificity and stability, *Plant Mol. Biol.* **25**, 141–157.

Takata, H., Takaha, T., Kuriki, T., Okada, S., Takagi, M., Imanaka, T. (1994) Properties and active center of the thermostable branching enzyme from *Bacillus stearothermophilus*, *Appl. Environ. Microbiol.* **60**, 3096–3104.

Takata, H., Takaha, T., Okada, S., Hizukuri, S. Takagi, M., Imanaka, T. (1996) Structure of the cyclic glucan produced from amylopectin by *Bacillus stearothermophilus* branching enzyme, *Carbohydr. Res.* **295**, 91–101.

Takata, H., Takaha, T., Shigetaka, O., Takagi, M., Imanaka, T. (1997) Characterization of a gene cluster for glycogen biosynthesis and a heterotetrameric ADP-glucose pyrophosphorylase from *Bacillus stearothermophilus*, *J. Bacteriol.* **179**, 4689–4698.

Uttaro, A. D., Ugalde, R. A. (1994) A chromosomal cluster of genes encoding ADP-glucose synthetase, glycogen synthase and phosphoglucomutase in *Agrobacterium tumefaciens*, *Gene* **150**, 117–122.

Uttaro, A. D., Ugalde, R. A., Preiss, J., Iglesias, A. A. (1998) Cloning and expression of the *glg* C gene from *Agrobacterium tumefaciens*. Purification and characterization of the ADPglucose synthetase, *Arch. Biochem. Biophys.* **357**, 13–21.

Wilkinson, J. F. (1959) The problem of energy-storage compounds in bacteria, *Exp. Cell Res.* Suppl. 7, 111–130.

Yang, H., Liu, M. Y., Romeo, T. (1996) Coordinate genetic regulation of glycogen catabolism and biosynthesis in *Escherichia coli* via the CsrA gene product, *J. Bacteriol.* **178**, 1012–1017.

3
Bacterial Cellulose

Prof. Dr. Eng. Stanislaw Bielecki[1], **Dr. Eng. Alina Krystynowicz**[2], **Prof. Dr. Marianna Turkiewicz**[3], **Dr. Eng. Halina Kalinowska**[4]

[1] Institute of Technical Biochemistry, Technical University of Lódz, Stefanowskiego 4/10, 90-924 Lódz, Poland; Tel: +48-4263-13442; Fax: +48-4263-402; E-mail: biochem@ck-sg.p.lodz.pl

[2] Institute of Technical Biochemistry, Technical University of Lódz, Stefanowskiego 4/10, 90-924 Lódz, Poland; Tel: +48-4263-13442; Fax: +48-4263-402; E-mail: biochem@ck-sg.p.lodz.pl

[3] Institute of Technical Biochemistry, Technical University of Lódz, Stefanowskiego 4/10, 90-924 Lódz, Poland; Tel: +48-4263-13442; Fax: +48-4263-402; E-mail: biochem@ck-sg.p.lodz.pl

[4] Institute of Technical Biochemistry, Technical University of Lódz, Stefanowskiego 4/10, 90-924 Lódz, Poland; Tel: +48-4263-13442; Fax: +48-4263-402; E-mail: biochem@ck-sg.p.lodz.pl

A-BC	bacterial cellulose from agitated culture
ATP	adenosine triphosphate
BC	bacterial cellulose
CBH	cellobiohydrolase
c-di-GMP	cyclic diguanosine monophosphate
Cel^-	cellulose-negative mutant
Cel6A, Cel7A	cellobiohydrolases belonging to 6A and 7A families, respectively
CM	carboxymethyl-
CS	cellulose synthase
CSL	corn steep liquor
D	aspartic acid

DMSO	dimethyl sulfoxide
DP	degree of polymerization
E	glutamic acid
FBP	fructose-1,6-biphosphate phosphatase
FK	fructokinase
Fru-bi-P	fructose-1,6-biphosphate
Fru-6-P	fructose-6-phosphate
G	guanine
GK	glucokinase
Glc	glucose
Glc-6(1)-P	glucose-6(1)-phosphate
G6PDH	glucose-6-phosphate dehydrogenase
H-S medium	Hestrin and Schramm medium (1954)
IS	insertion sequence
Lip	lipid
LP	UDPGPT: lipid pyrophosphate: UDPGlc- phosphotransferase
LPP	lipid pyrophosphate phosphohydrolase
Man	mannose
NMR	nuclear magnetic resonance
PC	plant cellulose
PDEA	phosphodiesterase A
PDEB	phosphodiesterase B
Pel^-	pellicle non-forming
1PFK	fructose-1-phosphate kinase
PGA	phosphogluconic acid
PGI	phosphoglucoisomerase
PMG	phosphoglucomutase
PTS	system of phosphotransferases
Q	glutamine
R	arginine
Rha	rhamnose
Rib	D-ribose
S	serine
S-BC	bacterial cellulose from static culture
TC	terminal complexe
U	uridine
UDP	uridine diphosphate
UDPGlc	uridine diphosphoglucose
UGP	pyrophosphorylase uridine diphosphoglucose
UMP	uridine monophosphate
v.v.m.	volume per volume per minute
W	tryptophan

1
Introduction

Cellulose is the most abundant biopolymer on earth, recognized as the major component of plant biomass, but also a representative of microbial extracellular polymers. Bacterial cellulose (BC) belongs to specific products of primary metabolism and is mainly a protective coating, whereas plant cellulose (PC) plays a structural role.

Cellulose is synthesized by bacteria belonging to the genera *Acetobacter, Rhizobium, Agrobacterium*, and *Sarcina* (Jonas and Farah, 1998). Its most efficient producers are Gram-negative, acetic acid bacteria *Acetobacter xylinum* (reclassified as *Gluconacetobacter xylinus*, Yamada et al., 1997; Yamada, 2000), which have been applied as model microorganisms for basic and applied studies on cellulose (Cannon and Anderson, 1991). Investigations have been focused on the mechanism of biopolymer synthesis, as well as on its structure and properties, which determine practical use (Legge, 1990; Ross et al., 1991). One of the most important features of BC is its chemical purity, which distinguishes this cellulose from that from plants, usually associated with hemicelluloses and lignin, removal of which is inherently difficult.

Because of the unique properties, resulting from the ultrafine reticulated structure, BC has found a multitude of applications in paper, textile, and food industries, and as a biomaterial in cosmetics and medicine (Ring et al., 1986). Wider application of this polysaccharide is obviously dependent on the scale of production and its cost. Therefore, basic studies run together with intensive research on strain improvement and production process development.

2
Historical Outline

Although synthesis of an extracellular gelatinous mat by *A. xylinum* was reported for the first time in 1886 by A. J. Brown, BC attracted more attention in the second half of the 20th century. Intensive studies on BC synthesis, using *A. xylinum* as a model bacterium, were started by Hestrin et al. (1947, 1954), who proved that resting and lyophilized *Acetobacter* cells synthesized cellulose in the presence of glucose and oxygen. Next, Colvin (1957) detected cellulose synthesis in samples containing cell-free extract of *A. xylinum*, glucose, and ATP. Further milestones in studies on BC synthesis, presented in this review, contributed to the elucidation of mechanisms governing not only the biogenesis of the bacterial polymer, but also that of plants, thus leading to the understanding of one of the most important processes in nature. The true historical outline is presented throughout all the paragraphs below, including the references.

3
Structure of BC

Cellulose is an unbranched polymer of β-1,4-linked glucopyranose residues. Extensive research on BC revealed that it is chemically identical to PC, but its macromolecular structure and properties differ from the latter (Figure 1). Nascent chains of BC aggregate to form subfibrils, which have a width of approximately 1.5 nm and belong to the thinnest naturally occurring fibers, comparable only to subelemental fibers of cellulose detected in the cambium of some plants and in quinee mucous (Kudlicka, 1989). BC subfibrils are crystallized into microfibrils (Jonas and Farah, 1998), these into bundles, and the latter into ribbons (Yamanaka et al.,

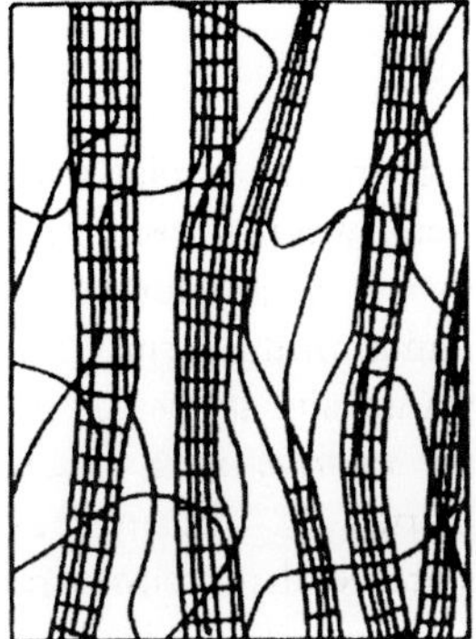
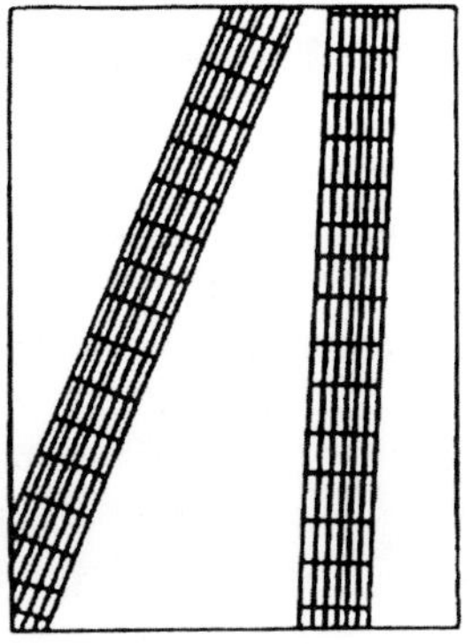

Fig. 1 Schematic model of BC microfibrils (right) drawn in comparison with the 'fringed micelles'; of PC fibrils (Iguchi et al., 2000; with kind permission).

2000). Dimensions of the ribbons are 3–4 (thickness) × 70–80 nm (width), according to Zaar (1977), 3.2 × 133 nm, according to Brown et al. (1976), or 4.1 × 117 nm, according to Yamanaka et al. (2000), whereas the width of cellulose fibers produced by pulping of birch or pine wood is two orders of magnitude larger ($1.4-4.0 \times 10^{-2}$ and $3.0-7.5 \times 10^{-2}$ mm, respectively). The ultrafine ribbons of microbial cellulose, the length of which ranges from 1 to 9 μm, form a dense reticulated structure (Figure 2), stabilized by extensive hydrogen bonding (Figure 3). BC is also distinguished from its plant counterpart by a high crystallinity index (above 60%) and different degree of polymerization (DP), usually between 2000 and 6000 (Jonas and Farah, 1998), but in some cases reaching even 16,000 or 20,000 (Watanabe et al., 1998b), whereas the average DP of plant polymer varies from 13,000 to 14,000 (Teeri, 1997).

Macroscopic morphology of BC strictly depends on culture conditions (Watanabe et al., 1998a; Yamanaka et al., 2000). In static conditions (Figure 4), bacteria accumulate cellulose mats (S-BC) on the surface of nutrient broth, at the oxygen-rich air–liquid interface. The subfibrils of cellulose are continuously extruded from linearly ordered pores at the surface of the bacterial cell, crystallized into microfibrils, and forced deeper into the growth medium. Therefore, the leather-like pellicle, supporting the population of *A. xylinum* cells, consists of overlapping and intertwisted cellulose rib-

a)

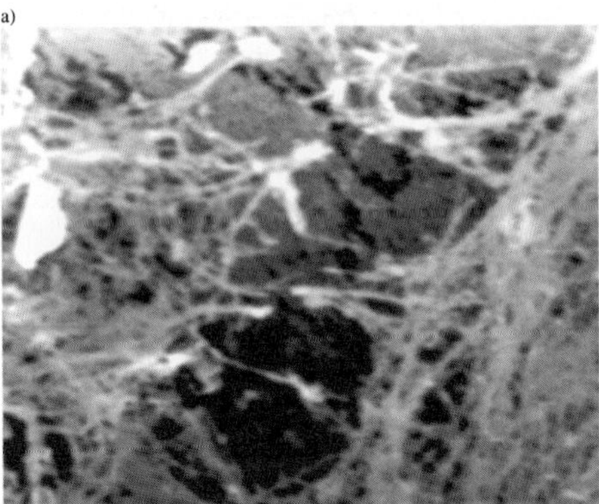

b)

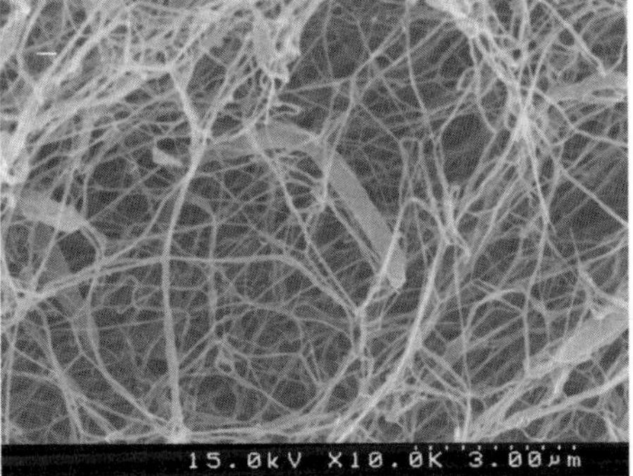

Fig. 2 Scanning electron microscopy images of BC membrane from static culture of *A. xylinum* (a) and bacterial cell with attached cellulose ribbons (b).

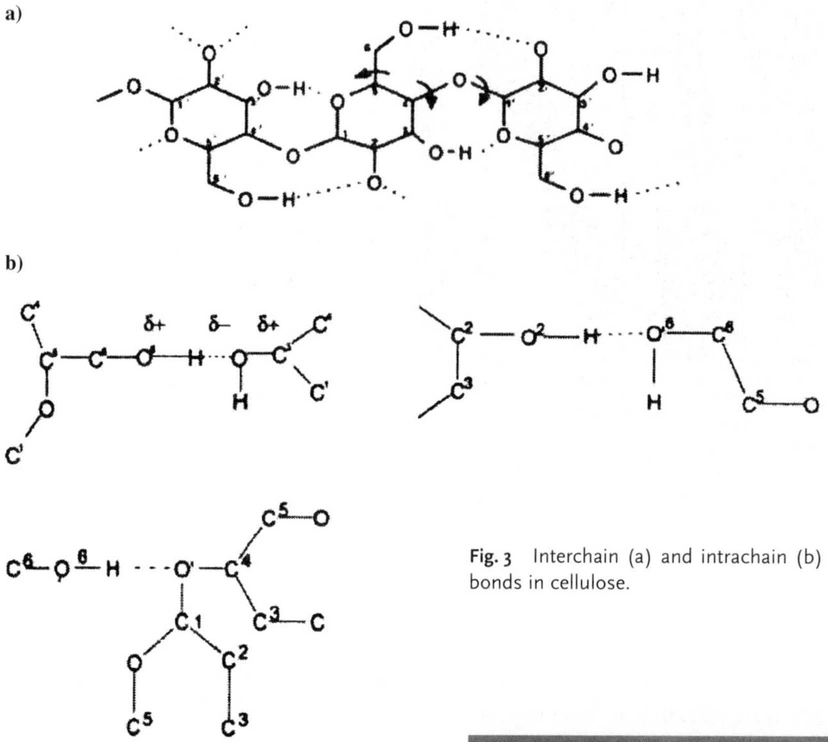

Fig. 3 Interchain (a) and intrachain (b) hydrogen bonds in cellulose.

bons, forming parallel but disorganized planes (Jonas and Farah, 1998). The adjacent S-BC strands branch and interconnect less frequently than these in BC produced in agitated culture (A-BC), in a form of irregular granules, stellate and fibrous strands, well-dispersed in culture broth (Figure 5) (Vandamme et al., 1998). The strands of reticulated A-BC interconnect to form a grid-like pattern, and have both roughly perpendicular and roughly parallel orientations (Watanabe et al., 1998a).

Differences in three-dimensional structure of A-BC and S-BC are noticeable in their scanning electron micrographs. The S-BC fibrils are more extended and piled above one another in a criss-crossing manner. Strands of A-BC are entangled and curved (Johnson and Neogi, 1989). Besides, they

Fig. 4 BC pellicle formed in static culture.

Fig. 5 BC pellets formed in agitated culture.

have a larger cross-sectional width (0.1–0.2 μm) than S-BC fibrils (usually 0.05–0.10 μm). Morphological differences between S-BC and A-BC contribute to varying degrees of crystallinity, different crystallite size and I_α cellulose content.

Two common crystalline forms of cellulose, designated as I and II, are distinguishable by X-ray, nuclear magnetic resonance (NMR), Raman spectroscopy, and infrared analysis (Johnson and Neogi, 1989). It is known that in the metastable cellulose I, which is synthesized by the majority of plants and also by *A. xylinum* in static culture, parallel β-1,4-glucan chains are arranged uniaxially, whereas β-1,4-glucan chains of cellulose II are arranged in a random manner. They are mostly antiparallel and linked with a larger number of hydrogen bonds that results in higher thermodynamic stability of the cellulose II.

A-BC has a lower crystallinity index and a smaller crystallite size than S-BC (Watanabe et al., 1998a). It was also observed that a significant portion of cellulose II occurred in BC synthesized in agitated culture. In nature, cellulose II is synthesized by only a few organisms (some algae, molds, and bacteria, such as *Sarcina ventriculi*) (Jonas and Farah, 1998); the industrial production of this kind of cellulose is based on chemical conversion of PC.

Using CP/MAS ^{13}C-NMR it is possible to reveal the presence of cellulose I_α and I_β – two distinct forms of cellulose (Watanabe et al., 1998a). These forms occur in algae-, bacteria-, and plant-derived cellulose. The latter one contains less I_α cellulose than BC (Johnson and Neogi, 1989). The irreversible crystal transformation from cellulose I_α to I_β shifts the X-ray and CP/MAS ^{13}C-NMR spectra because of the difference in the unit cell. S-BC contains more cellulose I_α than A-BC. It was reported that the difference in cellulose I_α content between A-BC and S-BC exceeded that in crystallinity index (Watanabe et al., 1998a), and the mass fraction of cellulose I_α was closely related to the crystallite size.

4 Chemical Analysis and Detection

For the detection of either crystalline or amorphous cellulose, several direct dyes, specific for the linear β-1,4-glucan, are used (Mondal and Kai, 2000). All of them are fluorescent brightening agents and form dye–cellulose complexes, stabilized by van der Waals and/or hydrogen bonding. One of these dyes is the fluorescent brightener Calcofluor. The direct dyes do not only enable visualization of cellulose chains, but also have been intensively applied for studies on nascent cellulose chains association and crystallization (see Section 7.3.2).

The weight-average DP of cellulose and the DP distribution are determined by high-performance gel-permeation chromatography (Watanabe et al., 1998a) of nitrated cellulose samples.

The differences between the reticulated structure of microbial cellulose, produced under agitated culture conditions, and the disorganized layered structure of cellulose pellicle, formed in static culture, are noticeable in scanning electron microscopy images (Johnson and Neogi, 1989).

To distinguish the parallel chain crystalline lattice of cellulose I from the antiparallel one of cellulose II, X-ray diffraction, Raman spectroscopy, infrared analysis, and NMR are applied (Johnson and Neogi, 1989). The crystallinity index and crystallite size are calculated based on X-ray diffraction measurements (Watanabe et al., 1998a).

Two distinct forms of cellulose I, i.e. cellulose I_α and I_β, are not distinguishable by X-ray diffraction, and therefore CP/MAS ^{13}C-NMR analysis, carried out on freeze-dried cellulose samples, has to be performed to determine their mass fractions (Watanabe et al., 1998a).

The physicochemical properties of cellulose such as water holding capacity, viscosity of disintegrated cellulose suspension, and the Young's modulus of dried sheets are determined using conventional methods (Watanabe et al., 1998a, Iguchi et al., 2000).

5 Occurrence

BC is synthesized by several bacterial genera, of which *Acetobacter* strains are best known. An overview of BC producers is presented in Table 1 (Jonas and Farah, 1998). The polymer structure depends on the organism, although the pathway of biosynthesis and mechanism of its regulation are probably common for the majority of BC-producing bacteria (Ross et al., 1991; Jonas and Farah, 1998).

Tab. 1 BC producers (Jonas and Farah, 1998, modified)

Genus	*Cellulose structure*
Acetobacter	extracellular pellicle composed of ribbons
Achromobacter	fibrils
Aerobacter	fibrils
Agrobacterium	short fibrils
Alcaligenes	fibrils
Pseudomonas	no distinct fibrils
Rhizobium	short fibrils
Sarcina	amorphous cellulose
Zoogloea	not well defined

A. xylinum (synonyms *A. aceti* ssp. xylinum, *A. xylinus*), which is the most efficient producer of cellulose, has been recently reclassified and included within the novel genus *Gluconacetobacter*, as *G. xylinus* (Yamada et al., 1998, 2000) together with some other species (*G. hansenii*, *G. europaeus*, *G. oboediens*, and *G. intermedius*).

6 Physiological Function

In natural habitats, the majority of bacteria synthesize extracellular polysaccharides, which form envelopes around the cells (Costeron, 1999). BC is an example of such a substance. Cells of cellulose-producing bacteria are entrapped in the polymer network, frequently supporting the population at the liquid–air interface (Wiliams and Cannon, 1989). Therefore, BC-forming strains can inhabit sewage (Jonas and Farah, 1998). The polymer matrix takes part in adhesion of the cells onto any accessible surface and facilitates nutrient supply, since their concentration in the polymer lattice is markedly enhanced due to its adsorptive properties, in comparison to the surrounding aqueous environment (Jonas and Farah, 1998; Costeron, 1999). Some authors suppose that cellulose synthesized by *A. xylinum* also plays a storage role and can be utilized by the starving microorganisms. Its decomposition would be then catalyzed by exo- and endoglucanases, the co-presence of which was detected in the culture broth of some cellulose-producing *A. xylinum* strains (Okamoto et al., 1994).

Because of the viscosity and hydrophilic properties of the cellulose layer, the resistance of producing bacterial cells against unfavorable changes (a decrease in water content, variations in pH, appearance of toxic substances, pathogenic organisms, etc.) in an habitat is increased, and they can further grow and develop inside the envelope. It was also found that cellulose placed over bacterial cells protects them from ultraviolet radiation. As much as 23% of the acetic acid bacteria cells covered with BC survived a 1 h treatment with ultraviolet irradiation. Removal of the protective polysaccharide brought about a drastic decrease in their viability (3% only) (Ross et al., 1991).

7 Biosynthesis of BC

Synthesis of BC is a precisely and specifically regulated multi-step process, involving a large number of both individual enzymes and complexes of catalytic and regulatory proteins, whose supramolecular structure has not yet been well defined. The process includes the synthesis of uridine diphosphoglucose (UDPGlc), which is the cellulose precursor, followed by glucose polymerization into the β-1,4-glucan chain, and nascent chain association into characteristic ribbon-like structure, formed by hundreds or even thousands of individual cellulose chains. Pathways and mechanisms of UDPGlc synthesis are relatively well known, whereas molecular mechanisms of glucose polymerization into long and unbranched chains, their extrusion outside the cell, and self-assembly into fibrils require further elucidation.

Moreover, studies on BC synthesis may contribute to better understanding of PC biogenesis.

7.1 Synthesis of the Cellulose Precursor

Cellulose synthesized by *A. xylinum* is a final product of carbon metabolism, which depending on the physiological state of the cell involves either the pentose phosphate cycle or the Krebs cycle, coupled with gluconeogenesis (Figure 6) (Ross et al., 1991; Tonouchi et al., 1996). Glycolysis does not operate in acetic acid bacteria since they do not synthesize the crucial enzyme of this pathway – phosphofructose kinase (EC 2.7.1.56) (Ross et al., 1991). In *A. xylinum*, cellulose synthesis is tightly associated with catabolic processes of oxidation and consumes as much as 10% of energy derived from catabolic reactions (Weinhouse, 1977). BC production does not interfere with other anabolic processes, including protein synthesis (Ross et al., 1991).

A. xylinum converts various carbon compounds, such as hexoses, glycerol, dihydroxyacetone, pyruvate, and dicarboxylic acids, into cellulose, usually with about 50% efficiency. The latter compounds enter the Krebs cycle and due to oxalacetate decarboxylation to pyruvate undergo conversion to hexoses via gluconeogenesis, similarly to glycerol, dihydroxyacetone, and intermediates of the pentose phosphate cycle (Figure 6).

The direct cellulose precursor is UDPGlc, which is a product of a conventional pathway, common of many organisms, including plants, and involving glucose phosphoryla-

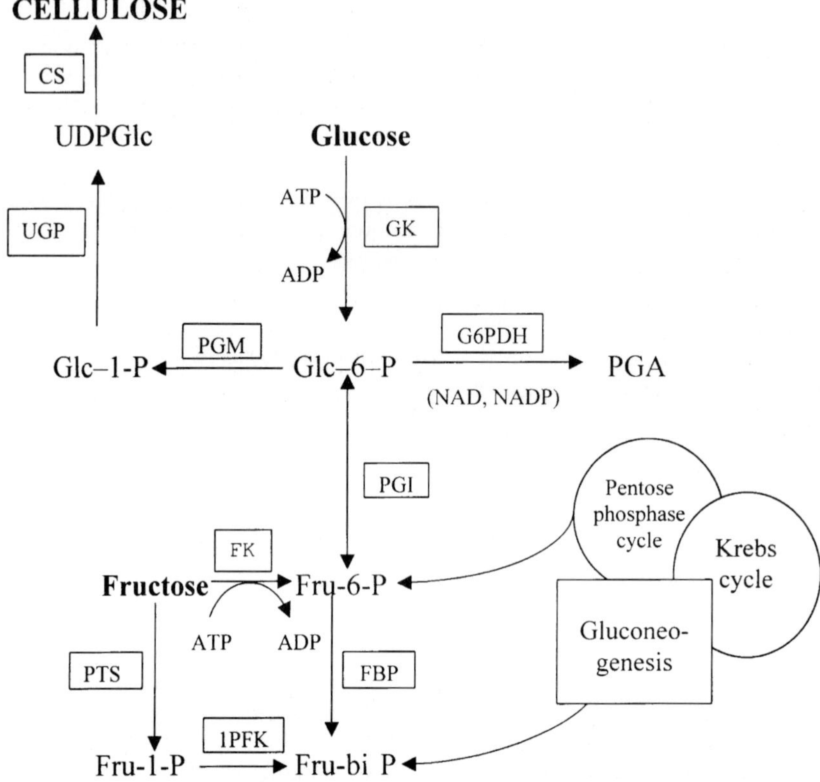

Fig. 6 Pathways of carbon metabolism in *A. xylinum*. CS, cellulose synthase (EC 2.4.2.12); FBP, fructose-1,6-biphosphate phosphatase (EC 3.1.3.11); FK, glucokinase (EC 2.7.1.2); G6PDH, glucose-6-phosphate dehydrogenase (EC 1.1.1.49); 1PFK, fructose-1-phosphate kinase (EC 2.7.1.56); PGI, phosphoglucoisomerase; PMG, phosphoglucomutase (EC 5.3.1.9); PTS, system of phosphotransferases; UGP, pyrophosphorylase UDPGlc (EC 2.7.7.9); Fru-bi-P, fructose-1,6-bi-phosphate; Fru-6-P, fructose-6-phosphate; Glc-6(1)-P, glucose-6(1)-phosphate; PGA, phosphogluconic acid; UDPGlc, uridine diphosphoglucose.

tion to glucose-6-phosphate (Glc-6-P), catalyzed by glucokinase, followed by isomerization of this intermediate to Glc-α-1-P, catalyzed by phosphoglucomutase, and conversion of the latter metabolite to UDPGlc by UDPGlc pyrophosphorylase. This last enzyme seems to be the crucial one involved in cellulose synthesis, since some phenotypic cellulose-negative mutants (Cel$^-$) are specifically deficient in this enzyme (Valla et al., 1989), though they display cellulose synthase (CS) activity, this was confirmed *in vitro* by means of observation of cellulose synthesis, catalyzed by cell-free extracts of Cel$^-$ strains (Saxena et al., 1989). Furthermore, the pyrophosphorylase activity varies between different *A. xylinum* strains and the highest activity was detected in the most effective cellulose producers, such as *A. xylinum* ssp. sucrofermentans BPR2001. The latter strain prefers fructose as a carbon source, displays high activity of phosphoglucoisomerase, and possesses a system of phosphotransferases, dependent on phosphoenolpyruvate. The system catalyzes conversion of fructose to fructose-1-phosphate and further to fructose-1,6-biphosphate (Figure 6).

7.2 Cellulose Synthase

Cellulose biosynthesis both in plants and in prokaryotes is catalyzed by the uridine diphosphate (UDP)-forming CS which is basically a processing 4-β-glycosyltransferase (EC 2.4.1.12, UDPGlc: 1,4-β-glucan 4-β-D-glucosyltransferase), since it transfers consecutive glucopyranose residues from UDPGlc to the newly formed polysaccharide chain and is all the time linked with this chain. Oligomeric CS complexes are frequently called terminal complexes (TCs). It is presumed that TCs are responsible first for β-1,4-glucan chains synthesis, extrusion through the outer membrane (if the globular, catalytic domain of CS is localized on the cytoplasmic side of the cell membrane), as well as association and crystallization into defined supramolecular structures, that follow the first two processes.

Cellulose synthase of *A. xylinum* is a typical membrane-anchored protein, having a molecular mass of 400–500 kDa (Lin and Brown, 1989), and is tightly bound to the cytoplasmic membrane. Because of this localization, purification of CS was extremely difficult, and isolation of the membrane fraction was necessary before CS solubilization and purification. Furthermore, *A. xylinum* CS appeared to be a very unstable protein (Lin and Brown, 1989). CS isolation from membranes was carried out using digitonin (Lin and Brown, 1989), or detergents (Triton X-100) and treatment with trypsin (Saxena et al., 1989), followed by CS entrapment on cellulose. According to Lin and Brown (1989), the purified CS preparation contained three different types of subunits, having molecular masses of 90, 67, and 54 kDa. Saxena et al. (1989) found only two types of polypeptides (83 and 93 kDa). The latter result seems to be more probable, since the mass of both subunits corresponds to the size of two genes *cesA* and *cesB*, detected in the CS operon by Saxena et al. (1990b, 1991), and the genes *bcsA* and *bcsB*, reported by Wong et al. (1990) and Ben-Bassat et al. (1993) for the same operon (see Section 7.4).

Photolabeling affinity studies indicate that the 83-kDa polypeptide is a catalytic subunit, displaying high affinity towards UDPGlc (Lin et al., 1990). According to the gene sequence, it contains 723 amino acid residues, of which the majority are hydrophobic. This subunit is probably synthesized as a proprotein, which contains a signal sequence, composed of 24 amino acid residues, which is cut off during the maturation process (Wong et al., 1990). The mature protein is anchored in the cell membrane by means of a transmembrane helix, close to the N-terminus of the polypeptide chain. The protein contains five more transmembrane helices, which can interact both with each other and with a large, catalytic site-comprising globular domain submerged in cytoplasma. Brown and Saxena (2000) claim that the catalytic subunit of *A. xylinum* CS operates in the same manner as processing glycosyltransferases of the second family, i.e. it catalyzes the direct glycoside bond synthesis in cellulose, assisted by simultaneous inversion of the configuration on the anomeric carbon, from the α-configuration in the UDPGlc, which is the monomer donor, to the β-configuration in the polysaccharide. A catalytically active aspartic acid residue (D), and short sequences DXD and QXXRW were found in the globular fragment of the numerous processing glycosyltransferases (Saxena and Brown, 1997). The same motives were detected in *A. xylinum* CS. This globular fragment has a two-domain character (i.e., comprises two domains A and B) in many of glycoside synthases (Saxena et al., 1995). The domain A includes the D residue, crucial for catalysis, and the DXD motive, which probably binds the nucleotide–sugar substrate. Participation of this motive in UDPGlc binding by *A. xylinum* CS was recently confirmed by Brown and Saxena (2000), who investigated enzyme mutants obtained using site-directed mutagenesis. It is not yet clear if the globular fragment of *A. xylinum* CS comprises one or two domains, although this second possibility is more probable. However, it is known that the catalytic subunit of *A. xylinum* CS is glycosylated. Two potential sites of glycosylation were found in its primary structure, deduced based on the gene sequence (Saxena et al., 1990a).

The 93-kDa polypeptide is tightly bound to the catalytic subunit. It contains the signal sequence close to its N-terminus and one transmembrane helix close to the C-terminus. The helix enables anchoring in the cell membrane. The polypeptide is probably a regulatory subunit, which interacts with the CS activator – cyclic diguanosine monophospahte (c-di-GMP) (see Section 7.5). The 93-kDa polypeptide does not combine with the 83-kDa subunit antibodies. However, it is not yet clear if CS is composed of these two subunits only, since this oligomeric complex plays multiple roles, i.e. β-1,4-glucan polymerization, extrusion of the subfibrils outside the cell, as well as self-assembly and crystallization of cellulose ribbons, composed of microfibrils. Therefore, the CS complex is probably comprised of other polypeptides involved in transmembrane pore formation. The analysis of microscopic images of the purified CS indicates that the enzyme tends to form tetrameric or octameric aggregates (Lin et al., 1989).

7.3
Mechanism of Biosynthesis

Synthesis of the metastable cellulose I allomorph, in *A. xylinum* and other cellulose-producing organisms, including plants, involves at least two intermediary steps, i.e. (1) polymerization of glucose molecules to the linear 1,4-β-glucan, and (2) assembly and crystallization of individual nascent polymer chains into supramolecular structures, characteristic for each cellulose-producing organism.

Assuming that the CS globular domain is localized on the cytoplasmic side of the cell membrane (see Section 7.2), the transfer of 1,4-β-glucan chains through the membrane, outside the cell is also required. All three steps are tightly coupled, and the rate of polymerization is limited by the rate of assembly and crystallization.

7.3.1
Mechanism of 1,4-β-Glucan Polymerization

Formation of BC is catalyzed by the CS complexes aligned linearly in the *A. xylinum* cytoplasmic membrane. Since these complexes can polymerize up to 200,000 molecules of Glc s^{-1} into the β-1,4-glucan chain (Ross et al., 1991), the process must run with high intensity. The mechanism of the reaction has not yet been definitely clarified. Currently, two different hypotheses for this mechanism in *A. xylinum* have been proposed.

The first of them, developed in Brown's laboratory (Brown, 1996; Brown and Saxena, 2000), assumes that 1,4-β-glucan polymerization does not involve a lipid intermediate, which transfers glucose from UDPGlc to the newly synthesized polymer chain. This hypothesis agrees with the fact that glycosyltransferases responsible for synthesis of unbranched homopolysaccharides are processing enzymes. The scheme of 1,4-β-glucan polymerization, proposed by Brown and his colleagues, is presented in Figure 7.

According to this model, there are three catalytic sites in the globular fragment of the CS catalytic subunit, similar to other proc-

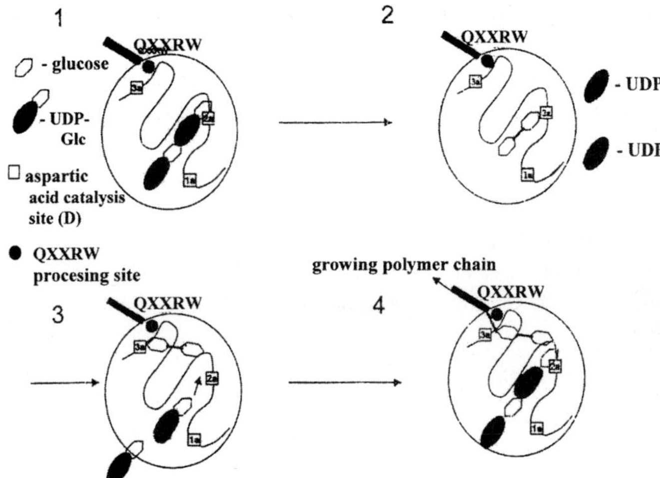

Fig. 7 Generalized concept for the polymerization reactions leading to β-1,4-glucan chain biosynthesis (Brown and Saxena, 2000; with kind permission).

essing glycosyltransferases. One of the catalytic sites (2a), comprising the DXD motive, binds two UDPGlc molecules (see Figure 7.1). The second catalytic site (1a), containing the crucial acidic aspartic acid (D) residue, catalyzes formation of the β-1,4-linkage between the two Glc residues docked in the pocket (see Figure 7.2), accompanied by releasing two UDP molecules. The third catalytic site (3a), containing the QXXRW motif, pulls the reducing end of the synthesized cellobiose (see Figure 7.3). The disaccharide leaves the area 1a–2a, which may bind two subsequent UDPGlc molecules (see Figure 7.4), and forms the second cellobiose molecule. In the next step, involving the QXXRW motif, the reducing end of the first cellobiose molecule is forced to the site of extrusion. At the same moment, the second cellobiose molecule is linked (with the aid of the D residue in the site 1a) to the nonreducing end of the first one, thus giving cellotetraose. The emptied 1a–2a area may bind two subsequent UDPGlc molecules, to repeat the cycle of reactions, attaching two more glucose residues to the chain. The simplified scheme of the proposed model of polymerization is depicted in Figure 8. According to this model, extrusion of the newly synthesized β-1,4-glucan chain starts from its reducing end.

Studies by Koyama et al. (1997) confirm this model. The authors proved that glucose residues are added to the nonreducing end of the polysaccharide and that reducing ends of nascent polymer chains are situated away from the cells. Furthermore, the torsion angle between two adjacent glucose residues in cellulose molecule is 180°, thus adding cellobiose residues (rather than single glucose moieties) to the growing chain favors maintaining the 2-fold screw axis of the β-1,4-glucan (Brown, 1996). Also Kuga and Brown (1988), who applied silver labeling, proved that cellulose chains were extruded outside the cell, starting from their reducing ends.

Han and Robyt (1998) who studied BC synthesis by the *A. xylinum* ATCC 10821 strain (resting cells and membrane preparations) using the ^{14}C pulse and chase reaction with D-glucose and UDPGlc, respectively,

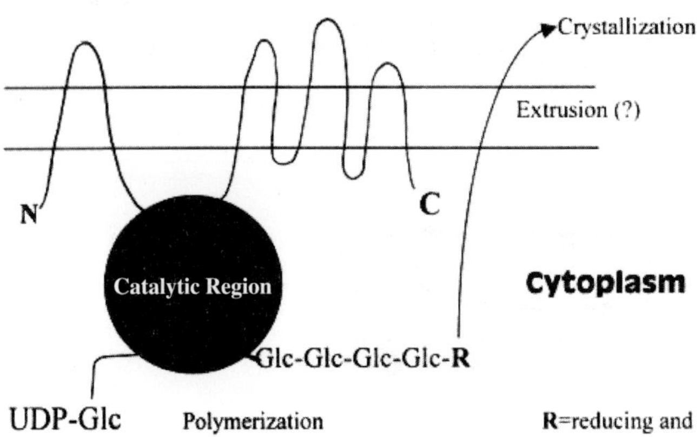

N and C – N and C-terminus of CS subunit of 83 kDa, respectively

Fig. 8 Simplified scheme of BC biosynthesis according to the Brown's model (Brown and Saxena, 2000; with kind permission).

proposed a second, different molecular mechanism of this process. They believe that the consecutive residues of the activated glucopyranoses (from UDPGlc) are linked to the reducing end of the growing cellulose chain, because the ratio of D-[^{14}C]glucitol obtained from the extruded reducing end of the cellulose chain to D-[^{14}C]glucose was decreasing with time. Han and Robyt assumed that a lipid pyrophosphate with a polyisoprenoid character takes part in the biosynthesis. According to them, formation of the lipid intermediate during BC synthesis was confirmed by an extraction of as much as 33% of the pulsed ^{14}C label with a mixture of chloroform and methanol. Its quantitative extraction was impossible, since when the polysaccharide chain length exceeded 8–10 sugar moieties, the complex with lipid became insoluble in the extraction mixture.

Han and Robyt proposed that BC biosynthesis involved three enzymes embedded in the cytoplasmic membrane: CS, lipid pyrophosphate: UDPGlc phosphotransferase (LipPP: UDPGlc-PT), and lipid pyrophosphate phosphohydrolase (LPP). The reaction mechanism, called by the authors the insertion reaction, is presented in the Figure 9.

The first enzyme transfers Glc-α-1-P from UDPGlc onto the lipid monophosphate (Lip-P), thus giving the lipid pyrophosphate-α-D-Glc (LipPP-α-Glc) (reaction 1, Figure 9). The α configuration on the anomeric carbon, involved in the Glc phosphoester bond, remains the same as in the substrate molecule. The second product of this reaction is UMP (according to Brown's model, UDP is released).

In the next reaction (reaction 2, Figure 9), catalyzed by CS, the glucose residue is transferred from one LipPP-α-Glc molecule onto another one and the β-1,4-glycosidic linkage between the two glucose residues is formed, due to the attack of the C-4 hydroxyl group of one of them onto C-1 hydroxyl group of the second Glc (from the second LipPP-α-Glc).

In the third reaction (reaction 3, Figure 9), hydrolysis of the lipid pyrophosphate formed in the previous step occurs and

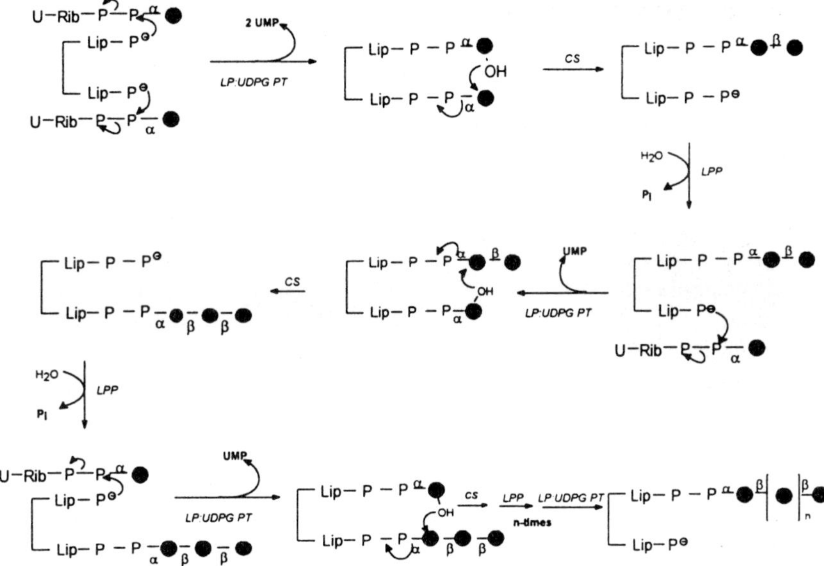

Fig. 9 Mechanism of BC biosynthesis involving lipid intermediates (Han and Robyt, 1998; with kind permission; modified). Lip, lipid; P, phosphoryl; Rib, D-ribose; U, uridine, •, glucosyl residue; LP: UDPGPT, lipid pyrophosphate: UDPGlc-phosphotransferase; CS; cellulose synthase; LPP, lipid pyrophosphate phosphohydrolase.

another Glc-α-1P from UDPGlc can be attached to the LipP, released in this reaction. The cycle of these three reactions is continued to give the β-1,4-glucan chain of an appropriate length. The mechanism includes the inversion of a configuration on the anomeric carbon in the Glc residue transferred either from one LipPP-α-Glc molecule to another, or to cellobiose, -triose, -tetraose, and subsequently to *n*-ose. An acceptor of the elongated chain of β-1,4-glucan is always a single α-D-Glc residue, linked with one of two LipPP molecules present in the polymerizing system. It means that the chain elongation occurs from its reducing end and nonreducing ends of nascent cellulose chains are situated away from the cells; this is contradictory to Brown's model. According to Han and Robyt's model, both the two cooperating LipPP molecules and the CS complex, which is not a processing glycosyltransferase, are embedded in the cytoplasmic membrane of the bacterial cell.

Participation of a lipid intermediate in cellulose biosynthesis was confirmed for *Agrobacterium tumefaciens* (Matthysse et al., 1995), as well as for synthesis of *Salmonella* O-antigen polysaccharide (Bray and Robbins, 1967) and *Xanthomonas campestris* xanthan (Ielpi et al., 1981). Furthermore, the lipid intermediate is probably also involved in the synthesis of acetan (De Lannino et al., 1988), which is a soluble polysaccharide produced together with cellulose by numerous *A. xylinum* strains (see Section 7.6). Further studies have to be carried out to elucidate whether the lipid intermediate plays in cellulose formation, the role postulated by Han and Robyt (1998). On the contrary, Brown's hypothesis (1996, 2000) is confirmed by cellulose synthesis *in vitro*, catalyzed by *A. xylinum* membrane preparations solubilized with digitonin (Lin et al., 1985; Bureau and Brown, 1987), and by polymer synthesis by purified CS subunits (Lin and Brown, 1989), which did not contain the lipid component.

According to both proposed hypotheses, BC biosynthesis does not require any primer, in agreement with earlier deductions (Canale-Parda and Wolte, 1964).

7.3.2
Assembly and Crystallization of Cellulose Chains

Polymerization of glucopyranose molecules to the β-1,4-glucan, whatever the mechanism, is the least complicated stage of cellulose synthesis, also in bacteria. The unique structure and properties of cellulose, which are dependent on its origin, result from the course of further stages: it means extrusion of the chains and their assembly outside the cell, giving supramolecular structures. Most important in these processes is the specific organization of multimeric complexes of CS, which are anchored in cytoplasmic membrane. In *A. xylinum* cells these complexes are ordered linearly along the long axis of the cell and one cell contains 50–80 TCs (Brown et al., 1976), whereas in vascular plants TCs form characteristic, 6-fold symmetrical rosettes (Brown and Saxena, 2000).

Stronger aeration, e.g., during agitated *A. xylinum* cultures, or the presence of certain substances, that cannot penetrate inside the cells, but can form competitive hydrogen bonds with the β-1,4-glucan chains (e.g., carboxymethyl-(CM)-cellulose, fluorescent brightener Calcofluor white), bring about significant changes in the supramolecular organization of cellulose chains, and instead of the ribbon-like polymer, i.e. the metastable cellulose I (Figure 10), which fibrils are colinear to CS complexes and gathered in a row, the second, thermodynamically more stable allomorph amorphous cellulose II is formed. This phenomenon is accompanied by disorganization of the TC' linear order. Discussing this problem, Brown and Saxena (2000) state that the conditions-dependent form of the final product may somehow determine the TC' arrangement in the cytoplasmic membrane, though it is believed that in *A. xylinum* these complexes have a stationary character, as opposed to some algae (Kudlicka, 1989). Worth mentioning is the well-known phenomenon of *A. xylinum* cell motion during BC synthesis, in the direction opposite to cellulose chain extrusion (Brown et al., 1976). Reversal movement of *A. xylinum* cells enabled calculation

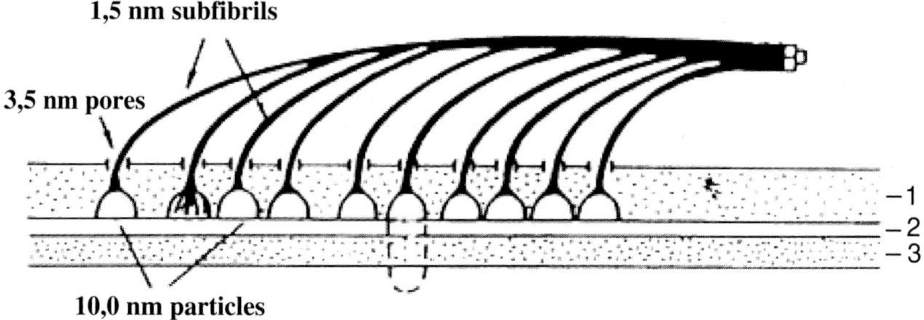

Fig. 10 Model of *A. xylinum* cellulose subfibrils formation: 1, lipopolysaccharide layer; 2, periplasmic space; 3, plasmalemma; 10 nm particles, CS subunits (Kudlicka, 1989; with kind permission).

of the typical chain elongation rate, which is equal to 2 μm min^{-1}, and corresponds to a polymerization of more than 10^8 glucose molecules into the β-1,4-glucan per hour and per single bacterial cell. Thus if forces associated with subfibrils assembly to bundles and ribbons, are strong enough to shift the motion of whole *A. xylinum* cells, these forces may as well influence the spatial arrangement of CS subunits, localized in the semi-liquid lipid bilayer of the membrane.

The spatial assembly of the β-1,4-glucan chains has a hierarchical character (Figure 10). In the first stage, 10–15 nascent β-1,4-glucan chains form a subfibril (also called a protomicrofibril), 1.5 nm in diameter, which is not a rod-like structure, but a left-handed triple helix (Ross et al., 1991). The subfibrils gather to form (also twisted) microfibrils composed of numerous parallel chains. The next structure in hierarchy is a bundle of microfibrils, followed by loosely wound ribbon, comprising about 1000 individual chains of the β-1,4-glucan (Haigler, 1985). In the presence of Calcofluor, which penetrates the outer membrane of the cell envelope, and forms complexes with protomicrofibrils, their aggregation to microfibrils is stopped. In the presence of much larger molecules such as CM-cellulose and xyloglucan (Yamamoto et al., 1996), which are too large to traverse the membrane, assembly of the bundles of microfibrils and ribbons is disrupted, thus indicating that formation of these two latter structures occurs exocellularly. Ross et al. (1991), who analyzed the mechanism of *A. xylinum* cellulose chain association, emphasized the important role of specific sites of adhesion in the inner and outer membranes of cell envelope, whose presence in bacterial cells was first reported by Bayer (1979). Their occurrence would enable an export of cellulose microfibrils without any interactions with the peptidoglycan gel, which fills the periplasmic space. Such interactions would prevent formation of correct hydrogen bonding between the protomicrofibrils.

The molecular organization of pores, through which *A. xylinum* cellulose microfibrils migrate outside the cell, remains unknown. The participation of expression products of the *bcsC* and *bcsD* genes, belonging to CS operon (see Section 7.4), in their formation cannot be excluded (Wong et al., 1990), all the more since these expression products appeared to be essential for cellulose synthesis *in vivo* (Ben-Bassat et al., 1993). Also the function of three proteins: CSAP20, 54, and 59 (CS-associated proteins), has to be scrutinized. They bind to cellulose together with the CS catalytic subunit, as revealed by experiments on CS purification (Benziman and Tal, 1995). It seems probable that isolation and purification of other components of the complex machinery responsible for cellulose synthesis, in concert with direct mutations of genes from *A. xylinum* CS operon, may help to elucidate the sequence of events during this process on the molecular level.

The process of assembly of nascent cellulose chains, not separated from parental *A. xylinum* cells, displays a unique and remarkable dynamics. Extrusion of these chains outside and their assembly into ordered structures is assisted by reversal of the motion of the cells. Throughout the whole process of the β-1,4-glucan synthesis and secretion, *A. xylinum* cell simultaneously turns around its own axis (uncoiling motion) (Brown et al., 1976). Also, this movement is driven by forming of twists of the ribbons, which are anchored in extrusion loci, localized on the side surface of the elongated cell. These twists are noticeable under the electron microscope.

7.4 Genetic Basis of Cellulose Biosynthesis

Cellulose biosynthesis involves several enzymes. Their function starts from the synthesis of the direct cellulose precursor UDPGlc, followed by polymerization of the monomer. It was found that the first reactions catalyzed by glucokinase, phosphoglucomutase and UDPGlc pyrophosphorylase, do not limit the final rate of BC synthesis, since *A. xylinum* synthesizes an excess of UDPGlc (Ben-Basat et al.,1993). The rate of BC synthesis is limited by diguanylate cyclase and oligomeric, regulatory complexes of CS. Therefore construction of more efficient cellulose-producing *A. xylinum* mutants, requires determination of organization of the genes coding for these enzymes. This has been studied by Wong et al. (1990) and Ben-Bassat et al. (1993), who proved that BC biosynthesis involves four coupled genes: *bcsA* (2261 bp), *bcsB* (2405 bp), *bcsC* (3956 bp), and *bcsD* (467 bp), forming the CS operon, which is 9217 bp in length, and is transcribed simultaneously as polycistron mRNA.

Although the function of the proteins encoded by each of these genes has not yet been precisely determined, some information on their role was derived from genetic complementation experiments (Wong et al., 1990). Results of these studies indicate that CS-deficient strains can be complemented by the gene *bcsB*, which codes for the 85-kDa protein (802 amino acid residues), and that *A. xylinum* mutants defective in both CS and diguanylate cyclase are complemented by the gene *bcsA*, coding for the 84-kDa protein (754 amino acid residues) (Ben-Bassat et al., 1993). Based on different mutations it was concluded that protein A takes part in the interaction of the CS complex with c-di-GMP, which is an allosteric activator of CS, earlier postulated by Saxena et al. (1991). The protein B is capable of binding UDPGlc and synthesizing β-1,4-glucan chains. A gene coding for the catalytic subunit (*cesA*) and analogous to the gene *bcsB* (Wong et al., 1990; Ben-Bassat et al., 1993) was also described by Saxena et al. (1990a), who one year later identified the gene coding for the regulatory subunit (*cesB*). However, Saxena et al. (1990a, 1991) suppose that the position of the first two genes in the CS operon is opposite and that the first gene codes for the CS catalytic subunit.

The role of the expression products of both *bcsC* and *bcsD* genes has not been determined so precisely. These genes probably also code for proteins (molecular mass of 141 and 17 kDa, respectively), anchored in the membrane, and both essential for cellulose synthesis *in vivo* and for protein secretion from the cells. Ben-Bassat et al. (1993) reported that splitting or deletion of the *bcsD* gene markedly reduced synthesis of cellulose. The presence of genes, which govern the assembly of cellulose chains, in the CS operon was also predicted by Saxena et al. (1990a) and Ross et al. (1991).

It is not clear if plasmid DNA participates in BC synthesis and, if so, what its role is. However it was proven that the composition and size of plasmids detected in 60% of the non-reverting Cel$^-$ mutants of the *A. xylinum* ATCC 10245 (currently 17005) strain, that were obtained by means of mutagenesis with *N*-methyl-*N'*-nitro-*N*-nitrosoguanidine, were markedly different from that of the wild strain. This observation suggests the plausible relationship between plasmid DNA structure and BC synthesis.

Mobile DNA fragments, including insertion sequences (IS), particularly widespread among prokaryotes, and 750–2500 bp in length (Galas and Chandler, 1989), are important factors of regulation of many genes. The ISs may activate or inactivate genes, and start DNA rearrangement such as deletions, inversions, and cointegrations. It was revealed that ISs participated in regulation of extracellular polysaccharides synthesis in *Pseudomonas atlantica* (Bartlett and Silverman, 1989), *Xanthomonas campestris* (Hotte et al., 1990), and *Zoogloea ramigera* (Easson et al., 1987). Their participation in BC synthesis regulation is also probable.

The best recognized insertion sequence in *A. xylinum* is the IS 1031 (950 bp, with known nucleotide sequence). Alteration in the IS 1031 profile was detected in the majority of Cel$^-$ mutants, in comparison to the wild strain. Some recombinants contained two or more IS 1031 fragments. Coucheron (1991) observed even more significant changes in the IS pattern, pointing to DNA rearrangement and tried to obtain revertants by splitting off the additional ISs from the inactivated gene and obtained pseudorevertants, which produced a wax-like substance on the surface of nutrient broth, but the capability of cellulose synthesis was not restored. These experiments indicate that the presence of ISs contributes to the genetic instability of *A. xylinum*.

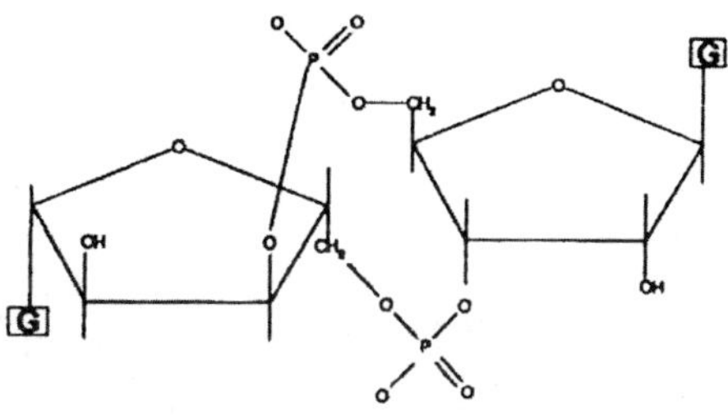

Fig. 11 Structure of c-di-GMP – the allosteric activator of *A. xylinum* CS.

7.5 Regulation of Bacterial Cellulose Synthesis

Cyclo-3,6′:3′6 diguanosine monophosphate (c-di-GMP, see Figure 11) is a reversible allosteric activator of *A. xylinum* CS and plays a crucial role in regulation of the whole β-1,4-glucan biogenesis (Ross et al., 1987).

This compound binds to the enzyme regulatory subunit, and induces conformational changes that facilitate association of the CS protomers and lead to the enhancement of its reactivity (Ross et al., 1987).

The concentration of c-di-GMP in *A. xylinum* cells is regulated by 3 enzymes: diguanyl cyclase (CDG), phosphodiesterase A (PDEA) and phosphodiesterase B (PDEB) (Figure 12) (Ross et al., 1987). PDEA and PDEB are anchored in the cytoplasmic membrane, and CDG has two forms. One of them is anchored in the cytoplasmic membrane and the second one operates in the cytoplasma (Ross et al., 1991). Recently, Weinhouse et al. (1997) reported on a new c-di-GMP-binding protein, which probably also participates in the intracellular regulation of free c-di-GMP concentration. Tal et al. (1998) proved that the cellular turnover of c-di GMP in *A. xylinum* is controlled by three operons.

CDG, the key regulatory enzyme in the *A. xylinum* cellulose synthesis system, is composed of two polypeptide chains and encoded by two genes (Nichols and Singletary, 1998). CDG is activated by Mg^{2+} ions and specifically inhibited by saponin (Ohana et al., 1998), a glycosylated terpenoid. CDG converts two GTP molecules at first into the linear (pppGpG) and then into the cyclic (c-di-GMP) diguanosine monophosphate, which activates CS. PDEA splits active c-di-GMP into a linear inactive dimer (pGpG, di-GMP), further decomposed by PDEB into two molecules of 5′GMP (Figure 12). PDEA is inhibited by Ca^{2+} ions, which do not influence PDEB. Therefore, the Ca^{2+} concentration indirectly affects the rate of cellulose synthesis. High concentration of these ions enhances this rate, since the

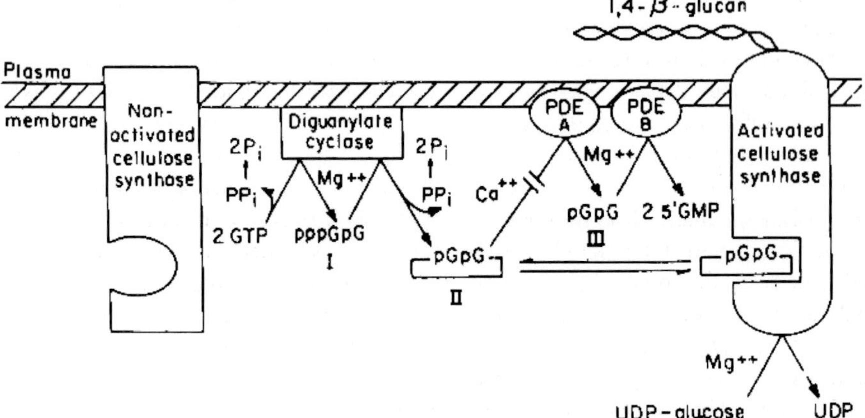

Fig. 12 Proposed model for cellulose synthesis in *A. xylinum*. For simplicity, the synthesis of a single β-1,4-glucan chain is depicted, although a more complex form of CS, polymerizing several chains simultaneously, might be the active enzyme unit in cellulose biogenesis (Ross et al., 1991; with kind permission). PDEA, phosphodiesterase A; PDEB, phosphodiesterase B; pppGpG, diphospho-di-GMP; pGpG (II), c-di-GMP (cyclic-di-GMP); pGpG, di-GMP (linear di-GMP).

conversion of c-di-GMP to the inactive linear form is inhibited.

It is believed that both the molecular background of the CS activation by c-di-GMP as well as the regulation of its synthesis, involving a positive influence of CDG, and a negative effect of PDEA and PDEB, have a unique character. To date, such a mechanism of cellulose synthesis regulation has only been detected in *A. xylinum*.

7.6 Soluble Polysaccharides Synthesized by *A. xylinum*

Apart from cellulose, some related soluble polysaccharides are also synthesized by *A. xylinum*. In 1970s, a soluble β-homoglucan was detected. Its main backbone was composed of β-1,4-linked glucose residues and every third glucose moiety of this chain was substituted with another glucose via β-1,2 linkage (Colvin and Leppard, 1977; Colvin et al., 1979). *A. xylinum* cellulose pellicle non-forming (Pel^-) strains synthesize α-glucan linked with the cell envelope (Dekker et al., 1977). Valla and Kjosbakken (1981) isolated a soluble polysaccharide containing Glc, Man, Rha, and glucuronic acid residues (molar ratio of 3:1:1:1, respectively) from the culture broth of the Cel^- strain. The related polymer, composed of the same moieties (altered molar ratio of 6:2:1:1, respectively), synthesized by another Cel^- strain, was detected by Minakami et al. (1984). Probably some of the residues in these polymers are acetylated (Tayama et al., 1985). Polysaccharides of this type have received the common name acetan (De Lannino et al., 1988). Moreover, some wild *A. xylinum* strains synthesize a similar soluble polysaccharide (Glc, Man, Rha, and glucuronic acid; 6:1:1:1, respectively) together with cellulose (Savidge and Colvin, 1985).

The question is whether cellulose-assisting soluble polysaccharides play any role in its biogenesis. It seems that they do not play any direct role, although kinetics of their biosynthesis are the same as that of cellulose (Marx-Figini and Pion, 1974) and, moreover,

UDPGlc was preferentially used for production of these soluble saccharides instead of the production of cellulose (Delmer, 1982). Some authors state that *in vivo* nascent cellulose microfibrils are coated with amorphous material (Leppard et al., 1975), which may be composed of the above-mentioned soluble polysaccharides. The absence of these soluble polymers in the liquid fraction of static culture broths and their presence in the cellulose mat confirms that assumption (Valla and Kjosbakken, 1981).

Much earlier, Ben-Hayyim and Ohad revealed in 1965 that CM-cellulose present in the medium, used for BC synthesis, resulted in incorporation of this soluble derivative of cellulose into microfibrils of the polymer. However, incorporation of the *A. xylinum* soluble exopolysaccharides into cellulose microfibrils has not been observed.

7.7 Role of Endo- and Exocellulases Synthesized by *A. xylinum*

The first reports on *A. xylinum* cellulases, detected independently by Okamoto et al. (1994) and Standal et al. (1994), included description of their genes. The first authors selected from *A. xylinum* IFO 3288 DNA gene libraries one gene coding for a 24-kDa protein (218 amino acid residues), which displayed CM cellulase activity. Standal et al. (1994) found another endocellulase gene in the *A. xylinum* ATCC 23769 Cel^- mutant, localized upstream the CS operon. They revealed that the loss of cellulase synthesis capability resulted from gene splitting. The conclusions of Tonouchi et al. (1997), who investigated the localization of the *A. xylinum* BPR 2001 endo-1,4-β-glucanase gene, were similar. The latter gene was localized upstream the CS operon, whereas the gene of the second cellulolytic enzyme, produced by this strain, i.e. an exo-1,4-β-glucosidase, was found downstream this operon.

A. xylinum BPR 2001 endo- and exocellulase were purified and characterized (Oikawa et al., 1997; Tahara et al., 1998). The studies of Tahara et al. (1997) also revealed that at pH 5.0, optimal for growth and BC synthesis, the activity of both cellulases is several times higher than at pH 4.0, at which the BC production is only slightly declined. It was also observed that at pH 5.0, the average DP was decreasing from 16,800 to 11,000 with the time of cultivation, whereas at pH 4.0 these changes were negligible. The cellulose obtained at pH 5.0 had inferior physical properties, i.e. a lower tensile strength (a lower Young's modulus value). These results and colocalization of genes of cellulases and CS operon, suggest that the endo-1,4-β-glucanases and exo-1,4-β-glucosidase are involved in *A. xylinum* cellulose biogenesis (Tahara, 1998).

More particularly, it relates to possible degradation of the nascent β-glucan chains, synthesized in the later phase of growth. Furthermore, the *A. xylinum* BPR 2001 exo-1,4-β-glucosidase displays both hydrolytic and transglycosylating activity (Tahara et al., 1998), similarly to other glycosidases, which do not cause inversion of configuration on the anomeric carbon. The latter activity may be responsible *in vivo* for changes in the average DP of the polymer; however, further research is necessary to explain this hypothesis. Tentative evidence for such a possibility might be the behavior of soybean cells cultured *in vitro*. The activity of their β-glucosidase, which also has transglycosylating activity, increases together with the length of the cells, since this enzyme probably participates in the transfer of glycosidic residues of hemicelluloses precursors to the growing terminus of the cell wall, in which intensive synthesis of these polysaccharides is taking place (Nari et al., 1983).

Whether *A. xylinum* cellulases are involved in releasing the energy stored in the form of cellulose, in periods of starvation, is still a question to answer, although some studies confirm this thesis (Okamoto et al., 1994). It is known that some *Sclerotium* endo-1,3-β-glucanases play such a role (Jones et al., 1974).

Current observations indicate that modification of the physical properties and DP of BC can be achieved either by using compounds that influence its biosynthesis, or by exploiting the activity and specificity of *A. xylinum* cellulases.

8
Biodegradation of BC

Independent of its origin, cellulose undergoes total biodegradation in nature. However, in comparison to PC, associated both with polymers susceptible to degradation, like hemicelluloses and pectin, and with lignin, which is the most resistant plant polymer, BC is relatively pure and *a priori* more susceptible to attack by cellulolytic enzymes, which are produced mainly by fungi and numerous bacterial species. In this respect, commercial application of BC is friendly for the environment.

Furthermore, the structure and properties of BC (large accessible surface area) make it the superior model substrate for studies on cellulases (1,4-β-glucan 4-glucanohydrolases, formerly endocellulases, EC 3.2.1.4) and cellobiohydrolases (cellulose 1,4-β-cellobiosidases, EC 3.2.1.91), that are the main components of cellulosomes, the specialized multienzymatic particles from cellulolytic bacteria, as well as fungal systems for cellulose decomposition.

Recently, Boisset et al. (1999) proved that *Clostridium thermocellum* cellulosomes completely decompose *A. xylinum* cellulose microfibrils faster than microcrystalline *Valonia ventricosa* cellulose. Ultrastructural observations of the hydrolysis process, using transmission electron microscopy, infrared spectroscopy, and X-ray diffraction analysis, indicated that the rapid hydrolysis of BC resulted from very efficient synergistic action of the various enzymic components, present in the cellulosome scaffolding structure. Further studies of Boisset et al. (2000) revealed that *Humicola insolvens* cellobiohydrolase Cel7A (previously CBH I) brought about thinning of dispersed BC ribbons, whereas the mixture of Cel6A (CBH II) and Cel7A (in the ratio of 2:1, respectively) cut the ribbons to shorter pieces, thus suggesting the partly endo-manner of CBH II attack. The phenomenon of low inherent endoactivity of some exoglucanases has been known for several years and explains more efficient synergistic action of cellobiohydrolases in comparison to endoglucanases (Teeri, 1997). According to current opinions, the system of multiple exo- and endocellulases, both in bacterial cellulosomes, and in fungal cellulolytic complexes, represents the perfectly balanced continuum of activity, providing efficient degradation of various cellulosic substrates.

BC susceptibility to cellulases was also observed by Samejima et al. (1997), who found that the mixture of *Trichoderma viride* CBH I and endoglucanase II, drastically disintegrated the twisted and bent ribbon-like structure of microfibril bundles to linear needle-like microcrystallites, and also caused rapid polymer fragmentation. Transformation of the coiled structure of BC ribbons to the needle-like one is driven by the remarkable twisting motion of the substrate, which is probably a result of a tension – released when the microfibrils are being decomposed to shorter fragments by cellulases. Samejima et al. (1997) also detected that the acid-treated BC, containing many microfibril aggregates, was not susceptible to attack of both enzymes.

Similar results were achieved by Srisodsuk et al. (1998), who digested microcrystalline BC with *Trichoderma reesei* CBH I and observed rapid solubilization of the polymer, but slow decrease in its DP. The endocellulase I from this organism attacked the substrate in the opposite manner. Both enzymes hydrolyzed cotton cellulose more slowly than BC, though cotton cellulose exhibits relatively high purity as compared to other plants.

One of the reasons of the susceptibility of BC to the attack of cellulases is its large accessible surface area, that even increases throughout hydrolysis and facilitates effective binding of cellulolytic enzymes, the crucial step in the process of degradation. It was concluded based on studies of Palonen et al. (1999), who found that the number of *T. reesei* CBH I and CBH II molecules linked to BC, increased during hydrolysis. Furthermore, the CBH II binding was markedly stronger (the adsorption/desorption of the enzyme from its substrate was assisted by 60–70% hysteresis), pointing to a distinctly different processing character of this enzyme.

Stalbrandt et al. (1998) compared the mode of attack of four *Cellulomonas fimi* endo- and exocellulases against microcrystalline BC and acid-swollen Avicel cellulose. The latter was decomposed more effectively by all the enzymes. In most cases 45–65% solubilization and a decrease in DP were observed, except for Cbh B, which caused 27% solubilization. A high degree of bacterial polymer solubilization (67%) was achieved for Cbh A, known as the processing enzyme. Results of Stalbrandt et al. (1998) proved that only the external surface of BC fibrils is accessible for *C. fimi* cellulases. Since Cbh A is the processing enzyme, it can remove external fibrils much faster than the other three cellulases and attack the deeper internal surface of the polymer. Therefore, the other enzymes (two nonprocessing endocellulases and Cbh B, which displays weaker processing properties) could not solubilize BC so effectively, whereas amorphous soluble cellulose was equally available to all of them.

Current results indicate that complete digestion of the highly crystalline bacterial polymer will not cause any problems, all the more since organisms responsible for BC biodegradation in natural environments produce a multitude of various endo- and exocellulases, whose activities complement each other. Furthermore, as opposed to cellulose of plant origin, which requires pretreatment to provide easier enzymatic conversion to sugars, BC can be directly digested by cellulases.

9
Biotechnological Production

To achieve high productivity and yields of BC and to reduce cost of its production, special emphasis should be given to the following aspects:

- development of screening methods providing selection of *A. xylinum* strains, which can efficiently produce cellulose from various inexpensive waste carbon sources – wild strains derived from the screening could be improved using genetic engineering methods;
- optimization of *A. xylinum* culture conditions (static or agitated culture, fermentor type), determining both a form (a pellicle or an amorphous gel) and properties (resilience, elasticity, mechanical strength, absorbency) of BC, which have to be tailored to the further polymer application;

- optimization of culture broth composition (carbon and nitrogen sources, biosynthesis stimulating compounds, microelements, etc.) and process conditions (pH, temperature, aeration).

These are discussed in the next sections.

9.1 Isolation from Natural Sources and Improvement of BC-producing Strains

One of the methods enabling selection of proper *A. xylinum* strains, is the screening for strains, which cannot oxidize glucose via gluconic acid to 2-, 5-, or 2,5-ketogluconate (Winkelman and Clark, 1984; Johnson and Neogi, 1989; De Wulf et al., 1996; Vandamme, 1998). In this respect, De Wulf et al. (1996) applied an agar nutrient medium containing Br^- and BrO_3^- ions, that combine at acidic pH to release molecular bromine, toxic to *A. xylinum* cells. Only those mutants, which do not convert glucose into gluconate or its derivatives, can survive in this medium. Vandamme et al. (1998) successfully used this method to isolate *A. xylinum* KJ33 strain, producing 3.3 g cellulose L^{-1}. *A. xylinum* strains, which do not metabolize glucose to gluconic acids, were also selected on $CaCO_3$ containing medium (Johnson and Neogi, 1989). Another approach to avoid conversion of glucose to organic acids, was screening for mutants defective in glucose-6-phosphate dehydrogenase. The mentioned increase in final concentration, volumetric productivity and total cellulose yield, was achieved in agitated cultures. For the same purpose, i.e. selection of the best cellulose producers, preparations of cellulases, added into the growth medium, have also been used (Brown, 1989c).

Mutation with chemical compounds, such as *N*-methyl-*N'*-nitro-*N*-nitrosoguanidine or ethylmethanesulphate, as well as with ultraviolet irradiation gave mutants that displayed higher cellulose productivity, reduced synthesis of the soluble polysaccharide acetane, and a much lower degree of glucose conversion into organic acids (Johnson and Noegi, 1989).

Screening of *A. xylinum* strains was also aimed at isolation of spontaneous or induced cellulose II-producing mutants. This cellulose is synthesized by bacteria that form smooth colonies and do not produce a pellicle on the surface of the nutrient broth. Instead, the polymer is dispersed in the whole volume of the medium.

Selection of the best BC producers has been also performed traditionally, by determination of the synthetic activity of a plethora of individual monocultures. For example, Toyosaki et al. (1995) isolated 2096 *Acetobacter* strains from a large number of plant samples (fruits, flowers) and 412 strains forming a mat on the surface of nutrient medium were subjected to further investigations. The procedure yielded *A. xylinum* ssp. sucrofermentans BPR2001 – the strain which efficiently synthesized BC in agitated culture.

9.1.1 Improvement of Cellulose-producing Strains by Genetic Engineering

Expression of genes coding for various carbohydrases from organisms other than *A. xylinum* enhances the scope of carbon sources available, that may increase BC yield and decrease the nutrient broth price. An example is the expression of sucrose phosphorylase (EC 2.4.1.7) gene from *Leuconostoc mesenteroides* in *Acetobacter* sp. (Tonouchi et al., 1998), that resulted in utilization of sucrose as a carbon source and increased cellulose production. In addition, Nakai et al. (1999) obtained double cellulose yield by means of expression of the mutant sucrose synthase (EC 2.4.1.13) gene from mung bean (*Vigna radiata*, Wilczek) in *A. xylinum* strain. In the mutant enzyme, the eleventh serine residue was replaced by glutamic acid (S11E); this caused higher affinity of the engineered enzyme toward sucrose and favored cleavage of this sugar for the synthesis of UDPGlc. Introduction of the mutant gene into *A. xylinum* not only changed sucrose metabolism in the recombinant strain, but also created a new metabolic pathway of direct UDPGlc synthesis (Figure 13). The energy driving this process is derived only from the cleavage of the glycosidic bond in sucrose without any other energy input required, which is the case when UDPGlc is synthesized via the conventional pathway. Furthermore, the UDP molecules resulting from glucose polymerization process can be directly recycled by the sucrose synthase activity, which increases the rate of the coupled reactions and a higher rate of cellulose synthesis is observed. It should be emphasized that higher plants have both systems of UDPGlc synthesis.

9.2 Fermentation Production

Growth on the surface of liquid media and synthesis of the gelatinous, leather-like mat, are natural properties of *A. xylinum* strains. Therefore static conditions seemed to be optimal for BC synthesis. However, the main drawbacks of static culture, such as the synthesis of the polymer only in the form of a sheet and relatively low productivity, have contributed to the development of new fermentation processes.

9.2.1 Carbon and Nitrogen Sources

In studies on factors affecting the BC production yield, much attention has been paid to carbon sources. Numerous mono-, di-, and polysaccharides, alcohols, organic acids, and other compounds were compared by Jonas and Farah (1998), who found out that the preferred carbon sources were D-arabitol and D-mannitol, which resulted in a 6.2- or 3.8-fold higher cellulose production, respectively, in comparison to glucose. Both sugar alcohols provided stabilization of the pH throughout the culture period, since they were not converted to gluconic acids.

Tonouchi et al. (1996), who used a strain of *A. xylinum* to obtain cellulose from glucose and fructose, found that fructose stimulated the activities of phosphoglucose isomerase and UDPGlc pyrophosphorylase, thus enhancing cellulose yield.

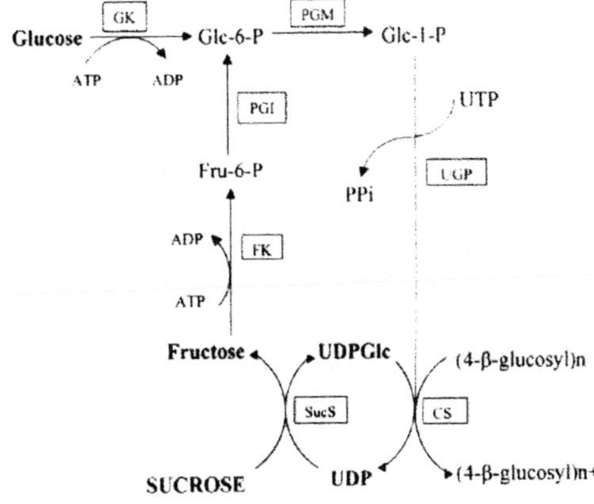

Fig. 13 UDPGlc biosynthesis pathways in a recombinant strain of *A. xylinum* containing the sucrose synthase (SucS) gene. Other abbreviations are the same as in Figure 6 (see Section 7.1).

When maltose was the sole carbon source in the nutrient medium (Masaoka, 1993) BC production was even 10 times lower than in glucose-containing medium and the polymer was markedly shorter (DP = 4000–5000) than in the presence of glucose (DP = 11,500).

Matsuoka et al. (1996) investigated BC synthesis by *A. xylinum* ssp. sucrofermentous BPR2001 in agitated culture and found out that the presence of lactate in the growth medium stimulated bacterial growth and enhanced the cellulose yield 4–5 times. The source of lactate is usually corn steep liquor (CSL), which is one of the main medium components applied for BC production, especially in agitated cultures. Contrary to glucose and fructose, lactate is not converted to UDPGlc, but is metabolized via pyruvate and oxalacetate in the Krebs cycle, being a source of energy for cellulose production. To provide lactate in the growth medium, cultivation of mixed cultures of lactic acid and acetic acid bacteria has been carried out. The best results gave strains of *Lactobacillus, Leuconostoc, Pediococcus,* and *Streptococcus.* The lactic and acetic acid bacteria were grown also in the presence of sucrose-hydrolyzing *Saccharomyces* yeast (β-fructofuranosidase producers). This principle provided a high BC yield, since after 14 days of agitated culture, as much as 8.1 g of cellulose L^{-1} was obtained, when in the absence of *Lactobacillus* strain, only 6.4 g L^{-1} (Seto et al., 1997).

One of the cellulose synthesis-stimulating compounds appeared to be ethanol (Naritomi et al., 1998). Continuous culture of *A. xylinum* in fructose-containing medium, revealed that 10 g ethanol L^{-1} enhanced a BC yield, but a concentration of 15 g L^{-1} prevented polymer synthesis. These results suggest that similarly to lactate, ethanol is also a source of energy (accumulated as ATP) and not a substrate for cellulose synthesis. ATP activates fructose kinase and inhibits glucose-6-phosphate dehydrogenase, thus halting conversion of 6-P-glucose to 6-phosphogluconate. It can be concluded that – due to these coupled reactions – the amount of fructose further isomerized to glucose-6-phosphate is increased.

Each cellulose-producing strain requires a specific complex nitrogen source, providing not only amino acids, but vitamins and mineral salts as well. These requirements are met with yeast extract, CSL as well as hydrolyzates of casein and other proteins. The preferred nitrogen sources are yeast extract and peptone, which are basic components of the model medium developed by Hestrin and Schramm (1954; H-S medium), applied in numerous studies on BC synthesis. However the most recommended nitrogen source for agitated cultures is CSL (Johnson and Noegi, 1989).

It was also found that a significant part of the expensive medium components, i.e., yeast extract and bactopeptone, can be replaced with CSL or even white cabbage juice. Also waste plant materials such as sugar beet molasses, spent liquors after glucose separation from starch hydrolyzates, as well as whey and some fermentation industry wastes (e.g., spent liquors after dextran precipitation with ethanol) were appropriate medium components (Krystynowicz et al., 2000).

Studies on the influence of vitamins on BC synthesis (Ishikawa et al., 1995, 1996b) revealed that the most stimulating ones were pyridoxine, nicotinic acid, *p*-aminobenzoic acid and biotin, and even CSL media should be fortified with these substances. Some other compounds, strongly stimulating cellulose production by *A. xylinum* strains, such as derivatives of choline, betaine, and fatty acids (salts and esters), were also identified (Hikawu et al., 1996).

High BC production (up to 25 g L^{-1}) can be achieved using optimum growth medium composition, designed by mathematical methods and computer analysis (Joris et al., 1990; Embuscado et al., 1994; Vandamme et al., 1998; Galas et al., 1999).

9.2.2 Effect of pH and Temperature

Analysis of the influence of pH on *A. xylinum* cellulose yield and properties, indicates that optimum pH depends on the strain, and usually varies between 4.0 and 7.0 (Johnson and Neogi, 1989; Galas et al., 1999). For instance, Ishikawa et al. (1996a) and Tahara et al. (1997), who applied two different *A. xylinum* strains, observed the highest polymer yield at pH 5.0. The same pH was found to be optimal by Krystynowicz et al. (1997). Comparison of adsorptive properties of the polymer obtained at different pH, learned that cellulose, accumulated in the S-H medium pH 4.8–6.0, displayed the highest water-binding capacity (Wlochowicz, 2001).

In addition to the pH of the nutrient broth, also the temperature influences BC yield and properties. In majority of reported experiments, the temperature ranged from 28 to 30 °C, and its variations caused changes of cellulose DP and water-binding capacity. For instance, BC synthesized at 30 °C had a lower DP (approximately 10,000) and a higher water-binding capacity (164%) in comparison to that produced at 25 and 35 °C (Wlochowicz, 2001).

9.2.3 Static and Agitated Cultures; Fermentor Types

Synthesis of BC is run either in static culture or in submerged conditions, providing proper agitation and aeration, necessary for medium homogeneity and effective mass transfer. The choice of culture conditions strictly depends on the polymer use and destination.

BC yield in static cultures is mostly dependent on the surface/volume ratio. Optimum surface/volume ratio protects from either too high (unnecessary) or too low aeration (cell growth and BC synthesis termination). Reported values of surface:volume ratio vary from 2.2 cm^{-1} (Joris et al., 1990) to 0.7 cm^{-1} (Krystynowicz et al., 1997). BC synthesis in static conditions can be achieved either in a one-step (medium inoculated with 5–10% cell suspension) or a two-step procedure. The latter one starts from agitated fermentation, followed by the static culture (Okiyama et al., 1992). Krystynowicz et al. (1997) modified the two-step procedure and applied two consecutive static cultures. Pellicles containing entrapped *A. xylinum* E_{25} cells, obtained in the first step (24 h), were cut to uniform pieces and used as an inoculum to start the next culture, run for 4–5 days. The method provided uniform bacterial growth and production of homogeneous pellicles.

The control of BC synthesis in static culture is very difficult since the pellicle limits access to the liquid medium. A particularly important parameter, which requires continuous control, is the pH. Accumulation of keto-gluconic acids in the culture broth brings about a decrease in pH far below the value that is optimum for growth and polysaccharide synthesis. Because conventional methods of pH adjustment cannot be used in static cultures, Vandamme et al. (1998) applied an *in situ* pH control via an optimized fermentation medium design, based on introducing acetic acid as an additional substrate for *Acetobacter* sp. LMG1518. Products of acetic acid catabolism counteracted the pH decrease caused by keto-gluconate formation, and provided a constant pH of 5.5 of the growth medium, throughout the whole process.

Fig. 14 *A. xylinum* cellulose formed on the surface of a roller in horizontal fermentor.

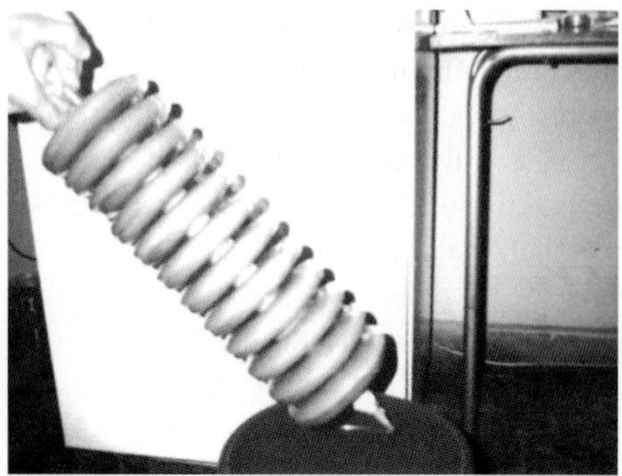

Fig. 15 *A. xylinum* cellulose deposited on disks in rotating disk fermentor.

Better control of BC synthesis can be achieved in special fermentors. Cellulose production in horizontal fermentors provides a combination of stationary and submerged cultures. The polymer is deposited on the surface of rollers or disks, rotating around the long axis (Figure 14). Part of their surface temporarily dips in the liquid medium or is above its surface (in the air). The advantages of this method include a larger polymer surface, synthesis of cellulose in a form of hollow fibers, different in diameter, as well as good process control, easy scale up, appropriate accessible surface for adhesion of bacteria and product deposition, better substrate utilization, higher rate of cellulose production, possibility of nutrient supply, additional aeration during the process, etc. (Sattler and Fiedler, 1990). Bungay and Serafica (1997) produced BC in a 1-L disk (12 cm in diameter) fermentor and revealed that optimum sugar (sucrose or glucose) concentration was 10 g L^{-1}, disks rotation rate 12 r.p.m., and constant pH 5.0. Krystynowicz et al. (1997) successfully applied a 11-L disk reactor containing 3 l of the H-S medium for BC synthesis by the *A. xylinum* E_{25} strain (Figure 15). The optimum rate of rotation of its 11 disks (12 cm in diameter each) appeared to be 3 r.p.m. As much as 4.2 g L^{-1} of dry cellulose was obtained after 7 d of growth.

BC production in submerged cultures usually requires common fermentors, equipped with some static parts (baffles, blades, etc.), enabling cell adhesion, since cellulose synthesis in the free aqueous phase usually drops. In analogy, various water-insoluble microparticles, such as sand, diatomaceous earth, or glass beads, added to the culture medium, enhanced BC productivity, since the biofilm formed by bacteria on the particles probably limited oxygen transfer and stopped glucose oxidation to gluconic acids (Vandamme et al., 1998).

Large-scale cellulose production, in fermentors with continuous agitation and aeration, encounters many problems, including spontaneous appearance of Cel^- mutants, which contribute to a decline in cellulose production. Optimized agitation and aeration prevent turbulence, which negatively effects cellulose polymerization and crystallization, thus reducing the polysaccharide yield. For instance, a rate of agitation equal to 60 r.p.m. and aeration of 0.6 volume per volume per minute (v.v.m.) were optimum for the *A. aceti* ssp. xylinum ATCC 2178 strain cultured in a 300-L fermentor for 45 h at 30 °C, and 10 g BC L^{-1} d^{-1} was obtained (Laboureur, 1988). Ben-Bassat et al. (1989) investigated BC production by the *A. xylinum* 1306–21 strain in 14-L Chemap fermentor, cultured in CSL medium containing glucose and CSL, 2% of each. The rate of agitation was equal to 900 r.p.m. and the dissolved oxygen concentration corresponded to 30% air saturation. The polymer yield amounted to 5.1 g L^{-1} d^{-1}. Chao et al. (2000) applied a 50-L internal loop airlift reactor for BC synthesis with the *A. xylinum* sp. BPR2001 strain. Aeration with oxygen-enriched air enlarged cellulose yield from 3.8 to 8 g L^{-1} after 67 h.

Production of BC in fermentors encounters similar agitation problems as cultivation of fungi or streptomycetes, since most of these organisms grow in pellet or filament form and their culture broths become non-Newtonian fluids. A high concentration of mycelium in the form of a dense suspension of diversely shaped particles limits agitation and gas transfer. Increasing the rate of agitation in order to improve aeration leads to damage to the product structure due to shearing forces. Bauer et al. (1992) took this aspect into consideration and introduced a polyacrylamide-protecting agent into the growth medium, applied for BC synthesis in fermentors. The protector reduced the shear damage and affected the specific productivity at high densities negatively since the cell growth rate was lower.

Accumulation of some metabolites can also affect production of cellulose. For instance, studies of Kouda et al. (1998) revealed that a high partial pressure of CO_2 negatively affected *A. xylinum* growth and reduced BC yield.

Laboureur (1998) developed a method for BC production by *A. xylinum* sp. ATCC 21780 strain in 300- and 500-L fermentors in a medium containing 5% sucrose, 0.05% yeast extract, 0.2% citric acid, nitrogen salts, Mg^{2+}, and phosphates. The medium was inoculated with 12% of inoculum, cultivated for 160 h. The process was run for 45 h at 30 °C and pH 4.8, and an aeration rate of 0.6 v.v.m. BC synthesis yield reached 18 g L^{-1} (10 g L^{-1} d^{-1}). Another strain, *Acetobacter* sp. ATCC 8303, was grown at 28 °C and pH 4.6, in a modified medium, containing 0.28% glucose, 0.07% maltose, 0.03% CSL, and 0.03% yeast extract. The aeration was 1 v.v.m. for the first 33 h and 0.5 v.v.m. for the last 12 h of the process, and the BC yield was 13 g L^{-1} d^{-1}.

Preparing an inoculum of appropriate cell density for large-volume fermentors is also a problem, mainly because the cells are entrapped in cellulose. To liberate the cells and increase their density, Brown (1989c) applied preparations of cellulases for partial cellulose hydrolysis. In the presence of these enzymes, the cell density reached 10^8 mL^{-1}, whereas in their absence it was only $1.12 \cdot 10^7$ mL^{-1}, after 30 h of growth.

In addition to the above-mentioned methods, the production of cellulose in the form of hollow fibrils of various diameters has also been tried (Yamanaka et al., 1990). Such material could be especially useful for the production of small-diameter blood vessels. The hollow BC fibril is obtained by growing the BC-producing bacteria on the inner and/or outer surface of an oxygen-permeable hollow carrier, produced from, for example, cellophane, Teflon, silicone, ceramics, etc.

9.2.4 Continuous Cultivation

Microbial cellulose can be also produced in a static continuous culture (Sakair et al., 1998). *A. xylinum* strain was cultured on trays, in the S-H medium, and after 2 d the pellicle produced on the surface was picked up, passed through a sodium dodecylsulfate bath to kill the cells, and set on a winding roller. The process was continued for a couple of weeks at a winding rate of 35 mm h^{-1} and fresh S-H medium was added into the trays every 8–12 h to keep its optimum level. A cellulose filament of more than 5 m was collected using this method, indicating its industrial application potential.

9.3 *In vitro* Biosynthesis

In vitro synthesis of cellulose has been one of the most difficult topics in the area of cellulose research. For the first time, cellulose was synthesized utilizing the 'reversed action' of cellulase as a catalyst, by Kobayashi et al. (1991), even though chemical synthetic methods have been applied before. Several monomers and catalysts have been used (Nakatsubo et al., 1989), but none of them gave the desired product, the stereoregular polymer of β-1,4-linked glucopyranoses. The idea of BC synthesis *in vitro* originated from experiments using *A. xylinum* cell-free extracts (Glaser, 1958), raw preparations of membranes (Colvin, 1980; Swissa et al., 1980; Aloni et al., 1982), and membranes solubilized with digitonin (Aloni et al., 1983). Such investigations proved that UDPGlc is the substrate for *A. xylinum* cellulose synthesis, c-di-GMP is the specific activator (Ross et al., 1987), and Ca^{2+} and Mg^{2+} ions are essential for this process (Swissa et al., 1980). Synthesis *in vitro*, carried out using *A. xylinum* membranes solubilized with digitonin (Lin et al., 1985), gave cellulose fibers 17 ± 2 Å in diameter. Their morphology and size resembled *A. xylinum* cellulose fibrils formed in the presence of factors disturbing crystallization of the nascent polymer (see Section 7.3.2). Further research, including X-ray diffraction analysis (Bureau and Brown, 1987) of cellulose produced by the *A. xylinum* membrane fractions, revealed that synthesis *in vitro* led to the amorphous cellulose II allomorph, whereas *in vivo* the highly crystalline cellulose I was obtained. Lin and Brown (1989) proved that purified CS also catalyzed cellulose synthesis *in vitro*. All these experiments were run on the microscale level and it is hard to believe today that commercial *in vitro* production of microbial polymer is possible, all the more since in the 1990s no progress in this area has been made.

9.4 Chemo-enzymatic Synthesis

Experiments on the chemical synthesis of cellulose have not given the expected results. Branched and low-molecular-weight glucans (Husemann and Muller, 1966) or polymers of β-1,4- and α-1,4-linked glucopyranoses (Micheel and Brodde, 1974) were obtained.

Successful experiments of Kobayashi et al. (1991) indicate that using coupled chemical and enzymatic methods may lead to the development of commercial *in vitro* synthesis of cellulose. The authors applied β-D-cellobiosyl fluoride, obtained by means of chemical synthesis, as a substrate and *T. reesei* cellulase as a catalyst in an aqueous–organic solvent system (a mixture of acetic buffer pH 5.0 and acetonitrile, 1:5 v/v) and achieved water-insoluble 'synthetic cellulose' having a DP > 22. X-ray and ^{13}C-NMR analyzes revealed that the product was the cellulose II. The authors stated that the DP of the product depends on reaction conditions, especially on the aqueous–organic solvent composition. They found that in the presence of higher concentrations of the substrate or acetonitrile, the cellulase – which acts as a glycosyltransferase – produces mainly soluble cellooligosaccharides (DP ≤ 8), which may also find numerous uses. Although these studies have not been continued, the method proposed by Kobayashi might be a promising alternative for total enzymatic *in vitro* cellulose production.

9.5 Production Processes Expected to be Applied in the Future

Recently, an original method of BC production on a large-scale was proposed by Nichols and Singletary (1998), who patented the idea of the construction of transgenic plants capable of this polymer synthesis. The authors intend to express three *A. xylinum* CS genes (*bcsA*, *bcsB*, and *bcsC*, see Section 7.4) and two *A. xylinum* diguanyl cyclase genes in storage tissues (roots, tubers, grains) of crops (potato, maize, oat, sorghum, millet, wheat, rice, sugarcane, etc.). They suppose to obtain masses of the pure polymer, easy and cheap to recover. According to their method, its production would be less expensive than by means of fermentation. The authors also emphasize the ecological aspect of their procedure, since an additional benefit of breeding of pure cellulose-producing transgenic plants would be a more economical use of forest resources.

9.6 Recovery and Purification

Microbial cellulose obtained via stationary or agitated culture is not completely pure and contains some impurities, such as culture broth components and whole *A. xylinum* cells. Prior to use in medicine, food production, or even the papermaking industry, all these impurities must be removed.

One of the most widely used purification methods is based on treatment of BC with a solution of hydroxides (mainly sodium and potassium), sodium chlorate or hypochlorate, H_2O_2, diluted acids, organic solvents, or hot water. The reagents can be used alone or in combination (Yamanaka et al., 1990). Immersing BC in such solutions (for 14–18 h, in some cases up to 24 h) at elevated temperature (55–65 °C) markedly reduces the number of cells and coloration degree.

Boiling in 2% NaOH solution after preliminary rinsing with running tap water was also reported (Yamanaka et al., 1989). Watanabe et al. (1998a) immersed the polymer in 0.1 NaOH at 80 °C for 20 min and then washed it with distilled water. Takai et al. (1997) treated BC with distilled water and 2% NaOH, and neutralized the mat with 2% acetic acid.

Krystynowicz et al. (1997) developed a procedure starting from washing crude BC in running tap water (overnight), followed by boiling in 1% NaOH solution for 2 h, washing in tap water to accomplish NaOH removal (1 d), neutralization with 5% acetic acid, and its removal with tap water. The final BC preparations obtained contained less than 3% protein, and were suitable for certain food and medical purposes.

Medical application of BC requires special procedures to remove bacterial cells and toxins, which can cause pyrogenic reactions. One of the most effective protocols begins with gentle pressing of the cellulose pellicle between absorbent sheets to expel about 80% of the liquid phase and then immersing the mat in 3% NaOH for 12 h. This procedure is repeated 3 times, and after that the pellicle is incubated in 3% HCl solution, pressed, and thoroughly washed in distilled water. The purified pellicle is sterilized in an autoclave or by ^{60}Co irradiation. It performs excellently as a wound dressing since it contains only 1–50 ng of lipopolysaccharide endotoxins, whereas BC purified using conventional methods usually contains 30 μg or more of these substances (Ring et al., 1986).

10
Properties

BC is an extremely insoluble, resilient, and elastic polymer, having high tensile strength. It has a reticulated structure, in which numerous ribbon-shaped fibrils are composed of highly crystalline and highly uniaxially oriented cellulose subfibrils. This three-dimensional structure, not found in plant-originating cellulose, brings about a higher crystallinity index (60–70%) of BC. Furthermore, fibers of the plant polymer are about 100 times thicker than BC microfibrils and therefore the bacterial polymer has an about 200 times larger accessible surface. In comparison to *A. xylinum* cellulose from agitated cultures, the polymer synthesized under static conditions has a higher DP (14,400 and 10,900, respectively), crystallinity index (71 and 63%, respectively), tensile strength (Young's modulus of 33.3 and 28.3, respectively), but lower water holding-capacity (45 and 170 g water g BC^{-1}, respectively) and suspension viscosity (0.04 and 0.52 Pa·s, respectively) in its disintegrated form (Watanabe et al., 1998a).

Microbial cellulose appears to be gelatinous, since its liquid component (usually water), present in voids among very fine ribbons, amounts to at least 95% by weight. The bacterial polysaccharide has a high water-binding capacity, but the majority of the water is not bound to the polymer and it can be squeezed out by gentle pressing. Drying of BC leads to paper-like sheets, having a thickness of 0.01–0.5 mm (Yamanaka et al., 1990; Krystynowicz et al., 1995, 1997) and good absorptive properties. In addition to a high Young's modulus, BC also displays high sonic velocity and, because of these unique mechanical properties, it can be applied as acoustic membranes (Vandamme et al., 1998).

The properties of BC can be modified either during its synthesis or when the culture is completed. Some compounds, like cellulose derivatives, sulfonic acids, alkylphosphates, or other polysaccharides (starch, dextran), introduced into the nutrient broth alter the macroscopic morphology, the tensile strength, the optical density, and the absorptive properties of the final product (Yamanaka et al., 2000). BC can also be combined with other substances, added either to wet or dried cellulose, thus giving composites of desired physicochemical properties (Yamanaka, 1990). The auxiliary materials used for this purpose comprise granules and fibers of various inorganic and organic compounds, such as alumina, glass, agar, alginates, carragenan, pullulan, dextran, polyacrylamide, heparin, polyhydroxylalcohols, gelatin, collagen, etc. They are combined with BC sheets by impregnation, lamination, or adsorption, or mixed with the disintegrated polymer. The composites are further subjected to a shaping treatment, thus giving various products.

Recently, Kim et al. (1999) reported an enzymatic method of BC modification using *L. mesenteroides* dextransucrase and alternansucrase. In their presence *A. xylinum* ATCC 10821 synthesized 'soluble cellulose', which was a glucan composed of 1,4-, 1,6-, and 1,3-linked monomers.

The basic BC properties are summarized in Table 2.

11
Applications

BC belongs to the generally recognized as safe (GRAS) polysaccharides and therefore it has already been put to a multitude of different uses. Commercial application of this polymer results from its unique properties and developments in effective technologies of production, based on growth of improved bacterial strains on cheap waste materials. The advantage of BC is its chem-

Tab. 2 Properties of BC

High purity
High degree of crystallinity
Greater surface area than that of conventional wood pulp
Sheet density from 300 to 900 kg m^{-3}
High tensile strength even at low sheet density (below 500 kg m^{-3})
High absorbency
High water-binding capacity
High elasticity, resilience, and durability
Nontoxicity
Metabolic inertness
Biocompatibility
Susceptibility to biodegradation
Good shape retention
Easy tailoring of physicochemical properties

ical purity and the absence of substances usually occurring in the plant polysaccharide, which requires laborious purification. In addition to the shape of BC sheets, their area and thickness can be tailored by means of culture conditions. Relatively easy BC modification during its biosynthesis enables regulation of such properties as molecular mass, elasticity, resilience, water-holding capacity, crystallinity index, etc. BC microfibrils may bind both low-molecular-weight substances and polymers, added for instance to the growth medium, thus giving novel commodities. BC can be also a raw material for further chemical modifications.

Based on the assumption that as much as 10,000 kg of the bacterial polymer can be obtained per year from static culture having 1 hectare surface area and only 600 kg of cotton is harvested in the same period of time from a field of the same area, the perspectives for a wider use of BC become more apparent (Kudlicka, 1989).

The main potential BC applications are summarized in the Table 3.

11.1 Technical Applications

Compared to PC sheets, BC has satisfactory tensile strength even at low sheet density (300–500 kg m^{-3}) (Johnson and Neogi,

Tab. 3 Applications of bacterial cellulose

Sector	*Application*
Cosmetics	Stabilizer of emulsions such as creams, tonics, nail conditioners, and polishes; component of artificial nails
Textile industry	Artificial skin and textiles; highly adsorptive materials
Tourism and sports	Sport clothes, tents, and camping equipment
Mining and refinery waste treatment	Spilt oils collecting sponges, materials for toxins adsorption, and recycling of minerals and oils
Sewage purification	Municipal sewage purification and water ultrafiltration
Broadcasting	Sensitive diaphragms for microphones and stereo headphones
Forestry	Artificial wood replacer, multi-layer plywood and heavy-duty containers
Paper industry	Specialty papers, archival documents repair, more durable banknotes, diapers, and napkins;
Machine industry	Car bodies, airplane parts, and sealing of cracks in rocket casings
Food production	Edible cellulose and 'nata de coco'
Medicine	Temporary artificial skin for therapy of decubitus, burns and ulcers; component of dental and arterial implants
Laboratory/research	Immobilization of proteins, chromatographic techniques, and medium component of *in vitro* tissue cultures

1989). Therefore, BC is an excellent component of papers, providing better mechanical properties. Microfibrils of the bacterial polymer form a great number of hydrogen bonds when the paper is subjected to drying, thus giving improved chemical adhesion and tensile strength. BC-containing paper not only shows better retention of solid additives, such as fillers and pigments, but is also more elastic, air permeable, resistant to tearing and bursting forces, and binds more water (Iguchi et al., 2000).

A beneficial effect such as an improved aging resistance was achieved by adding small amounts of BC to cotton fibers to obtain hand-made paper, used for old document repair. The paper displayed appropriate ink receptivity and specific snap (Krystynowicz et al., 1997). Paper containing 1% of BC meets the international standard ISO 9706:1994 for information and documentation papers, and also has a specific snap comparable to that of rag paper. It is possible to use BC for pressboard making, as well as production of paperboard used as an electro-insulating material and for bookbinding.

BC can also be used as a surface coating for specialty papers. For this purpose, a filtered suspension of BC (homogenized and mixed with other components) is added using special applicators, either within wet sheet formation, or to a partially or completely dried sheet. The coating improves properties such as gloss, brightness, smoothness, porosity, ink receptivity, and tensile strength. Other additives like starch, organic polymers, including CM-cellulose, organic, or inorganic pigments may also be used. Johnson and Neogi (1989) claim that paper coated with 3% of BC (solid matter) displays gloss properties and surface strength similar to rotogravure paper having 20% traditional coating. The authors also state that coating with a mixture of BC and CM-cellulose gives even better properties, since they act synergistically.

BC is also a valuable component of synthetic paper (Iguchi et al., 2000) since nonpolar polypropylene and polyethylene fibers, providing insulation, heat resistance, and fire-retarding properties, cannot form hydrogen bonds. The amount of wood pulp in this type of paper is usually from 20 to 50% to achieve good quality. Using BC enables us to decrease the amount of the additives without any effect on the synthetic paper properties.

BC also appeared to be a good binder in nonwoven fabric-like products (Yamanaka et al., 1990) commonly used in surgical drapes and gowns, and containing various hydrophilic and hydrophobic, natural and synthetic fibers, such as cellulose esters, polyolefin, nylon, acrylic glass, or metal fibers. Even small amounts of BC improve tensile and tear strength of the fabric, e.g. 10% of the bacterial polymer is equivalent to 20–30% of latex binder.

The scope of BC uses can be even wider since it is modifiable during synthesis (Brown, 1989a,b). For this purpose, CM-cellulose or copolymers of saccharides and dicarboxylic acids are added directly to the culture medium (Yamanaka et al., 1989). Cellulose obtained in the presence of CM-cellulose and dried with organic solvents has a resilient and elastic character, as well as higher water binding capacity (adsorbs faster more water). The optimum CM-cellulose concentrations range from 0.1 to 5% (w/v). The polarity of the organic solvent also influences the BC features. After treatment with acetone the polymer is elastic and rubber-like, whereas after drying with absolute ethanol, BC resembles leather.

Since BC is susceptible to enzymatic digestion, it is modified to obtain various composites of satisfactory biodegrability and strength. Their production is based either on chemical reaction between cellulose and the copolymer or on culturing the BC-producing strain in the copolymer-containing medium.

11.2 Medical Applications

A cellulose pad from a static culture is a ready-to-use, naturally 'woven' wound dressing material that meets the standards of modern wound dressings (Figure 16). It is sterilizable, biocompatible, porous, elastic, easy to handle and store, adsorbs exudation, provides optimum humidity, which is essential for fast wound healing, protects from secondary infection and mechanical injury, does not stick to the newly regenerated tissue, and alleviates pain by heat adsorption from burns. BC sheets are also excellent carriers for immobilization of medicinal preparations, which speed up the healing process.

Since BC pellicles can have various dimensions, it is relatively easy to produce dressings for extensive wounds. Because of recent problems with products of animal origin, collagen dressings can be replaced with BC ones. This thesis is additionally supported by undoubtedly positive results of clinical tests. For instance, the BC preparation, Prima Cel™, produced by Xylos Corp., according to the Rensselaer Polytechnic Institute (USA), has been applied as a wound dressing in clinical tests to heal ulcers. The results obtained were satisfactory, since after 8 weeks 54% of the patients recovered and the remaining ulcers were almost healed (Jonas and Farah, 1998). Other commercial preparations of *A. xylinum* cellulose, such as Biofil® and Bioprocess®, appeared to be excellent as skin transplants, and in the treatment of third-degree burns, ulcers, and decubitus. Another preparation, Gengiflex®, found application in recovering periodontal tissues (Jonas and Farah, 1998). Investigations on *A. xylinum* cellulose pads prepared by Krystynowicz et al. (2000) (Figure 17) indicate that general use as wound dressings is possible.

Results of investigations on BC hollow fibers use as artificial blood vessels and ureters are also promising (Yamanaka et al., 1990). Antithrombic BC property (blood

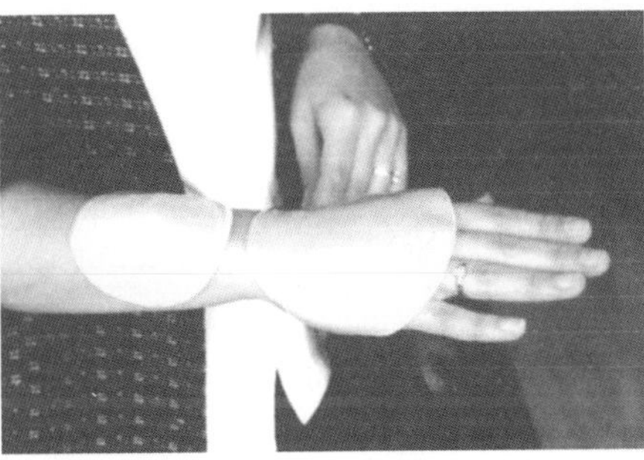

Fig. 16 Cellulose pellicle as a material for wound dressings.

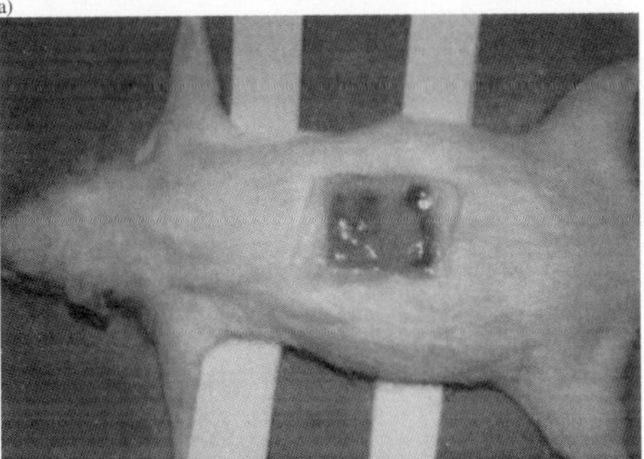

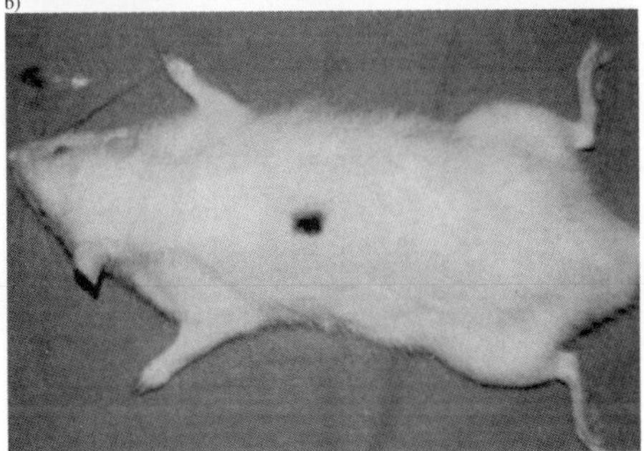

Fig. 17 Burns treated with a BC dressing, before (a) and after (b) healing (Krystynowicz et al., 2000).

compatibility) was evaluated by a test based on replacement of adult dog blood vessels (parts of the descending aorta and jugular vein) with the artificial counterpart made of BC. One month later the artificial vessel was removed and the state of adhesion of thrombi on its inner surface was examined. A good open state of the BC blood vessel was maintained.

Another successful experiment showing the application of biosynthetic cellulose was substitution of the dog dura matter in the brain (Jonas and Farah, 1998).

Because of its high tensile strength, elasticity, and permeability to liquids and gases, dried BC was applied as an additional membrane to protect immobilized glucose oxidase in biosensors used for assays of blood glucose levels. This BC membrane enhanced the electrode stability in 10 times diluted human blood solution, up to 200 h. Other commercial protecting membranes, such as cuprophan (AKZO, England), provided an electrode stability up to 30 h only. In undiluted human blood the biosensor coated with cuprophan was stable for 3–4 h, whereas the BC membrane prolonged its stability up to 24 h.

11.3
Food Applications

Chemically pure and metabolically inert BC has been applied as a noncaloric bulking and stabilizing agent in processed food. Similar to pure preparations of the plant polysaccharide and its derivatives, it is used for stabilization of foams, pectin and starch gels, emulsions, e.g. canned chocolate drinks and soups, texture modification, e.g. improvement of the consistency of pulps, enhancement of adhesion, replacement of lipids, including oils, and dietary fiber supplements (Ang and Miller, 1991; Kent et al., 1991; Krystynowicz et al., 1999).

The first successful commercial application of BC in food production is 'nata de coco' (Sutherland, 1998). It is a traditional dessert from the Philippines, prepared from coconut milk or coconut water with sucrose, which serves as a growth medium for BC-producing bacteria. Consumption of the pellicle is believed to protect against bowel cancer, artheriosclerosis, and coronary thrombosis, and prevents a sudden rise of glucose in the urine. Therefore 'nata de coco' is becoming increasingly popular, not only in Asia.

Another popular BC-containing food product is Chinese Kombucha (teakvass or tea-fungus), obtained by growing yeast and acetic acid bacteria on tea and sugar extract. The pellicle formed on the surface contains both cellulose and enzymes healthy for humans. Their abiotic activity is especially stimulating for the large bowel and the whole alimentary tract. Kombucha is believed to protect against certain cancers (Iguchi et al., 2000).

Results of studies carried out by the authors, who applied BC pellicles synthesized by *A. xylinum* E_{25} for wine and juice filtration, and for immobilization of polyphenols, are promising (Krystynowicz et al., 1999). Preparations of bioactive anthocyanins, enriched in dietary fiber, are excellent for functional food production. BC also appeared to be an attractive component of bakery products, since it plays a role of dietary fiber, is taste and odorless, and prolongs the shelf-life.

11.4
Miscellaneous Uses

The large accessible surface area, high durability, and superior adsorptive properties as well as possibility of modification by means of physical or chemical methods, means that BC can be applied as a carrier for immobilization of biocatalysts. Cellulose gels containing immobilized animal cells were used for their culture to produce interferon, interleukin-1, cytostatics, and monoclonal antibodies (Iguchi et al., 2000).

BC was also applied for adsorption of cells of *Gluconobacter oxydans, Acetobacter methanolyticus* and *Saccharomyces cerevisiae.* The immobilized strains appeared to be effective producers of gluconate (84–92% yield), dihydroxyacetone (90–98% yield), and ethanol (88–92% yield), respectively, and displayed better operational and thermal stability.

Purified BC can be a raw material for synthesis of cellulose acetate, nitrocellulose, CM-cellulose, hydroxymethylcellulose, methylcellulose, and hydroxycellulose (Yamanaka, 1990). If BC is produced in the presence of compounds that interfere with regular fibril assembly and which influence the β-1,4-glucan structure, such as CM-cellulose or other cellulose derivatives, and other carbohydrates (starch, dextran), sulfonates, and alkylphosphates, the resulting microbial polymer has novel, additive-dependent, and useful properties, including optical transparency or higher water-binding capacity, even after repeated soaking and drying.

The potential application of BC in the chemical, paper, and textile industries depends on its price and accessibility. To meet these demands, BC has to be produced by highly efficient strains, growing on cheap substrates, in sophisticated surface, solid-state, or submerged fermentors (Vandamme et al., 1998).

12 Patents

Growing rapidly, the interest in BC is reflected in a number of patents (annually about 20 since the 1980s) and publications (20–40 per year for the last 10 years) devoted to this unique polymer (Iguchi et al., 2000). Some basic patents concerning the biosynthesis, properties, and applications of BC are presented in Table 4. These patents that are cited in chapter are listed in the references.

13 Outlook and Perspectives

The first scientific paper reporting on an unusual substance formed by acetic acid bacteria and known for ages in many countries as 'vinegar plant' (Yamanaka et al., 1989) was published 115 years ago (Brown, 1886). Further studies revealed that the substance is a super-pure cellulose. Metabolic pathways and the complicated molecular machinery of the polysaccharide biosynthesis, as well as the intriguing dynamics of its nascent chains association into a structure with unique properties, has been elucidated. Although progress and limited commercialization of BC have undoubtedly taken place, the relevant biotechnology, competitive to modern industrial technologies for PC production, has not yet been developed.

So, what to do, to achieve success and accomplish commercialization of BC production? First of all, stable overproducer strains of the polymer have to be constructed, using recent achievements in molecular genetics and biology. These strains have to assimilate a wide range of carbon sources, as well as display lower tendency towards spontaneous mutation to $Cell^-$ mutants and effectively synthesize cellulose (10–15 $g\,L^{-1}\,d^{-1}$) under agitated culture conditions. Construction of new bioreactors for both stationary and submerged culture is necessary. The considerable reduction of BC production cost could be attained by replacing expensive nutrient media components

Tab. 4 Selected patents concerning BC

Patent number	Patent holder	Inventors	Title	Date of publi-cation	Major claims
WO 0125470	Novozymes A/S, Denmark	Herbert, W., Chanzy, H. D., Ernst, S., Schulein, M., Husum, T. L., Kongsbak, L.	Cellulose films for screening	2001	BC microfibril films containing fluorescein-labeled hemoglobin or galactomannan can be used to detect proteases and mannanases, respectively
WO 0105838	Pharmacia Corp., USA	Yang, Z. F., Sharma, S., Mohan, C., Kobzeff, J.	Process for drying reticulated bacterial cellulose without co-agents	2001	The reticulated *Acetobacter* cellulose subjected to dispersing in a solvent, e.g. hydrocarbon (hexane), aliphatic alcohol, and/or alkyl sulfoxide (DMSO), separation from the solvent, and drying, may be rehydrated to provide uniform dispersions having high viscosity
JP 11255806 A2	Bio-Polymer Research Co. Ltd, Japan	Watanabe, O.	Freeze drying method for microfibrous cellulose concentrate	1999	Freeze drying of *Acetobacter* cellulose gives dry microfibrous cellulose with good retention of its original properties after redispersing in water
WO 9940153	Monsanto	Smith, B. A., Colegrove, G. T., Rakitsky, W. G.	Acid-stable, cationic compatible cellulose compositions useful as rheology modifiers	1999	Cationic co-agents, acids, and/or cationic surfactants are used to form acid-stable *Acetobacter* cellulose compositions, which are useful as rheology modifiers
WO 9943748	Sony	Uryu, M., Tokura, K.	Biodegradable composite polymer material	1999	A new composite material has been obtained by drying the *A. xylinum* cellulose, its pulverization, and blending with a biodegradable polymer material
JP 0056669 A2	Canon Co., Japan	Minami, M., Mihara, C., Takeda, T., Kikuchi, Y.	Composite, for use in thermoformed articles, comprises cellulose and saccharide ester derivative	1999	A composite comprising BC and a saccharide ester can be used for production of biodegradable, thermoformed articles, which have improved processability, mechanical strength and flexibility
JP 11172115 A2	Ajinomoto Co. Inc., Japan	Suzuki, O., Kitamura, N., Matsumoto, R.	Bacterial cellulose-containing composite absorbents with high liquid absorption	1999	Dispersing highly water-absorbing polymer particles, dissolved in an organic solvent, in an aqueous solution of defibrillated BC, followed by organic solvent removal and partial drying, gives excellent and stable absorbents

Tab. 4 (cont.)

Patent number	*Patent holder*	*Inventors*	*Title*	*Date of publi-cation*	*Major claims*
JP 11246602 A2	Bio-Polymer Research Co. Ltd, Japan	Tahara, N., Hagamida, T., Miyashita, H., Watanabe, O.	Preservation of wet bacterial cellulose	1999	Wet BC can be preserved with alkyl sulfate salts or NaOH and/or KOH
JP 11187896 A2	Bio-Polymer Research Co. Ltd, Japan	Tabata, T., Toyosaki, H., Tsuchida, T., Yoshinaga, F.	A method for screening cellulose-producing bacteria using cellulase	1999	A rapid and convenient method for screening a large number of cellulose membrane-producing strains is presented; the metabolic peculiarities of the strains and conditions needed for them to produce cellulose pellicle are explained
JP 11092502 A2	Bio-Polymer Research Co. Ltd, Japan	Shoda, M., Kanno, Y., Koda, T., Yoshinaga, F.	Manufacture of bacterial cellulose under oxygen-enriched conditions	1999	Passing more than 21% oxygen-containing air through *A. xylinum* ssp. sucrofermentans culture in an air-lift fermentor, enables accumulation of 6.5 g L^{-1} of BC after 75 h
JP 11117120 A2	Toray Industries Inc., Japan	Hara, T., Amano, J.	Fibers made from blends of bacterial cellulose and polymers having flexible main backbone	1999	The blends contain polyvinyl alcohol-type polymers at weight ratio of 2–50%
JP 11181001 A2	Bio-Polymer Research Co. Ltd, Japan	Matsuoka, M., Toyosaki, H., Matsumura, T., Ougiya, H., Tsuchida, T., Yoshinaga, F.	Production of bacterial cellulose	1999	BC, useful for filler retention aids for paper-making, can be produced in agitated cultures of *Acetobacter* strains, in the presence of water-soluble polysaccharides, e.g. CM-cellulose
JP 11137163 A2	Shikishima Seipan Co. Ltd, Japan	Kondo, M., Yamada, M., Inoue, S.	Manufacture of bread from dough containing bacterial cellulose	1999	Addition of BC increases water absorption of dough, thus giving bread with high water-holding capacity
JP 11221072 A2	Bio-Polymer Research Co. Ltd, Japan	Ishikawa, A., Tsuchida, T., Yoshinaga, F.	Bacterial cellulose production enhancement	1999	An *A. xylinum* mutant having higher cellular levels of UDP, UTP, and UDPGlc as well as higher carbamoyl phosphate synthetase activity is an excellent cellulose producer

Tab. 4 (cont.)

Patent number	*Patent holder*	*Inventors*	*Title*	*Date of publi-cation*	*Major claims*
JP 11269797 A2	Toppan Printing Co. Ltd, Nakano Vinegar Co. Ltd, Japan	Yamawaki, K., Tomita, T., Harasawa, A., Kaminaga, J., Kawasaki, K., Matsuo, R., Sasaki, N., Fukagai, M., Tsukamoto, Y.	Impregnated paper with good water resistance and stiffness	1999	Paper impregnated with silane coupler-grafted BC derivatives is moisture resistant
PP 299907	Technical University of Lodz, Poland	Krystynowicz, A., Czaja, W., Bielecki, S.	Biosynthesis and aplication of bacterial cellulose	1999	Production of BC by wild and mutant *A. xylinum* strains, cultured in static or agitated cultures, and under various conditions as well as an influence of culture conditions on the cellulose properties are described
JP 11018758 A2	Bio-Polymer Research Co. Ltd, Japan	Yamamoto, T., Yano, H., Yoshinaga, F.	Horizontal type spinner culture vessel, having high oxygen-supplying efficiency	1999	BC can be produced in the spinner culture tank, equipped with mixing impellers, and providing high efficiency of oxygen supply
JP 10298204 A2	Ajinomoto Co. Inc., Japan	Ishikara, M., Yamanaka, S.	Bacterial cellulose with ribbon-like microfibril shape	1998	Cellulose fibrils having a short axis 10–100 nm and a long axis 160–1000 nm are produced extracellularly by cellulose-generating bacteria, e.g. *Acetobacter pasteurianus*, in a culture containing cell division inhibitor
US 005723764	Pioneer Hi-Bred International Inc., USA	Nichols, S. E., Singletary, G. W.	Cellulose synthesis in the storage tissue of transgenic plants	1998	Introducing the genes for cellulose biosynthesis from the species *A. xylinum* into a given plant provides a method of synthesizing cellulose in the storage tissue of transgenic plants
JP 10077302	Bio-Polymer Research Co. Ltd, Japan	Tabuchi, M., Watanabe, K., Morinaga, Y.	Solubilized bacterial cellulose and its compositions or composites for moldings and coatings.	1998	Cellulose synthesized by *A. xylinum* in stationary culture was solubilized by stirring in a mixture of DMSO and paraformaldehyde (25:5) at 100 °C for 3 h

Tab. 4 (cont.)

Patent number	*Patent holder*	*Inventors*	*Title*	*Date of publi-cation*	*Major claims*
WO 97/05271	Rensselaer Polytechnic Institute, USA	Bungay, H. R., Serafica, G.	Production of microbial cellulose using a rotating disc film bioreactor	1997	BC can be deposited by *A. xylinum*, cultured in a liquid medium inside the horizontal fermentor, on a plurality of disks, which rotate around the long axis of the fermentor and are partly submerged in the culture medium
JP 09025302 A2	Bio-Polymer Research Co. Ltd, Japan	Hioki, S., Watanabe, K., Ogya, H., Morinaga, Y.	Preparation of disintegrated bacterial cellulose for improved additive retention in paper manufacture	1997	BC which helps additive retention and causes no harm to freeness during paper manufacturing can be obtained by disintegration using a self-excited ultrasonic pulverizer
JP 09107892 A	Nakano Vinegar Co. Ltd, Japan	Furukawa, H., Maruyama, Y., Fukaya, M., Tsukamoto, Y., Kawamura, K.	Composition for stabilizing dispersion used in food	1997	A fine particulate (210 μm average particle diameter) BC, obtained by hydrolyzing *A. xylinum* cellulose with a mineral acid, is of low viscosity and is a sufficient dispersion stability agent in food products
WO 9744477	Bio-Polymer Research Co. Ltd, Japan	Naritomi, T., Kouda, T., Naritomi, M., Yano, H., Yoshinaga, F.	Continuous preparation of bacterial cellulose having a high production rate and yield	1997	BC obtained in continuous culture of *A. xylinum* (production rate at least 0.4 g L^{-1} h^{-1}) in a medium containing a substance enhancing the apparent substrate affinity to sugar (e.g. lactic acid) is useful as a thickener, humectant or stabilizer for production of food, cosmetics or paints, etc.
WO 9740135	Bio-Polymer Research Co. Ltd, Japan	Tsuchida, T., Tonouchi, N., Seto, A., Kojima, Y., Matsuoka, M., Yoshinaga, F.	Novel cellulose-producing bacteria and a process of producing it	1997	Production of cellulose with a new *A. xylinum* ssp. nonactoxidans, which lacks an ability to oxidize acetates and lactates, yields odor- and taste-less products, having excellent dispersability in water
WO 9712987	Bio-Polymer Research Co. Ltd, Japan	Kouda, T., Naritomi, T., Yano, H., Yoshinaga, F.	Production process for bacterial cellulose which is useful as material in various fields	1997	A new process for BC production involving maintaining a certain pressure inside a fermentor, reduces power required for agitation, and elevates production rate and yield
JP 09056392 A	Kikkoman Corp., Japan	Fukazawa, K., Imai, H., Kijima, T., Kikuchi, T.	Production of microorganism cellulose	1997	A cellulose pellicle synthesized by *A. xylinum* strain precultured in stationary conditions can be homogenized and used to inoculate the fresh culture medium

Tab. 4 (cont.)

Patent number	*Patent holder*	*Inventors*	*Title*	*Date of publi-cation*	*Major claims*
PL 171952 B1	Technical University of Lodz, Poland	Galas, E., Krystynowicz, A.	Method of obtaining bacterial cellulose	1997	A method of BC production in static culture using *A. xylinum* P23 strain is described
US 0824096	Bio-Polymer Research Co. Ltd, Japan	Kouda, T., Nagata, Y., Yano, H., Yoshinaga, F.	Production of bacterial cellulose through cultivation of cellulose-producing bacteria under specified conditions in aerated and agitated culture	1997	Cellulose produced by species of *Acetobacter, Agrobacterium, Rhizobium, Sarcina, Pseudomonas, Achromobacter, Alcaligenes, Aerobacter, Azotobacter,* and *Zoogloea*, in aerated and agitated cultures can be applied in production of food and cosmetics
JP 97–21905	Bio-Polymer Research Co. Ltd, Japan	Seto, H., Tsuchida, T., Yoshinaga, F.	Manufacture of bacterial cellulose by mixed culture of microorganisms	1997	To provide lactate and split sucrose, cellulose-producing *A. xylinum* strain can be grown together with lactic acid bacteria and *Saccharomyces* yeast; their presence in the culture broth markedly enhances BC yield
JP 08127601 A	Bio-Polymer Research Co. Ltd, Japan	Hioki, S., Watabe, O., Hori, S., Morinaga, Y., Yoshinaga, F.	Freeness regulating agent	1996	Production of a freeness regulating agent, comprising a defiberized BC, synthesized by *A. xylinum* ssp. sucrofermentans is described
JP 96316922	Bio-Polymer Research Co. Ltd, Japan	Hikawu, S., Hiroshi, T., Takayasu, T., Yoshinaga, F.	Manufacture of bacterial cellulose by addition of cellulose formation stimulators	1996	Compounds such as derivatives of choline, betaine, and fatty acids (salts and esters) appeared to stimulate cellulose production by *A. xylinum* strains
JP 08056689 A	Bio-Polymer Research Co. Ltd, Japan	Seto, H., Tsuchida, T., Yoshinaga, F.	Production of bacterial cellulose	1996	To obtain an edible BC, excellent in aqueous dispersibility, useful for retaining the viscosity of foods, cosmetics, coatings, etc., and enrichment of foods, *Acetobacter* strains can be cultured in saponin-containing medium
JP 08033494 A	Bio-Polymer Research Co. Ltd, Japan	Tawara, N., Koda, T., Hagamida, T., Morinaga, Y., Yano, H.	Method for circulating continuous production and separation of bacterial cellulose	1996	Circulating a culture liquid containing *A. xylinum* cells between a culture and a separation apparatus enables separation of the produced cellulose; also, a flotation separator or an edge filter can be applied for this purpose

Tab. 4 (cont.)

Patent number	*Patent holder*	*Inventors*	*Title*	*Date of publi-cation*	*Major claims*
JP 08034802 A	Gun Ei Chemical Industries Co. Ltd, Japan	Hirooka, S., Hamano, T., Miyashita, Y., Hanaue, K., Yamazaki, K., Shiichi, K., Shiichi, F.	Bacterial cellulose, production thereof and processed product made therefrom	1996	*A. xylinum* can synthesize cellulose in culture broths containing difficult to ferment, branched oligosaccharides
JP 08000260 A	Bio-Polymer Research Co. Ltd, Japan	Ishikawa, A., Tsuchida, T., Yoshinaga, F.	Production of bacterial cellulose with pyrimidine analogue-resistant strain	1996	A method of production of BC, in a high yield and at a low cost, using a pyrimidine analog-resistant *Acetobacter* mutant is presented
JP 08325301 A	Bio-Polymer Research Co. Ltd, Japan	Hori, S., Watabe, O., Morinaga, Y., Yoshinaga, F.	Cellulose having high dispersibility and its production	1996	BC synthesized by *A. xylinum* ssp. sucrofermentans, subjected after harvesting to partial hydrolysis with HCl, yields a fraction exhibiting high birefringence
JP 08276126	Bio-Polymer Research Co. Ltd, Japan	Ogiya, H., Watabe, O., Shibata, A., Hioki, S., Morinaga, Y., Yoshinaga, F.	Emulsification stabilizer	1996	The method of production of the emulsification stabilizer containing BC obtained from agitated culture of *A. xylinum* ssp. sucrofermentans and having low index of crystallization, is presented
JP 08009965 A	Bio-Polymer Research Co. Ltd, Japan	Ishikawa, A., Tsuchida, T., Yoshinaga, F.	Production of bacterial cellulose using microbial strain resistant to inhibitor of DHO-dehydrogenase	1996	A method of efficient production of BC, using *Acetobacter* strains resistant either to an inhibitor of DHO-dehydrogenase or to DNP is presented
US 005382656	Weyerhauser Co., USA	Benziman, M., Tal, R.	Cellulose synthase associated proteins	1995	CS-associated proteins, which have molecular weights of 20, 54, and 59 kDa are not encoded by CS operon genes *bcsA*, *B*, *C*, or *D*
JP 07313181 A	Bio-Polymer Research Co. Ltd, Japan	Takemura, H., Tsuchida, T., Yoshinaga, F., Matsuoka, M.	Production of bacterial cellulose using PQQ-unproductive strain	1995	Pyrroloquinolinequinone-unproductive *Acetobacter* mutant strain enables high-yield production of BC
JP 07184675 A	Bio-Polymer Research Co. Ltd, Japan	Matsuoka, M., Tsuchida, T., Yoshinaga, F.	Production of bacterial cellulose	1995	To obtain BC useful for retaining the viscosity of foods, cosmetics, or coatings, *Acetobacter* strains can be cultured in a methionine-containing medium

Tab. 4 (cont.)

Patent number	*Patent holder*	*Inventors*	*Title*	*Date of publi-cation*	*Major claims*
JP 07268128	Fujitsuko Co. Ltd, Japan	Kiriyama, S., Fukui, H., Toda, T., Yamagishi, H.	Dried material of cellulose derived from microorganism and its production	1995	Water-soluble stabilizer (preferably glucose or gelatin) added to cellulose gel obtained by culturing *A. xylinum* strain before drying of the material provides excellent and stable physical properties
WO 95/32279	Bio-Polymer Research Co. Ltd, Japan	Tonouchi, N., Tsuchida, T., Yoshinaga, F., Horinouchi, S., Beppu, T.	Cellulose-producing bacterium transformed with gene coding for enzyme related to sucrose metabolism	1995	*A. xylinum* transformant with a gene coding for invertase accumulates cellulose in a sucrose-containing medium
JP 07184677 A	Bio-Polymer Research Co. Ltd, Japan	Seto, H., Tsuchida, T., Yoshinaga, F.	Production of bacterial cellulose	1995	High-yield production of an edible BC useful for retaining the viscosity of cosmetics or coatings can be achieved by culturing an *Acetobacter* strain in an invertase-added medium containing sucrose as a carbon source
JP 07039386 A	Bio-Polymer Research Co. Ltd, Japan	Matsuoka, M., Takemura, H., Tsuchida, T., Yoshinaga, F.	Production of bacterial cellulose	1995	BC, useful for retaining the viscosity of a food, a cosmetic, a coating, etc., and usable as a food additive, an emulsion stabilizer, etc., can be obtained by culturing *A. xylinum* in a culture medium containing a carboxylic acid (e.g. lactic acid) salt
JP 07274987 A	Bio-Polymer Research Co. Ltd, Japan	Toda K., Asakura, T.	Production of bacterial cellulose	1995	To obtain BC in high yield, the cellulose-producing *Acetobacter* strain is cultured in a cylindrical container and oxygen is fed through an oxygen-permeable membrane at the bottom
JP 06125780	Nakano Vinegar Co. Ltd, Japan	Fukaya, M., Okumura, H., Kawamura, K.	Production of cellulosic substance of microorganism	1994	A method used to improve production efficiency of cellulose by an *Acetobacter* strain is presented; the method is based on adding a protein having affinity to the cellulose to the culture medium
JP 06248594 A	Mitsubishi Paper Mills Ltd, Japan	Katsura, T., Okafuro, K.	Low-density paper having high smoothness	1994	Excellent, low-density paper having high smoothness contains BC and the broad-leaved pulp in the ratio of 1/99:1/1

Tab. 4 (cont.)

Patent number	*Patent holder*	*Inventors*	*Title*	*Date of publi-cation*	*Major claims*
JP 06206904	Shin Etsu Chemical Co. Ltd, Japan	Horii, F., Yamamoto, H.	Bacterial cellulose, its production and method for controlling crystal structure thereof	1994	Xanthan gum or sodium CM-cellulose added to the culture broth of *A. xylinum* enable control of the crystal structure of BC
JP 06113873 A	Nippon Paper Industries Co. Ltd, Japan	Samejima, K., Mamoto, K.	Production of microbial cellulose	1994	Adding a sulfite pulp waste liquor and/or its permeate from ultrafiltration into a culture medium of a cellulose-producing microorganism enhances the cellulose yield and lowers the cost of its production
WO 94/20626	Bio-Polymer Research Co. Ltd, Japan	Beppu, T., Tonouchi, N., Horinouchi, S., Tsuchida, T.	*Acetobacter*, plasmid originating therein, and shuttle vector constructed from said plasmid	1994	Genetic recombination of cellulose-producing *Acetobacter* strains is performed using an *Acetobacter* strain, its endogenous plasmid, and a shuttle vector constructed from the latter plasmid and a plasmid from *Escherichia coli*
JP 06001647 A	Shimizu Corp., Japan	Yano, H., Narutomi, T., Okamura, K., Kawai, T., Minami, S.	Concrete and coating material	1994	The concrete or coating material containing disaggregated BC displays better dispersibility of cement or pigment particles and an enhanced fluidity
US 005268274 A	Cetus Corp., USA	Ben-Bassat, A., Calhoon, R. D., Fear, A. L., Gelfand, D. H., Meade, J. H., Tal, R., Wong, H., Benziman, M.	Methods and nucleic acid sequences for the expression of the cellulose synthase operon	1993	Nucleic acid sequences encoding the BC synthase from *A. xylinum*, and methods for isolating the genes and their expression in hosts are presented
EP 0396344 A2	Ajinomoto Co. Inc., Japan	Yamanaka, S., Ono, E., Watanabe, K., Kusakabe, M., Suzuki, Y.	Hollow microbial cellulose, process for preparation thereof, and artificial blood vessel formed of said cellulose	1990	BC prepared by culturing a cellulose-producing strain on one or both surfaces of an oxygen-permeable hollow carrier is useful as a substitute for a blood vessel or another internal hollow organ; the cellulose can be impregnated with a medium, cured, and cut if necessary

Tab. 4 (cont.)

Patent number	*Patent holder*	*Inventors*	*Title*	*Date of publi-cation*	*Major claims*
US 004950597	University of Texas, USA	Saxena, I. M., Roberts, E. M., Brown, R. M.	Modification of cellulose normally synthesized by cellulose-producing micro-organisms	1990	Mutants of *A. xylinum* that do not form a pellicle in liquid culture and synthesize cellulose almost exclusively as the allomorph cellulose II, that arise spontaneously or by nitrosoguanidine mutagenesis are described and the cellulose they produce is characterized
JP 02182194 A	Asahi Chemical Industries Co. Ltd, Japan	Matsuda, Y., Kamiide, K.	Production of cellulose with acetic acid bacterium	1990	Acetic acid bacteria having a synchronized cell cycle enable efficient production of BC, excellent in water holding properties, tensile strength, and purity, and displaying a relatively low DP
WO 89/12107	Brown, R. M.	Brown, R. M.	Microbial cellulose as a building block resource for specialty products and processes thereof	1989	A novel process for manufacturing BC using different bacterial species belonging to *Acetobacter, Rhizobium, Agrobacterium*, and *Pseudomonas*, and production of various articles from this polymer are described
EP 0258038 A3	Brown, R. M.	Brown, R. M.	Use of cellulase preparations in the cultivation and use of cellulose-producing micro-organisms	1989	To prepare an inoculum of appropriate cell density for large-volume fermentors, *A. xylinum* cells entrapped in cellulose can be liberated with cellulase preparation, which causes a partial cellulose hydrolysis
US 004863565	Weyerhauser Co., USA	Johnson, D. C., Neogi, A. M.	Sheeted products formed from reticulated microbial cellulose	1989	Strains of *Acetobacter* that are stable under agitated culture conditions and that exhibit reduced gluconic acids production, synthesize unique reticulated cellulose sheets, characterized by resistance to densification and great tensile strength
WO 89/11783	Brown, R. M.	Brown, R. M.	Microbial cellulose composites and processes for producing same	1989	Methods enabling production of various objects utilizing BC produced *in situ* or applied as a film are presented; a process for manufacturing currency from BC is described

Tab. 4 (cont.)

Patent number	*Patent holder*	*Inventors*	*Title*	*Date of publi-cation*	*Major claims*
EP 0323717 A3	ICI Plc, UK	Byrom, D.	Process for the production of microbial cellulose	1988	A process for the production of BC using a novel strain of the genus *Acetobacter* is described
EP 0 289993 A3	Weyerhaeuser Co., USA	Johnson, D. C., LeBlanc, H. A., Neogi, A. N.	Bacterial cellulose as surface treatment for fibrous web	1988	BC applied at relatively low concentrations, singularly or in combination, to at least one surface of a fibrous web gives excellent properties of gloss, smoothness, ink receptivity and holdout, and surface strength
WO 88/09381	Financial Union for Agricultural Development, France	Labourer, P. F.	Process for producing bacterial cellulose from material of plant origin	1988	Culturing of *A. xylinum* strain in a plant polysaccharide-containing medium enables efficient cellulose production
US 004655758	Johnson & Johnson products, Inc., USA	Ring, D. F., Nashed, W., Dow, T.	Microbial polysaccharide articles and methods of production	1987	After removal of excess liquid and bacterial cells, the cellulose pellicle can be impregnated and used for various purposes
WO 86/02095	Bio-Fill Industria e Comercio de Produtos Medico Hospitalares, Ltd, Brazil	Farah, L. F. X.	Process for the preparation of cellulose film, cellulose film produced thereby, artificial skin graft and its use	1986	BC film preparation, including optimal conditions of *A. xylinum* culturing, and methods of removing the formed film are described; the film appeared to be suitable for use as an artificial skin graft, a separating membrane, or artificial leather
EP 0200409 A3	Ajinomoto Co. Inc., Japan	Iguchi, M., Mitsuhashi, S., Ichimura, K., Nishi, Y., Uryu, M., Yamanake, S., Watanabe, K.	Molded material comprising bacteria-produced cellulose	1986	BC is an excellent component of molded materials having high dynamic strength as compared to conventional molded materials

with industrial wastes, rich in proper carbon sources, such as spent liquors from crystalline glucose production, etc. Simultaneously, a significant environmental benefit would be obtained. However, the major advantage of mass production of BC would be the protection of forests, which are presently disappearing at an alarming rate, thus leading to soil eutrophication and global climate changes.

BC has already been put to numerous uses, presented in this review. The polysaccharide can not only be replaced by some animal polymers (collagen), but also carriers of substances having a positive impact on human health (e.g. antioxidants and prebiotics). The usefulness of the bacterial polymer in medicine (wound, burn and ulcer dressing materials, component of implants) is not longer questioned. Moreover, cellulose granulates can be an excellent matrix for the immobilization of medicinal preparations. For example, if specific substances (receptors) are adsorbed on BC, the resulting molecules can be scavengers of either toxins or of the pathogenic microflora inhabiting the alimentary tract. Recently, unique nanocrystals (30×600–800 nm) of BC have been obtained, derived from its commercial preparation Prima Cel™ (Xylos). Selective modification (trimethyl silylation) of the surface of these nanocrystals while leaving their core intact has been achieved. Such modified crystals have great potential in several advanced technologies.

Deciphering of all the riddles of cellulose biosynthesis will lead to improvement and tailoring of BC supramolecular structure and properties; as a consequence, novel concepts for both its inexpensive production, and for its bulk and specialty applications will be developed.

14 References

Aloni, Y., Benziman, M. (1982) Intermediates of cellulose synthesis in *Acetobacter*, in: *Cellulose and Other Natural Polymers System* (Brown, R. M., Jr., Ed.), New York: Plenum Press, 341–361.

Aloni, Y., Cohen, Y., Benziman, M., Delmer, D. P. (1983) Solubilization of UDP-glucose: 1,4-β-glucan 4-β-D-glucosyl transferase (cellulose synthase) from *Acetobacter xylinum*, *J. Biol. Chem.* **258**, 4419–4423.

Ang, J. F., Miller, W. B. (1991) Multiple functions of powdered cellulose as a food ingredient, *J. Am. Ass. Cereal Chem.* **36**, 558–564.

Bartlett, D. H., Silverman, M. (1989) Nucleotide sequences of IS492, a novel insertion sequence causing variation of extracellular production in the marine bacterium *Pseudomonas atlantica*, *J. Bacteriol.* **171**, 1763–66.

Bauer, K., Codolington, K., Ben Bassat, A. (1992) Methods for improving production of bacterial cellulose, *Abstr. Paper Am. Chem. Soc.*, **203** Meet., Pt 1, Biot. 94.

Bayer, M. E. (1979) The fusion sites between outer membrane and cytoplasmic membrane in bacteria: their role in membrane assembly and virus infection, in: *Bacterial Outer Membranes* (Inouye, M., Ed.), New York: John Wiley & Sons, 167–202.

Ben-Bassat, A., Bruner, R., Wong, H., Shoemaker, S., Aloni, Y. (1989) Production of bacterial cellulose by *Acetobacter*, *Abstr. Pap. Am. Chem. Soc.*, **198** Meet., MBTD20.

Ben-Bassat, A., Calhoon, R. D., Fear, A. L., Gelfand, D. H., Mead, J. H., Tal, R., Wong, H., Benziman, M. (1993) Methods and nucleic acid sequences for expression of the cellulose synthase operon, US patent 5 268 274.

Ben-Hayyim, G., Ohad, I. (1965) Synthesis of cellulose by *Acetobacter xylinum*; VIII. On the formation and orientation of bacterial cellulose fibrils in the presence of acidic polysaccharides, *J. Cell Biol.* **25**, 191–207.

Benziman, M., Tal, R. (1995) Cellulose synthase associated proteins. US patent 5 382 656.

Boisset, C., Chanzy, H., Henrissat, B., Lamed, R., Shoham, Y., Bayer, E. A. (1999) Digestion of crystalline cellulose substrates by the *Clostridium thermocellum* cellulosome: structural and morphological aspects, *Biochem. J.* **340**, 829–835.

Boisset, C., Fraschini, C., Schulein, M., Henrissat, B., Chanzy, H. (2000) Imaging the enzymatic digestion of bacterial cellulose ribbons reveals the endo character of the cellobiohydrolase Cel6A from *Humicola insolens* and its mode of synergy with cellobiohydrolase Cel7A, *Appl. Environ. Microbiol.* **66**, 1444–1452.

Brown, A. J. (1886) An acetic ferment which forms cellulose, *J. Chem. Soc.* **49**, 432–439.

Brown R. M., Jr. (1989a) Microbial cellulose composites and processes for producing same. WO 89/11783.

Brown, R. M., Jr. (1989b) Microbial cellulose as a building block resource for specialty products and processes thereof, WO 89/12107.

Brown, R. M., Jr. (1989c) Use of cellulase preparations in the cultivation and use of cellulose-producing microorganisms, European patent 0258038A3.

Brown, R. M., Jr. (1996) The biosynthesis of cellulose, *Pure Appl. Chem.* **A33**, 1345–1373.

Brown, R. M., Jr., Saxena, I. M. (2000) Cellulose biosynthesis: a model for understanding the assembly of biopolymers, *Plant Physiol. Biochem.* **38**, 57–60.

Brown, R. M., Jr., Willison, J. H. M., Richardson, C. L. (1976) Cellulose biosynthesis in *Acetobacter xylinum*: visualisation of the site of synthesis and direct measurement of the *in vivo* process, *Proc. Natl. Acad. Sci. USA* **73**, 4565–4569.

Bungay, H. R., Serafica, G. (1997) Production of microbial cellulose using a rotating disc film bioreactor, WO 97/05271.

Bureau, T. E., Brown, R. M., Jr. (1987) *In vitro* synthesis of cellulose II from a cytoplasmic membrane fraction of *Acetobacter xylinum, Proc. Natl. Acad. Sci. USA* **84**, 6985–6989.

Canale-Parda, E., Wolfe, R. S. (1964) Synthesis of cellulose by *Sarcina ventriculi, Biochim. Biophys. Acta* **82**, 403–405.

Cannon, R. E., Anderson, S. M. (1991) Biogenesis of bacterial cellulose, *Crit. Rev. Microbiol.* **17**, 435–439.

Chao, Y., Ishida, T., Sugano, Y., Shoda, M. (2000) Bacterial cellulose production by *Acetobacter xylinum* in a 50 L internal-loop airlift reactor, *Biotechnol. Bioeng.* **68**, 345–352.

Colvin, J. R. (1957) Formation of cellulose microfibrils in a homogenate of *Acetobacter xylinum, Arch. Biochem. Biophys.* **70**, 294–295.

Colvin, J. R. (1980) The biosynthesis of cellulose, in: *Plant Biochemistry* (Priess, J., Ed.), New York: Academic Press, 543–570, Vol. 3

Colvin, J. R., Leppard, G. G. (1977) The biosynthesis of cellulose by *Acetobacter xylinum* and *Acetobacter acetigenus, Can. J. Microbiol.* **23**, 701–709.

Colvin, J. R., Sowden, L. C., Daoust, V., Perry, M. (1979) Additional properties of a soluble polymer of glucose from cultures of *Acetobacter xylinum, Can. J. Biochem.* **57**, 1284–1288.

Costeron, J. W. (1999) The role of bacterial exopolysaccharides in nature and disease, *J. Ind. Microbiol. Biotechnol.* **22**, 551–563.

Coucheron, D. H. (1991) An *Acetobacter xylinum* insertion sequence element associated with inactivation of cellulose production, *J. Bacteriol.* **173**, 5723–2731.

Dekker, R. F. H., Rietschel, E. T., Sandermann, H. (1977) Isolation of α-glucan and lipopolysaccharide fractions from *Acetobacter xylinum, Arch. Microbiol.* **115**, 353–357.

De Lannino, N., Cuoso, R. O., Dankert, M. A. (1988) Lipid-linked intermediates and the synthesis of acetan in *A. xylinum, J. Gen. Microbiol.* **134**, 1731–1736.

Delmer, D. P. (1982) Biosynthesis of cellulose, *Adv. Carbohydr. Chem. Biochem.* **41**, 105–153.

De Wulf, P., Joris, K., Vandamme, E. J. (1996) Improved cellulose formation by an *Acetobacter xylinum* mutant limited in (keto)gluconate synthesis, *J. Chem. Tech. Biotechnol.* **67**, 376–380.

Easson, D. D., Jr., Sinskey, A. J., Peoples, O. P. (1987) Isolation of *Zoogloea ramigera* I-16 M exopolysaccharides biosynthetic genes and evidence for instability within this region, *J. Bacteriol.* **169**, 4518–4524.

Embuscado, M. E., Marks, J. S., Miller, J. N. (1994) Bacterial cellulose. II. Optimization of cellulose production by *Acetobacter xylinum* through response surface methodology, *Food Hydrocolloids* **8**, 419–430.

Galas, D. J., Chandler, M. (1989) Bacterial insertion sequences, in: *Mobile DNA* (Berg, D. E., Howe, M. M., Eds.) Washington, DC: ASM, 109–162.

Galas, E., Krystynowicz, A., Tarabasz-Szymanska, L., Pankiewicz, T., Rzyska, M. (1999) Optimization of the production of bacterial cellulose using multivariable linear regression analysis, *Acta Biotechnol.* **19**, 251–260.

Glaser, L. (1958) The synthesis of cellulose in cell-free extracts of *Acetobacter xylinum, J. Biol. Chem.* **232**, 627–636.

Haigler, C. H. (1985) The function and biogenesis of native cellulose, in: *Cellulose Chemistry and its Applications* (Nevel, R. P., Zeronian, S. H., Eds.), Chichester: Ellis Horwood, 30–83.

Han, N. S., Robyt, J. F. (1998) The mechanism of *Acetobacter xylinum* cellulose biosynthesis: direction of chain elongation and the role of lipid pyrophosphate intermediates in the cell membrane, *Carbohydr. Res.* **313**, 125–133.

Hestrin, S., Aschner, M., Mager J. (1947) Synthesis of cellulose by resting cells of *Acetobacter xylinum, Nature* **159**, 64–65.

Hestrin, S., Schramm, M. (1954) Synthesis of cellulose by *Acetobacter xylinum*, II. Preparation of freeze-dried cells capable of polymerizing glucose to cellulose, *Biochem. J.* **58**, 345–352.

Hikawu, S., Hiroshi, T., Takayasu, T., Yoshinaga, F. (1996) Manufacture of bacterial cellulose by addition of cellulose formation stimulators, Japanese patent 96316922.

Hotte, B., Roth-Arnold, I., Puhler, A., Simon, R. (1990) Cloning and analysis of a 35.3 kb DNA region involved in exopolysaccharide production by *Xanthomonas campestris, J. Bacteriol.* **172**, 2804–2807.

Husemann, E., Muller, G. J. M. (1966) Synthesis of unbranched polysaccharides, *Macromol. Chem.* **91**, 212–230.

Ielpi, L., Couso, R., Dankert, M. A. (1981) Lipid-linked intermediates in the biosynthesis of xanthan gum, *FEBS Lett.* **130**, 253–256.

Iguchi, M., Yamanaka, S., Budhioko, A. (2000) Bacterial cellulose – a masterpiece of nature's arts, *J. Mater. Sci.* **35**, 261–270.

Ishikawa, A., Matsuoka, M., Tsuchida, T., Yoshinaga, F. (1995) Increasing of bacterial cellulose production by sulfoguanidine-resistant mutants derived from *Acetobacter xylinum* subsp. sucrofermentans BPR2001, *Biosci. Biotechnol. Biochem.* **59**, 2259–2263.

Ishikawa, A., Tsuchida, T., Yoshinaga, F. (1996a) Production of bacterial cellulose using microbial strain resistant to inhibitor of DHO-dehydrogenase, Japanese patent 08009965A.

Ishikawa, A., Tsuchida, T., Yoshinaga, F. (1996b) Production of bacterial cellulose with pyrimidine analogue-resistant strain, Japanese patent 08000260.

Johnson, D. C., Neogi, A. N. (1989) Sheeted products formed from reticulated microbial cellulose, US patent 4 863 565.

Jonas, R., Farah, L. F. (1998) Production and application of microbial cellulose, *Polym. Degrad. Stabil.* **59**, 101–106.

Jones, D., Gordon, A. H., Bacon, J. S. D. (1974) β-1,3-Glucanases from *Sclerotium rolfsii*, *Biochem. J.* **140**, 47–55.

Joris, K., Billiet, F., Drieghe, S., Brachx, D., Vandamme, E. (1990) Microbial production of β-1,4-glucan, *Meded. Fac. Landbouwwet Rijksuniv. Gent* **55**, 1563–1566.

Kenji, K., Yukiko, M., Hidehi, L., Kunihiko, O. (1990) Effect of culture conditions of acetic acid bacteria on cellulose biosynthesis, *Br. Polym. J.* **22**, 167–171.

Kent, R. A., Stephens, R. S., Westland, J. A. (1991) Bacterial cellulose fiber provides an alternative for thickening and coating, *Food Technol.* **45**, 108.

Kim, D., Kim, Y. M., Park, D. H. (1999) Modification of *Acetobacter xylinum* bacterial cellulose using dextransucrase and alternansucrase, *J. Microbiol. Biotechnol.* **9**, 704–708.

Kobayashi, S., Kashiwa, K., Kawasaki, T., Shoda, S. (1991) Novel method for polysaccharide synthesis using an enzyme: the first *in vitro* synthesis of cellulose via nonbiosynthetic path utilising cellulase as catalyst, *J. Am. Chem. Soc.* **113**, 3079–3084.

Kouda, T., Naritomi, T., Yano, H., Yoshinaga, F. (1998) Inhibitory effect of carbon dioxide on bacterial cellulose production by *Acetobacter* in agitated culture, *J. Ferment. Bioeng.* **85**, 318–321.

Koyama, M., Helbert, W., Imai, T., Sugiyama, J., Henrissat, B. (1997) Parallel-up structure evidence the molecular directionality during biosynthesis of bacterial cellulose, *Proc. Natl. Acad. Sci. USA* **94**, 9091–9095.

Krystynowicz, A., Turkiewicz, M., Drynska, E., Galas, E. (1995) Bacterial cellulose – biosynthesis and application, *Biotechnologia* **30**, 120–132.

Krystynowicz, A., Galas, E., Pawlak, E. (1997) Method of bacterial cellulose production. Polish patent P-299907.

Krystynowicz, A., Czaja, W., Bielecki, S. (1999) Biosynthesis and application of bacterial cellulose, *Zywnosc* **3**, 22–33.

Krystynowicz, A., Czaja, W., Pomorski, L., Kolodziejczyk, M., Bielecki, S. (2000) The evaluation of usefulness of microbial cellulose as a wound dressing material, 14th Forum for Applied Biotechnology, Gent, Belgium, *Meded. Fac. Landbouwwet Rijksuniv. Gent*, Proceedings Part I, 213–220.

Kudlicka K. (1989) Terminal complexes in cellulose synthesis, *Postepy biologii komórki* **16**, 197–212 (abstract in English).

Kuga, G., Brown, R. M., Jr. (1988) Silver labeling of the reducing ends of bacterial cellulose, *Carbohydr. Res.* **180**, 345–350.

Laboureur, P. (1988) Process for producing bacterial cellulose from material of plant origin, WO 88/09381.

Legge R. L. (1990) Microbial cellulose as a specialty chemical, *Biotechnol. Adv.* **8**, 303–319.

Leppard G. G., Sowden, L. C., Ross, C. J. (1975) Nascent stage of cellulose biosynthesis, *Science* **189**, 1094–1095.

Lin, F. C., Brown, R. M., Jr. (1989) Purification of cellulose synthase from *Acetobacter xylinum*, in: *Cellulose and Wood Chemistry and Technology* (Schmerck, C., Ed.), New York: John Wiley & Sons, 473–492.

Lin, F. C., Brown, R. M., Jr., Cooper, J. B., Delmer, D. P. (1985) Synthesis of fibrils *in vitro* by a solubilized cellulose synthase from *Acetobacter xylinum*, *Science* **230**, 822–825.

Lin, F. C., Brown, R. M., Jr., Drake, R. R., Jr., Haley. B. E. (1990) Identification of the uridine-5′-diphosphoglucose (UDPGlc) binding subunit of cellulose synthase in *Acetobacter xylinum* using the photoaffinity probe 5-azido-UDPGlc, *J. Biol. Chem.* **265**, 4782–4784.

Marx-Figini, M., Pion, B. G. (1974) Kinetic investigations on biosynthesis of cellulose by *Acetobacter xylinum*, *Biochim. Biophys. Acta* **338**, 382–393.

Masaoka, S., Ohe, T., Sakota, N. (1993) Production of cellulose from glucose by *Acetobacter xylinum*, *J. Ferment. Bioeng.* **75**, 18–22.

Matsuoka, M., Tsuchida, T., Matsuchita, K., Adachi, O., Yoshinaga, F. (1996) A synthetic medium for bacterial cellulose production by *Acetobacter xylinum* subsp. sucrofermentans, *Biosci. Biotechnol. Biochem.* **60**, 575–579.

Matthyse, A., Thomas, D. I., White, A. R. (1995) Mechanism of cellulose synthesis in *Agrobacterium tumefaciens*, *J. Bacteriol.* **177**, 1076–1081.

Micheel, F., Brodde, O. E. (1974) Polymerization of 1,4-anhydro-2,3,6-tri-*O*-benzyl-α-D-glucopyranose, *Liebigs Ann. Chem.* **124**, 702–708.

Minakami, H., Entani, K., Tayama, S., Fujiyama, S., Masai, H. (1984) Isolation and characterization of a new polysaccharide-producing *Acetobacter* sp. *Agric. Biol. Chem.* **48**, 2405–2414.

Mondal, I. H., Kai, A. (2000) Control of the crystal structure of microbial cellulose during nascent stage, *J. Appl. Polym. Sci.* **79**, 1726–1734.

Nakai, T., Tonouchi, N., Konishi, T., Kojima, Y., Tsuchida, T., Yoshinaga, F., Sakai, F., Hayashi, T. (1999) Enhancement of cellulose production by expression of sucrose synthase in *Acetobacter xylinum*, *Proc. Natl. Acad. Sci. USA* **96**, 14–18.

Nakatsubo, F., Takano, T., Kawada, T., Murakami, K. (1989) Toward the synthesis of cellulose: synthesis of cellooligosaccharides, in: *Cellulose: Structural and Functional Aspects* (Kennedy, J. F., Philips, G. O., Williams, P. A., Eds.), New York: Ellis Horwood, 201–206.

Nari, J., Noat, G., Richard, J., Franchini, E., Monstacas, A. M. (1983) Catalytic properties and tentative function of a cell wall β-glucosyltransferase from soybean cells cultured *in vitro*, *Plant Sci. Lett.* **28**, 313–320.

Naritomi, T., Kouda, T., Yano, H., Yoshinaga, F. (1998) Effect of ethanol on bacterial cellulose production from fructose in continuous culture, *J. Ferment Bioeng.* **85**, 598–603.

Nichols, S. E., Singletary, G. W. (1998) Cellulose synthesis in the storage tissue of transgenic plants, US patent 5 723 764.

Ohana, P., Delmer, D. P., Volman, G., Benziman, M. (1998) Glycosylated triterpenoid saponin: a specific inhibitor of diguanylate cyclase from *Acetobacter xylinum*, *Plant Cell Physiol.* **39**, 153–159.

Oikawa, T., Kamatani, N., Kaimura, T., Ameyama, M., Soda, K. (1997) Endo-β-glucanase from *A. xylinum* – purification and characterization, *Curr. Microbiol.* **34**, 309–313.

Okamoto, T., Yamano, S., Ikeaga, H., Nakamura, K. (1994) Cloning of the *A. xylinum* cellulase gene and its expression in *E. coli* and *Zymomonas mobilis*, *Appl. Microbiol. Biotechnol.* **42**, 563–568.

Okiyama, A., Shirae, H., Kano, H., Yamanaka, S. (1992) Two-stage fermentation process for cellulose production by *Acetobacter aceti*, *Food Hydrocolloids* **6**, 471–477.

Palonen, H., Tenkanen, M., Linder, M. (1999) Dynamic interaction of *Trichoderma reesei* cellobiohydrolases Cel6A and Cel7A and cellulose at equilibrium and during hydrolysis, *Appl. Environ. Microbiol.* **65**, 5229–5233.

Ring, D. F., Nashed, W., Dow, T. (1986) Liquid loaded pad for medical applications, US patent 4 588 400.

Ross, P., Weinhouse, H., Aloni, Y., Michaeli, D., Ohana, P., Mayer, R., Braun, S., de Vroom, E., van der Marel, G. A., van Boom, J. H., Benziman, M. (1987) Regulation of cellulose synthesis in *Acetobacter xylinum* by cyclic diguanylic acid, *Nature* **325**, 279–281.

Ross, P., Mayer, R., Benziman, M. (1991) Cellulose biosynthesis and function in bacteria, *Microbiol. Rev.* **55**, 35–58.

Sakair, N., Asamo, H., Ogawa, M., Nishi, N., Tokura, S. (1998) A method for direct harvest of bacterial cellulose filaments during continuous cultivation of *Acetobacter xylinum*, *Carbohydr. Polym.* **35**, 233–237.

Samejima, M., Sugiyama, J., Igarashi, K., Eriksson, K. E. L. (1997) Enzymatic hydrolysis of bacterial cellulose, *Carbohydr. Res.* **305**, 281–288.

Sattler, K., Fiedler, S. (1990) Production and application of bacterial cellulose. II. Cultivation in a rotating drum fermentor, *Zbl. Microbiol.* **145**, 247–252.

Savidge, R. A., Colvin, J. R. (1985) Production of cellulose and soluble polysaccharides by *Acetobacter xylinum*, *Can. J. Microbiol.* **31**, 1019–1025.

Saxena, I. M., Brown, R. M., Jr. (1989) Cellulose biosynthesis in *Acetobacter xylinum*: a genetic approach, in: *Cellulose and Wood Chemistry and Technology* (Schnerck, C., Ed.). New York: John Wiley & Sons, 537–557.

Saxena, I. M., Brown, R. M., Jr. (1997) Identification of cellulose synthase(s) in higher plants: Sequence analysis of processive β-glycosyltransferases with common motif 'D, D, D 35Q (RQ)XRW', *Cellulose* **4**, 33–49.

Saxena, I. M., Brown, R. M., Jr. (2000) Cellulose synthases and related proteins, *Curr. Opin. Plant Biol.* **3**, 523–531.

Saxena, I. M., Lin, F. C., Brown, R. M., Jr. (1990a) Cloning and sequencing of the cellulase synthase catalytic subunit gene of *Acetobacter xylinum*, *Plant Mol. Biol.* **15**, 673–683.

Saxena, I. M., Roberts, E. M., Brown, R. M., Jr. (1990b) Modification of cellulose normally synthesised by cellulose-producing microorganisms, US patent 4 950 597.

Saxena, I. M., Lin, F. C., Brown, R. M., Jr. (1991) Identification of a new gene in an operon for cellulose biosynthesis in *Acetobacter xylinum*, *Plant Mol. Biol.* **16**, 947–954.

Saxena, I. M., Brown, R. M., Jr., Fevre, M., Geremia, R. A., Henrissat, B. (1995) Multidomain architecture of β-glycosyl transferase: Implications for mechanism of action, *J. Bacteriol.* **177**, 1419–1424.

Seto, H., Tsuchida, T., Yoshinaga, F., Beppu, T., Horinouchi, S. (1996) Characterization of the biosynthetic pathway of cellulose from glucose and fructose in *Acetobacter xylinum*, *Biosci. Biotechnol. Biochem.* **60**, 1377–1379.

Seto, H., Tsuchida, T., Yoshinaga, F. (1997) Manufacture of bacterial cellulose by mixed culture of microorganisms, Japanese patent 9721905.

Srisodsuk, M., Kleman-Leyer, K., Keranen, S., Kirk, T. K., Teeri, T. T. (1998) Modes of action on cotton and bacterial cellulose of a homologous endoglucanase-exoglucanase pair from *Trichoderma reesei*, *Eur. J. Biochem.* **62**, 185–187.

Stalbrandt, H., Mansfield, S. D., Saddler, J. N., Kilburn, D. G., Warren, R. A. J., Gilkes, N. R. (1998) Analysis of molecular size of cellulose by recombinant *Cellulomonas fimi* β-1,4-glucanase, *Appl. Environ. Microbiol.* **64**, 2374–2379.

Standal, R., Iversen, T. G., Coucheron, D. H., Fjaervik, E., Blatny, J. M., Valla, S. (1994) A new gene required for cellulose production and gene encoding cellulolytic activity in *A. xylinum* are localized with *bcs* operon, *J. Bacteriol.* **176**, 665–672.

Sutherland, I. W. (1998) Novel and established applications of microbial polysaccharides, *TIBTECH* **16**, 41–46.

Swissa, M., Aloni, Y., Weinhouse, H., Benziman, M. (1980) Intermediary steps in cellulose synthesis in *Acetobacter xylinum*: studies with whole cells and cell-free preparation of the wild type and a cellulose-less mutant, *J. Bacteriol.* **143**, 1142–1150.

Tahara, N., Tabuchi, M., Watanabe, K., Yano, H., Morinaga, Y., Yoshinaga, F. (1997) Degree of polymerisation of cellulose from *A. xylinum* BPR 2001 decreased by cellulase producing strain, *Biosci. Biotechnol. Biochem.* **61**, 1862–1865.

Tahara, N., Tonouchi, M., Yano, H., Yoshinaga, F. (1998) Purification and characterization of exo-1,4-β-glucosidase from *A. xylinum* BPR 2001, *J. Ferment. Bioeng.* **85**, 589–594.

Takai, M., Tsuta, Y., Watanabe, S. (1997) Biosynthesis of cellulose by *Acetobacter xylinum* and characterization of bacterial cellulose, *Polym. J.* **7**, 137–146.

Tal, R., Wong, H. C., Calhon, R., Gelfand, D., Fear, A. L., Volman, G., Mayer, R., Ross, P., Amikam, D., Weinhouse, H., Cohen, A., Sapir, S., Ohana, P., Benziman, M. (1998) Three *cdg* operons control cellular turnover of c-di-GMP in *Acetobacter xylinum*: genetic organization and occurrence of conserved domain in isozymes, *J. Bacteriol.* **180**, 4416–4425.

Tayama, K., Minakami, H., Entani, E., Fujiyama, S., Masai, H. (1985) Structure of an acidic polysaccharide from *Acetobacter*, sp. NBI 1022, *Agric. Biol. Chem.* **49**, 959–966.

Teeri, T. T. (1997) Crystalline cellulose degradation: new insight into the function of cellobiohydrolases, *TIBTECH* **15**, 160–167.

Tonouchi, N., Tsuchida, T., Yoshinaga, F., Beppu, T. (1996) Characterization of the biosynthetic pathway of cellulose from glucose and fructose in *Acetobacter xylinum*, *Biosci. Biotechnol. Biochem.* **60**, 1377–1379.

Tonouchi, N., Tahara, N., Kojima, Y., Nakai, T., Sakai, F., Hayashi, T., Tsuchida, T., Yoshinaga, F. (1997) A β-glucosidase gene downstream of the cellulase synthase operon in cellulase producing *Acetobacter*, *Biosci. Biotechnol. Biochem.* **61**, 1789–1790.

Tonouchi, N., Hirinouchi, S., Tsuchida, T., Yoshinaga, F. (1998) Increased cellulose production by *Acetobacter* after introducing the sucrose phosphorylase gene, *Biosci. Biotechnol. Biochem.* **62**, 1778–1780.

Toyosaki, H., Naritomi, T., Seto, A., Matsuoka, M., Tsuchida, T., Yoshinga, F. (1995) Screening of bacterial cellulose- producing *Acetobacter* strains suitable for agitated culture, *Biosci. Biotechnol. Biochem.* **59**, 1498–1502.

Valla, S., Kjosbakken, J. (1981) Isolation and characterization of a new extracellular polysaccharide from a cellulose-negative strain of *Acetobacter xylinum*, *Can. J. Microbiol.* **27**, 599–603.

Valla, S., Coucheron, D. H., Fjaervik, E., Kjosbakken, J., Weinhose, H., Ross, P., Amikam, D., Benziman, M. (1989) Cloning of a gene involved in cellulose biosynthesis in *Acetobacter xylinum*: complementation of cellulose-negative mutant by the UDPG pyrophosphorylase structure gene, *Mol. Gen. Genet.* **217**, 26–30.

Vandamme, E. J., De Baets, S., Vanbaelen, A., Joris, K., De Wulf P. (1998) Improved production of bacterial cellulose and its application potential, *J. Polymer Degrad. Stabil.* **59**, 93–99.

Watanabe, K., Tabuchi, M., Morinaga, Y., Yoshinaga, F. (1998a) Structural features and properties of bacterial cellulose produced in agitated culture, *Cellulose* **5**, 187–200.

Watanabe, K., Tabuchi, M., Ischikawa, M., Takemura, H., Tsuchida, T., Morinaga, Y. (1998b)

Acetobacter xylinum mutant with high cellulose productivity and ordered structure, *Biosci. Biotechnol. Biochem.* **62**, 1290–1292.

Weinhouse, R. (1977) Regulation of carbohydrate metabolism in *Acetobacter xylinum*, PhD thesis, Hebrew University of Jerusalem. Jerusalem, Israel.

Weinhouse, H., Sapir, S., Amikam, D., Shiko, Y., Volman, G., Ohana, P., Benziman, M. (1997) c-di-GMP-binding protein, a new factor regulating cellulose synthesis in *Acetobacter xylinum*, *FEBS Lett.* **416**, 207–211.

Wiliams, W. S., Cannon, R. E. (1989) Alternative environmental roles for cellulose produced by *Acetobacter xylinum*, *Appl. Environ. Microbiol.* **55**, 2448–2452.

Winkelman, J. W., Clark, D. P. (1984) Proton suicide: general method for direct selection of sugar transport and fermentation-defective mutants, *J. Bacteriol.* **11**, 687–690.

Wlochowicz, A. (2001) Personal communication.

Wong, H. C., Fear, A. L., Calhoon, R. D., Eichinger, G. M., Mayer, R., Amikam, D., Benziman, M., Gelfand, D. H., Meade, J. H., Emerick, A. W., Bruner, R., BenBassat, A., Tal, R. (1990) Genetic organization of cellulose synthase operon in *A. xylinum*, *Proc. Natl. Acad. Sci. USA* **87**, 8130–8134.

Yamada, Y., Hoshino, K., Ishikawa, T. (1997) The phylogeny of acetic acid bacteria based on the partial sequences of 16 S ribosomal RNA: the elevation of the subgenus *Gluconacetobacter* to the generic level, *Biosci. Biotechnol. Biochem.* **61**, 1244–51.

Yamada, Y., Hoshino, K., Ishikawa, T. (1998) *Gluconacetobacter corrig.* (*Gluconoacetobacter* [sic]), in: Validation of publication of new names and new combinations previously effectively published outside the IJSB, List no 64, *Int. J. Syst. Bacteriol.* **48**, 327–328.

Yamada, Y. (2000) Transfer *Acetobacter oboediens* and *A. intermedius* to the genus *Gluconacetobacter* as *G. oboediens* comb. nov. and *G. intermedius* comb. nov., *Int. J. System. Evolut. Microbiol.* **50**, 2225–2227.

Yamamoto, H., Horii, F., Hirai, A. (1996) *In situ* crystallization of bacterial cellulose. 2. Influence of different polymeric additives on the formation of celluloses I_α and I_β at the early stage of incubation, *Cellulose* **3**, 229–242.

Yamanaka, S., Watanabe, K., Kitamura, N., Iguchi, M., Mitsuhashi, S., Nishi, Y., Uryu, M. (1989) The structure and mechanical properties of sheets prepared from bacterial cellulose, *J. Mater. Sci.* **24**, 3141–3145.

Yamanaka, S., Watanabe, K., Suzuki, Y. (1990) Hollow microbial cellulose, process for preparation thereof, and artificial blood vessel formed of said cellulose, European patent 0396344A2.

Yamanaka, S., Ishihara, M., Sugiyama, J. (2000) Structural modification of bacterial cellulose, *Cellulose* **7**, 213–225.

Zaar, K. (1977) The biogenesis of cellulose by *Acetobacter xylinum*, *Cytobiologie* **16**, 1–15.

4
Bioemulsans: Surface-active Polysaccharide-containing Complexes

Prof. Dr. Eugene Rosenberg[1], **Prof. Dr. Eliora Z. Ron**[2]

[1] Department of Molecular Microbiology and Biotechnology, Tel Aviv University, Ramat Aviv, Israel 69978; Tel.: +972-3-640 9838; Fax: +972-3-642 9377; E-mail: eros@post.tau.ac.il

[2] Department of Molecular Microbiology and Biotechnology, Tel Aviv University, Ramat Aviv, Israel 69978; Tel.: +972-3-640 9879; Fax: +972-3-641 4138; E-mail: eliora@post.tau.ac.il

CMC critical micelle concentration
OmpA outer membrane protein A
PAHs polyaromatic hydrocarbons
pgi phosphoglucoisomerase gene
SCP single-cell protein

1 Introduction

In general, microorganisms are specialists, and in any particular ecological niche one microorganism or a limited number of strains will dominate. These microorganisms have adapted through evolution to be able to survive in this niche during the long periods when growth is impossible, but when nutrients do become available they can outgrow their competitors. The rapid growth of microorganisms depends largely on their high surface-to-volume ratio, which allows for the efficient uptake of nutrients and release of waste products. The price that the microorganism pays for the high surface-to-volume ratio is that it is totally exposed. All of the components on the outside of the cell must be able to function under the specific conditions of the ecological niche. It is probably for this reason that the "diversity of the microbial world" is best expressed on the outside of the cell. Because of their high surface-to-volume ratio and the diversity of their exocellular polymers, microorganisms are a rich source of potentially useful polymers. One group of such polymers, referred to as emulsans, is amphipathic polysaccharides and/or proteins that stabilize oil-in-water emulsions.

2 Historical Outline

Research on microbial bioemulsifiers can be traced to pioneering attempts to: (1) solve practical problems in the petroleum industry; and (2) understand how microorganisms grow on water-insoluble hydrocarbons.

2.1 The Petroleum Industry Connection

The two major problems of the petroleum industry that involved surface-active agents (surfactants) were the secondary recovery of oil, and the production of single-cell protein (SCP) from hydrocarbons. One of the earliest studies of the microbial production of surfactants was carried out by La Riviere (1955a,b), who showed that eight different microbial cultures, including *Aspergillus ni-*

ger, *Psuedomonas aeruginosa*, *Candida lipolytica*, *Desulfovibrio desulfuricans* and *Mycobacterium phlei*, lowered surface tension by 14 to 34 mN m^{-1}. Updegraff and Wren (1954) patented a process for the release of oil from petroleum-bearing materials using sulfate-reducing bacteria. However, subsequent studies demonstrated that only small increases (ca. 5%) in oil recovery were obtained with microbial surfactants (Dostalek and Spurny, 1958). The early work on the use of microorganisms for secondary recovery of oil has been reviewed by Davis (1967) and Zajic and Panchal (1976).

During the early 1960s, British Petroleum announced that it intended to manufacture microbial protein from hydrocarbons on a large scale. This stimulated many oil companies to support in-house as well as university research on the optimization of the microbial conversion of n-paraffins to what came to be called SCP. It soon became clear that the growth of microorganisms at high cell density in fermentors on n-paraffins was limited by the low solubility of hydrocarbons in water (Johnson, 1968; Erickson and Nakahara, 1975). For example, the solubilities of decane, dodecane and tetradecane in water at 25 °C are 3.1×10^{-7} M, 1.7×10^{-8} M, and 9.8×10^{-10} M, respectively (Goldberg, 1985). Accordingly, the growth rate of microorganisms on hydrocarbons is limited by the transfer rate of the substrate into solution, and then across the cell membrane. Numerous observations were reported indicating that the dispersion of hydrocarbons in oil increased during fermentation as a result of increased interfacial area between the oil drops and the aqueous culture broth. This led to the discovery that microorganisms, in this case yeast, produce extracellular bioemulsifiers, such as fatty acids and various polymers (Iguchi et al., 1969; Abbot and Gledhill, 1971).

2.2 The Emulsan Story

By means of the enrichment culture technique, a mixed population of microorganisms was obtained which catalyzed the emulsification of crude oil in supplemented seawater (Reisfeld et al., 1972). From this mixed culture, one strain (RAG-1) was isolated that degraded ca. 60% of the oil and efficiently dispersed the nondegraded oil into small droplets (2–5 μm diameter). The initial studies indicated that the material responsible for emulsification was present in the extracellular fluid of cultures grown on crude oil as the carbon and energy source. The emulsifier-producing RAG-1 strain was initially mis-classified as a member of the genus *Arthrobacter*. Subsequently, the strain was identified as an *Acinetobacter* (Pines and Gutnick, 1981) and referred to as *A. calcoaceticus* RAG-1 (ATCC 31012).

RAG-1 was initially isolated with the goal of using it to treat oil pollution in the sea. However, it soon became clear that this goal was impracticable because growth of this bacterium, as with any known microorganism, requires utilizable nitrogen (N) and phosphorus (P) sources in addition to the crude oil. The oceans are “deserts” when it comes to N and P compounds, and crude oil contains very little N and P. The addition of utilizable N and P compounds (e.g., urea, nitrates and phosphates) to an open system such as an ocean is of no use because these water-soluble compounds diffuse away from the oil and are not available to the bacteria adhering to the oil spill. Consequently, Rosenberg et al. (1974) used RAG-1 in a closed system, the cargo compartment of an oil tanker during its ballast voyage, in order to prevent oil pollution resulting from discharge of oily ballast water.

The oil tanker/RAG-1 experiment was performed in January, 1973 on a ballast voyage from Eilat in the Gulf of Aqaba to Khargh Island in the Persian Gulf. One of the slop tanks containing 100 m^3 of oily ballast water over a layer of sludge was supplemented with urea and phosphate. After inoculation with RAG-1, the slop tank was oxygenated with air at 3 $m^3 min^{-1}$. A nonaerated slop tank served as a control. Samples removed from the tanks during the first 3 days of the voyage showed that the number of bacteria in the tank that was supplemented with urea and phosphate, aerated and inoculated with RAG-1, reached over 10^8 bacteria per mL, whereas the nonaerated control tank contained ca. 10^4 bacteria per mL.

Furthermore, the nondegraded oil in the experimental tank was thoroughly emulsified. After 4 days, the tanks were emptied, whereupon the experimental tank was completely free of the thick layer of oily sludge that remained in the control tank. In this experiment, removal of the sludge was made possible by providing conditions that favored bioemulsification rather than optimum cell growth. This microbiological cleaning method has the advantage that cleaning takes place while the tank is full of ballast water, thereby decreasing the danger of explosion. Moreover, all submerged components of the tank are in contact with the cleaning agent.

The potential of using RAG-1 to clean the cargo compartments of oil tankers during their ballast voyage (Gutnick and Rosenberg, 1977) led to an initial patent on the process (U.S. patent No. 3,941,692) and more than 20 subsequent patents throughout the world on RAG-1 and emulsans. The oil tanker experiment reached the attention of Leslie Misrock, a patent attorney and pioneer biotechnology entrepreneur, who obtained the rights from Tel Aviv University and established a company, Petroferm USA, to exploit these discoveries. Between 1975 and 1990, most of the research on emulsans was supported by Petroferm. RAG-1 emulsan was the first of a number of emulsans that are produced by microorganisms. During the past few years, several reviews have been published on different aspects of emulsans: their role in bacterial adhesion (Neu, 1996); growth on hydrocarbons (Rosenberg and Ron, 1996); surface-active polymers from the genus *Acinetobacter* (Rosenberg and Ron, 1998a); production (Wang and Wand, 1990); molecular genetics (Sullivan, 1998); commercial applications (Fiechter, 1992); enhanced oil recovery (Banat, 1995); bioemulsans (Rosenberg and Ron, 1997); high and low molecular mass microbial surfactants (Rosenberg and Ron, 1999); and their natural roles (Ron and Rosenberg, 2001).

3 Occurrence and Chemical Properties

A large number of bacterial species from different genera produce exocellular polymeric surfactants composed of polysaccharides, protein, lipopolysaccharides, lipoproteins, or complex mixtures of these biopolymers (Table 1). Xanthan, although it has some surfactant properties, is primarily a thickener, and as such, will not be discussed in this chapter.

3.1 RAG-1 Emulsan

The best-studied polymeric bioemulsifiers are the emulsans produced by different species of *Acinetobacter* (Rosenberg and Ron, 1998a). RAG-1 emulsan is a complex of an anionic heteropolysaccharide and protein (Rosenberg and Kaplan, 1987). Its surface activity is due to the presence of fatty acids, comprising 15% of the emulsan dry weight, that are attached to the polysaccha-

Tab. 1 Selected microbially produced emulsans

Emulsan	*Producing microorganisms*	*References*
RAG-1 emuslan	*A. calcoaceticus* RAG-1	Rosenberg et al. (1979a,b)
BD4 emulsan	*A. calcoaceticus* BD413	Kaplan and Rosenberg (1982)
Alasan	*A. radioresistens* KA53	Navon-Venezia et al. (1995)
Biodispersan	*A. calcoaceticus* A2	Rosenberg et al. (1988a,b)
Mannan-lipid-protein	*C. tropicalis*	Kaeppeli et al. (1984)
Liposan	*C. lipolytica*	Cirigliano and Carman (1984)
Emulsan 378	*P. fluorescens*	Persson et al. (1988)
Protein complex	*M. thermoautotrophium*	De Acevedo and McInerney (1996)
Insecticide emulsifier	*P. tralucida*	Appaiah and Karanth (1991)
Thermophilic emulsifier	*B. stearothermophilus*	Gunjar et al. (1995)
Acetylheteropolysaccharide	*S. paucimobilis*	Ashtaputre and Shah (1995)
Food emulsifier	*C. utilis*	Shepherd et al. (1995)
Sulfated polysaccharide	*H. eurihalinia*	Calvo et al. (1998)
Undefined complex	*Rhodococcus* sp.	Bredholt and Eimhjellen (1999)

ride backbone via O-ester and N-acyl linkages (Belsky et al., 1979). The chemical and physical properties of RAG-1 emulsan are summarized in Table 2. RAG-1 cells excrete maximum amounts of emulsan in shake flasks when grown in a minimal medium containing 2% ethanol as the sole carbon and energy source. Under these conditions, approximately 80% of the emulsan produced is released when the cells are in stationary phase (Goldman et al., 1982). Emulsan is an effective emulsifier at low concentrations (0.01–0.001%), representing emulsan-to-hydrocarbon ratios of 1:100 to 1:1000, and exhibits considerable substrate specificity (Rosenberg et al., 1979a). RAG-1 emulsan does not emulsify pure aliphatic, aromatic, or cyclic hydrocarbons; however, all mixtures that contain an appropriate mixture of an aliphatic and an aromatic (or cyclic alkane) are emulsified efficiently. Maximum emulsifying activity occurs over a wide pH range and requires the presence of divalent cations (Rosenberg et al., 1979b).

Tab. 2 Chemical and physical properties of RAG-1 emulsan (Data from Rosenberg et al., 1979b)

Measurement	*Result*
Chemical composition	D-galactosamine, 25% L-galactosaminouronic acid, 25% Dideoxy-diaminohexose, 25% 3-Hydroxydodecanoic acid, 10% 2-Hydroxydodecanoic acid, 10% Water and ash, 10%
Intrinsic viscosity ($cm^3 g^{-1}$)	550
Diffusion constant ($cm^3 g^{-1}$)	5.3×10^{-8}
Partial molar volume ($cm^3 g^{-1}$)	0.71
Molecular weight (kDa)	980
Dimensions (nm)	3×200

3.2 BD4 Emulsan

A. calcoaceticus BD4, which was initially isolated and characterized by Taylor and Juni (1961), produces a large polysaccharide capsule. Under certain growth conditions (e.g., 2% ethanol as the carbon and energy source), an enhanced release of the capsules was obtained (Sar and Rosenberg, 1983). However, in contrast to RAG-1, no decrease in capsular polysaccharide was observed; rather, capsular polysaccharide levels remained constant when enhanced levels of extracellular polysaccharide were obtained. When the crude capsular material was applied to a Sepharose 4B column, it eluted in a single peak containing a polysaccharide–protein complex. The polysaccharide component was obtained by deproteinization and its chemical structure elucidated (Figure 1). The protein component was obtained from the extracellular supernatant fluid of strain BD4-R7, a capsule-negative mutant of BD4 that produces no extracellular polysaccharide. The polysaccharide and protein components had no emulsifying activity by themselves. However, mixing the protein and polysaccharide fractions led to a reconstitution of the emulsifying activity (Kaplan et al., 1987). Apparently, the protein (which is hydrophobic) binds to the hydrocarbon initially in a reversible fashion. The polysaccharide then attaches to the protein and stabilizes the oil-in-water emulsion.

3.3 Alasan

Alasan, produced by a strain of *Acinetobacter radioresistens*, is a complex of an anionic polysaccharide and protein with a molecular weight of approximately 10^3 kDa (Navon-Venezia et al., 1995). However, the polysaccharide component of alasan is unusual in that it contains covalently bound alanine. The protein component of alasan appears to play an important role in both the structure and activity of the complex (Navon-Venezia et al., 1998). Deproteinization of alasan with hot phenol or treatment with specific proteinases caused a loss in most of the emulsifying activity. When a solution of alasan was exposed to increasing temperatures, there were large changes in the viscosity and emulsifying activity of the complex (Figure 2). Between 30 and 50 °C, the viscosity increased 2.6-fold, but with no change in activity. However, between 50 and 90 °C the viscosity decreased 4.8-fold while the activity increased 5-fold. None of these changes occurred with the protein-free polysaccharide, indicating that they were due to the interactions of the protein and polysaccharide portions of the complex. Alasan lowers interfacial tension from 69 to 41 dynes cm^{-1} at 20 °C and has a critical micelle concentration (CMC) of 200 μg/ml (Barkay et al., 1999). Although alasan, as well as the other emulsans, is not as effective as some of the lower molecular-weight bioemulsifiers at lowering interfacial tension, the emulsans are extremely effective at stabilizing oil-in-water emulsions.

```
→3)-a-L-Rha-(1 → 3)-α-D-Man-(1 → 3)-α-L-Rha-(1 → 3)-α-L-Rha-(1 → 3)-β-D-Glc-(1→
                      2
                      ↑
                      1
  α-L-Rha-(1 →4)-β-D-GlcUA
```

Fig. 1 The chemical structure of the *Acinetobacter calcoaceticus* BD4 capsular polysaccharide (Kaplan et al., 1985).

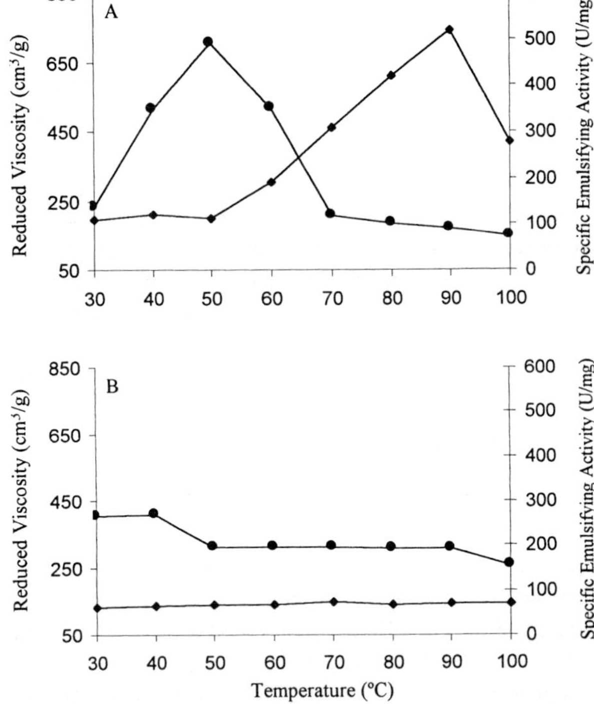

Fig. 2 Effect of temperature on viscosity and emulsifying activity of alasan (A) and apo-alasan (B). Solutions of alasan and apo-alasan (2 mg mL^{-1} in 20 mM Tris–HCl buffer, pH 7.0) were placed in a thermoblock and incubated for 10 min at 30 °C. After treatment, samples were taken for measuring emulsifying activity (squares) and reduced viscosity (circles). The solutions were then heated at the next highest temperature for 10 min. The procedure was repeated up to 100 °C. All measurements of viscosity and emulsifying activity were carried out at 30 °C.

Recently, alasan has been fractionated into three proteins (of 16, 31 and 45 kDa) and the alasan polysaccharide (Toren et al., 2001). Each of the three alasan proteins showed emulsifying activity: the 45-kDa protein had the highest specific activity, 11% higher than the intact alasan complex. The N-terminal amino acid sequence of the 45-kDa protein showed high similarity to the OmpA protein of several Gram-negative bacteria. The function of the alasan polysaccharide is not clear, but it may play a role in the release of the proteins into the medium and in protecting the protein complex against proteolytic activities. It was interesting to note that the purified 45-kDa protein was readily hydrolyzed by trypsin, whereas the protein that was bound to the polysaccharide was resistant.

3.4 Biodispersan

A. calcoaceticus A2 produces an extracellular anionic polysaccharide surfactant of molecular mass 51,400 Daltons that effectively disperses limestone and titanium dioxide (Rosenberg et al., 1988a,b). The biopolymer, referred to as biodispersan, binds to powdered calcium carbonate and changes its surface properties in a way that allows for better dispersion in water. In addition to being a surfactant, biodispersan acts also as a surfactant, aiding in the fracturing of limestone during the grinding process (Rosenberg et al., 1989).

3.5 Other High Molecular-mass Bioemulsifiers

Sar and Rosenberg (1983) reported that the majority of *Acinetobacter* strains produce extracellular nondialyzable emulsifiers. These strains included both soil and hospital isolates. Marin et al. (1996) have reported the isolation of a strain of *A. calcoaceticus* from contaminated heating oil which emulsifies that substrate. Neufeld and Zajic (1984) demonstrated that whole cells of *A. calcoaceticus* 2CA2 have the ability to act as emulsifiers, in addition to producing an extracellular emulsifier.

A large number of high molecular-weight complex bioemulsifiers have been reported, though in general little is known about these other than the producing organism and the overall chemical composition of the crude mixture. An alkane-oxidizing *Rhodococcus* sp. produces a high molecular-weight complex that when released into the medium stabilizes oil-in-water emulsions (Bredholt and Eimhjellen, 1999). *Halomonas eurihalina* produces an extracellular sulfated heteropolysaccharide (Calvo et al., 1998), while *Pseudomonas tralucida* produced an extracellular acetylated polysaccharide that was effective in emulsifying several insecticides (Appaiah and Karanth, 1991). Several recently reported bioemulsifiers are effective at high temperature, including the protein complex from *Methanobacterium thermoautotrophium* (De Acevedo and McInerney, 1996) and the protein–polysaccharide–lipid complex of *Bacillus stearothermophilus* ATCC 12980 (Gunjar et al., 1995).

Yeasts produce a number of emulsifiers, which is particularly interesting because of the food-grade status of several yeasts. Liposan is an extracellular emulsifier produced by *Candida lipolytica* (Cirigliano and Carman, 1985) and is composed of 83% carbohydrate and 17% protein. Mannanprotein emulsifiers are produced by *Saccharomyces cerevisiae* (Cameron et al., 1988), while a variety of polymeric bioemulsifiers for potential use in foods was reported by Shepard et al. (1995).

4 Natural Role of Emulsans

Although bioemulsifiers are produced by a large number of microorganisms and are clearly significant in many aspects of growth, it is difficult to generalize on their role in microbial physiology. To begin with, relevant experiments have been performed only with few emulsifier-producing microbes. In addition, with increasing numbers of identified microbial emulsifiers it has become clear that microbial surfactants have very different structures, are produced by a wide variety of microorganisms, and have very different surface properties. Thus, it is expected that bioemulsifiers have various roles, some of which are unique to the physiology and ecology of the producing microorganisms. At this stage it is impossible to draw any generalizations or to identify one or more functions that are clearly common to all microbial surfactants. Thus, most of the concepts have been derived from a consideration of the surface properties of the biosurfactants and experiments in which biosurfactants are added to microorganisms growing on water-insoluble substrates. In the following discussion, we will present a few natural roles for emulsans that have been suggested or demonstrated.

4.1 Increasing the Surface Area of Hydrophobic Water-insoluble Substrates

For bacteria growing on hydrocarbons the growth rate can be limited by the interfacial surface area between water and oil. When the surface area becomes limiting, biomass increases arithmetically rather than exponentially. The evidence that emulsification is a natural process brought about by extracellular agents is indirect, and there are certain conceptual difficulties in understanding how emulsification can provide an (evolutionary) advantage for the microorganism producing the emulsifier. Stated briefly, emulsification is a cell density-dependent phenomenon; that is, the greater the number of cells, the higher the concentration of extracellular product. The concentration of cells in an open system, such as an oil-polluted body of water, never reaches a high enough value to emulsify oil effectively. Furthermore, any emulsified oil would disperse in the water and not be any more available to the emulsifier-producing strain than to competing microorganisms. One way to reconcile the existing data with these theoretical considerations is to suggest that the emulsifying agents do play a natural role in oil degradation, but not in producing macroscopic emulsions in the bulk liquid. If emulsion occurs at or very close to the cell surface, and no mixing occurs at the microscopic level, then each cell creates its own micro-environment and no cell density dependence would be expected.

4.2 Increasing the Bioavailability of Hydrophobic Water-insoluble Substrates

One of the major reasons for the prolonged persistence of high molecular-weight hydrophobic compounds is their low water solubility that increases their sorption to surfaces and limits their availability to biodegrading microorganisms. When organic molecules are bound irreversibly to surfaces, biodegradation is inhibited (Van Delden et al., 1998). Biosurfactants can enhance growth on bound substrates by desorbing them from surfaces or by increasing their apparent water solubility (Deziel et al., 1996). Surfactants that lower interfacial tension are particularly effective in mobilizing bound hydrophobic molecules and making them available for biodegradation. Recently, it has been demonstrated that alasan increases the apparent solubilities of polycyclic aromatic hydrocarbons (PAHs) between 5- and 20-fold, and also significantly increases their rate of biodegradation (Barkay et al., 1999; Rosenberg et al., 1999).

4.3 Binding of Toxic Heavy Metals

Emulsans, like most anionic polysaccharides, bind cations. In some cases, the uronic acid residues of the polysaccharide are arranged so that divalent cations are bound avidly. RAG-1 emulsan is particularly efficient at binding uranyl ions, partially because of its negative charge and partially due to hydrophobic interactions (Zosim et al., 1983).

4.4 Regulating the Attachment-detachment of Microorganisms to and from Surfaces

One of the most fundamental survival strategies of microorganisms is their ability to locate themselves in an ecological niche where they can multiply. This is true not only for microbes that live in or on animals and plants, but also for those that inhabit soil and aquatic environments. The key elements in this strategy are cell surface structures that are responsible for the attachment of the microbes to the proper surface. Neu (1996) has reviewed how surfactants can affect the interaction between bacteria and interfaces. If a biosurfactant is excreted it can form a conditioning film on an interface, thereby stimulating certain microorganisms to attach to the interface while inhibiting the attachment of others. In the case where the substratum is also a water-insoluble substrate (e.g., sulfur and hydrocarbons), the biosurfactant stimulates growth (Beebe and Umbreit, 1971; Bunster et al., 1989). If the biosurfactant is cell-bound it can cause the microbial cell surface to become more hydrophobic, depending on its orientation. For example, the cell- surface hydrophobicity of *P. aeruginosa* was greatly increased by the presence of cell-bound rhamnolipid (Zhang and Miller, 1994), whereas the cell-surface hydrophobicity of *Acinetobacter* strains was reduced by the presence of its cell-bound emulsifier (Rosenberg et al., 1983). These data suggest that microorganisms can use their biosurfactants to regulate their cell-surface properties in order to attach or detach from surfaces according to need. This has been demonstrated for *A. calcoaceticus* RAG-1 growing on crude oil (Rosenberg and Ron, 1998a) when, during exponential growth, emulsan is cell-bound in the form of a minicapsule. This bacterium utilizes only relatively long-chain n-alkanes for growth. When these compounds have been utilized, RAG-1 becomes starved, despite still being attached to the oil droplet that is enriched in aromatics and cyclic paraffins. Starvation of RAG-1 causes release of the minicapsule of emulsan. It was shown that this released emulsan forms a polymeric film on the n-alkane-depleted oil droplet, thereby desorbing the starved cell (Rosenberg et al., 1983). In effect, the "emulsifier" frees the cell to find fresh substrate. At the same time, the depleted oil droplet has been "marked" as having been used, because it now has a hydrophilic outer surface to which the bacterium cannot attach.

5
Genetics and Regulation of Bioemulsan Production

The genetics of bioemulsifier production has been studied using mutants, both naturally occurring or those induced by transposition. The screening of such mutants is made difficult by the fact that the loss of ability to produce the emulsifier usually does not result in an easily selectable phenotype. In addition, the genetics of many emulsifier-producing bacteria has not been adequately elucidated, and important genetic tools (plasmids, transposons, gene libraries) are still to be developed.

The synthesis of high molecular-weight heteropolysaccharides requires a large number of genes, and the genetics is even more complex for polysaccharide–protein complexes. From the genetics point of view, the best-studied polysaccharide bioemulsifier is that produced by *A. calcoaceticus* BD4. The genes involved in its synthesis were identified in a cosmid library that was used to complement nonproducing mutants (Stark, 1996). These biosynthetic genes are organized in a cluster of about 60 kilobases. The first gene in the biosynthesis of the BD4 emulsifier (Accession no. X89900) was identified as a homologue of genes coding for phosphoglucoisomerase (*pgi*). The product of this gene is a protein of about 60 kDa molecular weight and carries out the bidirectional conversion of glucose-6-phosphate to fructose-6-phosphate. The gene is highly conserved from bacteria to mammals, with about 40% homology in amino acids. Additional genes in the cluster (*epsX* and *epsM*; Accession no. X81320) show homology to the genes coding for GDP-mannose pyrophosphorylase and phosphomannose isomerase of enteric bacteria (Stark, 1996). It is interesting to note that mutants in the *epsX* and *epsM* grow poorly under conditions that favor formation of the bioemulsifier, and these deleterious mutations can be overcome by an additional (suppressor) mutation in the first gene in the pathway – *pgi*. These results suggest that mutants in the biosynthesis of polysaccharide accumulate a toxic intermediate, probably fructose-6-phosphate (Fraenkel, 1992). The accumulation of the toxic substance, as well as the inhibition of growth, can be overcome by a mutation in a previous metabolic reaction, since it blocks the synthesis of this toxic intermediate. The finding that some mutants in capsule synthesis grow poorly under conditions that favor capsule synthesis may explain the difficulty often encountered in obtaining such mutants (as an example, mutants unable to synthesize alasan were not obtained after screening more than 7000 transposants of *A. radioresistens* KA53; Dahan, 1998). As screening for the mutants is usually performed on media that maximize the contrast between the capsule-producing wild-type and the mutant, i.e., media that favor capsule formation, it is possible that these conditions are strongly inhibitory – or even lethal – for many of the mutants.

In *Acinetobacter*, the production of polysaccharide bioemulsifiers is concurrent with stationary phase. It has also been suggested that UDP-glucose – one of the precursors in the synthesis of polysaccharide bioemulsifiers – is a signal molecule in the control of sigma S and sigma S-dependent genes (Bohringer et al., 1995). The bioemulsifier of *A. calcoaceticus* BD4 (Kaplan and Rosenberg, 1982), emulsan of *A. calcoaceticus* RAG-1 (Rubinovitz et al., 1982), biodispersan of *A. calcoaceticus* A2 (Rosenberg et al., 1988a), alasan of *A. radioresistens* (Navon Venezia et al., 1995; Dahan, 1998), and an uncharacterized bioemulsifier from *A. junii* (Goldenberg-Dvir, 1998) can be detected in cultures only after more than 10 h of growth, and maximal production occurs when the cultures have progressed well into the stationary phase. These results suggest the possibility that the high molecular-weight bioemulsifiers are also controlled by quorum sensing, although there is, as yet, no direct proof.

The results presented here suggest that production of bioemulsifiers by bacteria is correlated with high bacterial density. This finding may be fortuitous, or may reflect an indirect correlation with one or more physiological factors affected by high bacterial density such as availability of energy, nitrogen, or oxygen. However, it is possible that the production of bioemulsifiers at high bacterial density has a selective advantage. For emulsifiers produced by pathogens, it has been suggested (Sullivan, 1998) that – being virulence factors – they are produced when the cell density is high enough to cause a localized attack on the host. It is easier to explain the need for bioemulsifiers in bacteria growing on hydrocarbons. As these bacteria are growing at the oil–water interface, production of emulsifiers when the density is high will increase the surface area of the drops, allowing more bacteria to feed. Furthermore, when the utilizable fraction of the hydrocarbon is consumed, as in the case of oil that consists of many types of hydrocarbons, the production of the emulsifiers allows the bacteria to detach from the "used" droplet and find a new one (Rosenberg et al., 1983).

6 Biodegradation

Although it is generally assumed that bioemulsifiers are biodegradable, few studies have been conducted on the biodegradation of bioemulsans. A mixed bacterial culture was obtained by enrichment culture that was capable of degrading RAG-1 emulsan and using it as a carbon source (Shoham et al., 1983). From this mixed culture an aerobic, Gram-negative, rod-shaped bacterium was isolated which formed translucent plaques on *A. calcoaceticus* lawns. The plaques were due to the solubilization of the emulsan capsule of RAG-1. A pure culture of the capsule-degrading *Bacillus* depolymerized RAG-1 emulsan, and then used the breakdown products for growth. Subsequently it was shown that depolymerization of RAG-1 emulsan was due to the secretion of an extracellular emulsan depolymerase of molecular weight 89 000 daltons (Shoham and Rosenberg, 1983).

Emulsan depolymerase activity was the result of an eliminase reaction which split glycosidic linkages within the heteropolysaccharide backbone of emulsan to generate reducing groups and α,β-unsaturated uronides with an absorbance maximum of 233 nm. Deesterified emulsan was degraded by emulsan depolymerase at only 27% of the rate of the native polymer. The treatment of emulsan solutions with emulsan depolymerase for brief periods caused a rapid and parallel drop in viscosity and emulsifying activity. More than 75% of the viscosity and emulsifying activity was lost at a time when less than 0.5% of the glycosidic linkages were broken. These data indicate that: (1) emulsan depolymerase is an endoglycosidase; and (2) the higher the molecular weight of emulsan, the greater its emulsifying activity.

To study alasan degradation, an enrichment culture was used to isolate a bacterial strain that grew on alasan as the sole source of carbon and energy. A strain was obtained that degraded alasan, causing the loss of the protein portion of alasan, as well as the emulsifying activity (Navon-Venezia et al., 1998). The degradation was mediated by extracellular proteinases/alasanases. One of these enzymes, referred to as alasanase II, was purified to homogeneity. Alasanase II, as well as pronase, inactivated alasan, whereas a polysaccharide-degrading enzyme mixture, snail juice, had no effect on emulsifying activity.

7 Production

The production of both high and low molecular-weight bioemulsifiers was reviewed by Desai and Banat (1997).

7.1 Shake-flask Experiments

Initial studies on the production of RAG-1 emulsan was carried out using hexadecane and ethanol as the carbon sources (Rosenberg et al., 1979b). In both cases, extracellular emulsifier was produced during the exponential growth phase, but continued to accumulate during stationary phase. Using antibodies specific for emulsan, it was subsequently shown that cells in the early exponential growth phase exhibited relatively large amounts of cell-associated emulsan in the form of a mini-capsule (Goldman et al., 1982). During late exponential phase and early stationary phase, the cell-bound emulsan was released and was accompanied by a rise in the extracellular bioemulsifier concentration.

When exponentially growing cultures of *A. calcoaceticus* RAG-1 were either treated with inhibitors of protein synthesis or starved of a required amino acid, there was a stimulation in the production of emulsan (Rubinowitz et al., 1982). Emulsan synthesis in the presence of chloramphenicol was dependent on utilizable sources of carbon and nitrogen, and was inhibited by cyanide or azide or anaerobic conditions. Radioactive tracer experiments indicated that the enhanced production of emulsan after the addition of chloramphenicol was due to both the release of material synthesized before the addition of the antibiotic (40%) and to *de novo* synthesis of the polymer (60%). Chemical analysis of RAG-1 cells demonstrated the presence of large amounts of polymeric amino sugars; it was estimated that cell-associated emulsan comprised about 15% of the dry weight of growing cells. These data are consistent with the hypothesis that a polymeric precursor of emulsan accumulates on the cell surface during the exponential growth phase; in the stationary phase or during inhibition of protein synthesis, the polymer is released as a potent emulsifier.

The heavily encapsulated *A. calcoaceticus* BD4 and the "mini-encapsulated" single-step mutant *A. calcoaceticus* BD413 produced extracellular polysaccharides in addition to the capsular material (Kaplan and Rosenberg, 1982). The molar ratio of rhamnose to glucose (3:1) in the extracellular BD413 polysaccharide fraction was similar to the composition of the capsular material. In both strains, the increase in capsular polysaccharide was parallel to cell growth and remained constant in stationary phase. The extracellular polysaccharides were detected starting from mid-logarithmic phase and continued to accumulate in the growth medium for 5–8 h after the onset of stationary phase. Strain BD413 produced one-fourth the total rhamnose exopolysaccharide per cell that strain BD4 did. Depending on the growth medium, 32 to 63% of the rhamnose polysaccharide produced by strain BD413 was extracellular, whereas in strain BD4 only 7–14% was extracellular. In all cases, strain BD413 produced more extracellular rhamnose polysaccharide than did strain BD4. In glucose medium, strain BD413 also produced approximately 10-fold more extracellular emulsifying activity than did strain BD4. The isolated capsular polysaccharide obtained after shearing of BD4 cells showed no emulsifying activity. Thus, strain BD413 either produces a modified extracellular polysaccharide or excretes an additional substance(s) that is responsible for the emulsifying activity.

7.2 Production in Small Fermentors

The growth of *A. calcoaceticus* RAG-1 and emulsan production was studied in a 5-L jar fermentor at 30 °C, operating at 500 r.p.m. and 0.5 v.v.m. (volume air per volume medium per min) (Choi et al., 1996). Optimum emulsan production was achieved at an ethanol concentration of 6.5 g L^{-1} and a phosphate concentration of 12.1 g L^{-1}. In a fed-batch culture, high volumetric emulsan production was achieved by continuous ethanol and phosphate feeding to maintain their concentrations at optimum levels. During the exponential growth phase, *A. calcoaceticus* RAG-1 accumulates emulsan on the cell surface in the form of a minicapsule. When the cells are starved, the cell-bound emulsan is released into the extracellular medium. An exocellular esterase is required for the release of emulsan (Shabtai and Gutnick, 1985).

Mechanisms for RAG-1 emulsan accumulation in immobilized *A. calcoaceticus* RAG-1 cells were studied using a Celite support matrix (Wang and Wand, 1990). The ratio of cell-bound emulsan to cell dry weight was much higher for immobilized cells grown at low shear forces than for free cells or immobilized cells grown at shear stress values above 5 dynes cm^{-2}. The most efficient production of emulsan appears to be via self-cycling fermentation (Brown and Cooper, 1991). The specific emulsan productivity achieved by self-cycling fermentation was about 50-fold greater than that of a batch culture, and more than twice that of a chemostat.

The production of alasan was studied in a batch-fed 2.5-L fermentor (Navon-Venezia et al., 1995). *A. radioresistens* KA53 had an approximate doubling time of 2.4 h during the initial 20-h exponential phase of growth, while the viable count decreased slowly after reaching a maximum of 8.0×10^9 cells per mL at 68 h. The total dry weight (cell biomass and extracellular material) at the end of the fermentation was 23.6 g L^{-1}. The ratio of emulsifying activity to cell biomass increased from 5.3 to 7.3 to 11.6 at 24, 64, and 87 h, respectively. The extracellular activity was 190 U mL^{-1} after 87 h, indicating that the majority of emulsification activity was extracellular. The extracellular biopolymer as a fraction of the cell biomass increased during growth from 9.8% at 20 h to 24% at 87 h.

The production and secretion of biodispersan by *A. calcoaceticus* A2 was studied in protein secretion mutants (Elkeles et al., 1994), the active component being an extracellular anionic polysaccharide. Extracellular fluid also contains a high concentration of secreted proteins that create problems in the purification and application of biodispersan. Strains that were defective in protein secretion were selected following chemical mutagenesis. These mutants produced equal, or even higher, levels of extracellular biodispersan than the parent strain, suggesting that the production and release of polysaccharides proceeded independently of protein secretion. Thus, protein secretion mutants are potentially useful in the production of extracellular polysaccharides.

7.3 Large-scale Production of Emulsans

Several tons of RAG-1 emulsan was produced by Petroferm USA in California during the 1980s. No information is available on the fermentation conditions, other than that denatured ethanol was used as the carbon source. Part of the emulsan that was produced was used for cleaning the decks of aircraft carriers and oil separators by the U.S. Navy (J. Jones-Meehan, personal communication).

Alasan was produced in a 20 000-L fermentor at 30 °C using ethanol as the carbon source and maintaining the pH at 6.8–7.0 (by addition of NH_3). The ethanol concentration was maintained at 0.5–1.5% (v/v) by periodic addition, and the foam was controlled by Dow Corning 1520 Silicone antifoam. The yield was 120 kg. The alasan produced was used in grinding limestone for the production of paper by American-Israeli Paper Mills Ltd. (Rosenberg et al., 1989).

7.4 Patents

A number of material, process and production patents were issued on RAG-1 emulsan; the U.S. patents are listed in Table 3. Patents have also been issued on biodispersan (Rosenberg and Ron, 1990) and alasan (Rosenberg and Ron, 1998b).

8 Potential Applications

Surfactants are used widely in industry, agriculture and medicine. The materials currently in use commercially as surfactants are produced mainly by chemical synthesis or as a relatively inexpensive by-product of an industrial process (e.g., lignin sulfates from

Tab. 3 U.S. patents issued on RAG-1 emulsan, alasan and emulcyan

Authors	*Title*	*U.S. Patent No.*
Gutnick, D.L., Rosenberg, E. (1976)	Cleaning of cargo compartments	3,941,692
Gutnick, D.L., Rosenberg, E. (1980)	Production of emulsans	4,230,801
Gutnick, D.L., Rosenberg. E., Shoham, Y. (1980)	Production of α-emulsans	4,234,689
Gutnick, D.L., Rosenberg, E. (1981)	Cleaning oil-contaminated vessels with α-emulsans	4,276,094
Gutnick, D.L., Rosenberg, E., Belsky. I., Zosim, Z. (1982)	Apo-β-emulsans	4,311,829
Gutnick, D.L., Rosenberg, E., Belsky, I., Zosim, Z. (1982)	Apo-β-emulsans	4,311,830
Gutnick, D.L., Rosenberg, E., Belsky, I., Zosim, Z. (1982)	Apo-γ-emulsans	4,311,831
Gutnick, D.L., Rosenberg, E., Belsky, I., Zosim, Z. (1982)	Proemulsans	4,311,832
Gutnick, D.L., Rosenberg, E., Belsky, I., Zosim, Z. (1983)	γ-emulsans	4,380,504
Gutnick, D.L., Rosenberg, E., Belsky, I., Zosim, Z. (1983)	Polyanionic heteropolysaccharide biopolymers	4,395,353
Gutnick, D.L., Rosenberg, E, Belsky, I., Zosim, Z. (1983)	Emulsans	4,395,354
Shoham, Y., Gutnick, D.L., Rosenberg, E. (1987)	Enzymatic degradation of lipopolysaccharide bioemulsifiers	4,704,360
Sar, N., Rosenberg, E., Gutnick, D.L., Nestaas, E. (1984)	Bioemulsifier production by *Acinetobacter calacoaceticus* strain	4,676,916
Hayes, M.E. et al. (1990)	Bioemulsifier-stabilized hydrocarbosols	4,943,390
Gutnick, D.L. et al. (1989)	Bioemulsifier production by *Acinetobacter* strains	4,883,757
Hayes, M.E. (1989)	Bioemulsified-containing personal care products	4,870,010
Shilo, M., Fattom, A. (1986)	Cyanobacterial produced bioemulsifier composition	4,693,842
Rosenberg, E., Ron, E.Z. (1998)	Alasan	5,840,847

paper manufacturing). In order for a microbial polysaccharide surfactant to penetrate the market, it must provide a clear advantage over the existing competing materials. Special properties of polysaccharide surfactants that may provide added incentive for their commercialization include:

1) Biodegradability and controlled inactivation. Several chemically synthesized, commercial available surfactants (e.g., perfluorinated anionics) resist biodegradation, accumulate in nature, and cause ecological problems. Microbial polysaccharides, like all natural products, are susceptible to degradation by microorganisms in water and soil (Zajic et al., 1977; Shoham et al., 1983). Hydrocarbon-in-water emulsions stabilized by emulsan can be broken by minute quantities of the enzymes. In principle, it should be possible to isolate a specific enzyme that will reverse the effect of any particular polysaccharide surfactant. The ability rapidly to break a specific emulsion or dispersion under mild conditions by the addition of catalytic quantities of an enzyme may lead to new applications.
2) Selectivity for specific interfaces. Compared with chemically synthesized materials, one of the general features of polysaccharide molecules is their remarkable specificity. Two examples of microbial surfactants that exhibit substrate specificity are emulsan toward a mixture of aliphatic and aromatic hydrocarbons (Rosenberg et al., 1979a), and the solubilizing factor of *Pseudomonas* PG-1 toward pristane. Some of the best examples of specificity involve polysaccharide–lectin interactions.
3) Surface modifications. A polysaccharide emulsifying or dispersing agent not only causes a reduction in the average particle size, but also changes the surface properties of the particle in a fundamental manner (Neu and Poralla, 1990). For example, oil droplets that have been emulsified with emulsan contain an outer hydrophilic shell of spatially oriented carboxyl and hydroxyl groups. These emulsan-stabilized oil droplets bind large quantities of uranium (Zosim et al., 1983). In an elegant experiment, Pines et al. (1983) demonstrated that emulsan on the surface of hexadecane droplets acts as a receptor for *A. calcoaceticus* bacteriophage ap3. Clearly, each polysaccharide alters, in a characteristic manner, the interface to which it adheres. Usually, 0.1 – 1.0% of a polysaccharide bioemulsifier or biodispersant is sufficient to saturate the surface of the dispersed material. Thus, small quantities of a dispersant can alter dramatically a material's surface properties, such as its surface charge, its hydrophobicity and, most interestingly, its pattern recognition based on the three-dimensional structure of the adherent polymer.

At present, the use of bioemulsifiers is limited by the cost of production and insufficient experience in applications. However, since there is increasing awareness of water quality and environmental conservation, as well as an expanding demand for natural products, it appears inevitable that high-quality, microbially produced, bioemulsifiers will replace the currently used chemical emulsifiers in many applications.

9 Outlook and Perspectives

Although a large number of bioemulsans have been isolated, much of the research has been quite descriptive: characterization of the producing microorganism; flask conditions for producing the bioemulsifier; and a very superficial description of the macromolecules. With the exception of the BD4 polysaccharide (Kaplan et al., 1985) and one of the alasan proteins (Toren et al., 2001), none of the bioemulsan macromolecules has been purified to homogeneity, and neither has their chemical structure been elucidated.

The requirements for bioemulsans to find their rightful position in microbial physiology and biotechnology are threefold. First, the chemical structures of bioemulsans must be elucidated. In the case of polysaccharides, structures can be determined using modern NMR studies coupled to classical carbohydrate chemistry. With surface-active proteins, powerful molecular genetic techniques are now available to determine the precise amino acid sequence of the protein. Second, when structures have been determined, then it becomes possible to perform structure–function studies. The fundamental questions remaining to be answered include: (1) How do bioemulsans differ from other macromolecules? (2) What aspect of their three-dimensional structure makes them such effective stabilizers of oil-in-water emulsions, or what other surface-active properties do they express? (3) Are their activities a direct outcome of their monomeric sequences, or are they modified after polymerization? Using the power of modern structural biology, these questions should be answered within the next decade – at least for some bioemulsifiers.

The third question relates to the reason why bioemulsans have not made a greater penetration into the market place, despite their special properties. The two major problems are cost and lack of the necessary toxicity testing. The cost of existing commercial emulsifiers ranges from US$5 to US$15 per kg, compared with the current cost of producing RAG-1 emulsan of about US$50 per kg. It should be borne in mind however, that for many applications bioemulsans are 5- to 10-fold more effective on a weight basis than commercial emulsifiers. Furthermore, up-scaling and strain improvement should reduce the production cost of bioemulsans considerably. The greater obstacle to overcome before bioemulsans achieve their commercial potential is the cost of the toxicity testing required in order to introduce any new product. This is especially true in the food industry, but is equally applicable to the cosmetics and home care industries. Consequently, it is predicted that the company which chooses the correct bioemulsan and invests in lowering its production costs and obtaining the necessary product approvals, will reap the benefits of introducing this new biotechnology product.

10
References

Abbot, B. J., Gledhill, W. E. (1971) The extracellular accumulation of metabolic products by hydrocarbon-degrading microorganisms, *Adv. Appl. Microbiol.* **14**, 249–388.

Appaiah, A. K. A., Karanth, N. G. K. (1991) Insecticide specific emulsifier production by hexachlorocyclohexane-utilizing *Pseudomonas tralucida* Ptm + strain, *Biotechnol. Lett.* **13**, 371–374.

Ashtaputre, A. A., Shah, A. K. (1995) Emulsifying property of a viscous exopolysaccharide from *Sphingomonas paucimobilis, World J. Microbiol. Biotechnol.* **11**, 219–222.

Banat, I. M. (1995) Biosurfactants production and possible use in microbial enhanced oil recovery and oil pollution remediation: a review, *Biosource Technol.* **51**, 1–12.

Barkay, T., Navon-Venezia, S., Ron, E. Z., Rosenberg, E. (1999) Enhancement of solubilization and biodegradation of polyaromatic hydrocarbons by the bioemulsifier alasan, *Appl. Environ. Microbiol.* **65**, 2697–2702.

Beebe, J. L., Umbreit, W. W. (1971) Extracellular lipid of *Thiobacillus thiooxidans, J Bacteriol.* **108**, 612–615.

Belsky, I., Gutnick, D. L., Rosenberg, E. (1979) Emulsifier of *Arthrobacter* RAG-1: determination of emulsifier-bound fatty acids, *FEBS Lett.* **101**, 175–178.

Bohringer, J., Fischer, D., Mosler, G., Hengge-Aronis, R. (1995) UDP-glucose is a potential intracellular signal molecule in the control of expression of sigma S and sigma S-dependent genes in *Escherichia coli, J. Bacteriol.* **177**, 413–422.

Bredholt, H., Eimhjellen, K. (1999) Induction and development of the oil emulsifying system in an alkane oxidizing *Rhodococcus* species, *Can. J. Microbiol.* **45**, 700–708.

Brown, W. A., Cooper, D. G. (1991) Self-cycling fermentation applied to *Acinetobacter calcoaceticus* RAG-1, *Appl. Environ. Microbiol.* **57**, 2901–2906.

Bunster, L., Fokkema, N. J., Shippers, B. (1989) Effect of surface-active *Pseudomonas* spp. on leaf wettability, *Appl. Environ. Microbiol.* **55**, 1434–1435.

Calvo, C., Martinez-Checa, F., Mota, A., Bejar, V., Quesada, E. (1998) Effect of cations, pH and sulfate content on the viscosity and emulsifying activity on the *Halomonas eurihalina, J. Ind. Microbiol. Biotechnol.* **20**, 205–209.

Cameron, D. R., Cooper, D. G., Neufeld, R. J. (1988) The mannoprotein of *Saccharomyces cerevisiae* is an effective bioemulsifier, *Appl. Environ. Microbiol.* **54**, 1420–1425.

Choi, J. W., Choi, H. G., Lee, W. H. (1996) Effects of ethanol and phosphate on emulsan production by *Acinetobacter calcoaceticus* RAG-1, *J. Biotechnol.* **45**, 217–225.

Cirigliano, M. C., Carman, G. M. (1984) Purification and characterization of liposan, a bioemulsifier from *Candida lipolytica, Appl. Environ. Microbiol.* **50**, 846–850.

Dahan, O. (1998) Isolation and characterization of alasan mutants in *Acinetobacter radioresistens*, MSc Thesis, Tel Aviv University.

Davis, J. B. (1967) *Petroleum Microbiology.* New York: Elsevier.

De Acevedo, G. T., McInerney, M. J. (1996) Emulsifying activity in thermophilic and extremely thermophilic microorganisms, *J. Ind. Microbiol.* **16**, 17–22.

Desai, J., Banat, I. (1997) Microbial production of surfactants and their commercial potential, *Microbiol. Mol. Biol. R.* **61**, 47–48.

Deziel, E., Paquette, G., Villemur, R., Lepine, F., Bisaillon, J. G. (1996) Biosurfactant production by a soil *Pseudomonas* strain growing on polycyclic aromatic hydrocarbons, *Appl. Environ. Microbiol.* **62**, 1908–1912.

Dostalek, M., Spurny, M. (1958) Bacterial release of oil, *I. Folia Biol.* (Prague), **IV**, 166–171.

Elkeles, A., Rosenberg, E., Ron, E. Z. (1994) Production and secretion of the polysaccharide biodispersan of *Acinetobacter calcoaceticus* A1 in protein secretion mutants, *Appl. Environ. Microbiol.* **60**, 4642–4645.

Erickson, L. E., Nakahara, T. (1975) Growth in cultures with two liquid phases: hydrocarbon uptake and transport, *Process Biochem.* **10**, 9–13.

Fiechter, A. (1992) Biosurfactants: moving towards industrial application, *Trends Biotechnol.* **10**, 208–217.

Fraenkel, D. G. (1992) Genetics and intermediary metabolism, *Annu. Rev. Genet.* **26**, 159–177.

Goldberg, I. (1985) *Single Cell Protein.* Heidelberg: Springer-Verlag.

Goldenberg-Dvir, V. (1998) A new bioemulsifier produced by the oil-degrading *Acinetobacter junii* V-26, MSc thesis, Tel-Aviv University.

Goldman, S., Shabtai, Y., Rubinovitz, C., Rosenberg, E., Gutnick, D. L. (1982) Emulsan in *Acinetobacter calcoaceticus* RAG-1: distribution of cell-free and cell-associated cross-reacting materials, *Appl. Environ. Microbiol.* **44**, 165–170.

Gunjar, M., Khire, J. M., Khan, M. I. (1995) Bioemulsifier production by *Bacillus stearothermophilus* VR8 isolate, *Lett. Appl. Microbiol.* **21**, 83–86.

Gutnick, D. L., Rosenberg, E. (1977) Oil tankers and pollution: a microbiological approach, *Annu. Rev. Microbiol.* **31**, 379–396.

Iguchi, T., Takeda, I., Ohsawa, H. (1969) Emulsifying factor of hydrocarbon produced by a hydrocarbon-assimilating yeast, *Agric. Biol. Chem.* **33**, 1657–1658.

Johnson, M. J. (1968) Utilization of hydrocarbon by microorganisms, *Chem. Ind.* (London) **36**, 1532–1537.

Kaplan, N., Rosenberg E. (1982) Exopolysaccharide distribution and bioemulsifier production in *Acinetobacter calcoaceticus* BD4 and BD413, *Appl. Environ. Microbiol.* **44**, 1335–1341.

Kaplan, N., Jann, B., Jann, K. (1985) Structural studies on the capsular polysaccharide of *Acinetobacter calcoaceticus* BD4, *Eur. J. Biochem.* **152**, 453–458.

Kaplan, N., Zosim, Z., Rosenberg, E. (1987) *Acinetobacter calcoaceticus* BD4 emulsan: reconstitution of emulsifying activity with pure polysaccharide and protein, *Appl. Environ. Microbiol.* **53**, 440–446.

Kaeppeli, O., Walther, P., Mueller, M., Fiechter, A. (1984) Structure of cell surface of the yeast *Candida tropicalis* and its relation to hydrocarbon transport, *Arch. Microbiol.* **138**, 279–282.

La Riviere, J. W. M. (1955a) The production of surface active compounds by microorganisms and its possible significance in oil recovery. I. Some general observations on the change of surface tension in microbial cultures, *Antonie van Leeuwenhoek, J. Microbiol. Serol.* **21**, 1–8.

La Riviere, J. W. M. (1955b) The production of surface active compounds by microorganisms and its possible significance in oil recovery. II. On the release of oil from oil-sand mixtures with the aid of sulphate-reducing bacteria. *Antonie van Leeuwenhoek, J. Microbiol. Serol.* **21**, 9–16.

Marin, M., Pedregosa, A., Laborda, F. (1996) Emulsifier production and microscopical study of emulsions and biofilms formed by the hydrocarbon-utilizing bacteria *Acinetobacter calcoaceticus* MM5, *Appl. Microbiol. Biotechnol.* **44**, 660–667.

Navon-Venezia, S., Zosim, Z., Gottlieb, A., Legmann, R., Carmeli, S., Ron, E. Z., Rosenberg, E. (1995) Alasan, a new bioemulsifier from *Acinetobacter radioresistens*, *Appl. Environ. Microbiol.* **61**, 3240–3244.

Navon-Venezia, S., Banin, E., Ron, E. Z., Rosenberg, E. (1998) The bioemulsifier alasan: role of protein in maintaining structure and activity. *Appl. Microbiol. Biotechnol.* **49**, 382–384.

Neu, T. R. (1996) Significance of bacterial surface-active compounds in interaction of bacteria with interfaces, *Microbiol. Rev.* **60**, 151–166.

Neu, T. R., Poralla, K. (1990) Emulsifying agent from bacteria isolated during screening for cells with hydrophobic surfaces, *Appl. Microbiol. Biotechnol.* **32**, 521–525.

Neufeld, R. J., Zajic, J. E. (1984) The surface activity of *Acinetobacter calcoaceticus* sp. 2CA2, *Biotechnol. Bioeng.* **26**, 1108–1114.

Pines, O., Bayer, E.A., Gutnick, D. L. (1983) Localization of emulsan-like polymers associated with the cell surface of *Acinetobacter calcoaceticus*, *J. Bacteriol.* **154**, 893–905.

Pines, P., Gutnick, D. L. (1981) Relationship between phase resistance and emulsan production, interaction of phases with the cell-surface of *Acinetobacter calcoaceticus*, *Arch. Microbiol.* **130**, 129–133.

Reisfeld, A., Rosenberg, E., Gutnick, D. (1972) Microbial degradation of crude oil: factors affecting the dispersion in sea water by mixed and pure cultures, *Appl. Environ. Microbiol.* **24**, 363–368.

Ron, E.Z., Rosenberg, E. (2001) Natural roles of biosurfactants, *Environ. Microbiol.* **3**, 229–236.

Rosenberg, E., Kaplan, N. (1987) Surface-active properties of *Acinetobacter* expolysaccharides, in: *Bacterial Outer Membranes as Model Systems* (Inouye, M., Ed.), New York: John Wiley & Sons, 311–342.

Rosenberg, E., Ron, E. Z. (1990) Bacterial process for the production of biodispersants. Israel-Patent Application No. 76981.

Rosenberg, E., Ron, E. Z. (1996) Bioremediation of petroleum contamination, in: *Bioremediation: Principles and Applications* (Crawford, R. L., Crawford, D. L., Eds.), Cambridge University Press, 100–124.

Rosenberg, E., Ron, E. Z (1997) Bioemulsans: microbial polymeric emulsifiers, *Curr. Opin. Biotechnol.* **8**, 313–316.

Rosenberg, E., Ron, E. Z. (1998a) Surface active polymers from the genus *Acinetobacter*, in: *Biopolymers from Renewable Resources* (Kaplan, D. L., Ed.), New York: Springer-Verlag, 281–291.

Rosenberg, E., Ron, E. Z. (1998b) Novel bioemulsifiers. Israeli Patent Application No. 112,254, 5/1/95: U.S. Patent No. 5,840,547 (November 24, 1998).

Rosenberg, E., Ron, E. Z. (1999). High and low molecular mass microbial surfactants, *Appl. Microbiol. Biotechnol.* **52**, 154–162.

Rosenberg, E., Horowitz, A., Englander, E., Gutnick, D. L. (1974) Bacterial emulsion of crude oil in seawater, Symposium on Impact of Microorganisms on the Aquatic Environment, *USEPA Publication*, 157–168.

Rosenberg, E., Perry, A., Gibson, D. T., Gutnick, D. (1979a) Emulsifier of *Arthrobacter* RAG-1: specificity of hydrocarbon substrate, *Appl. Environ. Microbiol.* **37**, 409–413.

Rosenberg, E., Zuckerberg, A., Rubinovitz, C., Gutnick, D. L. (1979b) Emulsifier of *Arthrobacter* RAG-1: Isolation and emulsifying properties, *Appl. Environ. Microbiol.* **37**, 402–408.

Rosenberg, E., Gottlieb, A., Rosenberg, M. (1983) Inhibition of bacterial adherence to hydrocarbons and epithelial cells by emulsan, *Infect. Immun.* **39**, 1024–1028.

Rosenberg, E., Rubinovitz, C., Gottlieb, A., Rosenhak, S., Ron, E. Z. (1988a) Production of biodispersan by *Acinetobacter calcoaceticus* A2, *Appl. Environ. Microbiol.* **54**, 317–322.

Rosenberg, E., Rubinovitz, C., Legmann, R., Ron, E. Z. (1988b) Purification and chemical properties of *Acinetobacter calcoaceticus* A2 biodispersan, *Appl. Environ. Microbiol.* **54**, 323–326.

Rosenberg, E., Schwartz, Z., Tenenbaum, A., Rubinovitz, C., Legmann, R., Ron, E. Z. (1989) A microbial polymer that changes the surface properties of limestone: effect of biodispersan in grinding limestone and making paper, *J. Dispersion Sci. Technol.* **10**, 241–250.

Rosenberg, E., Barkay, T., Navon-Venezia, S., Ron, E. Z. (1999) Role of *Acinetobacter* bioemulsans in petroleum degradation, in: *Novel Approaches for Bioremediation of Organic Pollution* (Fass, R., Flasher, Y., Eds), New York: Kluwer Academic/Plenum Publishers.

Rubinovitz, C., Gutnick, D. L., Rosenberg, E. (1982) Emulsan production by *Acinetobacter calcoaceticus* in the presence of chloramphenicol, *J. Bacteriol.* **152**, 126–132

Sar, N., Rosenberg, E. (1983) Emulsifier production by *Acinetobacter calcoaceticus* strains, *Curr. Microbiol.* **9**, 309–314.

Shabtai, Y., Gutnick, D. L. (1985) Exocellular esterase and emulsan release from the cell surface of *Acinetobacter calcoaceticus, J. Bacteriol.* **161**, 1176–1181.

Shepard, R., Rockey, J., Sutherland, I. W., Roller, S. (1995) Novel bioemulsifiers from microorganisms for use in foods, *J. Biotechnol.* **40**, 207–217.

Shoham, Y., Rosenberg, E. (1983) Enzymatic depolymerization of emulsan, *J. Bacteriol.* **156**, 161–167.

Shoham, Y., Rosenberg, M., Rosenberg, E. (1983) Bacterial degradation of emulsan, *Appl. Environ. Microbiol.* **46**, 573–579.

Stark, M. (1996) Analysis of the exopolysaccharide gene cluster from *Acinetobacter calcoaceticus* BD4, PhD thesis, Tel-Aviv University.

Sullivan, E. R. (1998) Molecular genetics of biosurfactant production, *Curr. Opin. Biotechnol.* **9**, 263–269.

Taylor, W. H., Juni, E. (1961) Pathways for biosynthesis of a bacterial capsular polysaccharide. I. Characterization of the organism and polysaccharide, *J. Bacteriol.* **81**, 688–693.

Toren, A., Navon-Venezia, S., Ron, E. Z., Rosenberg, E. (2001) Emulsifying activities of purified alasan proteins from *Acinetobacter radioresistens* KA53, *Appl. Environ. Microbiol.* **67**, 1102–1106.

Updegraff, D. M., Wren, G. B. (1954) The release of oil from petroleum bearing materials by sulfate reducing bacteria, *Appl. Microbiol.* **2**, 309–316.

Van Delden, C., Pesci, E. C., Pearson, J. P., Iglewski, B. H. (1998) Starvation selection restores elastase and rhamnolipid production in a *Pseudomonas aeruginosa* quorum-sensing mutant, *Infect. Immun.* **66**, 4499–4502.

Wang, S. D., Wand, D. I. C. (1990) Mechanisms for biopolymer accumulation in immobilized *Acinetobacter calcoaceticus* system, *Biotechnol. Bioeng.* **36**, 402–410.

Zajic, J. E., Panchal, C. J. (1976) Bioemulsifiers, *CRC Crit. Rev. Microbiol.* **5**, 39–66.

Zajic, J. E., Guignard, H., Gerson, D. F. (1977) Properties and biodegradation of a bioemulsifier from *Corynebacterium hydrocarboclastus*, *Biotechnol. Bioeng.* **19**, 1303–1320.

Zhang, Y., Miller, R. M. (1994) Effect of a *Pseudomonas* rhamnolipid biosurfactant on cell hydrophobicity and biodegradation of octadecane, *Appl. Environ. Microbiol.* **60**, 2101–2106.

Zosim, Z., Gutnick, D. L., Rosenberg, E. (1983) Uranium binding by emulsan and emulsanosols, *Biotech. Bioeng.* **25**, 1725–1735.

5 Other Bacterial Glycolipids

Dr. Jitendra D. Desai[1], Prof. Dr. Anjana J. Desai[2]

[1] Department of Environment and Ecology, Indian Petrochemicals Corp. Ltd., P.O. Petrochemicals, Vadodara, Gujarat 391 346, India; Fax: +91-265-270414; E-mail: desaijd@ipclmail.com

[2] Department of Microbiology and Biotechnology Center, Faculty of Science, M. S. University of Baroda, Vadodara, Gujarat 390 002, India; Tel: +91-265-794396; Fax: +91-265-792508; E-mail: desai_aj@yahoo.com

CMC critical micelle concentration
LPS lipopolysaccharide
MEOR microbial enhanced oil recovery
RL rhamnolipid
TDP thymidine diphosphate
UDP uridine diphosphate

What is perfect, this is perfect,
What comes out from such perfection is also perfect;
When the part is taken-out of the perfection,
The remaining part yet remains perfect.
Isa Upanishad. Chapter V.1.1

1 Introduction

A wide spectrum of glycolipids is synthesized by bacteria, yeasts, and fungi during growth on various carbon sources under appropriate environmental and nutritional conditions. Recent interest in bacterial glycolipids can be attributed to their diversity, environmental friendly and biocompatible nature, low toxicity, high biodegradability, high selectivity, activity at extreme environmental conditions, cost-effective production through fermentation, etc. Due to surfactant properties and other characteristics, they have promising commercial applications in the oil, health care, agriculture, cosmetic, paper, metal, leather, and food-processing industries. This aspect has been the subject of many recent reviews (Rosenberg, 1993; Banat, 1995; Miller, 1995; Deziel et al., 1996; Desai and Banat, 1997; Lang and Wullbrandt, 1999). The structure, properties, biosynthesis, and chemistry of microbial glycolipid biosurfactants have been reviewed (Lang and Wagner, 1987; Desai and Desai, 1993; Desai and Banat, 1997; Lang, 1999).

This chapter describes the current status of research on the structure, properties, function, biosynthesis, genetics, production, and potential commercial applications of bacterial glycolipids. Cell-wall-associated exopolysaccharides and complex glycolipids such as emulsan are exclusively covered elsewhere in this volume.

2 Historical Outline

Historically, studies on bacterial glycolipids are associated with hydrocarbon degradation in the environment. For 90 years it has been known that certain microorganisms are able to use petroleum hydrocarbons as the sole source of carbon and energy (Atlas, 1981). Jarvis and Johnson (1949) first described the production of rhamnose-containing glycolipids in *Pseudomonas aeruginosa* and its role in hydrocarbon emulsification. Soon after the biosynthetic pathway of rhamnolipids (RLs) in *P. aeruginosa* was investigated (Hauser and Karnovsky, 1954). Interest in trehalolipids as general surfactants is linked to the discovery of trehalose dimycolate in the emulsion layer of culture broth of *Arthrobacter paraffineus* during growth on hydrocarbon (Suzuki et al., 1969). One of the most powerful emulsifiers known today, emulsan, a complex glycolipid produced by *A. calcoaceticus*, was isolated and characterized by Rosenberg et al. (1979). Subsequently, several bacterial strains producing glycolipids both on water-soluble and -insoluble substrates have been isolated, and the production has been optimized. The genes for RL synthesis have been isolated and characterized (Ochsner and Reiser, 1995). Breakthroughs have been achieved in the field of rapid and reliable assessment of potential glycolipid-producing bacteria, cost-effective glycolipid production, and the enzymatic synthesis of tailor-made glycolipids and sugar esters.

3 Occurrence

Glycolipids are widely distributed among microorganisms and are broadly classified on the basis of the producer organism as bacterial, yeast, or fungal glycolipids. Although standard enrichment culture and the estimation of surface tension, interfacial tension, and critical micelle concentration (CMC) values are the most widely employed techniques for the assessment of glycolipid-producing bacteria, recently developed quick and reliable methods listed below have played an important role in the advancement of this area.

Axisymmetric drop shape analysis by the profile technique of Van der Vegt et al. (1991) in which the profile of culture drops placed on a fluoroethylene–propylene surface is determined and surface tension is calculated.

The drop-collapsing test of Jain et al. (1991), in which a drop of a surface active cell suspension placed on an oil-coated surface collapses.

A direct thin layer chromatography method as described by Finnerty and Singer (1985) and Matsuyama et al. (1991).

An agar plate diffusion technique using a dye reagent described by Sigmond and Wagner (1991) and Koch et al. (1991).

Estimation of the emulsification index value of the culture broth on mixing with an equal volume of kerosene as per Cooper and Goldenberg (1987).

The most common and well-studied glycolipids and their producer organisms are listed in Table 1.

4 Chemical Structure

Glycolipids are carbohydrate lipids in which carbohydrates are attached to long chain aliphatic acids or hydroxyaliphatic acids either by glycosidic or ester linkages. The glycolipids have α,α or β,β configurations but do not appear in the α,β form. The elucidation of the chemical structure of glycolipids involves isolation, purification, and hydrolysis, followed by structure elucidation of fatty acid and carbohydrate moieties. Most glycolipids are excreted in the growth medium from which they can be concentrated by precipitation or by solvent extraction after the removal of cells. Chromatographic techniques (thin-layer or high-performance liquid chromatography) are employed for their purification. The glyco-

Tab. 1 Common microbial glycolipids and their origin

Glycolipids	*Producing organism*
Rhamnolipids	*Pseudomonas* spp.
Trehalolipids	*R. erythropolis, Mycobacterium* spp., *Corynebacterium* spp., *A. paraffineus, N. erythropolis*
Sophorolipids	*Torulopsis* spp.
Cellobiolipids	*Ustilago* spp.
Glucose, sucrose, and fructose lipids	*Arthrobacter paraffineus, Corynebacterium* spp., *R. erythropolis*
Penta- and disaccharide lipids	*Nocardia corynebacteroides*
Phenolic glycolipid	*Mycobacterium leprae*
Diglycosyl diglycerides	*Lactobacillus fermentii*
Mannosylerythritol lipid	*Candida* spp., *Sch. melanogramma*
Lipo-heteropolysaccharide (emulsan)	*Acinetobacter calcoaceticus* RAG-1
Biodispersan	*A. calcoaceticus*
Polysaccharide–fatty acid–protein	*Candida tropicalis*
Carbohydrate–protein–lipids	*P. fluorescens, D. polymorphis, Corynebacterium* spp.
Polysaccharide–protein	*C. hydrocarboclastus, C. lipolytica*

sidic linkage between the two sugars and the hydroxyl groups of the lipid moiety in the purified product can be broken by acid hydrolysis. The ester linkage between the fatty acid group and the hydroxyl group of the sugar is broken by alkaline hydrolysis. The structures of mono-, di-, and oligosaccharides are determined by thin-layer, gas, or high-performance liquid chromatography; or colorimetric or enzymatic techniques. Mass, infrared, Fourier transform infrared and nuclear magnetic resonance spectroscopy complementary to other techniques are frequently used in the determination of complex carbohydrates. The structure elucidation of fatty acids involves separation of the mixture of fatty acids by preparative liquid chromatography, converting them into methyl esters before their subjection to gas chromatography and gas chromatography–mass spectroscopy. Some of these techniques have been recently reviewed by Brandenburg and Seydel (1998), Bush et al. (1999), Harvey (1999), and Mercade et al. (1997).

The major and best-studied bacterial glycolipids include glycosyl diglycerides, RLs, trehalolipids, and complex glycolipids.

4.1 Glycosyl Diglycerides

Glycosyl diglycerides containing monosaccharides (Figure 1a) are distributed widely in Gram-positive bacteria, but have so far not been found in Gram-negative bacteria. Production of glucose lipids by *Alcanivorax borkumensis* (Yakimov et al., 1998), glycoglycerolipids containing mannose by *Microbacterium* spp. (Wicke et al., 2000), and sucrose and fructose lipids from *Arthrobacteria, Corynebacteria,* and *Nocardia* (Suzuki et al., 1974; Itoh and Suzuki, 1974) have been reported. Brennan et al. (1970) have isolated acyl-glucose (Figure 1b) from *Corynebacteria* and *Mycobacteria* during growth on glucose. Li et al. (1984) and Vollbrecht et al. (1998, 1999) demonstrated synthesis of various mono-, di-, or trisaccharide lipids in *Arthrobacter* sp. and *Tsukamurella* sp. nov, respectively.

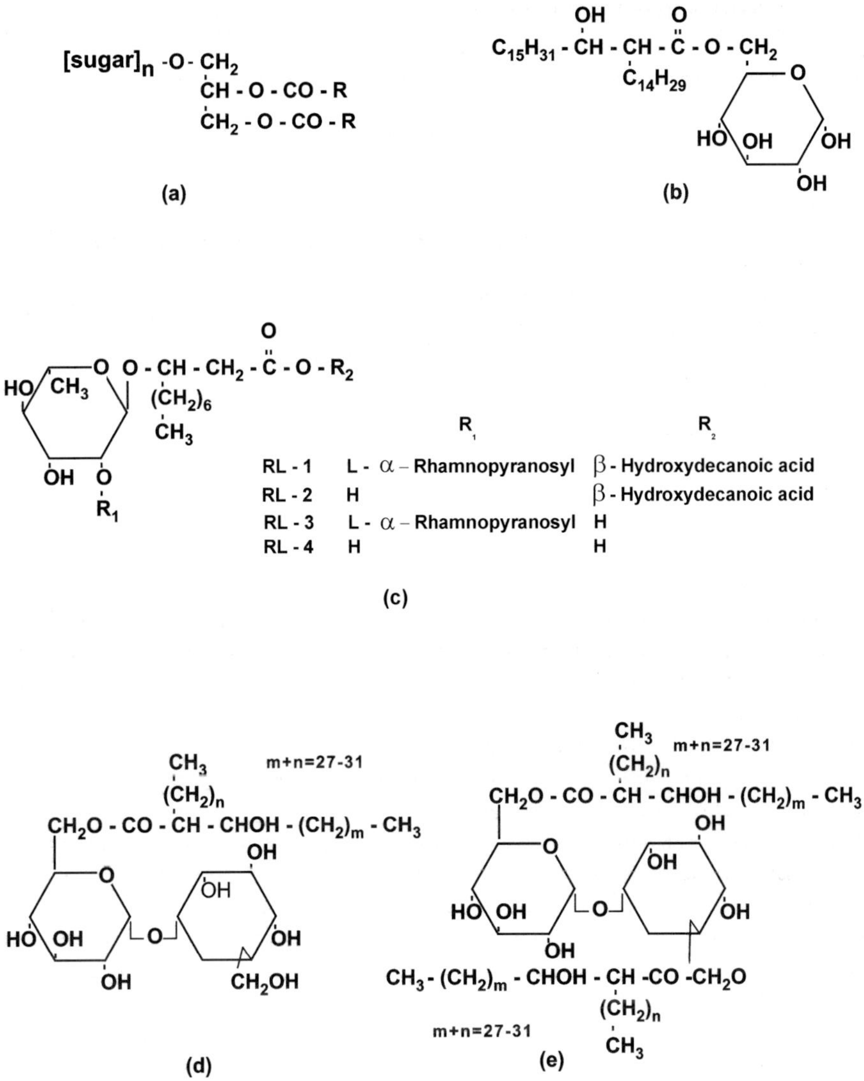

(a)

(b)

	R_1	R_2
RL - 1	L - α – Rhamnopyranosyl	β - Hydroxydecanoic acid
RL - 2	H	β - Hydroxydecanoic acid
RL - 3	L - α – Rhamnopyranosyl	H
RL - 4	H	H

(c)

m+n=27-31

(d)

m+n=27-31

m+n=27-31

(e)

Fig. 1 Structures of major bacterial glycolipids. (a) Glycosyl diglyceride in which commonly found sugar residues are disaccharides containing glucose and/or galactose, while the fatty acid component (R) has a composition similar to that of phospholipids of the cell. (b) Glucose-6-monocorynomycolate. (c) RL1–4. (d) Trehalose monomycolate. (e) Trehalose dimycolate.

4.2
Rhamnolipids

Most glycolipids have a carbohydrate moiety linked to the glycerol component of the glyceride. However, in RLs, one or two molecules of rhamnose is/are linked to one or two molecules of β-hydroxydecanoic acid. Jarvis and Johnson (1949) first described the production of rhamnose-containing glycolipids in *P. aeruginosa*. Four types of RLs (Figure 1c) containing two rhamnose and two β-hydroxydecanoic acid (RL-1), two β-hydroxydecanoic acid and one rhamnose (RL-2), one β-hydroxydecanoic acid and two rhamnose (RL-3), and one β-hydroxydecanoic acid and one rhamnose (RL-4) have been reported from *Pseudomonas* spp. (Jarvis and Johnson, 1949; Edward and Hayashi, 1965; Syldatk et al., 1985). Hirayama and Kato (1982) demonstrated the production of methyl ester derivatives of RL-1 and -2 by *P. aeruginosa*. Production of RLs similar to RL-2 and -1 with additional acylation by α-decenoic acid, known as RL-A and -B, respectively, have been shown by Yamaguchi et al. (1976). Novel RLs with alternative fatty acid chains and hydroxy fatty acids with different chain lengths have also been reported (Lang and Wullbrandt, 1999). The predominant type of RLs is strain specific, and depends on the environmental and cultivation conditions, especially the medium composition.

4.3 Trehalolipids

The α-branched β-hydroxy fatty acids esterified with the C-6 and C-6′ hydroxy group of the trehalose unit are found to be associated with the cell wall structure of most species of *Mycobacterium, Brevibacterium, Rhodococcus, Nocardia, Corynebacterium*, etc. Trehalolipids from different organisms differ in the size and structure of mycolic acid, the number of carbon atoms, and the degree of unsaturation (Asselineau and Asselineau, 1978; Lang and Wagner, 1987; Lang and Philp, 1998). Mycolic acids possess 60–90 carbon atoms, nocardiomycolic acids have 40–60 carbon atoms, while corynomycolic acids have 20–40 carbon atoms. The trehalolipid associated with the virulence of *M. tuberculosis* contains two molecules of mycolic acid attached to trehalose. In human and bovine strains of *M. tuberculosis*, the side chain is $C_{24}H_{49}$, while in avian and saprophytic mycobacteria it is $C_{22}H_{45}$. The most extensively studied trehalose lipids are trehalose dimycolate containing long-chain α-branched β-hydroxy saturated fatty acids ranging from C_{32} to C_{38}, and trehalose in the molar ratio of 2:1 (Figure 1d) and trehalose monomycolate containing trehalose and mycolic acid in the molar ratio of 1:2 (Figure 1e) produced by *R. erythropolis* (Rapp et al., 1979; Kretschmer et al., 1982). Trehalose monomycolates with a long carbon chain, trehalose esters of straight chain mono hydroxylated C_{16-18} fatty acids, and acylated polyunsaturated fatty acids from *Mycobacterium* spp. have also been isolated (Asselineau and Asselineau, 1978).

Two novel trehalose lipids, i.e. nonionic 6-mycolyl-acyl trehalose and anionic 2-octanoyl-di-decanoyl-succinoyl trehalose, have been isolated from *M. paraffinicum* by Batrakov et al. (1981). Modified trehalolipids such as anionic trehalose esterified with one succinoyl, one octadecanoyl and two decanoyl residues at the 2, 3, 4, and 2′ position of rhamnose; mono- and di-succinoyl trehalose lipids and anionic succinoyl-alkanoyl trehalose lipid have been reported in *R. erythropolis* (Uchida et al., 1989; Ristau and Wagner, 1993). Powalla et al. (1989) and Kim et al. (1990) have also observed production of penta- and disaccharide possessing trehalose units in *N. corynebacteroides*.

4.4 Complex Glycolipids

Lipo-heteropolysaccharide–protein complexes produced by *Acinetobacter* spp. have been extensively studied. Only a brief account is given here as they are covered in detail elsewhere in this volume. *A. calcoaceticus* RAG-1 produces polyanionic amphipathic heteropolysaccharides called "emulsan". The heteropolysaccharide backbone contains a repeating trisaccharide of *N*-

acetyl-D-galactosamine, *N*-acetylgalactosaminuronic acid, and *N*-acetyl-diamino-dideoxy-glucosamine. Fatty acids are covalently linked to the polysaccharides through *O*-ester linkages (Rosenberg, 1993).

Biodispersan, an extracellular anionic heteropolysaccharide produced by *A. calcoaceticus* A2, has an average molecular weight of 51,400 and contains glucosamine, 6-methylaminohexose, galactosamine uronic acid, and an unidentified amino sugar (Rosenberg et al., 1988a,b).

The production of complex glycolipids containing proteins, carbohydrates, and lipids has been reported in *Pseudomonas* spp. (Desai et al., 1988; Chameotra and Singh, 1990). Recently, Arino et al. (1998) demonstrated the synthesis of unusual extracellular glycolipids composed of 11 compounds with fatty acids and hydroxy fatty acids in the range of C_{10} to C_{18}, and a glycosidic moiety consisting of glucose, rhamnose, and ribose.

5 Physiological Functions

Bacterial glycolipids, amphipathic molecules either secreted extracellularly or attached to the cell, are synthesized predominantly during the growth on water- immiscible substrates and under extreme conditions. The diversity of glycolipids in bacteria indicates their diverse physiological functions. Several investigators have shown poor growth of RL-negative mutants of *P. aeruginosa* on hydrocarbons compared to their parent strain and indicate that the addition of RL to the medium restored their growth (Itoh and Suzuki, 1972; Koch et al., 1991). Thus, glycolipids play an important role in the survival of the producer organism and in the biodegradation of water-immiscible substrates. Recent work suggests that glycolipid-enhanced biodegradation of water-immiscible substrates may be attributed to either solubilization of hydrophobic compounds within micelle structures, which increases their effective solubility for uptake by cells, or to increased cell-surface hydrophobicity, which increases the affinity between the cell and the water-immiscible substrate (Oberbremer et al., 1990; Hommel, 1994; Zhang and Miller, 1994, 1995). Ragheb et al. (2000) have demonstrated that glycolipid-induced release of lipopolysaccharide (LPS) is a possible mechanism of enhanced cell-surface hydrophobicity in *P. aeruginosa*. Bacterial glycolipids lower the interfacial tension of the water/oil system to below 1 mN m^{-1}, the surface tension of water to 25–30 mN m^{-1} and stabilize the oil-in-water emulsion. Many glycolipids are produced under nutrient-limiting conditions, a common hostile environmental condition to which bacteria need to adapt. Glycolipids provide stability to the producer cells by excreting antimicrobial effects, solubilization of cell components of competing organisms, increasing cell adherence to susceptible hosts, prevention of phagocytic ingestion, and cell desorption to find a new habitat for survival (McClure and Schiller, 1992; Lang and Wagner, 1993; Roger and de Bentzman, 1998). Mycolic acid, the cord factor, has been found to be associated with the virulence of *Mycobacteria*. It imparts negative effects on migration of leukocytes and mitochondrial membrane permeability of many drugs. Recently, Ozeki et al. (1997) have shown induction of thymic atrophy via apoptosis by mycobacterial cord factor, which is closely linked with granuloma formation.

6 Biosynthesis

Glycolipids contain hydrophilic and hydrophobic moieties. Various pathways for the synthesis of these two groups of precursors are known and these depend on the carbon source in the growth medium. Despite this diversity, there are some common features in their synthesis and regulation. The detailed biosynthetic paths for major hydrophilic and hydrophobic moieties have been documented in standard books and hence are not covered here. However, a brief account by Hommel and Ratledge (1993) may be referred to. Various possibilities for the synthesis of glycolipids include:

De novo synthesis of both hydrophobic and hydrophilic moieties by two independent pathways.

De novo synthesis of the hydrophilic moiety and substrate induced synthesis of the hydrophobic moiety.

De novo synthesis of the hydrophobic moiety and substrate dependent synthesis of the hydrophilic moiety.

Substrate-dependent synthesis of both the moieties.

The biosynthesis of RLs in *P. aeruginosa* was first studied *in vivo* using radioactive precursors in resting cells by Hauser and Karnovsky (1954). Ochsner *et al.* (1994a,b) have confirmed the above results through genetic analysis and now it is known that the biosynthesis of RLs includes two sequential glycosyl transfers catalyzed by specific rhamnosyltransferases from thymidine diphosphate (TDP)-rhamnose as depicted in Figure 2. TDP-glucose is synthesized from glucose through enzymatic steps, which on dehydration, epimerization, and reduction of glycosyl moiety is converted to TDP-rhamnose (Glaser and Konfeld, 1966; Hommel and Ratledge, 1993).

The synthesis of trehalolipids follows a similar pattern, except that corynemycolic acid and trehalose-6-phosphate are synthesized independently, and are subsequently esterified to give trehalose mono- and dimycolates. Trehalose-6-phosphate formation is catalyzed by trehalose-6-phosphate synthase which links two α-D-glucopyranosyl units at C1 and C1′. Uridine diphosphate (UDP)-glucose and trehalose-6-phosphate act as intermediate precursors (Kretschmer and Wagner, 1983) in the synthesis as described in below.

$$\text{UDP-glucose} + \text{glucose-6-phosphate} \rightarrow \text{trehalose-6-phosphate} + \text{UDP} \quad (1)$$

$$\text{Trehalose-6-phosphate} + \text{corynomycolic acid} \rightarrow \text{trehalose monomycolate} \quad (2)$$

$$\text{Trehalose monomycolate} + \text{corynomycolic acid} \rightarrow \text{trehalose dimycolate} \quad (3)$$

The precursor of the hydrophobic moiety, β-hydroxydecanoic acid, is synthesized as an intermediate through fatty acid β-oxidation or by *de novo* fatty acid synthesis. Dimer formation of this precursor takes place by an esterification reaction (Boulton and Ratledge, 1987). For the detail account one may refer to Lang and Philp (1998) and Kretschmer and Wagner (1983).

Among bacterial glycolipids, the biochemistry of RL synthesis has been extensively studied. The enzyme rhamnosyltransferase-1 has been found to be in both soluble and membrane bound fractions. It is a complex of a 47-kDa protein and LPS. Removal of LPS from the protein results in a loss of enzyme activity. Rhamnosyltransferase-2 has been found mainly in the particulate fraction and purification attempts have been unsuccessful. A detailed description of these enzymes can be found in Oschner et al. (1995a).

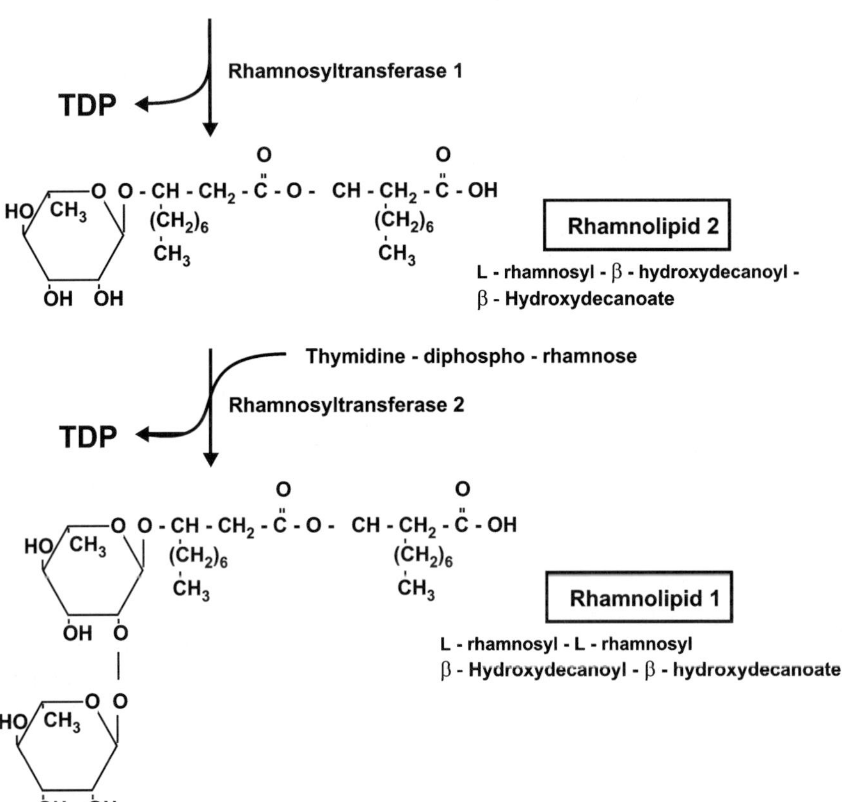

Fig. 2 Biosynthesis of RLs in *P. aeruginosa* (Ochsner et al., 1995; Lang and Wulbrandt, 1999).

7 Biodegradation

There have been numerous studies on the biosynthesis and production of glycolipids due to their commercial applications. On the other hand, only limited information is available on their biodegradation. In general, bacterial glycolipid biosurfactants are less toxic and easily biodegradable in natural conditions than synthetic surfactants. This has been confirmed by a study of Poremba et al. (1991) on the biodegradation of nine biosurfactants and eight synthetic surfactants. The biodegradation of glycolipids is initiated by the enzymatic cleavage of glycosidic or ester linkages yielding aliphatic or hydroxyaliphatic fatty acid and carbohydrate moieties.

Complete degradation of the glycolipid is accomplished by a series of enzymatic steps in the fatty acid and carbohydrate degradative pathways, and are very well documented in standard biochemistry text books.

8 Genetics and Regulation

Very little information is available on the genetics of bacterial glycolipid synthesis, except for the RLs in *Pseudomonas* spp. by Ochsner et al. (1994a, 1995b) and Ochsner and Reiser (1995). They adopted a strategy to isolate large numbers of RL-negative mutants followed by their genetic complementation by wild-type genes. Their findings led to the identification of four complementation groups consistent with four putative genes in the *rhl* cluster and involvement of an *rhlR–rhlI* system plus an autoinducer in the synthesis of RLs (Figure 3). The *rhlABR* gene cluster is responsible for the synthesis of a regulatory protein (RhlR) and rhamnosyltransferase, both essential for RL synthesis. The regulatory genes, *rhlR* and *rhlI*, code for the 28-kDa regulatory protein and the autoinducer synthetase, respectively. The regulatory protein, RhlR, is converted to its active form by binding on to autoinducer, a product of the autoinducer synthetase. The binding of active RhlR to the *rhlA* promoter initiates the transcription of the *rhlAB* operon encoding for rhamnosyltransferase. RhlR also controls the expression of the *rhlI* gene (Ochsner et al., 1994b).

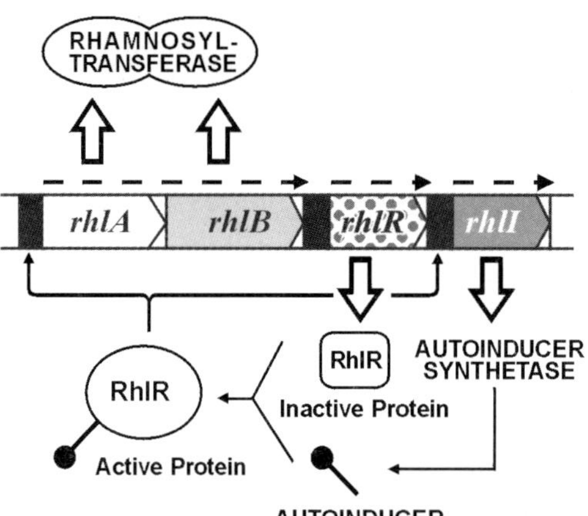

Fig. 3 Genes involved in the synthesis and regulation of RLs in *P. aeruginosa*. Map position at 47 min (Ochsner et al., 1994a,b; Ochsner and Reiser, 1995).

In general, the regulation of glycolipid synthesis in bacteria is governed by either induction, repression, or nitrogen/multivalent ion-mediated effects (Desai and Banat, 1997). The inductions of trehalolipid synthesis in *R. erythropolis* by hydrocarbons and of glycolipid synthesis in *P. aeruginosa* with alkanes are well documented (Rapp et al., 1979). Repression of glycolipid production in many bacteria by glucose and organic acids has been reported. Many investigators have observed RL production in *P. aeruginosa* on exhaustion of nitrogen and the initiation of the stationary phase of growth. In the event of nitrogen limitation cell metabolism changes from nitrogen to carbon resulting in the overproduction of glycolipids. The expression of the genes for RL synthesis has been observed only under the nitrogen-limiting condition in *Pseudomonas* spp. (Suzuki et al., 1974). The higher production of RLs has also been demonstrated under the limiting concentration of multivalent cations and iron in *Pseudomonas* spp. (Lang and Philp, 1998; Lang and Wullbrandt, 1999).

9 Production

The production of bacterial glycolipids has been found to take place either by fermentation or biotransformation. Details of these routes and product recovery are briefly described in the following subsections.

9.1 Fermentative Production

Industrial-scale fermentative production of biomolecules demands strain improvement and optimization of various process parameters including the medium constituents. No generalization on this aspect could be made due to the large variation in the producer bacteria and the types of glycolipids. Case by case, the engineering concept, either batch/fed-batch or continuous and with or without cell recycling, can be drawn to maintain the overproduction cycle for longer duration. For convenience, kinetic parameters can be grouped into the following types:

Growth associated glycolipid production, e.g. RL by *Pseudomonas* spp. (Yamaguchi et al., 1976; Syldatk et al., 1985; Robert et al., 1989).

Production under the limitation of medium components, e.g. production of glycolipids (Guerra-Santos et al., 1986; Ochsner et al., 1995a) by *Pseudomonas* spp. under nitrogen limitation.

Production by resting cells, such as glycolipid production by *Pseudomonas* spp. and *Arthrobacter* spp. and trehalose tetraester by *R. erythropolis* (Li et al., 1984; Syldatk *et al.*, 1985).

Production with precursor supplementation such as, production of mono-, di-, or trisaccharides in *A. paraffineus, Corynebacterium* sp., and *Nocardia* sp. on supplementation of the corresponding sugar in the growth medium (Itoh and Suzuki, 1974; Suzuki et al., 1974; Li et al., 1984).

Production of bacterial glycolipids has been achieved on water soluble carbon sources such as sugars, alcohols, etc. (Jarvis and Johnson, 1949; Syldatk et al., 1985; Palejwala and Desai, 1989) as well as on water-immiscible substrates such as hydrocarbons, olive oil, etc. (Desai et al., 1988; Robert et al., 1989; Mercade et al., 1997; Haba et al., 2000). The carbohydrate in the medium has a great bearing on the type of glycolipid synthesized by *Pseudomonas* spp. (Syldatk et al., 1985), *Rhodococcus* sp. (Espuny et al., 1996; Lang and Philp, 1998), and *A. corynebacteria* and *Nocardia* (Itoh and Suzuki, 1974; Suzuki et al., 1974). During growth of *A. paraffineus* on glucose, supplementation of hexadecane in the medium caused a significant increase in glycolipid yield (Duvnjak et al., 1983). Finnerty and Singer (1985) and Rapp et al. (1979) showed qualitative variation in glycolipid synthesis reflective to the chain length of the alkane.

The nitrogen concentration in the medium has a profound effect on glycolipid synthesis. Ammonium salts and urea are preferred for glycolipid production by *A. paraffineus* (Duvnjak et al., 1983), whereas nitrate supported maximum production in *P. aeruginosa* (Guerra-Santos et al., 1986; Kim et al., 1990), in *Pseudomonas* 44T1 (Robert et al., 1989), and in *Rhodococcus* spp. (Espuny et al., 1996; Lang and Philp, 1998). In *A. paraffineus*, glycolipid production significantly increased by the addition of L-amino acids to the medium (Duvnjak et al., 1983).

Syldatk et al. (1985) showed that nitrogen limitation not only causes over-production of glycolipids in *Pseudomonas* sp. DSM 2874, but also changes its composition. Guerra-Santos et al. (1986) and Haba et al. (2000) observed maximum RL production in *P. aeruginosa* at a C:N ratio in the range of 8–18. According to Hommel (1994), the absolute quantity of nitrogen is important for optimum biomass yield, while the concentration of carbon determines the yield of glycolipids.

Rhamnolipid production in *Pseudomonas* sp. is maximum in a pH range of 6.0–6.5 (Guerra-Santos et al., 1986). Powalla *et al.* (1989) showed penta- and disaccharide lipid production by *N. corynebacteroides* in the pH range of 6.5–8.0. Addition of EDTA results in an increase in glycolipid production (Vollbrecht et al., 1998). Other factors influencing glycolipid production include temperature, aeration rate, and salt concentration in the medium.

9.2 Production by Biotransformation and Chemical Synthesis

Recently, considerable attention has been directed towards the *in vitro* production of glycolipids through biotransformation using whole cells and enzymatic synthesis. Hydrophobic and hydrophilic domains are synthesized by fermentation, which can be joined enzymatically to produce commercial glycolipids. Such enzyme systems are highly specific and reactions take place at normal temperature and pressure. In biotransformation of lipids, lipases and phospholipases are commercially most important as they perform nonspecific or specific transesterification reactions (Totani and Hara, 1991; Mutua and Akoh, 1993). The recent interest in monoacylation of sugars with fatty acids is due to the synthesis of pure nonionic surfactants with a definite structure that could be used as adjuvants in the pharmaceutical industry.

Ikada and Klibanov (1993) succeeded in the acylation of glucose at the 6 position with vinylacrylate on a preparative scale. Scheckermann et al. (1995) reported the synthesis of lipase-catalyzed monoacylated fructose fatty acid esters. Sarney and Vulfson (1995) and Sieman and Wagner (1993) have presented a detailed account on the prospects and limitations of enzymatic synthesis of glycolipids.

Complete chemical synthesis of complex molecules of glycolipids is not common and commercially feasible. However, chemical conversion of the carboxylic domain of RLs to the nonionic methyl ester with improved properties (Ishigami et al., 1993) and synthesis of RL-2 from ethyl dimethoxybutane-thio-rhamnopyranoside and phenacyl hydroxydecanoate involving six steps (Duynstee et al., 1998) have been demonstrated. Fatty acid esters of sorbitol are produced by methods involving acid catalysis at 225–250°C. However, numerous byproducts are formed (Fregapane et al., 1994). Selective mono-acylation of hexoaldoses has been reported using immobilized *C. antarctica* lipase in acetone, an accepted solvent by the European Economic Community in food applications (Arcos et al., 1998). Recently, Torres et al. (2000) reported the selective enzymatic transesterification of ethyl lactate with octyl-β-D-glucopyranoside in acetone with 90% conversion, which is commercially very attractive in cosmetic applications.

9.3 Recovery and Purification

The recovery of glycolipids may account for as high as 60% of the total production costs. The product is isolated by precipitation, solvent extraction, or crystallization as a mixture of glycolipids. However, the predominant type would depend on the producing organism and on culture conditions. Glycolipids are recovered and concentrated after the removal of cells by centrifugation or filtration. Cross-filtration is more cost-effective than centrifugation.

The most commonly employed recovery method is solvent extraction using chloroform:methanol, dichloromethane:methanol, butanol, ethyl acetate, pentane, hexane, ether, etc. Trehalose lipids of *Mycobacterium* sp. and *A. paraffineus* (Suzuki et al., 1974; Asselineau and Asselineau, 1978; Li et al., 1984), trehalose corynomycolates and tetraesters of *R. erythropolis* (Rapp et al., 1979;

Kretschmer *at al.*, 1982; Ristau and Wagner, 1993, Lang and Philp, 1998), mono-, di- and pentasaccharide lipids of *A. paraffineus* and *N. corynebacterioides* (Li et al., 1984; Powalla et al., 1989), and RLs of *Pseudomonas* spp. (Yamaguchi et al., 1976; Desai et al., 1988) are some of the well-known examples of product recovery by solvent extraction.

Rhamnolipids of *P. aeruginosa* are also extracted by chilled acetone with or without adsorption on charcoal or by acidification followed by extraction in chloroform:methanol mixtures (Robert et al., 1989; Chameotra and Singh, 1990; Mercade et al., 1997; Lang and Wullbrandt, 1999). Ammonium sulfate precipitation has been successfully employed in isolating emulsan (Rosenberg et al., 1979) and glycolipid bioemulsifier (Palejwala and Desai, 1989).

Continuous removal of the glycolipid during fermentation eliminates the product inhibition with a net increase in yield (Mattei et al., 1986; Ochsner et al., 1995a). High-purity RLs have been obtained by (1) Continuous ultrafiltration of culture broth followed by precipitation or foam fractionation (Mulligan and Gibbs, 1990; Ochsner et al., 1995a), (2) continuous adsorption on Amberlite XAD-2 followed by purification and freeze drying (Fiechter 1992), and (3) application of Amicon XM-50 (a 5000 Da cut-off) and YM-10 (10, 000 Da cut-off) membranes followed by isopropanol precipitation (Bryant, 1990).

10 Applications

Currently bacterial glycolipids are not able to compete with chemical products with similar properties on cost considerations. However, a number of industrial applications have been envisaged. Due to their selectivity, effectiveness to broad ranges of oil and reservoir conditions, and low toxicity and biodegradable nature, most potential uses are in bioremediation of contaminated soils, microbially enhanced oil recovery (MEOR), and marine oil spill management (Fiechter, 1992; Banat, 1995; Finnerty, 1994; Desai and Banat, 1997).

Banat et al. (1991) applied bacterial glycolipids to increase oil recovery from crude storage tank sludge. Rhamnolipids from *P. aeruginosa* removed substantial quantities of oil from contaminated Alaskan gravel from the *Exxon Valdez* oil spill (Harvey at al., 1990), and recovered hydrocarbons from contaminated sandy-loam and silt-loam soil (Van Dyke et al., 1993; Scheibenbogen et al., 1994). Glycolipid biosurfactants have been shown to enhance the removal and mineralization of hydrocarbons (Oberbremer et al., 1990; Muller-Hurtig et al., 1993; Miller and Zhang, 1997, Ragheb et al., 2000). In another study, Bai et al. (1997) showed that RLs had positive effects on the physicochemical removal of oil from sand, but failed to assist microbial oil degradation. Microbial remediation of hydrocarbons and industrial waste contaminated soils is an emerging technology involving the application of bacterial glycolipids (Harvey at al., 1990; Banat, 1995; Ghosh et al., 1995; Ragheb et al., 2000). Zhang and Miller (1994, 1995) and Sharve et al. (1995) have demonstrated a stimulatory effect of RLs on the degradation of hexadecane and octadecane by *Pseudomonas* strains. Pretreatment with hydrogen peroxide is a common practice in bioremediation technology to overcome oxygen limitation for aerobic biodegradation.

RLs are also very useful in the bioremediation of sites contaminated with toxic heavy metals like uranium, cadmium, lead, zinc, etc. (Miller, 1995; Betts, 1997). Recently, sandrin et al. (2000) have reported the use of the RLs to reduce cadmiun toxicity during the biodegradation of naphthalene by *Burk-*

holderia sp. in a cocontaminated system. Mulligan et al. (1999) have also shown application of RLs for the removal of copper from oil-contaminated soil. The efficiency of glycolipid biosurfactants has been reported in remediation of the phenanthrene (Providenti et al., 1995), polycyclicaromatic hydrocarbons (Deziel et al., 1996), and polychlorinated biphenyl (Van Dyke et al., 1993) soil contamination.

Pollution of the seas and coasts with aromatic-containing crude oils resulting from oil tanker discharges and accidents is a worldwide problem. The potential application of bacterial glycolipid biosurfactants in oil spill management is attributed to the emulsification of hydrocarbon–water mixtures and increased hydrocarbon degradation (Harvey et al., 1990; Oberbremer et al., 1990; Bertrand et al., 1994; Banat, 1995).

Large numbers of glycolipid compounds for cosmetic applications are prepared by enzymatic conversion of hydrophobic molecules by various lipases and whole cells (Totani and Hara, 1991; Sieman and Wagner, 1993). The cosmetic industry demands surfactants with a minimum shelf life of 3 years, which gives preference to a saturated acyl group over the unsaturated compounds.

Other applications of bacterial glycolipids which have a bearing in the agriculture area include improvement of the wettability of leaf surfaces (Bunster et al., 1989). Haferburg et al. (1987) successfully used 1% emulsions of RLs for the treatment of leaves of *Nicotiana glutinosa* infected with the tobacco mosaic virus and for the control of disease on *N. tabacum* by potato X virus. The zoosporicidal activity on phytopathogens *Pythium aphanidermatum, Phytophtora capsici,* and *Plasmopara radicis* has also been reported by Stanghellini and Miller (1997).

11
Patents

Many glycolipid biosurfactants and their production procedures have been patented. However, only few have been commercialized. There are number of patents on the enzymatic synthesis of glycolipids and their applications in the cosmetic and health care products. Table 2 lists some of the important patents on glycolipids.

12
World Market and Cost Economics

The global surfactant market has now become extremely competitive and mature. Over the last 4 years it has grown at a rate of 10%. The current demand is almost entirely met by chemical surfactants. Applications of glycolipids have been reported in the oil, cosmetics, food, pharmaceuticals, agriculture, and environmental sectors. However, the current target is in marine pollution abatement, bioremediation of heavy metals, industrial waste and oil contaminated sites, MEOR, and in biocosmetics (Fiechter, 1992; Finnerty, 1994; Miller, 1995; Bai et al., 1997; Betts, 1997; Desai and Banat, 1997; Mercade et al., 1997; Mulligan et al., 1999; Sandrin et al., 2000).

Glycolipid fermentation processes hold the key to improving the overall process economics. The major cost centers are the product recovery and raw material costs, which contribute about 50–60 and 10–30%, respectively, to the overall production costs (Mulligan and Gibbs; 1993, Desai and Banat, 1997). Glycolipids are produced on both water-soluble and -insoluble substrates. From the process economics point of view, water-insoluble substrates are expensive.

Complete chemical synthesis of complex molecules of glycolipids is not commercially

Tab. 2 List of important patents on bacterial glycolipids

Patent No.	*Patent holder*	*Inventors*	*Title*	*Year*
US 4374735	Wintershall AG and GBF, Germany	Lindorfer, W., Schulz, W., Wanger, F., Jahn-Held, W.	Method for removal of oil from adsorbents using glycolipids	1983
US 4328030	Petrotec Forschungs AG, Switzerland	Kaeppeli, O., Guerra-Santos, L.	Process for the production of rhamnolipids	1986
US 4663039	Wintershall AG and GBF, Germany	Lindorfer, W., Schulz, W., Wagner, F., Jahn-Held, W.	Apparatus for removal of oil from water and adsorbents using glycolipids	1987
US 4720456	Wintershall AG, Germany	Wanger, F., Ristau, E., Li Zu-yi, Lang, S., Schulz, W., Hofmann, H. J., Sewe, K. U., Lindorfer, W.	Trehalose lipid tetraesters	1988
US 4814272	Wintershall AG, Germany	Wanger, F., Syldatk, C., Matulowic, U., Hofmann, H. J., Sewe, K. U., Lindorfer, W.	Process for the biotechnical production of rhamnolipids including rhamnolipids with only one β-hydroxydecanoic acid residue in the molecule	1989
US 5440028	Hoechst AG, Germany	Buchholz, R., Fricke, U., Mixich, J.	Process for preparing purified glycolipids by membrane separation process	1995
US 5455232	–	Piljac, G., Piljac, V.	Pharmaceutical preparation based on rhamnolipid	1995
Spanish 9602091	CSIC, Madrid, Spain	Arcos, J., Otero, C.	Method for enzymatic preparation of acyl glucose esters	1996
US 5514661	–	Piljac, G., Piljac, V.	Immunological activity of rhamnolipids	1996
US 5656747	Hoechst AG, Germany	Mixich, J., Rothert, R., Wullbrandt, D.	Process for the quantitative purification of glycolipids	1997
US 5767090	Regent, University of Arizona, Tucson, USA	Stanghellini, M. E., Miller, R. M., Rasmussen, S. L., Kim, D. H., Zhang, Y.	Microbially produced rhamnolipids (biosurfactants) for the control of plant pathogenic zoosporic fungi	1998
US 5866376	Universidad Simon Bolivar, Venezuela	Rocha, C. A., Gonzalez, D., Iturralde, M. L., Lacoa, U. L., Morales, F. A.	Production of oily emulsions mediated by a microbial tenso-active agent	1999
US 5998344	Novo-Nordisk A/S, Denmark	Christenson, P.N., Kakum, B., Anderson, O.	Detergent composition comprising a glycolipid and anionic surfactant for cleaning hard surfaces	1999

feasible. Recently, chemical synthesis of RL-2 from ethyl dimethoxybutane-thio-rhamnopyranoside and phenacyl hydroxydecanoate involving six steps has been reported (Duynstee et al., 1998). Reports by Scheckermann et al. (1995) on the synthesis of fructose fatty acid ester with 40% yield and by Torres et al. (2000) on selective enzymatic transesterification of ethyl lactate with octyl-β-D-glucopyranoside in acetone with 90% conversion are commercially very attractive.

Rhamnolipid yields of more than 100 g L^{-1} on 160 g L^{-1} soybean oil at the rate of 0.4 g L^{-1} h^{-1} (Lang and Wullbrandt, 1999) and 2 g L^{-1} h^{-1} with nearly 0.48 g g^{-1} corn oil (Ochsner et al., 1995a) during growth of *Pseudomonas* spp. and trehalolipid yield of 0.35 g L^{-1} on *n*-alkane by *R. erythropolis* (Kim et al., 1990) have been achieved. Currently only sophorolipid from *C. bombicola* has been produced at commercially viable yields of 300 g L^{-1} (Davila et al., 1997). Rhamnolipid production by *Pseudomonas* species has great commercial potential, but *Pseudomonas* are opportunistic human pathogens and hence not classified as *generally regarded as safe* (GRAS). The production of trehalose lipids using *Rhodococcus* is economically nonviable as their recovery from the cells needs expensive solvent extraction processes. Attempts have also been made to use RLs as the source of L-rhamnose, a high-quality flavor compound.

For enhanced oil recovery applications, lignin-based sulfonates cost in the range of US$2.5–3.5 kg^{-1} (Rosenberg, 1993). Fermentative production of RLs costs US$3–5 kg^{-1} (Ochsner et al., 1995b; Lang and Wulbrandt, 1999). In order to compete in the surfactant market, the cost of glycolipids should be below US$2 kg^{-1} (Mulligan and Gibbs, 1993). The costs of chemical surfactants, e.g. ethoxylates and alkyl polyglycosides, are in the range of US$1–3 kg^{-1}. Thus, future applications of bacterial glycolipids will be in the areas which demand enhanced properties and could absorb the associated higher costs, such as pharmacology and cosmetics. The global market for cosmetic industry in 1999 is estimated to be US$116 billion (Reisch, 2000).

13 Outlook and Perspectives

Bacterial glycolipids possess great diversity, selectivity and stability. They lower the surface tension of water from 72 to 25–30 mN m^{-1} at as low as 100 mg L^{-1} concentration. They are less toxic and easily biodegradable. As a result, bacterial glycolipids have assumed a significant place in expanding the existing range of surfactants.

Attempts have been made to increase glycolipid yields by manipulating physiological conditions, medium composition, and genes responsible for glycolipids synthesis. The development of rapid and simple screening methods for glycolipid-producing bacteria has helped to obtain biochemical and genetic information on glycolipid production. Mutants with enhanced glycolipid production and genes responsible for the synthesis of RLs have been isolated, characterized (Ochsner and Reiser, 1995; Ochsner et al., 1995b), and cloned in heterogeneous hosts (Hardegger et al., 1994; Ochsner et al., 1995b). These efforts have resulted in several fold higher yields and a change in raw material requirement for production (Koch et al., 1988; Ochsner et al., 1995b).

The development of continuous recovery processes for glycolipids has replaced expensive extraction procedures with an improved recovery and increased overall output by relieving product inhibition (Reiling et al., 1986; Bryant, 1990; Ochsner et al., 1995a). Another recent development towards cost reduction is the use of agro-industrial wastes as substrates for glycolipid production (Mercade et al. 1993; Mercade and Manresa, 1994; Babu et al., 1996; Patel and Desai, 1997; Daniel et al. 1998; Vollbrecht et al., 1999).

There are numerous applications of bacterial glycolipids and each demands a different set of properties. Therefore, glycolipid product development with unique properties will be the major future research activity. It is possible to modify the properties of glycolipids genetically, biologically, enzymatically, or chemically for specific applications. From cost considerations in the oil sector, most applications would include either whole bacterial cell cultures producing glycolipids or crude preparations. Their lower CMC value (10- to 40-fold) and the high crude oil price have reduced the actual gap in cost to their chemical counterparts, and the economics are slowly turning in favor of glycolipid surfactants. Further reductions in the cost of bacterial glycolipids to make them economically attractive will largely depend on the development of cheaper and more efficient product recovery processes, use of low-cost raw materials, increased product yields through super-active mutants and genetically engineered bacteria, and efficient chemical or enzymatic glycolipid synthesis

Acknowledgments

J. D. D. acknowledges the permission granted by the Indian Petrochemicals Corp., Vadodara (India) for publication of this work. We thank Dr. Ibrahim M. Banat (University of Ulster, UK) and Dr. Siegmund Lang (Technical University of Braunschweig, Germany) for constructive suggestions, and for providing preprints and unpublished work.

14
References

Arcos, J. A., Bernabe, M., Otero, C. (1998) Different strategies for selective monoacylation of hexoaldoses in acetone, *J. Surf. Detergents* **1**, 345–352.

Arino, S., Marchal, R., Vandecasteele, J. P. (1998) Production of new extracellular glycolipids by a strain of *Cellulomonas cellulans* (*Oerskovia xenthineolytica*) and their structural characterization, *Can J. Microbiol.* **44**, 238–243.

Atlas, R. M. (1981) Microbial degradation of petroleum hydrocarbons: an environmental perspective, *Microbiol. Rev.* **45**, 180–209.

Asselineau, C., Asselineau, J. (1978) Trehalose containing glycolipids, *Prog. Chem. Fats Lipids* **16**, 59–99.

Babu, P. S., Vaidya, A. N., Bal, A. S., Kapur, R., Juwarkar, A., Khanna, P. (1996) Kinetics of biosurfactant production by *Pseudomonas aeruginosa* strain BS2 from industrial wastes, *Biotechnol. Lett.* **18**, 263–268.

Bai, G., Brusseau M. L., Miller, R. M. (1997) Biosurfactant-enhanced removal of residual hydrocarbon from soil, *J. Contamin. Hydrol.* **25**, 157–170.

Banat, I. M., Samarah, N., Murad, M., Horne, R., Benerjee, S. (1991) Biosurfactant production and use in oil tank clean-up, *World J. Microbiol. Biotechnol.* **7**, 80–84.

Banat, I. M. (1995) Biosurfactants production and possible uses in microbial enhanced oil recovery and oil pollution remediation – a review, *Bioresource Technol.* **51**, 1–12.

Batrakov, S. G., Rozynov, B. V., Koronelli, T. V., Bergelson, L. D. (1981) Two novel types of trehaloselipids, *Chem. Phys. Lipids* **29**, 241–266.

Bertrand, J. C., Bonin, P., Goutx, M., Gauthier, M., Mille, G. (1994) The potential application of biosurfactants in combating hydrocarbon pollution in marine environment, *Res. Microbiol.* **145**, 53–56.

Betts, K. S. (1997) Biosurfactants remove metals from soil, *Environ. Sci. Technol.* **31**, 547–548.

Boulton, C. A., Ratledge, C. (1987) Biosynthesis of lipid precursors to surfactant production, in: *Biosurfactants and Biotechnology* (Kosaric, N., Cairns, W. L., Gray, N. C. C., Eds.), Marcel Dekker, New York, 47–87.

Brandenburg, K., Seydel, U. (1998) Infrared spectroscopy of glycolipids, *Chem. Phys. Lipids* **96**, 23–40.

Brennan, P. J., Lehane, D. P., Thomas, D. W. (1970) Acylglucoses of the *Corynebacteria* and *Mycobacteria*, *Eur. J. Biochem.* **13**, 117–12.

Bryant, F. O. (1990) Improved method for the isolation of biosurfactant glycolipids from *Rhodococcus* sp. strain H13A, *Appl. Environ. Microbiol.* **56**, 1494–1496.

Bunster, L. Fokkema, N. J., Schippers, B. (1989) Effect of surface-active *Pseudomonas* spp. on leaf wettability, *Appl. Environ. Microbiol.* **55**, 1340–1345.

Bush, C. A., Martin Pastor, M., Imborty, A. (1999) Structure and conformation of complex carbohydrates of glycoproteins, glycolipids and bacterial polysaccharides, *Annu. Rev. Biophys. Biomol. Struct.* **28**, 269–278.

Chameotra, S. S., Singh, H. D. (1990) Purification and characterization of alkane solubilizing factor produced by *Pseudomonas* sp. PG-1, *J. Ferment. Bioeng.* **69**, 341–344.

Cooper, D. G., Goldenberg, B. G. (1987) Surface active agents from two *Bacillus* species, *Appl. Environ. Microbiol.* **53**, 224–229.

Daniel, H. J., Otto, T. T., Binder, M., Reuss, M., Syldatk, C. (1998) Sophorolipid production with high yields on whey concentrated and rapeseed oil without consumption of lactose, *Biotechnol. Lett.* **20**, 805–807.

Davila, A. M., Marchal, R., Vandecasteele, J. P. (1997) Sophorose lipid fermentation with differentiated substrate supply for growth and production phases, *Appl. Microbiol. Biotechnol.* **47**, 496–501.

Desai, A. J., Patel, K. M., Desai, J. D. (1988) Emulsifier production by *Pseudomonas fluorescens* during the growth on hydrocarbons, *Curr. Sci.* **57**, 500–501.

Desai, J. D., Desai, A. J. (1993) Production of biosurfactants, in: *Biosurfactants, Production, Properties, Applications* (Kosaric, N., Ed.), Marcel Dekker, New York, 65–97.

Desai, J. D., Banat, I. M. (1997) Microbial production of surfactants and their commercial potential, *Microbiol. Mol. Biol. Rev.* **61**, 47–64.

Deziel, E., Paquette, G., Villemur, R., Lepine, F., Bisaillon, J. (1996) Biosurfactant production by soil *Pseudomonas* strain growing on polycyclic aromatic hydrocarbons, *Appl. Environ. Microbiol.* **62**, 1908–1912.

Duvnjak, Z., Cooper, D. G., Kosaric, N. (1983) Effect of nitrogen source on surfactant production by *Arthrobacter paraffines* ATCC 19558, in: *Microbial Enhanced Oil Recovery* (Zajic, J. E., Cooper, D. G., Jack, T. R., Kosaric, N., Eds.), Pennwell, Tulsa, OK, 66–72;

Duynstee, H. I., Vliet, M. J., Marel, G. A., Boom, J. H. (1998) An efficient synthesis of (*R*)-3-{(*R*)-3-[2-*O*-(α-L-rhamnopyranosyl)-α-L-rhamnopyranosyl] oxydecanoyl} oxydecanoic acid, a rhamnolipid from *Pseudomonas aeruginosa*, *Eur. J. Org. Chem.* **1998**, 303–307.

Edward, J. R., Hayashi, J. A. (1965) Structure of a rhamnolipid from *Pseudomonas aeruginosa*, *Arch. Biochem. Biophys.* **111**, 415–421.

Espuny, M. J., Ejido, S., Rodon, I., Manresa, A., Mercade, M. E. (1996) Nutritional requirements of a biosurfactant producing strain *Rhodococcus* sp. 51T7, *Biotechnol. Lett.* **18**, 521–526.

Fiechter, A. (1992) Biosurfactants, moving towards industrial application, *Trends Biotechnol.* **10**, 208–217.

Finnerty, W. R. (1994) Biosurfactants in environmental biotechnology, *Curr. Opin. Biotechnol.* **5**, 291–295.

Finnerty, W. R., Singer M. E. (1985) A microbial biosurfactant physiology, Biochemistry and applications, *Dev. Ind. Microbiol.* **25**, 31–46.

Fregapane, G., Sarney, D. B., Greenberg, S. G., Knight, D. J., Vulfson, E. N. (1994) Enzymatic synthesis of monosaccharide fatty acid esters and their comparison with conventional products, *J. Am. Chem. Soc.* **71**, 87–91.

Glaser, L., Konfeld, S. (1966) Formation of dTDP-L-rhamnose from dTDP-D-glucose, *Methods Enzymol.* **8**, 303–306.

Guerra-Santos, L. H., Kappeli, O., Fiechter, A. (1986) Dependence of *Pseudomonas aeruginosa* continuous culture biosurfactant production on nutritional and environmental factors, *Appl. Microbiol. Biotechnol.* **24**, 443–448.

Haba, E., Espuny, M. J., Busquets, M., Manresa, A. (2000) Screening and production of rhamnolipids by *Pseudomonas aeruginosa* 47 T2 NCB 40044 from waste frying oils, *J. Appl. Microbiol.* **88**, 379–387.

Haferburg, D., Hommel, R., Kleber, H. P., Kluge, S., Schuster, G., Zschiegner, H. J. (1987) Antiphytovirale Aktivitat von Rhamnolipid aus *Pseudomonas aeruginosa*, *Acta Biotechnol.* **7**, 353–356.

Hardegger, M., Koch, A. K., Ochsner, U. A., Fiechter, A., Reiser, J. (1994) Cloning and heterologous expression of a gene encoding an alkane induced extracellular protein involved in alkane assimilation from *Pseudomonas aeruginosa*, *Appl. Environ. Microbiol.* **60**, 3679–3687.

Harvey, D. J. (1999) Matrix-assisted laser-desorption ionization mass spectroscopy of carbohydrates, *Mass Spectrosc. Rev.* **18**, 349–450.

Harvey, S., Elashvili, I., Valdes, J. J., Kamely, D., Chakrabarty, A. M. (1990) Enhanced removal of *Exxon Valdez* spilled oil from Alaskan gravel by a microbial surfactant, *Bio/Technology* **8**, 228–230.

Hauser, G., Karnovsky, M. L. (1954) Studies on the production of glycolipid by *Pseudomonas aeruginosa*, *J. Bacteriol.* **68**, 645–654.

Hirayama, T., Kato, I. (1982) Novel rhamnolipids from *Pseudomonas aeruginosa*, *FEBS Lett.* **139**, 81–85.

Hommel, R. K. (1994) Formation and function of biosurfactants for degradation of water insoluble substrates, in: *Biochemistry of Microbial Degradation* (Ratledge. C., Ed.), Kluwer, London, 63–87.

Hommel, R. K., Ratledge, C. (1993) Biosynthetic mechanisms of low molecular weight surfactants and their precursor molecules, in: *Biosurfactants, Production, Properties, Applications* (Kosaric, N., Ed.), Marcel Dekker, New York, 3–63.

Ikada, I., Klibanov, A. M. (1993) Lipase catalyzed acylation of sugars solubilized in hydrophobic solvents by complexation, *Biotechnol. Bioeng.* **42**, 788–791.

Ishigami, Y., Gama, Y., Ishli, F., Kook, C. Y. (1993) Colloid chemical effect of polar head moieties of a rhamnolipid type biosurfactant, *Langmuir* **9**, 1634–1636.

Itoh, S., Suzuki, T. (1972) Effect of rhamnolipids on growth of *Pseudomonas aeruginosa* mutant deficient in *n*-paraffin utilizing ability, *Agric. Biol. Chem.* **36**, 2233–2235.

Itoh, S., Suzuki, T. (1974) Fructose lipids of *Arthrobacter, Corynebacteria, Nocardia* and *Mycobacteria* grown on fructose, *Agric. Biol. Chem.* **38**, 1443–1449.

Jain, D. K., Thompson, D. L. C., Lee, H., Trevors, J. T. (1991) A drop-collapsing test for screening surfactant producing microorganisms, *J. Microbiol. Methods* **13**, 271–279.

Jarvis, F. G., Johnson, M. J. (1949) A glycolipid produced by *Pseudomonas aeruginosa*, *J. Am. Chem. Soc.* **71**, 4124–4126.

Kim, J. S., Powalla, M., Lang, S., Wagner, F., Lunsdorf, H., Wrey, V. (1990) Microbial glycolipid production under nitrogen limitation and resting cell conditions, *J. Biotechnol.* **13**, 257–266.

Koch, A. K., Reiser, J., Kappeli, O., Fiechter, A. (1988) Genetic construction of lactose-utilizing stains of *Pseudomonas aeruginosa* and their application in biosurfactant production, *Bio/Technology* **6**, 1335–1339.

Koch, A. K., Kappeli, O., Fiechter, A., Reiser, J. (1991) Hydrocarbon assimilation and biosurfactant production in *Pseudomonas aeruginosa* mutants, *J. Bacteriol.* **173**, 4212–4219.

Kretschmer, A., Wagner, F. (1983) Characterization of biosynthetic intermediates of trehalose dimycolate from *Rhodococcus erythropolis* grown on *n*-alkanes, *Biochim. Biophys. Acta* **753**, 306–313.

Kretschmer, A., Bock, H., Wagner, F. (1982) Chemical and physical characterisation of interfacial-active lipids from *Rhodococcus erythropolis* grown on *n*-alkane, *Appl. Environ. Microbiol.* **44**, 864–870.

Lang, S. (1999) Production of microbial glycolipids, in: *Methods in Biotechnology* (Bucke, C., Ed.), Humana Press, Totowa, NJ, 103–118.

Lang, S., Philp, J. (1998) Surface active lipids in *Rhodococcus, Antonie van Leeuwenhoek.* **74**, 59–70.

Lang, S., Wagner, F. (1987) Structure and properties of biosurfactants, in: *Biosurfactants and biotechnology* (Kosaric, N., Cairns, W. L., Gray, N. C. C., Eds.), Marcel Dekker, New York, 21–47.

Lang, S., Wagner, F. (1993) Biological activities of biosurfactants, in: *Biosurfactants, Production, Properties, Applications* (Kosaric, N., Ed.), Marcel Dekker, New York, 251–268.

Lang, S., Wullbrandt, D. (1999) Rhamnose lipids – biosynthesis, microbial production and application potential, *Appl. Microbiol. Biotechnol.* **51**, 22–32.

Li, Z. Y., Lang, S., Wagner, F., Witte, L., Wray, V. (1984) Formation and identification of interfacial-active glycolipids from resting microbial cells of *Arthrobacter* sp. and potential use in tertiary oil recovery, *Appl. Environ. Microbiol.* **48**, 610–617.

Matsuyama, T., Sogawa, M., Yano, I. (1991) Direct colony thin layer chromatography and rapid characterization of *S. marcescens* mutants defective in production of wetting agents, *Appl. Environ. Microbiol.* **53**, 1186–1188.

Mattei, G., Rambeloarisoa, E., Giusti, G., Rontani, J. F., Bertrand, J. C. (1986) Fermentation procedure of a crude oil in continuous culture on seawater, *Appl. Microbiol. Biotechnol.* **23**, 302–304.

McClure, C. D., Schiller, N. L. (1992) Effect of *Pseudomonas aeruginosa* rhamnolipids on human monocyte derived macrophages, *J. Leukoc. Biol.* **51**, 97–102.

Mercade, M. E., Espuny, M. J., Manresa, A. (1997) The use of oil substrate for biosurfactant production, *Recent Res. Dev. Oil. Chem.* **1**, 177–185.

Mercade, M. E., Manresa, M. A. (1994) The use of agro industrial byproducts for biosurfactant production, *J. Am. Oil. Chem. Soc.* **71**, 61–64.

Mercade, M. E., Manresa, M. A., Robert, M., Espuny, M. J., de Andres, C., Guinea, J. (1993) Olive oil mill effluent (OOME) new substrate for biosurfactant production, *Bioresource Technol.* **43**, 1–6.

Miller, R. M. (1995) Biosurfactant-facilitated remediation of metal-contaminated soils, *Environ. Health Perspect.* **103**, 59–62.

Miller, R. M., Zhang, Y. (1997) Measurement of biosurfactant enhanced solubilization and biodegradation of hydrocarbons, *Methods Biotechnol.* **2**, 59–66.

Muller-Hurtig, R., Wagner, F., Blaszczyk, R., Kosaric, N. (1993) Biosurfactants for environmental control, in: *Biosurfactants: Production, Properties, Applications* (Kosaric, N., Ed.), Marcel Dekker, New York, 447–469.

Mulligan C. N., Gibbs, B. F. (1990) Recovery of biosurfactants by ultrafiltration, *J. Chem. Tech. Biotechnol.* **47**, 23–29.

Mulligan, C. N., Gibbs, B. F. (1993) Factors influencing the economics of biosurfactants, in: *Biosurfactants, Production, Properties, Applications* (Kosaric, N., Ed.), Marcel Dekker, New York, 329–371.

Mulligan, C. N., Yong, R. N., Gibbs, B. F. (1999) On the use of biosurfactants for the removal of heavy metals from oil contaminated soil, *Process Safety Prog.* **18**, 50–54.

Mutua, L. N., Akoh, C. C. (1993) Lipase catalyzed modification of phospholipids, incorporation of *n*-3-fatty acids into biosurfactants, *J. Am. Oil Chem. Soc.* **70**, 125–128.

Oberbremer, A., Muller-Hurtig, R., Wagner, F. (1990) Effect of the addition of microbial surfactant on hydrocarbon degradation in a soil population in a stirred reactor, *Appl. Microbiol. Biotechnol.* **32**, 485–489.

Ochsner, U. A., Reiser, J. (1995). Autoinducer-mediated regulation of rhamnolipid biosurfactant synthesis in *Pseudomonas aeruginosa, Proc. Natl Acad. Sci. USA* **92**, 6424–6428.

Ochsner, U. A., Fiechter, A., Reiser, J. (1994a). Isolation, characterization and expression in *Escherichia coli* of the *Pseudomonas aeruginosa rhlAB* genes encoding a rhamnosyltransferase involved in rhamnolipid biosurfactant synthesis, *J. Biol. Chem.* **269**, 19787–19795.

Ochsner, U. A., Koch, A. K., Fiechter, A. Reiser, J. (1994b) Isolation and characterization of a regulatory gene affecting rhamnolipid biosurfactant synthesis in *Pseudomonas aeruginosa, J. Bacteriol.* **176**, 2044–2054.

Ochsner, U. A., Hembach, T., Fiechter, A. (1995a) Production of rhamnolipid biosurfactants, *Adv. Biochem. Eng. Biotechnol.* **53**, 89–118.

Ochsner, U. A., Reiser, J., Fiechter, A., Witholt, B. (1995b). Production of *Pseudomonas aeruginosa* rhamnolipid biosurfactants in heterogeneous host, *Appl. Environ. Microbiol.* **61**, 3503–3506.

Ozeki, Y., Kaneda, K., Fujiwara, N., Morimoto, M., Oka, S., Yano, I. (1997) *In vivo* induction of apoptosis in the thymus by administration of *Mycobacterial* cord factor (trehalose-6,6′-dimycolate), *Infect. Immun.* **65**, 1793–1799.

Palejwala, S., Desai, J. D. (1989) Production of extracellular emusifier by a Gram negative bacterium, *Biotechnol. Lett.* **11**, 115–118.

Patel, R. M., Desai, A. J. (1997) Biosurfactant production by *Pseudomonas aeruginosa* GS3 from molasses, *Lett. Appl. Microbiol.* **25**, 91–94.

Poremba, K., Lang, S., Wagner, F, Gunkel, W. (1991) Marine biosurfactants. III. Toxicity testing with marine microorganisms and comparison with synthetic surfactants, *Z. Naturforsch.* **46c**, 210–216.

Powalla, M., Lang, S., Wray, V. (1989) Penta- and disaccharide lipid formation by *Nocardia corynebacteroides* grown on *n*-alkanes, *Appl. Microbiol. Biotechnol.* **31**, 473–479.

Providenti, M. A., Flemming, C. A., Lee, H., Trevors, J. T. (1995) Effect of addition of rhamnolipid biosurfactants or rhamnolipid-producing *Pseudomonas aeruginosa* on phenanthrene mineralization in soil slurries, *FEMS Microbiol. Ecol.* **17**, 15–26.

Ragheb, A. T., Al-Tahhan Sandrin, T. R., Bodour, A. A., Maier, R. M. (2000) Rhamnolipid induced removal of lipopolysaccharide from *Pseudomonas aeruginosa*: effect in cell surface properties and interaction with hydrophobic substrates, *Appl. Environ. Microbiol.* **66**, 3262–3268.

Rapp, P., Bock, H., Wray, V., Wagner, F. (1979) Formation, isolation and characterization of trehalose dimycolates from *Rhodococcus erythropolis* grown on *n*-alkanes, *J. Gen. Microbiol.* **115**, 491–503.

Reiling, H. E., Wyass, U. T., Guerra-Santos, L. H., Hirt, R., Kappali, O., Fiechter, A. (1986) Pilot plant production of rhamnolipid biosurfactant by *Pseudomonas aeruginosa, Appl. Environ. Microbiol.* **51**, 985–989.

Reisch, M. C. (2000) Hair care products, *Chem. Eng. News* **78** (12), 15–26.

Ristau, E., Wagner, F. (1993) Formation of novel anionic trehalose-tetraesters from *Rhodococcus erythropolis* under growth limiting conditions, *Biotechnol. Lett.* **5**, 95–100.

Robert, M., Mercade, M. E., Bosch, M. P., Parra, J. L., Espuny, M. J., Manresa, M. A., Guinea, J. (1989) Effect of the carbon source on biosurfactant production by *Pseudomonas aeruginosa* 44T, *Biotechnol. Lett.* **11**, 871–874.

Roger, P., de Bentzmann, S. (1998) *Pseudomonas aeruginosa* adherence to the respiratory mucosa, *Medicine Maladies Infect.* **28**, 119–125.

Rosenberg, E. (1993) Exploiting microbial growth on hydrocarbon – new markets, *Trends Biotechnol.* **11**, 419–424.

Rosenberg, E., Zuckerberg, A., Rubinovitz, C., Gutnick, D. L. (1979) Emulsifier *Arthrobacter* RAG-1, isolation and emulsifying properties, *Appl. Environ. Microbiol.* **37**, 402–408.

Rosenberg, E., Rubinovitz, C., Gottlieb, A., Rosenhak, S., Ron, E. Z. (1988a) Production of biodispersan by *Acinetobacter calcoaceticus* A2, *Appl. Environ. Microbiol.* **54**, 317–322.

Rosenberg, E., Rubinovitz, C., Legmann, R., Ron, E. Z. (1988b) Purification and chemical properties of *Acinetobacter calcoaceticus* A2 biodispersan, *Appl. Environ. Microbiol.* **54**, 323–326.

Sandrin, T. R., Chech, A. M., Maier, R. M. (2000) A rahmnolipd biosurfactant reduces cadmium toxicity during naphthalene biodegradation, *Appl. Environ. Microbiol.* **66**, 4585–4588.

Sarney, D. B., Vulfson, E. N. (1995) Application of enzymes to the synthesis of surfactants, *Trends Biotechnol.* **13**, 164–175.

Scheckermann, C., Schlotterbeck, A., Schmidt, M., Wray, V., Lang, S. (1995) Enzymatic monoacylation of fructose by two procedures, *Enz. Microbial. Technol.* **17**, 157–162.

Scheibenbogen, K., Zytner, R. G., Lee, H., Trevors, J. T. (1994) Enhanced removal of selected hydrocarbons from soil by *Pseudomonas aeruginosa* UG2 biosurfactants and some chemical surfactants, *J. Chem. Technol. Biotechnol.* **59**, 53–59.

Sharve, G. S., Ingura, S., Gunnam, C. (1995) Rhamnolipid biosurfactant enhancement of hexadecane biodegradation by *Pseudomonas aeruginosa, Mol. Mar. Biol. Biotechnol.* **4**, 331–337

Sieman, M., Wagner, F. (1993) Prospects and limits for the production of biosurfactants using immobilized biocatalysts, in: *Biosurfactants, Production, Properties, Applications* (Kosaric, N., Ed.), Marcel Dekker, New York, 99–133.

Sigmond, I., Wagner. F. (1991) New Method for detecting rhamnolipids excreted by *Pseudomonas* species grown on mineral agar, *Biotechnol. Tech.* **5**, 265–268.

Stanghellini, M. E., Miller, R. M. (1997) Biosurfactants – their identity and potential efficacy in the biological control of Zoosporic plant pathogens, *Plant Dis.* **81**, 4–12

Suzuki, T., Tanaka, H., Itoh, S. (1974) Sucrose lipids of *Arthrobacteria, Corynebacteria* and *Nocardia* grown on sucrose, *Agric. Biol. Chem.* **38**, 557–563.

Suzuki T., Tanaka K., Matsubara, I., Kinoshita S. (1969) Trehalose lipids and alpha branched beta hydroxyfatty acids formed by bacteria grown on *n*-alkanes, *Agric. Biol. Chem.* **33**, 1619- 1627.

Syldatk, C., Lang, S., Wagner, F. (1985) Chemical and physical characterization of four interfacial-active rhamnolipids from *Pseudomonas* sp. DSM 2874 grown on *n*-alkanes, *Z. Naturforsch.* **40c**, 51–60.

Torres, C., Bernabe, M., Otero, C. (2000). Enzymatic synthesis of lactic acid derivatives with emulsifying properties, *Biotechnol. Lett.* **22**, 331–334.

Totani, Y., Hara, S. (1991) Preparation of polyunsaturated phospholipids by lipase-catalysed transesterification, *J. Am. Oil Chem. Soc.* **68**, 848–851.

Uchida Y., Tsuchiya, R., Chino, M., Hirano, J., Tabuchi, T. (1989) Extracellular accumulation of mono- and di-succinyl trehalose lipids by a strain of *Rhodococcus erythropolis* grown on *n*-alkanes, *Agric. Biol. Chem.* **53**, 757–763.

Van der Vegt, W., Vander Mei, H. C., Noordmans, J. Busscher, H. J. (1991) Assessment of bacterial biosurfactant production through axisymmetric drop shape analysis by profile, *Appl. Microbiol. Biotechnol.* **35**, 766–770.

Van Dyke, M. I., Counture, P., Brauer, M., Lee, H., Trevors, J. T. (1993) *Pseudomonas aeruginosa* UG2 rhamnolipid biosurfactants: structural characterization and their use in removing hydrophobic compounds from soil, *Can. J. Microbiol.* **39**, 1071–1078.

Vollbrecht, E., Heckmann, R., Wray, V., Nimtz, M., Lang, S. (1998) Production and structure elucidation of di- and oligosaccharide lipids (biosurfactants) from *Tsukamurella* sp. Nov, *Appl. Microbiol. Biotechnol.* **50**, 530–537.

Vollbrecht, E., Rau, U., Lang, S. (1999) Microbial conversion of vegetable oils into surface active di-, tri-, and tetrasaccharide lipids (biosurfactants) by the bacterial strain *Tsukamurella sp., Fats Lipids* **101**, 389–394.

Wicke, C., Huners, M., Wray, V., Nimtz, M., Bilitewsky, U., Lang, S. (2000) Production and structure elucidation of glycoglycerolipids from marine sponge associated *Microbacterium* sp., *J. Nat. Prod.* **63**, 621–626.

Yakimov, N. M., Golyshin, P. N., Lang, S., Moore, E. R. B., Abraham, W. R., Lunsdorf, H., Timmis, K. N. (1998) *Alcanivorax borkumensis* gen. nov., Sp. nov., a new, hydrocarbon degrading and surfactant producing marine bacterium, *Int. J. Syst. Bacteriol.* **48**, 339–348.

Yamaguchi, M., Sato, A., Yukuyama, A. (1976) Microbial production of sugar lipids, *Chem. Ind.* **17**, 741–742.

Zhang, Y., Miller, R. M. (1994) Effect of a *Pseudomonas* rhamnolipid biosurfactant on cell hydrophobicity and biodegradation of octadecane, *Appl. Environ. Microbiol.* **60**, 2101–2106.

Zhang, Y., Miller, R. M. (1995) Effect of rhamnolipid (biosurfactant) structure on solubilization and biodegradation of *n*-alkanes, *Appl. Environ. Microbiol.* **61**, 2247–2251.

6
Curdlan

Dr. In-Young Lee
Korea Research Institute of Bioscience and Biotechnology and DawMaJin Biotech Corp., P.O. Box 115, Daeduk Valley, Daejon 305-600, Korea; Tel.: +82-428620722; Fax: +82-428620702; E-mail: leeiy@dmj-biotech.com

AIDS	acquired immunodeficiency syndrome
AMP	adenosine 5′-monophosphate
ATCC	American Type Culture Collection
DMSO	dimethylsulfoxide
DP	degree of polymerization
FDA	Food and Drug Administration
HIV	human immunodeficiency virus
MNNG (also NTG)	*N*-methyl-*N*-nitro-*N*-nitrosoguanidine
NMR	nuclear magnetic resonance
TEM	transmission electron microscopy
UDP-glucose	uridine 5′-diphosphate glucose
UMP	uridine 5′-monophosphate

1 Introduction

Curdlan, an insoluble microbial exopolymer is composed almost exclusively of β-(1,3)-glucosidic linkages. One of the unique features of curdlan is that aqueous suspensions can be thermally induced to produce high-set gels, which will not return to the liquid state upon reheating (Harada et al., 1968), and this has attracted the attention of the food industry. In addition to this, curdlan offers many health benefits, as the beta-glucan family is well known among the scientific community to have immunestimulatory effects. Many informative reviews have been produced on this subject (Harada, 1977; Harada et al., 1993); however, apart from offering a brief review of β-(1,3)-glucan, the present chapter article provides an updated overview of the production, properties, and application of curdlan.

2 Historical Outline

Curdlan was discovered in 1966 by Professor Harada and coworkers, and given its name because of its ability to "curdle" when heated (Harada et al., 1966). At this time, Harada and his colleagues were working on the identification of organisms capable of utilizing petrochemical materials, and isolated *Alcaligenes faecalis* var. *myxogenes* 10C3 from soil. This organism was found to be capable of growing on a medium containing 10% ethylene glycol as the sole carbon source

(Harada et al., 1965), and also produced a new β-(1,3)-glucan that contained about 10% succinic acid, and which was named succinoglucan (Harada 1965; Harada and Yoshimura, 1965). They were also able to derive a spontaneous mutant that mainly produced a water-insoluble neutral polysaccharide, β-(1,3)-glucan, and which did not contain succinoglucan.

Scientists at Takeda Chemical Industries Ltd. (Osaka, Japan) have played a pioneering role in both the research and development of curdlan. Thus, as early as 1989, curdlan was approved and commercialized for food usage in Korea, Taiwan, and Japan. Upon obtaining approval in December 1996, Pureglucan™ – the tradename of curdlan – was launched in the US market as a formulation aid, processing aid, stabilizer, and thickener or texture modifier for food use (Spicer et al., 1999). No evidence of any toxicity nor carcinogenicity of Pureglucan has been observed.

3 Structure

3.1 Beta-1,3-glucan

Both chemical and enzymatic analyses have confirmed that curdlan is a homopolymer of D-glucose linked in β-(1,3) fashion (Saito et al., 1968) (Figure 1). Curdlan has an average degree of polymerization (DP) of approximately 450, and is unbranched (Naganishi et al., 1976). Nakata et al. (1998) reported that the average molecular weight of curdlan in 0.3 N NaOH is in the range of 5.3×10^4 to 2.0×10^6 daltons. Within the class of polysaccharides classified as β-(1,3)-glucans, there are a number of structural variants. The sources of glucans and their structural differences are listed in Table 1. Mycelial fungi are an abundant source of β-(1,3)-glucans; grifolan, which is produced from *Grifola frondosa* and stimulates cytokine production from macrophages, is a β-(1,3)-glucan with a molecular weight of $> 4.5 \times 10^5$ daltons (Okazaki et al., 1995), while lentinan, from *Lentinus edodes*, has a molecular weight of 5×10^5 daltons and two glucose branches for every five β-(1,3)-glucosyl units in the backbone (Jong and Birmingham, 1993). The structure of schizophyllan is very similar to that of lentinan, but it has one glucose branch for every third glucose in the β-1,3-backbone and its molecular weight is 4.5×10^5 daltons (Misaki et al., 1993). Scleroglucan from *Sclerotium rolfsii* has one glucose branch for every third glucose unit (Farina et al., 2001), whereas SSG from *Sclerotinia sclerotiorum* is a highly branched β-(1,3)-glucan (Sakurai et al., 1991). Pachyman, from *Poria cocos*, has an average of 3.2 branch points per molecule of β-(1,3)-glucan (Okuyama et al., 1996; Zhang et al., 1997), whilst krestin is a protein-linked β-glucan with a molecular weight of c. 100,000 daltons, which can be extracted from the mycelia of *Coriolus versicolor* (Azuma, 1987). β-(1,3)-Glucan is also present in the inner cell wall of the bakers' yeast *Saccha-*

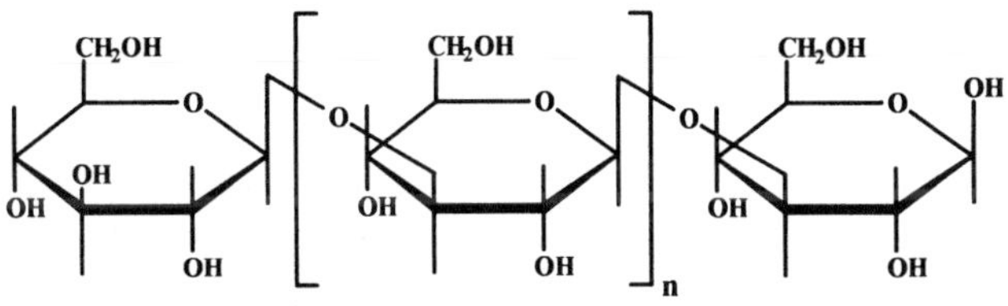

Fig. 1 Structure of curdlan.

Tab. 1 A variety of glucans having β-1,3 linkage in their backbones

Source	*Branch*	M_w	*Reference*
Bacteria			
Curdlan (*Agrobacterium sp. Alcaligens sp.*)	Exclusively β-(1,3)-glucosidic linkages	5.3×10^4 -2.0×10^6	Nakata et al. (1998)
Fungi			
Grifolan (*Grifola frondosa*)	Branched β-1,3-gulcan	4.5×10^5	Okazaki et al. (1995)
Lentinan (*Lentinus eeodes*)	Two glucose branches for every five glucose unit	5×10^5	Jong and Birmingham (1993)
Schizophyllan (*Schizophyllum commune*)	One glucose branch for every third glucose unit	4.5×10^5	Misaki et al. (1993)
Scleroglucan (*Sclerotium glucanum*)	One glucose branch for every third glucose unit	$1.6-5.0\times10^6$	Farina et al. (2001)
SSG (*Sclerotinia sclerotiorum*)	Highly branched β-1,3-glucan	$2\times10^5-2\times10^6$	Okazaki et al. (1995)
Pachyman (*Poria cocos*)	Several β-1,6-linked branch points per molecule	2.06×10^4, 8.93×10^4	Zhang et al. (1997)
Krestin (*Coriolus versicolor*)	β-1,3-glucan	1.0×10^6	Azuma (1987)
Yeast (*Saccharomyces cerevisiae*)			
Soluble glucan	β(1,6) linkage to β-(1,3) backbone	2×10^5–2×10^6	Janusz et al. (1986)
Insoluble glucan	β-(1,6) linkage to β-(1,3) backbone	3.53×10^4, 4.57×10^6	Williams et al. (1994)
Brown algea			
Laminarin (*Laminaria digitata*)	β-1,3-glucan and β-1,6-glucan		Read et al. (1996)

romyces cerevisiae to support the structural strength of its cell wall. Unlike lentinan, schizophyllan and scleroglucan, the side branches of yeast β-(1,3)-glucans are chains of glucose molecules, and not single glucose residues. Depending on the extraction procedure used and their subsequent treatment, the yeast glucans may be either particulate water-insoluble or water-soluble macromolecules.

3.2 Conformation in Solution

Many researchers have investigated the molecular structures of curdlan in aqueous system. Three conformers of soluble curdlan have been reported, including single-helix, triple-helix, and random coil. Ogawa et al. (1972) studied the conformational behavior of curdlan in alkaline solution by measuring the optical rotatory dispersion, intrinsic viscosity and flow birefringence. At low concentrations of sodium hydroxide, curdlan has a helical (ordered) conformation, but a significant conformational change occurs at a NaOH concentration of 0.19–0.24 N NaOH. In alkaline solution >0.2 N NaOH, curdlan is completely soluble and exists as random coils, but upon neutralization the polymer adopts an 'ordered state', which is composed of a mixture of single and triple helices. A ^{13}C-NMR study supported this finding (Saito et al., 1977). Increasing the salt concentration shifts the point of conformational transition to a higher alkali concentration (Ogawa et al., 1973a), and addition of nonsolvents such as 2-chloroethanol, dioxan or water to dimethylsulfoxide (DMSO) solution also changes the conformation of curdlan to a rigid, ordered structure (Ogawa et al., 1973b). These workers also showed that the optical rotation was dependent upon the DP of the curdlan in 0.1 N sodium hydroxide (Ogawa et al., 1973c), and concluded that the content of the ordered form increases with DP until becoming constant at DP values of about 200. Electron microscopic comparison of the molecular structures of curdlan with different DPs showed that only curdlan with higher DP can form a gel when heated (Koreeda et al., 1974).

The conversion between triple-helix and single-helix conformers is mediated by different chemical or physical treatments. Treatment of the triple-helix schizophyllan with NaOH has been used to prepare single helix-rich forms (Ohno et al., 1995). Young et al. (2000) proposed a transition mechanism after an investigation using fluorescence resonance energy transfer spectroscopy, which showed that a partially opened triple-helix conformer was formed on treatment with NaOH, and that increasing degrees of strand opening were associated with increasing concentrations of NaOH. After neutralizing the NaOH, the partially opened conformers gradually reverted to the triple-helix.

3.3 Gel Structure

Some clarification of the fine structure of dispersed molecules and networks is necessary to understand the viscoelastic properties of curdlan, which forms two distinct types of gel. For both gels – described as low-set and high-set gels – transmission electron microscopy (TEM) showed them to be composed of three curdlan molecules that are associated to form a triple helix. Tada et al. (1997) proposed a mechanism of formation of the low-set gel using static light-scattering measurements. The molecular associates are formed at a NaOH concentration of 0.01–0.1 N at 25 °C, and this association progresses with as the NaOH concentration decreases. Consequently, the average molecular weight for the molecular associate in 0.01 N NaOH is higher than that in 0.1 N NaOH aqueous solution. The molecular associates consist of a dense core and hydrophilic surface at low NaOH concentrations. By contrast, Kasai and Harada (1980) proposed an annealing model to form a high-set, resilient gel upon heating to >80 °C. This annealing is associated with the irreversible loss of water, and resulted in a more tightly coiled triple helix. The structure crystallizes as a triplex of right-handed, six-fold helical chains in a hexagonal unit cell with a fiber repeating length of 18.78 Å (Chuah et al., 1983). Further removal of water from this structure, by drying under vacuum, results in further tightening of the six-fold triple helix and a decrease in the fiber period to only 5.87 Å (Deslandes et al., 1980).

4 Occurrence

As described in Section 2, β-(1-3)-D-glucans are present in a variety of living systems, including fungi, yeasts, algae, bacteria and higher plants. However, until now only bacteria belonging to the *Alcaligenes* and *Agrobacterium* species have been reported to produce the linear β-(1,3)-glucan type of homopolymer, curdlan. Figure 2 illustrates

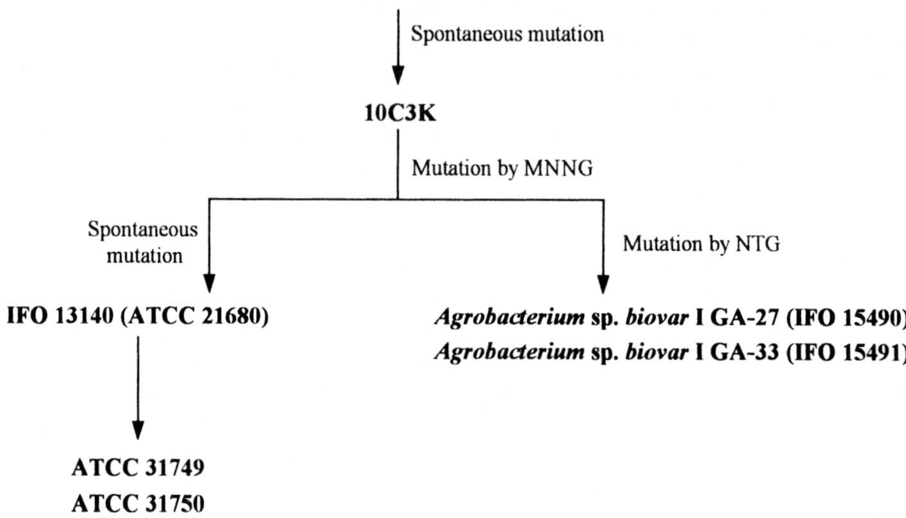

Fig. 2 Lineage of representative strains for curdlan production.

the lineage of the curdlan-producing strains since Harada et al. first isolated *Alcaligenes faecalis* var. *myxogenes* 10C3 during the screening of soil bacteria capable of metabolizing various petroleum fractions (Harada et al., 1965). The parent strain produced two different types of exopolysaccharide; a water-insoluble neutral homoglucan called 'curdlan', and a water-soluble acidic heteroglucan containing about 10% succinic acid, and referred to as 'succinoglucan' (Harada, 1965; Harada and Yoshimura 1965). Moreover, a mutant strain 10C3K was isolated from a stock culture of 10C3, which produced only curdlan. Strain 10C3K is a spontaneous mutant with a stable ability to produce exocellular polysaccharide. By inducing mutagenesis with *N*-methyl-*N*-nitro-*N*-nitrosoguanidine (MNNG), Takeda Chemical Industries Ltd. later isolated a uracil auxotrophic mutant from strain 10C3K, which was named *Alcaligenes faecalis* var. *myxogenes* IFO 13140 (ATCC 21680) and had improved gel-forming β-(1,3)-glucan-producing ability. Phillips and Lawford (1983) isolated a mutant strain from strain ATCC 21680 in a nitrogen-limited chemostat culture (the accession number was ATCC 31749). Unlike its auxotrophic parent strain, ATCC 31749 does not require uracil for its growth, and is not a revertant as it can be distinguished from 10C3K by its inability to hydrolyze starch and its ability to grow on citric acid as sole carbon source. ATCC 31750, which arose as a spontaneous variant of the parent ATCC 31749, produces only the water-insoluble curdlan-type glucan, while the parent strain produces both soluble and insoluble polysaccharides. All of these strains, which were formerly regarded as *Alcaligenes* species, have now been taxonomically reclassified as *Agrobacterium* species (IFO Research Communications, Vol. 15, pp. 57–75 (1991)). Takeda Chemical Industries Ltd. derived further mutant strains from strain 10C3K, which reduced the activity of the enzyme, phosphoenol pyruvic acid carboxykinase (Kanegae et al., 1996).

Naganishi et al. (1974) examined the occurrence of curdlan-type polysaccharides in microorganisms by using the water-soluble dye aniline blue, with which curdlan forms a blue complex. It was also shown that the rate of color complex formation was dependent both on the polymer concentration and DP; hence these findings provided an excellent tool for the screening of curdlan-producing bacteria. Naganishi et al. (1976) tested 687 strains of different genus of bacteria using the aniline blue staining technique. Among those examined, some strains of *Alcaligenes* and *Agrobacterium* species turned blue on agar plates containing aniline blue, and these have been used widely in the production of curdlan-type polysaccharides. Some strains of *Bacillus* formed blue complexes with aniline blue, but their polymeric constitution has not yet been studied.

5 Biosynthesis

Sutherland (1977, 1993) generalized the biosynthesis of extracellular polysaccharides into three major steps: (1) substrate uptake; (2) intracellular formation of polysaccharide; and (3) extrusion from cell. A metabolic pathway for exopolysaccharide biosynthesis is shown schematically in Figure 3. First, a carbohydrate substrate enters the cell by active transport and group translocation involving substrate phosphorylation. The substrate is then directed along either catabolic pathways, or those leading to polysaccharide synthesis. UDP-glucose, a key precursor, is synthesized by the UDP-glucose pyrophosphorylase-induced conversion of glucose-1-phosphate to UDP-glucose. Subsequently, polymer construction occurs together with the transfer of monosaccharides from UDP-glucose to a carrier lipid.

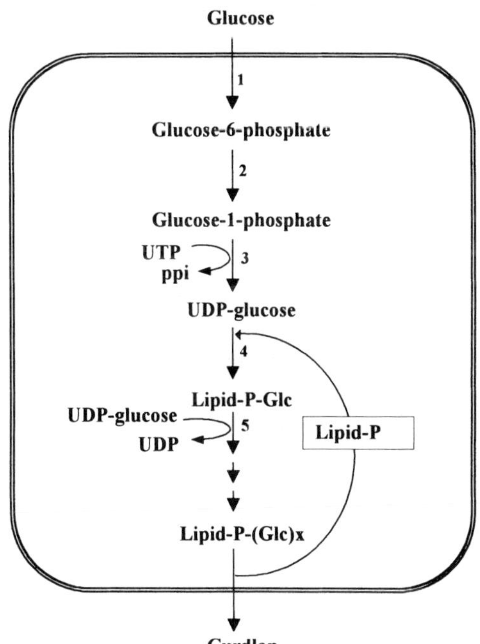

Fig. 3 Metabolic pathway for the synthesis of curdlan. 1, hexokinase; 2, phosphoglucomutase; 3, UDP-glucose pyrophosphorylase; 4, transferase; 5, polymerase. Lipid-P represents isoprenoid lipid phosphate.

After further chain elongation, the polymer is extruded from the cells.

Nitrogen-limited culture has also been employed for the production of curdlan, and has been generally explained by the roles of the carrier lipids (Sutherland, 1977; Ielpi et al., 1993). The availability of isoprenoid lipid may provide a way of regulating polysaccharide synthesis. Since curdlan biosynthesis takes place most extensively after cell growth has been stopped due to nitrogen exhaustion, isoprenoid lipids would be more available for carrying oligosaccharides instead of cellular liposaccharide and peptidoglycan.

The synthesis of precursor molecules is also of considerable importance in polysaccharide synthesis, in terms of the metabolic driving force. UDP-glucose serves as an activated precursor for glycosyl moieties in the synthesis of curdlan, a homopolysaccharide composed exclusively of β-1,3-linked glucose residues. In addition, cellular nucleotides not only play an important role in the synthesis of sugar nucleotides, but also have a widespread regulatory potential in cellular metabolism. Kim et al. (1999) examined the change of intracellular nucleotide levels and their stimulatory effects on curdlan synthesis in *Agrobacterium* species under different culture conditions. Under nitrogen-limited conditions where curdlan synthesis was stimulated, intracellular levels of UMP and AMP were at least twice as high as those occurring under nitrogen-sufficient conditions, though UDP-glucose levels were similar. The time profiles of curdlan synthesis and cellular nucleotide levels showed that curdlan synthesis is positively related with intracellular levels of UMP and AMP. *In vitro* enzyme reactions involved in the synthesis of UDP-glucose showed that a higher UMP concentration promotes the synthesis of UDP-glucose, while AMP neither inhibits nor facilitates the activity of UDP-glucose pyrophosphorylase. The addition of UMP to the medium also increased curdlan synthesis. From these results, these workers concluded that the higher intracellular UMP levels caused by nitrogen limitation enhance the metabolic flux of curdlan synthesis by promoting cellular UDP-glucose synthesis.

Kai et al. (1993) reported a study of the biosynthetic pathway of curdlan using ^{13}C-labeled glucose in *Agrobacterium* sp. ATCC 31749. By analyzing the labeled products, the biosynthesis of curdlan was interpreted as involving five routes: direct polymerization from glucose; rearrangement; isomerization of cleaved trioses; from fructose-6-phosphate; and from fructose fragments produced in various pathways of glycolysis. However, it was noted that more than 60% of curdlan is synthesized by direct polymerization, and that curdlan biosynthesis via glycolysis is comparatively low. This analysis also indicated that glycolysis occurs mainly via the pentose cycle and the Entner–Doudoroff pathway rather than the Embden–Meyerhof pathway.

6 Molecular Genetics

Little is known about the molecular genetics of bacterial curdlan biosynthesis, while there is growing information about the genes required for β-(1,3)-glucan synthesis in yeasts and filamentous fungi. Recently, Stasinopoulos et al. (1999) cloned genes that were essential for the production of curdlan and, by using comparative sequence analysis, identified them as putative curdlan synthase genes. Further genetic investigations will open up new avenues for curdlan synthesis, and these will doubtlessly be exploited to produce curdlan in higher yields.

7 Production

7.1 Carbon Source

Many crucial factors that affect curdlan production, including carbon, nitrogen, phosphate, oxygen supply, and pH have been investigated. High productivity using cheap carbon sources is important for the industrial production of curdlan. Lee, I.-Y. et al. (1997) reported that maltose and sucrose were efficient carbon sources for the production of curdlan by a strain of *Agrobacterium* species, with maximal production (60 g L^{-1}) being obtained from sucrose, with a productivity of 0.5 g L^{-1} h^{-1} when nitrogen was limited at a cell concentration of 16 g L^{-1}. Molasses, which contains large amounts of sucrose, might also be the substrate of choice, with up to 42 g L^{-1} of curdlan, at a yield of 0.35 g curdlan per gram total sugar, being obtained in 5-day cultivation. Sucrose is a less expensive substrate than glucose, and as sugar beet or sugar cane molasses are cheap byproducts widely available from the sugar industry, they are very attractive carbon sources from an economic point of view.

7.2 Nitrogen Effect

As previously described, relatively few strains of *Agrobacterium* and *Alcaligenes* species are known to produce curdlan. In such strains, curdlan production is associated with the poststationary phase of nitrogen depletion; thus, the operation involves an initial production of biomass, which is followed by curdlan production. Therefore, it is important to determine the initial concentration of the nitrogen source because it provides the limiting factor for cell growth during batch fermentation. Kim, M.-K. et al. (2000) reported that the cell growth rate decreased as the ammonium concentration increased. However, since higher cell concentrations produce more curdlan, an optimal ammonium concentration should be determined to provide an appropriate cell concentration while minimizing the inhibitory effect of the ammonium ion.

7.3 Oxygen Supply

Curdlan-producing stains are highly aerobic, and an adequate oxygen supply is therefore a key factor in production. Since curdlan is insoluble in water, the fermentation broth is of relatively low viscosity, and there is little resistance to oxygen transfer from gas to the liquid. However, a layer of insoluble exopolymer surrounding the cell mass offers resistance to oxygen transfer from the liquid into the cell, and therefore a high dissolved oxygen concentration is required for maximal productivity. Shake-flask fermentation results have shown that the specific production rate decreases as the volume of medium is increased, indicating that these cultures are limited by the relatively low oxygen transfer capacity of the system. Several investigations have been made into developing the process for curdlan production, especially with respect to reactor design (Lawford et al., 1986; Lawford and Rousseau, 1991, 1992). These workers employed two different types of impeller: a radial-flow, flat-blade impeller; and an axial-flow impeller. The radial-flow impeller was effective at providing high oxygen transfer rates to increase the production of curdlan, but the high shear characteristics of this design yielded a product of inferior quality in terms of tensile strength of the thermally induced gel. An axial-flow impeller typically produces less shear and more pumping. High volumetric oxygen transfer can be achieved by low shear designs equipped with sparging devices, which consist of microporous materials through which oxygen-enriched air is dispersed. The maximal specific production rate was 90 mg per g cells h^{-1} when 30% oxygen-enriched air was supplied in the low-shear system.

7.4 Phosphate Effect

Phosphate concentration must also be considered because it significantly influences cell growth and product formation. The production of rhamnose-containing polysaccharide by a *Klebsiella* strain was enhanced by a reduction in the phosphate content of the medium (Farres et al., 1997). In contrast, a sufficiency of phosphate resulted in good alginate yields in *Pseudomonas* strains and showed growth-associated production (Conti et al., 1994). Thus, the effect of phosphate on the production of polysaccharides is variable. Kim, M.-K. et al. (2000) investigated the influence of inorganic phosphate concentration on the production of curdlan by *Agrobacterium* species. Under nitrogen-limited conditions which allow curdlan production, the concentration of phosphate remains constant as it is not further utilized for cell growth. The optimal residual phosphate concentration for curdlan production was in the range 0.1–0.5 g L^{-1}. Relatively low concentrations appeared to be optimal for curdlan production, although without phosphate, curdlan production was extremely low. However, on increasing the cell phosphate concentration from 0.42 to 1.68 g L^{-1}, curdlan production increased from 0.44 to 2.80 g L^{-1}. Moreover, the optimal phosphate concentration range was not dependent upon cell concentration, and the specific production rate was about 70 mg curdlan per g cells h^{-1}, irrespective of cell concentration.

7.5 pH Effect

The pH of the culture is one of the most important factors because it significantly influences rates of both cell growth and product formation. High viscosity of the culture broth is often a critical problem in polysaccharide production. However, the viscosity problem can be obviated by operating the fermentation at a slightly acidic pH, since curdlan is insoluble under these conditions. Moreover, there appears to be more than one single optimal pH because fermentation of the culture for curdlan production is divided into two phases – the cell growth phase and the curdlan production phase. Lee, J.-H. et al. (1999) sought an optimal pH profile to maximize curdlan production in a batch fermentation of *Agrobacterium* species. The cell growth rate was maximal at pH 7.0, while curdlan production was maximal at pH 5.5. The pH profile provided a strategy to shift the culture pH from the optimal growth condition (pH 7.0) to the optimal production one (pH 5.5) at the time of ammonium exhaustion. By adopting the optimal pH profile in a batch process, these workers obtained a significant improvement in curdlan production (64 g L^{-1}) compared with that obtained using a constant-pH operation (36 g L^{-1}).

7.6 Batch Production

A high curdlan production was attempted by employing the optimal operation strategy with an *Agrobacterium* strain (Lee, I.-Y. et al., 1997, 1999a; Lee, J.-H. et al., 1999; Kim, M.-K. et al., 2000) that was tolerant of high concentrations of sucrose. This made batch operation possible, with an initial sucrose concentration of 140 g L^{-1}. An initial ammonium concentration 0.8 g L^{-1} was chosen, which produced 6.4 g L^{-1} of cells, the cell yield from ammonium being 8.0 g cells per g ammonium. A typical batch fermentation profile in a 300-L stirred tank reactor is shown in Figure 4. To produce a volumetric oxygen transfer coefficient of 146.5 h^{-1}, the agitation speed was set at 200 r.p.m. over the whole fermentation period, with an inner pressure of 0.2 bar. When the ammonium was exhausted in the culture broth at 20 h, the cell concentration had reached 6.8 g L^{-1}. The pH was controlled at 7.0 with 4 N NaOH/KOH at the cell growing stage. The culture pH was then shifted from 7.0 to 5.5 by adding 3 N HCl at the time of nitrogen limitation. Aeration rates were maintained at 0.5 vvm (volume volume^{-1} min^{-1}). Curdlan production began from the onset of nitrogen exhaustion, and a maximum concentration of 58 g L^{-1} was obtained in 120 h cultivation. The dissolved oxygen level fell rapidly during cell growth, and increased immediately after nitrogen exhaustion, as cell growth ceased. Subsequently, as the curdlan concentration increased, the dissolved oxygen level became limited due to the increased viscosity of the culture broth.

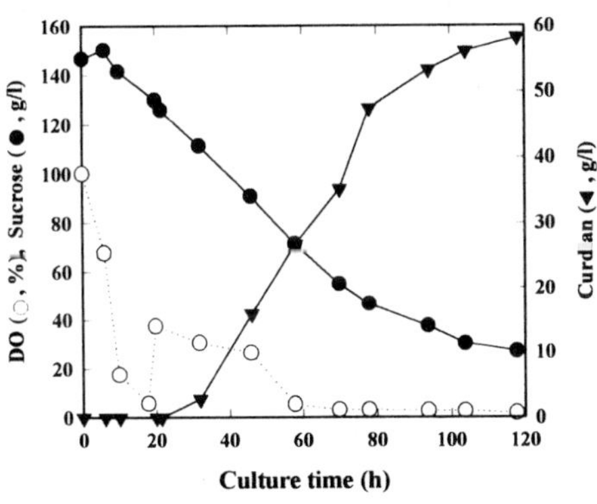

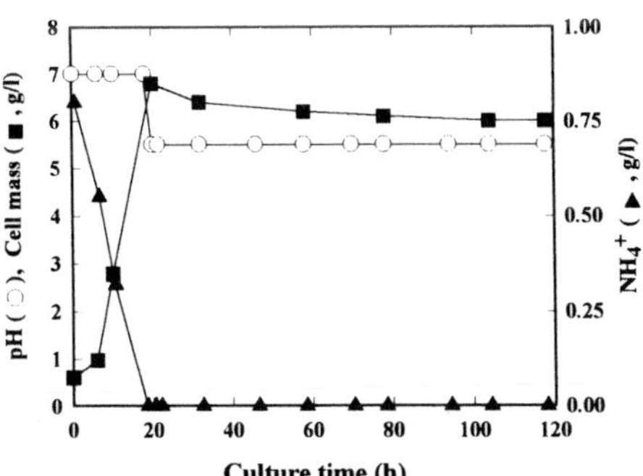

Fig. 4 Batch production of curdlan with *Agrobacterium sp.* ATCC 31750 in a 300-L jar fermenter (Lee, I.-Y. et al., 1999a).

7.7 Continuous Production

Continuous production of curdlan was attempted in a two-stage continuous process (Phillips and Lawford, 1983). In the first stage, the curdlan-producing strain *Alcaligenes* sp. ATCC 31749 was grown aerobically in a medium containing carbon and nitrogen. However, the amount of nitrogen in the first stage was so limited that the effluent contained substantially no inorganic nitrogen. The effluent was fed into the second stage in a constant-volume fermenter, where it was mixed with a nitrogen-free medium in order to induce curdlan production without further cell growth. Curdlan production was 7 g L^{-1} at a dilution rate of 0.02 h^{-1} in a steady-state culture, providing a curdlan productivity of 0.14 g L^{-1} h^{-1}.

7.8 Isolation Process

The recovery procedure is based on the conformational transition which occurs when the concentration of alkali exceeds 0.2 N (Ogawa et al., 1972). Under alkaline conditions, the biomass can be separated from the dissolved curdlan, which remains in the supernatant. Upon neutralization of the alkaline supernatant, the polymer forms an insoluble gel that can be recovered by centrifugation, the polymer being subsequently washed free of contaminating salt. Procedures for preserving curdlan in the dry state include both dehydration with organic solvents and spray drying.

8 Properties

8.1 Gel Formation

Curdlan is water-insoluble but soluble in alkali, and forms two-types of gels: high-set or low-set gels. The high-set gel is formed upon heating to around 80 °C, depending on its concentration in water (Maeda et al., 1967). The high-set curdlan gels are thermally irreversible and, unlike agar gel, further heating does not cause them to revert to the liquid state; they are also very elastic and resilient, whereas agar gels are brittle and relatively fragile. Curdlan forms a low-set reversible gel when its suspension is heated to 55–65 °C and then cooled. When a low-set gel is heated to 80 °C, however, it turns into a high-set gel. At the same curdlan concentration, a low-set gel has a weaker strength than a high-set gel. Changes in external factors such as pH, temperature, and ionic strength greatly affect gelation ability. Curdlan also forms a reversible gel when its alkaline solution is neutralized, and the addition of calcium or magnesium ions to a weakly alkaline solution of curdlan produces a gel with a bridged structure (Aizawa et al., 1974). Other methods to prepare curdlan gels include cooling DMSO solutions, or dialyzing its alkaline or DMSO solutions in water (Ogawa et al., 1972). The strength of the curdlan gel increases with temperature, concentration, and heating time (Maeda et al., 1967; Takeda technical report, 1997) (Figure 5). Curdlan, unlike other gelling agents, has a unique ability to form gels over a wide range of pH values (3.0–10.0).

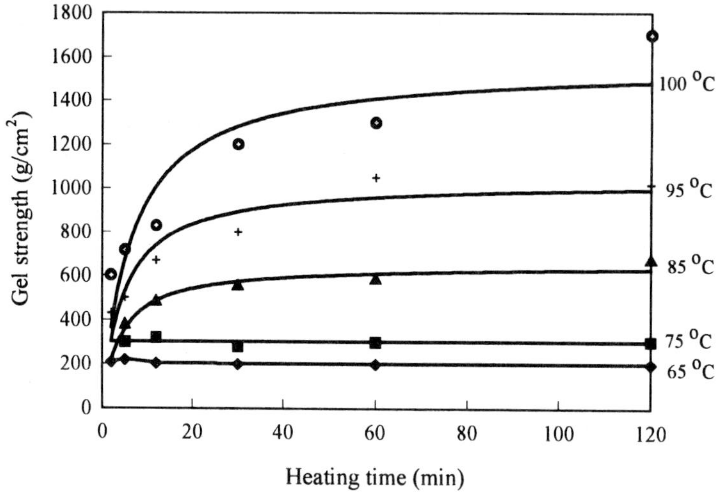

Fig. 5 Effect of heating temperature and time on curdlan gel strength (2% gel) (Takeda technical report, 1997).

8.2 Immunestimulatory Activity

β-(1,3)-Glucan has been found to be effective in stimulating macrophages (leukocytes or white blood cells). Macrophages form the immune system's first line of defense against foreign invaders, and can recognize and kill tumor cells, remove foreign debris resulting from oxidative and radiation damage, speed up the recovery of damage tissue, further activate cytokines, and initiate an immune cascade system to mobilize B and T cells (Di Luzio et al., 1979; Di Luzio, 1983; Suzuki et al., 1992; Hadden, 1993). Curdlan is a bacterial polysaccharide composed entirely of β-(1,3)-D-glucosidic linkages, and has been reported to show antitumor activity (Sasaki and Takasuka, 1976; Yoshioka et al., 1985). However, differences in biological activities seem to be dependent upon the degree of branching, molecular conformation, and molecular weight. Sasaki et al. (1978) examined the effect of chain length on antitumor activity by using different molecular weight glucans, which were obtained by the acid hydrolysis of curdlan. The results show that insoluble glucans with a number-average DP >50 have strong antitumor activity. As described in Section 2, β-glucans exist in three conformers: single-helix, triple-helix, and random coil. Among these, the ordered (helical) conformations are considered to be biologically active forms (Bohn and BeMiller, 1995). Some researchers reported that the triple-helical structure of schizophyllan, a β-glucan obtained from *Schizophyllum commune*, is essential for its anti-tumor activity (Yanaki et al., 1983; Kojima et al., 1986). However, other studies have suggested that the single-helix is more potent in this respect (Saito et al., 1991; Aketagawa et al., 1993). Ohno et al. (1995, 1996) showed that both the triple- and single-helix conformers of schizopyllan are active against tumors and leukopenia. In addition, the triple-helix conformer had a significant antagonistic effect upon zymosan-mediated hydrogen peroxide synthesis in peritoneal macrophages, whereas the single-helix conformer showed strong activity on the synthesis of tumor necrosis factor, nitric oxide, and hydrogen peroxide. It remains unclear as to which of the helical conformers is the most active, but it is likely that the structure–activity relationship of the β-glucan-mediated immunopharmacological activities vary, and are dependent upon the assay systems adopted.

9 Applications

9.1 Food Applications

Since the time of curdlan's original discovery, it has been mainly proposed for use in the food industry as a possible extender or substitute for natural plant gums in food preparations requiring a thickening or bodying additive. The food applications of curdlan are summarized in Table 2. These food additives and essential ingredients are generally divided in terms of how much curdlan is used; food additives require less than 1% curdlan, and essential ingredients more than 1%. Curdlan is useful as a gelling material to improve the textural quality, water-holding capacity and thermal stability of various foods. The polymer can be added during the production process, before heating, either as a powder, or as a suspension or slurry in either water or aqueous alcohol. The foods employing these functions include soy-bean curd (tofu), sweet bean paste jelly, boiled fish paste, noodles, sausages, jellies, and jams (Masayuki and Yukihiro, 1990; Taguchi et al., 1991; Ken and Akirou, 1992; Masanori et al., 1992; Shunsuke and Masatoshi, 1992; Hideaki et al., 1993; Hiroki and Etsuko, 1993; Masahiro and Masaru, 1993; Masatoshi et al., 1993; Yukihiro and Takahiro, 1993; Yasuhiro, 1994; Masaaki and Takaaki, 1995; Masataka et al., 1997; Akihiro and Takaaki, 1998; Hiroshi et al., 1999; Kazushi, 1999; Masao and Toshimi, 1999; Takaaki, 1999). Various tests have shown that this polymer is safe. The gel of curdlan has properties intermediate between the brittleness of agar gel and the elasticity of gelatin (Kimura et al., 1973). Furthermore, since the β-(1,3)-glucan polymer is not readily degraded by human digestive enzymes, it offers the possibility of new calorie-reduced products, such as those often referred to as "dietetic" foods. The fact that curdlan set-gels are known efficiently to absorb high concentrations of sugars from syrups and are relatively resistant to syneresis, suggests a use in sweet jellies and various other dessert-type foods. The resistance of curdlan gels to degradation by freezing and thawing also indicates potential in frozen food products. The pseudoplastic flow behavior of curdlan-containing fluids points to a possible role as thickeners and stabilizers in liquefied foods, such as salad dressings and spreads. Curdlan is also used to produce an edible casting film for foods with other water-soluble polymeric substances having heat-sealability (Akira and Atsushi, 1989).

9.2 Pharmaceutical Applications

Curdlan sulfate having a β-(1,3)-glucan backbone showed high anti-AIDS (acquired immunodeficiency syndrome) virus activity, with few adverse side effects; therefore, curdlan sulfate has a growing potential as an anti-AIDS drug (Jagodzinski et al., 1994; Takeda-Hirokawa et al., 1997). The Phase I/II trials (toxicity testing) of curdlan sulfate for human immunodeficiency virus (HIV) carriers have been carried out in the United States under the auspices of the Food and Drugs Administration (FDA) since 1992. A sulfuric acid ester of curdlan low molecular polymer, with an average DP ranging from 10 to 400 was found to be an active anti-retrovirus ingredient useful for preventing and treating infection because of its high water solubility, low toxicity and relatively high inhibitory activity upon human retrovirus multiplication (Takashi and Junji, 1991). Mikio et al. (1995) disclosed that a

Tab. 2 Application of curdlan as food additives and essential ingredient (Spicer et al., 1999)

Function	*Objective foods*
Essential ingredient	
Gelling agent	Dessert, jelly, pudding, dry mixes
Bulking agent	Dietetic foods, diabetic foods
Low-calorie foods	Edible film, edible fiber casings
Food additive	
Improving viscoelasticity	Noodle, hamburger, sausage
Binding agent	Hamburger, starch jelly
Water-holding agent	Noodle, sausage, ham, starch jelly
Prevention of deterioration	Frozen egg products
Masking of malodors or aromas	Boiled rice
Retention of shape	Starch jelly, dry desert mixes
Thickeners and stabilizers	Salad dressing, low-calorie foods, frozen foods
Coating agent	Flavors

curdlan derivative modified by reaction with glycidol developed excellent antiviral activity whilst showing extremely low toxicity. Evans et al. (1998) subsequently reported that the low toxicity of curdlan and its marked anti-invasion activity on merozoites makes it a potential auxiliary treatment for severe malaria. Moreover, linear β-(1,3)-glucan sulfate has potential as a blood coagulation inhibitor useful for both the prevention and treatment of thrombosis, because linear β-(1,3)-glucan sulfate has been found to have a strong inhibitory action on blood coagulation (Tsuneo, 1990). Curdlan is also used as a sustained release suppository containing drug-active components, such as indomethacin, diclofenac sodium and ibuprofen (Toshiko, 1992). Kanke et al. (1992) investigated the *in vitro* release of a curdlan tablet containing theophylline; drug release from the tablet was the lowest tested, and was unaffected by pH and various other ions. Kim, B.-S. et al. (2000) proposed that hydroxyethyl derivatives of curdlan can be used as protein drug delivery vehicles. A curdlan hydrolysate with a number average molecular weight ranging from 340 to 4000 was used as an immunoactivator and did not cause any adverse side effects (Masahiro et al., 1998a). Hiroshi et al. (1997) have also reported that linear or branched β-(1,3)-glucans such as lentinan and curdlan are effective in the treatment of dementia.

9.3 Agricultural Applications

Fumio et al. (1989) reported that polysaccharides with the β-(1,3)-glucan structure, such as lentinan, schizophyllan, curdlan or laminarin are useful for the multiplication of *Bifidobacterium* bacteria and, when given to animals in the food, also inhibit the intestinal putrefying bacteria, thereby preventing the animals from aging. In addition, curdlan is incorporated into fish feed mixtures to improve immune activity (Masahiro et al., 1998b).

9.4 Other Industrial Applications

Curdlan has been studied as a support for immobilized enzymes (Murooka et al., 1977). For example, in the immobilization of an enzyme or a mold using curdlan gel, the curdlan is gelatinized by heating an aqueous suspension containing curdlan, an enzyme or a mold (Toshio, 1986). Other potential industrial applications include the use of curdlan films and fibers which are water-insoluble, biodegradable and impermeable to oxygen. Curdlan is also used to prevent nozzle clogging, and the electrostatic damage of electronic components when used as a seal in an ink-jet recoding head (Yoichi et al., 1996). Furthermore, curdlan functions as a concrete admixture to produce concrete with high fluidity but low separating properties (Toshiyuki et al., 1995; Lee, I.-Y. et al., 1999b). Applications of curdlan other than in the food industry are shown in Figure 6.

10 Outlook and Perspectives

Curdlan has been well accepted in the meat processing and noodle industries for its extremely unusual property of forming gels that are resistant to freezing or retorting. Takeda has been marketing this innovative product in Japan for several years, and sales are promising. Indeed, production has continued to grow by more than 10% each year during the past eight years such that, in recent years, Takeda Chemical Industries, Ltd. has produced 600–700 tonnes of curdlan annually.

Pharmaceutical applications

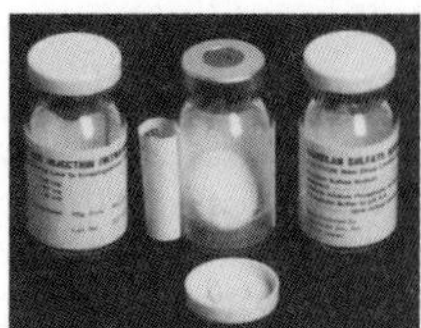

Anti-HIV agent
(www.iis.u-tokyo.ac.jp)

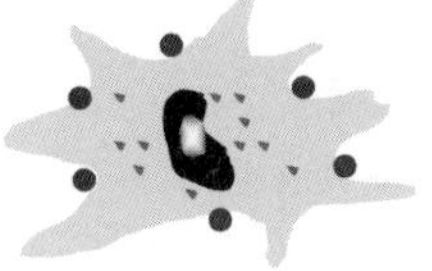

Immunoactivator
(Macrophage activation)

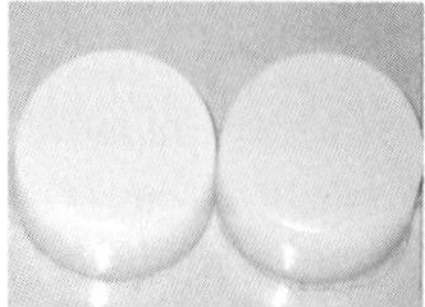

Drug delivery vehicle
(Curdlan gel)

Agricultural applications

Feed for domestic animals

Feed for fish animals

Plant fertilizer

Other industrial applications

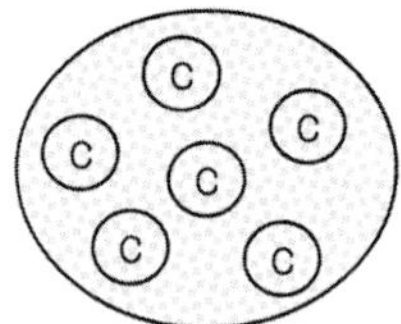

Immobilization support
(Gel entrapment)

Edible film

Concrete admixture
(Segregation reducing agent)

Fig. 6 Applications of curdlan in fields other than the food industry.

In addition, it should be noted that a variety of glucans with the β-1,3 linkage have immunestimulatory effects. Krestin, a β-glucan extracted from the mycelium of the basidiomycete *Coriolus versicolor*, has a general immunestimulatory effect in humans and has also been used as an anticancer agent (Kamisato and Nowakowski, 1988). Lentinan from *Lentinus edodes* is another glucan which inhibits tumor growth and increases resistance to infections by bacteria, viruses, and parasites (Jong and Birmingham, 1993). Schizophyllan (Numasaki et al., 1990) from *Schizophyllum commune* has the property of inducing the formation of cytotoxic macrophages and generating antitumor activity in the human body. In their preparation, these fungal glucans require extensive extraction operations to be performed in order to obtain comparatively small amounts of pure glucans, which makes the process very expensive. By contrast, curdlan as an exotype of β-(1,3)-glucan can be readily produced on a large scale by a submerged bacterial culture. Thus, if the material is correctly designed on a molecular

Tab. 3 Patent holders and publication year

Company	***Year of publication***													
	1988	***1989***	***1990***	***1991***	***1992***	***1993***	***1994***	***1995***	***1996***	***1997***	***1998***	***1999***	***2000***	***Total***
Ajinomoto	2		1							2				5
Asai							1							2
Chem Reizou	1													1
Dainippon	1	1							1	1	1			5
Daiichi		1												1
Daito										1				1
Endo Akira	1													1
Ezaki Glico														1
Fuji Oil	1				2	1							1	5
Hinoshoku												1	1	2
House foods	1							1						2
Hodaka						1								1
Ina						1					1		4	6
Japan Organo						2		2	2	7	2	2	2	19
Kanai Masako												1		1
Kanebo					2			1			1			4
Kanegafuchi				1			1				1			3
Kiteii					1									1
Kibun foods									1					1
Kyokuto				1										1
Meiji			2		1					1	1			5
Morigana												1		1
Nagai				1	1									2
Nikken									1					1
Nippon					2	1							1	4
Nitta Gelatin							1							1
Nokyo									1					1
Okada					2									2
Sanei		4		1			1		1					7
Sanyo	1													1
Shiseido							1		2					3
Snow Brand					1						2	1	2	6
Sumisho												1		1
Takeda		1	12	8	6	5	1	5	3	5	9	3	4	62
Taiyo Kagaku							1							1
Tsubuki								1						1
Tsukishima						1								1
Unie Colloid							2							2
Wako	5													5
Yamamoto										1				1
Total	13	7	15	12	18	12	8	13	9	19	19	11	15	171

basis to enhance its immunestimulatory activity, there is a large potential market in both the nutraceutical and pharmaceutical industries.

11 Patents

The Japanese patent literature between 1988 and 2000 was reviewed, as most of the patents relating to the application of curdlan were first filed in Japan. The companies holding patents, together with the annual number of patents filed, are listed in Table 3. Takeda and Japan Organo each hold about one-half of total 171 patents filed, indicating their major role in the development of curdlan application. It should also be noted that 122 patents relate to the application of curdlan as food additives or ingredients. The patents described in Section 8 are listed in Table 4.

Acknowledgments
The author thanks Mi-Kyoung Kim for her help in the preparation of the manuscript.

Tab. 4 A list of the patents of curdlan application

Patent no.	*Inventor(s)*	*Title*	*Year*
JP-61015688	Toshio, O.	Immobilized enzyme, or the like and preparation thereof	1986
JP-01289457	Akira, K., Atsushi, I.	Edible film	1989
JP-01137990A	Fumio, I., Kimikazu, I., Satoshi, S.	Polysaccharide having activity for multiplying *Bifidobacterium*	1989
JP-02124902	Tsuneo, A., Koichi, K., Junji, K.	Blood coagulation inhibitor	1990
JP-02249466	Masayuki, T., Yukihiro, N.	Noodle made of rice powder and producing method thereof	1990
JP-03218317	Takashi, T., Junji, K.	Anti-retrovirus agent	1991
JP-03157401	Taguchi, T., Hiroshi, K., Yukihiro, N.	Production of linear gel	1991
JP-04210507	Masanori, T., Tetsuya, T., Yukihiro, N.	Noodles	1992
JP-04330256	Shunsuke, O., Masatoshi, S.	Production of uncooked Chinese noodle and boiled Chinese noodle	1992
JP-04197148	Ken, O., Akirou, M.	Soybean protein-containing solid food	1992
JP-04074115	Toshiko, S.	Sustained release suppository	1992
JP-05184310	Masatoshi, H., Masami, F., Kenichi, H., Hiromi, K.	Devil's tongue for frozen food and its production	1993
JP-05207859	Hideaki, Y., Chieko, M., Takeshi, A.	Ganmodoki and its production	1993
JP-05023143	Yukihiro, N., Takahiro, F.	Ground fish meat or fish or cattle paste product and composition therefor	1993
JP-05076290	Hiroki, O., Etsuko, M.	Noodle-like soybean protein-containing food	1993
JP-05288909	Masahiro, K., Masaru, N.	Preparation of processed edible meat	1993
JP-06335370	Yasuhiro, S.	Formed jelly-containing liquid food and its production	1994
JP-07010624	Toshiyuki, U., Yoshio, T., Minoru, Y.	Additive for concrete	1995

Tab. 4 (cont.)

Patent no.	***Inventor(s)***	***Title***	***Year***
JP-07274876	Masaaki, A., Takaaki, I.	Quality improver for wheat flour product and wheat flour product having improved texture	1995
JP-07228601	Mikio, K., Yoshiro, O., Hirotomo, O., Yoshikazu, K., Hiroshi, I., Hisashi, M., Takeshi, K.	Water soluble β-1,3-glucan derivative and antiviral agent containing the derivative	1995
JP-0818787	Yoichi, T., Hiroshi, S., Makiko, K.	Biodegradable component for protecting ink jet recording head	1996
USP-5508191	Kanegae, Y. Yutani, A., Nakatsui, I.	Mutant strains of *Agrobacterium* for producing beta-1,3-glucan	1996
JP-09075017	Masataka, M., Hideo, Y., Yukihiro, N.	Devil's-tongue jelly and its production	1997
Patent number	Inventors	Title	Year
JP-09255579	Hiroshi, S., Nobuyoshi, N., Yutaro, K., Toru, M.	Medicine for treating dementia	1997
JP-10194977	Masahiro, K., Masaaki, K., Shinji, M., Hiroaki, K.	Immunoactivator	1998a
JP-10313794	Masahiro, K., Mikitomo, A., Akitomo, A., Tomonori, Y.	Feed composition for fish kind cultivation	1998b
JP-10014541	Akihiro, S., Takaaki, I.	Quality improving agent for fishery paste product and production of fishery paste product	1998
JP-11187819	Hiroshi, Y., Yuichi, S., Norio, I., Nana, I.	Frozen dessert food and its production	1999
JP-11075726	Masao, K., Toshimi, T.	Jelly-like food	1999
JP-11178533	Kazushi, M.	Production of bean curd including ingredient	1999
JP-11056246	Takaaki, I	Quality improver for bean jam product and production of bean jam product	1999

12
References

Aizawa, M., Takahashi, M., Suzuki, S. (1974) Gel formation of curdlan-type polysaccharide in DMSO-H_2O mixed solvents, *Chem. Lett.* 193–196.

Aketagawa, J., Tanaka, S., Tamura, H., Shibata, Y., Saito, H. (1993) Activation of *Limulus* coagulation factor G by several (1→3)-β-D-glucans: comparison of the potency of glucans with identical degree of polymerization but different conformations, *J. Biochem.* **113**, 683–686.

Akihiro, S., Takaaki, I. (1998) Quality improving agent for fishery paste product and production of fishery paste product. Japanese patent 10014541.

Akira, K., Atsushi, I. (1989) Edible film. Japanese patent 01289457.

Azuma, I. (1987) Development of immunostimulants in Japan, in: *Immunostimulants: Now and Tomorrow* (Azuma, I., Jolles, G., Eds.), Tokyo: Japan Sci. Soc. Press/Berlin: Springer-Verlag, 41–45.

Bohn, J. A., BeMiller, J. N. (1995) (1→3)-β-D-glucans as biological response modifiers: a review of structure–functional activity relationships, *Carbohydr. Res.* **28**, 3–14.

Chuah, C. T., Sarko, A., Deslandes, Y., Marchessault, R. H. (1983) Triple-helical crystalline structure of curdlan and paramylon hydrates, *Macromolecules* **16**, 1375–1382.

Conti, E., Flaibani, A., O'Regan, M., Sutherland, I. W. (1994) Alginate from *Pseudomonas fluorescens* and *P. putida*: production and properties, *Microbiology* **140**, 1125–1132.

Deslandes, Y., Marchessault, R. H., Sarko, A. (1980) Triple-helical structure of (1→3)-β-D-glucan, *Macromolecules* **13**, 1466–1471.

Di Luzio, N. R. (1983) Immunopharmacology of glucan: a broad spectrum enhancer of host defence mechanisms, *Trends Pharmacol. Sci.* **4**, 344–347.

Di Luzio, N. R., Williams, D. L., McNamee, R. B., Edwards, B. F., Kitahama, A. (1979) Comparative tumor-inhibitory and anti-bacterial activity of soluble and particulate glucan, *Int. J. Cancer* **24**, 773–779.

Evans, S. G., Morrison, D., Kaneko, Y., Havlik, I. (1998) The effect of curdlan sulfate on development *in vitro* of *Plasmodium falciparum*, *Trans. R. Soc. Trop. Med. Hyg.* **92**, 87–89.

Farina, J. I., Sineriz, F., Molina, O. E., Perotti, N. I. (2001) Isolation and physico-chemical characterization of soluble scleroglucan from *Sclerotium rolfsii*. Rheological properties, molecular weight and conformational characteristics, *Carbohydr. Res.* **44**, 41–50.

Farres, J., Caminal, G., Lopez-Santin, J. (1997) Influence of phosphate on rhamnose-containing exopolysaccharide rheology and production by *Klebsiella* I-714, *Appl. Microbiol. Biotechnol.* **48**, 522–527.

Fumio, I., Kimikazu, I., Satoshi, S. (1989) Polysaccharide having activity for multiplying *Bifidobacterium*. Japanese patent 01137990A.

Hadden, J. W. (1993) Immunostimulants, *Immunology Today* **14**, 275–280.

Harada, T. (1965) Succinoglucan 10C3: a new acidic polysaccharide of *Alcaligenes faecalis* var. *myxogenes*, *Arch. Biochem. Biophys.* **112**, 65–69.

Harada, T. (1977) Production, properties, and application of curdlan, in: *Extracellular Microbial Polysaccharides* (Sanford, P. A. Laskin, A., Eds.), Washington, DC: American Chemical Society, 265–283.

Harada, T., Yoshimura, T. (1965) Rheological properties of succinoglucan 10C3 from *Alcaligenes faecalis* var. *myxogenes*, *Agr. Biol. Chem.* **29**, 1027–1032.

Harada, T., Yoshimura, T., Hidaka, H., Koreeda, A. (1965) Production of a new acidic polysaccharide, succinoglucan by *Alcaligenes faecalis* var. *myxogenes*, *Agr. Biol. Chem.* **29**, 757–762.

Harada, T., Masada, M., Fujimori, K., Maeda, I. (1966) Production of a firm, resilient gel-forming polysaccharide by a mutant of *Alcaligenes faecalis* var. *myxogenes* 10C3, *Agr. Biol. Chem.* **30**, 196–198.

Harada, T., Misaki, A., Saito, H. (1968) Curdlan: a bacterial gel-forming β-1,3-glucan, *Arch. Biochem. Biophys.* **124**, 292–298.

Harada, T., Terasaki, M., Harada, A. (1993) Curdlan, in: *Industrial Gums* (Whistler, R. L., BeMiller, J. N., Eds.), San Diego, CA: Academic Press, Inc., 427–445.

Hideaki, Y., Chieko, M., Takeshi, A. (1993) Ganmodoki and its production. Japanese patent 05207859.

Hiroki, O., Etsuko, M. (1993) Noodle-like soybean protein-containing food. Japanese patent 05076290.

Hiroshi, S., Nobuyoshi, N., Yutaro, K., Toru, M. (1997) Medicine for treating dementia. Japanese patent 09255579.

Hiroshi, Y., Yuichi, S., Norio, I., Nana, I. (1999) Frozen dessert food and its production. Japanese patent 11187819.

Ielpi, L., Couso, R. O., Dankert, M. A. (1993) Sequential assembly and polymerization of the polyphenol-linked pentasaccharide repeating unit of the xanthan polysaccharide in *Xanthomonas campestris*, *J. Bacteriol.* **175**, 2490–2500.

Jagodzinski, P. P., Wiaderkiewicz, R., Kurzawski, G., Kloczewiak, M., Nakashima, H., Hyjek, E., Yamamoto, N., Uryu, T., Kaneko, Y., Posner, M. R., Kozbor, D. (1994) Mechanism of the inhibitory effect of curdlan sulfate on HIV-1 infection *in vitro*, *Virology* **202**, 735–745.

Janusz, M. J., Austen, K. F., Czop, J. K. (1986) Isolation of soluble yeast β-glucans that inhibit human monocyte phagocytosis mediated by β-glucan receptors, *J. Immunol.* **137**, 3270–3276.

Jong, S. C., Birmingham, J. M. (1993) Medicinal and therapeutic value of the Shiitake mushroom, *Adv. Appl. Microbiol.* **39**, 153–184.

Kai, A., Ishino, T., Arashida, T., Hatanaka, K., Akaike, Y., Matsuzaki, K., Kaneko, Y., Mimura, T. (1993) Biosynthesis of curdlan from culture media containing ^{13}C-labeled glucose as the carbon source, *Carbohydr. Res.* **240**, 153–159.

Kamisato, J. K., Nowakowski, M. (1988) Morphological and biochemical alterations of macrophages produced by a glucan, PSK, *Immunopharmacology* **16**, 88–96.

Kanegae, Y. Yutani, A., Nakatsui, I. (1996) Mutant strains of *Agrobacterium* for producing beta-1,3-glucan. US patent 5508191.

Kanke, M., Koda, K., Koda, Y., Katayama, H. (1992) Application of curdlan to controlled drug delivery. I. The preparation and evaluation of theophylline-containing curdlan tablets, *Pharmaceut. Res.* **9**, 414–418.

Kasai, N., Harada, T. (1980) Ultrastructure of curdlan. in: *Fiber Diffraction Methods*, (French, A. D., Gardner, K. H., Eds.), Washington, DC: American Chemical Society Symposium Series, 363–383.

Kazushi, M. (1999) Production of bean curd including ingredient. Japanese patent 11178533.

Ken, O., Akirou, M. (1992) Soybean protein-containing solid food. Japanese patent 04197148.

Kim, B.-S., Jung, I.-D., Kim, J.-S., Lee, J.-H., Lee, I.-Y., Lee, K.-B. (2000) Curdlan gels as protein drug delivery vehicles, *Biotechnol. Lett.* **22**, 1127–1130.

Kim, M.-K., Lee, I.-Y., Ko, J.-H., Rhee, Y.-H., Park, Y.-H., (1999) Higher intracellular levels of uridine monophosphate under nitrogen-limited conditions enhance metabolic flux of curdlan synthesis in *Agrobacterium* species, *Biotechnol. Bioeng.* **62**, 317–323.

Kim, M.-K., Lee, I.-Y., Lee, J.-H, Kim, K.-T., Rhee, Y.-H., Park, Y.-H. (2000) Residual phosphate concentration under nitrogen-limiting conditions regulates curdlan production in *Agrobacterium* species, *J. Ind. Microbiol. Biotechnol.* **25**, 180–183.

Kimura, H., Moritaka, S., Misaki, M. (1973) Polysaccharide 13140: a new thermo-gelable polysaccharide, *J. Food Sci.* **38**, 668–670.

Kojima, T., Tabata, K., Itoh, W., Yanaki, T. (1986) Molecular weight dependence of the antitumor activity of schizophyllan, *Agr. Biol. Chem.* **50**, 231–232.

Koreeda, A., Harada, T., Ogawa, K., Sato, S. Kasai, N. (1974) Study of the ultrastructure of gel-forming (1→3)-β-D-glucan (curdlan-type polysaccharide) by electron microscopy, *Carbohydr. Res.* **33**, 396–399.

Lawford, H. G., Rousseau, J. D. (1991) Bioreactor design considerations in the production of high-quality microbial exopolysaccharide, *Appl. Biochem. Biotechnol.* **28/29**, 667–684.

Lawford, H. G., Rousseau, J. D. (1992) Production of β-1,3-glucan exopolysaccharide in low shear systems, *Appl. Biochem. Biotechnol.* **34/35**, 597–612.

Lawford, H., Keenan, J., Phillips, K., Orts, W. (1986) Influence of bioreactor design on the rate and amount of curdlan-type exopolysaccharide production by *Alcaligenes faecalis*, *Biotechnol. Lett.* **8**, 145–150.

Lee, J.-H, Lee, I.-Y., Kim, M.-K., Park, Y.-H. (1999) Optimal pH control of batch processes for production of curdlan by *Agrobacterium* species, *J. Ind. Microbiol. Biotechnol.* **23**, 143–148.

Lee, I.-Y., Seo, W.-T., Kim, K.-G., Kim, M.-K., Park, C.-S., Park, Y.-H. (1997) Production of curdlan using sucrose or sugar cane molasses by two-step fed-batch cultivation of *Agrobacterium* species, *J. Ind. Microbiol. Biotechnol.* **18**, 255–259.

Lee, I.-Y, Kim, M.-K., Lee, J.-H., Seo, W.-T., Jung, J.-K., Lee, H.-W., Park, Y.-H., (1999a) Influence of agitation speed on production of curdlan by *Agrobacterium* species, *Bioprocess Eng.* **20**, 283–287.

Lee, I.-Y., Kim, S.-W., Lee, J.-H., Kim, M.-K., Cho, I.-S., Park, Y.-H. (1999b) A high viscosity of curdlan at alkaline pH increases segregational resistance of concrete, *Korean J. Biotechnol. Bioeng.* **14**, 1–5.

Maeda, I., Saito, H., Masda, M., Misaki, A., Harada, T. (1967) Properties of gels formed by heat treatment of curdlan, a bacterial β-1,3 glucan, *Agr. Biol. Chem.* **31**, 1184–1188.

Masaaki, A., Takaaki, I. (1995) Quality improver for wheat flour product and wheat flour product having improved texture. Japanese patent 07274876.

Masahiro, K., Masaru, N. (1993) Preparation of processed edible meat. Japanese patent 05288909.

Masahiro, K., Masaaki, K., Shinji, M., Hiroaki, K. (1998a) Immunoactivator. Japanese patent 10194977.

Masahiro, K., Mikitomo, A., Akitomo, A., Tomonori, Y. (1998b) Feed composition for fish kind cultivation. Japanese patent 10313794.

Masanori, T., Tetsuya, T., Yukihiro, N. (1992) Noodles. Japanese patent 04210507.

Masao, K., Toshimi, T. (1999) Jelly-like food. Japanese patent 11075726.

Masataka, M., Hideo, Y., Yukihiro, N. (1997) Devil's-tongue jelly and its production. Japanese patent 09075017.

Masatoshi, H., Masami, F., Kenichi, H., Hiromi, K. (1993) Devil's tongue for frozen food and its production. Japanese patent 05184310.

Masayuki, T., Yukihiro, N. (1990) Noodle made of rice powder and producing method thereof. Japanese patent 02249466.

Mikio, K., Yoshiro, O., Hirotomo, O., Yoshikazu, K., Hiroshi, I., Hisashi, M., Takeshi, K. (1995) Water soluble β-1,3-glucan derivative and antiviral agent containing the derivative. Japanese patent 07228601.

Misaki, A., Kishida, E., Kakuta, M., Tabata, K. (1993) Antitumor fungal (1→3)-β-D-glucans: structural diversity and effects of chemical modification. In: *Carbohydrates and Carbohydrate Polymers* (Yalpani, M., Ed.), Mount Prospect, IL: ATL Press, 116–129.

Murooka, Y., Yamada, T., Harada, T. (1977) Affinity chromatography of *Klebsiella* arylsulfatase on tyrosyl-hexamethylenediamine-β-1,3-glucan and immunoadsorbent, *Biochim. Biophys. Acta* **485**, 134–140.

Naganishi, I., Kimura, K., Kusui, S., Yamazaki, E. (1974) Complex formation of gel-forming bacterial (1→3)-β-D-glucan (curdlan type polysaccharide) with dyes in aqueous solution, *Carbohydr. Res.* **32**, 47–52.

Naganishi, I., Kimura, K., Suzuki, T., Ishikawa, M., Banno, I., Sakene, T., Harada, T. (1976) Demonstration of curdlan-type polysaccharide and some other β-1,3-glucan in microorganisms with aniline blue, *J. Gen. Appl. Microbiol.* **22**, 1–11.

Nakata, M., Kawaguchi, T., Kodama, Y., Konno, A. (1998) Characterization of curdlan in aqueous sodium hydroxide, *Polymer* **39**, 1475–1481.

Numasaki, Y., Kikuchi, M., Sugiyama, Y., Ohba, Y. (1990) A glucan Sizofiran: T cell adjuvant property and antitumor and cytotoxic macrophage inducing activities, *Oyo Yakuri Pharmacometrics* **39**, 39–48.

Ogawa, K., Watanabe, T., Tsurugi, J., Ono, S. (1972) Conformational behavior of a gel-forming (1→3)-β-D-glucan in alkaline solution, *Carbohydr. Res.* **23**, 399–405.

Ogawa, K., Tsurugi, J., Watanabe, T. (1973a) Effect of salt on the conformation of gel-forming β-1,3-D-glucan in alkaline solution, *Chem. Lett.* 95–98.

Ogawa, K., Miyagi, M., Fukumoto, T., Watanabe, T. (1973b) Effect of 2-chloroethanol, dioxane, or water on the conformation of a gel-forming β-1,3-glucan in DMSO, *Chem. Lett.* 943–946.

Ogawa, K., Tsurugi, J., Watanabe, T. (1973c) The dependence of the conformation of a (1→3)-β-D-glucan on chain length in alkaline solution, *Carbohydr. Res.* **29**, 397–403.

Ohno, N., Miura, N. N., Chiba, N., Adachi, Y., Yadomae, T. (1995) Comparison of the immunopharmacological activities of triple and single-helical schizophyllan in mice, *Biol. Pharm. Bull.* **18**, 1242–1247.

Ohno, N., Hashimoto, T., Adachi, Y., Yadomae, T. (1996) Conformation dependency of nitric oxide synthesis of murine peritoneal macrophages by β-glucans *in vitro*, *Immunol. Lett.* **52**, 1–7.

Okazaki, M., Adachi, Y., Ohno, N., Yadomae, T. (1995) Structure–activity relationship of (1→3)-β-D-glucans in the induction of cytokine production from macrophages, *in vitro*, *Biol. Pharm. Bull.* **18**, 1320–1327.

Okuyama, K., Obata, Y., Noguchi, K., Kusaba, T., Ito, Y., Ohno, S. (1996) Single structure of curdlan triacetate, *Biopolymers* **38**, 557–566.

Phillips, K. R., Lawford, H. G. (1983) Curdlan: Its properties and production in batch and continuous fermentations, in: *Progress Industrial Microbiology* (Bushell, D. E., Ed.), Amsterdam: Elsevier Scientific Publishing Co., 201–229.

Read, S. M., Currie, G., Bacic, A. (1996) Analysis of the structural heterogeneity of laminarin by electrospray-ionisation-mass spectrometry, *Carbohydr. Res.* **281**, 187–201.

Saito, H., Misaki, A., Harada, T. (1968) Comparison of structure of curdlan and pachyman, *Agr. Biol. Chem.* **32**, 1261–1269.

Saito, H., Ohki, T., Sasaki, T. (1977) A ^{13}C Nuclear Magnetic Resonance study of gel-forming (1 → 3)-β-D-glucans. Evidence of the presence of single-helical conformation in a resilient gel of a curdlan-type polysaccharide 13140 from *Alcaligenes faecalis* var. *myxogenes* IFO 13140, *Biochemistry* **16**, 908–914.

Saito, H., Yoshioka, Y., Uehara, N., Aketagawa, J., Tanaka, S., Shibata, Y. (1991) Relationship between conformation and biological response for (1 → 3)-β-D-glucans in the activation of coagulation factor G from limulus amebocyte lysate and host-mediated antitumor activity. Demonstration of single-helix conformation as a stimulant, *Carbohydr. Res.* **217**, 181–190.

Sakurai, T., Suzuki, I., Kinoshita, A., Oikawa, S., Masuda, A., Ohsawa, M., Tadomae, T. (1991) Effect of intraperitoneally administered β-1,3-glucan, SSG, obtained from *Sclerotinia sclerotiorum* IFO 9395 on the functions of murine alveolar macrophages, *Chem. Pharm. Bull.* **39**, 214–217.

Sasaki, T., Takasuka, N. (1976) Further study of the structure of lentinan, an anti-tumor polysaccharide from *Lentinus edodes*, *Carbohydr. Res.* **47**, 99–104.

Sasaki, T., Abiko, N., Sugino, Y., Nitta, K. (1978) Dependence on chain length of antitumor activity of (1 → 3)-β-D-glucan from *Alcaligenes faecalis* var. *myxogenes*, IFO 13140, and its acid-degraded products, *Cancer Res.* **38**, 379–383.

Shunsuke, O., Masatoshi, S. (1992) Production of uncooked Chinese noodle and boiled Chinese noodle. Japanese patent 04330256.

Spicer, E. J. F., Goldenthal, E. I., Ikeda, T. (1999) A toxicological assessment of curdlan. *Fd. Chem. Toxicol.* **37**, 455–479.

Stasinopoulos, S. J., Fisher, P. R., Stone, B. A., Stanisich, V. A. (1999) Detection of two loci involved in (1 → 3)-β-glucan (curdlan) biosynthesis by *Agrobacterium* sp. ATCC 31749, and comparative sequence analysis of the putative curdlan synthase gene, *Glycobiology* **9**, 21–31.

Sutherland, I. W. (1977) Microbial exopolysaccharide synthesis, in: *Extracellular Microbial Polysaccharides* (Sanford, P. A., Laskin, A., Eds.), Washington, DC: American Chemical Society, 40–57.

Sutherland, I. W. (1993) Biosynthesis of extracellular polysaccharides, in: *Industrial Gums* (Whistler, R. L., BeMiller, J. N., Eds.), San Diego, CA: Academic Press, Inc., 69–85.

Suzuki, T., Ohno, N., Saito, K., Yamdomae, T. (1992) Activation of the complement system by (1,3)-β-D-glucan having different degrees of branching and different ultrastructures, *J. Pharmacobiodyn.* **15**, 277–285.

Tada, T., Matsumoto, T., Masuda, T. (1997) Influence of alkaline concentration on molecular association structure and viscoelastic properties of curdlan aqueous systems, *Biopolymers* **42**, 479–487.

Taguchi, T., Hiroshi, K., Yukihiro, N. (1991) Production of linear gel. Japanese patent 03157401.

Takaaki, I. (1999) Quality improver for bean jam product and production of bean jam product. Japanese patent 11056246.

Takashi, T., Junji, K. (1991) Anti-retrovirus agent. Japanese patent 03218317.

Takeda technical report. (1997) Pureglucan: basic properties and food applications. Takeda Chemical Industries, Ltd. Japan.

Takeda-Hirokawa, N., Neoh, L. P., Akimoto, H., Kaneko, H., Hishikawa, T., Sekigawa, I., Hashimoto, H., Hirose, S.-I., Murakami, T., Yamamoto, N., Mimura, T., Kaneko, Y. (1997) Role of curdlan sulfate in the binding of HIV-1 gp120 to CD4 molecules and the production of gp120-mediated THF-α, *Microbiol. Immunol.* **41**, 741–745.

Toshio, O. (1986) Immobilized enzyme, or the like and preparation thereof. Japanese patent 61015688.

Toshiko, S. (1992) Sustained release suppository. Japanese patent 04074115.

Toshiyuki, U., Yoshio, T., Minoru, Y. (1995) Additive for concrete. Japanese patent 07010624.

Tsuneo, A., Koichi, K., Junji, K. (1990) Blood coagulation inhibitor. Japanese patent 02124902.

Williams, D. L., Pretus, H. A., Ensley, H. E., Browder, I. W. (1994) Molecular weight analysis of a water-insoluble, yeast-derived (1 → 3)-β-D-glucan by organic-phase size-exclusion chromatography, *Carbohydr. Res.* **253**, 293–298.

Yanaki, T., Ito, W., Tabata, K., Kojima, T., Norysuye, T., Takano, N., Fujita, H. (1983) Correlation between the antitumor activity of a polysaccharide schizophyllan and its triple-helical conformation in dilute aqueous solution, *Biophys. Chem.* **17**, 337–342.

Yasuhiro, S. (1994) Formed jelly-containing liquid food and its production. Japanese patent 06335370.

Yoichi, T., Hiroshi, S., Makiko, K. (1996) Biodegradable component for protecting ink jet recording head. Japanese patent 0818787.

Yoshioka, Y., Tabeta, R., Saito, R., Uehara, N., Fukuoka, F. (1985) Antitumor polysaccharides from *P. ostreatus* (Fr.) Quel: isolation and structure of a beta-glucan, *Carbohydr. Res.* **140**, 93–100.

Young, S.-H., Dong, W.-J., Jacobs, R. R. (2000) Observation of a partially opened triple-helix conformation in (1 → 3)-β-glucan by fluorescence resonance energy transfer spectroscopy, *J. Biol. Chem.* **275**, 11874–11879.

Yukihiro, N., Takahiro, F. (1993) Ground fish meat or fish or cattle paste product and composition therefor. Japanese patent 05023143.

Zhang, L., Ding, Q., Zhang, P., Zhu, R., Zhou, Y. (1997) Molecular weight and aggregation behaviour in solution of β-D-glucan from *Poria cocos* sclerotium, *Carbohydr. Res.* **303**, 193–197.

7 Succinoglycan

Dr. Miroslav Stredansky
Polytech, Area Science Park, Padriciano 99, 34012 Trieste, Italy;
Tel: +39-040-3756621; Fax: +39-040-9220016;
E-mail: miro@polytech3.area.trieste.it

CoA	coenzyme A
EPS	exopolysaccharide
HMW	high molecular weight
LMW	low molecular weight
NMR	nuclear magnetic resonance
T_m	melting temperature
UDP	uridine 5′-diphosphate

1 Introduction

Succinoglycan is an acidic extracellular heteropolysaccharide produced by several strains of bacteria belonging to the genera *Rhizobium, Agrobacterium, Alcaligenes,* and *Pseudomonas* (Harada and Harada, 1996; Zevenhuizen, 1997). Succinoglycan chains are made up of octasaccharide repeating units modified with pyruvate, succinate, and acetate substituents. The repeat unit is branched, and contains one galactose and seven β-linked glucoses (Jansson et al., 1977; Chouly et al., 1995). The biosynthetic pathway of succinoglycan and the identity of the genes encoding the enzymes involved in it have been elucidated on the model symbiotic strain *R. meliloti* Rm 1201 and its mutants (Reuber and Walker, 1993a; Becker et al., 1993a; Gonzalez et al., 1996). The biosynthesis consists of the synthesis of nucleotide sugar precursors, stepwise assembly of the octasaccharide subunit, addition of decorations (succinate, pyruvate, and acetate), and finally polymerization followed by a transport to the cell surface.

Succinoglycan was found to play an important role in bacterium–plant symbiotic interactions (Long, 1989; Gonzalez et al., 1996). It was the first exopolysaccharide (EPS) to be recognized as being crucial for inducing the formation of nitrogen-fixing nodules on the roots of their leguminous host plants and it was considered to be a likely signal molecule. On the other hand, interesting rheological properties of succinoglycan together with satisfactory production processes led to its industrial applications as a thickening, suspending, emulsion-stabilizing, gel-forming, and precipitation agent (Sutherland, 1990).

2 Historical Outline

During studies on bacteria utilizable petrochemicals, Harada and Yoshimura (1964) isolated the strain *Alcaligenes faecalis* var. myxogenes, which grew on ethylene glycol and produced an acidic EPS containing succinate. The EPS was firstly named succinoglucan (Harada et al., 1965). Later the name was changed to succinoglycan because this polymer was found to contain also a minor portion of galactose besides glucose (Misaki et al., 1969). Subsequently some strains of the genera *Rhizobium, Agrobacterium,* and *Pseudomonas* were found to produce succinoglycans (Björndal et al., 1971; Zevenhuizen, 1971, 1997). The structure of the biopolymer was intensively studied and elucidated (Jansson et al., 1977; Åman et al.,

1981; Matulova et al., 1994; Chouly et al., 1995). Succinoglycans from all mentioned sources are composed of polymerized branched octasaccharide subunits, each of which consists of one galactose and seven glucoses (β-linked), and carries succinyl, pyruvyl, and acetyl modifications. Pyruvate is present in a stoichiometric ratio; the number and type of succinyl and acetyl groups vary, and depend on the bacterial origin and culture conditions. Only the biopolymer with this structure is called succinoglycan, although there are other polysaccharides containing succinate, e.g. marginalan (Osman and Fett, 1989). The biosynthetic pathway of succinoglycan and its genetic base have been fully determined (Reuber and Walker, 1993a; Gonzalez et al., 1996).

Succinoglycan is one of the best-understood symbiotically important EPSs. It is required for the successful invasion of nodules by bacteria in several *Rhizobium*–legume symbioses (Leigh and Walker, 1994). In addition to its great biological importance, succinoglycan has been produced industrially, because of its useful applications as an emulsion-stabilizing, suspending, and thickening agent (Sutherland, 1990). Various fermentation processes, such as batch, fed-batch, continuous, and solid state, have been developed for its production (Linton et al., 1987a; Stredansky and Conti, 1999a; Stredansky et al., 1999a; Knipper et al., 2000).

3 Chemical Structure

Succinoglycans from different sources have a common structure unit, a branched octasaccharide consisting of one galactose and seven glucoses with succinyl, pyruvyl, and acetyl modifications. The structure (Figure 1) was elucidated by various methods, such as sequential degradation of methylated saccharides followed by gas–liquid chromatography or high-performance liquid chromatography-mass spectroscopy (Jansson et al., 1977; Åman et al., 1981), enzymatic hydrolysis followed by paper or gas chromatography (Amemura and Harada,

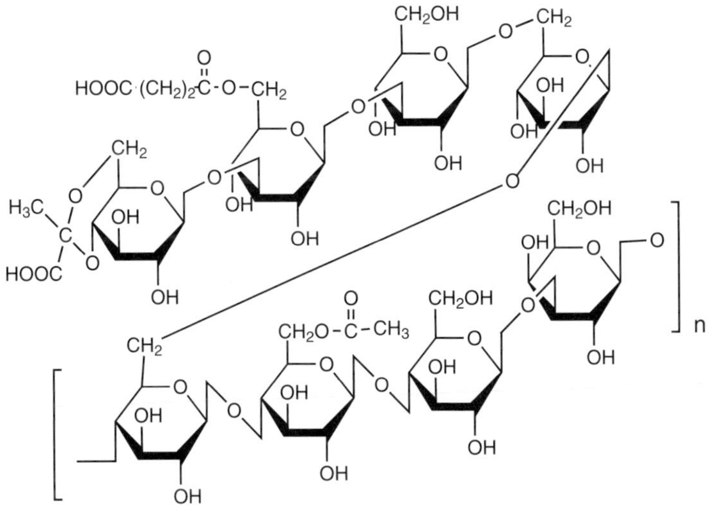

Fig. 1 Chemical structure of succinoglycan from *R. meliloti*.

1983), one- or two-dimensional nuclear magnetic resonance (NMR) studies (Matulova et al., 1994; Chouly et al., 1995; Evans et al., 2000), tandem mass spectroscopy with electrospray ionization, and collision-induced dissociation (Reinhold et al., 1994). The polymer chains have a neutral backbone with a regularly positioned tetrasaccharide side chain composed of β-D-glucose residues. The side chain, which contains pyruvate acetal linked at positions O^4 and O^6 on the terminal glucose residue, is connected to the main chain via a $\beta(1 \rightarrow 6)$ glycosidic linkage. Other glycosidic bonds in succinoglycan are $\beta(1 \rightarrow 3)$ and $\beta(1 \rightarrow 4)$. All hexosyl residues were determined to be β-pyranosidic from the low optical rotation and ^{1}H-NMR spectrum (Jansson et al., 1977). Each octasaccharide repeating unit (Figure 1) is further substituted by *O*-succinyl (side chain) and *O*-acetyl (main chain) groups. Their location was determined by detailed NMR studies (Matulova et al., 1994; Chouly et al., 1995) and by mass spectroscopy (Reinhold et al., 1994). The content of succinate and acetate per repeating unit is less than 1, and varies with the source. Molecular weight of succinoglycan is in the range from several hundred thousand to several million Daltons, and strongly depends on the used strain and culture conditions (Meade et al., 1995; Burova et al., 1996; Ridout et al., 1997; Stredansky et al., 1998).

4 Chemical Analysis and Detection

The detection and quantification of succinoglycan is important for both research work and monitoring production processes. However, there are no specific methods for succinoglycan determination. The only exception is a Calcofluor method used for the detection of succinoglycan-producing colonies of *Rhizobium* growing on agar plates (Leigh et al., 1985). The ability of succinoglycan to bind laundry whitener Calcofluor and fluoresce under UV light greatly facilitated subsequent genetic analysis of succinoglycan's biosynthesis, and its symbiotic roles, regulation, and degradation (Leigh and Walker, 1994). This method has never been applied for quantitative succinoglycan determination. Thus classical methods based on the EPS precipitation from cell-free supernatant with organic solvents (isopropanol, acetone, and ethanol) or quaternary ammonium compounds followed by gravimetry have been most frequently used (Meade et al., 1995; Linton et al., 1987a; Navarini et al., 1997; Stredansky et al., 1998). The precipitation with quaternary ammonium compounds is more specific for acidic EPS. The colorimetric determination of succinoglycan after the anthrone–sulfuric acid reaction was also reported (Zevenhuizen and Faleschini, 1991; Glucksmann et al., 1993b). The quantitative determination of nonsaccharide substituents was performed either directly by colorimetry (Zevenhuizen and Faleschini, 1991) and by NMR analysis (Matulova et al., 1994) or after mild hydrolysis of the EPS. Acetate and succinate are liberated by alkali hydrolysis and pyruvate by acidic hydrolysis (Reuber and Walker, 1993a). Subsequently, they could be determined by well-known enzymatic, spectrophotometric, or chromatographic methods.

5 Occurrence

Succinoglycan is produced and secreted by soil bacteria belonging to the genera *Rhizobium, Agrobacterium, Pseudomonas,* and *Alcaligenes,* which also often produce other EPSs (Harada and Harada, 1996; Zevenhuizen, 1997). Harada and Yoshimura (1964) isolated the soil bacterium *A. faecalis* var. myxogenes 10C3, which grew on ethylene glycol and produced succinoglycan. The production was significantly improved when sugars were used as a carbon source (Harada et al., 1965). The strain was later reclassified as belonging to the genera *Agrobacterium* (Yokota and Sakane, 1991). Thus the ability of *Alcaligenes* strains to produce succinoglycan is questionable, because no other strains of this genera were reported to synthesize it. On the other hand, a number of strains of *Rhizobium*, *Agrobacterium*, and *Pseudomonas* were found to produce succinoglycan. In the genera *Rhizobium* these include *R. meliloti* (Björndal et al., 1971), *R. trifolii*, *R. phaseoli*, *R. lupini* (Amemura and Harada, 1983), and *R. hedysari* (Navarini et al., 1997). Succinoglycan of *Rhizobium* strains is involved in symbiotic processes with legumes and is also termed EPS I or EPS a. In the genera *Agrobacterium* these include *A. radiobacter* (Harada et al., 1979; Linton et al., 1987a) and *A. tumefaciens* (Zevenhuizen, 1973; Stredansky et al., 1998). The *Pseudomonas* strains producing succinoglycan have been not fully classified (Cripps et al., 1981; Meade et al., 1995; Zevenhuizen, 1997).

6 Biological Function

A number of *Rhizobium* strains are capable of inducing the formation of nitrogen-fixing nodules on the roots of their leguminous host plants. This process involves specific recognition and progressive differentiation of both bacterial and host cells, where bacterial EPS play an important role. The *Rhizobium*–legume symbiosis is of a great economic importance, since legumes are widely cultivated as food or forage crops and actively used in agricultural rotations.

Nodulation of the leguminous plants by *Rhizobium* is a complex developmental process that requires a series of signal exchanges between the host and bacteria (Long, 1989). Plant flavonoids and bacterial Nod factors are among the best characterized signals involved in this interorganismal communication process. These specific flavonoids are secreted from the roots and induce Nod factor production by bacteria (van Rhijn and Vanderleyden, 1995). The Nod factor then induces root hair curling and initiates nodule development (Denarie and Cullimore, 1993). These are colonized by *Rhizobium* and form infection threads filled with bacterial cells. The infection threads elongate inside the root hairs, enter the root cortex, and release bacterial cells into newly formed plant cells in the nodule primordia. The bacterial cells then differentiate into bacteroids and fix nitrogen inside those plant cells (Kijne, 1992). In order to invade the nodules they elicit on alfalfa roots, *R. meliloti* cells must be able to synthesize at least one of the following polysaccharides: succinoglycan (EPS I), galactoglucan (EPS II), or a particular capsular polysaccharide (Leigh and Walker, 1994; Reuhs et al., 1995).

Succinoglycan was the first such EPS to be recognized as being crucial for bacterial invasion of nodules and was considered to be a likely signal. Although nodules formed in response to *exo* mutants (not producing succinoglycan), indicating that succinoglycan was neither the initiating signal nor the primary determinant of host specificity, they remained uncolonized, indicating that succinoglycan was required for nodule invasion (Leigh et al., 1985). Invasion appeared to be blocked at the stage where the infection thread penetrates beyond the epidermal cell layer. The structural features of the EPS required for nodule invasion suggest that its role is more specific than previously attributed to bacterial EPS. The first clue to this specificity was the observation that *exoH* mutants, which produce succinoglycan that lacks the succinyl moiety, have the same invasion phenotype as *exo* mutants that produce no succinoglycan: they fail to invade nodules (Leigh and Walker, 1994). The succinoglycan produced by *exoH* mutants also lacks the low-molecular-weight (LMW) fraction produced by the wild strain. This might be because the succinyl moiety is required for degradation of the EPS to LMW forms (York and Walker, 1998a). Thus either succinylated or LMW succinoglycan is required for stimulating the infection thread in the plant (Battisti et al., 1992; Cheng and Walker, 1998a). On the contrary, the acetyl substituent is not required (Reuber and Walker, 1993b). The specificity is further supported by the fact that the deficiency of *exo* mutants in nodule invasion can be partially suppressed by adding homologous EPS to the plant root at the time of inoculation (Djordjevic et al., 1987; Battisti et al., 1992; Urzainqui and Walker, 1992). It has been shown for three different *Rhizobium*–plant pairs that wild-type EPS from the usual nodulating strain will partially restore invasion of an *exo* mutant strain, while EPS from another species will not. Furthermore, invasion of alfalfa nodules by *exo* mutants of *R. meliloti* requires a particular LMW succinoglycan which is a trimer of the repeating subunit having succinyl and pyruvyl substituents (Gonzalez et al., 1998). The LMW succinoglycan is produced either by its direct biosynthesis or by expressing glycanases that cleave nascent HMW succinoglycan (York and Walker, 1998a).

The biological function of succinoglycan in *Pseudomonas* and *Agrobacterium* is unknown. It is apparently unnecessary in *A. tumefaciens*–plant pathogenic processes (Leigh and Coplin, 1992). However, the production and function of succinoglycan in bacterial ecosystems should not be restricted to a mere specific role during symbiosis of *Rhizobium* bacteria with their appropriate host plant; it points to a much more general functioning in bacterial systems, such as water-holding capacity, adhesive properties in bacterial flocs and biofilms of water organisms, and in aiding soil aggregation and stability (Zevenhuizen, 1997).

Tab. 1 The genes involved in the succinoglycan biosynthesis

Exo genes	*Function of encoded protein(s)*	*Reference*
C	phosphoglucomutase	Uttaro et al., 1990
N	UDP-pyrophosphorylase	Glucksmann et al., 1993b
B	UDP-glucose-4-epimerase	Buendia et al., 1991
Y, F	transfer of galactose onto the lipid carrier	Gray et al., 1990; Müller et al., 1993
A, L, M, O, U, W	glucosyl transferases	Glucksmann et al., 1993a
Z	acetylation	Buendia et al., 1991
H	succinylation	Becker et al., 1993b
V	pyruvylation	Glucksmann et al., 1993b
P, Q, T	polymerization/export of succinoglycan	Glucksmann et al., 1993b, Müller et al., 1993
K	cleavage of HMW succinoglycan	Becker et al., 1993b
X, R, S	regulation of biosynthesis	Gray et al., 1990; Reed et al., 1991a; Cheng and Walker, 1998b

7 Biosynthesis

The biosynthetic pathway of succinoglycan and the identity of the genes encoding the enzymes involved in it (Table 1) have been studied intensively on the model symbiotic strain *R. meliloti* Rm 1201 and its mutants (Reuber and Walker, 1993a; Becker et al., 1993a; Gonzalez et al., 1996). The pathway in strains of *A. tumefaciens* looks to be very similar (Cangelosi et al., 1987). Moreover many biosynthetic steps, enzymes, and gene sequences involved in succinoglycan synthesis showed a similarity to their counterparts in the biosynthesis of other bacterial EPSs (Reuber and Walker, 1993a; Coplin and Cook, 1990). Succinoglycan biosynthesis consists of the synthesis of sugar precursors, stepwise assembly of the octasaccharide subunit, addition of substituents, and finally polymerization followed by transport to the cell surface.

7.1 Biosynthetic Pathway

Succinoglycan is synthesized by a set of more than 20 *exo* gene products (Glucksmann et al., 1993a,b; Becker et al., 1995). The assembly of polymer subunits takes place on lipid carriers on the cytoplasmic face of the cytoplasmic membrane and polymerization takes place on the periplasmic side (Tolmasky et al., 1982; Reuber and Walker, 1993a). Sugar precursors, acetyl-coenzyme A (CoA), succinyl-CoA, and phosphoenolpyruvate are derived from the glucose catabolism pathway (Entner–Doudoroff and Krebs cycle) (Dussap et al., 1991). The scheme of succinoglycan synthesis is illustrated in Figure 2. The *exoC* gene encoding for phosphoglucomutase is the only one located on the chromosome. *exoN* protein is 42% identical to the *Acetobacter xylinum celA* gene product, which codes for a uridine 5′-diphosphoglucose (UDP)-pyro-

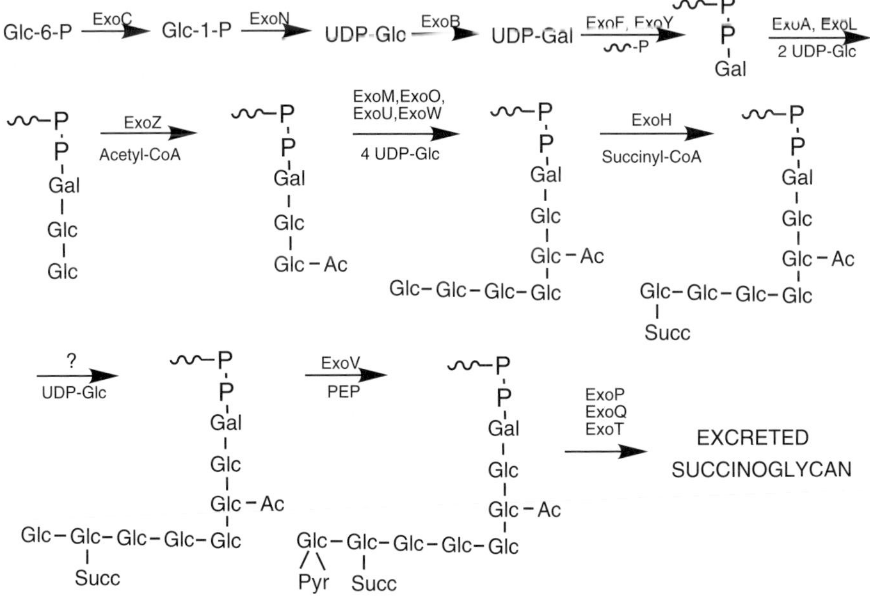

Fig. 2 Proposed model of the succinoglycan biosynthetic pathway (Reuber and Walker, 1993a; Gonzalez et al., 1996).

phosphorylase catalyzing formation of UDP-glucose (Glucksmann et al., 1993b). *exoB* functions as a UDP-glucose-4-epimerase (Buendia et al., 1991). The *exoY*, galactosyl-1-P-transferase, and *exoF* gene products are required for the transfer of galactose onto the lipid carrier. The *exoA, exoL, exoM, exoO, exoU,* and *exoW* gene products showed glucosyltransferase activities (Glucksmann et al., 1993a). The *exoZ* and *exoH* genes code for transmembrane proteins, which are needed for acetyl and succinyl modification, respectively. The *exoV* product is a ketalase responsible for transferring pyruvate to the terminal glucose (Reuber and Walker, 1993a; Glucksmann et al., 1993b). The membrane-associated proteins *exoP, exoQ,* and *exoT* are involved in polymerization and export of succinoglycan, where the *exoQ* produces high-molecular-weight (HMW) and the *exoT* produces LMW polymer (Gonzalez et al., 1996; Becker and Pühler, 1998). Another way to produce LMW succinoglycan, which is important in *Rhizobium*–legume interactions, is the cleavage of HMW succinoglycan by the *exoK* gene product showing endo-β-glycanase activity (Becker et al., 1993b).

7.2 Genetic Basis of Biosynthesis

The work of Gonzalez et al. (1996) showed that succinoglycan is synthesized by a set of 22 *exo* gene products, with 19 of them clustered in a 25-kb region on the *R. meliloti* megaplasmid 2 (Figure 3). Becker et al. (1995) extended it and showed that 21 *exo* and two *exs* genes are located in a 27-kb gene cluster on this megaplasmid. The *exp* gene cluster, responsible for the synthesis of EPS II, (galactoglucan) is localized on the same megaplasmid, separated from the *exo* cluster by about 200 kb (Charles and Finan, 1991). The DNA sequence of the *exo* region has been determined. Functional and evolutionary relatedness of genes for EPS synthesis in various *Rhizobium* strains were found (Zhan et al., 1990; Leigh and Coplin, 1992), and similarity to some genes encoding EPS in other bacteria is also apparent (Coplin and Cook, 1990).

7.3 Regulation of Biosynthesis

The succinoglycan synthesis in *Rhizobium* is regulated at both the transcriptional and posttranscriptional levels. The EPS production is regulated in part by the *exoS, chvI,* and *exoR* (a negative regulatory locus) genes located on the chromosome (Cheng and Walker, 1998b; Reed et al., 1991a). *exoS* and *chvI* proteins constitute a two-component regulatory system involved in controlling *exo* gene expression. *exoS* is the membrane sensor and *chvI* is the response regulator. The negative regulatory nature of *exoR* is supported by the existence of several independent insertion mutations, all of which caused increased EPS synthesis, and by the recessive nature of the mutations. Alkaline phosphatase activities of *exo–phoA* fusions indicate that *exoR* regulates the transcription or translation of most of the genes of the *exo*

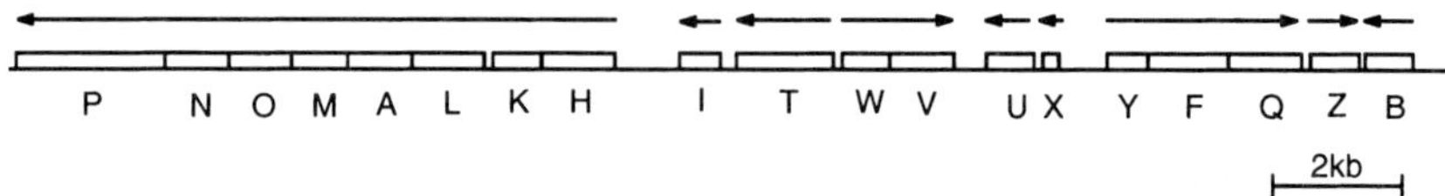

Fig. 3 Map of the *exo* gene cluster (Glucksmann et al., 1993b; Gonzalez et al., 1996).

cluster, with the exception of *exoB* (Reuber et al., 1991).

Another regulatory mechanism, mediated by the *exoX* and *exoY* genes, is conserved among several species of *Rhizobium* (Reed et al., 1991b; Zhan and Leigh, 1990). A multicopy of the small protein encoded by *exoX* inhibits EPS synthesis posttranscriptionally, because the gene dosage of *exoX* does not alter the expression of translational fusions to other *exo* genes. The inhibitory effect of *exoX* is counterbalanced in a copy number-dependent manner by *exoY*, which stimulates EPS synthesis.

The recently identified chromozomal *mucR* gene from *Rhizobium* codes for the regulatory protein that plays a key role in the control of the biosynthesis of both EPSs important in the nodulation process, i.e. succinoglycan and galactoglucan (Keller et al., 1995). The encoded protein exerts a positive effect on the transcription of the *exoK*, *exoY*, and *exoF* genes.

Succinoglycan synthesis has been shown to be regulated both in the free-living stage and during symbiosis. In the free-living stage, EPS production by *R. meliloti* is sensitive to the concentration of ammonia, phosphate, and sulfate present in the growth media. For the symbiotic stage, gene fusions have been used to show that the *exo* genes are expressed only in the invasion zone inside the nodule, not in the more mature nodule regions (Reuber et al., 1991; Mendrygal and Gonzalez, 2000).

8 Biodegradation

Succinoglycan is a fully biodegradable biopolymer. A partial cleavage of HMW to LMW succinoglycan in *R. meliloti* plays an important role in the nodulation process (York and Walker, 1998b). The *exoK* gene encodes a protein employed in this cleavage. This gene was sequenced and its high homology to β(1,3)–β(1,4)-glucanases genes present in various bacteria was found (Becker et al., 1993b; Glucksmann et al., 1993b).

In addition to succinoglycan-producing bacteria, other microorganisms also showed ability to degrade this polymer and enzymes from various sources were used in the structure elucidation (Harada et al., 1979; Amemura and Harada, 1983). *Cytophaga arvensicola* was capable growing on succinoglycan as the sole carbon source by means of extracellular succinoglycan depolymerase and intracellular endo-(1 → 6)β-D-glucanase. The former enzyme cleaves the polymer to octasaccharide repeating units and the latter enzyme hydrolyzes the octasaccharide to two tetrasaccharides. These could be further hydrolyzed by other enzymes of this bacterium or by glycosidases from other various sources to form oligo- and monosaccharides (Harada and Harada, 1996).

9 Solution and Rheological Properties

Succinoglycan is a polyelectrolyte and exhibits reversible pseudoplastic behavior in aqueous solutions. It does not form gels, but gives rise to extremely viscous solutions characterized under proper conditions by a viscoelastic behavior typical of 'weak gels'. This behavior is exhibited by worm-like polysaccharides in aqueous solution at sufficiently high concentration and/or molecular weight and under conditions of an ordered conformation. The rheological properties are sensitive to the bacterial origin of the EPS, and hence to the level and type of noncarbohydrate substituents (Gravanis et al., 1990b). The viscosity of succinoglycan solutions is affected by the ionic strength, shear rate, nature of counterions, concentration, and molecular weight of the polymer (Gravanis et al., 1990b; Ridout et al., 1997). Succinoglycans show a sharp order–disorder transition on heating and cooling. The conformational transition was studied by various techniques, such as NMR, optical rotation, differential scanning calorimetry, conductivity and viscosity measurements, and atomic force microscopy (Dentini et al., 1989; Gravanis et al., 1990a; Burova et al., 1996; Ridout et al., 1997; Balnois et al., 2000). The ordered form is promoted by lowering the temperature and/or raising the ionic strength. Variation of the transition temperature (T_m, melting temperature) with ionic strength is consistent with the behavior expected for a single helix to stretched coil transition. Burova et al. (1996) showed the single helix–coil type of transition mechanism at low EPS concentrations in salt-free solution, and with increasing EPS and/or salt concentrations the transition included two stages: the cooperative dissociation of the helix dimer and subsequent two-state melting of the helix monomer. The reported T_ms are in the range from 60 to 75 °C depending on the ionic strength, EPS composition (presence or absence of acyl groups), and molecular weight. T_m could also be controlled by addition of various salts and alcohols, in particular polyols (Lau and Davies, 1992). The viscosity of succinoglycan samples falls on heating to above the T_m and increases on subsequent recooling. The first thermal cycle results in an irreversible decrease in viscosity as measured under ambient conditions and it is accompanied by a decrease in molecular weight (Dentini et al., 1989; Gravanis et al., 1990b; Stredansky et al., 1999a). Such changes have been attributed to chain breakage and/or disruption of aggregates. No further reductions occur on subsequent thermal recycling.

10
Biotechnological Production

Succinoglycan is a subject of commercial interest, thus various processes, such as batch, fed-batch, continuous, and solid-state fermentation, have been developed for its production using strains of the genera *Rhizobium*, *Agrobacterium*, and *Pseudomonas* (Cripps et al., 1981; Azoulay, 1983; Linton et al., 1987a; Stredansky and Conti, 1999a; Stredansky et al., 1999a; Knipper et al., 2000).

During the course of fermentation, the excretion of EPS results in a highly viscous shear-thinning broth. The rheological behavior of the viscous broth causes serious problems with regard to mixing, heat transfer, and oxygen supply, thus limiting both the maximum polymer concentration achievable and product quality. As in the case of many microbial EPS production processes, a specific design of bioreactors, in particular impeller systems, is necessary for improving the bioprocess performance. The design of the agitation system requires special attention to the correct distribution of power to ensure good culture homogeneity and turbulence, to minimize bubble coalescence, to promote small bubble formation, and to achieve adequate fluid movement at heat transfer surfaces (Margaritis and Pace, 1985).

Control of the cultivation medium composition, its addition, and other environmental parameters are critical in achieving the desired rates of synthesis and yields of succinoglycan. The media suited for succinoglycan production are relatively poor with a high carbon/nitrogen ratio (Linton et al., 1987a; Stredansky et al., 1999a; Knipper et al., 2000). The best carbon sources are sugars (glucose, sucrose, and maltose) and alditols (mannitol and sorbitol) in the concentration range from 1 to 5% with 50–70% conversion into polymer. A high sugar concentration had an inhibitory effect (Stredansky et al., 1998). Nitrogen sources, such as ammonium salts, nitrates, glutamate, and lysine, are used mostly as growth-limiting factors, thus the excess energy deriving from carbon substrate metabolism is committed to EPS synthesis. A low level of minerals and other growth factors is also important. The optimum cultivation temperature for succinoglycan production by most strains is around 30 °C. Maintenance of a neutral pH during the process is required. Oxygen availability depends on the performance of aeration and agitation systems (discussed above), and showed a significant effect. Under nonlimited oxygen conditions a better yield of succinoglycan with a HMW was reached (Stredansky et al., 1998, 1999a).

10.1
Continuous Process

The major advantages of the continuous process are a high production rate and overall bioreactor productivity. On the other hand, some problems arise connected with poor long-term stability of the strain, higher contamination risk, and diluted product, which increases downstream costs. The process is easily controlled through a single limiting nutrient and the dilution rate. Detailed studies of the continuous production of succinoglycan by *A. radiobacter* were reported by Linton and his coworkers (Linton et al., 1987a,b; Cornish et al., 1987). The maximum observed yield of EPS was 3.5 g/g O_2 and 0.67 g/g glucose in nitrogen-limiting chemostat culture at dilution rate 0.047 h^{-1} (steady-state) with a succinoglycan concentration of 9.8 g/L. When the observed yields were corrected for cell production, they came very close to the theoretical values. The yields of EPS decreased sharply when the carbon source was either more reduced or oxidized than glucose.

10.2
Batch and Fed-Batch Fermentation

The batch process is simpler than the continuous one; in particular, when performed on a large scale, the risk of strain instability and culture contamination is lower, and a higher final product concentration could be reached. The batch process looks to be favorable in the case of succinoglycan, which is formed prevailingly in the non-growth phase of cultivation. A rapid period of cell growth is followed by a stationary phase, during which the product is generated (Linton et al., 1985; Stredansky et al., 1998). Draw fill techniques may be applied to the batch process to increase overall bioreactor productivity. The best reported succinoglycan yields per substrate in batch process are high, more than 60%. On the other hand, the highest volumetric yields obtained in laboratory bioreactors are about 16 g/L (Dasinger and McArthur, 1992; Meade et al., 1995; Knipper et al., 2000), i.e. lower than those reported for some other EPS (Margaritis and Pace, 1985). However, improved yields could be expected in large-scale bioreactors with a suitable design and/or substrate feeding. Stredansky et al. (1999a) reported an increase of the succinoglycan yield of 47% using a simple fed-batch process, as compared to the batch one. The batch production allows also utilization of less-defined cheap substrates, e.g. wood hydrolyzates (Meade et al., 1995). The reported cultivation times are 3–4 days.

10.3
Solid-state Fermentation

Solid-state fermentation has been used for many centuries in Asian and African countries for the production of fermented food. It is a process in which microorganisms grow on a moist substrate in the absence of free water and which allows utilization of cheap raw materials and residues of agro and food industries as substrates (Pandey, 1992). Production of succinoglycan by *Agrobacterium* and *Rhizobium* strains and production of other EPS, e.g. xanthan, on spent malt grains impregnated with a nutrient solution have been recently reported (Stredansky et al., 1999b,c, Stredansky and Conti, 1999a,b). The solid-state fermentation process overcomes the problems connected to the highly viscous broth typical of the submerged cultivation. The highest succinoglycan yield was achieved with *A. tumefaciens* in a 4-day static cultivation, reaching 42 g/L of impregnating solution, corresponding to 30 g/kg of total wet solid substrate (Stredansky and Conti, 1999a). The increased EPS production in a solid-state fermentation system could be partially accounted for by a higher tolerance of cells to a high sugar concentration. The enhanced metabolic activity observed in microorganisms growing in solid-state fermentation might be stimulated by the formation of a concentration gradient of nutrients, causing a local drop in substrate concentration, which significantly minimizes catabolite repression. However, many scale-up problems of the solid-state fermentation process have to be solved before a laboratory scale process can be transferred to the large-scale level.

10.4 Recovery and Purification

The cost of recovery of EPS is a significant part of the total production cost due to the dilute nature of the stream leaving the bioreactor, the presence of cells and solutes in the stream, and its high viscosity. Cell removal is essential if succinoglycan is to be used not only in cosmetics or foods, but also in enhanced oil recovery where the presence of solids in the polymer flood causes plugging of the pores of the oil-bearing rock. The high viscosity of the fermentation broth results in conventional separation steps such as centrifugation, filtration, and flocculation being less effective and slow. However, some advantages may be derived during processing by heating the broth above the T_m of succinoglycan (60–75 °C) to lower its viscosity, but care must be taken not to degrade the product (Gravanis et al., 1990b, Stredansky et al., 1999a).

To obtain a low-grade product, the cell-free supernatant could be concentrated and/or dried; to obtain a high-grade product, the EPS must be precipitated. Suitable precipitation agents are organic solvents, such as isopropanol (Linton et al., 1987a), acetone (Navarini et al., 1997) and ethanol (Stredansky et al., 1998), and or quaternary ammonium compounds (Meade et al., 1995). Increasing the concentration of EPS and salts prior to precipitation will decrease the amount of solvent processed per unit weight of polymer recovered. The precipitate is subsequently dewatered, dried, and milled to a suitable particle size. A simple and effective isolation procedure was described by Stredansky et al. (1999a). The whole fermentation broth was heated to 90 °C after fermentation. Cell removal, deodorization, and decoloration of the broth were simultaneously achieved by addition of activated charcoal (1% w/v), followed by stirring and filtration through a layer of washed kieselguhr. The warm filtrate could be directly dried or concentrated without cooling to save additional energy costs. As the fermentation process led to almost complete substrate utilization and the medium was poor in salts, succinoglycan of a high purity was obtained.

10.5 Patents and Commercial Products

The production of succinoglycan with various strains of genera *Rhizobium*, *Agrobacterium*, and *Pseudomonas* is covered by a number of patents (Table 2).

Several succinoglycan preparations have been produced industrially and introduced into the market. The company Shell marketed two products under the trademarks Shellflo-S and Enorflo-S produced with the strains *Pseudomonas* sp. NCIB 11592 and *A. radiobacter* NCIB 11883 (Clarke-Sturman et al., 1986; Linton et al., 1987a). Succinoglycan (a powder form) from *A. tumefaciens* I-736 modified by partial hydrolysis of acidic groups is actually marketed by the company Rhodia under the trademark Rheozan (around US$25/kg). The Halliburton Company offers the liquid concentrate of succinoglycan under the trademark FLO-PAC (about US$90/gallon). All these products are claimed for applications mainly in the oil industry.

11 Applications

Succinoglycan is used as a thickening, suspending, emulsion-stabilizing, gel-forming, and precipitation agent. The main field of its application is in the oil industry, where various synthetic and natural polymers are employed in various phases of the oil production. Succinoglycan possesses a combination of desirable properties for the use in this field: shear-thinning rheology, high stability at acidic and neutral pH, compatibility with anionic and nonionic surfactants, ease of mixing, temperature-insensitive viscosity below its T_m, and adjustable T_m over a wide range of temperature. It is often applied as a component of drilling fluids in the drilling of exploratory or production wells, when its lower T_m than that of xanthan is required (Linton et al., 1985; Lau and Davies, 1992; Aubert et al., 1999). Succinoglycan applications in replacement fluids used in enhanced oil recovery have been also patented (Cripps et al., 1981; Azoulay, 1983; Linton et al., 1985; Lau and Davies, 1992; Sydansk, 1992). The biopolymer-containing fluids and foam fluids are utilized for completion, workover, and kill operations in wells, and in profile modifications (Dasinger and McArthur, 1992; Rae and Johnston, 1995; Sydansk, 1998; Carpenter et al., 1998). Succinoglycan is particularly suitable in completing operations such as gravel packing, because it is effective in situations where cheaper hydroxyethylcellulose and xanthan are not (Lau, 1993; Lau and Bernardi, 1993). Gravel packing is the most commonly used method of sand control.

In addition to oil production, succinoglycan is used also in other fields. It has been employed in the manufacture of cosmetics, e.g. shampoos, creams, emulsions, and hair

Tab. 2 List of patents on the succinoglycan production

Patent	*Publication date*	*Patent holder*	*Inventors*
EP 040445A1	25 November 1981	Shell	R. E. Cripps, R. N. Ruffel and A. J. Sturman
US 4400466	23 August 1983	Dumas & Inchnuspe	E. E. Y. Azoulay
EP 0138255A2	24 April 1985	Shell	J. D. Linton, M. W. Evans and A. R. Godley
EP 0251638B1	22 April 1992	Pfizer	B. Dasinger and , H. A. I. McArthur
IT 1293620	8 March 1999	Polytech	M. Stredansky, E. Conti and C. Bertocchi
EP 0527061B1	10 May 2000	Rhodia Chimie	M. Knipper, A. Senechal and M. Raffart

dye compositions, as a thickener, stabilizer, and humectant (Guerin et al., 1998; Yoshida, 1998; Yoshida and Suzuki, 1998; Yoshida, 1999). Succinoglycan use in the production of detergents (Guerin and Knipper, 1991; Guillou, 1995), inks (Kondo and Matsubara, 1995; Kitaoka, 1998; Sugimoto and Yamamoto, 1999), and agricultural compositions (Philips et al., 1991; Macri et al., 1996) was described. Its application as a precipitation or stabilizing agent in the manufacturing processes of solid catalysts, adsorbents, and ceramics was also claimed (Chopin et al., 1991; Besnard et al., 1994). Although a potential use of succinoglycan in food has been described (Fung et al., 1992; Fuisz, 1993), it is probably not used for these purposes.

12
References

Åman P., McNeil, M., Franzen, L.-E., Darvill, A. G., Albersheim, P. (1981) Structural elucidation, using HPLC-MS and GLC-MS, of the acidic polysaccharide secreted by *Rhizobium meliloti* strain 1021, *Carbohydr. Res.* **95**, 263–282.

Amemura, A., Harada, T. (1983) Structural studies on extracellular acidic polysaccharides secreted by three non-nodulating *Rhizobia*, *Carbohydr. Res.* **112**, 85–93.

Aubert, D., Frouin, L., Morvan, M., Vincent, M.-M. (1999) Gel of an apolar medium, its use for the preparation of water-based drilling fluids, US patent 5 858 928.

Azoulay, E. E. Y. (1983) Process for the preparation of viscous water by action of microorganisms, US patent 4 400 466.

Balnois, E., Stoll, S., Wilkinson, K. J., Buffle, J., Rinaudo, M., Milas, M. (2000) Conformation of succinoglycan as observed by atomic force microscopy, *Macromolecules* **33**, 7440–7447.

Battisti, L., Lara, J. C., Leigh, J. A. (1992) Specific oligosaccharide form of the *Rhizobium meliloti* exopolysaccharide promotes nodule invasion in alfalfa, *Proc. Natl. Acad. Sci. USA* **89**, 5625–5629.

Becker, A., Pühler, A. (1998) Specific amino acid substitution in the proline-rich motif of the *Rhizobium meliloti exoP* protein results in enhanced production of low-molecular-weight succinoglycan at the expense of high-molecular-weight succinoglycan, *J. Bacteriol.* **180**, 395–399.

Becker, A., Kleickmann, A., Küster, H., Keller, M., Arnold, W., Pühler, A. (1993a) Analysis of the *Rhizobium meliloti* genes *exoU, exoV, exoW, exoT,* and *exoI* involved in exopolysaccharide biosynthesis and nodule invasion: *exoU* and *exoW* probably encode glucosyltransferases, *Mol. Plant–Microbe Interact.* **6**, 735–744.

Becker, A., Kleickmann, A., Arnold, W., Pühler, A. (1993b) Analysis of the *Rhizobium meliloti exoH, exoK, exoL* fragment: *exoK* shows homology to excreted endo-β-1,3–1,4 glucanases and *exoH* resembles membrane proteins, *Mol. Gen. Genet.* **238**, 145–154.

Becker, A., Küster, H., Niehaus, K., Pühler, A. (1995) Extension of the *Rhizobium meliloti* succinoglycan biosynthesis gene cluster: identification of *exsA* gene encoding an ABC transporter protein, and the *exsB* gene which probably codes for a regulator of succinoglycan biosynthesis, *Mol. Gen. Genet.* **249**, 487–497.

Besnard, M.-M., David, C., Knipper, M. (1994) Mixed polysaccharide precipitating agents and insulating articles shaped therefrom, US patent 5 350 524.

Björndal, H., Erbing, C., Lindberg, B., Fahraeus, G., Ljunggren, H. (1971) Studies on an extracellular polysaccharide from *Rhizobium meliloti*, *Acta Chem. Scand.* **25**, 1281–1286.

Buendia, A. M., Enenkel, B., Köplin, R., Niehaus, K., Arnold, W., Pühler, A. (1991) The *Rhizobium meliloti exoZ/exoB* fragment of megaplasmid 2: *exoB* functions as a UDP-glucose-4-epimerase and *exoZ* shows homology to *NodX* of *Rhizobium leguminosarum* biovar. vicia strain TOM, *Mol. Microbiol.* **5**, 1519–1530.

Burova, T. V., Golubeva, I. A., Grinberg, N. V., Mashkevich, A. Y., Grinberg, V. Y., Usov, A. I., Navarini, L., Cesàro, A. (1996) Calorimetric study of the order-disorder conformational transition in succinoglycan, *Biopolymers* **39**, 517–529.

Cangelosi, G. A., Hung, L., Puvanesarajah, V., Stacey, G., Ozga, D. A., Leigh, J. A., Nester, E. W. (1987) Common loci for *Agrobacterium tumefaciens* and *Rhizobium meliloti* exopolysaccharide synthesis and their roles in plant interactions, *J. Bacteriol.* **169**, 2086–2091.

Carpenter, R. B., Wilson, J. M., Loughridge, B. W., Ravi, K. M., Jones, R. R. (1998) Well com-

pletion spacer fluids and methods, US patent 5 789 352.
Charles, T. C., Finan, T. M. (1991) Analysis of a 1600-kilobase *Rhizobium* megaplasmid using defined deletions generated *in vivo*, *Genetics* **127**, 5–20.
Cheng, H.-P., Walker, G. C. (1998a) Succinoglycan is required for initiating and elongation of infection threads during nodulation of alfalfa by *Rhizobium meliloti*, *J. Bacteriol.* **180**, 5183–5191.
Cheng, H.-P., Walker, G. C. (1998b) Succinoglycan production by *Rhizobium meliloti* is regulated through the *exoS–chvI* two-component regulatory system, *J. Bacteriol.* **180**, 20–26.
Chopin, T., Quemere, E., Nortier, P., Schuppiser, J.-L., Segaud, C. (1991) Mechanically improved shaped articles, US patent 4 983 563.
Chouly, C., Colquhoun, I. C., Jodelet, A., York, G., Walker, G. C. (1995) NMR studies of succinoglycan repeating-unit octasaccharides from *Rhizobium meliloti* and *Agrobacterium radiobacter*, *Int. J. Biol. Macromol.* **17**, 357–363.
Clarke-Sturman, A. J., Pedley, J. B., Sturla, P. L. (1986) Influence of anions on the properties of microbial polysaccharides in solution, *Int. J. Biol. Macromol.* **8**, 355–360.
Coplin, D. L., Cook, D. (1990) Molecular genetics of extracellular polysaccharide synthesis in vascular phytopathogenic bacteria, *Mol. Plant–Microbe Interact.* **3**, 271–279.
Cornish, A., Linton, J. D., Jones, C. W. (1987) The effect of growth conditions on the respiratory system of a succinoglucan-producing strain *Agrobacterium radiobacter*, *J. Gen. Microbiol.* **133**, 2971–2978.
Cripps, R. E., Ruffel, R. N., Sturman, A. J. (1981) Fluid displacement with heteropolysaccharide solutions, and the microbial production of heteropolysaccharides, European patent 040445A1.
Dasinger, B., McArthur, H. A. I. (1992) Aqueous gel composition derived from succinoglycan, European patent 0251638B1.
Denarie, J., Cullimore, J. (1993) Lipo-oligosaccharide nodulation factors: a new class of signalling molecules mediating recognition and morphogenesis, *Cell* **74**, 951–954.
Dentini, M., Crescenzi, V., Fidanza, M., Coviello, T. (1989) On the aggregation and conformational state in aqueous solution of a succinoglycan polysaccharide, *Macromolecules* **22**, 954–959.
Djordjevic, S. P., Chen, H., Batley, M., Redmond, J. W., Rolfe, B. G. (1987) Nitrogen fixation ability of exopolysaccharide mutants of *Rhizobium sp.* strain NGR234 and *Rhizobium trifolii* is restored by the addition of homologous exopolysaccharides, *J. Bacteriol.* **169**, 53–60.
Dussap, C. G., De Vita, D., Pons, A. (1991) Modeling growth and succinoglycan production by *Agrobacterium radiobacter* NCIB 9042 in batch cultures, *Biotechnol. Bioeng.* **38**, 65–74.
Evans, L. R., Linker, A., Impallomeni, G. (2000) Structure of succinoglycan from an infectious strain of *Agrobacterium radiobacter*, *Int. J. Biol. Macromol.* **27**, 319–326.
Fuisz, R. C. (1993) Method of preparing a frozen comestible, US patent 5 238 696.
Fung, F.-N., Miller, J. W., Wuesthoff, M. T. (1992) Low-calorie fat substitute, US patent 5 158 798.
Glucksmann, M. A., Reuber, T. L., Walker, G. C. (1993a) Family of glycosyl transferases needed for the synthesis of succinoglycan by *Rhizobium meliloti*, *J. Bacteriol.* **175**, 7033–7044.
Glucksmann, M. A., Reuber, T. L., Walker, G. C. (1993b) Genes needed for modification, polymerization, export, and processing of succinoglycan by *Rhizobium meliloti*: a model for succinoglycan biosynthesis, *J. Bacteriol.* **175**, 7045–7055.
Gonzalez, J. E., York, G. M., Walker, G. C. (1996) *Rhizobium meliloti* exopolysaccharides: synthesis and symbiotic function, *Gene* **179**, 141–146.
Gonzalez, J. E., Semino, C. E., Wang, L.-X., Castellano-Torres, L. E., Walker, G. C. (1998) Biosynthetic control of molecular weight in the polymerization of the octasaccharide subunits of succinoglycan, a symbiotically important exopolysaccharide of *Rhizobium meliloti*, *Proc. Natl. Acad. Sci. USA* **95**, 13477–13482.
Gravanis, G., Milas, M., Rinaudo, M., Clarke-Sturman, A. J. (1990a) Conformational transition and polyelectrolyte behaviour of a succinoglycan polysaccharide, *Int. J. Biol. Macromol.* **12**, 195–200.
Gravanis, G., Milas, M., Rinaudo, M., Clarke-Sturman, A. J. (1990b) Rheological behaviour of a succinoglycan polysaccharide in dilute and semi-dilute solutions, *Int. J. Biol. Macromol.* **12**, 201–206.
Gray, J. X., Djordjevic, M. A., Rolfe, B. G. (1990) Two genes that regulate exopolysaccharide production in *Rhizobium* sp. strain NGR234: DNA sequences and resultant phenotypes, *J. Bacteriol.* **172**, 193–203.
Guerin, G., Knipper, M. (1991) Stable zeolite/succinoglycan suspensions, US patent 5 104 566.
Guerin, G., Larrey, M.-D., Nabavi, M., Ricca, J.-M. (1998) Aqueous cosmetic compositions with base of non-volatile insoluble silicon, stabilized by a succinoglycan, WO 98/24408.

Guillou, V. (1995) Sulfamic acid cleaning/stripping compositions comprising heteropolysaccharide thickening agents, US patent 5 431 839.

Harada, T., Yoshimura, T. (1964) Production of new acidic polysaccharide containing succinic acid by a soil bacterium, *Biochim. Biophys. Acta* **83**, 374.

Harada, T., Harada, A. (1996) Curdlan and succinoglycan, in: *Polysaccharides in Medicinal Applications* (Dumitriu, S., Ed.), New York: Marcel Dekker, 21–58.

Harada, T., Yoshimura, T., Hidaka, H., Koreda, A. (1965) Production of new acidic polysaccharide, succinoglucan, by *Alcaligenes faecalis* var. myxogenes, *Agric. Biol. Chem.* **29**, 757.

Harada, T., Amemura, A., Jansson, P.-E., Lindberg, B. (1979) Comparative studies of polysaccharides elaborated by *Rhizobium, Alcaligenes,* and *Agrobacterium, Carbohydr. Res.* **77**, 285–288.

Jansson, P.-E., Kenne, L., Lindberg, B., Ljunggren, H., Lönngren, J., Ruden, U., Svensson, S. (1977) Demonstration of an octasaccharide repeating unit in the extracellular polysaccharide of *Rhizobium meliloti* by sequential degradation, *J. Am. Chem. Soc.* **99**, 3812–3815.

Keller, M., Roxlau, A., Weng, W. M., Schmidt, M., Quandt, J., Niehaus, K., Jording, D., Arnold, W., Puhler, A. (1995) Molecular analysis of the *Rhizobium meliloti MucR* regulating the biosynthesis of the exopolysaccharides succinoglycan and galactoglucan, *Mol. Plant–Microbe Interact.* **8**, 267–277.

Kijne, J. W. (1992) The *Rhizobium* infection process, in: *Biological Nitrogen Fixation* (Stacey, G., Burris, R. H., Evans H. J., Eds), New York: Chapman & Hall, 349–398.

Kitaoka, N. (1998) Aqueous ink for ball point pen, Japanese patent 10130563A.

Knipper, M., Besnard, M.-M., David, C. (1995) Composition comportant un succinoglycane, European patent 0469953B1.

Knipper, M., Senechal, A., Raffart, M. (2000) Composition dérivant d'un succinoglycane, son procédé de préparation et ses applications, European patent 0527061B1.

Kondo, M., Matsubara, N. (1995) Aqueous ink composition for writing instrument, US patent 5 466 283.

Lau, H. C. (1993) Gravel packing process, US patent 5184679.

Lau, H. C., Bernardi, L. A. (1993) Low-viscosity gravel packing process, US patent 5 251 699.

Lau, H. C., Davies, D. R. (1992) Aqueous polysaccharide compositions and their use, GB patent 2 250 761A.

Leigh, J. A., Coplin, D. L. (1992) Exopolysaccharides in plant–bacterial interactions, *Annu. Rev. Microbiol.* **46**, 307–346.

Leigh, J. A., Walker, G. C. (1994) Exopolysaccharides of *Rhizobium*: synthesis, regulation and symbiotic function, *Trends Genet.* **10**, 63–67.

Leigh, J. A., Signer, E. R., Walker, G. C. (1985) Exopolysaccharide deficient mutants of *Rhizobium meliloti* that form ineffective nodules, *Proc. Natl. Acad. Sci. USA* **82**, 6231–6235.

Linton, J. D., Evans, M. W., Godley, A. R. (1985) Process for preparing a heteropolysaccharide obtained thereby, its use, and strain NCIB 11883, European patent 0138255A2.

Linton, J. D., Evans, M. W., Jones, D. S., Gouldney, D. N. (1987a) Exocellular succinoglycan production by *Agrobacterium radiobacter* NCIB 11883, *J. Gen. Microbiol.* **133**, 2961–2969.

Linton, J. D., Jones, D. S., Woodard, S. (1987b) Factors that control the rate of exopolysaccharide production by *Agrobacterium radiobacter* NCIB 11883, *J. Gen. Microbiol.* **133**, 2979–2987.

Long, S. R. (1989) *Rhizobium*–legume nodulation: life together in the underground, *Cell* **56**, 203–214.

Macri, C. A., Miller, J. W., Sarges, D. A., Sprott, D. J. (1996) Agricultural foam composition, European patent 0400914B1.

Margaritis, A., Pace, G. W. (1985) Microbial polysaccharides, in: *Comprehensive Biotechnology* (Moo Young, M., Ed.), Oxford: Pergamon Press, 1005–1044, Vol. 3.

Matulova, M., Toffanin, R., Navarini, L., Gilli, R., Paoletti, S., Cesaro, A. (1994) NMR analysis of succinoglycans from different microbial sources: partial assignment of their ^{1}H and ^{13}C NMR spectra and location of the succinate and the acetate groups, *Carbohydr. Res.* **265**, 167–179.

Meade, M. J., Tanenbaum, S. W., Nakas, J. P. (1995) Production and rheological properties of a succinoglycan from *Pseudomonas* sp. 31260, grown on wood hydrolysates, *Can. J. Microbiol.* **41**, 1147–1152.

Mendrygal, K. E., Gonzalez, J. E. (2000) Environmental regulation of exopolysaccharide production in *Sinrhizobium meliloti*, *J. Bacteriol.* **182**, 599–606.

Misaki, A., Saito, H., Ito, T, Harada, T. (1969) Structure of succinoglycan, an extracellular acidic polysaccharide from *Alcaligenes faecalis* var. myxogenes, *Biochemistry* **8**, 4645–4650.

Müller, P., Weng, M. W. M., Quandt, J., Arnold, W., Pühler, A. (1993) Genetic analysis of *Rhizobium meliloti exoYFQ* operon: exoY is homologous to sugar transferases and ExoQ represents a transmembrane protein, *Mol. Plant–Microbe Int.* **6**, 55–65.

Navarini, L., Stredansky, M., Matulova, M., Bertocchi, C. (1997) Production and characterization of an exopolysaccharide from *Rhizobium hedysari* HCNT1, *Biotechnol. Lett.* **19**, 1231–1234.

Osman, S. F., Fett, W. F. (1989) Structure of an acidic exopolysaccharide from *Pseudomonas marginalis* HT041B, *J. Bacteriol.* **171**, 1760–1762.

Pandey, A. (1992) Recent process development in solid state fermentation, *Process Biochem.* **27**, 109–117.

Philips, J. C., Hoyt IV, H. L., Macri, C. A., Miller, W. L., O'Neilll, J. J. (1991) Agricultural gel-forming compositions, US patent 5 077 314.

Rae, P., Johnston, N. (1995) Storable liquid cementitious slurries for cementing oil and gas wells, US patent 5 447 197.

Reed, J. W., Glazebrook, J., Walker, G. C. (1991a) The *exoR* gene of *Rhizobium meliloti* affects RNA levels of other *exo* genes but lacks homology to known transcriptional regulators, *J. Bacteriol.* **173**, 3789–3794.

Reed, J. W., Capage, M., Walker, G. C. (1991b) *Rhizobium meliloti exoG* and *exoJ* mutations affect the exoX–exoY system for modulation of exopolysaccharide production, *J. Bacteriol.* **173**, 3776–3788.

Reinhold, B. B., Chan, S. Y., Reuber, T. L., Marra, A., Walker, G. C., Reihold, V. N. (1994) Detailed structural characterization of succinoglycan, the major exopolysaccharide of *Rhizobium meliloti* Rm1201, *J. Bacteriol.* **176**, 1997–2002.

Reuber, T. L., Walker, G. C. (1993a) Biosynthesis of succinoglycan, a symbiotically important exopolysaccharide of *Rhizobium meliloti*, *Cell* **74**, 269–280.

Reuber, T. L., Walker, G. C. (1993b) The acetyl substituent of succinoglycan is not necessary for alfalfa nodule invasion by *Rhizobium meliloti* Rm1201, *J. Bacteriol.* **175**, 3653–3655.

Reuber, T. L., Long, S., Walker, G. C. (1991) Regulation of *Rhizobium meliloti exo* genes in free-living cells and in plants examined by using Tn*phoA* fusions, *J. Bacteriol.* **173**, 426–434.

Reuhs, B. L., Williams, M. N. V., Kim, J. S., Carlson, R. W., Côté, F. (1995) Suppression of the Fix$^-$ phenotype of *Rhizobium meliloti exoB* mutants by *lpsZ* is correlated to a modified expression of the K polysaccharide, *J. Bacteriol.* **177**, 6231–6235.

Ridout, M. J., Brownsey, G. J., York, G. M., Walker, G. C., Morris, V. J. (1997) Effect of *o*-acyl substituent on the functional behaviour of *Rhizobium meliloti* succinoglycan, *Int. J. Biol. Macromol.* **20**, 1–7.

Stredansky, M., Conti, E. (1999a) Succinoglycan production by solid-state fermentation with *Agrobacterium tumefaciens*, *Appl. Microbiol. Biotechnol.* **52**, 332–337.

Stredansky, M., Conti, E. (1999b) Xanthan production by solid state fermentation, *Process Biochem.* **34**, 581–587.

Stredansky, M., Conti, E., Bertocchi, C., Matulova, M., Zanetti, F. (1998) Succinoglycan production by *Agrobacterium tumefaciens*, *J. Ferm. Bioeng.* **85**, 398–403.

Stredansky, M., Conti, E., Bertocchi, C., Navarini, L., Matulova, M., Zanetti, F. (1999a) Fed-batch production and simple isolation of succinoglycan from *Agrobacterium tumefaciens*, *Biotechnol. Tech.* **13**, 7–10.

Stredansky, M., Conti, E., Navarini, L., Bertocchi, C. (1999b) Production of bacterial exopolysaccharides by solid substrate fermentation, *Process Biochem.* **34**, 11–16.

Stredansky, M., Conti, E., Bertocchi, C. (1999c) Processo per la produzione di esopolisaccaridi batterici mediante fermentazione aerobica su substrato solido, Italian patent 1 293 620.

Sutherland, I. W. (1990) *Biotechnology of Microbial Polysaccharides*. Cambridge: Cambridge University Press.

Sugimoto, Y., Yamamoto, T. (1999) Ink composition for water base ballpoint, and water-base ballpoint, JP 11256090A.

Sydansk, R. D. (1992) Enhanced liquid hydrocarbon recovery process, US patent 5 129 457.

Sydansk, R. D. (1998) Polymer enhanced foam workover, completion, and kill fluids, US patent 5 706 895.

Tolmasky, M. E., Staneloni, R. J., Leloir, L. F. (1982) Lipid-bound saccharides in *Rhizobium meliloti*, *J. Biol. Chem.* **257**, 6751–6757.

Urzainqui, A., Walker, G. C. (1992) Exogenous suppression of the symbiotic deficiencies of *Rhizobium meliloti exo* mutants, *J. Bacteriol.* **174**, 3403–3406.

Uttaro, A. D., Cangelosi, G. A., Geremia, R. A., Nester, E. W., Ugalde, R. A. (1990) Biochemical characterization of avirulent *exoC* mutants of *Agrobacterium tumefaciens*, *J. Bacteriol.* **172**, 1640–1646.

van Rhijn, P., Vanderleyden, J. (1995) The *Rhizobium*–plant symbiosis, *Microbiol. Rev.* **59**, 124–142.

Yokota, A., Sakane, T. (1991) Taxonomic significance of fatty acid compositions in whole cells and lipopolysaccharides in *Rhizobiaceae, Inst. Ferment. Osaka, Res. Commun.* **15**, 57.

York, G. M., Walker, G. C. (1998a) The succinyl and acetyl modifications of succinoglycan influence susceptibility of succinoglycan to cleavage by the *Rhizobium meliloti* glycanases *exoK* and *exsH*, *J. Bacteriol.* **180**, 4184–4191.

York, G. M., Walker, G. C. (1998b) The *Rhizobium meliloti exoK* and *exsH* glycanases specifically depolymerize nascent succinoglycan chains, *J. Bacteriol.* **180**, 4912–4917.

Yoshida, M. (1998) Cosmetic material, Japanese patent 10203953A.

Yoshida, M. (1999) Composition for acid hair dye, Japanese patent 11240824A.

Yoshida, M., Suzuki, K. (1998) Hairdye compositions, WO 98/31330A1.

Zevenhuizen, L. P. T. M. (1971) Chemical composition of exopolysaccharides of *Rhizobium* and *Agrobacterium*, *J. Gen. Microbiol.* **68**, 239–243.

Zevenhuizen, L. P. T. M. (1973) Methylation analysis of acidic exopolysaccharides of *Rhizobium* and *Agrobacterium*, *Carbohydr. Res.* **26**, 409–419.

Zevenhuizen, L. P. T. M. (1997) Succinoglycan and galactoglucan, *Carbohydr. Polymers.* **33**, 139–144.

Zevenhuizen, L. P. T. M., Faleschini, P. (1991) Effect of the concentration of sodium chloride in the medium on the relative proportions of poly- and oligo-saccharides excreted by *Rhizobium meliloti* strain YE-2SL, *Carbohydr. Res.* **209**, 203–209.

Zhan, H., Leigh, J. A. (1990) Two genes that regulate exopolysaccharide production in *Rhizobium meliloti*, *J. Bacteriol.* **172**, 5254–5259.

Zhan, H., Gray, J. X., Levery, S. B., Rolfe, B., Leigh, J. A. (1990) Functional and evolutionary relatedness of genes for exopolysaccharide synthesis in *Rhizobium meliloti* and *Rhizobium* sp. strain NGR234, *J. Bacteriol.* **172**, 5245–5253.

8
Alginates from Bacteria

Dr. Bernd H. A. Rehm
Institut für Mikrobiologie, Westfälische Wilhelms-Universität Münster, Corrensstraβe 3, 48149 Münster, Germany; Tel.: +49-251-8339848; Fax: +49-251-8338388; E-mail: rehm@uni-muenster.de

CF cystic fibrosis
EPS extracellular polysaccharide (exopolysaccharide)
HSL homoserine lactone
LPS lipopolysaccharide
PMI-GMP phosphomannose isomerase/guanosine diphosphomannose pyrophosphorylase
PMI-GMP phosphomannose isomerase-GDP-mannose pyrophosphorylase
PMM phosphomannomutase
PMM phosphomannomutase
SCLM scanning laser confocal microscopy
TCA cycle tricarboxylic acid cycle

1 Introduction

Alginates are a family of unbranched, nonrepeating copolymers consisting of variable amounts of (1-4)-linked β-D-mannuronic acid and its epimer α-L-guluronic acid. The monomers are distributed in blocks of continuous mannuronate residues (M-blocks), guluronate residues (G-blocks), or alternating residues (MG-blocks) (Figure 1). Alginates isolated from different natural sources vary in the length and distribution of the different block types. Alginates are produced by bacteria and brown seaweeds, and the mannuronate residues of the bacterial–but not those of the seaweed polymers–are acetylated to a variable extent at positions O-2 and/or O-3 (Skjåk-Bræk et al., 1986). The variability in monomer block structures and acetylation strongly affects the physicochemical and rheological properties of the polymer, and the biological basis for the variability is therefore of both scientific and

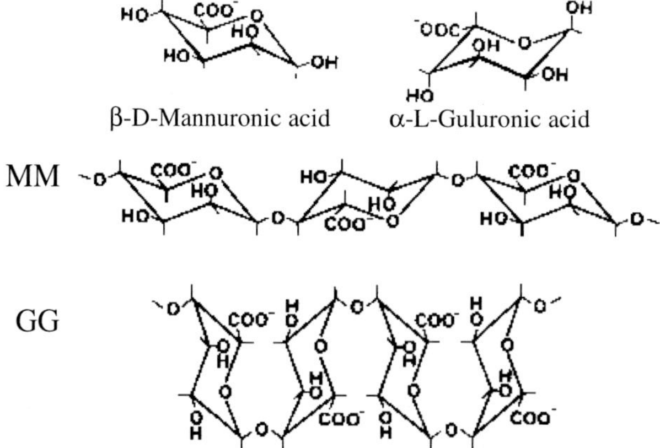

Fig. 1 The structure of alginate. MM: polymannuronate; GG: polyguluronate.

applied importance (Smidsrød and Draget, 1996). Alginate is used for a variety of industrial purposes, for example as a stabilizing, thickening and gelling agent in food production, or to immobilize cells in pharmaceutical and biotechnology industries (see below) (Onsøyen, 1996). The production is currently based exclusively on the harvesting of brown seaweeds. Several bacteria belonging to the genera *Pseudomonas* and *Azotobacter* also produce alginate (Gorin and Spencer, 1966; Linker and Jones, 1966; Govan et al., 1981; Cote and Krull, 1988), and the structures of the blocks of monomer residues are similar in alginates produced by seaweeds and those synthesized by *Azotobacter vinelandii*. In contrast, all *Pseudomonas* alginates lack G-blocks (Skjåk-Bræk et al., 1986). Most of our knowledge of the genetics of alginate biosynthesis originates from studies of *Pseudomonas aeruginosa*, mainly because of the medical relevance of this organism as an opportunistic human pathogen, particularly for patients suffering from cystic fibrosis (CF) (Govan and Harris, 1986; May and Chakrabarty, 1994). Alginate plays an important role as a virulence factor during the infectious process (Gacesa and Russell, 1990). The reason for this appears to be related to the alginate-mediated mode of biofilm growth, which causes resistant colonization of the lung. *A. vinelandii* and *P. aeruginosa* produce alginate as an extracellular polysaccharide (EPS) in vegetatively growing cells, whereas in *A. vinelandii* alginate is also involved in the differentiation process leading to a so-called cyst (Sadoff, 1975).

2 Historical Outline

The polysaccharide alginate was firstly isolated from marine macroalgae in the 19th century, but it was approximately 80 years later that a bacterial source (from *P. aeruginosa*) of the polysaccharide was identified (Linker and Jones, 1966). This alginate was later found to be similar to the commercially useful polymer obtained from marine algae (Lin and Hassid, 1966a; Linker and Jones, 1966) and also to the polysaccharide produced as a capsule by *Azotobacter vinelandii* (Gorin and Spencer, 1966). The association of mucoid *P. aeruginosa*, i.e., alginate-overproducing strains, with chronic lung infections in patients who suffer from the inherited disease CF has been well established, and is recognized as a major cause of pathogenesis in these individuals. Mucoid *P. aeruginosa* have also been isolated, albeit less frequently, from other patients, for example bronchiotatics and those with urinary tract or middle-ear infections (McAvoy et al., 1989), although not normally from individuals with infected burn sites. Although most of the mucoid isolates of *P. aeruginosa* have been obtained from clinical samples, it is clear that alginate production is important in a much wider context. Ten of 81 *P. aeruginosa* isolates from technical water systems showed a mucoid phenotype, which implies a more widespread occurrence of ecological niches for mucoid forms (Grobe et al., 1995). Alginate biosynthesis is a key factor in the establishment of stable mature biofilms of *P. aeruginosa* in a wide range of environmental situations (Nivens et al., 2001). Alginate production is fairly widespread amongst rRNA homology group I pseudomonads, as indicated by Southern hybridization experiments using various alginate biosynthesis genes as probes (Fialho et al., 1990; Fett et al., 1992; Rehm, 1996). Genomic DNA from representatives of groups II–IV gave very weak or no hybridization with the probes, except for *algC*, indicating that the ability to produce alginate is restricted to members of rRNA homology group I. This substantiates earlier physiological studies in which alginate was isolated from *Pseudomonas fluorescens, Pseudomonas putida, Pseudomonas mendocina* (Govan et al., 1981) and *P. syringae* (Fett et al., 1986; Gross and Rudolph, 1987). Alginate is also synthesized by *A. vinelandii* as part of the encystment process (Gorin and Spencer, 1966). The mature cysts are surrounded by two discrete alginate-rich layers (exine and intine) which enable the dormant cells to survive long periods of desiccation. Strains of *Azotobacter chroococcum* also produce alginate (Cote and Krull, 1988). In *Azotobacter*, abnormalities in alginate production resulted in impaired encystment, which indicated that alginate biosynthesis is required to survive and adapt to famine conditions, as was supposed for the establishment of biofilms by pseudomonads. Enzymological evidence for the biosynthesis pathway of alginate was obtained from the brown algae *Fucus gardneri* by Lin and Hassid (1966a) and from *A. vinelandii* about ten years later by Pindar and Bucke (1975). These studies indicated that fructose-6-phosphate is the first alginate precursor, which was derived from the Entner–Doudoroff pathway and from the fructose-1,6-bisphosphate aldolase reaction. The presence of the first alginate biosynthetic enzymes, phosphomannose isomerase, GDP-mannose pyrophosphorylase and GDP-mannose dehydrogenase, was demonstrated by Piggot et al. (1981) in *P. aeruginosa*. Later, Padgett and Phibbs (1986) detected another alginate biosynthesis enzyme, phosphomannomutase. The medical relevance of mucoid *P. aeruginosa* stimulated research on the genetics of alginate biosynthesis. A stable mucoid mutant of *P. aeruginosa* (8830) had to be generated by chemical mutagenesis in order to obtain nonmucoid mutants (Darzins and Chakrabarty, 1984). Complementation studies of the various mutants allowed the identification and characterization of the alginate biosynthesis genes, as well as the respective regulatory genes (May and Chakrabarty, 1994). The identification and biochemical characterization of regulatory alginate biosynthesis proteins enabled a deeper understanding of the rather complex regulatory network (Govan and Deretic, 1996; Rehm and Valla, 1997). The extracellular Ca^{2+}-dependent C-5-epimerases were the first alginate-related enzymes from *A. vinelandii*, which were biochemically and genetically characterized (Ertesvåg et al., 1994, 1999). Meanwhile, hybridization of a lambda gene library of *A. vinelandii* with the outer membrane encoding the *algE* gene from *P. aeruginosa* as a probe, led to the identification of the second alginate biosynthesis gene cluster and the first gene cluster of a biotechnologically relevant microorganism (Rehm, 1996; Rehm et al., 1996). Comparative analysis of the genetics, biochemistry and regulation of alginate biosynthesis revealed strong similarities (Rehm and Valla, 1997; Gacesa, 1998). Moreover, the plant-pathogenic *P. syringae* pv. *syringae* alginate biosynthesis gene cluster was identified and characterized, exhibiting a virtually identical alginate gene arrangement (Penaloza-Vazquez et al., 1997).

3
Chemical Structures

Alginate is composed of the uronic acid β-D-mannuronate and its C-5 epimer α-L-guluronate. These monomers may be arranged in homopolymeric (poly-mannuronate or poly-guluronate) or heteropolymeric block structures (see Figure 1). In addition, bacterial alginates are normally O-acetylated on the 2 and/or 3 position(s) of the β-D-mannuronate residues. Consequently, bacteria produce a range of alginates with different block structures and degrees of O-acetylation.

The high molecular mass of bacterial alginate and the negative charge ensure that the polysaccharide is highly hydrated and viscous. It is well established that alginates from *P. aeruginosa* do not contain polyguluronate blocks (Sherbrock-Cox et al., 1984) but those from *A. vinelandii* may do so. The block structure and degree of O-acetylation, as well as the molecular weight, determine the physico-chemical properties of alginate. Alginates which contain polyguluronate form rigid gels in the presence of Ca^{2+}, and are therefore important in structural roles, for example the outer cyst wall (exine) of *A. vinelandii*. Conversely, an absence of polyguluronate, as in *P. aeruginosa*, produces relatively flexible gels in the presence of Ca^{2+}. Extensive O-acetylation of alginate increases the water-binding capacity of the polysaccharide, which may be significant in enhancing survival under desiccating conditions.

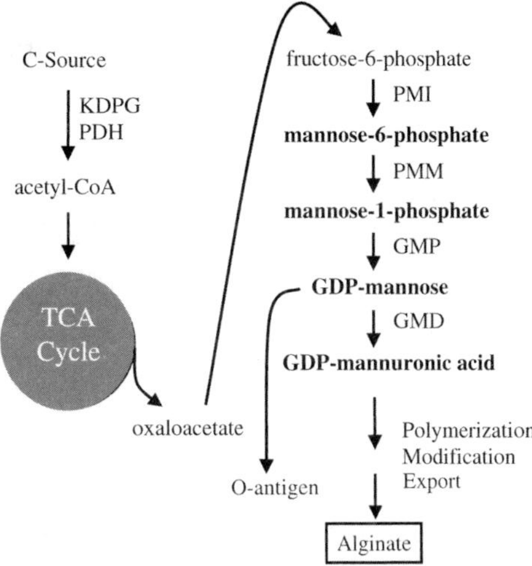

Fig. 2 Biosynthesis pathway of alginate. Oxaloacetate is converted to fructose-6-phosphate via gluconeogenesis. KDPG: ketodeoxyphosphogluconate pathway (Entner–Doudoroff pathway); PDH: pyruvate dehydrogenase; PMI-GMP: phosphomannose isomerase-GDP-mannose pyrophosphorylase; PMM: phosphomannomutase; GMD: GDP-mannose dehydrogenase. Intermediates shown in bold type are precursors of alginate.

4 Biosynthetic Pathway of the Alginate Precursor, GDP-Mannuronic Acid

A convincing pathway for alginate biosynthesis in *A. vinelandii* was first proposed by Pindar and Bucke (1975) based on the assay of individual enzyme activities. However, the corresponding enzymes in *P. aeruginosa* have proved more difficult to assay directly, and the pathway has been elucidated using a combination of complementation analyses and gene overexpression studies. Although the initial steps in the pathway are indisputable (Figure 2) (and the same applies to *A. vinelandii* and pseudomonads), there is still considerable debate about the final stages of biosynthesis and the export of alginate.

The alginate biosynthesis starts from fructose-6-phosphate in the cytosol. Radiolabeling studies have established that six-carbon growth substrates are oxidized via the Entner–Doudoroff pathway, and that the resultant pyruvate [1 mol (mol hexose)$^{-1}$] is ultimately channeled into alginate biosynthesis (Lynn and Sokatch, 1984). More detailed analyses of the labeling patterns in alginate indicate that the pyruvate derived from the oxidation of sugars is fed into the tricarboxylic acid (TCA) cycle prior to synthesis of fructose-6-phosphate and alginate (Narbad et al., 1988). Recent experiments using ^{13}C-NMR have clearly established the key role of the Entner–Doudoroff pathway in the oxidation of hexoses, and also the obligatory requirement of triose intermediates in alginate biosynthesis (Beale and Foster, 1996). The latter authors concluded that the pyruvate derived from the oxidation of hexoses feeds into alginate biosynthesis via the formation of oxaloacetate and subsequent gluconeogenesis (see Figure 2). This supports earlier data employing ^{13}C-labeled precursors, which suggested an obligatory role for the TCA cycle in the conversion of glucose to alginate (Narbad et al., 1988). This important role of the TCA cycle in alginate biosynthesis is supported by more recent genetic studies using nonmucoid mutants of *P. aeruginosa*. Levels of only the phosphorylated (active) form of the key TCA cycle enzyme succinyl-CoA synthetase are reduced in *algQ* mutants (Schlictman et al., 1994). The normally rare occurrence of mucoid forms of *P. aeruginosa* in culture can be significantly increased by growth on energy-poor media (Terry et al., 1991). However, there is a recent report that glucose can stimulate *algD* transcription and alginate production (Ma et al., 1997), though this contradicts earlier findings which proposed that glucose repression of *algD* occurs (Devault et al., 1991). Interestingly, alginate biosynthesis occurs in response to energy deprivation, despite biosynthesis of the polysaccharide being an energy-consuming process. It has been suggested that the AlgQ-controlled expression of succinyl-CoA synthetase and nucleoside-diphosphate kinase may result in a decreased pool size of GTP and hence less available GDP-mannose for alginate biosynthesis (Schlictman et al., 1994; Kim et al., 1998; Kapatral et al., 2000). However, quantification of the nucleotide sugars indicates that GDP-mannose is present in great excess (Tatnell et al., 1993), even when the GDP-mannose dehydrogenase gene was overexpressed (Tatnell et al., 1994). The initial steps in the alginate biosynthesis are related to general carbohydrate metabolism, and the intermediates are widely utilized. In particular, the intermediate GDP-mannose serves not only as a precursor for alginate biosynthesis but also for lipopolysaccharide (LPS) biosynthesis (Goldberg et al., 1993). Accordingly, the GDP-mannose dehydrogenase (AlgD) exhibits a key role in the biosynthesis of alginate. The *algD* gene is proximal to the promoter on the alginate biosynthesis gene cluster and expression is tightly controlled (Schurr et al., 1993). Analyses of nucleotide sugar pools and exopolysaccharide production clearly indicate that GDP-mannose dehydrogenase is the kinetic control point in the alginate pathway (Tatnell et al., 1994). However, the alginate biosynthesis enzyme phosphomannose isomerase/ guanosine-diphosphomannose pyrophosphorylase (PMI-GMP (AlgA)) is a bifunctional protein catalyzing the initial and third steps of alginate synthesis (see Figure 2). The PMI reaction pulls the fructose-6-phosphate out of the metabolic pool, leading to the first intermediate, mannose-6-phosphate. Phosphomannomutase (AlgC) then catalyzes the second step, resulting in the formation of mannose-1-phosphate. This is not the only reaction catalyzed by AlgC, which also exhibits phosphoglucomutase activity and which is evidently involved in rhamnolipid, LPS and alginate biosynthesis (Olvera et al., 1999). The GMP activity of PMI-GMP (AlgA) then, with concomitant GTP hydrolysis, converts mannose-1-phosphate to GDP-mannose. The enzyme favors the reverse reaction, but because of the efficient removal of the GDP-mannose in the next step, the entire pathway proceeds efficiently in the direction of alginate synthesis. The almost irreversible oxidation of GDP-mannose to GDP-mannuronic acid involves the enzyme guanosine-diphosphomannose dehydrogenase, and the reaction product is the immediate precursor for polymerization (see Figure 2). For further details on this pathway, readers are referred to the review on *P. aeruginosa* alginate synthesis by May and Chakrabarty (1994). Interestingly, the intracellular activities of key biosynthesis enzymes appear to be very low, even in extracts prepared from highly mucoid *P. aeruginosa* cells. It has been speculated that the enzymes phosphomannose isomerase-GDP-mannose pyrophosphorylase (PMI-GMP) and phosphomannomutase (PMM) may exist as an enzyme complex ("metabolon"), which allows the coupling of enzymatic reactions, as described for many metabolic pathways (Mathews, 1993).

5 Genetics of Alginate Biosynthesis

At least 24 genes have been directly implicated in alginate biosynthesis in *P. aeruginosa* (Figure 3; Table 1), and there is good evidence that others may also be involved, e.g., *glpM* (Schweizer et al., 1995). It is not possible to assign all *alg* genes identified so far as solely functioning in alginate biosynthesis, as it is now evident for example that some of the regulator genes act globally and encode proteins such as alternative sigma factors (Yu et al., 1995). Other "*alg*" genes, e.g., *algA, C,* are also required for LPS biosynthesis (Goldberg et al., 1993). The genes involved in the synthesis of the precursor GDP-mannuronic acid have all been identified and characterized, and they have been assigned the designations *algA* (encoding PMI-GMP), *algC* (encoding phosphomannomutase), and *algD* (encoding guanosine-diphosphomannose dehydrogenase). With the exception of *algC* (located at 10 min on the chromosome map; between 5,992,000 and 5,994,000 bp of the *P. aeruginosa* genome sequence), the other two genes and all other known structural genes involved in alginate biosynthesis in *P. aeruginosa* are clustered at 34 min (between positions 3,962,000 bp and 3,980,000 bp of the *P. aeruginosa* genome sequence) (Figure 3; Table 1). The biological functions of many of the gene products putatively involved in the polymerization process are poorly understood, mainly because the polymerase has

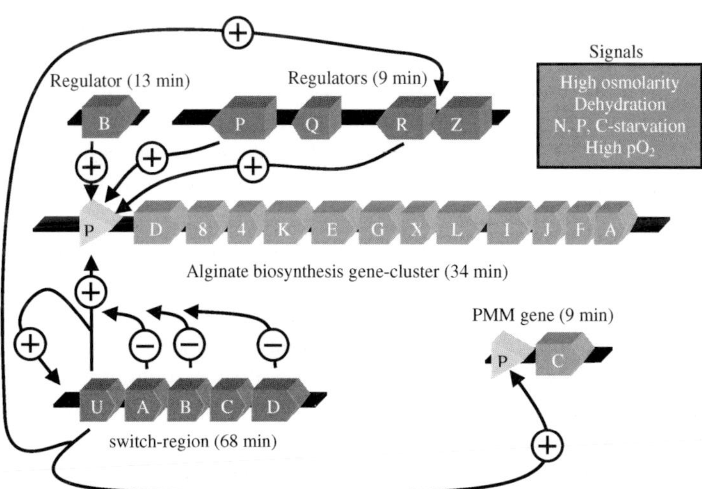

Fig. 3 Genetic organization of alginate genes and their regulation. Signals which induce alginate biosynthesis gene expression are given in the gray box. +: induction (positive effector); – : inhibition (negative effector); P: promoter; switch-region: muc-gene region (genotypic switch to alginate overproduction by mutation in *mucA*, *mucB* or *mucD*, respectively).

Tab. 1a Alginate genes from *Pseudomonas aeruginosa*

Gene	*Location*	*Gene product*
algD	34	GDP mannose dehydrogenase
alg8	34	Polymerase/export function?
alg44	34	Polymerase/export function?
*algK**	34	Polymerase/export function?
algE	34	Outer-membrane porin?
algG	34	Mannuronan C-5-epimerase
algX	34	Unidentified function but high sequence similarity to *algJ*
algL	34	Alginate lyase
algI	34	*O*-Acetylation
algJ	34	*O*-Acetylation
algF	34	*O*-Acetylation
algA	34	Phosphomannose isomerase/GDP mannose pyrophosphorylase
algB	13	Member of *ntrC* subclass of two-component regulators
algC	10	Phosphomannomutase
algH	?	Unknown function
algR1	9	Regulatory component of two-component sensory transduction system
algR2 (algQ)	9	Protein kinase or kinase regulator
algR3(algP)	9	Histone-like transcription regulator
algZ	9	AlgR cognate sensor
algU (algT)	68	Homologue of *E. coli* σ^E global stress response factor
algS (mucA)	68	Anti σ factor
algN (mucB)	68	Anti σ factor?
algM (mucC)	68	Regulator?
algW (mucD)	68	Homologue of serine protease (HtrA)

Modified according to Rehm and Valla (1997)

not been purified and no *in vitro* alginate synthesis assay has been established. However, the gene products of *alg8, alg44, algX* (formerly *alg60*) and *algK* are candidates for being subunits of the alginate polymerase. The deduced amino acid sequences of these proteins contain hydrophobic regions, suggesting a localization in the cytoplasmic membrane (Wang et al., 1987; Maharaj et al., 1993; Rehm and Valla, 1997). The *algK* gene has been identified located directly downstream of *alg44* in *P. aeruginosa* (Aarons et al., 1997). Evidence was obtained that AlgK is entirely periplasmic and is probably anchored in the cytoplasmic membrane. AlgK might be also involved in polymerization and/or export of alginate. Although most of the genes are essential for alginate biosynthesis, those encoding the epimerase and O-acetyltransferase(s) can be inactivated, provided that essential genes downstream of the mutation are expressed in trans, without abolishing alginate biosynthesis. The role of a gene (*algL*) encoding an alginate lyase in the biosynthesis operon (Boyd et al., 1993; Schiller et al., 1993; Monday and Schiller, 1996) is unknown. Contradictory data were published regarding the role of AlgL in alginate biosynthesis. Boyd and coworkers (1993) showed that a mutation in the *algL* gene had no effect on alginate biosynthesis. However, Monday and Schiller (1996) demonstrated that an *algL* mutation strongly impaired alginate biosynthesis, as

Tab. 1b Alginate genes from *Azotobacter vinelandii*

Gene	*Gene product*
alg8	Polymerase?
alg44	Polymerase/export function?
algA	Phosphomannose isomerase GDP mannose pyrophosphorylase
algD	GDP mannose dehydrogenase
algE1-7	Mannuronan C-5-epimerases
algG	Mannuronan C-5-epimerase
algJ	Export of alginate?
algL	Alginate lyase
algU	Homologue of *E. coli* σ^E global stress response factor
mucA	Anti σ factor
mucB	Anti σ factor
mucC	Regulator?
mucD	Homologue of serine protease (HtrA)

Modified according to Rehm and Valla (1997).

was recently confirmed by an *algL* mutant of *P. syringae* pv. *syringae* (Penaloza-Vazquez et al., 1997). The *P. aeruginosa* gene cluster at 34 min also contains the *algE* gene, encoding an outer membrane protein (Chu et al., 1991; Rehm et al., 1994a). Production of this protein is strictly correlated with the mucoid phenotype of *P. aeruginosa* (Grabert et al., 1990; Rehm et al., 1994a).

In *A. vinelandii*, our understanding of the genetics of alginate synthesis has improved greatly over the past few years. The first genes involved in alginate synthesis to be cloned and characterized encode a set of seven strongly related Ca^{2+}-dependent mannuronan C-5-epimerases, designated AlgE1 to AlgE7 (Ertesvåg et al., 1999). These proteins are structurally unusual, since they can all be described as repeats of two types of protein modules, designated A (385 amino acids) and R (153 amino acids). Each protein contains a short motif designated S at the carboxy-terminal end. The A modules are present once or twice in each protein, while the R modules are represented one to seven times. Each R module contains four to seven repeats of a nine-amino-acid sequence repeat putatively involved in the binding of Ca^{2+} ions (Ertesvåg et al., 1999). These epimerase genes (*algE1–algE7*) are clustered in the *A. vinelandii* chromosome, and they share no significant sequence homology to the *P. aeruginosa algG* gene.

It was shown recently that the *A. vinelandii* genome encodes a gene (*algD*) corresponding to the *P. aeruginosa algD* gene, and these two genes share 73% identity with each other at the protein level (Campos et al., 1996). An *A. vinelandii* gene (*algJ*) apparently corresponding to the *P. aeruginosa algE* gene was also recently identified (Rehm, 1996). AlgJ shares about 52% sequence identity with AlgE from *P. aeruginosa*, and topological modeling suggests that this protein is structurally similar to AlgE. It is also believed to be functionally equivalent, forming a pore which is involved in alginate export (Rehm, 1996). Surprisingly, the *A. vinelandii* genome also encodes a mannuronan C-5-epimerase (AlgG), which belongs to a different class from AlgE1–E7, and this epimerase seems to represent the equivalent of AlgG in *P. aeruginosa* (Rehm et al., 1996). In addition *A. vinelandii* encodes a protein

(AlgY) containing one A and one R module, but this protein appears to display no epimerase activity after expression in *Escherichia coli* (Svanem et al., 1999). Recently, the entire alginate biosynthesis gene cluster of *A. vinelandii* was identified, which revealed a similar physical organization of the *alg* genes as has been found in *P. aeruginosa* (Lloret et al., 1996; Rehm et al., 1996). A similar arrangement of *alg* genes has been also described in *Pseudomonas syringae* pv. *syringae* (Penaloza-Vazquez et al., 1997), and is also evident for *P. fluorescens* based on genome sequence analysis (B. H. A. Rehm, unpublished results). These data suggested that the physical arrangement of alginate genes in bacteria is conserved.

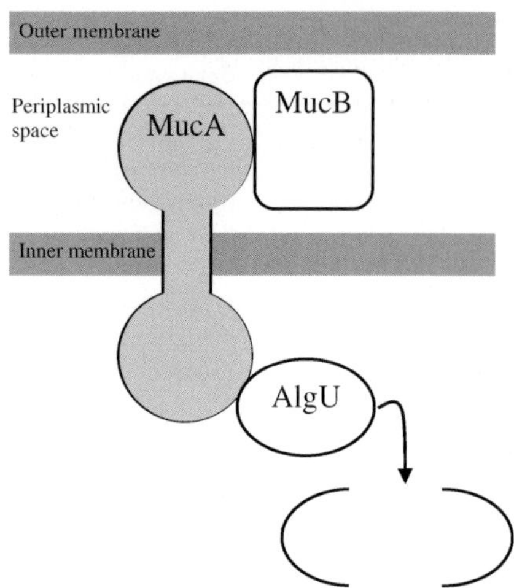

Fig. 4 Model of the action of anti-sigma factors MucA and MucB. AlgU is the alternative sigma-factor, which is required for *alg* gene expression. Binding of AlgU to MucA might expose AlgU to proteolytic digestion by MucD (Mathee et al., 1997).

6 Regulation of Alginate Biosynthesis

The regulation of alginate biosynthesis is complex, and involves specific gene products and those that act more globally (Figure 3 and Figure 4). Expression of the entire alginate biosynthesis gene cluster of *P. aeruginosa* is under the control of the *algD* promoter, and in essence this region (*algD-algA*) acts as an operon (Chitnis and Ohman, 1993), although there is sequence-based evidence for weak promoters within the gene cluster. One of the alginate biosynthesis genes, *algC*, is located at 10 min on the PAO1 map and is transcribed independently of the 34-min region, but is coordinately induced with the biosynthesis gene cluster. The regulatory genes in *P. aeruginosa* map at 9 min and 13 min, and genes responsible for a genotypic switch to alginate overproduction are found at 68 min (see Figure 3). The alginate structural genes are controlled via the positively regulated *algD* promoter (Deretic et al., 1989). In *A. vinelandii*, on the other hand, there seems to exist two additional *algD*-independent promoters, from which *alg8-algJ* and *algG-algA* are transcribed, respectively. These promoters were found to be independently regulated (Lloret et al., 1996; Vazquez et al., 1999). The genotypic switch region comprising *algU, mucA, mucB, mucC* and *mucD* (831,000–835,000, positions relative to the *P. aeruginosa* genome sequence) is found in both bacteria, and the corresponding gene sequences are highly homologous (see Figure 3). The genes have also been found to be biologically active and play a similar role in the two species (Martinez-Salazar et al., 1996). All the known *alg* genes and their corresponding proteins are listed in Table 1.

6.1 Environmentally Induced Activation of *alg* Genes

Conditions of high osmolarity, N, P or carbon starvation, dehydration, and the presence of phosphorylcholine activate a cascade of regulatory proteins in *P. aeruginosa* involved in the activation of the *algD*-promoter (Gacesa and Russell, 1990; Terry et al., 1991). Genes located at a region spanning 9 and 13 min on the *P. aeruginosa* chromosome, *algR(algR1)*, *algQ(algR2)* and *algP(algR3)*, and *algB* modulate the production of alginate and have been described as auxiliary regulators of mucoidy (Govan and Deretic, 1996). A two-component signal-transduction pathway comprising the putative sensor proteins AlgQ (kinase) and AlgZ, interacting with the cognate response regulator proteins such as AlgR and AlgB, were identified (see Figure 3) (May and Chakrabarty, 1994; Yu et al., 1997). Analysis of sequence data indicates that *algZ* encodes a sensory component of a signal transducer system, but that it lacks several expected motifs typical of histidine protein kinases (Yu et al., 1997). The best characterized of these regulators is AlgR, which gene is transcribed in response to the protein AlgU. AlgR binds to three sites upstream of the *algD* promoter and, in conjunction with AlgU, up-regulates transcription of *algD* and the downstream genes. AlgR also promotes expression of *algC*. The efficiency of AlgR is increased by phosphorylation by the cognate kinase AlgQ. AlgB also modulates *algD* expression and, based on sequence analysis, is a member of the NtrC subclass of two-component regulators (Wozniak and Ohman, 1991); however, the *algB* and *algR* regulatory systems appear to operate independently of each other (Wozniak and Ohman, 1994). Interestingly, AlgB and AlgR showed phosphorylation-independent activity on the induction of alginate biosynthesis (Ma et al., 1998). Binding of these positive regulators upstream of the *algD* promoter, presumably leads to the formation of a suprahelical structure with the aid of the histone-like AlgP protein, causing activation of transcription (Deretic and Konyecsni, 1989; Konyecsni and Deretic, 1990; Deretic et al., 1994). A comprehensive account of the inter-relationships of the regulators has been reviewed (Govan and Deretic, 1996). Furthermore, the recently identified sigma-like factor AlgU (AlgT) is responsible for the initiation of *algD* transcription (se Figure 3) (Hershberger et al., 1995). On the basis of sequence analysis (Martin et al., 1994), AlgU is a member of the σ^E class of sigma factors, i.e., analogous to RpoE of *Escherichia coli*, and is essential for alginate production. Subsequent studies have established that AlgU and RpoE are functionally equivalent (Yu et al., 1995), and that AlgU forms complexes with RNA polymerase (Schurr et al., 1995). AlgU causes an increase in alginate biosynthesis by a direct action on the *algD* promoter (see below), and indirectly by up-regulating transcription of another regulatory gene, *algR* (Martin et al., 1994). This environmentally induced transcription of the *alg* cluster and the resulting production of alginate occur only at a rather low level.

6.2 Genotypic Switch

Copious amounts of alginate are only produced in combination with inactivation (mutations) of the negative regulators of the AlgU activity (anti-sigma factors) MucA (AlgS) or MucB (AlgN) (Martin et al., 1993a,b). Mutational inactivation (genotypic switch) of MucA, MucB (Schurr et al., 1996) or MucD (Boucher et al., 1996) leads to full activity of AlgU, which allows strong transcription of the *alg* operon (Figure 4). MucA is supposed to be located in the cytoplasmic membrane interacting with MucB in the periplasm upon perception of an unknown stimulus, and transducing a signal to the cytoplasm which mediates degradation of

AlgU presumably due to the action of MucD (Mathee et al., 1997) (see Figure 4). MucD is orthologous to the *E. coli* periplasmic protease and chaperone DegP. DegP homologues are known virulence factors that play a protective role in stress responses in various species. Recently, negative regulation of AlgU by anti-sigma factors MucA and MucB and the transcriptional regulation of the *algD* gene have been also described for *A. vinelandii* (Campos et al., 1996; Martinez-Salazar et al., 1996).

7
Polymerization and Export of the Alginate Chain

Since no undecaprenol-linked intermediate has been identified in either *P. aeruginosa* or *A. vinelandii*, the polymerase–which presumably is localized as a protein complex in the cytoplasmic membrane–might synthesize alginate by an undecaprenol-independent mechanism (Sutherland, 1982). Alginate synthesis might occur similarly to bacterial cellulose synthesis. The cellulose synthase of *Acetobacter xylinum* is localized in the cytoplasmic membrane, and appears to be a protein complex of 420 kDa. This enzyme catalyzes the processive polymerization of glucose residues (from UDP-glucose), and the nascent β-(1,4)-linked glucosan chains appear to remain attached to the synthase during polymerization (Ross et al., 1991). Correspondingly, the alginate polymerase in the cytoplasmic membrane might accept the GDP-mannuronic acid at the cytosolic site while simultaneously translocating the nascent alginate chain through the cytoplasmic membrane (Rehm and Winkler, 1996). Preliminary studies using ^{14}C-GDP-mannuronic acid as substrate and defined oligomannuronates as primer have revealed that the envelope fraction of mucoid *P. aeruginosa* exhibited *in vitro* alginate polymerase activity (B. H. A. Rehm, unpublished results). Biochemical and electrophysiological characterization of AlgE revealed that it forms an anion-selective pore in the outer membrane (Rehm et al., 1994b). This pore could be partially blocked by GDP-mannuronic acid in lipid bilayer experiments. In addition, a topological model of AlgE has been developed and, according to this model, the protein is a β-barrel consisting of 18 β-strands (Rehm et al., 1994b) (Figure 5). A three-dimensional model of AlgE was developed by homology modeling, indicating pore diameters eligible for alginate export (see Figure 5). These data are consistent with the hypothesis that AlgE forms an alginate-specific pore that enables export of the nascent alginate chain through the outer membrane. Figure 6 summarizes all the findings regarding polymerization, modification and export in a model.

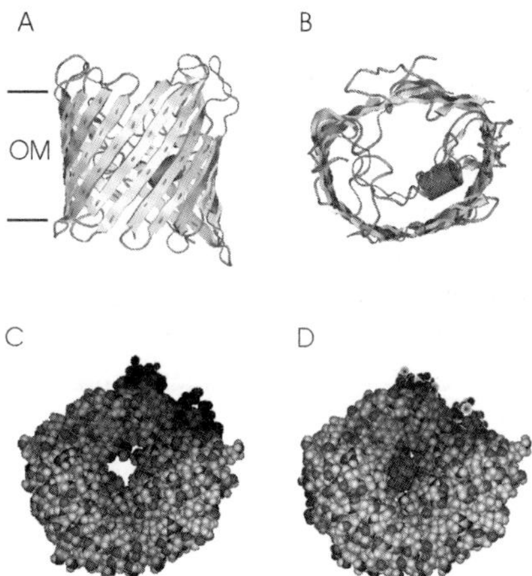

Fig. 5 Topological model of the outer membrane protein AlgE from *P. aeruginosa*. Experimental evidence was obtained that AlgE is involved in export of alginate through the outer membrane. (A) Side-view of the AlgE model (the bottom is exposed to periplasm, whereas the top is cell-surface-exposed. OM: outer membrane. (B) Top view of the AlgE model from outside the cell. (C) Top view of the AlgE model (in CPK format) from outside the cell. D, C with inserted alginate chain (dark gray).

8 Alginate-Modifying Enzymes

The alginate-modifying enzymes (the transacetylase and the C-5-mannuronan epimerase) carry N-terminal signal sequences and are mainly found in the periplasm in *P. aeruginosa* (Franklin et al., 1994). The corresponding alginate-modification reactions occur at the polymer level, presumably in the periplasm (see Figure 6). The genes encoding the transacetylase and other proteins involved in transacetylation (*algI, algJ, algF*), the epimerase (*algG*), and the lyase (*algL*) have been cloned, and the gene products have been characterized (Franklin and Ohman, 1993, 1996; Shinabarger et al., 1993; Franklin et al., 1994; Monday and Schiller, 1996). Transacetylation occurs at position(s) O-2 and/or O-3 of the mannuronic acid residue, preventing these residues from being epimerized to guluronic residues by AlgG and from degradation by AlgL (Franklin and Ohman, 1993; Franklin et al., 1994; Wong et al., 2000).

Thus, the periplasmic acetylase indirectly controls the periplasmic epimerase and lyase activity on the alginate polymer. The increasing degree of acetylation also causes the alginate polymer to have an enhanced water-binding capacity. This might be particularly important under dehydrating conditions, for example during infection and colonization of the lungs of patients with CF. The alginate lyase presumably functions as an editing protein to control the length of the polymer, or it might also serve the polymerase with alginate oligomers to prime synthesis. The lyase is not involved in providing a carbon source (Boyd et al., 1993).

8.1 Mannuronan C-5-epimerases

The G residues in alginates originate from a post-polymerization reaction catalyzed by mannuronan C-5-epimerases. In *P. aeruginosa*, and presumably also in other species belonging to this genus, there appears to be only one such enzyme encoded by the gene *algG* (Franklin et al., 1994). Like the proteins necessary for acetylation (Franklin and Ohman, 1996), the Ca^{2+}-independent AlgG is also probably located in the periplasm. The *A. vinelandii* genome also contains an *algG* homologue (Rehm et al., 1996), but this species in addition modifies its alginates by a family of extracellular Ca^{2+}-dependent epimerases that according to sequence-alignment studies are unrelated to AlgG. Seven

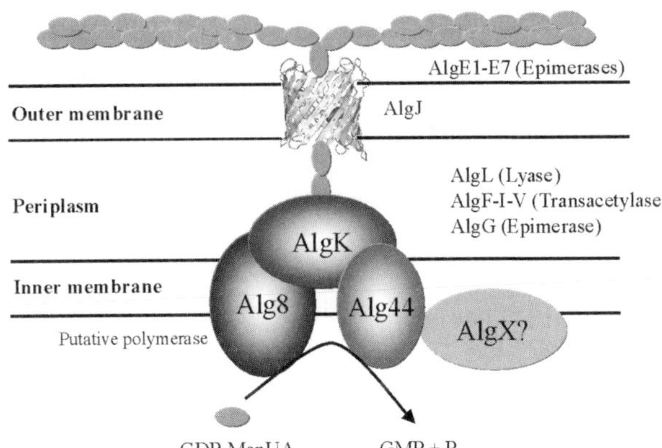

Fig. 6 Model of alginate polymerization, modification and export in *A. vinelandii* (Modified according to Rehm and Valla, 1997).

such enzymes (AlgE1–7) are now known (Ertesvåg et al., 1994, 1995; Svanem et al., 1999), and they can all be seen as composites of two structurally distinct modules, designated A and R. The A modules (about 385 amino acids) are present in one or two copies in each enzyme, while the R modules (about 155 amino acids) are present in one (AlgE4) and up to seven (AlgE3) copies. The N-terminal ends of each R module contain four to seven copies of a nine-amino-acids motif putatively involved in the binding of Ca^{2+}. In addition, *A. vinelandii* encodes a protein (AlgY) containing one A and one R module, but this protein appears to display no epimerase activity after expression in *E. coli* (Svanem et al., 1999).

8.1.1 Functional Differences

The A modules alone appeared to be sufficient both for catalyzing the epimerization reaction and for determining the epimerization pattern (Ertesvåg and Valla, 1999). It is, therefore, particularly important to understand the structure–function relationships in these modules. The epimerization patterns generated by all seven AlgE epimerases can be divided into two main groups: those which almost exclusively generate MG-blocks; and those which form G- blocks. From NMR spectroscopy analyses of the reaction products of all the enzymes, it is immediately obvious that only AlgE4 belongs to the first group. In addition, the C-terminal parts of AlgE1 and AlgE3 (AlgE1-2 and AlgE3-2) display this property, but it is not reasonable that these truncated forms are made *in vivo* in *A. vinelandii*. It is not known why the *A. vinelandii* genome encodes a specialized enzyme like AlgE4, but the physical properties of the alternating structure of its reaction product can be predicted to be quite different from the gel-forming G-block alginates (Smidsrød and Draget, 1996). All Ca^{2+}-dependent epimerases except AlgE4 are involved in the formation of G-blocks. AlgE3 and AlgE1 are composite enzymes, and due to the properties of each part they can both make long G-blocks and presumably put alternating structures between them. Comparison of the epimerization patterns of these two enzymes indicates that they share similar properties, but it was observed that less AlgE3 (measured as initial activity) compared to AlgE1, is needed to obtain high degrees of epimerization. For AlgE1 (Ertesvåg et al., 1998b) and AlgE3 the relative amount of G-blocks also increases with decreasing concentration of Ca^{2+}. This property, which was not observed for AlgE2 (Ramstad et al., 1999), probably reflects that the C-terminal part of AlgE3 displays less activity at low concentrations of Ca^{2+} than the N-terminal part, similar to what has been observed earlier for AlgE1 (Ertesvag et al., 1998b). It is also known that whole AlgE1 needs only 0.8 mM $CaCl_2$ for full activity (Ertesvåg et al., 1998b), while AlgE3 requires 3 mM $CaCl_2$, and displays less than 40% of full activity at 1 mM concentration of this cation. At low or moderate levels of epimerization, AlgE6 introduces more alternating structures than AlgE2 and AlgE5, although less than the composite enzymes AlgE1 and AlgE3. The average lengths of the G-blocks at about 40% epimerization, on the other hand, are similar to those made by AlgE2 and AlgE5. Alginate may, however, be epimerized to 90% G by AlgE6, and such highly epimerized alginate contain very long G-blocks and can be predicted to form very strong gels. This degree of epimerization has so far not been achieved using AlgE2 or AlgE5. These different epimerase specificities may indicate the requirement of the organism to generate alginate structures of importance for formation of the metabolically dormant and alginate-containing cysts which are generated under conditions of environmental stress in *A. vinelandii* (Sadoff, 1975). Furthermore, the significant alginate lyase activity of AlgE7 might be needed for the germination of the cysts (Wyss et al., 1961).

8.1.2 The Biotechnological Potential

Commercially, the alginates are harvested from different species of brown algae (Smidsrød and Draget, 1996), and it seems very likely that the composition of these products will not always meet industrial and biotechnological needs. In addition, alginate structures that are not available from seaweeds might have properties that would open the possibilities for totally new applications. By epimerizing algal alginates with the recombinantly produced epimerases, one might therefore be able either to upgrade their value or to generate completely new products. Since it is already clear that this is feasible, the questions are rather what prices are acceptable for each particular application, what the potential uses are, and how the market will react to products made using recombinant enzymes. A further extension of alginate modifications *in vitro* would be to design cells that directly synthesize the alginates of interest *in vivo*. This might lead to products of lower price, but it may prove to be more difficult to obtain the same level of control as achieved by in-vitro epimerization. In any case, the mannuronan C-5-epimerization system raises many interesting questions for basic science and applied biotechnology, and consequently the enzymes are likely to be the subject of active studies for many years to come.

8.2 O-Transacetylases

The products of *algI*, *algJ*, and *algF* from *P. aeruginosa* are required for the addition of O-acetyl groups to the alginate polymer, and mutations in *algI*, *algJ*, or *algF* resulted in production of an alginate polymer that was not O-acetylated (Shinabarger et al., 1993; Franklin and Ohman, 1996; Nivens et al., 2001). The mannuronate residues undergo modification by C-5 epimerization to form the L-guluronates and by the addition of acetyl groups at the O-2 and O-3 positions. By using genetic analysis, *algF* was identified and located upstream of *algA* in the 18-kb alginate biosynthesis operon, as a gene required for alginate acetylation (Franklin and Ohman, 1993; Shinabarger et al., 1993). An *algI*::*Tn501* mutant, which was defective in *algIJFA* because of the polar nature of the transposon insertion, produced alginate when *algA* was provided in *trans*. This indicated that the *algIJF* gene products were not required for polymer biosynthesis. To examine the potential role of these genes in alginate modification, mutants were constructed by gene replacement in which each gene (*algI*, *algJ*, or *algF*) was replaced by a polar gentamicin resistance cassette (Franklin and Ohman, 1996). Proton nuclear magnetic resonance (NMR) spectroscopy showed that polymers produced by strains deficient in *algIJF* still contained a mixture of D-mannuronate and L-guluronate, indicating that C-5 epimerization was not affected. Alginate acetylation was evaluated by a colorimetric assay and Fourier transform-infrared (FTIR) spectroscopy, and this analysis showed that strains deficient in *algIJF* produced nonacetylated alginate. Plasmids that supplied the downstream gene products affected by the polar mutations were introduced into each mutant. The strain defective only in *algF* expression produced an alginate that was not acetylated, confirming previous results. Strains missing only *algJ* or *algI* also produced nonacetylated alginates. Providing the respective missing gene (*algI*, *algJ*, or *algF*) *in trans* restored alginate acetylation. Mutants defective in *algI* or *algJ*, obtained by chemical and transposon mutagenesis, were also defective in their ability to acetylate alginate. Therefore, *algI* and *algJ* represent newly identified genes that, in addition to *algF*, are required for alginate acetylation. Once in the periplasmic space, alginate is O-acetylated by the combined action of the *algF*, *algI* and *algJ* gene products. Mutants deficient in any one of these three genes are unable to produce O-acetylated alginate, but the epimerization process and overall yields of alginate appear to be unaffected (Shinabarger et al., 1993; Franklin and Ohman, 1996). Acetyl-CoA is almost certainly the primary donor of O-acetyl groups for alginate modification; however, this metabolite is localized in the cytoplasm, whereas the O-acetylation process occurs in the periplasm. Therefore, at least one of the *algF*, *algI* or *algJ* gene products is likely to be involved in transport of O-acetyl groups across the cell membrane into the periplasmic space. Sequence data indicate that AlgI is probably a membrane-bound protein and may fulfil this role (Franklin and Ohman, 1996). AlgF has a signal peptide, which is processed by *E. coli*, indicating that it is a periplasmic protein (Shinabarger et al., 1993) and therefore could be the O-acetyltransferase. AlgJ shows remarkable similarity (30% identity, 69% similarity) to another gene product, AlgX (Monday and Schiller, 1996), which is of unknown function but is essential for alginate biosynthesis. However, neither AlgJ nor AlgX shows any significant similarity to other proteins in the databases and therefore their function in the O-acetylation process remains unresolved at this stage. The observation that a cell suspension of *P. syringae* is able to O-acetylate seaweed alginate suggests that this event is either periplasmic or extracellular (Lee and Day, 1995). It is likely that the O-acetylation of mannuronate is catalyzed by at least two enzymes each specific for either the 2- or 3-hydroxyl on the sugar ring. However, at this stage it is not known which gene products might be involved in determining these specific modifications.

8.3 Alginate Lyases

It is not clear why a degradative enzyme should be expressed concurrently with the enzymes involved in biosynthesis. One suggestion has been that the lyase may be involved in excising polysaccharide fragments from the biosynthesis complex, although there is no real evidence to support this contention. It is clear though that overexpression of the *algL* gene product (May and Chakrabarty, 1994) or of other alginate lyases (Gacesa and Goldberg, 1992) results in the release of planktonic bacteria from biofilms.

8.3.1 Reaction Mechanism

Alginate lyase catalyzes the degradation of alginate by a β-elimination mechanism, targeting the glycosidic 1,4 O-linkage between monomers. A double bond is formed between the C4 and C5 carbons of the six-membered ring from which the 4-O-glycosidic bond is eliminated, depolymerizing alginate and simultaneously yielding a product containing 4-deoxy-L-*erythro*-hex-4-eno-pyranosyluronic acid as the nonreducing terminal moiety (Haug et al., 1967). Gacesa (1987, 1992) proposed a catalytic mechanism for alginate lyase that described a three-step reaction to depolymerize alginate. This mechanism may also be shared with epimerase, another enzyme that acts on the alginate polymer. The two reactions differ only in the last step of the three-stage transformation of alginate. The three steps include: (1) removal of the negative charge on the carboxyl anion – essentially neutralizing the charge by a salt bridge (lysine or arginine may be the candidate residue); (2) a general base-catalyzed abstraction of the proton on C5 (aspartic acid, glutamic acid, histidine, lysine, and cysteine have been suggested for this role), where one residue may be required as the proton abstractor and another as the proton donor, although the proton may be derived from the solvent environment; and (3) a transfer of electrons from the carboxyl group to form a double bond between C4 and C5, resulting in the β-elimination of the 4-O-glycosidic bond. In the proposed mechanism for epimerase, the replacement of the proton at C5 (epimerization) takes place in step 3. The principle of the catalytic mechanism was confirmed by the identification of putative catalytic residues by structural analysis of the *Sphingomonas* lyase ALY1-III complexed with a trisaccharide (Yoon et al., 2001).

8.3.2
Function in Alginate-Producing Bacteria

Very few bacteria synthesize polysaccharides as well as the specific degrading enzymes (Kennedy et al., 1992; Sutherland, 1995). Periplasmically localized alginate lyases have been found in various species of *Pseudomonas* and *Azotobacter* that can synthesize an extracellular alginate but are unable to use alginate as a carbon or energy source. The extracellular poly(M)-rich alginate produced by these bacteria is O-acetylated at the C2 and/or C3 positions on the M residues to various degrees, which makes this polymer more resistant to degradation by the endogenously produced lyase (Nguyen and Schiller, 1989; Kennedy et al., 1992). Interestingly, alginate lyase genes in *Pseudomonas* and *Azotobacter* species are localized within their respective alginate biosynthesis gene clusters. This genetic organization has been reported in *P. aeruginosa* (Chitnis and Ohman, 1993), *P. syringae* pv. *syringae* (Penaloza-Vazquez et al., 1997), *A. vinelandii* (Rehm et al., 1996), and *A. chroococcum* (Pecina et al., 1999). This localization of the alginate lyase gene (*algL*) raises questions about the function of AlgL in alginate biosynthesis. It is possible that the lyase works as part of a polymerization complex within the periplasm, assembling the alginate exopolysaccharide for transport to the cell surface, or it may provide oligomeric alginate, which might serve as primer. Accordingly, the lyase-negative *P. aeruginosa* produced only small amounts of alginate (Monday and Schiller, 1996). However, Boyd et al. (1993) reported that lyase was not required for alginate production by *P. aeruginosa*. A similar study demonstrated that the absence of lyase activity reduced alginate production by *P. syringae* pv. *syringae* by ~50% (Penaloza-Vazquez et al., 1997). Hence, alginate lyase seems not to be essential for alginate biosynthesis, but for maximum production. May and Chakrabarty (1994) proposed that the lyase may function either as an editing protein to control the length of the alginate polymer or to provide short pieces of alginate to prime the polymerization reaction. The nascent polymannuronate in the periplasmic space would be most sensitive to endogenous lyase before acetylation and epimerization, whereas after modification, the polymer would be ready for export to the cell surface. The levels of Ca^{2+} and Mn^{2+} ions have been shown selectively to activate or inhibit alginate lyase and epimerization activity in algae (Madgwick et al., 1978). Since various cations have a strong influence on bacterial alginate lyase activity (Wong et al., 2000), this observation suggests that periplasmic ionic conditions could also regulate alginate modification in *P. aeruginosa*.

The production of lyase by *P. aeruginosa* may also be important in facilitating dissemination of the bacteria (Boyd and Chakrabarty, 1994). Alginate synthesis is increased upon attachment of the bacteria to a cell surface, resulting in stronger bacterial adhesion to the surface and colonization. However, overexpression of the lyase gene within a mucoid strain of *P. aeruginosa* led to a decrease in the length of alginate polymers and increased bacterial detachment from the surface (Boyd and Chakrabarty, 1994). Thus, cleavage of the alginate polymer within *P. aeruginosa* biofilms could enhance detachment of the bacteria, allowing them to spread and colonize new sites. In *Azotobacter* sp., alginate is produced by vegetatively growing cells as capsule, and by cells in the metabolically dormant-state in the cyst coat (Page and Sadoff, 1975, Sadoff, 1975). Both *A. vinelandii* and *A. chroococcum* strains produce M-specific endolytic lyases that are localized in the periplasmic space. These lyases may be biologically important in the differentiation of *Azotobacter* cells during encystment, when they most likely play a role in concert with epimerases to form the desiccation-resistant cyst capsule. It has now been reported that *A. vinelandii* has multicopy epimerase genes, making a gene family that is highly likely to be responsible for the synthesis of complex alginates of various polymeric composition in the cyst capsule (Svanem et al., 1999). One of these epimerases, AlgE7, also exhibited lyase activity. Although alginate-negative strains of *A. vinelandii* fail to encyst, it is not yet clear whether alginate lyase-negative strains can differentiate. An increase in alginate lyase activity just before cyst germination (Kennedy et al., 1992) suggests that lyase expression might support this process.

From the collective information on alginate and alginate lyases, it is concluded that alginate lyases are important enzymes in a broad spectrum of biological roles and applications. The lyases maintain a balance in the cell physiology of alginate-producers that efficiently use alginate as functional biopolymers and also in the natural environment, where the recycling of alginate is achieved through metabolic breakdown (for review, see Wong et al., 2000). In addition, the ability of lyases selectively to depolymerize alginate–which has become a very useful industrial polysaccharide–makes them important tools with great potential for advanced biotechnological uses.

8.3.3 Structure–Function Analysis

The cloning and sequencing of many alginate lyase genes have now enabled investigators to focus on structure–function analysis of this enzyme. DNA sequences and primary amino acid sequences are available for 23 alginate lyases (Wong et al., 2000). Various alginate lyases have been characterized with respect to enzyme properties (Wong et al., 2000; Table 2). Meanwhile, the coordinates of the three-dimensional structure of ALY1-III *Sphingomonas* sp. can be found in the Protein Data Bank (Brookhaven, NY) (Yoon et al., 1999). Based on sequence information, most alginate lyases appear to fall into three major classes according to their molecular mass: 20–35 kDa; ~40 kDa; and ~60 kDa. Analysis of amino acid sequences of alginate lyases revealed that they contain a hydrophilic central region and a hydrophobic sequence in the C terminus, with a slightly charged end. All alginate lyases share amino acid sequence similarities from 18 to 95%. Although several regions of similarity exist between these lyases, the core region residues of the lyases within the 40-kDa class

Tab. 2 Alginate lyases from Gram-negative bacteria: localization, characteristics, and sequence accession numbers (modified according to Wong et al., 2000)

Source	*Localization*	*Substrate specificity*	*Endo/exolytic*	*Cleavage site*	*Molecular mass (kDa)*	*pI*	*Opt. pH*	*Opt. T*	*Cations needed*	*GenBank accession No.*
A. chroococcum	Periplasmic	AlgL: M	Endolytic	–	43	–	–	30°C, pH 7.5	Na^+, K^+ Mg^{2+}	AJ223605
	Periplasmic	M			–	–	6.8	30°C		N/A
A. chroococcum 4A1M	Extracellular	PolyM	Endolytic	–	23–24	5.6	6	60°C	Ca^{2+}	N/A
A. vinelandii	Intracellular	M	–	–	~50		7.5	–	–	N/A
	Periplasmic	M			–	–	7.2	30°C		N/A
	Periplasmic	AlgL: M, acetyl'd (nonconsecutive M)	Endolytic not G-M	M-M and M-G,	39	5.1	8.1–8.4		Na^+, divalents not needed	AF037600
	Periplasmic	AlgE7 (epimerase and lyase): M and G	Endolytic	G-GM and G-MM	(384aa)	–	–	–	–	AF099800
A. vinelandii	N/A	M	N/A	N/A	(375aa)	N/A	N/A	N/A	N/A	AF027499
Enterobacte cloacae M-1	Extracellular	G	Endolytic	–	32–38	8.9	7.8	30°C	Ca^{2+}, Al^{3+}, Mn^{2+}	N/A
	Intracellular	G	Endolytic	7 subsites, cleaves G-G between sub-sites 2 and 3	31–39	8.9	7.5	40°C	–	N/A
K. aerogenes type 25	Intracellular	PolyG	Endolytic	–	28–31.6	–	7	37°C	Na^+ and K^+ (0.1–0.3 M)	N/A
	Extracellular	G	Endolytic	G-G, G-M	–	–	7	–	Mg^{2+} (0.05–0.1 M)	N/A
K. pneumoniae subsp. aerogenes	Extra/intra-celluar (R)	AlyA: polyG	–	–	28 8.9	–	–	–	N/A	25
	Extracellular (R)	AlyA: polyG	Endolytic		31.4 9.39 (calc.)	7.0 –7.6 h		50°C	Na^+	L19657
Pseudomonas sp. W7	N/A (R)	N/A	N/A	N/A	(345aa)	N/A	N/A	N/A	N/A	AF050114
P. aeruginosa	Intracellular	M, nonacetyl'd	–	–	–	–	6.2	20–40°C	–	N/A
P. aeruginosa	Intracellular	AlgL: M nonace-tyl'd	–	–	53	–	8	–	Mg^{2+}, K^+, Na^+	N/A
P. aeruginosa CFl/Ml	Periplasmic	AlgL :M,	Endolytic	6 subsites, nonacetyl'd cleaves M-M between sub-sites 3 and 4	–	–	–	–	–	N/A
P. aeruginosa	Intracellular (R)	AlgL: M, nonace-tyl'd	Endolytic	–	43.5	–	–	–	–	L14597
P. aeruginosa FRD1	Peripl.(N)/intracell.(R)	AlgL :M, nonace-tyl'd	Endolytic	M-M	39	9 (calc.)	7.0	–	Mg^{2+} , Na^+	U27829/L09724
P. aeruginosa	N/A (R)	AlgY :M	N/A	N/A	(685aa)	N/A	N/A	N/A	N/A	Z54213

Tab. 2 (cont.)

Source	*Localization*	*Substrate specificity*	*Endo/exolytic*	*Cleavage site*	*Molecular mass (kDa)*	*pI*	*Opt. pH*	*Opt. T*	*Cations needed*	*GenBank accession No.*
P. maltophilia and *P. putida*	Intracellular	M, acetyl'd	Endolytic	–	–	–	7.7–7.8	28–30°C	–	N/A
P. syringae pv. syringae	Periplasmic	PolyM (prefers deacetylated)	Endolytic		40	8.2	7	42°C	Cations not needed	AF22020
Pseudomonas	Intracellular	Multiple		–	–	90, 72, 60, 54	–	–	–	– N/A
sp. OS-ALG-9	Intra/ extracellular (R)	ALY or AlyP.M	Endolytic (preferred), G		46.3	–	–	–	AlyP:—	D10336
	Intracellular	(ALY or AlyP:—)	Endolytic	–	45	–	7.5	45°C	AlyP:—	N/A
	Intracellular (R)	ALYII : M	Endolytic	–	79.8 (calc.)	8.3	7	30°C	ALYII: None required; EDTA stimulates activity	AB003330
Sphingomonas sp.	Cytoplasmic	ALY1-I :M, nonacetyl'd, acetyl'd		Endolytic	–	60	9.03	7.5–8.5	70°C	ALY1-I, -II, -III: 2009330A
	Cytoplasmic	ALY1-II :G	Endolytic	–	25	6.82	7.5–8.5	70°C	None required;	sequenced
	Cytoplasmic	ALY1-III : M, acetyl'd (highly active) and nonacetyl'd	Endolytic	M-M, hetero MG	38	10.16	7.5–8.5	70°C	EDTA: no effect	1QAZ

N/A, not available

that share significant alignment are very well conserved; especially notable is the highly conserved six-amino-acid hydrophilic sequence "NNHSYW" in the center of the protein sequences. This region is also conserved in the alginate lyase ALY1-III from a *Sphingomonas* sp., which otherwise differs significantly in amino acid sequence from the other lyases in this grouping. Yoon et al. (1999) recently solved the three-dimensional structure for ALY1-III, revealing a structure with 12 α-helices, organized in a twisted α/α helix barrel composed of six inner and five outer helices. Analysis of this conformation identified a deep tunnel-like cleft, which was proposed as the catalytic site. The highly conserved NNHSYW sequence is located in the center of this cleft. As described below, studies suggest that alteration of the histidine residue in this site inactivates the lyase, providing evidence that this region is critical for enzyme activity. There is also a well-conserved nine-amino-acid hydrophobic sequence, "WLEPYCALY," in the C terminus of the lyases from *P. aeruginosa* and *Azotobacter* sp. This nine-amino-acid block is weakly conserved in ALY1-III from *Sphingomonas* sp. Sequence and structural information indicates that this nine-amino-acid region is in H11, an inner α-helix of ALY1-III (Yoon et al., 1999).

The three-dimensional structure for alginate lyase from *Sphingomonas* sp. revealed an interesting feature of the enzyme, a disulfide bridge between Cys188 and Cys189, giving the structure a tight turn at the edge of the proposed active-site cleft (Yoon et al., 1999). This disulfide bond immediately precedes the conserved sequence of "NNHSYW" and is suggested to be very important in the maintenance of the active site conformation. This cysteine-cysteine pair in the alginate lyase of *Sphingomonas* sp. is a structural feature that is unique to this lyase, which is not surprising because the primary protein sequence of ALY1-III departs significantly from the general consensus of the majority of alginate lyases in the 40-kDa group. M-specific ALY1-III has an α-helix-rich structure and has no β-strand or sheet conformation (Yoon et al., 1999). The ALY1-III structure was used as a template to develop a threading model of the alginate lyase of *P. aeruginosa*, which exhibited 22% similarity to ALY1-III (B. H. A. Rehm, unpublished results). This model showed the tunnel-like cleft with the potential catalytic residues N171, H172, R219, Y226 closely arranged inside the cleft. Residues N171 and H172 are located in the conserved motif "NNHSYW" (Figure 7). Considering the many differences observed for the alginate lyases (primary sequence, M_W, substrate specificities, hydrophobicity profiles, etc.), it is important that additional enzymes will be biochemically analyzed as well as by X-ray crystallography. This will allow complete comparisons of structure and function between these enzymes, as well as related enzymes, and define their relationship with the structure of ALY1-III from *Sphingomonas* sp. (Yoon et al., 1999).

8.3.4 Substrate-Binding and Catalytic Sites

Experiments with defined alginate oligomers have indicated the specific substrate recognition sequences for alginate lyase action, highlighting the optimal capacity of the catalytic site for substrate units. Using a series of oligomannuronates (n D 2–9), Rehm (1998) reported that the poly(M) lyase from *P. aeruginosa* CF1/M1 would not accept oligomers that were smaller than five mannuronates for β-elimination, and that the highest enzyme activity was noted with the hexameric mannuronate oligomer, whereas the trimer was the most abundantly accumulating product. This study suggested that the catalytic site of alginate lyases probably

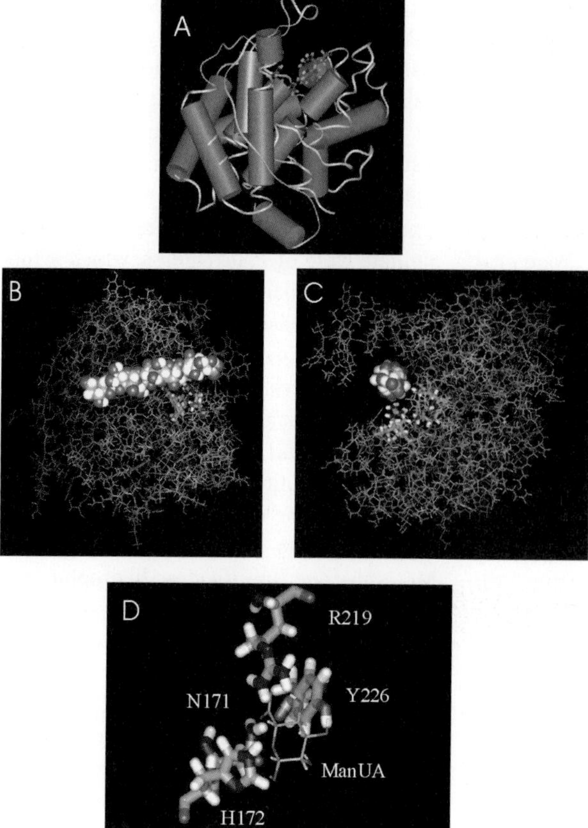

Fig. 7 Threading model of the alginate lyase from *P. aeruginosa* based on the structure of the *Sphingomonas* sp. ALY1-III alginate lyase. (A) Alginate lyase threading model indicating the secondary structure composition (cylinders represent α-helical segments) and amino acid side chain of the putative catalytic residues were demonstrated (see D). (B) The alginate lyase model in stick format complexed (docked) with the hexameric oligomannuronate and emphasis of the putative catalytic residues by thicker sticks in the tunnel-like cleft. (C) Model (B) turned around by 90°. (D) Spatial arrangement of the putative catalytic residues complexed (docked) with a mannuronic acid monomer (ManUA).

accommodates five to six residues. In the crystal structure of ALY1-III (Yoon et al., 1999), His192 is located in the highly conserved region NNHSYW, which is in the center of the proposed active-site cleft. Chemical modification of histidine residues in ALY1-III inactivated the enzyme (Yoon et al., 1999), which suggests strongly the role of histidine in the catalytic activity of ALY1-III. In ALY1-III, four conserved tryptophan residues, together with other aromatic residues, are located along the cleft of the active site, and two conserved arginine residues flank the entrance to the cleft. The charged arginine and lysine residues on both sides of the cleft and the aromatic side chains lining the active site have been suggested to be substrate-binding molecules. This active cleft can accommodate at least five residues of poly(M) along the curved surface (Yoon et al., 1999). The highly conserved sequence "INNHSY" located in the central region of the 40-kDa lyases is also highly conserved as "E/FNNVSY" in mannuronan C5-epimerases (Ertesvag et al., 1998b). This motif, found in alginate lyases and epimerases from *P. aeruginosa* (Boyd et al., 1993; Schiller et al., 1993; Franklin et al., 1994), *A. chroococcum* (Pecina et al., 1999), and *A. vinelandii* (Rehm et al., 1996; Ertesvag et al., 1998a; Svanem et al., 1999), appears to be mainly located within an average of 200 residues from the encoded N terminus of the proteins. However, this pattern of residues is not found in the G-lyases or the 30-kDa alginate lyases. Because both the alginate lyases and epimerases from *P. aeruginosa* and *A. vinelandii* are active on mannuronate units of alginate, it is highly possible that this pattern could be a binding motif for the poly(M) or the mannuronate and its glycosidic bond. The AlgG epimerases of the alginate biosynthesis operon in *A. vinelandii* (Rehm et al., 1996) and *P. aeruginosa* (Franklin et al., 1994) differ slightly from the epimerases found in the *A. vinelandii* epimerase gene family cluster (Rehm et al., 1996; Ertesvag et al., 1998a; Svanem et al., 1999). The AlgG proteins are smaller than most of the other epimerases (AlgE or AlgY), and the conserved "INNHSY" sequences from both AlgGs have greater homology between themselves than with the other epimerases. Also, the conserved sequences are located ~350 residues (instead of 200) from the encoded N termini of the respective AlgG. Through comparisons of the common "INNHSY" and "ENNVSY" motifs, it is clear that asparagine (N), serine (S), and tyrosine (Y) residues are the essential residues and are conserved throughout. A valine (V) or arginine (R) residue in the epimerases substitutes for the histidine (H) residue position in the M lyases. In the M lyases, the histidine (H) residue is highly important for catalytic activity, and this may be the residue that contributes to the main difference in catalytic mechanism between a lyase and an epimerase. However, only recently the crystal structure of ALY1-III complexed with trimeric mannuronate was obtained (Yoon et al., 2001). The binding of this substrate in the tunnel-like cleft, strongly suggested that the four residues – N191, H192, R239 and Y246 – are directly involved in substrate binding and catalysis. The following catalytic mechanism has been proposed: (1) The C5 carboxylate group is neutralized by R239 and N191; (2) subsequently, the C5 proton can more easily removed by Y246, the catalytic nucleophile, resulting in the formation of the carboxylate dianion intermediate; (3) Y246 then donates the proton to the oxygen of the glycosidic bond, which results in bond cleavage and formation of a C4–C5 double bond. H192 presumably stabilizes the carboxylate dianion intermediate during catalysis (Figure 8). A similar arrangement of the catalytic residues was obtained in the threading model of the alginate lyase from *P. aeruginosa*, supporting their role in catalysis (see Figure 7).

8.3.5
Future Applications

The therapeutic use of alginate lyase for the treatment of alginate in biofilms of *P. aeruginosa* colonizing the lungs of CF patients remains one of the most important goals of studying alginate lyase. Mrsny et al. (1996) described the complex distribution of DNA and alginate within the mucin matrix of CF sputa. The combination of alginate lyase and deoxyribonuclease demonstrated an additive reduction of the sputum viscoelasticity, which suggests that this approach deserves further study (Mrsny et al., 1994). Alginate lyases may also be used to generate defined products with potential applications in various fields. Alginates with low molecular weights act like oligosaccharides in their ability to regulate physiological processes in

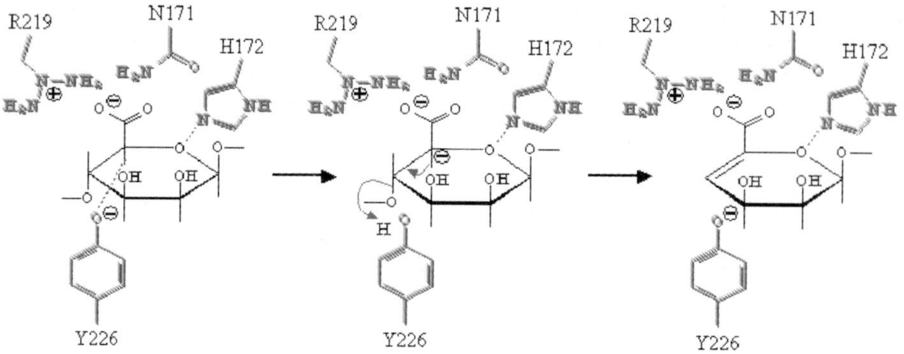

Fig. 8 The postulated catalytic mechanism of alginate lyases (see text for detailed description).

plants (Albersheim and Darvill, 1985). Oligomeric alginate (average M_W 2000 Daltons) obtained from lyase degradation of high-molecular weight alginate, can promote growth of *Bifidobacteria* sp. and thus has been proposed for use as a physiological food source (Murata et al., 1993; Akiyama et al., 1992). Furthermore, alginate lyase-degraded products (average M_W 1800 Daltons) greatly enhanced germination and shoot elongation in plants, despite repressing the growth of *Chlamydomonas* sp. and HeLa cells (Yonemoto et al., 1993). Trisaccharides (Natsume et al., 1994) or alginate lyase-lysate (Tomoda et al., 1994) has also been found to promote root growth in barley. In the presence of epidermal growth factor, dimers, trimers, and tetramers that possessed guluronic acid at the reduced ends highly induced the proliferation of keratinocytes (Kawada et al., 1997). Other alginate polymers (average M_W 230,000 Daltons) have antitumor effects and enhance the phagocytic activity of macrophages (Fujihara and Nagumo, 1993). Poly(M) block-rich alginate exhibited high antitumor activity (Fujihara and Nagumo, 1992) and could stimulate production of cytokines by human monocytes (Otterlei et al., 1991). The conformational properties of the macromolecules and the anionic character and molecular weight of the polysaccharide were suggested to be important in the effectiveness of their antitumor activity (Fujihara et al., 1984). Therefore, alginate lyases are crucial in the generation of such useful oligomeric products. The promising use of calcium alginate beads as a biomaterial in wide-ranging applications (Skjåk-Bræk and Martinsen, 1991) extends further the potential for alginates and alginate lyases. Calcium alginate, with or without a polylysine, polyarginine, or chitosan protective coating, has been used for the encapsulation of a variety of materials including drugs for controlled delivery; DNA and oligonucleotides for tumor development studies, gene delivery, gene therapy, and antisense oligonucleotide therapeutic agents; yeast cells coentrapped with lipase for the production of flavor esters; plant tissue, with the aim of developing artificial seed technology; *P. fluorescens* as a biocatalyst for phenolic-compound removal from wastewater; and entomopathogenic nematodes for agricultural biocontrol (for review, see Wong et al., 2000). Alginate in combination with other biomaterials such as hyaluronate and chitosan can be highly effective in many medical applications, including use in wound dressings impregnated with antibiotics and encapsulation of chondrocytes to engineer cartilage tissues *in vitro* for cartilage transplant and repair. Alginate with a G content >70% and average G-block length of >15, and with low polyphenol contamination, provides the most suitable characteristics for immobilization beads. Apart from generating alginate with a predictable monomeric sequence from native sources, it is possible to use a combination of D-mannuronan C5-epimerase and alginate lyase on a particular alginate substrate to engineer novel alginate polymers of defined composition (Skjåk-Bræk et al., 1986). The composition and properties of alginate gels are still being extensively studied, and the collective information will provide a guideline for the design of these novel polymers.

9
The Role of Alginate in Biofilm Formation

It has been proposed that contact with a surface may induce changes in gene expression, and there is evidence to support this idea in *P. aeruginosa*. Studies by Davies and coworkers showed that one of the genes required for the synthesis of the exopolysaccharide (EPS) alginate (*algC*) is up-regulated three- to five-fold in recently attached cells compared to their planktonic counterparts (Davies et al., 1993; Davies and Geesey, 1995). This result is not surprising because alginate – the regulation of which has been studied in depth (Govan and Fyfe, 1978; May et al., 1991) – has long been implicated as the extracellular matrix in biofilms of *P. aeruginosa*. These experiments were among the first to show surface contact-induced gene expression in *P. aeruginosa*. Recent studies in the laboratory of Wozniak have taken this observation a step further. Wozniak and colleagues noted that isolates of *P. aeruginosa* from the CF lung that produced large quantities of alginate (mucoid strains) were also nonmotile, and these authors suspected a link between the two phenotypes. In a series of genetic experiments, they showed that expression of a sigma factor (AlgU or σ^{22}) required for alginate synthesis resulted in down-regulation of a key flagellar biosynthetic gene (Garrett et al., 1999). These data suggest that, on contacting the surface, flagellar synthesis is down-regulated and alginate synthesis is up-regulated. Moreover, a recent study by Whiteley and coworkers (2001) employing DNA microarray analysis demonstrated that only about 1% of the *P. aeruginosa* genes were differentially regulated in mature biofilms as compared to planktonic cells. Interestingly, none of the *alg* genes appeared to be differentially regulated, whereas biosynthesis genes for pili IV and flagella were repressed. Scanning laser confocal microscopy (SCLM) analysis confirmed this result, demonstrating that nonmucoid strains formed densely packed biofilms that were generally less than 6 μm in depth. In contrast, *P. aeruginosa* FRD1 produced microcolonies that were approximately 40 μm in depth. An *algJ* mutant strain that produced alginate lacking O-acetyl groups produced only small microcolonies. After 44 h, the *algJ* mutant switched to the nonmucoid phenotype and formed uniform biofilms, similar to biofilms produced by the nonmucoid strains (Nivens et al., 2001). Results of both the ATR/FT-IR (attenuated total reflexion/Fourier transform infrared) and SCLM analyses of nonmucoid strains demonstrated that alginate was not required for *P. aeruginosa* biofilm formation, and therefore alginate did not act as a primary adhesin for the *P. aeruginosa* cells to these surfaces. Biofilm formation by *P. aeruginosa* FRD1131 that had a *Tn501* insertion in the alginate biosynthesis gene *algD* was still possible. Strain FRD1131 showed a delay in biofilm formation but ultimately formed biofilms, indicating that alginate biosynthesis is not essential for biofilm formation.

These results demonstrate that alginate, although not required for *P. aeruginosa* biofilm development, plays a role in the biofilm structure and may act as intercellular material, required for formation of thicker three-dimensional biofilms. The results also demonstrate the importance of alginate O-acetylation in *P. aeruginosa* biofilm architecture.

The portion of the biofilm developmental pathway that concerns detachment represents an important area of future research. One possible signal for detaching may be starvation, although this has not been investigated in detail. However, Boyd and Chakrabarty (1994) reported that the enzyme alginate lyase may play a role in the detachment phase in *P. aeruginosa*. These authors showed that overexpression of alginate lyase could speed detachment and cell sloughing from biofilms (Boyd and Chakrabarty, 1994). A recent study by Allison et al. (1998) showed that a *P. fluorescens* biofilm decreased after extended incubation, which was attributed – at least in part – to the loss of EPS. Furthermore, these authors presented evidence showing that acyl-HSLs and/or another factor present in stationary-phase culture supernatants mediated this effect (Allison et al., 1998). Little else is known about the functions or regulatory pathways involved in the release of bacteria from the biofilm.

10
The Applied Potential of Bacterial Alginates

At present, all alginates used for commercial purposes are produced from harvested brown seaweeds. The prices of such alginates are generally low, and it seems to be a difficult task to establish a competitive bacterial production process in this price range. However, there are at least two factors that now make it more probable that bacterial alginates may become commercial products. The first is related to the environmental concerns associated with seaweed harvesting and processing, and the second is related to the quality of the final polymer product. The environmental impact is not the topic of this review, but the possibility of producing bacterial alginates with improved qualities will be discussed on the basis of recent scientific discoveries. Alginates with unique qualitative properties have the obvious advantage that they may potentially be sold at higher prices than the bulk materials, and such new polymer products may at least theoretically also open new kinds of markets for this polymer. The properties of pure alginates are determined by their degree of polymerization and acetylation, and by their monomer composition and sequence (Moe et al., 1995). It appears likely that it will become possible to control these three parameters in bacterial alginates. The corresponding possibility does not seem realistic for alginates obtained by the harvesting of oceanic seaweeds, and producers of such products will therefore be limited by the need to fractionate the polymer mixtures produced by these organisms in their natural environments. The degree of polymerization affects the viscosity of alginates, and it will most likely be possible – at least to some extent – to control this parameter in bacteria by strict control of the fermentation conditions and/or genetic manipulations. One possible target for genetic modifications might be the alginate lyase found in both *Pseudomonas* and *Azotobacter* species (Kennedy et al., 1992; Lloret et al., 1996; Monday and Schiller, 1996; Rehm et al., 1996). Hence, bacterial alginates might in principle provide a particular viscosity if reduced quantities of material (on a weight basis) are used. Extreme viscosities would, on the other hand, also create problems with oxygen transfer in fermentors. From a commercial point of view it is not clear if this is a fruitful approach. It now seems obvious that the most interesting possibilities offered by bacterial alginates relate to the control of their monomer composition and sequential structures. Both seaweed and bacterial alginates are heterogeneous mixtures of molecules, and their sequential monomer composition can be described by only statistical models (Stokke et al., 1991), in contrast to accurately defined protein and nucleic acid sequences. If alginate structures are to be controlled, one must therefore first understand the origin of the structural variability observed in nature. On the basis of current knowledge it appears that *Pseudomonas* species have only one mannuronan C-5-epimerase, encoded by the *algG* gene, and this enzyme is able to introduce only single guluronic acid residues into the mannuronan chain. Consequently, these alginates cannot form the divalent-cation-dependent (typically Ca^{2+}) gels formed by alginates which contain G-blocks. The potential for in-vivo manipulations of monomer structures in *Pseudomonas* alginates therefore appears to be limited. However, by knocking out the *algG* gene, it may be possible to produce pure mannuronan, and evidence for this has been presented (Franklin et al., 1994). Mannuronan is known to be a strong immunostimulant, and might have a commercial potential in applications where this property is of interest (Skjåk-Bræk and Espevik, 1996). In *A. vinelandii*, the epimerization system (AlgE1–E7) is complicated, as described above. The known secreted epimerases and their different substrate specificities might allow the *in vitro* design of alginates (see above). Specialized alginates might be made *in vitro* by first producing deacetylated poly(mannuronic acid), possibly from a *Pseudomonas* species. Deacetylated alginates are needed because early studies indicate that acetylation seems to protect them from epimerization (Skjåk-Bræk et al., 1985). The acetyl groups might be prevented from being introduced by the use of an acetylation-deficient mutant of the production strain (Franklin and Ohman, 1993, 1996; Shinabarger et al., 1993), or they might be removed later using standard chemical deacetylation procedures. Mannuronan produced in this way could then be epimerized by one particular recombinantly produced epimerase, or by combinations of such enzymes. Alternatively, one might homogenize commercially produced alginates from brown seaweeds using recombinant epimerases. It is now very probable that almost any alginate structure of applied interest can be produced by such procedures. The problems of marketing these products are therefore most likely more related to parameters such as price, and to the technical advantages of the products compared with those already available from brown seaweeds. For food and pharmaceutical applications, approval by legal authorities may also be a major obstacle. In order to evaluate the technological properties of the alginates modified *in vitro*, the enzymes must be produced in sufficient quantities to allow physical and functional studies of the modified polymers. Alginates produced as described above would probably be quite expensive and, if the price were to become a hindrance to certain applications, then the production of similar products *in vivo* using metabolic engineering techniques might well be considered.

The first applications of bacterial alginates are likely to involve their use either as immunostimulants (e.g., mannuronan, see above) or as gel-forming agents for the immobilization of cells (e.g., in tissue engineering) (Gutowska et al., 2001). Immobilized cells might be used for a variety of biotechnological production processes (see

Skjåk-Bræk and Espevik, 1996 for mini review), or in medical transplantation technologies. A variety of proposals have been suggested by different groups (Skjåk-Bræk and Espevik, 1996), and one of the most interesting examples involves the reversal of type I diabetes by immobilizing insulin-producing cells in alginate capsules. These capsules have been implanted into the body of whole animals and even humans, and the biological effects of this kind of approach are currently being evaluated (Soon-Shiong, 1995).

Acknowledgments

These studies were supported by the Deutsche Forschungsgemeinschaft, the Deutsche Fördergesellschaft der Mukoviszidose-Forschung e.V. and the Minister für Wissenschaft und Forschung des Landes Nordrhein-Westfalen.

11 References

Aarons, S. J., Sutherland, I. W., Chakrabarty, A. M., Gallagher, M. P. (1997) A novel gene, *algK*, from the alginate biosynthesis cluster of *Pseudomonas aeruginosa*, *Microbiology* **143**, 641–652.

Akiyama, H., Endo, T., Nakakita, R., Murata, K., Yonemoto, Y., Okayama, K. (1992) Effect of depolymerized alginates on the growth of bifidobacteria, *Biosci. Biotechnol. Biochem.* **56**, 355–356.

Albersheim, P., Darvill, A. G. (1985) Oligosaccharins, *Sci. Am.* **253**, 58–64.

Allison, D. G., Ruiz, B., SanJose, C., Jaspe, A., Gilbert, P. (1998) Extracellular products as mediators of the formation and detachment of *Pseudomonas fluorescens* biofilms, *FEMS Microbiol. Lett.* **167**, 179–184.

Beale, J. M., Foster, J. L. (1996) Carbohydrate fluxes into alginate biosynthesis in *Azotobacter vinelandii* NCIB 8789 – nmr investigations of the triose pools, *Biochemistry* **35**, 4492–4501.

Boucher, J. C., Martinez-Salazar, J., Schurr, M. J., Mudd, M. H., Yu, H., Deretic, V. (1996) Two distinct loci affecting conversion to mucoidy in *Pseudomonas aeruginosa* in cystic fibrosis encode homologs of the serine-protease HtrA, *J. Bacteriol.* **178**, 511–523.

Boyd, A., Chakrabarty, A. M. (1994) Role of alginate lyase in cell detachment of *Pseudomonas aeruginosa*, *Appl. Environ. Microbiol.* **60**, 2355–2359.

Boyd, A., Gosh, M., May, T. B., Shinabarger, D., Keogh, R., Chakrabarty, A. M. (1993) Sequence of the *algL* gene from *Pseudomonas aeruginosa* and purification of its alginate lyase product, *Gene* **131**, 1–8.

Campos, M.-E., Martinez-Salazar, J. M., Lloret, L., Moreno, S., Nunez, C., Espin, G., Soberon-Chavez, G. (1996) Characterization of the gene coding for GDP-mannose dehydrogenase (*algD*) from *Azotobacter vinelandii*, *J. Bacteriol.* **178**, 1793–1799.

Chitnis, C. E., Ohman, D. E. (1993) Genetic analysis of the alginate biosynthetic gene cluster of *Pseudomonas aeruginosa* shows evidence of an operonic structure, *Mol. Microbiol.* **8**, 583–590.

Chu, L., May, T. B., Chakrabarty, A. M., Misra, T. K. (1991) Nucleotide sequence and expression of the *alg*E gene involved in alginate biosynthesis by *Pseudomonas aeruginosa*, *Gene* **107**, 1–10.

Cote, G. L., Krull, L. H. (1988) Characterization of the exocellular polysaccharides from *Azotobacter chroococcum*, *Carbohydr. Res.* **181**, 143–152.

Darzins, A., Chakrabarty, A. M. (1984) Cloning of genes controlling alginate biosynthesis from a mucoid cystic fibrosis isolate of *Pseudomonas aeruginosa*, *J. Bacteriol.* **159**, 9–18.

Davies, D. G., Geesey, G. G. (1995) Regulation of the alginate biosynthesis gene *algC* in *Pseudomonas aeruginosa* during biofilm development in continuous culture, *Appl. Environ. Microbiol.* **61**, 860–867.

Davies, D. G., Chakrabarty, A. M., Geesey, G. G. (1993) Exopolysaccharide production in biofilms – substratum activation of alginate gene-expression by *Pseudomonas aeruginosa*, *Appl. Environ. Microbiol.* **59**, 1181–1186.

Deretic, V., Konyecsni, W. M. (1989) Control of mucoidy in *Pseudomonas aeruginosa*: transcriptional regulation of *alg*R and identification of the second regulatory gene *alg*Q, *J. Bacteriol.* **171**, 3680–3688.

Deretic, V., Dikshit, R., Konyecsni, W. M., Chakrabarty, A. M., Misra, T. K. (1989) The *algR* gene, which regulates mucoidy in *Pseudomonas aeruginosa*, belongs to a class of environmentally response genes, *J. Bacteriol.* **171**, 1278–1283.

Deretic, V., Schurr, M. J., Boucher, J. C., Martin, D. W. (1994) Conversion of *Pseudomonas aeruginosa* to mucoidy in cystic fibrosis: environmental stress and regulation of bacterial virulence by alternative sigma factors, *J. Bacteriol.* **176**, 2773–2780.

Devault, J. D., Hendrickson, W., Kato, J., Chakrabarty, A. M. (1991) Environmentally regulated *algD* promoter is responsive to the cAMP receptor protein in *Escherichia coli, Mol. Microbiol.* **5**, 2503–2509.

Ertesvåg, H., Valla, S. (1999) The A modules of the *Azotobacter vinelandii* mannuronan-C-5-epimerase AlgE1 are sufficient for both epimerization and binding of Ca^{2+}, *J. Bacteriol.* **181**, 3033–3038.

Ertesvåg, H., Doseth, B., Larson, B., Skjåk-Bræk, G., Valla, S. (1994) Cloning and expression of an *Azotobacter vinelandii* mannonuran C-5-epimerase gene, *J. Bacteriol.* **176**, 2846–2853.

Ertesvåg, H., Hoidal, H. K., Hals, I. K., Rian, A., Doseth, B., Valla, S. (1995) A family of modular type mannuronan C-5-epimerase genes controls alginate structure in *Azotobacter vinelandii, Mol. Microbiol.* **9**, 719–731.

Ertesvåg, H., Frode, E., Skjåk-Bræk, G., Rehm, B. H. A., Valla, S. (1998a) Biochemical properties and substrate specificities of a recombinantly produced *Azotobacter vinelandii* alginate lyase, *J. Bacteriol.* **180**, 3779–3784.

Ertesvåg, H., Hoidal, H. K., Skjåk-Bræk, G., Valla, S. (1998b) The *Azotobacter vinelandii* mannuronan C-5-epimerase AlgE1 consists of two separate catalytic domains, *J. Biol. Chem.* **273**, 30927–30932.

Ertesvåg, H., Hoidal, H. K., Schjerven, H., Svanem, B. I., Valla, S. (1999) Mannuronan C-5-epimerases and their application for *in vitro* and *in vivo* design of new alginates useful in biotechnology, *Metab. Eng.* **1**, 262–269.

Fett, W. F., Osman, S. F., Fishman, M. L., Siebles, T. S. (1986) Alginate production by plant-pathogenic Pseudomonads, *Appl. Environ. Microbiol.* **52**, 466–473.

Fett, W. F., Wijey, C., Lifson, E. R. (1992) Occurrence of alginate gene-sequences among members of the Pseudomonad ribosomal-RNA homology groups I-IV, *FEMS Microbiol. Lett.* **99**, 151–157.

Fialho, A. M., Zielinski, N. A., Fett, W. F., Chakrabarty, A. M., Berry, A. (1990) Distribution of alginate gene-sequences in the *Pseudomonas* ribosomal-RNA homology group I-*Azomonas-Azotobacter* lineage of superfamily-B procaryotes, *Appl. Environ. Microbiol.* **56**, 436–443.

Franklin, M. J., Ohman, D. E. (1993) Identification of *algF* in the alginate biosynthetic gene cluster of *Pseudomonas aeruginosa* which is required for alginate acetylation, *J. Bacteriol.* **175**, 5057–5065.

Franklin, M. J., Ohman, D. E. (1996) Identification of *algI* and *algJ* in the *Pseudomonas aeruginosa* alginate biosynthetic gene cluster which are required for alginate acetylation, *J. Bacteriol* **178**, 2186–2195.

Franklin, M. J., Chitnis, C. E., Gacesa, P., Sonesson, A., White, D. C., Ohman, D. E. (1994) *Pseudomonas aeruginosa* AlgG is a polymer level alginate C5-mannuronan epimerase, *J. Bacteriol.* **176**, 1821–1830.

Fujihara, M., Nagumo, T. (1992) The effect of the content of D-mannuronic acid and L-guluronic acid blocks in alginates on antitumor activity, *Carbohydr. Res.* **224**, 343–347.

Fujihara, M., Nagumo, T. (1993) An influence of the structure of alginate on the chemotactic activity of macrophages and the antitumor activity, *Carbohydr. Res.* **243**, 211–216.

Fujihara, M., Iizima, N., Yamamoto, I., Nagumo, T. (1984) Purification and chemical and physical characterization of an antitumor polysaccharide from the brown seaweed *Sargassum fulvellum, Carbohydr. Res.* **125**, 97–106.

Gacesa, P. (1987) Alginate-modifying enzymes: a proposed unified mechanism of action for the lyases and epimerases, *FEBS Lett.* **212**, 199–202.

Gacesa, P. (1992) Enzymic degradation of alginates, *Int. J. Biochem.* **24**, 545–552.

Gacesa, P. (1998) Bacterial alginate biosynthesis – recent progress and future prospects, *Microbiology* **144**, 1133–1143.

Gacesa, P., Goldberg, J. B. (1992) Heterologous expression of an alginate lyase gene in mucoid and non-mucoid strains of *Pseudomonas aeruginosa, J. Gen. Microbiol.* **138**, 1665–1670.

Gacesa, P., Russell, N. J. (1990) The structure and property of alginate, in: *Pseudomonas* Infection and Alginates (Gacesa, P., Russell, N. J., Eds.), London: Chapman & Hall, 29–49.

Garrett, E. S., Perlegas, D., Wozniak, D. J. (1999) Negative control of flagellum synthesis in *Pseudomonas aeruginosa* is modulated by the alternative sigma factor AlgT (AlgU), *J. Bacteriol.* **181**, 7401–7404.

Goldberg, J. B., Hatano, K., Pier, G. B. (1993). Synthesis of lipopolysaccharide-O side-chains by *Pseudomonas aeruginosa* PAO1 requires the enzyme phosphomannomutase, *J. Bacteriol.* **175**, 1605–1611.

Gorin, P. A., Spencer, J. F. T. (1966) Exocellular alginic acid from *Azotobacter vinelandii, Can. J. Chem.* **44**, 993–998.

Govan, J. R. W., Deretic, V. (1996) Microbial pathogenesis in cystic fibrosis: mucoid *Pseudo-*

monas aeruginosa and *Burkholderia cepacia*, *Microbiol. Rev.* **60**, 539–574.

Govan, J. R. W., Fyfe, J. F. M. (1978) Mucoid *Pseudomonas aeruginosa* and cystic fibrosis: resistance of the mucoid form to carbenicillin, flucloxacillin and tobramycin and the isolation of mucoid variants *in vitro*, *J. Antimicrob. Chemother.* **4**, 233–240.

Govan, J. R. W., Harris, G. S. (1986) *Pseudomonas aeruginosa* and cystic fibrosis: unusual bacterial adaptation and pathogenesis, *Microbiol. Sci.* **3**, 302–308.

Govan, J. R. W., Fyfe, J. F. M., Jarman, T. R. (1981) Isolation of alginate-producing mutants of *Pseudomonas fluorescens, Pseudomonas putida*, and *Pseudomonas mendocina*, *J. Gen. Microbiol.* **125**, 217–220.

Grabert, E., Wingender, J., Winkler, U. K. (1990) An outer membrane protein characteristic of mucoid strains of *Pseudomonas aeruginosa*, *FEMS Microbiol. Lett.* **68**, 83–88.

Grobe, S., Wingender, J., Trüper, H. G. (1995) Characterization of mucoid *Pseudomonas aeruginosa* strains isolated from technical water systems, *J. Appl. Bacteriol.* **79**, 94–102.

Gross, M., Rudolph, K. (1987) Demonstration of levan and alginate in bean-plants (*Phaseolus vulgaris*) infected by *Pseudomonas syringae* pv. *phaseolicola*, *J. Phytopathol.* **120**, 9–19.

Gutowska, A., Jeong, B., Jasionowski, M. (2001) Injectable gels for tissue engineering, *Anat. Rec.* **263**, 342–349.

Haug, A., Larsen, B., Smidsrød, O. (1967) Studies on the sequence of uronic acid residues in alginic acid, *Acta Chem. Scand.* **21**, 691–704.

Hershberger, C. D., Ye, R. W., Parsek, M. R., Xie, Z. D., Chakrabarty, A. M. (1995) The *algT* (*algU*) gene of *Pseudomonas aeruginosa*, a key regulator involved in alginate biosynthesis, encodes an alternative sigma factor (sigma^{E}), *Proc. Natl. Acad. Sci. USA* **92**, 7941–7945.

Kapatral, V., Bina, X., Chakrabarty, A. M. (2000) Succinyl coenzyme A synthetase of *Pseudomonas aeruginosa* with a broad specificity for nucleoside triphosphate (NTP) synthesis modulates specificity for NTP synthesis by the 12-kilodalton form of nucleoside diphosphate kinase, *J. Bacteriol.* **182**, 1333–1339.

Kawada, A., Hiura, N., Shiraiwa, M., Tajima, S., Hiruma, M. (1997) Stimulation of human keratinocyte growth by alginate oligosaccharides, a possible co-factor for epidermal growth factor in cell culture, *FEBS Lett.* **408**, 43–46.

Kennedy, L., McDowell, K., Sutherland, I. W. (1992) Alginases from *Azotobacter* species, *J. Gen. Microbiol.* **138**, 2465–2471.

Kim, H. Y., Schlictman, D., Shankar, S., Xie, Z., Chakrabarty, A. M., Kornberg, A. (1998) Alginate, inorganic polyphosphate, GTP and ppGpp synthesis co-regulated in *Pseudomonas aeruginosa*: implications for stationary phase survival and synthesis of RNA/DNA precursors, *Mol. Microbiol.* **27**, 717–725.

Konyecsni, W. M., Deretic, V. (1990) DNA sequence and expression analysis of *alg*P and *alg*Q, components of the multigene system transcriptionally regulating mucoidy in *Pseudomonas aeruginosa*: *algP* contains multiple direct repeats, *J. Bacteriol.* **172**, 2511–2520.

Lee, J. W., Day, D. F. (1995) Bioacetylation of seaweed alginate, *Appl. Environ. Microbiol.* **61**, 650–655.

Lin, T.-Y., Hassid, W. Z. (1966a) Pathway of alginic acid synthesis in the marine brown alga, *Fucus gardneri* Silva, *J. Biol. Chem.* **241**, 5284–5297.

Lin, T.-Y., Hassid, W. Z. (1966b) Isolation of guanosine diphosphate uronic acids from a marine brown alga, *Fucus gardneri* Silva, *J. Biol. Chem.* **241**, 3283–3293.

Linker, A., Jones, R. S. (1966) A new polysaccharide resembling alginic acid isolated from pseudomonads, *J. Biol. Chem.* **241**, 3845–3851.

Lloret, L., Barreto, R., Leon, R., Moreno, S., Martínez-Salazar, J., Espín, G., Soberón-Chávez, G. (1996) Genetic analysis of the transcriptional arrangement of *Azotobacter vinelandii* alginate biosynthetic genes: identification of two independent promoters, *Mol. Microbiol.* **21**, 449–457.

Lynn, A. R., Sokatch, J. R. (1984) Incorporation of isotope from specifically labelled glucose into alginates of *Pseudomonas aeruginosa* and *Azotobacter vinelandii*, *J. Bacteriol.* **158**, 1161–1162.

Ma, J. F., Phibbs, P. V., Hassett, D. J. (1997) Glucose stimulates alginate production and *algD* transcription in *Pseudomonas aeruginosa*, *FEMS Microbiol. Lett.* **148**, 217–221.

Ma, S., Selvaraj, U., Ohmann, D. E. Quarless, R., Hassett, D. J., Wozniak, D. J. (1998) Phsophorylation-independent activity of the response regulators AlgB and AlgR in promoting alginate biosynthesis in mucoid *Pseudomonas aeruginosa*, *J. Bacteriol.* **180**, 956–968.

Madgwick, J., Haug, A., Larsen, B. (1978) Ionic requirements of alginate-modifying enzymes in the marine alga *Pelvetia canaliculata* (L.) Dcne. et Thur, *Bot. Mar.* **21**, 1–3.

Maharaj, R., May, T. B., Wang, S. K., Chakrabarty, A. M. (1993) Sequence of the *alg8* and alg44 genes involved in the synthesis of alginate by *Pseudomonas aeruginosa*, *Gene* **136**, 267–269.

Martin, D. W., Schurr, M. J., Mudd, M. H., Deretic, V. (1993a) Differentiation of *Pseudomonas aeruginosa* into the alginate-producing form: inactivation of *mucB* causes conversion to mucoidy, *Mol. Microbiol.* **9**, 495–506.

Martin, D. W., Schurr, M. J., Mudd, M. H., Govan, J. R. W., Holloway, B. W., Deretic, V. (1993b) Mechanism of conversion to mucoidy in *Pseudomonas aeruginosa* infecting cystic fibrosis patients, *Proc. Natl. Acad. Sci. USA* **90**, 8377–8381.

Martin, D. W., Schurr, M. J., Yu, H., Deretic, V. (1994) Analysis of promoters controlled by the putative sigma-factor AlgU regulating conversion to mucoidy in *Pseudomonas aeruginosa* relationship to sigma(e) and stress-response, *J. Bacteriol.* **176**, 6688–6696.

Martínez-Salazar, J. M., Moreno, S., Najera, R., Boucher, J. C., Espín, G., Soberón-Chávez, G., Deretic, V. (1996) Characterization of genes coding for the putative sigma factor AlgU and its regulators MucA, MucB, MucC, and MucD in *Azotobacter vinelandii* and evaluation of their roles in alginate biosynthesis, *J. Bacteriol.* **178**, 1800–1808.

Mathee, K., McPherson, C. J., Ohmann, D. E. (1977) Posttranslational control of the algT (algU)-encoded sigma22 for expression of the alginate regulon in *Pseudomonas aeruginosa* and localization of its antagonist proteins MucA and MucB (AlgN), *J. Bacteriol.* **179**, 3711–3720.

Mathews, C. K. (1993) Enzyme organization in DNA precursor biosynthesis, *Prog. Nucleic Acid Res. Mol. Biol.* **44**, 167–203

May, T. B., Chakrabarty, A. M. (1994) *Pseudomonas aeruginosa*: genes and enzymes of alginate biosynthesis, *Trends Microbiol.* **2**, 151–157.

May, T. B., Shinabarger, D., Maharaj, R., Kato, J., Chu, L., Devault, J. D., Roychoudhury, S., Zielinski, N. A., Berry, A., Rothmel, R. K., Misra, T. K., Chakrabarty, A. M. (1991) Alginate synthesis by *Pseudomonas aeruginosa*: a key pathogenic factor in chronic pulmonary infections of cystic fibrosis patients, *Clin. Microbiol. Rev.* **4**, 191–206.

McAvoy, M. J., Newton, V., Paull, A., Morgan, J., Gacesa, P., Russell, N. J. (1989) Isolation of mucoid strains of *Pseudomonas aeruginosa* from non-cystic-fibrosis patients and characterization of the structure of their secreted alginate, *J. Med. Microbiol.* **28**, 183–189.

Moe, S. T., Draget, K. I., Skjåk-Bræk, G., Smidsrød, O. (1995) Alginates, in: *Food Polysaccharides and Applications* (Stephen, A. M., Ed.), New York: Marcel Dekker, Inc., 245–286.

Monday, S. R., Schiller, N. L. (1996) Alginate synthesis in *Pseudomonas aeruginosa*: the role of AlgL (alginate lyase) and AlgX, *J. Bacteriol.* **178**, 625–632.

Mrsny, R. J., Lazazzera, B. A., Daugherty, A. L., Schiller, N. L., Patapoff, T. W. (1994) Addition of a bacterial alginate lyase to purulent CF sputum *in vitro* can result in the disruption of alginate and modification of sputum viscoelasticity, *Pulm. Pharmacol.* **7**, 357–366.

Mrsny, R. J., Daugherty, A. L., Short, S. M., Widmer, R., Siegel, M. W., Keller, G.-A. (1996) Distribution of DNA and alginate in purulent cystic fibrosis sputum: implications to pulmonary targeting strategies, *J. Drug Target.* **4**, 233–243.

Murata, K., Inose, T., Hisano, T., Abe, S., Yonemoto, Y. (1993) Bacterial alginate lyase: enzymology, genetics and application, *J. Ferment. Bioeng.* **76**, 427–437.

Narbad, A., Russell, N. J., Gacesa, P. (1988) Radiolabelling patterns in alginate of *Pseudomonas aeruginosa* synthesized from specifically-labelled ^{13}C monosaccharide precursors, *Microbios* **54**, 171–179.

Natsume, M., Kamo, Y., Hirayama, M., Adachi, T. (1994) Isolation and characterization of alginate-derived oligosaccharides with root growth-promoting activities, *Carbohydr. Res.* **258**, 187–197.

Nivens, D. E., Ohman, D. E., Williams, J., Franklin, M. J. (2001) Role of alginate and its O acetylation in formation of *Pseudomonas aeruginosa* microcolonies and biofilms, *J. Bacteriol.* **183**, 1047–1057.

Nguyen L. K., Schiller, N. L. (1989) Identification of a slime exopolysaccharide de-polymerase in mucoid strains of *Pseudomonas aeruginosa*, *Curr. Microbiol.* **18**, 323–329.

Olvera, C., Goldberg, J. B., Sanchez, R., Soberon-Chavez, G. (1999) The *Pseudomonas aeruginosa algC* gene product participates in rhamnolipid biosynthesis, *FEMS Microbiol. Lett.* **179**, 85–90.

Onsøyen, E. (1996) Commercial applications of alginates. *Carbohydr. Eur.* **14**, 26–31.

Otterlei, M., Østgaard, K., Skjåk-Bræk, G., Smidsrød, O., Soon-Shoing, Espevik, T. (1991) Induction of cytokine production from human monocytes stimulated with alginate, *J. Immunother.* **10**, 286–291.

Padgett, P. J., Phibbs, P. V., Jr. (1986) Phosphomannomutase activity in wild-type and alginate-producing strains of *Pseudomonas aeruginosa*, *Curr. Microbiol.* **14**, 187–192.

Page, W. J., Sadoff, H. L. (1975) Relationship between calcium and uronic acids in the encystment of *Azotobacter vinelandii*, *J. Bacteriol.* **122**, 145–151.

Pecina, A., Pascual, A., Paneque, A. (1999) Cloning and expression of the *algL* gene, encoding the

Azotobacter chroococcum alginate lyase: purification and characterization of the enzyme, *J. Bacteriol.* **181**, 1409–1414.

Penaloza-Vazquez, A., Kidambi, S. P., Chakrabarty, A. M., Bender, C. L. (1997). Characterisation of the alginate biosynthetic gene cluster in *Pseudomonas syringae* pv. *syringae*, *J. Bacteriol.* **179**, 4464–4472.

Piggott, N. H., Sutherland, I. W., Jarman, T. R. (1981) Enzymes involved in the biosynthesis of alginate by *Pseudomonas aeruginosa*, *Eur. J. Appl. Microbiol. Biotechnol.* **13**, 179–183.

Pindar, D. F., Bucke, C. (1975) The biosynthesis of alginic acid by *Azotobacter vinelandii*, *Biochem. J.* **152**, 617–622.

Ramstad, M. V., Ellingsen, T. E., Josefsen, K. D., Hoidal, H. K., Valla, S., Skjåk-Bræk, G., Levine, D. W. (1999) Properties and action pattern of the recombinant mannuronan C-5-epimerase AlgE2, *Enzyme Microbiol. Technol.* **24**, 636–646.

Rehm, B. H. A. (1996) The *Azotobacter vinelandii* gene *algJ* encodes an outer membrane protein presumably involved in export of alginate, *Microbiology* **142**, 873–880.

Rehm, B. H. A. (1998) The alginate lyase from *Pseudomonas aeruginosa* CF1/M1 prefers the hexameric oligomannuronate as substrate, *FEMS Microbiol. Lett.* **165**, 175–180.

Rehm, B. H. A., Valla, S. (1997) Bacterial alginates: biosynthesis and applications, *Appl. Microbiol. Biotechnol.* **48**, 281–288.

Rehm, B. H. A., Winkler, U. K. (1996) Alginatbiosynthese bei *Pseudomonas aeruginosa* und *Azotobacter vinelandii*: Molekularbiologie und Bedeutung, *BIOspektrum* **4**, 31–36.

Rehm, B. H. A., Grabert, G., Hein, J., Winkler, U. K. (1994a) Antibody response of rabbits and cystic fibrosis patients to alginate-specific outer membrane protein of a mucoid strain of *Pseudomonas aeruginosa*, *Microb. Pathog.* **16**, 43–51.

Rehm, B. H. A., Boheim, G., Tommassen, J., Winkler, U. K. (1994b) Overexpression of *alg*E in *Escherichia coli*: subcellular localization, purification, and ion channel properties, *J. Bacteriol.* **176**, 5639–5647.

Rehm, B. H. A., Ertesvåg, H., Valla, S. (1996) A new *Azotobacter vinelandii* mannuronan C-5-epimerase gene (*alg*G) is part of an *alg* gene cluster physically organized in a manner similar to that in *Pseudomonas aeruginosa*, *J. Bacteriol.* **178**, 5884–5889.

Ross, P., Meyer, R., Benziman, M. (1991) Cellulose biosynthesis and function in bacteria, *Microbiol. Rev.* **55**, 35–58.

Sadoff, H. L. (1975) Encystment and germination in *Azotobacter vinelandii*, *Bacteriol. Rev.* **39**, 516–539.

Schiller, N. L., Monday, S. R., Boyd, C. M., Keen, N. T., Ohman, D. E. (1993) Characterization of the *Pseudomonas aeruginosa* alginate lyase gene (*algL*) – cloning, sequencing, and expression in *Escherichia coli*, *J. Bacteriol.* **175**, 4780–4789.

Schlictman, D., Kavanaughblack, A., Shankar, S., Chakrabarty, A. M. (1994) Energy metabolism and alginate biosynthesis in *Pseudomonas aeruginosa* – role of the tricarboxylic acid cycle, *J. Bacteriol.* **176**, 6023–6029.

Schurr, M. J., Martin, D. W., Mudd, M. H., Hibler, N. S., Boucher, J. C., Deretic, V. (1993) The *algD* promoter – regulation of alginate production by *Pseudomonas aeruginosa* in cystic fibrosis, *Cell. Mol. Biol. Res.* **39**, 371–376.

Schurr, M. J., Yu, H., Martinez-Salazar, J. M., Hibler, N. S., Deretic, V. (1995) Biochemical characterisation and posttranslational modification of AlgU, a regulator of stress response in *Pseudomonas aeruginosa*, *Biochem. Biophys. Res. Commun.* **216**, 874–880.

Schurr, M. J., Yu, H., Martinez-Salazar, J. M., Boucher, J. C., Deretic, V. (1996) Control of *alg*U, a member of the sigma(e)-like family of stress sigma factors, by the negative regulators MucA and MucB and *Pseudomonas aeruginosa* conversion to mucoidy in cystic fibrosis, *J. Bacteriol.* **178**, 4997–5004.

Schweizer, H. P., Po, C., Bacic, M. K. (1995) Identification of *Pseudomonas aeruginosa glp*M, whose gene-product is required for efficient alginate biosynthesis from various carbon sources, *J. Bacteriol.* **177**, 4801–4804.

Sherbrock-Cox, V., Russell, N. J., Gacesa, P. (1984) The purification and chemical characterization of the alginate present in extracellular material produced by mucoid strains of *Pseudomonas aeruginosa*, *Carbohydr. Res.* **135**, 147–154.

Shinabarger, D., May, T. B., Boyd, A., Ghosh, M., Chakrabarty, A. M. (1993) Nucleotide sequence and expression of the *Pseudomonas aeruginosa algF* gene controlling acetylation of alginate, *Mol. Microbiol.* **9**, 1027–1035.

Skjåk-Bræk, G., Espevik, T. (1996) Application of alginate gels in biotechnology and medicine, *Carbohydr. Eur.* **14**, 19–25.

Skjåk-Bræk, G., Martinsen, A. (1991) Applications of some algal polysaccharides in biotechnology, in: *Seaweed Resources in Europe: Uses and Potential* (Guiry, M.D., Blunden, G., Eds.), New York: John Wiley & Sons, 219–257.

Skjåk-Bræk G., Larsen, B. Grasdalen, H., (1985) The role of O-acetyl groups in the biosynthesis of alginate by *Azotobacter vinelandii*, *Carbohydr. Res.* **145**, 169–174.

Skjåk-Bræk G., Grasdalen, H., Larsen, B. (1986) Monomer sequence and acetylation pattern in some bacterial alginates, *Carbohydr. Res.* **154**, 239–250.

Smidsrød, O., Draget, K. I. (1996) Chemistry and physical properties of alginates, *Carbohydr. Eur.* **14**, 6–13.

Svanem, B. J., Skjåk-Bræk G., Ertesvåg, H., Valla, S. (1999) Cloning and expression of three new *Azotobacter vinelandii* genes closely related to a previously described gene family encoding mannuronan C-5-epimerases, *J. Bacteriol.* **181**, 68–77

Soon-Shiong, P. (1995) Encapsulated islet cell therapy for the treatment of diabetes: intraperitoneal injection of islets, *J. Controlled Release* **39**, 399–409.

Stokke, B. T., Smidsrød, O., Bruheim, P., Skjåk-Bræk, B. (1991) Distribution of uronate residues in alginate chains in relation to alginate gelling properties, *Macromolecules* **24**, 4637–4645.

Sutherland, J. W. (1982) Biosynthesis of microbial exopolysaccharides, *Adv. Microb. Physiol.* **23**, 79–150.

Sutherland, I. W. (1995) Polysaccharide lyases, *FEMS Microbiol. Rev.* **16**, 323–347.

Tatnell, P. J., Russell, N. J., Gacesa, P. (1993) A metabolic study of the activity of GDP-mannose dehydrogenase and concentrations of activated intermediates of alginate biosynthesis in *Pseudomonas aeruginosa*, *J. Gen. Microbiol* **139**, 119–127.

Tatnell, P. J., Russell, N. J., Gacesa, P. (1994). GDP-mannose dehydrogenase is the key regulatory enzyme in alginate biosynthesis in *Pseudomonas aeruginosa*: evidence from metabolite studies, *Microbiology* **140**, 1745–1754.

Terry, J. M., Pina, S. E., Mattingly, S. J. (1991) Environmental conditions which influence mucoid conversion in *Pseudomonas aeruginosa* PAO1, *Infect. Immun.* **59**, 471–477.

Tomoda, Y., Umemura, K., Adachi, T.(1994) Promotion of barley root elongation under hypoxic conditions by alginate lyase-lysate, *Biosci. Biotechnol. Biochem.* **58**, 202–203.

Vazquez, A., Soledad, M., Guzman, J., Alvarado, A., Espin, G. (1999) Transcriptional organization of the *Azotobacter vinelandii algGXLVIFA* genes: characterization of *algF* mutants, *Gene* **232**, 217–222.

Wang, S. K., Sa-Correia, I., Darzins, A., Chakarabarty, A. M. (1987) Characterization of *Pseudomonas aeruginosa* alginate (*alg*) gene region II, *J. Gen. Microbiol.* **133**, 2303–2314.

Whiteley, M., Bangera, M. G., Bumgarner, R. E., Parsek, M. R., Teitzel, G. M., Lory, S., Greenberg, E. P. (2001) Gene expression in *Pseudomonas aeruginosa* biofilms, *Nature* **413**, 860–864.

Wong, T. Y., Preston, L. A., Schiller, N. L. (2000) Alginate lyase: review of major sources and enzyme characteristics, structure–function analysis, biological roles, and applications, *Annu. Rev. Microbiol.* **54**, 289–340.

Wozniak, D. J., Ohman, D. E. (1991) *Pseudomonas aeruginosa* AlgB, a 2-component response regulator of the NtrC family, is required for *algD* transcription, *J. Bacteriol.* **173**, 1406–1413.

Wozniak, D. J., Ohman, D. E. (1994) Transcriptional analysis of the *Pseudomonas aeruginosa* genes *algR*, *algB*, and *algD* reveals a hierarchy of alginate gene expression which is modulated by *algT*, *J. Bacteriol.* **176**, 6007–6014.

Wyss, O., Neumann, M. G., Socolofsky, M. D. (1961). Development and germination of the *Azotobacter* cyst, *J. Biophys. Biochem. Cytol.* **10**, 555–565.

Yonemoto, Y., Tanaka, H., Yamashita, T., Kitabatake, N., Ishida, Y. (1993) Promotion of germination and shoot elongation of some plants by alginate oligomers prepared with bacterial alginate lyase, *J. Ferment. Bioeng.* **75**, 68–70.

Yoon, H.-J., Mikami, B., Hashimoto, W., Murata, K. (1999) Crystal structure of alginate lyase A1-III from *Sphingomonas* species A1 at 1.78 Å resolution, *J. Mol. Biol.* **290**, 505–514.

Yoon, H.-J., Hashimoto, W., Miyake, O., Murata, K., Mikami, B. (2001) Crystal structure of alginate lyase A1-III complexed with trisaccharide product at 2.0 Å resolution, *J. Mol. Biol.* **307**, 9–16.

Yu, H., Schurr, M. J., Deretic, V. (1995) Functional equivalence of *Escherichia coli* sigma(e) and *Pseudomonas aeruginosa* AlgU. *Escherichia coli* RpoE restores mucoidy and reduces sensitivity to reactive oxygen intermediates in AlgU mutants of *Pseudomonas aeruginosa*, *J. Bacteriol.* **177**, 3259–3268.

Yu, H., Mudd, M., Boucher, J. C., Schurr, M. J., Deretic, V. (1997) Identification of the *algZ* gene upstream of the response regulator *algR* and its participation in control of alginate production in *Pseudomonas aeruginosa*, *J. Bacteriol.* **179**, 187–193.

9
Poly-(1 → 4)-β-D-glucuronan

Prof. Josiane Courtois[1], Prof. Bernard Courtois[2]
[1] Laboratoire des Polysaccharides Microbiens et Végétaux, Université de Picardie Jules Verne, IUT/GB, Avenue des Facultés le Bailly, 80025 Amiens Cedex 1, France; Tel.: ++33-322-534092; Fax: ++33-322-956254; E-mail: josiane.courtois@u-picardie.fr
[2] Laboratoire des Polysaccharides Microbiens et Végétaux, Université de Picardie Jules Verne, IUT/GB, Avenue des Facultés le Bailly, 80025 Amiens Cedex 1, France; Tel.: ++33-322-534094; Fax: ++33-322-956254; E-mail: bernard.courtois@u-picardie.fr

bv	biovar
CFU	colony forming unit
DS	degree of substitution
EPS	exopolysaccharide
HMW	high-molecular weight
LMW	low-molecular weight
M_w	molecular weight
Glc*p*A	glucopyranosyluronic acid residue
2 (3)-*O*-Ac-Glc*p*A	glucopyranosyluronic acid residue acetylated on the carbon 2 (or 3)
2,3-di-*O*-Ac-Glc*p*A	glucopyranosyluronic acid residue acetylated on the carbons 2 and 3
RC	Rhizobium Complete medium
RCS	Rhizobium Complete medium with sucrose
TNF	tumor necrosis factor

1 Introduction

With several hundred known polymers, polysaccharides offer a great diversity of chemical structures, ranging from simple linear homopolymers to branched heteropolymers containing neutral and/or acidic sugars. Moreover, such polymers may contain noncarbohydrate substituents such as acetate, pyruvate, succinate, phosphoglycerol, amine, or sulfate functions. Numerous uses are proposed for these biopolymers, and new polysaccharides are constantly being discovered. Many carbohydrate polymers may be derived from eukaryotes: algae, plants, or animals; however, the widest diversity of polysaccharides is produced by prokaryotic cells, both Gram-negative and Gram-positive bacteria. Bacteria belonging to the Rhizobiacae family deserve special attention. Rhizobia strains are able to produce various polysaccharides, from insoluble extracellular cellulosic microfibrils to soluble viscous exopolysaccharides (EPS) and cyclic β-D-glucans. The EPS produced by Rhizobia wild-type strains can be linear, such as the galactoglucan produced by *Sinorhizobium* (*Rhizobium*) *meliloti* strains, or branched such as the succinoglycan (Figure 1) produced by almost all the *S. meliloti* wild-type strains. While *S. meliloti* strains produce principally polymers containing neutral sugars (galactose, glucose) substituted by acetate, succinate and pyruvate residues, other *Rhizobium* species, i.e., *Rhizobium leguminosarum* biovars *trifolii*, *phaseoli* and *viciae*, *Bradyrhizobium japonicum* and *Bradyrhizobium elkanii*, excrete various polysaccharides containing both neutral and acidic sugars. In many cases, the acidic sugar corresponds to glucopyranosyluronic acid residues (Glc*p*A). Some particular *S. meliloti* strains excrete polysaccharides that contain both neutral and acidic sugars, such as *S. meliloti* IFO 13336 (Amemura et al., 1981) (Figure 2) or *S. meliloti* 201 (Figure 3) (Yu et al., 1983).

Only one *S. meliloti* mutant strain (*S. meliloti* M5N1CS (NCIMB 40472)) has been

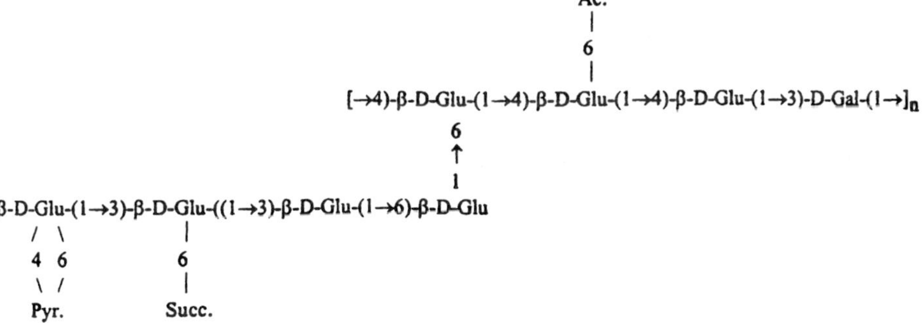

Fig. 1 Primary structure of the succcinoglycan produced by *Sinorhizobium meliloti* strains.

[→4)-β-D-Glu-(1→4)-β-D-Glu-(1→4)-β-D-Glu-(1→3)-β-D-Gal-(1→]n
6
1
α-D-RibA-(1→4)-α-D-GluA-(1→4)-β-D-Glu-(1→6)-β-D-Glu

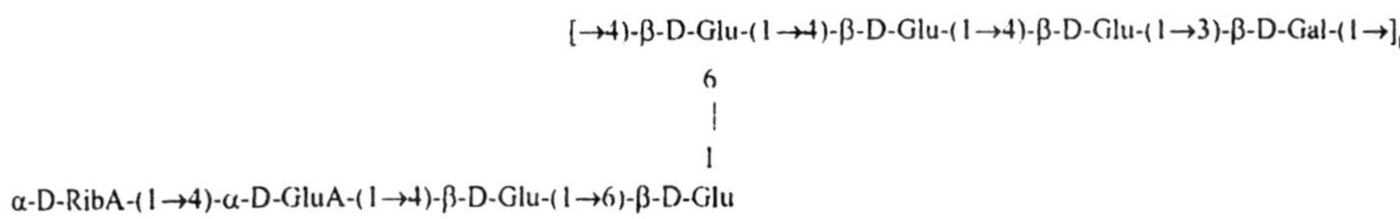

Fig. 2 Primary structure of the polysaccharide produced by the *Sinorhizobium meliloti* IFO 13336.

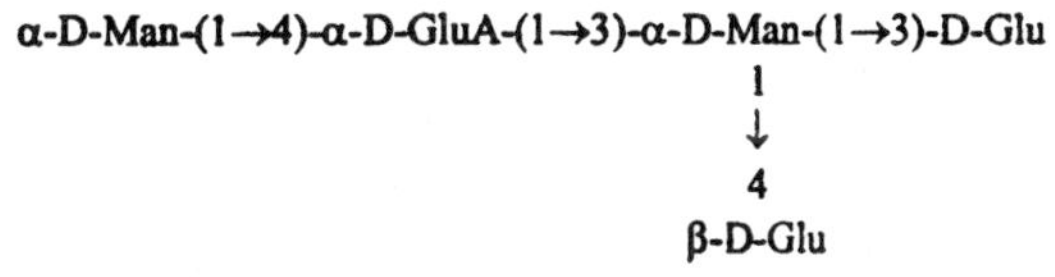

α-D-Man-(1→4)-α-D-GluA-(1→3)-α-D-Man-(1→3)-D-Glu
1
↓
4
β-D-Glu

Fig. 3 Primary structure of the polysaccharide produced by the *Sinorhizobium meliloti* 201.

described to produce a linear homopolysaccharide containing β-(1 → 4)-linked glucopyranosyluronic acid residues. This new polysaccharide, named glucuronan, contains acetyl residues (Figure 4). It displays physico-chemical and biological properties that allow its application in medicine, cosmetics, and agriculture.

2 Historical Outline

Polysaccharides from plants as starch and pectins, or from algae as agar or alginate, have a long history of industrial applications, and are widely used in various industries as texturing agents. In fact, the application domains of polysaccharides are equally diverse as polymers, mainly due to their structure–property relationships. In order to produce new polymers, chemical modifications have been applied to well-known polymers, with the best example probably being cellulose. However, studies on bacterial polysaccharides showed that they were equally good candidates for industrial applications as were gellan (from *Pseudomonas elodea*; O'Neil et al., 1983) or xanthan (from *Xanthomonas campestris*; Jansson et al., 1975). The wide diversity of microorganisms corresponds to a possible source of new polymers that present interesting biological or physico-chemical properties. Moreover, these specific microorganisms can be easily modified, whether by classical mutations or by gene transfer. By using this knowledge, bacteria belonging to the Rhizobiaceae family–which are known to produce a wide diversity of polysaccharides–were considered as good candidates for mutations, in order to produce new polysaccharides. In addition, using the knowledge of biological and physico-chemical properties of uronic acid-containing polymers from plants, bacteria or animals as pectins, alginic acid and dermatan or heparan sulfate (Yalpani, 1988), particular interest was focused on Rhizobia mutants producing uronic acid-containing polymers. A mutant of *S. meliloti* (*S. meliloti* M5N1CS (NCIMB 40472)) was selected as it produces a homopolymer made of β-(1 → 4)-linked glucuronic acid residues that are partially acetylated. This polymer, which is a new natural polysaccharide, was studied for its physico-chemical properties, as well as being subjected to a battery of biological tests on both animals and plants.

3 Chemical Structure

The glucuronan produced by *S. meliloti* M5N1CS is a linear homopolysaccharide that contains β-(1 → 4)-linked glucopyranosyluronic acid residues. The polymer has an average molecular weight of 250,000 ±

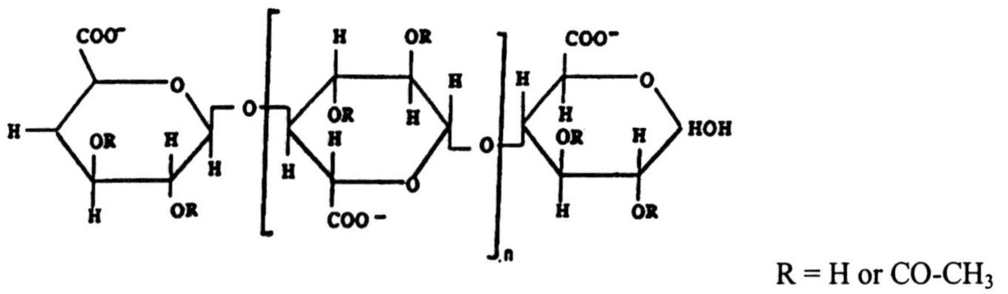

R = H or $CO\text{-}CH_3$

Fig. 4 Primary structure of the glucuronan produced by the *Sinorhizobium meliloti* M5N1CS strain.

100,000 daltons, and its chemical structure reveals an important degree of heterogeneity due to the acetyl residues linked on the glucuronic acid units. Three acetylated residues have been identified: β-D-2-*O*-Ac-Glc*p*A, β-D-3-*O*-Ac-Glc*p*A, β-D-2,3-di-*O*-Ac-Glc*p*A units, and unacetylated β-D-Glc*p*A units. Unacetylated residues may be linked to acetylated residues, or belong to a set of unacetylated residues (Courtois et al., 1994). A persistent length (which characterizes the stiffness of the chain) corresponding to 105 Å for a glucuronan with a low degree of acetate substitution (DS = 36%) indicates that glucuronans are semi-rigid polymers (Dantas et al., 1994a).

4
Occurrence

Bacteria belonging to the Rhizobiaceae family produce EPS from simple glucans, i.e. cyclic glucans, to complex heteropolysaccharides, i.e. succinoglycan. In this particular family, the *S. meliloti* strains produce mainly a succinoglycan containing only glucose and galactose as carbohydrate units, but with succinate, acetate, and pyruvate as substituents. However, some atypical strains, for example *S. meliloti* IFO 13336 (Amemura et al., 1981), excrete polymers that contain not only both glucose and galactose, but also glucuronic and riburonic acid residues. In order to select strains producing new polysaccharides, chemical mutagenesis using *N*-methyl-*N'*-nitro-*N*-nitrosoguanidine was performed on the succinoglycan-producing strain *S. meliloti* M5N1. The mutant *S. meliloti* M5N1CS (NCIMB 40472) was selected since it was able to induce nodule formation on the roots of lucerne (Courtois et al., 1993). The polymer produced by this selected strain was characterized as a homopolymer containing glucuronic acid where the residues were β-(1 → 4)-linked, and could be considered as a natural "acidic cellulose". Such a polymer shows promise with regard to its biological applications, for example as an immunostimulating agent (Espevik et al., 1993). The chemical production of a polymer comprising only Glc*p*A residues by oxidation of cellulose, remains difficult due to degradation of the polymer during the oxidation procedure; moreover, the presence of Glc*p* residues in the oxidized polymer indicates that oxidation is incomplete (Berntzen et al., 1998). Only a naturally occurring low molecular-weight (LMW) (5500–10,000 daltons) unacetylated glucuronan has been described from Mucorales (De Ruiter et al., 1992); however, due to the low level of LMW glucuronan obtained, no further studies were undertaken.

5
Physiological Function

Bacterial polysaccharides are crucial for the establishment of nitrogen-fixing symbioses between Rhizobia strains and plants belonging to the Fabacea family. In the case of *S. meliloti* strains that infect alfalfa, the succinoglycan polysaccharide was shown to be involved in the nodulation processes (Chakravorty et al., 1982; Leigh et al., 1985; Müller et al., 1988; Leigh and Walker, 1994). However, *S. meliloti* also has the capacity to produce a second exopolysaccharide, which is a galactoglucan (noted EPSII) (Glazebrook and Walker, 1989; Keller et al., 1990), while atypical *S. meliloti* IFO 13336 and 201 strains are described that produce EPS containing respectively Glc, Gal, GlcA, RibA and Man, GlcA, and Glc (Amemura et al., 1981; Yu et al., 1983). The production of high molecular-weight (HMW) EPS by Rhizobia strains was first shown to be necessary for the establishment of symbiosis between

Rhizobium and the corresponding host plant; more recently, the production of specific LMW EPS was described as being crucial for nodule invasion (Pellock et al., 2000). One class corresponds to cyclic glucans (Leigh and Walker, 1994), and a second corresponds to LMW succinoglycan and galactoglucan (Battisti et al., 1992; Gonzales, J. H. et al., 1996; Wang et al., 1999). The *S. meliloti* M5N1CS (NCIMB 40472) strain which is deficient in the production of succinoglycan, is able to produce not only HMW glucuronan, but also LMW polysaccharides composed of glucuronans with various degrees of acetylation (Michaud et al., 1997; Pirlet et al., 1998), as well as cyclic glucans (Michaud et al., 1995). The glucuronan-producing strain is able to induce the formation of protuberances on the root system of alfalfa, from typical nodule-like structures to calli corresponding to small excrescences on the root surface (Gonzales, M. L. et al., 1996). Electron microscopic examination (Figure 5) reveals the presence of bacteria only in typical nodules.

Typical infection threads containing the glucuronan-producing strain are observed: these bacteria contain inclusion bodies, characteristic for polyhydroxybutyrate. Moreover, the glucuronan-producing *Rhizobium* is also present in nontypical infection threads characterized by knobs, which are papillae-like protrusions. Few bacteria are observed in plant cells infected by the glucuronan-producing strain. These results, which were obtained with the glucuronan-producing strain, revealed a possible infection of alfalfa by the strain deficient in succinoglycan production. However, the infection process leading to nontypical infection threads may indicate that glucuronan induces such modifications, or that oligosuccinoglycans are required (as proposed by Wang et al., 1999), and cannot be replaced by oligoglucuronans.

6 Chemical Analysis and Detection

The structure of the polymer was deduced using classical chemical and physico-chemical analysis.

6.1 Chemical Hydrolysis

The glucuronan produced by *S. meliloti* M5N1CS strain (NCIMB 40472) is not susceptible to classical acid hydrolysis (H_2SO_4 1 M, 12 h, 100 °C). However, under drastic conditions (70% H_2SO_4, at room temperature, before dilution with water to reduce the H_2SO_4 concentration to 1 M and heating overnight at 100 °C), a brown coloration is observed. Analysis of the neutralized hydrolysate by high-pressure liquid chromatography (HPLC) (CHO-692 column, water as eluant at 85 °C) showed an absence of neutral sugars. Such behavior is characteristic of uronic acid-rich polysaccharides, and reflects the need for polymer reduction before hydrolysis.

The polymer, when reduced by action of a carbodiimide {*N*-cyclohexyl-*N'*[β(*N*-methylmorpholino)ethyl]carbodiimide-*p*-toluenesulfonate} and $NaBH_4$ at pH 7.0 (to prevent alkaline hydrolysis) (Bouffar and Heyraud, 1987), was subsequently hydrolyzed (H_2SO_4 1 M, 12 h, 100 °C), and only glucose was detected on HPLC analysis of the hydrolyzate (Heyraud et al., 1993). Structural determination was carried out on the reduced polymer, which was first *O*-methylated before hydrolysis with CF_3CO_2H (Hakomori, 1964). The resulting sugars, when converted into partially *O*-methylated alditol acetates and analyzed by gas-liquid chromatography (GLC), revealed the presence of a product with a retention time similar to that of 1,4,5-tri-*O*-methyl-glucitol. Confirmation by GLC-mass spectrometry (MS) indicated

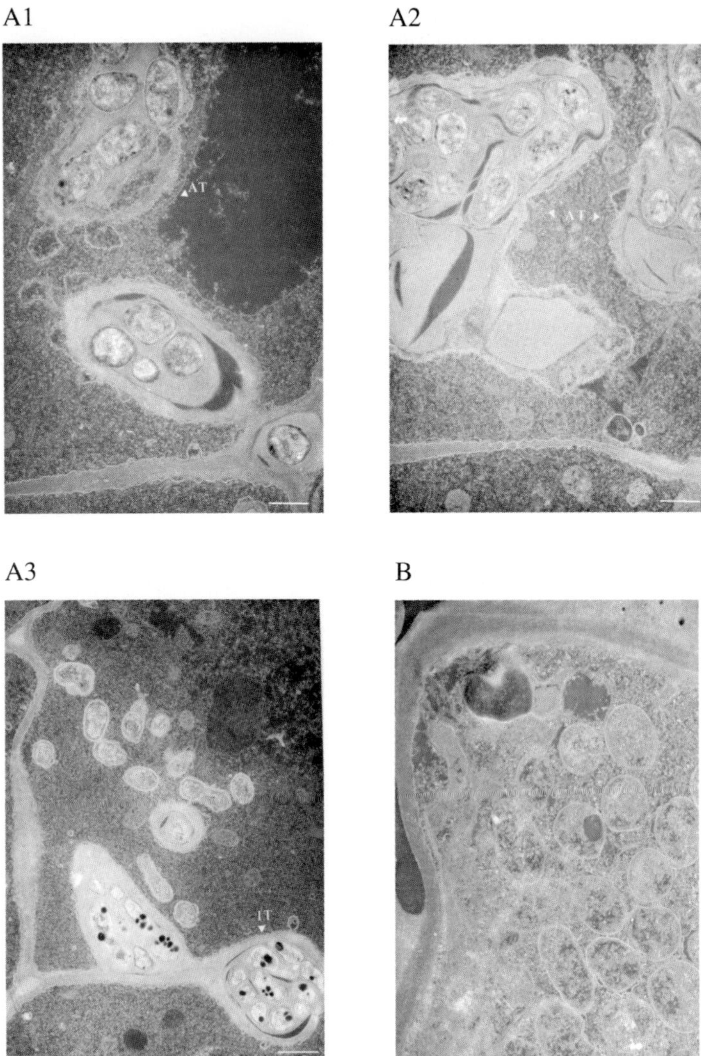

Fig. 5 Electron micrographs of alfalfa nodule cells colonized by the *S. meliloti* M5N1CS strain (NCIMB 40472) (A1–3), and by a wild-type strain *S. meliloti* (B). IT: typical infection threads; AT: abnormal infection threads. Scale bar = 1 μm.

that the new polymer contained either 4-linked glucopyranosyluronic acid residues or 5-linked glucofuranosyluronic acid residues (Heyraud et al., 1993).

6.2 Enzymatic Hydrolysis

Enzymatic hydrolysis performed with cellulase (Celluclast; Novo) on the reduced polymer led to the liberation of monosaccharides, disaccharides and trisaccharides

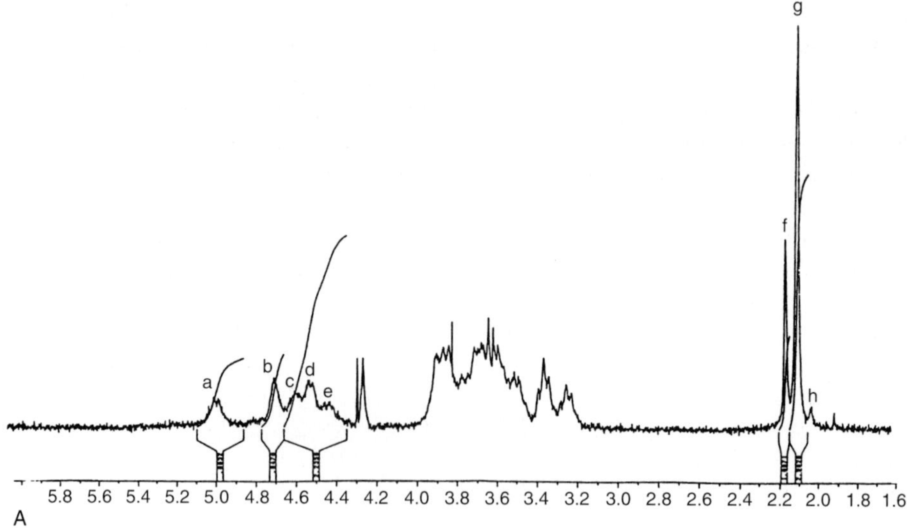

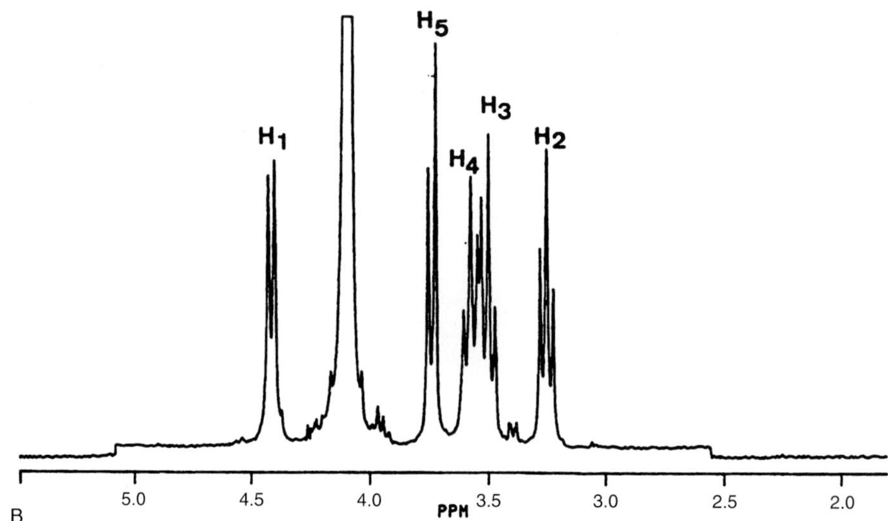

Fig. 6 ^{1}H-NMR spectra of native (A) and deacetylated (B) glucuronans in D_2O (at 300 MHz; T = 85 °C) (a) H3 of 3-*O*-acetylated residues; (b) H1 + H2 of 2-*O*-acetylated residues; (c) H1 of 3-*O*-acetylated residues; (d) H1 in a set of unacetylated residues; (e) H1 of unacetylated residues before or after acetylated residues; (f) protons of the acetyl group at C2 in the 2-*O*-acetylated residues; (g) protons of the acetyl group at C3 in the 3-*O*-acetylated residues and at C2 in the 2,3-di-*O*-acetylated residues; (h) protons of the acetyl group at C3 in the 2,3-di-*O*-acetylated residues. H1–5: protons in residues from a deacetylated glucuronan.

that could be separated by HPLC (CHO-682 column using water as eluant at 85 °C). The oligosaccharides were identified by ^{13}C-NMR as glucose, cellobiose, and cellotriose respectively. From its dextrorotatory characteristics, the glucopyranose unit was seen to have D-chirality.

6.3 NMR Studies

The ^{1}H-NMR spectrum of a native glucuronan (Figure 6A) revealed signals at 2 p.p.m. that were characteristic of *O*-acetyl groups, while a complex system in the ring proton region was due to acetyl groups linked to glucuronic residues. The ^{1}H-NMR spectrum of the deacetylated polymer (Figure 6B) (deacetylation was achieved with NaOH, pH >8.0 at ambient temperature for ≥ 2 h) revealed only five protons corresponding to glucuronic residues. The chemical shift of the anomeric proton (4.41 p.p.m.) with a $J_{1,2}$ coupling constant of 7.7 Hz indicated the presence of a β linkage between the glucopyranosyluronic acid residues. From ^{1}H- and ^{13}C-NMR (Figure 7) analyses, the structure of (1→4)-β-D-glucuronan substituted by acetyl residues was proposed for the polymer produced by the *S. meliloti* M5N1CS strain (NCIMB 40472).

Data from *O*-deacetylated and native glucuronans ^{1}H-NMR one- and two-dimensional spectra (Figure 8; Table 1) were used to determine the proportion of the different species of residues in the native glucuronan (Table 2).

While the molar proportion of the different species of residues is constant for glucuronans produced under similar conditions, it was noted that glucuronans produced by the *S. meliloti* M5N1CS strain in media supplemented with magnesium were more acetylated. Glucuronans with a higher degree of acetylation contain more diacetylated residues, while the molar ratio of 2-*O*-acetylated residues appears to be constant in the different glucuronans.

7 Biosynthesis

The biosynthesis of glucuronan by the *S. meliloti* M5N1CS strain can be achieved in

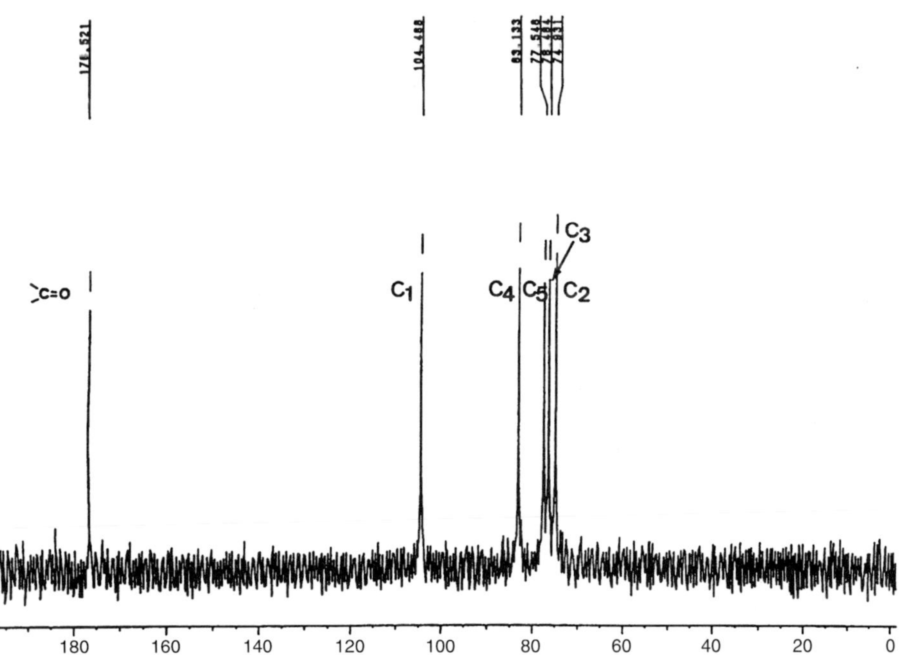

Fig. 7 ^{13}C-NMR spectrum of a deacetylated glucuronan (at 75 MHz, T = 80 °C).

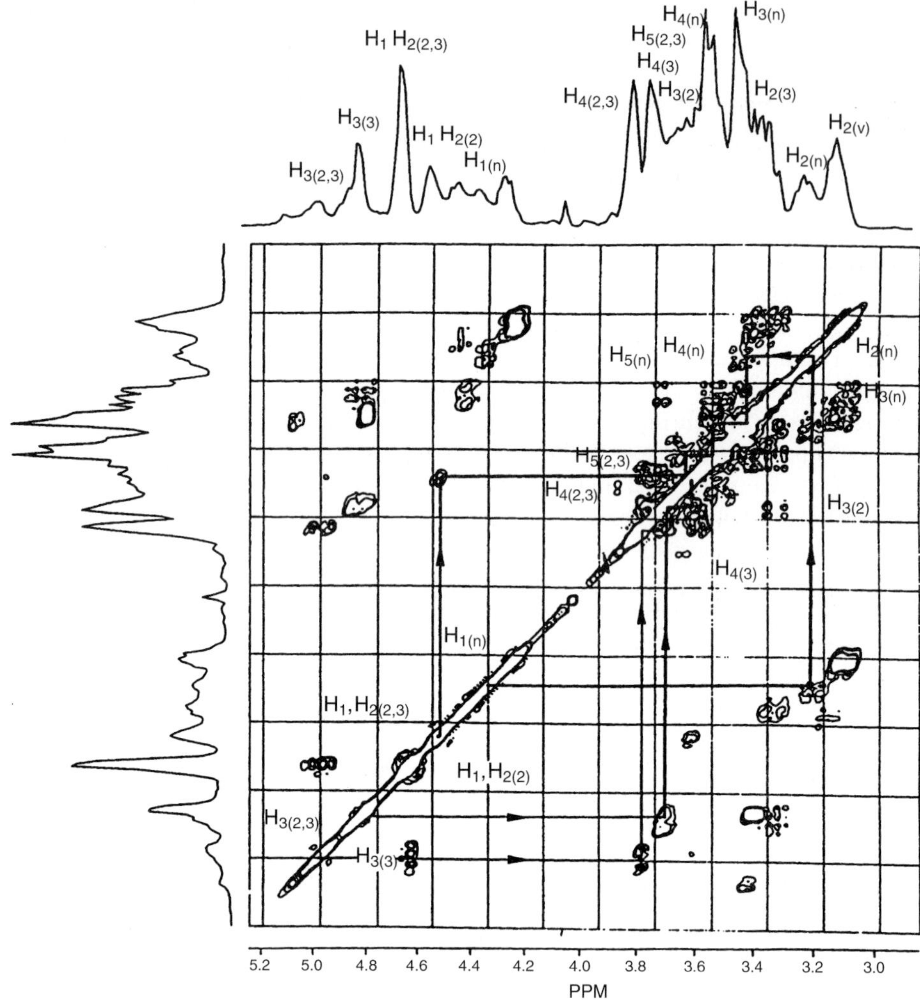

Fig. 8 300 MHz homonuclear relayed COSY plot at 85 °C of a native glucuronan.

Tab. 1 Experimental ^{1}H chemical shifts of the ring protons in native and deacetylated glucuronans (at 85 °C)

Residues in native glucuronan	*H-1*	*H-2*	*H-3*	*H-4*	*H-5*
β-D-Glc*p*A[a]	4.39	3.22	3.53	3.66	3.84
β-D-Glc*p*A[a]	4.44	3.25	3.56	3.73	3.84
β-D-Glc*p*A[b]	4.53	3.35	3.60	3.71	3.86
β-D-2-*O*-Ac-Glc*p*A	4.69	4.69	3.78	3.89	3.92
β-D-3-*O*-Ac-Glc*p*A	4.61	3.50	4.98	3.89	3.92
β-D-2,3-di-*O*-Ac-Glc*p*A	4.80	4.80	5.10	3.96	3.96
Residues in deacetylated glucuronan	4.53	3.36	3.61	3.69	3.85

[a] The β-D-Glc*p*A residue is situated behind or after acetylated residues. [b] The β-D-Glc*p*A residue belongs to a set of unacetylated residues.

Tab. 2 Molar proportion of the different species of β-D-glucuronic acid residues in native glucuronans[a]

	Glucuronan produced in RCS medium	*Glucuronan produced in RCS medium + $MgSO_4$, $7H_2O$*
Degree of substitution [%][b]	46	59
β-D-GlcpA[c]	12	8
β-D-GlcpA[c]	16	16
β-D-GlcpA[d]	29	23
β-D-2-*O*-Ac-GlcpA	14.5	12
β-D-3-*O*-Ac-GlcpA	26	16
β-D-2,3-di-*O*-Ac-GlcpA	2.5	24

[a] The glucuronans are produced in a RCS medium at pH 7.2 without or with addition of $MgSO_4 \cdot 7H_2O$ (0.6 g L^{-1} per day). [b] Degree of substitution (DS) per residue determined from integration of the resonances in the downfield, upfield and acetyl region. [c] The β-D-GlcpA residue is situated behind or after acetylated residues. [d] The β-D-GlcpA residue belongs to a set of unacetylated residues.

complex media containing yeast extracts supplemented with fructose, sucrose, mannitol, or glucose; a small decrease in the production of HMW glucuronan is observed after 80 h in media supplemented with $MgSO_4 \cdot 7H_2O$ (Figure 9). Mineral media supplemented with the above carbohydrate substrates are convenient for glucuronan production, as well as for succinoglycan production by the *S. meliloti* M5N1 strain. However, EPS biosynthesis by both strains under resting-cell conditions is not comparable. In nonproliferating media (NPM) supplemented with different carbohydrate substrates, only the *S. meliloti* M5N1 cells show long-term production of EPS (Figure 10) (Heyraud et al., 1985; Courtois et al., 1986). Cessation of glucuronan biosynthesis by the *S. meliloti* M5N1CS strain when incubated for 40 h in a medium deficient in nitrogen, in contrast to the regular production of succinoglycan by the *S. meliloti* M5N1, indicates that the stability of enzymes catalyzing glucuronan biosynthesis is less than that of those catalyzing succinoglycan biosynthesis.

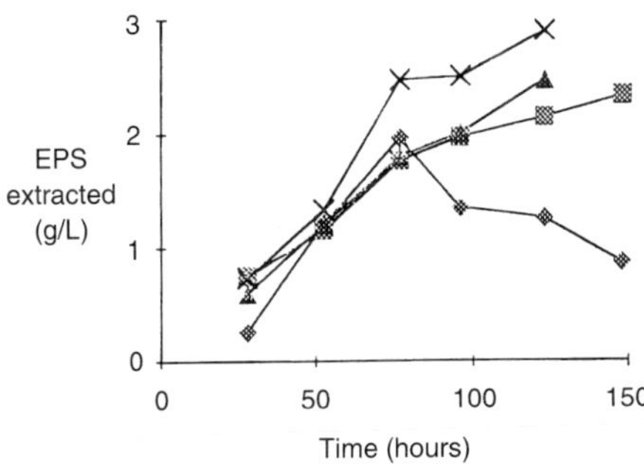

Fig. 9 Glucuronan production by the *S. meliloti* M5N1CS strain during fermentation in RCS medium with $MgSO_4 \cdot 7H_2O$, 0.2 g L^{-1} (▲—▲), 0.6 g L^{-1} (■—■), 0.8 g L^{-1} per day (◆—◆), or without (×—×).

The degree of substitution by acetate of glucuronan excreted after 150 h in RCS medium without added $MgSO_4 \cdot 7H_2O$ or supplemented with $MgSO_4 \cdot 7H_2O$ at 0.2, 0.6, and 0.8 g L^{-1} per day was respectively 52%, 55%, 65%, and 74%. A small increase in 2,3-di-*O*-acetyl residues and a decrease in unacetylated species in glucuronans produced in media supplemented with $MgSO_4 \cdot 7H_2O$ at 0.2 g L^{-1} per day was noted; these variations were amplified when the magnesium concentration increased (Figure 11). Hence, under specific conditions of production, various species of native glucuronans with various degrees of acetylation can be obtained, from 52% to 74%.

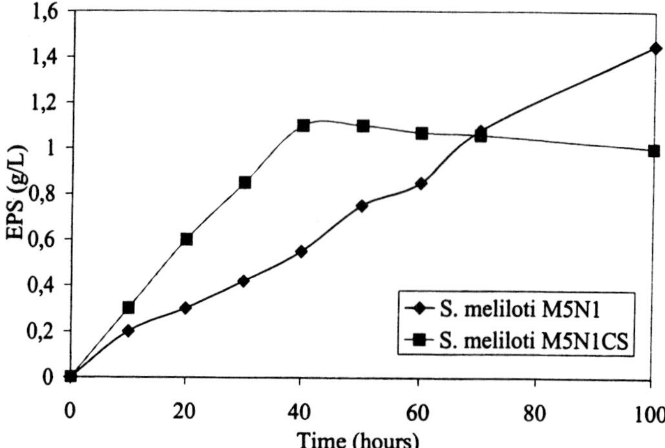

Fig. 10 Synthesis of succinoglycan and glucuronan by respectively *S. meliloti* M5N1 and *S. meliloti* M5N1CS strains (10^9 CFU mL^{-1}) suspended in nonproliferating media supplemented with fructose (1%, w/v).

8 Biodegradation

A weak glucuronan biodegradation was observed during production in specific fermentation media. The extraction of HMW glucuronan by ultrafiltration through a 100,000 normal-molecular-weight cut-off (NMWCO) cellulose acetate membrane (Sartorius) from cell-free broth samples taken during the *S. meliloti* M5N1CS growth, revealed two phases. First, the amount of HMW glucuronan produced was increased to about 80 h of fermentation, after which the production of HMW glucuronan (M_w > 100,000 daltons) decreased (see Figure 9). While glucuronan production during the first period was quite similar in the three media, the amount of HMW glucuronan produced during 87 h of fermentation in the RCS medium supplemented with $MgSO_4 \cdot 7H_2O$ at 0.8 g L^{-1} per day was half that obtained in a standard RCS medium (Table 3).

The decrease in the amount of HMW glucuronan produced was accompanied by an increase in LMW and medium molecular-weight (MMW) EPS (Table 4).

Tab. 3 Amount and molecular weight (Mw) of EPS (g L^{-1}) extracted during fermentation by ultrafiltration on a 100,000 cut-off membrane

	M1	*M2*	*M3*
Amount of EPS extracted	2.1	1.93	1.08
Mw	4.5×10^5	9.4×10^4	7.9×10^4

Ultrafiltration on the 100,000 cut off membrane (Sartorius) was carried out on microfiltered RCS medium without addition of magnesium after 87 h (M1) and 96 h (M2) of fermentation and after 96 h in RCS medium supplemented with 0.8 g L^{-1} $MgSO_4 \cdot 7H_2O$ per day (M3).

^{1}H-NMR spectra of the glucuronan produced during 96 h in RCS media without (see Figure 6A) or with addition of magnesium ($MgSO_4 \cdot 7H_2O$ 0.8 g L^{-1} per day) (Figure 12) were similar, except for the signals at 5.85, 5.25, and 5.15 p.p.m.

The signal at 5.85 p.p.m. was attributed to the H4 of an unsaturated unit corresponding to the nonreducing terminus unit of the glucuronan, while the signals at 5.25 and 5.15 p.p.m. were attributed to the H1α of the reducing terminus and the H1 of the

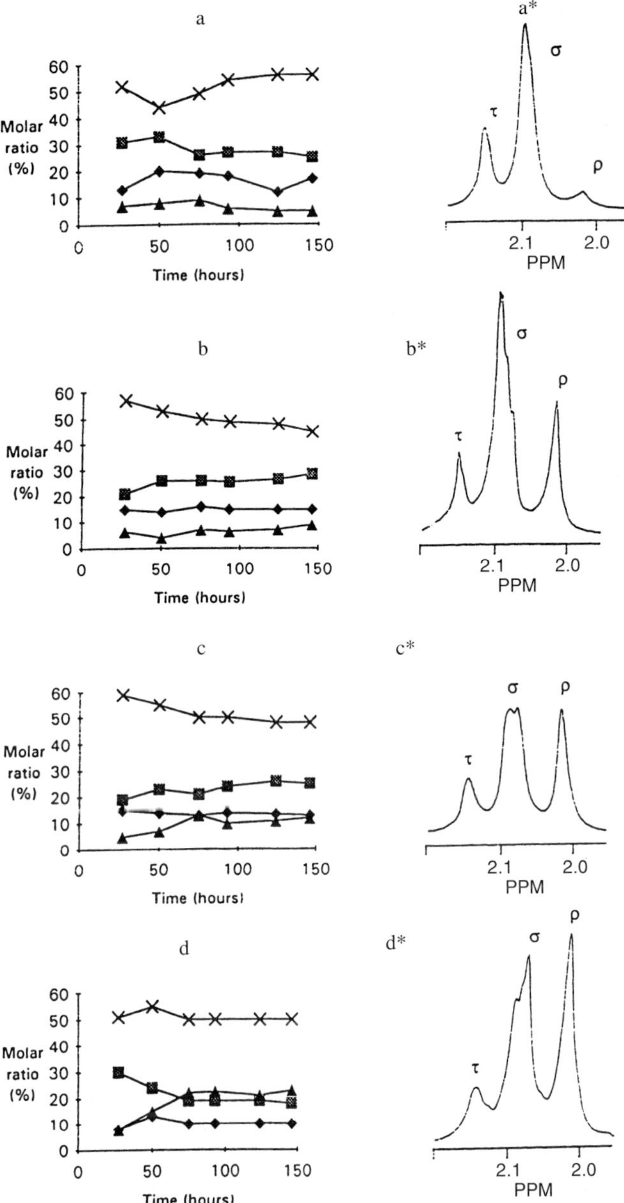

Fig. 11 Analysis of glucuronans produced in RCS medium (a, a*) and RCS medium with added $MgSO_4 \cdot 7H_2O$: 0.2 g L^{-1} (b, b*), 0.6 g L^{-1} (c, c*), 0.8 g L^{-1} (d, d*) per day. a, b, c, d: molar ratio of the different species of residues in glucuronans during production: 2-*O*-Ac-Glc*p*A (◆—◆), 3-*O*-Ac-Glc*p*A (■—■), 2,3-di-*O*-Ac-Glc*p*A (▲—▲), and unacetylated (**x—x**). a*, b*, c*, d*: ^{1}H-NMR spectra of acetyl signal of glucuronans obtained after 150 h of fermentation. τ: CH_3 of acetyl group in C-2 in the 2-*O*-Ac-Glc*p*A residues; σ: CH_3 of acetyl group in C-3 in the 2-*O*-Ac-Glc*p*A residues and in C-2 in the 2,3-di-*O*-Ac-Glc*p*A residues; ρ: CH_3 of acetyl group in C-3 in the 2,3-di-*O*-Ac-Glc*p*A residues.

Tab. 4 Weight average molecular weight (Mw̄) of glucuronans extracted during fermentation

	*M1**	*M2**	*M3**
Mw̄ (XM50-Diaflo)[a]	6.8×10^4	7.1×10^4	3.8×10^4
Mw̄ (YM1-Diaflo)[b]	2.3	9.2×10^3	6.0×10^3
Ratio (%) of EPS extracted (XM50-Diaflo)/(YM1-Diaflo)[c]	20	16	

[a] Extraction by ultrafiltration on a 50,000 XM50-Diaflo membrane (Amicon) of microfiltered broth after 87 h (M1*) and 96 h (M2*) of fermentation in RCS medium without addition of magnesium, or with 0.8 g L^{-1} $MgSO_4 \cdot 7H_2O$ per day (M3*). [b] Extraction by ultrafiltration on a 1000 YM-Diaflo membrane (Amicon) of the filtrate collected after ultrafiltration of M1*, M2*, and M3* fractions on a 50,000 XM50-Diaflo membrane. [c] Ratio of the amount of glucuronan extracted after filtration on a 1000 YM1-Diaflo membrane of M1*, M2*, and M3* fractions first ultrafiltered on a 50,000 XM50-Diaflo membrane to the amount of glucuronan extracted after filtration on the 50,000 XM50-Diaflo membrane.

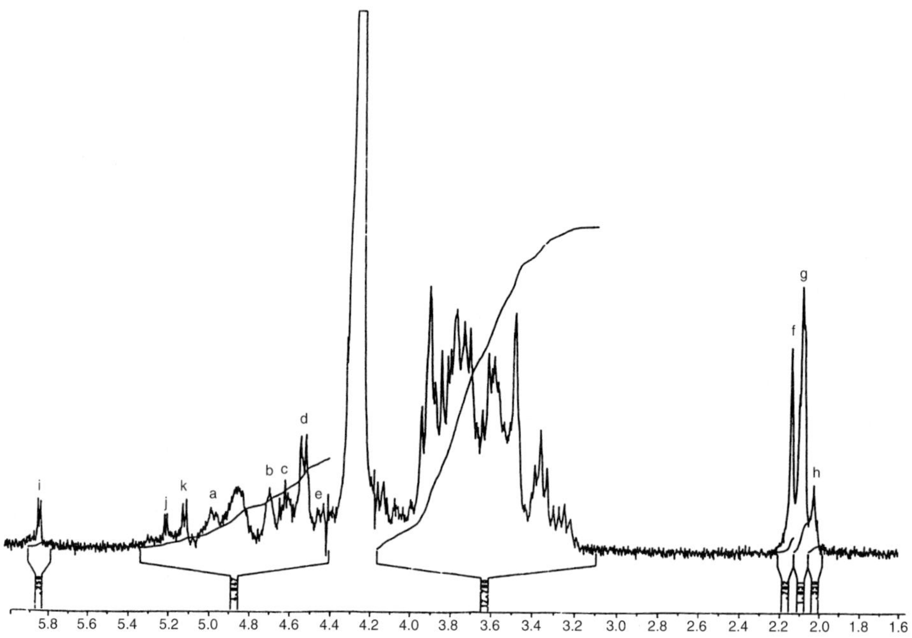

Fig. 12 ^{1}H-NMR spectrum (at 300 MHz, T = 60 °C) of LMW glucuronans. (i) H4 of the unsaturated nonreducing terminus; (j) H1α of the reducing terminus; (k) H1 of the unsaturated nonreducing terminus.

unsaturated nonreducing terminus, respectively (Michaud et al., 1994) (Figure 13).

LMW glucuronans can be obtained by degradation of the native glucuronan with *S. meliloti* M5N1CS cell extracts (Michaud et al., 1997). A glucuronan lyase with a molecular mass of 20,000 daltons was characterized (Da Costa et al., 2001). The degree of polymerization (DP) of the ultimate enzyme substrate was estimated from ^{1}H-NMR spectra by comparison of the integral of the H1 of the unsaturated unit with the integral of the H1 signal of the central unit. Oligoglucuronans with a DP of 4 were obtained. These results agreed with those described in the literature, where it was

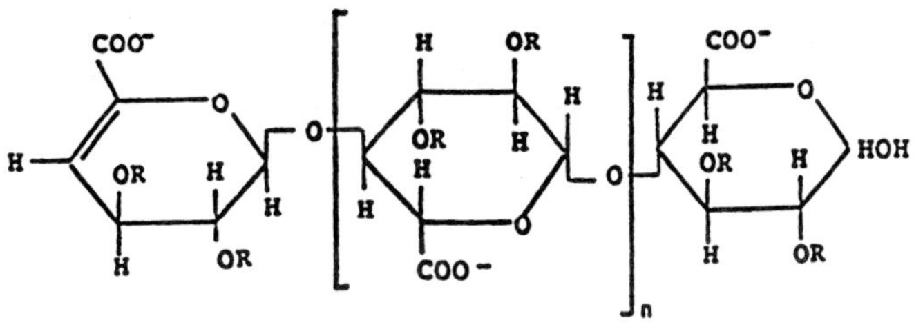

Fig. 13 Schematic representation of a glucuronan containing a 4,5-unsaturated terminal unit.

proposed that lyases cleave anionic polymers and form oligosaccharides with a DP between 2 and 5 (Waffanschmidt and Jaenicke, 1987). The glucuronan lyase activity is reduced on glucuronans which are highly acetylated (Michaud et al., 1997). However, small amounts of oligoglucuronans with a DP of 2 have been identified, as well as oligoglucuronans of DP > 3 with an average substitution degree of 24% and 38% (Table 5).

Due to the degree of acetylation of the polymer (54%), few degradations (0.4% of the total glucuronan production) occurred; however, the production of LMW glucuronan with a substitution degree of 38% was dominant due to the weak yield of glucuronan sequences that were poorly acetylated. Hence, by using a combination of culture conditions and enzymatic degradation by glucuronan lyase (from the *S. meliloti* M5N1CS strain), it was possible to obtain a wide diversity of LMW glucuronans with specific degrees of acetylation.

Tab. 5 Substitution degree and molar ratio of different species of residues in oligoglucuronans (DP>3) obtained by degradation of the polymer during fermentation

Oligoglucuronans		
Average substitution degree [%]	38	24
2-*O*-acetylated Glc*p*A [%]	6.5	3.5
3-*O*-acetylated Glc*p*A [%]	12	9
2,3-*O*-acetylated Glc*p*A [%]	15.5	9
Unacetylated [%]	65.5	79

9
Molecular Genetics

Although genetic manipulations related to glucuronan production have not yet been developed extensively, such studies will clearly be necessary in order to improve the bacterial production of the polysaccharide. Neither have any studies been developed on glucuronan-like polysaccharide production by mutants (unlike the case for xanthan-like polymers by *Xanthomonas campestris* mutants; Betlach et al., 1987), mainly due to the absence of any side chains on the glucuronan polymer. Glucuronans with various degrees of substitution by acetyl residues can be easily obtained by bacterial fermentations under specific conditions (see Chapter 10.2). Indeed, several genetic studies on glucuronan degradation are currently in progress, more precisely on a glucuronidase gene coding for an endo polyglucuronate lyase (Da Costa et al., 2001).

10
Biotechnological Production

HMW glucuronan is produced only by the *S. meliloti* strain NCIMB 40472; a LMW (5500–10,000 daltons) unacetylated glucuronan has been described by De Ruiter et al. (1992), but no further studies have been conducted on these glucuronan molecules due to the difficulty of preparing large amounts of pure polysaccharide from the fungal (Mucorales) source.

10.1 Fermentative Production

For production studies, the *S. meliloti* NCIMB 40472 is cultivated in a 20-L bioreactor containing 15 L of Rhizobium Complete (RC) medium (Courtois et al., 1983) supplemented with sucrose (1%, w/v) (RCS). The inoculum corresponds to 1.5 L of a *Rhizobium* culture in RCS medium grown in a flask at 30 °C to a cell density of $\sim 6 \times 10^8$ colony forming units (CFU) mL^{-1}. During cultivation at 30 °C, the pH in the reactor is maintained at 7.2 with KOH; the Po_2 is first stabilized to 80% during exponential growth, and then to 50%. The sucrose concentration in the broth is controlled by HPLC analysis and maintained in the range of 5 to 10 g L^{-1}. The HMW glucuronans produced during fermentation are extracted, purified by ultrafiltration on a 100,000 cut-off membrane (Sartorius), and then dried by lyophilization.

10.2 Specific Production Conditions for Specific Glucuronan

As for many Rhizobia strains producing EPS, the production of glucuronan by the *S. meliloti* NCIMB 40472 in a RCS medium is almost linear during 120 h of fermentation (see Figure 9). The yield of glucuronan (M_w > 100,000 daltons) produced in medium supplemented with $MgSO_4 \cdot 7H_2O$ is dependent upon the concentration of magnesium added. In fact, a high concentration of $MgSO_4 \cdot 7H_2O$ (0.8 g L^{-1} per 24 h) led to the production of glucuronan with a reduced M_w, the value of the EPS falling from 7.5×10^5 to 5×10^5 between 27 and 75 h of fermentation (see Figure 9). The reduction of the glucuronan molecular weight correlated with degradation of the HMW glucuronan by glucuronan lyase; the lyase was probably more active at a high concentration of $MgSO_4 \cdot 7H_2O$ (see Figure 9) (Michaud et al., 1994).

Since magnesium ions induce the production of an EPS which is more extensively acetylated and contains a higher molar ratio of 2,3-di-*O*-acetyl residues, glucuronans with a high or a low degree of substitution can be obtained for specific applications.

10.3 Chemical Routes as an Alternative?

Deacetylated glucuronan may be compared with an oxidized cellulose. Oxidation of cellulose by $NaNO_2$ has been widely described (Cesaro et al., 1985), although the polymer thus obtained is a partially C6 oxidized cellulose.

10.4 Recovery and Purification

The recovery of all glucuronans produced during fermentation can be achieved by centrifugation of the fermentation broth (13,880 × *g*, 20 min), followed by the addition of 1 volume of ethanol or isopropanol to the bacteria-free medium. The precipitate formed is then solubilized in the same initial volume of water and re-precipitated in order to purify the glucuronan. The polysaccharide is then dried either by lyophilization or under vacuum, and stored.

Cross-flow filtration using a 7 mm (internal diameter) tubular ceramic membrane of 0.5-μm pore size under fixed transmembrane pressure or fixed permeate flux and dynamic filtration with a 0.2 nylon membrane using a 16-cm rotating disc, gave the best results compared with classical membranes (Harscoat et al., 1999a). However, the sieving coefficient was dependent on the polymer characteristics (Figure 14). During all filtration conditions tested, the glucuronan molecular weight remained constant (Harscoat et al., 1999b). The most interesting filtration results were obtained with dynamic filtration as it allowed operation at high-shear rates and low transmembrane pressure (Figure 15). The glucuronan extracted can then be precipitated with alcohol, or concentrated and dried.

The recovery of specific HMW ($\geq$ 100,000 daltons), MMW (20,000–100,000 daltons) or LMW (< 20,000 daltons) glucuronans was achieved by ultrafiltration of bacterial-free fermentation media on specific molecular-weight cut-off polysulfone or cellulose triacetate membranes.

Due to the interesting biological properties of LMW glucuronans, specific procedures have been investigated for the purification of oligoglucuronans with specific degrees of acetylation. Anion-exchange chromatography on a DEAE Sepharose column (Figure 16) is useful in the separation of LMW glucuronans (5000–20,000 daltons) (Table 6), but glucuronans with similar DPs can be separated only on the basis of the degree of acetylation (Figure 16) (Pirlet et al., 1999a).

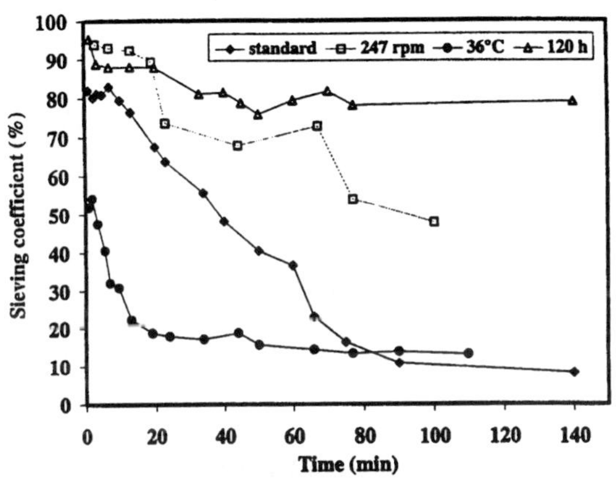

Fig. 14 Polysaccharide-sieving coefficient plotted versus time for glucuronans with various molecular weights obtained in RCS medium at 30 °C during 90 h (standard) ($M_w = 4.07 \times 10^5$ g mol^{-1}), 120 h ($M_w = 2.33 \times 10^5$ g mol^{-1}, and at 36 °C ($M_w = 2.68 \times 10^5$ g mol^{-1})*, and with the standard glucuronan under faster stirring conditions (247 r.p.m.); *the glucuronan produced at 36 °C provides viscous solutions and contains exclusively monoacetylated residues.

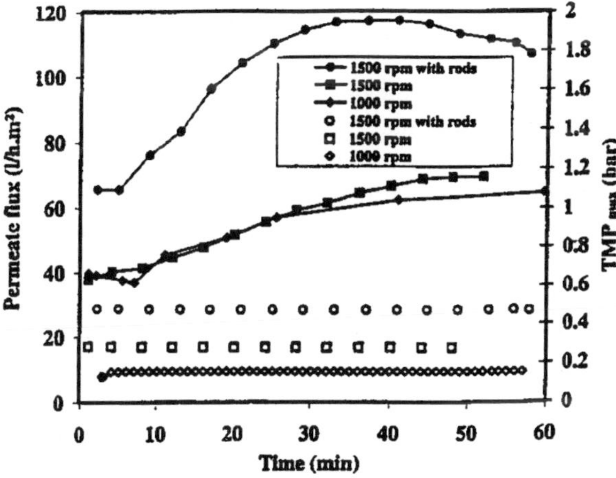

Fig. 15 Permeate-flux and transmembrane pressures (TMP) at 1000 and 1500 r.p.m. for smooth and modified discs (closed symbols, flux; open symbols, TMP).

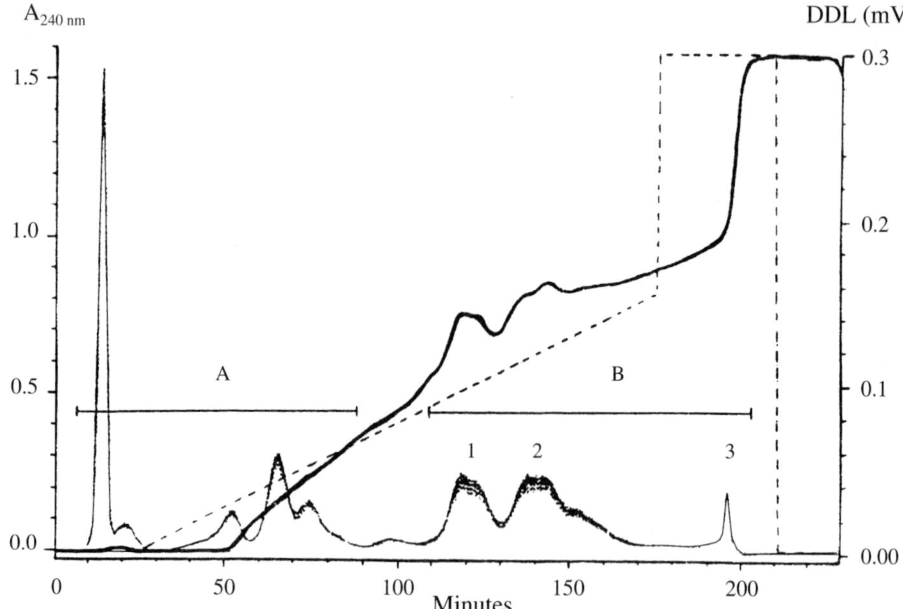

Fig. 16 Fractionation on a DEAE-Sepharose column of the low molecular-weight saccharidic fraction obtained by fermentation of the *S. meliloti* M5N1CS strain. Detection with an UV detector at 240 nm (—) and with a light-scattering detector (—). A, fraction containing cyclic glucans; B1, 2, 3, LMW glucuronan fractions.

Tab. 6 Average degree of substitution (DS) and degree of polymerization (DP) of LMW glucuronans fractionated on a DEAE-Sepharose column

	Peak B1	*Peak B2*	*Peak B3*
DS [%]	38	24	17
DP	2–10	3–>10	>10

Membranes with immobilized histidine, which are used for the purification of proteins and based on dipole interaction, are also useful for the selective adsorption of LMW glucuronans, with discrimination of acetylation (Pirlet et al., 1998, 1999b).

LMW glucuronans (DP >15) which are poorly acetylated (substitution degree 15.5) and without 2,3-di-*O*-acetylated residues, are selectively extracted from a complex mixture corresponding to the whole $M_w < 20{,}000$ dalton fraction, obtained by ultrafiltration (20,000 NMWCO membrane) of a *S. meliloti* M5N1CS fermentation medium. From the first chromatographed mixture, it is possible to extract LMW glucuronans of increasing degree of acetylation. This selective purification allows biological tests to be performed with pure acetylated glucuronan fractions.

11 Properties

Physico-chemical studies were approached with great enthusiasm, as the glucuronan represented a new polysaccharide that contained only uronic acid residues, as alginate.

11.1 Physical and Chemical Properties

In order for glucuronan to be used as a texturizing agent, it was necessary to conduct a series of preliminary rheological studies on polymer solutions.

11.1.1
Rheological Behavior

Based on data obtained from size-exclusion chromatography studies with glucuronans in the range of 60,000–400,000 daltons molecular weight, a viscosity relationship was obtained whereby: $[\eta]=2\times10^{-2}\ M_w^{0.9}$, where $[\eta]$ is the intrinsic viscosity (in mL g^{-1}) and M_w is the molecular weight (Heyraud et al., 1993).

HMW glucuronan ($M_w > 300{,}000$ daltons) solutions of concentration > 5 g L^{-1} produce thermoreversible gels at room temperature. The same phenomenon is observed with glucuronan solutions at < 5 g L^{-1}, when the ionic strength is increased (Dantas et al., 1994a). The stability and strength of the gels depends on the counterion used, and the degree of acetylation of the polymer. Changes in optical rotation are observed when the temperature is between 15 °C and 90 °C, for glucuronan solutions with NaCl (1 M) or NH_4Cl (1 M) (Figure 17) (Courtois et al., 1993). In the presence of divalent or trivalent cations, cross- linkages between glucuronan molecules are formed instantaneously, and with trivalent cations, higher-strength gels are formed. The Young's modulus (E) of gels obtained with glucuronans (10 g L^{-1}) and $FeCl_3$ or $CrCl_3$ (0.34 M) is seven-fold higher than that obtained with Ca^{2+}-gels (Table 7).

The Ca-gels obtained are thermally stable in 0.34 M $CaCl_2$ (Table 8); in pure water, the gel became swollen after heating 5 h at 100 °C and, after 24 h, was completely destroyed due to osmotic effects.

Tab. 7 Determination of the Young's modulus (E)* of polymer gels formed in the presence of divalent or trivalent cations (0.34 M)

Cation	*Cr^{3+}*	*Fe^{3+}*	*Ca^{2+}*
E	12.63	13.70	1.87

* $E\times10^4$: expressed in N m^{-2}.

Tab. 8 Evolution of the Young's modulus (E)[a] on heating at 100 °C a Ca-gel obtained by dialysis of a 10 g L^{-1} glucuronan[b] gel against water alone and 0.34 M $CaCl_2$

	Time [h]				
	1	*2*	*3*	*4*	*24*
Ca-gel in H_2O	1.20	1.10	1.10	0.90	[c]
Ca gel in $CaCl_2$ (0.34 M)	1.30	1.30	1.30	1.25	1.20

[a] $E\times10^{-4}$; expressed in N m^{-2}. [b] Mw: 2.2×10^5, 16% acetate. [c] Gel destroyed.

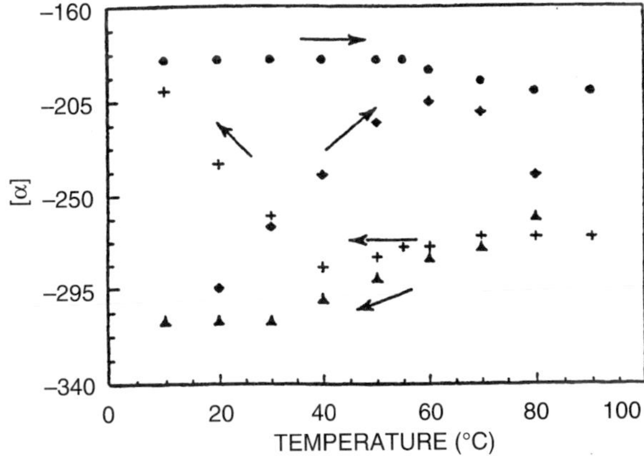

Fig. 17 Effect of temperature on the optical rotation [α]300 of glucuronan solutions (0.5 g L^{-1}) in 1 M NaCl (◆, ▲, glucuronan-NaCl solutions; ●, +, glucuronan-NH_4Cl solutions; →= rise; ←= fall in temperature).

11.1.2
Crystallization and X-ray Analysis

Gels from glucuronan solutions supplemented with saturated $CaCl_2$ solutions, and pulled rapidly from solution, allow the formation of fibers after drying (Heyraud et al., 1994).

X-ray fiber diagrams of acetylated and deacetylated (1→4)-β-D-glucuronan fibers show certain differences (Figure 18). The pattern of the deacetylated glucuronan displays a better resolution than the acetylated sample; moreover, it shows a good correspondence with fiber diagrams obtained with other β-(1→4)-D-linked polysaccharides such as cellulose (Figure 19), chitin or poly-β-D-mannuronic acid. X-ray patterns obtained with the reduced glucuronan are similar to those obtained with cellulose II.

11.2
Biological Properties

Before any biological test could be conducted with glucuronan, it was necessary to investigate any possible biodegradation of the polymer by microbial enzymes.

11.2.1
Biodegradation

The deacetylated glucuronan is degraded by a commercial cellulase mixture (Celluclast 1.5 L; Novo) at 35 °C (Dantas et al., 1994b). Oligoglucuronans with a DP of 2 are obtained (Figure 20), though the substrate limit corresponds to a DP 5 glucuronan. Similar results are obtained with glucuronan lyase obtained from the *S. meliloti* M5N1CS strain.

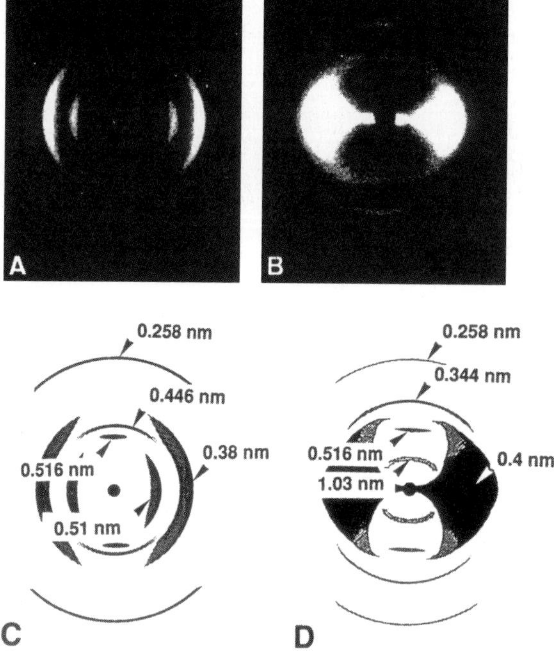

Fig. 18 (A) X-ray fiber diagram of deacetylated (1→4)-β-D-glucuronan. (B) X-ray fiber diagram of acetylated (1→4)-β-D-glucuronan. (C) Schematic diagram corresponding to the pattern A. (D) Schematic diagram corresponding to the pattern B.

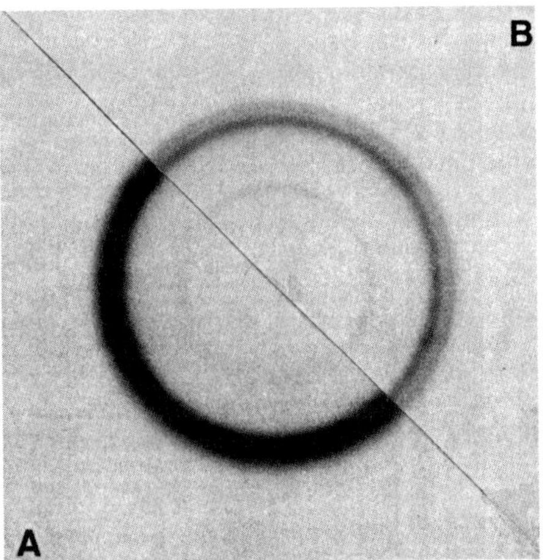

Fig. 19 (A) Powder diagram corresponding to reduced (1→4)-β-D-glucuronan. (B) Powder diagram of hydrolyzed fortisan, taken as a cellulose II crystalline standard.

11.2.2 Activity on Plant Cells

The native glucuronan and oligoglucuronans (DP average of 8) obtained from enzymatic degradation of the deacetylated glucuronan, when tested on protoplasts from *Rubus fructicosus*, induced a 1.5-fold increase in the 1,3-β-D-glucanase activity in 20 min, and a 1.2-fold increase in 1,4-β-D-glucanase activity (see Table 9; Lienart et al., 2000).

11.2.3 Activity on Animal Cells

LMW glucuronans tested on animal cells such as monocytes, were shown to induce the production of tumor necrosis factor (TNF) (Berntzen et al., 1998; Courtois et al., 1998).

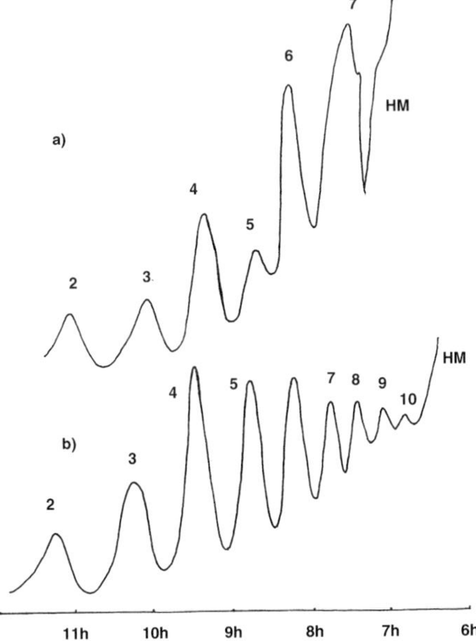

Fig. 20 Fractionation of oligoglucuronans (obtained by enzymatic hydrolysis of a (1→4)-β-D-glucuronan) on a Bio-Gel P6 column (detection by refractometry). The degree of polymerization is indicated by numbers above the peaks.

12 Applications

Applications requiring large amounts of the glucuronan have not been investigated due to the high cost of the polymer. Consequently, the following areas relate only to glucuronan applications requiring small amounts of the polymer, where cost is not a limiting factor.

12.1 Cosmetics

Glucuronan is used in cosmetic preparations in association with extracts from a microalgae (*Haematococcus pluvialis*). It is used in products for skin care, including anti-aging, anti-wrinkle, and anti-free radical treatments. Cosmetic preparations containing glucuronan are also suitable for the treatment of acne as they improve the suppleness of the skin.

12.2
Agriculture

Due to their activity on plant cells, glucuronans and oligoglucuronans may be used in agriculture as activators of plant natural defense mechanisms. In this way they may lead to a reduction in the amounts of chemicals and pesticides required to limit plant diseases.

12.3
Medicine

Glucuronan, such as mannuronan, can be used as an immunostimulating agent.

13
Patents

A patent related to glucuronan production and its main physico-chemical properties was published (Courtois et al., 1993) by the University of Picardie (Amiens, France). Most investigations on the *S. meliloti* mutant strain and on glucuronan have been carried out at the University of Picardie, though some analyses were developed in association with CERMAV-CNRS at the University of Grenoble, France (Table 9). Studies of the biological activities of the HMW polymer and oligoglucuronates led to the publication of three patents related to: (1) the immunostimulating activities of LMW glucuronans (Courtois et al., 1998); (2) the stimulation of plant defenses (Centre National de la Recherche Scientifique, Paris, France) (see Table 9; Lienart et al., 2000); and (3) applications of glucuronan in cosmetic preparations (SEDERMA Company, France) (Lintner, 1999).

14
Outlook and Perspectives

The development of glucuronan applications will clearly require an increased bacterial production of the polysaccharide by the Rhizobia strain. Genetic engineering will undoubtedly provide the opportunity to

Tab. 9 List of patents relating to glucuronan

Number of patent (date of publication)	***Patent holder***	***Inventors***	***Title***
WO-A-93/18174 (1993)	University of Picardie (Amiens France)	Courtois, J. Courtois, B. Heyraud, A. Colin-Morel, P. Rinaudo, M.	Polymers of glucuronic acid, preparation procedure and uses as gelling or thickening agent, moisturising, stabilizing, chelating or flocculating agent
FR 2 781 673-A1 (1998)	University of Picardie (Amiens France)	Courtois, J. Courtois, B.	Glucuronan uses as immunostimulating agent, preparation procedure
WO 9913855 (1999)	SEDERMA Company Le perray en Yvelines France	Lintner, K.	Compositions for cosmetic or dermopharmaceutical uses containing a mixture of algae extracts and an exopolysaccharide
WO 0100025 (2001)	Centre National de la Recherche Scientifique (CNRS)(Paris France)	Lienart, Y. Heyraud, A. Sevenou, O.	Glycuronic polysaccharides and oligosaccharides uses as plant health products and/or as fertilizing agents

study the biosynthesis of glucuronan, and also to create modifications of the regulation of its biosynthesis. A number of applications is envisaged that include uses in human healthcare, and with glucuronan either modified chemically or used in association with other polysaccharides. In fact, an association with polysaccharides may well provide a pathway by which the copolymers physico-chemical properties can not only be identified, but also used more beneficially with the greatest range of applications.

15
References

Amemura, A., Hisamatsu, M., Ghai, S., Harada, T. (1981) Structural studies on a new polysaccharide containing D-riburonic acid from *Rhizobium meliloti* IFO 13336, *Carbohydr. Res.* **91**, 59–65.

Battisti, L., Lara, J. C., Leigh, J. A. (1992) Specific oligosaccharide form of the *Rhizobium meliloti* EPS promotes nodule invasion in alfalfa, *Proc. Natl. Acad. Sci. USA* **89**, 5625–5629.

Berntzen, G., Flo, T. H., Medvedev, A., Kilaas, L., Skjak-Braek, G., Sundan, A., Espevik, T. (1998) The tumor necrosis factor-inducing potency of lipopolysaccharide and uronic acid polymers is increased when they are covalently linked to particles, *Clin. Diagn. Lab. Immunol.* **5**, 355–361.

Betlach, M. R., Capage, M. A., Doherty, D. H., Hassler, R. A., Henderson, N. M., Vanderslice, R. W., Marelli, J. D., Ward, M. B.(1987) Genetically engineered polymers: Manipulation of xanthan biosynthesis, in: *Progress in Biotechnology: Industrial Polysaccharides* (Yalpani M., Ed.), Amsterdam: Elsevier Science Publishers, vol. 3, 35–50.

Bouffar, C., Heyraud, A. (1987) Determination of relative proportion of D-mannuronic and D-guluronic acid in alginic acid by HPLC, *Carbohydr. Res.* **1**, 559–561.

Cesaro, A., Delben, F., Painter, T. J., Paoletti, S., Crescenzi, V., Dea, I. C. M., Stivala, S. S. (1985) *New Developments in Industrial Polysaccharides*, New York: Gordon and Breach Science Publishers.

Chakravorty, A. K., Zurkowski, W., Shine, J., Rolfe, B. G. (1982) Symbiotic nitrogen fixation: molecular cloning of *Rhizobium* genes involved in exopolysaccharide synthesis and effective nodulation, *J. Mol. Appl. Genet.* **1**, 585–596.

Courtois, B., Hornez, J. P., Courtois, J., Derieux, J. C. (1983) Classification of *Rhizobium meliloti* by a metabolic property, *Ann. Microbiol.* **134B**, 141–147.

Courtois, B., Courtois, J., Heyraud, A., Rinaudo, M. (1986) Effect of biosynthesis conditions on the chemical composition of the water-soluble polysaccharides of fast-growing *Rhizobia*, *J. Gen. Appl. Microbiol.* **32**, 519–526.

Courtois, J., Seguin, J. P., Declosmesnil, S., Heyraud, A., Colin-Morel, P., Dantas, L., Barbotin, J. N., Courtois, B. (1993) A β-(1 → 4)-D-glucuronan excreted by a mutant of the *Rhizobium meliloti* M5N1 strain, *J. Carbohydr. Chem* **12**, 441–448.

Courtois, J., Seguin, J. P., Roblot, C., Heyraud, A., Gey, C., Dantas, L., Barbotin, J. N., Courtois, B. (1994) Exopolysaccharide production by the *Rhizobium meliloti* M5N1CS strain. Location and quantitation of the sites of O-acetylation, *Carbohydr. Polym.* **25**, 7–12.

Da Costa, A., Michaud, P., Petit, E., Heyraud, A., Colin-Morel, P., Courtois, B., Courtois, J. (2001) Purification and properties of a glucuronan lyase from *Sinorhizobium meliloti* M5N1CS (NCIMB 40472), *Appl. Environ. Microbiol.*, **67**, 000–000.

Dantas, L., Heyraud, A., Courtois, B., Courtois, J., Milas, M. (1994a) Physico-chemical properties of "exogel" exocellular β-(1 → 4)-D-glucuronan from *Rhizobium meliloti* strain M5N1 CS (NCIMB 40472), *Carbohydr. Polym.* **24**, 185–191.

Dantas, L., Courtois, J., Courtois B., Seguin J. P., Gey, C., Heyraud, A. (1994b) NMR spectroscopic investigation of oligoglucuronates prepared by enzymic hydrolysis of a β-(1 → 4)-D-glucuronan, *Carbohydr. Res.* **265**, 303–310.

De Ruiter, G. A., Josso, S. L., Colquhoun, I. J., Voragen, A. G. J., Rombouts, F. M. (1992) Isolation and characterization of beta β-(1 → 4)-D-glucuronans from extracellular polysaccharides of moulds belonging to Mucorales, *Carbohydr. Pol.* **18**, 1–7.

Espevik, T., Otterlei, M., Skjak-Braek, G., Ryan, L., Wright, S. D., Sundan, A. (1993) The involvement of CD14 in stimulation of cytokine production by uronic polymers, *Eur. J. Immunol.* **23**, 255–261.

Glazeborook, J., Walker, G. C. (1989) A novel exopolysaccharide can function in place of the Calcofluor-binding exopolysaccharide in nodulation of alfalfa by *Rhizobium meliloti*, *Cell* **56**, 661–672.

Gonzales, J. H., Reuhs, B. L., Walker, G. C. (1996a) Low molecular weight EPSII of *Rhizobium meliloti* allows nodule invasion in *Medicago sativa*, *Proc. Natl. Acad. Sci. USA* **93**, 8636–8641.

Gonzales, M. L., Courtois, J., Colin-Morel, P., Michaud, P., Barbotin, J. N., Courtois, B. (1996b) Selection of a succinoglycan-deficient *Rhizobium meliloti* mutant producing a partially acetylated β-(1→4)-D-glucuronan, *Ann. N.Y. Acad. Sci.* **782**, 53–60.

Hakomori, S. (1964) A rapid permethylation of glycolipid and polysaccharide catalysed by methylsulfinyl-carbanion in dimethyl sulfoxide, *J. Biochem.* **55**, 205–208.

Harscoat, C., Jaffrin, M., Paullier, P., Courtois, B., Courtois, J. (1999a) Recovery of microbial polysaccharides from fermentation broths by microfiltration on ceramic membranes, *J. Chem. Tech. Biotechnol.* **74**, 571–579.

Harscoat, C., Jaffrin, M., Bouzerar, R., Courtois, J. (1999b) Influence of fermentation conditions and microfiltration processes on membrane fouling during recovery of glucuronan polysaccharides from fermentation broths, *Biotechnol. Bioeng.* **65**, 500–511.

Heyraud, A., Rinaudo, M., Courtois, B. (1985) Comparative studies of extracellular polysaccharide elaborated by *Rhizobium meliloti* strain M5N1 in defined medium and in non-growing cell suspensions, *Int. J. Biol. Macromol.* **8**, 85–88.

Heyraud, A., Courtois, J., Dantas, L., Colin-Morel, P., Courtois, B. (1993) Structural characterisation and rheological properties of an extracellular glucuronan produced by a *Rhizobium meliloti* M5N1 mutant strain, *Carbohydr. Res.* **240**, 71–78.

Heyraud, A., Dantas, L., Courtois, J., Courtois, B., Helbert, W., Chanzy, H. (1994) Crystallographic data on bacterial β-(1→4)-D-glucuronan, *Carbohydr. Res.* **258**, 275–279.

Jansson, P. E., Kenne, L., Lindberg, B. (1975) Structure of the extracellular polysaccharide from *Xanthomonas campestris*, *Carbohydr. Res.* **45**, 275–282.

Keller, M., Arnold, W., Müller, P., Niehaus, K., Shmidt, M., Quandt, J., Weng, W. M., Pühler, A. (1990) *Rhizobium meliloti* genes involved in exopolysaccharide production and infection of alfalfa nodules, in: Pseudomonas: *Biotransformations, Pathogenesis, and Evolving Biotechnology* (Silver, S., Chakrabarty, A. M., Iglewski, B., Kaplan, S., Eds.). Washington, DC: American Society for Microbiology Press, 91–97.

Leigh, J. A., Walker, G. C. (1994) Exopolysaccharides of *Rhizobium*: synthesis, regulation and symbiotic function, *Trends Genet.* **10**, 63–67.

Leigh, J. A., Signer, E. R., Walker, G. C. (1985) Exopolysaccharide-deficient mutants of *R. meliloti* that form ineffective nodules, *Proc. Natl. Acad. Sci. USA* **82**, 6231–6235.

Michaud, P., Courtois, B., Courtois, J., Heyraud, A., Colin-Morel, P., Seguin, J. P., Barbotin, J. N. (1994) Physicochemical properties of extracellular (1→4)-β-D-glucuronan produced by the *Rhizobium meliloti* M5N1CS strain during fermentation: evidence of degradation by an exoenzyme activated by Mg^{2+}, *Int. J. Biol. Macromol.* **16**, 301–305.

Michaud, P., Courtois, J., Courtois, B., Heyraud, A., Colin-Morel, P., Seguin, J. P., Barbotin, J. N. (1995) Cyclic (1→2)-β-D-glucans excreted by the glucuronan-producing strain *Rhizobium meliloti* M5N1CS (NCIMB 40472) and by the succinoglycan-producing strain *Rhizobium meliloti* M5N1, *Int. J. Biol. Macromol.* **17**, 369–372.

Michaud, P., Pheulpin, P., Petit, E., Seguin, J. P., Barbotin, J. N., Heyraud, A., Courtois, B., Courtois, J. (1997) Identification of a glucuronan lyase from a mutant strain of *Rhizobium meliloti*, *Int. J. Biol. Macromol.* **21**, 3–9.

Müller, P., Hynes, M., Kapp, D., Niehaus, K., Pühler, A. (1988) Two classes of *Rhizobium meliloti* infection mutants differ in exopolysaccharide production and coinoculation properties with nodulation mutants, *Mol. Gen. Genet.* **211**, 17–26.

O'Neil, M. A., Selvendran, R. R., Morris, V. J. (1983) Structure of the extracellular gelling polysaccharide produced by *Pseudomonas elodea*, *Carbohydr. Res.* **124**, 123–133.

Pellock, B. J., Cheng, H. P., Walker, G. C. (2000) Alfalfa root nodule invasion efficiency is dependent on *Sinorhizobium meliloti* polysaccharides, *J. Bacteriol.* **182**, 4310–4318.

Pirlet, A. S., Pitiot, O., Guentas, L., Heyraud, A., Courtois, B., Courtois, J., Vijalayakshmi, M. A. (1998) Separation of low-molecular-mass acetylated glucuronans on L-histidine immobilized onto poly(ethylene-vinyl alcohol) hollow-fiber membranes, *J. Chromatogr.* A, **826**, 157–166.

Pirlet, A. S., Guentas, L., Heyraud, A., Pheulpin, P., Vijalayakshmi, M., Barbotin, J. N., Courtois, B.,

Courtois J. (1999a) Influence of acetyl substituent on oligoglucuronans separation by anion exchange chromatography, *Carbohydr. Polym.* **38**, 155–160.

Pirlet, A. S., Guentas, L., Pitiot, O., Heyraud, A., Vijalayakshmi, M., Courtois, B., Courtois, J. (1999b) Selective affinity of L-histidine immobilized onto poly(ethylene-vinyl alcohol) hollow-fiber membranes for various oligoglucuronans. Influence of the degree of polymerisation and the degree of substitution by acetate, *J. Chromatogr.* A, **841**, 1–8.

Waffanschmidt, S., Jaenicke, L. (1987) Assay of reducing sugars in the nanomole rang with 2,2′-bicinchoninate, *Ann. Biochem.* **165**, 337–340.

Wang, L. X., Wang, Y., Pellock, B., Walker, G. C. (1999) Structural characterisation of the symbiotically important low-molecular-weight succinoglycan of *Sinorhizobium meliloti*, *J. Bacteriol.* **181**, 6788–6796.

Yalpani, M. (1988) Polysaccharides: syntheses, modifications and structure/property relations, in: *Studies in Organic Chemistry 36*, Amsterdam: Elsevier Science Publishers.

Yu, N. X., Hisamatsu, M., Amemura, A., Harada, T. (1983) Structural studies on an extracellular acidic polysaccharide (APS-I) of *Rhizobium meliloti* 201, *Agric. Biol. Chem.* **47**, 491–498.

10
Sphingan Group of Exopolysaccharides (EPS)

Dr. Thomas J. Pollock
Shin-Etsu Bio, Inc., 6650 Lusk Boulevard, Suite B106, San Diego, CA 92121-2775, USA; Tel.: + 1-8584558506; Fax: + 1-8585872716; E-mail: tompollock@attglobal.net

DNA	deoxyribonucleic acid
EPS	exopolysaccharide
Glc	glucose
GlcA	glucuronic acid
HPLC	high-performance liquid chromatography
IP	C-55 isoprenylphosphate
Man	mannose
PGM	phosphoglucomutase
PHB	poly-hydroxybutyrate
Rha	rhamnose
Sps	sphingan polysaccharide synthesis
TDP	thymidine-5′-diphosphate
UDP	uridine-5′-diphosphate

1 Introduction

Several members of the bacterial genus *Sphingomonas* secrete acidic capsular heteropolysaccharides with similar, but not identical, structures (Pollock, 1993). These related polymers are referred to as "sphingans" after the common genus. The sphingans are used to control the rheology of aqueous solutions in a variety of commercial applications (Baird et al., 1983; Moorhouse, 1987).

2 Historical Outline

Many of the sphingans were discovered by the Kelco Company (now CP-Kelco) during widespread searches for useful water-soluble polymers, beginning in the 1970s. Commercialization began in the late 1980s, and this group is still growing by accretion as more polysaccharide structures are determined and polysaccharide-producing bacteria are subjected to more accurate taxonomic classification. Future searches can take advantage of newly recognized common identifiers for *Sphingomonas* bacteria (Pollock, 1993), and it is reasonable to expect the discovery of new polysaccharide structures within this group which will likely have unique and useful physical properties.

3
Chemical Structures

The main chain of a sphingan contains a repeat unit which is largely, but not completely, conserved. The common main chain repeat is [→4) α-L-Rha or α-L-Man (1→3) β-D-Glc (1→4) β-D-GlcA (1→4) β-D-Glc (1→]. This order of sugars matches the monomeric repeat unit that is assembled in the cell attached to an isoprenylphosphate (IP) lipid carrier. The IP carrier is linked to the Glc which is on the reducing side of GlcA.

3.1
Structural Variations

Specific repeat units of the sphingans are shown in Figure 1. In two sphingans, S-88 and S-198, L-mannose substitutes for about one-half of the L-rhamnose in the main chain (Jansson et al., 1986a; Chowdhury et al., 1987a). A similar substitution occurs for the side chain of welan (Jansson et al., 1985; O'Neill et al., 1986; Stankowski and Zeller, 1992; Jansson and Widmalm, 1994). NW-11 is unique among the sphingans in lacking L-Rha entirely and having only L-Man in the main chain (O'Neill et al., 1990). L-Mannose is a partially oxidized form of L-rhamnose and is exceedingly rare in nature. Its biosynthetic origin in *Sphingomonas* is unknown. Two sphingans, S-7 and I-886, contain 2-deoxy-β-D-arabino-hexuronic acid (2-deoxy glucuronic acid) in the main chain in place of the more common glucuronic acid (Falk et al., 1996; Thorne et al., 2000; Gulin et al., 2001). The 2-deoxy glucuronic acid is also rare and, because it is especially labile to acid, these two sphingans are very sensitive to acid hydrolysis.

Excluding the acetyl groups, there are six types of side chains depicted in Figure 1 which are attached to one or the other of the two glucose residues: L-glyceric acid, L-rhamnose, L-mannose, L-rhamnose-L-rhamnose, D-glucose-α-D-glucose, and D-glucose-β-D-glucose. Sphingans S-88 and S-198 differ only in the position of the L-rhamnose side chain (Jansson et al., 1986a; Chowdhury et al., 1987a). S-88 and S-657 are different due to the additional L-rhamnose in the side chain of S-657 (Jansson et al., 1986b; Chowdhury et al., 1987b). Rhamsan and I-886 differ only in the oxidation state of the uronic acid residue (Jansson et al., 1986; Falk et al., 1996). It is the combination of these common and distinct structural features which make the sphingans a rich source of polysaccharides for structural and functional studies, and which confer the unique and useful physical properties to each member of the group.

A single acetyl group is generally attached to about one-half of the repeat units of the sphingans, but the position has only been determined for welan and gellan. In welan the acetyl group is attached to the 2-position of L-rhamnose (Stankowski and Zeller, 1992), and in gellan it is attached to the 6-position of the glucose located to the immediate reducing side of rhamnose (Kuo et al., 1986). Gellan also has an L-glyceric acid group attached to the same glucose residue at the 2 position (Jansson et al., 1983; Kuo et al., 1986).

3.2
Analytical Methods

The sphingans and associated cells are easily separated from the culture medium by precipitation of the biomass with one to three volumes of isopropyl alcohol or etha-

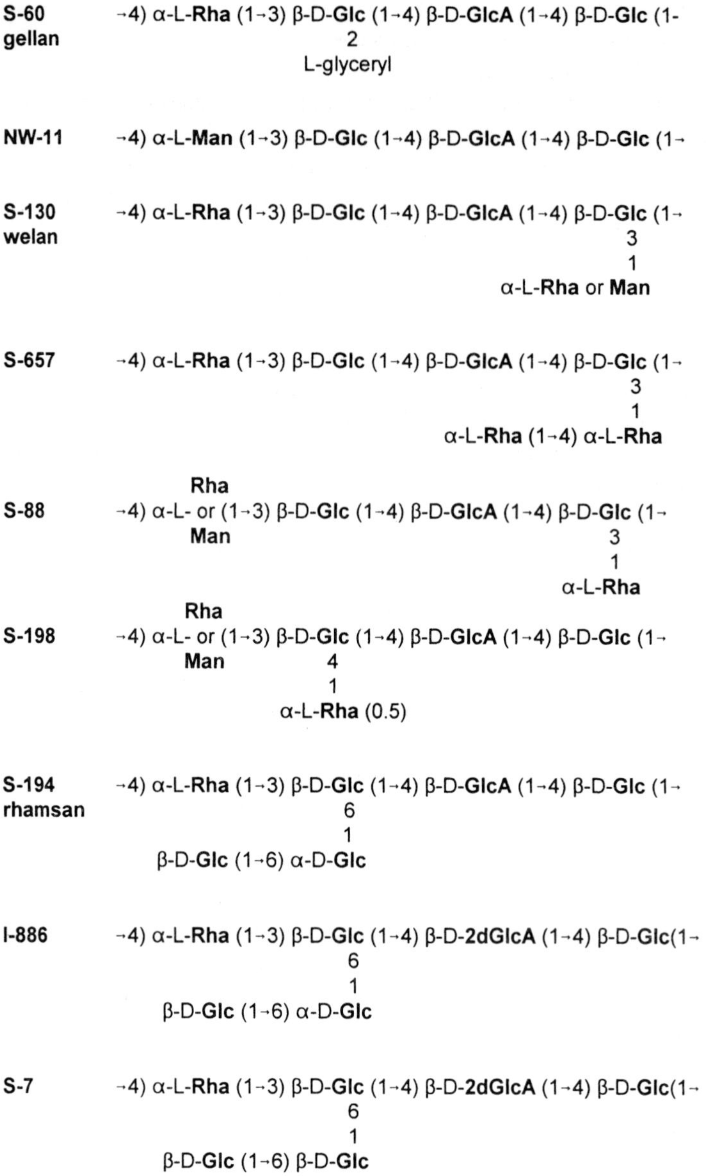

S-60 gellan
→4) α-L-**Rha** (1→3) β-D-**Glc** (1→4) β-D-**GlcA** (1→4) β-D-**Glc** (1-
2
L-glyceryl

NW-11
→4) α-L-**Man** (1→3) β-D-**Glc** (1→4) β-D-**GlcA** (1→4) β-D-**Glc** (1→

S-130 welan
→4) α-L-**Rha** (1→3) β-D-**Glc** (1→4) β-D-**GlcA** (1→4) β-D-**Glc** (1→
3
1
α-L-**Rha** or **Man**

S-657
→4) α-L-**Rha** (1→3) β-D-**Glc** (1→4) β-D-**GlcA** (1→4) β-D-**Glc** (1→
3
1
α-L-**Rha** (1→4) α-L-**Rha**

S-88
Rha
→4) α-L- or (1→3) β-D-**Glc** (1→4) β-D-**GlcA** (1→4) β-D-**Glc** (1→
Man
3
1
α-L-**Rha**

S-198
Rha
→4) α-L- or (1→3) β-D-**Glc** (1→4) β-D-**GlcA** (1→4) β-D-**Glc** (1→
Man
4
1
α-L-**Rha** (0.5)

S-194 rhamsan
→4) α-L-**Rha** (1→3) β-D-**Glc** (1→4) β-D-**GlcA** (1→4) β-D-**Glc** (1→
6
1
β-D-**Glc** (1→6) α-D-**Glc**

I-886
→4) α-L-**Rha** (1→3) β-D-**Glc** (1→4) β-D-**2dGlcA** (1→4) β-D-**Glc**(1→
6
1
β-D-**Glc** (1→6) α-D-**Glc**

S-7
→4) α-L-**Rha** (1→3) β-D-**Glc** (1→4) β-D-**2dGlcA** (1→4) β-D-**Glc**(1→
6
1
β-D-**Glc** (1→6) β-D-**Glc**

Fig. 1 Structures of repeating subunits of sphingan polymers. Acetyl groups are not shown. Abbreviations: Rha, rhamnose; Glc, glucose; Man, mannose; GlcA, glucuronic acid; and 2dGlcA, 2-deoxy-glucuronic acid.

nol. The cells usually represent between 20–40% of the dry weight of the precipitated biomass. Samples can be hydrolyzed with 2 M trifluoroacetic acid for a few hours at 100 °C, followed by washing and resuspension in water. Samples and monosaccharide standards are then separated on an anion-exchange (HPLC) column and quantified (Clarke et al., 1991). Rhamnose, glucose, mannose and to a lesser extent glucuronic acid are easily detected and measured. However, the 2-deoxy-glucuronic acid residues found in sphingans S-7 and I-886 are destroyed by these hydrolytic conditions. Assays using hydroxylamine to measure the degree of acetylation are routine (Hestrin, 1949). Special precipitation and washing methods are used to detect and measure gellan in foods (Baird and Smith, 1989; Graham, 1991).

Gellan (Kang et al., 1982a; Baird et al., 1983) and an exopolysaccharide produced by *Sphingomonas paucimobilis* GS1 (Ashtaputre and Shah, 1995) form thermoreversible gels. The properties of these gels are measured by standard methods. Gelation of the former is optimal in the presence of divalent cations, while the latter forms gels more readily with sodium ions. The structure of the polymer secreted by *S. paucimobilis* strain GS1 is not known.

4 Occurrence

Many of the sphingan-producing bacteria were isolated from pond waters or in association with moist plant surfaces. The gellan strain was isolated from a common freshwater plant, *Elodea* (Kang et al., 1982b), and the strains secreting rhamsan (Peik et al., 1983) and S-198 (Peik et al., 1985) were obtained from freshwater mountain ponds. The NW-11 strain was found in a waste water pond (O'Neill et al., 1990) and the S-657 strain was isolated from an algal sample taken from a marsh (Peik et al., 1992). Strain S-7 was discovered in a subsurface organic-rich soil after rain (Kang and McNeely, 1975a) and I-886 was found in the rhizosphere of Pearl millet (Hebbar et al., 1992).

5 Physiology

The sphingan polysaccharides form a capsule surrounding each bacterium (Pollock and Armentrout, 1999). The capsules appear to be bound physically to the cells, although the mechanism of attachment is unknown. *Sphingomonas* cells cannot be removed from the surrounding capsule simply by dilution with water and centrifugation, although following a limited acid hydrolysis, centrifugation will yield a cell pellet and a relatively cell-free supernatant.

When sphingan-producing cells are growing in liquid suspension they are aggregated into large sheets containing from hundreds to tens of thousands of cells. The multicellular sheets are reminiscent of biofilms formed by sessile communities of bacteria. In still cultures without agitation, *Sphingomonas* cells form a pellicle which adheres preferentially to the glass vessel, near the liquid surface. A minority of cells do not join in the aggregate and remain highly motile, presumably seeking out more advantageous concentrations of nutrients. The occurrence of sessile communities producing water-soluble capsular polysaccharides and the option for planktonic movement suggest that these cells might be specially adapted for growth on surfaces periodically exposed to water, and that the capsule may protect the biofilm from desiccation.

6 Properties

The sphingans have related structures, but the different locations and types of carbohydrate and acyl side groups and the alternative of L-rhamnose or L-mannose give rise to an array of physical properties which are unique for each member of the group (Baird et al., 1983; Kang et al., 1983; Moorhouse, 1987; Nussinovitch, 1997). Gellan (S-60) produces a thermoreversible gel when heated and cooled in the presence of cations (Nussinovitch, 1997). Divalent cations such as magnesium and calcium give stronger gels compared with monovalent cations such as sodium and potassium. For example, a gel strength of 700–800 g cm^{-2} is obtained for 1% gellan with 0.1% calcium or magnesium, and a gel strength of 500–600 g cm^{-2} for 1% KCl or NaCl. A model for gelation was proposed based on light-scattering studies with gel-promoting cations and gel-restricting bulky tetramethylammonium cations (Gunning and Morris, 1990). In this model gelation is due to localized lateral associations or crystallization among double helical fibrils. X-ray diffraction patterns for native and deacylated gellan show an extended interwound three-fold, left-handed, half-staggered, double helix (Chandrasekaran et al., 1992), the double helices being stabilized by interchain hydrogen bonds. Although the glyceric acid groups have no effect on the formation of the double helices, they mask the carboxylate groups of glucuronic acid and would be expected to alter its ion-binding properties. This in turn would reduce the propensity for aggregation and should cause native gellan to form weak gels. Indeed, the physical properties of a gellan gel depend on the degree of acylation of the polymer, such that a weak elastic gel is formed with fully acylated gellan, whilst a strong brittle gel is formed with deacylated gellan. By contrast, the acetyl groups do not affect the structure of gellan.

Welan (S-130) is an excellent suspending agent, imparting a high viscosity to aqueous solutions at low concentration and low shear rate. Although welan is shear-thinning for shear rates between 0.1 to 1000 per second, it is relatively resistant to breakage by shearing, and able to recover nearly all internal structure within minutes. It is thermally stable up to 150 °C, and the viscosity and solubility of aqueous welan solutions are relatively compatible with high concentrations of divalent ions, as in salt water or brine. Welan is also resistant to pH from 2 to 12. X-ray diffraction studies indicate that deacylation does not affect its three-dimensional structure (Attwool et al., 1986). Rhamsan (S-194) displays rheological properties similar to those of welan, but it is more viscous at low concentration and low shear, and has a greater resistance to shear forces, such as are generated in mills. Rhamsan loses its viscosity more than welan at temperatures above 100 °C.

Polysaccharides S-198 and S-88 are similar in rheological properties to welan and rhamsan. However, S-198 is even more resistant to shear forces such as those in high-pressure pumps. S-198 and S-88 both retain their viscosity at temperatures up to 150 °C.

Sphingan S-7 shows the broad pH and thermal stabilities of xanthan gum and is shear-thinning, but S-7 generates a two-fold higher viscosity in aqueous solutions and is a better suspending agent. The viscosity of S-7 is unaffected by salt water or brine, and is compatible with high concentrations of most salts. It can be cross-linked with polyvalent cations (Cr^{3+}) at high pH.

Polysaccharide S-657 is noteworthy as being the most thermally stable of all the known sphingans. Aqueous solutions of S-657 are also more viscous at low concentrations than either S-194 or xanthan gum.

7 Biosynthesis

Copious secretion of EPS is usually most noticeable when bacteria are supplied with an abundant carbon source and minimal nitrogen. Colonies of *Sphingomonas* bacteria growing on the surface of nutrient agar plates containing high levels of carbohydrates (1% glucose) are raised, opaque, and rubbery, the cells adhering tightly one to another. This phenotype is referred to as Sps^+ or positive for sphingan polysaccharide synthesis. Nonproducing mutant colonies (Sps^-) are more flat, translucent, and watery. The Sps^+ phenotype in liquid cultures is detected by observing or measuring the readily apparent increase in viscosity of the culture medium. Other known *Sphingomonas* species which are not obviously mucoid might also have the capacity for sphingan synthesis if presented with more advantageous conditions.

7.1 Synthesis of Sugar-nucleotide Substrates

Monosaccharides are transported into bacterial cells, phosphorylated, and then attached to nucleotide monophosphates to form the sugar-nucleotide substrates for polysaccharide synthesis (Sutherland, 1982). Consistent with the carbohydrate composition of gellan, several biosynthetic enzymes required for the synthesis of UDP-Glc, dTDP-Rha, and UDP-GlcA were detected in crude cell extracts prepared from *Sphingomonas* ATCC 31461 (Martins and Sá-Correia, 1991, 1993; Martins et al., 1996).

7.2 Assembly of the Repeat Unit, Polymerization and Secretion

The activated sugars are sequentially and specifically transferred to a membrane-associated IP carrier to assemble an oligosaccharide repeat unit. Sphingan assembly follows the same mechanisms established in detail for Gram-negative bacteria, such as *Xanthomonas campestris*, which secretes xanthan (Ielpi et al., 1993), and *Sinorhizobium meliloti*, which secretes succinoglycan (Reuber and Walker, 1993). The transferase-catalyzed steps for *Sphingomonas* have been studied mainly with strain S-88 as a model (Pollock et al., 1994; Yamazaki et al., 1996). Assembly of the S-88 repeat unit – and probably all of the other sphingans – begins with the transfer of glucose-1-phosphate from UDP-glucose to IP. Availability of the IP carrier in Gram-negative bacteria is reduced by the antibiotic bacitracin, which selectively binds to and sequesters isoprenylpyrophosphate (Stone and Strominger, 1971), the immediate precursor of IP. Among the *Sphingomonas* mutants which are resistant to bacitracin, the majority have lost their ability to produce nonessential sphingans. The mutants seem to survive by committing all of the remaining IP carrier to the synthesis of essential peptidoglycans of the cell wall (Pollock et al., 1994).

Subsequent additions of monosaccharides to the nonreducing end of the growing repeat unit by other specific glycosyl transferases give rise to a complete oligosaccharide repeat unit linked to the carrier. The sphingan repeat units are shown diagrammatically in Figure 1, where the carrier is attached to one of the glucose residues at the reducing end. In order to deduce which of the two glucose residues in the main chain repeat unit was initially transferred to the IP carrier it was necessary to determine that glucuronic acid was the second sugar added to the assembly (Pollock et al., 1998). The mechanism of attachment of the side chain sugars is not known; they are either added after the main chain repeat is completed in the form of a branch, or they may be added prior to completion of the main chain tetrasaccharide.

The enzymes and genes involved in polymerization have also not yet been identified in *Sphingomonas*, but they are expected to function as in other Gram-negative bacteria (Sutherland, 1982). The growing chains are thought to remain attached to the lipid carriers as additional repeat units are inserted at the reducing end of the polymer adjacent to the lipid carrier.

At least five gene products have been identified which are likely to act during the secretion of the polymer from the cell (Yamazaki et al., 1996).

7.3 Genetics

Many of the genes involved in polysaccharide biosynthesis in Gram-negative bacteria have been identified and cloned by using genetic complementation of polysaccharide-negative mutants. Spontaneous bacitracin-resistant, sphingan-negative (Sps⁻) mutants were used as recipients to test DNA fragments cloned from the chromosome of *Sphingomonas* S88 for their ability to restore sphingan synthesis to the Sps⁻ mutants (Pollock et al., 1994; Yamazaki et al., 1996). Because the DNA segments included about 25,000 base pairs each, and because the biosynthetic genes were clustered in the chromosome, it was possible to isolate many of the necessary genes on a single cloned segment. Figure 2 shows the organization of a single contiguous gene cluster from strain S88, and the functions of many of the genes. The cluster includes four genes arranged in an operon which code for the synthesis of one of the sugar-nucleotide precursors, dTDP-rhamnose. Also present are four genes which code for the glycosyl-transferases required to assemble the main chain of the repeat unit: *spsB* (transferase 1), *spsK* (transferase 2), *spsL* (transferase 3), and *spsQ* (transferase 4). Secretion is the likely function for *spsS*, *spsJ*, *spsD*, *spsC*, and *spsE*. However, all of the essential genes for synthesis are not located in this cluster.

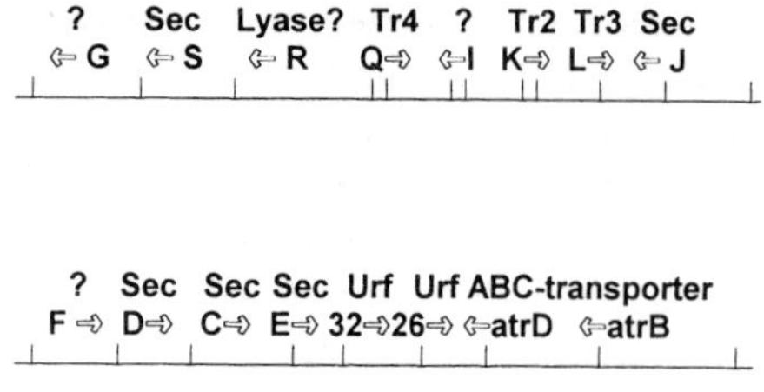

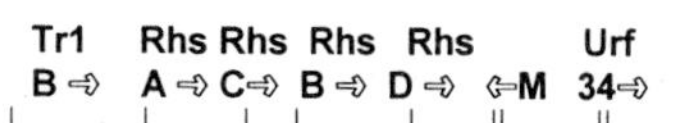

Fig. 2 Gene cluster for sphingan S-88 synthesis. The map spans about 30,000 base pairs and is divided into three contiguous segments, left to right, top to bottom. The direction of transcription is shown by the arrows. Abbreviations: sec, secretion; Tr1, 2, 3, and 4, glycosyl transferases for glucose, glucuronic acid, glucose and rhamnose, respectively; Urf, unassigned reading frame; Rhs, rhamnose synthetic operon.

According to the complete DNA sequence of the S88 cluster, the genes coding for the synthesis of UDP-glucose and UDP-glucuronic acid appear to be elsewhere in the chromosome. Analogous segments from the chromosomes of other sphingan-producing bacteria including *Sphingomonas* S7, S60, S198, S194 and S130 have similar (but not identical) gene arrangements (Thorne et al., 2000).

The SpsB gene product was identified as transferase 1 since its amino acid sequence was very similar to a distinctive family of enzymes which recognize both the IP carrier and UDP-glucose. Although a comparable analysis of the SpsK, SpsL, and SpsQ gene products indicated that they were likely to be transferases catalyzing subsequent steps in assembly, it was not possible to deduce the sugar specificities simply from the nucleotide and amino acid sequences. Rather, genetic complementation was used instead. Glycosyl transferases which recognize identical substrates (nucleotide-sugars and lipid-linked carbohydrates) can often substitute one for another in bacterial polysaccharide biosynthesis, even if the enzymes originate in different genera of bacteria. Intergeneric substitution was carried out between *Sphingomonas* and *Rhizobium meliloti*, and this allowed the assignment of the substrate specificities of three of the transferase genes: *spsK* (glucuronic acid), *spsL* (glucose), and *spsQ* (rhamnose) (Pollock et al., 1998). This genetic analysis also suggested the order of assembly of the sugars, and distinguished between the two glucose residues in the main chain as to which was the initiating sugar.

7.4
Regulation

Sphingomonas bacteria inhabit many different environments and display a phenotypic dimorphism. They can adopt either a planktonic or a sessile behavior in liquid culture (Pollock and Armentrout, 1999). The sessile morphology is marked by the presence of a sphingan polysaccharide capsule. Sensing of environmental stimuli and genetic control over synthesis of the capsule are key events in alternating between these two phenotypes. The level of oxygen appears to be an important determinant for switching between the two forms. A shift down in oxygen level results in the disaggregation of sheets of *Sphingomonas* cells in liquid culture, and the onset of planktonic motility.

A special class of *Sphingomonas* mutants arise in nonagitated liquid cultures. They do not synthesize sphingan polysaccharides, and they are highly motile. Two tightly linked genes were isolated from these strains which, when introduced back into the mutants in multiple copies, restored the wild-type phenotype of sphingan production and nonmotility. The genes, *spsAZ*, belong to a large family of structurally similar two-component sensory and regulatory systems. The identity of the sensory signal and which genes might be directly or indirectly regulated by this sensory-regulatory system is unknown.

7.5
Use of Classical Genetics and Recombinant DNA Technologies to Modify Production

There have been three efforts to improve sphingan production during fermentation by modifying specific genes involved in biosynthesis (Baird and Cleary, 1994; Vartak et al., 1995; Thorne et al., 2000). The objective of each effort was to increase the conversion of the sugar substrate to polysaccharide. An undesirable characteristic of gellan fermentation with *Sphingomonas* S60 is the abundant evolution of CO_2 and the simultaneous synthesis of poly-β-hydroxybutyrate (PHB). As much as 50% of the carbon source may be diverted to CO_2 production, and PHB may represent as much as 15–25% by weight of commercial nonclarified gellan. Microbial growth conditions with high sugar levels and low nitrogen favor the synthesis of both polymers. In addition, PHB is an insoluble contaminant which contributes to the turbidity of gellan solutions.

In the first study, random mutagenesis was used to disrupt the synthesis of PHB during the production of gellan. Individual mutants were analyzed directly for PHB accumulation, and those which were at least partially defective were mutagenized again in an iterative process until the progeny accumulated no PHB. However, the PHB-defective strains did not accumulate increased amounts of gellan and continued to divert 50% of the carbon source to CO_2 production (Vartak et al., 1995). A conclusion about whether or not the two polymers actually compete for the available carbon source will require a comparison between isogenic strains, one of which carries a well-defined mutation.

The second attempt to increase gellan production was based on measurements of the activities of specific enzymes which were reasonably expected to be involved in sugar metabolism (Vartak et al., 1995). The strategy was to divert the carbon flow away from the pentose phosphate and Entner–Doudoroff pathways, which evolve considerable CO_2. Site-specific mutagenesis was used to inactivate the *zwf* gene which codes for glucose-6-phosphate dehydrogenase. However, the parent and Zwf-negative strains showed the same conversion of glucose to gellan, and CO_2 evolution was unchanged. There appeared to be another metabolic route which bypassed the absent ZWF enzyme.

The third attempt at metabolic engineering of sphingan synthesis was with *Sphingomonas* S7 (Thorne et al., 2000). The strategy was two-fold: first, to increase the intracellular level of the enzyme phosphoglucomutase (PGM) to provide more glucose-1-phosphate for synthesis of the nucleotide sugars; and second, to increase the levels of the glycosyl transferases required for assembly of the repeat subunit for the S-7 polysaccharide. The normal complement of chromosomal genes was augmented with additional genes carried on multicopy plasmid vectors. However, augmentation of PGM resulted mainly in increased cell growth with a small percentage increase in the accumulation of gellan. Synthesis of cell wall polysaccharides and precursors such as UDP-glucosamine seemed to compete with gellan synthesis. Augmentation with additional sugar transferase genes gave a larger increase in conversion of glucose to sphingan S-7. To be commercially meaningful any such increase in the laboratory must be verified at a much larger scale. Unexpectedly, this particular scheme of transferase augmentation caused the synthesis of a modified S-7 polysaccharide, which had a lower amount of glucose relative to rhamnose. Structural analysis showed that this was due to reduced attachment of glucose side chains, possibly because the added genes did not include those required for attachment of the side chain sugars to the repeat unit.

8 Biodegradation

Aspects of degradation are discussed in Volume 9 of this book series.

9 Production

The sphingans which have been produced commercially include gellan, welan, and rhamsan. Each is produced by aerobic submerged fermentation of an inexpensive carbon source (Kang et al., 1983). The key factors to economic production are the yield from the sugar source as a function of time, the scale of production, and the method of recovery from an extremely viscous broth.

9.1 Fermentation

Sphingomonas bacteria are not fastidious, and generally grow to high cell densities on a variety of inexpensive and readily available medium components, usually at 30–32 °C. A bacterial inoculum is developed in stages, with each stage requiring 18–28 h. *Sphingomonas* doubling times during logarithmic growth are usually from 1.5 to 2.5 h. Starting from a 1% (v/v) inoculum, a production vessel grows exponentially until the ammonium is depleted (about 9–12 h), after which growth gradually decelerates to give a final absorbance (600 nm) between 16 and 24.

A fermentation medium that gives rapid bacterial growth and conversion of a carbon source to sphingan polymers includes: ammonium nitrate (0.75–1.5 g L^{-1}) and hydrolyzed soy protein (0.5–1 g L^{-1}) together as sources of nitrogen; dipotassium phosphate, magnesium sulfate and trace minerals; and either corn syrup, high-fructose corn syrup, or sucrose as inexpensive carbon sources. Deionized water can replace mineral-rich water when a calcium-free product is required. If the medium contains 5 g L^{-1} of dipotassium phosphate as a buffer then no additional pH control is usually required. Otherwise, potassium hydroxide can be added during a fermentation to control pH. Even in a buffered medium with the initial pH adjusted to about 7–7.5, consumption of the ammonium causes the pH to decrease naturally to around 6. The pH then increases to about 6.5 as the nitrate is taken up. When all of the inorganic nitrogen is consumed the pH slowly decreases until the carbon source is exhausted. Maintenance of a constant pH is not necessary, and is difficult once the broth becomes highly viscous. In addition, some of the sphingan-producing bacteria require chemical control of foaming, especially during the late exponential growth phase.

Thickening of the culture broth resulting from polysaccharide accumulation becomes visible and physically measurable when the cell density reaches an absorbance at 600 nm of about 4–5. Due to a dramatic increase in viscosity during these fermentations, aeration and agitation are key parameters which affect the overall productivity. Agitation and aeration must be increased during a typical fermentation to satisfy the oxygen demand. Nevertheless, despite supplying more oxygen it is not unusual to observe no detectable dissolved oxygen in the vessel at later times when the viscosity exceeds 10,000 centipoise (Brookfield LVTDV-II viscometer; shear rate of 2.5 s^{-1}). To achieve a maximum yield of polymer per liter and conversion of substrate to product (grams of polymer per gram of carbon source), the type of carbon source, the amount of nitrogen added, and the level of dissolved oxygen must be optimized together for each bacterial strain. Furthermore, each strain has its own characteristic metabolic evolution of CO_2 and capacity to accumulate intracellular PHB. It is possible to obtain 50–70% conversion of glucose to total biomass in a gellan culture after about 50 h, when 50–75% of the total biomass is the gellan capsule itself. The overall productivity for gellan from *Sphingomonas* is considerably less than for xanthan gum synthesis from *X. campestris*, where the overall conversion to xanthan gum can be 70%. The welan-producing *Sphingomonas* bacteria convert about 60% of the substrate carbon source to polysaccharide within 60–70 h. Although not produced commercially at this time, fermentation of *Sphingomonas* strains S7 and S88 on a laboratory scale gives similar yields under similar conditions (Thorne et al., 2000). Of course, actual productivities obtained at very large commercial scales of operation (up to 150 m^3 liquid fermentation broth or culture volume) may differ from those indicated here.

9.2 Recovery and Purification

Gellan is produced in multiple forms by different treatment and recovery processes. Native gellan (Kelcogel™ LT100) retains all of its acetyl and glyceric acid groups, and is not clarified to remove cells and cell debris. For native gellan the culture broth is heated to 90–95 °C for a brief period, and then mixed with about two volumes of isopropyl alcohol. The heating step not only kills the cells, which remain with the capsular polysaccharide, but also greatly reduces the viscosity of the broth, and this facilitates mixing during the precipitation. The precipitated fibrous biomass is then dried in a vacuum and milled. Native gellan is used to make elastic or weak gels for applications that do not require clarification. A deacylated gellan is produced in either unclarified (Kelcogel™ LT) or clarified (Kelcogel™) forms. Deacylated gellan forms rigid or strong gels. After heating the broth to 90–95 °C, potassium hydroxide is added to raise the pH to around 10, and then after about 15 min the pH is adjusted to neutral with sulfuric acid. The hot deacylated gellan is either precipitated immediately with isopropyl alcohol or is clarified by filtration to remove cells and cell debris and then precipitated with the alcohol. Until the precipitation is complete the temperature must be kept above 60 °C to maintain fluidity and to prevent gel formation. Gelrite™ is a special grade that is deacylated and clarified like Kelcogel, and has gel properties and consistency comparable with those of the agar used in solidified microbiological and plant tissue culture media. Welan and rhamsan do not form gels, and are simply recovered by alcohol precipitation, drying and milling.

9.3 Producers

To date, only CP-Kelco has produced and sold either gellan, welan or rhamsan. Since the composition-of-matter patents covering these polymers have recently expired, other companies may soon begin production, and new grades of these polymers with modified properties are anticipated.

9.4 World Market

Although accurate world market figures are unavailable, the gellan market for year 2000 is estimated at 1000 metric tons and is valued at approximately US$50,000,000. The gellan market in Japan for 2000 is estimated at about 200 metric tons, and consists mainly of food applications. The total market for welan is believed to be considerably smaller, and regular production of rhamsan by CP-Kelco was recently discontinued. However, a new sphingan polymer (S-657) from CP-Kelco may be produced in the future.

Annual growth in these markets is probably less than 5%. Two impediments to market growth in the past might have been the higher cost of production of the sphingan polymers relative to xanthan gum and the existence of only one supplier. In 2000, the selling price for the Kelcogel grade of gellan was about US$50–55 kg^{-1} in 25-kg drums, whilst Gelrite was priced at US$76 kg^{-1} in 50-kg drums. Welan was priced at US$14–18 kg^{-1}, and rhamsan became available only by special arrangement at a negotiated price of up to about US$100 kg^{-1}. The different prices reflect not only the value of the unique functional attributes of each polymer to the user for specific targeted applications, but also the cost of production arising from different yields and processing schemes.

9.5 Applications

The existing and potential uses for the sphingan polysaccharides are diverse and reflect the different structures and functional properties within the group. The most widely used sphingan is gellan.

9.5.1 Solid Culture Media for Growth of Microorganisms and Plants

A clarified grade of gellan, Gelrite™, is used as an agar substitute to solidify nutrient media for the growth of microorganisms and plants (Kang et al., 1982b). The product withstands several autoclaving cycles, is resistant to degradation by a variety of enzymes, supports the growth of diverse microorganisms, and has a texture that resembles that of agar. Depending on its purity, agar for culture media sells for US$100–300 kg^{-1} (year 2000), and is used at about 15 g L^{-1} of medium. By contrast, Gelrite sells for US$76 kg^{-1} and is used at only 6 g L^{-1} of medium. Phytagel™, a proprietary grade of gellan exhibiting high clarity and solubility, is produced by CP-Kelco and sold by Sigma Chemical especially for use in plant culture medium.

9.5.2 Food Hydrocolloids

Gellan has been approved in the United States for use in foods as a stabilizer or thickener under certain prescribed conditions (Section 172.665 of Title 21 of the Code of Federal Regulations). It is always present in a complex food mixture, often with additional hydrocolloids. Some of the polymers used in foods that may be replaced with gellan include pectin, starch, gelatin, agar, xanthan gum, locust bean gum, and carageenan. Applications include confectioneries, jams and jellies, pie fillings and puddings, bakery fillings, meat-like structures, pet foods, reduced-fat batters and coatings for cooking meats and vegetables, icings and frostings, beverages, and dairy products (Nussinovitch, 1997). Gellan can also be used in the fining of alcoholic beverages such as beer and wine. A potentially large future market for gellan may be as a replacement for animal-derived gelatin in foods, soft capsules, microcapsules and photographic films (Moorhouse, 1987).

Gellan, as a replacement for agar and carageenan in processed foods, is often used at concentrations which are only 33–50% of the substituted polymers. Texture and mouthfeel of the final products with gellan are satisfactory or better than with other hydrocolloids, and the gelation properties for gellan are consistent over a broad pH range. Although mixing of food components in the presence of gellan is simplified because gellan imparts no significant initial viscosity to the mixture, there are solubility issues that need testing for every food process in which gellan is used. A potential problem is the reduced solubility of gellan produced from cultures having elevated levels of calcium, though chelating agents such as citrate may alleviate some of these problems.

9.5.3 Gel Electrophoresis in Biotechnological Research

Gels of gellan can be used as a solid matrix for separating DNA fragments on the basis of size by electrophoresis (Cole, 2000). Gel electrophoresis is a widely practiced and key procedure in molecular biology. Gellan-based electrophoresis gels must include a second polymer such as hydroxyethylcellulose or polyethylene oxide to reduce electro-osmosis, an undesirable flow of cations which distorts the electrophoretic separations and which results from the many negatively charged glucuronic acids in gellan. In this application gellan can replace highly refined agarose which costs about US$700–3500 kg^{-1} and which is used at about 1% concentration. By contrast, the gellan costs less than US$100 kg^{-1} and is used at only 0.125% concentration.

9.5.4 Cement-based Materials and Oil-field Applications

Welan is used to thicken cement-based concretes and mortars, and related manufactured or extruded products. It is compatible with cement, traps a relatively small volume of air, prevents water loss, and imparts a high resistance to repeated freezing and thawing. However, its flow properties are inferior to those of alternative additives such as cellulose derivatives. A method was developed to prepare a low-viscosity welan for use in the construction and oil industries at concentrations of 0.1–0.5%, and especially at temperatures above 90 °C (Allen et al., 1990). Welan is also used to control the viscosity of aqueous oil drilling muds and fluids at high temperatures in deep wells (Moorhouse, 1987), and a cross-linked welan can improve the sweep efficiency during fluid flooding in oil recovery (Hoskin et al., 1991).

9.5.5 Applications for Sphingan S-7

S-7 has an unusually high viscosity, and its suspending ability exceeds that of most other polymers (Kang and McNeely, 1975a,b). Many years after its original discovery by CP-Kelco, the producing bacterium was shown to be a member of the *Sphingomonas* group (Pollock, 1993) and its structure as a sphingan was established (Thorne et al., 2000; Gulin et al., 2001). Since S-7 is very pseudoplastic, its high viscosity at low shear makes this hydrocolloid one of the best suspending agents. Furthermore, its low viscosity at high shear confers the ability to be transported by pumping, and to not impede the rotation of a drill. Thus it could be used as a key component of an oil drilling fluid (Kang and McNeely, 1976), a dripless water-based paint (Kang and McNeely, 1975a), or an environmentally friendly anti-icing composition for airplane surfaces (Lockyer et al., 1998).

9.5.6 Other Potential Applications for Sphingans

Although the productivity of sphingan S-657 is reportedly low, it has promise as an oil drilling fluid due to its very high viscosity and thermal resistance, exceeding that of either xanthan gum or welan (Peik et al., 1992). Rhamsan has been used as a suspending agent for fertilizers and pesticides because it is compatible with high concentrations of ammonium phosphate and salts (Baird et al., 1983). Sphingan S-198 exhibits an exceptional stability to shear, and a suspending ability and thermal resistance comparable with that of the other sphingans. It has been evaluated for use as a lubricant or hydraulic fluid (Kang et al., 1983).

10 Outlook and Perspectives

The bacterial genus *Sphingomonas* is a rich source for useful commercial heteropolysaccharides. The properties of the capsular polysaccharides from *Sphingomonas* suggest many applications where the control of aqueous viscosity is important, especially in extreme physical conditions. The commercial potential of each sphingan will be judged on the basis of the cost per unit of function delivered.

Bacterial fermentation provides a reliable and consistent source for specialized biopolymers, and the technologies and know-how for submerged stirred culturing and product recovery are widespread. However, the relatively high costs and current availability of product from only one manufacturing source limits the applications of these polymers. The productivities for the capsular sphingans from *Sphingomonas* are generally only about 50–75% of the higher productivity for xanthan gum from *X. campestris*, and the processing of fermentation broths containing capsular sphingans to provide clarified products lacking cellular contaminants is more difficult than for xanthan gum. Therefore, increased use of these microbial polysaccharides in existing and new applications will require improvements in fermentation yield and in downstream processes. Classical genetic and recombinant DNA technologies might be used to develop bacterial strains which at least partially overcome these productivity and processing problems.

It is reasonable to expect increased interest in the sphingans for studies of the relation between polysaccharide structure and physical properties in solution. Existing members of the *Sphingomonas* genus have been isolated in almost every environment, from surface waters to deep subterranean sediments. Additional candidates for structural/functional studies may become available from the application of simple microbiological screening methods to expand the library of sphingan producers.

11 Patents

Key patents covering novel compositions of matter for the sphingans and methods of production, as well as selected patents which represent the diverse applications for sphingans are listed in Table 1. The number of patents issued in the United States in this field exceeds 600 for gellan alone, of which over 400 were issued between 1996 and 2001. These can be accessed by search routines at the "uspto.gov" website. Some of the patents have recently expired and are now in the public domain.

Tab. 1 Key patents for sphingans

Patent no.	Holder	Inventors	Date of Publication	Title and (subject)
U.S. 3,894,976	Kelco Co.	Kenneth Suk Kang William H. McNeely	July 15, 1975	Pseudoplastic water base paint containing a novel heteropolysaccharide (EPS S-7)
U.S. 3,915,800	Kelco Co.	Kenneth Suk Kang William H. McNeely	Oct. 28, 1975	Polysaccharide and bacterial fermentation Process for its preparation (EPS S-7)
U.S. 3,979,303	Merck and Co.	Kenneth Suk Kang William H. McNeely	Sept. 7, 1976	Oil well drilling fluid (EPS S-7)
U.S. 4,326,052	Merck and Co.	Kenneth S. Kang George T. Colegrove George T. Veeder	April 20, 1982	Deacetylated polysaccharide S-60
U.S. 4,326,053	Merck and Co.	Kenneth S. Kang George T. Veeder	April 20, 1982	Polysaccharide S-60 and bacterial fermentation process for its preparation
U.S. 4,331,440	Merck and Co.	Joseph S. Racciato	May 25, 1982	Use of S-88 in printing paste systems
U.S. 4,342,866	Merck and Co.	Kenneth S. Kang George T. Veeder	Aug. 3, 1982	Heteropolysaccharide S-130
U.S. 4,401,760	Merck and Co.	Jerry A. Peik Suzanna M. Steenbergen Harold R. Hayden	Aug. 30, 1983	Heteropolysaccharide S-194
U.S. 4,462,836	Gulf Oil Corp.	Wilford S. Baker James J. Harrison	July 31, 1984	Cement composition and method of cement casing a well (EPS S-7)
U.S. 4,503,084	Merck and Co.	John K. Baird Jaewon L. Shim	March 5, 1985	Non-heated gellan gum gels
U.S. 4,529,797	Merck and Co.	Jerry A. Peik Suzanna M. Steenbergen Harold R. Hayden	July 16, 1985	Heteropolysaccharide S-198
U.S. 4,535,153	Merck and Co.	Kenneth S. Kang George T. Veeder	Aug. 13, 1985	Heteropolysaccharide S-88
U.S. 4,869,916	Merck and Co.	Ross C. Clark Daniel R. Burgum	Sept. 26, 1989	Blends of high acyl gellan gum with starch
U.S. 4,874,044	Texaco Inc.	Peter D. Robison Arthur J. Stipanovic	Oct. 17, 1989	Method for oil recovery using a modified heteropolysaccharide (NW-11)

Tab. 1 (cont.)

Patent no.	Holder	Inventors	Date of Publication	Title and (subject)
U.S. 4,898,810	Ciba-Geigy AG	Ute Eggert Andreas Engel Ekkehard Kramp	Feb. 6, 1990	Layers for photographic materials (gellan)
U.S. 4,963,668	Merck and Co.	Floyd L. Allen Glen H. Best Thomas A. Lindroth	Oct. 16, 1990	Welan gum in cement compositions
U.S. 4,981,520	Mobil Oil Co.	Dennis H. Hoskin Thomas O. Mitchell Paul Shu	Jan. 1, 1991	Oil reservoir permeability profile control with crosslinked welan gum biopolymers
U.S. 5,112,445	Merck and Co.	Philip E. Winston, Jr. Harold D. Dial Kenneth Clare Theresa M. Ortega	May 12, 1992	Gellan gum sizing (paper manufacture)
U.S. 5,175,278	Merck and Co.	Jerry A. Peik Suzanna M. Steenbergen George T. Veeder	Dec. 29, 1992	Heteropolysaccharide S-657
U.S. 5,300,429	Kelco Co.	John K. Baird Joseph M. Cleary	April 5, 1994	PHB-free gellan gum
U.S. 5,534,286	Monsanto Co.	William F. Chalupa George R. Sanderson	July 9, 1996	No-fat gellan gum spread (in food)
U.S. 5,772,912	NASA, USA	Robert T. Lockyer John Zuk Leonard A. Haslim	June 30, 1998	Environmentally friendly anti-icing (EPS S-7)
U.S. 5,855,826	Pacific Co. (Korea)	Chung Nam Lee Ok Sob Lee Hak Hae Kang Dong Soon Park	Jan. 5, 1999	Matrix-double encapsulation method and a cosmetic composition containing matrix-double capsules (gellan)
U.S. 6,203,680	Commerce, USA	Kenneth D. Cole	March 20,2001	Electrophoresis gels

12
References

Allen, F. L., Best, G. H., Lindroth, T. A. (1990) Welan gum in cement compositions, U.S. Patent 4,963,668.

Ashtaputre, A. A., Shah, A. K. (1995) Studies on a viscous, gel-forming exopolysaccharide from *Sphingomonas paucimobilis* GS1, *Appl. Environ. Microbiol.* **61**, 1159–1162.

Attwool, P. T., Atkins, E. D. T., Miles, M. J., Morris, V. J. (1986) Microbial polysaccharide diffraction results from *Alcaligenes* (ATCC 31555) microbial polysaccharide (S-130) and a comparison with gellan gum, *Carbohydr. Res.* **138**, c1–c4.

Baird, J. K., Cleary, J. M. (1994) PHB-free gellan gum broth, U.S. Patent 5,300,429.

Baird, J. K., Shim, J. L. (1985) Non-heated gellan gum gels, U.S. Patent 4,503,084.

Baird, J. K., Smith, W. W. (1989) An analytical procedure for gellan gum in food gels, *Food Hydrocolloids* **3**, 407–411.

Baird, J. K., Sandford, P. A., Cottrell, I. W. (1983) Industrial applications of some new microbial polysaccharides, *Bio/Technology* **1**, 778–783.

Baker, W. S., Harrison, J. J. (1984) Cement composition and method of cement casing in a well, U.S. Patent 4,462,836.

Chalupa, W. F., Sanderson, G. R. (1996) No-fat gellan gum spread, U.S. Patent 5,534,286.

Chandrasekaran, R., Radha, A., Thailambal, V. G. (1992) Roles of potassium ions, acetyl and L-glyceryl groups in native gellan double helix: an X-ray study, *Carbohydr. Res.* **224**, 1–17.

Chowdhury, T. A., Lindberg, B., Lindquist, U., Baird, J. (1987a) Structural studies of an extracellular polysaccharide (S-198) elaborated by *Alcaligenes* ATCC 31853, *Carbohydr. Res.* **161**, 127–132.

Chowdhury, T. A., Lindberg, B., Lindquist, U., Baird, J. (1987b) Structural studies of an extracellular polysaccharide, S-657, elaborated by *Xanthomonas* ATCC 53159, *Carbohydr. Res.* **164**, 117–122.

Clark, R. C., Burgum, D. R. (1989) Blends of high acyl gellan gum with starch, U.S. Patent 4,869,916.

Clarke, A. J., Sarabia, V., Keenleyside, W., MacLachlan, P. R., Whitfield, C. (1991) Compositional analysis of bacterial extracellular polysaccharides by high performance anion-exchange chromatography, *Anal. Biochem.* **199**, 68–74.

Cole, K. D. (2000) Reversible gels for electrophoresis and isolation of DNA, *BioTechniques* **26**, 748–756.

Eggert, U., Engel, A., Kramp, E. (1990) Layers for photographic materials, U.S. Patent 4,898,810.

Falk, C., Jansson, P.-E., Rinaudo, M., Heyraud, A., Widmalm, G., Hebbar, P. (1996) Structural studies of the exocellular polysaccharide from *Sphingomonas paucimobilis* strain I-886, *Carbohydr. Res.* **285**, 69–79.

Graham, H. D. (1991) Isolation of gellan gum from foods by use of monovalent cations, *J. Food Sci.* **56**, 1342–1346.

Gulin, S., Kussak, A., Jansson, P., Widmalm, G. (2001) Structural studies of S-7, another exocellular polysaccharide containing 2-deoxy-*arabino*-hexuronic acid, *Carbohydr. Res.* **331**, 285–290.

Gunning, A. P., Morris, V. J. (1990) Light scattering studies of tetramethyl ammonium gellan, *Int. J. Biol. Macromol.* **12**, 338–341.

Hebbar, K. P., Gueniot, B., Heyraud, A., Colin-Morel, P., Heulin, T., Balandreau, J., Rinaudo, M. (1992) Characterization of exopolysaccharides produced by rhizobacteria, *Appl. Microbiol. Biotechnol.* **38**, 248–253.

Hestrin, S. (1949) The reaction of acetylcholine and other carboxylic acid derivatives with hydroxylamine, and its analytical application, *J. Biol. Chem.* **180**, 249–261.

Hoskin, D. H., Mitchell, T. O., Shu, P. (1991) Oil reservoir permeability profile control with cross-linked welan gum biopolymers, U.S. patent 4,981,520.

Ielpi, L., Couso, R. O., Dankert, M. A. (1993) Sequential assembly and polymerization of the polyprenol-linked pentasaccharide repeating unit of the xanthan polysaccharide in *Xanthomonas campestris*, *J. Bacteriol.* **175**, 2490–2500.

Jansson, P.-E., Widmalm, G. (1994) Welan gum (S-130) contains repeating units with randomly distributed L-mannosyl and L-rhamnosyl terminal groups, as determined by FABMS, *Carbohydr. Res.* **256**, 327–330.

Jansson, P.-E., Lindberg, B., Sandford, P. A. (1983) Structural studies of gellan gum, an extracellular polysaccharide elaborated by *Pseudomonas elodea*, *Carbohydr. Res.* **124**, 135–139.

Jansson, P.-E., Lindberg, B., Widmalm, G., Sandford, P. A. (1985) Structural studies of an extracellular polysaccharide (S-130) elaborated by *Alcaligenes* ATCC 31555, *Carbohydr. Res.* **139**, 217–223.

Jansson, P.-E., Kumar, N. S., Lindberg, B. (1986a) Structural studies of a polysaccharide (S-88) elaborated by *Pseudomonas* ATCC 31554, *Carbohydr. Res.* **156**, 165–172.

Jansson, P.-E., Lindberg, B., Lindberg, J., Maekawa, E., Sandford, P. A. (1986b) Structural studies of a polysaccharide (S-194) elaborated by *Alcaligenes* ATCC 31961, *Carbohydr. Res.* **156**, 157–163.

Kang, K. S., McNeely, W. H. (1975a) Pseudoplastic water base paint containing a novel heteropolysaccharide, U.S. patent 3,894,976.

Kang, K. S., McNeely, W. H. (1975b) Polysaccharide and bacterial fermentation process for its preparation, U.S. patent 3,915,800.

Kang, K. S., McNeely, W. H. (1976) Oil well drilling fluid, U.S. patent 3,979,303.

Kang, K. S., McNeely, W. H. (1977) PS-7 – A new bacterial heteropolysaccharide, in: *Extracellular Microbial Polysaccharides* (Sandford, P. A., Laskin, A., Eds.), Washington D.C.: American Chemical Society, 220–230.

Kang, K. S., Veeder, G. T. (1982a) Heteropolysaccharide S-130, U.S. Patent 4,342,866.

Kang, K. S., Veeder, G. T. (1982b) Polysaccharide S-60 and bacterial fermentation process for its preparation, U.S. Patent 4,326,053.

Kang, K. S., Veeder, G. T. (1985) Heteropolysaccharide S-88, U.S. Patent 4,535,153.

Kang, K. S., Colegrove, G. T., Veeder, G. T. (1982a) Deacetylated polysaccharide S-60, U.S. patent 4,326,052.

Kang, K. S., Veeder, G. T., Mirrasoul, P. J., Kaneko, T., Cottrell, I. W. (1982b) Agar-like polysaccharide produced by a *Pseudomonas* species: production and basic properties, *Appl. Environ. Microbiol.* **43**, 1086–1091.

Kang, K. S., Veeder, G. T., Cottrell, I. W. (1983) Some novel bacterial polysaccharides of recent development, *Prog. Ind. Microbiol.* **18**, 231–253.

Kuo, M.-S., Mort, A. J., Dell, A. (1986) Identification and location of L-glycerate, an unusual acyl substituent in gellan gum, *Carbohydr. Res.* **156**, 173–187.

Lee, C. N., Lee, O. S., Kang, H. H., Park, D. S. (1999) Matrix-double encapsulation method and a cosmetic composition containing matrix-double capsules, U.S. Patent 5,855,826.

Lockyer, R. T., Zuk, J., Haslim, L. A. (1998) Environmentally friendly anti-icing, U.S. patent 5,772,912.

Martins, L. O., Sá-Correia, I. (1991) Gellan gum biosynthetic enzymes in producing and non-producing variants of *Pseudomonas elodea*, *Biotechnol. Appl. Biochem.* **14**, 357–364.

Martins, L. O., Sá-Correia, I. (1993) Temperature profiles of gellan gum synthesis and activities of biosynthetic enzymes, *Biotechnol. Appl. Biochem.* **20**, 385–395.

Martins, L. O., Fialho, A. M., Rodrigues, P. L., Sá-Correia, I. (1996) Gellan gum production and activity of biosynthetic enzymes in *Sphingomonas paucimobilis* mucoid and non-mucoid variants, *Biotechnol. Appl. Biochem.* **24**, 47–54.

Moorhouse, R. (1987) Structure/property relationships of a family of microbial polysaccharides, in: *Industrial polysaccharides: genetic engineering, structure/property relations and applications* (Yalpani, M., Ed.), Amsterdam, Elsevier Science Publishers B.V., 187–206.

Nussinovitch, A. (1997) Gellan gum, in: *Hydrocolloid Applications*, London: Blackie Academic and Professional, 63–82.

O'Neill, M., Selvendran, R. R., Morris, V. J., Eagles, J. (1986) Structure of the extracellular polysaccharide produced by the bacterium *Alcaligenes* (ATCC 31555) species, *Carbohydr. Res.* **147**, 295–313.

O'Neill, M. A., Darvill, A. G., Albersheim, P., Chou, K. J. (1990) Structural analysis of an acidic polysaccharide secreted by *Xanthobacter* sp. (ATCC 53272), *Carbohydr. Res.* **206**, 289–296.

Peik, J. A., Steenbergen, S. M., Hayden, H. R. (1983) Heteropolysaccharide S-194, U.S. patent 4,401,760.

Peik, J. A., Steenbergen, S. M., Hayden, H. R. (1985) Heteropolysaccharide S-198, U.S. patent 4,529,797.

Peik, J. A., Steenbergen, S. M., Veeder, G. T. (1992) Heteropolysaccharide S-657, U.S. patent 5,175,278

Pollock, T. J. (1993) Gellan-related polysaccharides and the genus *Sphingomonas, J. Gen. Microbiol.* **139**, 1939–1945.

Pollock, T. J., Armentrout, R. W. (1999) Planktonic/sessile dimorphism of polysaccharide-encapsulated sphingomonads, *J. Ind. Microbiol. Biotechnol.* **23**, 436–441.

Pollock, T. J., Thorne, L., Yamazaki, M., Mikolajczak, M. J., Armentrout, R. W. (1994) Mechanism of bacitracin resistance in gram-negative bacteria that synthesize exopolysaccharides, *J. Bacteriol.* **176**, 6229–6237.

Pollock, T. J., van Workum, W. A. T., Thorne, L., Mikolajczak, M. J., Yamazaki, M., Kijne, J. W., Armentrout, R. W. (1998) Assignment of biochemical functions to glycosyl transferase genes which are essential for biosynthesis of exopolysaccharides in *Sphingomonas* strain S88 and *Rhizobium leguminosarum, J. Bacteriol.* **180**, 586–593.

Racciato, J. S. (1982) Use of gum S-88 in printing paste systems, U.S. patent 4,331,440.

Reuber, T. L., Walker, G. C. (1993) Biosynthesis of succinoglycan, symbiotically important exopolysaccharide of *Rhizobium meliloti, Cell* **74**, 269–280.

Robison, P. D., Stipanovic, A. J. (1989) Method for oil recovery using a modified heteropolysaccharide, U.S. Patent 4,874,044.

Stankowski, J. D., Zeller, S. G. (1992) Location of the O-acetyl group in welan by the reductive-cleavage method, *Carbohydr. Res.* **224**, 337–341.

Stone, K. J., Strominger, J. L. (1971) Mechanism of action of bacitracin: complexation with metal ion and C55-isoprenyl pyrophosphate, *Proc. Natl. Acad. Sci. USA* **68**, 3223–3227.

Sutherland, I. W. (1982) Biosynthesis of microbial exopolysaccharides, *Adv. Microb. Physiol.* **24**, 79–150.

Thorne, L., Mikolajczak, M. J., Armentrout, R. W., Pollock, T. J. (2000) Increasing the yield and viscosity of exopolysaccharides secreted by *Sphingomonas* by augmentation of chromosomal genes with multiple copies of cloned biosynthetic genes. *J. Ind. Microbiol. Biotechnol.* **25**, 49–57.

Vartak, N. B., Lin, C. C., Cleary, J. M., Fagan, M. J., Saier, M. H., Jr. (1995) Glucose metabolism in *Sphingomonas elodea*: pathway engineering via construction of a glucose-6-phosphate dehydrogenase insertion mutant, *Microbiology* **141**, 2339–2350.

Winston, P. E., Jr., Dial, H. D., Clare, K., Ortega, T. M. (1992) Gellan gum sizing, U.S. Patent 5,112,445.

Yamazaki, M., Thorne, L., Mikolajczak, M., Armentrout, R. W., Pollock, T. J. (1996) Linkage of genes essential for synthesis of a polysaccharide capsule in *Sphingomonas* strain S88, *J. Bacteriol.* **178**, 2676–2687.

11
Xanthan

Dr. Karin Born[1], Dr. Virginie Langendorff[2], Dr. Patrick Boulenguer[3]
[1] DEGUSSA Texturant Systems France SAS, Research Center,
F-50500 Baupte, France; Tel.: +33-2-33-713433; Fax: +33-2-33-713492;
E-mail: karin.born@degussa.com
[2] DEGUSSA Texturant Systems France SAS, Research Center,
F-50500 Baupte, France; Tel.: +33-2-33-713433; Fax: +33-2-33-713492;
E-mail: virginie.langendorff@degussa.com
[3] DEGUSSA Texturant Systems France SAS, Research Center,
F-50500 Baupte, France; Tel.: +33-2-33-713433; Fax: +33-2-33-713492;
E-mail: patrick.boulenguer@degussa.com

ADI	acceptable daily intake
AFFF	asymmetric flow field fractionation
AFM	atomic force microscopy
EPS	exopolysaccharide
GM	genetically modified
GPC	gel-permeation chromatography
LALLS	low-angle laser light scattering
LBG	locust bean gum
MALLS	multi-angle laser light scattering
M_w	molecular weight
O/W	oil in water
SEC	size-exclusion chromatography
W/O	water in oil

1 Introduction

During the second half of the 20th century, many new and useful polysaccharides of scientific and commercial interest have been discovered which can be obtained from microbial fermentations. Microorganisms such as bacteria and fungi produce three distinct types of carbohydrate polymers: (1) extracellular polysaccharides, which can be found either as a capsule that envelops the microbial cell or as an amorphous mass secreted into the surrounding medium; (2) structural polysaccharides, which can be part of the cell wall; or (3) intracellular storage polysaccharides. The scientific and industrial success of polysaccharides of microbial origin is due to several factors. First, they can be produced under controlled conditions with selected species; second, they usually present a high structural regularity; and third, different microorganisms can synthesize a wide range of very specific ionic and neutral polysaccharides with widely varying compositions and properties. Such variety is not found among plant polysaccharides and, perhaps more importantly, it cannot be imitated by means of synthetic chemistry.

The usefulness of water-soluble carbohydrate polymers in industry relies on their wide range of functional properties. The most important characteristic is their ability to modify the properties of aqueous environments, that is their capacity to thicken, emulsify, stabilize, flocculate, swell and suspend or to form gels, films and membranes. Another very important aspect is that polysaccharides obtained from natural, renewable sources are both biocompatible and biodegradable.

Xanthan, a microbial biopolymer produced by the *Xanthomonas* bacterium, has provoked great scientific and industrial interest since its discovery in the late 1950s. In 1999 alone, more than 300 references of articles or patents dealing with xanthan are listed in *Chemical Abstracts.* Since 1990, more than 2000 patents have been listed in *Derwent World Patents Index.* This interest is due to the extraordinary properties of xanthan as well as to the successful establishment of an industrial process for its production.

2 Historical Outline

Xanthan gum was discovered in the 1950s by scientists of the Northern Regional Research Laboratory of the U.S. Department of Agriculture in the course of a screening which aimed at identifying microorganisms that produced water-soluble gums of commercial interest. The first industrial production of xanthan was carried out in 1960, and the product first became available commercially in 1964. Toxicology and safety studies showed that xanthan caused no acute toxicity, had no growth-inhibiting activity, and did not alter organ weights, hematological values or tumors when fed to rats or dogs, neither in short-term, nor in long-term feeding studies. The approval for food use was given by the U.S. Food and Drug administration in 1969, and the FAO/OMS specification followed in 1974. Authorization in France was given in March 1978, and approval in Europe was obtained in 1982, under the E number E415. The official definition of the EU food regulations for E415 is: "Xanthan gum is a high molecular-weight polysaccharide gum produced by a pure culture fermentation of a carbohydrate with natural strains of *Xanthomonas campestris,* purified by recovery with ethanol or propane-2-ol, dried and milled. It contains D-glucose and D-mannose as the dominant hexose units, along with D-glucuronic acid and pyruvic acid, and is prepared as the sodium, potassium or calcium salt. Its solutions are neutral. The molecular weight must be approximately 1 MDa and the color must be cream". Xanthan gum is approved as food additive with an acceptable daily intake (ADI) "not specified"; that is, no limit for ADI is defined and the gum may be used *quantum satis,* which means with just the quantity useful for the application. Today, xanthan is produced commercially by several companies, such as Monsanto/Kelco, Rhodia, Jungbunzlauer, Archer Daniels Midland, and SKW Biosystems. For the past few years xanthan gum has also been produced in China. Annual volumes worldwide are estimated to be about 35,000 tons in 2001.

3 Structure

3.1 Chemical Structure

Xanthan is a heteropolysaccharide with a very high molecular weight, consisting of repeating units (Figure 1). The sugars present in xanthan are D-glucose, D-mannose, and D-glucuronic acid. The glucoses are linked to form a β-1,4-D-glucan cellulosic backbone, and alternate glucoses have a short branch consisting of a glucuronic acid sandwiched between two mannose units. The side chain consists therefore of β-D-mannose-(1,4)-β-D-glucuronic acid-(1,2)-α-D-mannose. The terminal mannose moiety may carry pyruvate residues linked to the 4- and 6-positions. The internal mannose unit is acetylated at C-6. Acetyl and pyruvate substituents are linked in variable amounts to the side chains, depending upon which *X. campestris* strain the xanthan is isolated from. The pyruvic acid content also varies with the fermentation conditions. On average, about half of the terminal mannoses carry a pyruvate, with the number and positioning of the pyruvate and acetate residues conferring a certain irregularity to the otherwise very regular structure. Usually, the degree of substitution for pyruvate varies between 30 and 40%, whereas for acetate the degree of substitution is as high as 60–70%. Some of the repeating units may be devoid of the trisaccharide side chain.

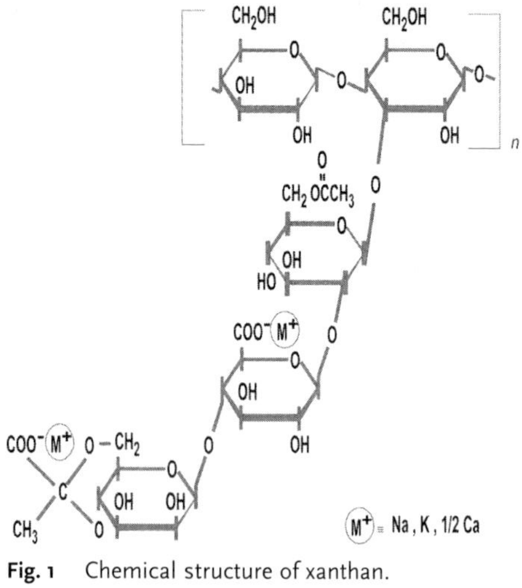

Fig. 1 Chemical structure of xanthan.

3.2 Superstructure/Secondary Structure

The secondary structure of xanthan depends on the conditions under which the molecule is characterized. The molecule may be in an ordered or in a disordered conformation. The ordered confirmation can be either native or renatured; in the native form the conformation is present at temperatures below the melting point of the molecule, a temperature which depends on the ionic strength of the medium in which xanthan is dissolved. The secondary structure of xanthan and the methods to analyze it have been recently reviewed in detail by Stokke et al. (1998). X-ray scattering results indicate that native xanthan in the ordered conformation exists as a right-hand helix with five-fold symmetry with a pitch of 4.7 nm and a diameter of 1.9 nm (Moorhouse, 1977). Two models, a single-strand helix and a double-strand or multi-strand helix, have been proposed, though most authors currently support the idea of a double helix. The helix is stabilized by noncovalent bonds, such as hydrogen bonds, electrostatic interaction, and steric effects; its structure can be described as rigid rod. In aqueous solution, the molecule may undergo a conformational transition which can be driven by changes in temperature and ionic strength, and which depends on the degree of ionization of the carboxyl groups and acetyl contents. The temperature-induced transition from an ordered to a disordered conformation is generally attributed to a complete or partial separation of the double-strand form. Renaturation may occur under favorable conditions, which means temperatures below the transition temperature and high salt concentrations. The transition from the denatured to the renatured state is reversible, whereas that from the native to the denatured state is irreversible. The model of a double-strand structure has been supported by several studies. Capron et al. (1997) demonstrated that upon heating to temperatures above the order–disorder transition temperature, denaturation of the native ordered conformation occurred, together with a reduction in molecular weight. The molecular weight is roughly halved, which supports the model of a double-strand conformation for the native form. The molecular weight was found to be invariant after renaturation on cooling. The renaturation probably occurs as an intramolecular process, which means that the restoration of the ordered form of xanthan seems to take place within a single molecule. Most likely, the conformation for the renatured form of xanthan is that of an anti-parallel, double-stranded structure consisting of one chain folded as a hairpin loop (Liu et al., 1987). The viscosity of renatured xanthan is higher than the viscosity of native xanthan however, thereby supporting the hypothesis that single-stranded xanthan molecules associate during renaturation to form supramolecular structures.

4
Occurrence

Xanthan is produced by *X. campestris*, a plant-associated bacterium that is generally pathogenic for plants belonging to the family Brassicaceae. *Xanthomonas* causes a variety of disease symptoms such as necrosis, gummosis and vascular parenchymatous diseases on leaves, stems or fruits; an example is "black rot" of crucifers such as cabbage, cauliflower or broccoli. *Xanthomonas* does not form spores, but it is very resistant to desiccation during relatively long periods. Survival at room temperature for 25 years has been reported by Leach et al. (1957). The resistance is usually due to the protective effect of the xanthan gum produced and exuded by the bacteria. Xanthan also protects the bacteria from the effects of light, and generally causes wilting of the leaves by blocking water movements (Leach et al., 1957). Exopolysaccharides, like xanthan, are also known to provide protection against bacteriophages by building a physical barrier against the phage attack (MacNeely, 1973). The polysaccharide is not a reserve energy source because in general the bacterium is not able to catabolize its own extracellular polysaccharide.

5
Physiological Function

The physiological function of the exopolysaccharide xanthan has received little attention compared with investigations into the molecule's production, properties, and applications. The bacteria (*Xanthomonas* sp.) that produce xanthan gum as a secondary metabolite are usually phytopathogenic, or may live in asymptomatic association with plant tissues or epiphytes. *Xanthomonas* infections have been observed in over 120 monocotyledonous and over 150 dicotyledonous plant species.

Xanthan is the predominant component of the bacterial slime. The physiological function of xanthan has been deduced as being analogous to the functions of other exopolysaccharides (Yang and Tseng, 1988; De Crecy et al., 1990; Daniels and Leach, 1993; Chan and Goodwin, 1999).

Enclosure of bacterial cells in the exopolysaccharide (EPS) results in prolonged survival, and increased resistance to both temperature and ultraviolet (UV) light (Leach et al., 1957). In rice, the wilting induced by EPS seems to play a role in pathogenesis (Kuo et al., 1970). EPS may increase cell membrane leakage, which in turn leads to wilting (Vidhyasekaran et al., 1989). In general, bacterial EPS induce water-soaking of the intercellular space which is important for bacterial colonization. It is also possible that xanthan forms a gel-like slime in the intercellular space as a result of synergy with other plant polysaccharides. This gel may then promote bacterial colonization of plant tissue by retarding the desiccation of the bacterial colony, by protecting the bacteria from bacteriostatic compounds, and by preventing close morphological contact of the colony with the cell wall, thus preventing the triggering of plant defense reactions. The amount of EPS produced by *Xanthomonas* is correlated to the organism's virulence; strains with attenuated virulence usually produce less EPS (Ramirez et al., 1988) and the distribution of the polysaccharide in infected leafs coincides with that of bacteria. This was seen in a study in rice, where EPS and bacteria were distributed in both the xylem and transverse veins (Watabe et al., 1993).

6
Analysis and Detection

In order to characterize xanthan, different parameters must be taken into consideration, such as chemical structure, acetate and pyruvate contents, molecular weight, secondary structure, and rheological behavior.

6.1
Chemical Characterization

By using chemical analysis, the sugar composition of the molecule as well as the nature and the degree of substituent content can be ascertained.

6.1.1
Sugar Composition

The sugar composition of xanthan is difficult to obtain as the cellulosic backbone is highly resistant to hydrolysis. Moreover, in the side chains the presence of uronic acid prevents complete hydrolysis of the aldobiouronic (β-D-GlcAp-(1 → 2)-D-Manp) acid without degradation of the glucuronic acid. Well-documented reports of this situation have been made (Tait and Sutherland, 1989; Tait et al., 1990) in which the most suitable conditions to determine the neutral sugars, the aldobiouronic (β-D-GlcAp-(1 → 2)-D-Manp) acid and the substituents are described. A single condition with one form of hydrolysis is insufficient to characterize all the constituents of xanthan quantitatively. Due to these problems, the official description of xanthan, e.g., by the JECFA (Joint Expert Committee for Food Additives), does not refer to its chemical composition, but only to its ability to gellify in the presence of locust bean gum (LBG). In the official description, there is no reference to acetyl groups, only to pyruvic acid.

6.1.2 Pyruvic Acid Determination

After hydrolysis (Cheetham and Punruckvong, 1985; Tait et al., 1990), the pyruvic acid content of xanthan can be determined in several ways. The oldest method described is a colorimetric procedure using 2,4-dinitrophenylhydrazine (DNPH) (Slonecker and Orentas, 1962), and this is still the reference method for the JECFA. Duckworth and Yaphe (1970), have developed an enzymatic method using lactate dehydrogenase (LDH), the reaction being as follows:

$$\text{Pyruvate} + \text{NADH} + \text{H}^+ \xrightarrow{\text{LDH/LD}} \text{lactate} + \text{NAD}^+ \quad (1)$$

The amount of NAD released is measured at 340 nm. More recent determinations use high-pressure liquid chromatography (HPLC) (Cheetham and Punruckvong, 1985; Tait et al., 1990) or nuclear magnetic resonance (NMR) methods, both of which permit the simultaneous detection of both pyruvate and acetate.

6.1.3 Acetate Determination

In 1949, Hestrin published a method for the determination of acetate which uses hydroxamic acid, but today (see Section 6.1.2) NMR and HPLC methods are more often used (Cheetham and Punruckvong, 1985; Tait et al., 1990).

6.2 Physical Characterization

Besides the chemical composition of xanthan, its physical characteristics such as molecular weight, secondary structure and rheological properties are the most important determinants of the behavior of this molecule in its final application.

6.2.1 Molecular Weight

Values reported in the literature for the molecular weight (M_w) of xanthan are usually between 4 and 12×10^6 g mol^{-1}. Accurate determination of the M_w of xanthan is difficult for several reasons, including: (1) the very high molecular weight; (2) the stiffness of the molecule; and (3) the presence of aggregates. Nonetheless, several techniques have been used to determine the molecular weight of xanthan, including GPC-MALLS (gel- permeation chromatography with multi-angle laser light scattering), AFFF-MALLS (asymmetrical flow field fractionation combined with multiangle laser light scattering) and electron microscopy.

GPC-MALLS

GPC-MALLS is a technique that permits the estimation of absolute molecular weight and gyration radius of polysaccharides, without the need for column calibration methods or standards. Often used in the field of polymer analysis, this technique is constituted by a GPC system which allows molecules to be separated as a function of their molecular size, and also by MALLS, which allows information to be obtained on the molecular weight of the fraction eluted from the column. With xanthan, the GPC technique presents several difficulties: first, the high molecular weight of the xanthan combined with its rod-like structure causes the xanthan molecules to have a very high hydrodynamic volume. The columns which are used today are unable to separate molecules with such high hydrodynamic volumes, and so xanthan appears as a monodisperse molecule. Second, for the same reason – and also due to the tight stationary phase of the column – the xanthan is submitted to a very high shear rate when eluted across the column, and this can degrade the molecule. In addition, the MALLS detection for xanthan analysis is problematic as, due to its high molecular weight, extrapolation to zero angle is not easy.

There are two classical methods to determine the molecular weight: the Zimm method; and the Debye method. In the Zimm method, we express $Kc/\Delta R_\theta = f(\sin \theta/2)$, whereby K is an optical parameter and R is the Rayleigh ratio. Since $1/M$ is obtained by extrapolation to zero angle, this method is not suitable for xanthan. The values obtained for $1/M$ are in the order of 10^{-7}, which leads to a significant variation of the value calculated for the molecular weight. In the Debye method, we express $\Delta R_\theta/Kc = f(\sin \theta/2)$; hence, by extrapolation to zero angle, M can be determined directly. This method is preferable for xanthan, but even in this case the very great angular dependence prevents the linear extrapolation and a polynomial of 3rd or 4th order is needed in order to obtain reproducible results. Some authors (Capron et al., 1997) prefer to use LALLS (low-angle laser light scattering), which provides a measure at a very low angle (5 °) and avoids this problem, but this type of apparatus does not provide any information on the gyration radius. Another problem is the presence of aggregates in the xanthan solution, and these are probably responsible for the very high molecular weight values reported in the literature. Such aggregates, even when present in very low quantities only, give very important signals in light scattering.

AFFF-MALLS

In order to avoid the problems which occur with GPC, recent investigations have used AFFF (Janca, 1988). AFFF uses a narrow channel in which a solvent flows, and a field is applied perpendicularly to this channel. Usually, the perpendicular field is created by a perpendicular flow. The sample is injected into the inlet of the channel and eluted by the solvent. At the same time, the field applied across the channel presses the sample against the wall (accumulation wall) of the channel. Due to the gradient of velocity across the channel coming from the laminar flow, the particles – depending on their distances from the wall – have different velocities. In AFFF, the separation is governed by the diffusion coefficient, and so this technique allows the separation of molecules on the basis of their hydrodynamic radius up to a size of several micrometers. An advantage of this method is that no packing material is needed, and this also avoids the problem of shear. The problem is that the different parameters – the two flows, the injected volume, and the solvent – must be carefully adjusted in order to obtain good results.

Electron Microscopy

This technique allows direct measurements of the xanthan molecule to be made after vacuum drying in the presence of glycerol and covering the molecule with a platinum film (Stokke et al., 1998). The contour length L of individual xanthan chains can be visualized by electron microscopy, and the average value of L reflects the molecular weight. Electron microscopy can also be used to detect the formation of microgels.

6.2.2 Secondary Structure

The physico-chemical properties of xanthan in aqueous solutions can be studied by means of various experimental techniques, such as light scattering measurements, hydrodynamic measurements, thermodynamic properties such as ion activity, dependence of the transition temperature on the ionic strength and calorimetric measurements. A relatively new method for studying the superstructure of xanthan is atomic force microscopy (AFM) (Capron et al., 1998a; Morris et al., 1999). AFM allows visualization of the surface of a sample, with a resolution close to the atomic scale. The mechanical properties of both native and denatured xanthan can be measured on the molecular scale (Li et al., 1999; Morris et al., 1999). AFM measures the interaction between the sample and the tip of the measuring device. A force exists between the atom of the tip of the microscope and those of the sample, separated by only a few Angstroms. By moving the tip, it is possible to follow the variation of this force on the surface of the sample and so obtain an image of the sample and estimate its shape. No modification of the sample is needed for the measurement, and this technique can be applied to both conducting and nonconducting samples. AFM avoids the drying step of the sample, and can provide images of individual xanthan molecules as well as of aggregated molecules. Molecules and molecular interactions can be imaged in the liquid environment.

Capron et al. (1997) have studied the size and conformation changes associated with the temperature-induced denaturation and renaturation of native xanthan under different salt conditions. The different methods used were LALLS, size- exclusion chromatography coupled with multi-angle light scattering (SEC-MALLS), low shear intrinsic viscosity measurement, and circular dichroism. The conformational transition can be monitored using NMR, optical rotation or calorimetric measurements. The specific optical rotation changes suddenly at the melting temperature of the molecule, from about −120 to almost zero. Circular dichroism studies near 200 nm show a decrease in overall ellipticity when passing through the transition region.

6.2.3 Rheology

Depending on the medium conditions, that is polymer concentration, salts or addition of other hydrocolloids, xanthan systems can be a Newtonian or pseudo-plastic solution or a gel. The rheological behavior can be determined using viscometers by applying shear rate and measuring shear stress and viscosity or using controlled shear stress or deformation rheometers to perform dynamic viscoelastic or flow measurements. Xanthan solutions can be characterized by classical rheological parameters, such as intrinsic viscosity $[\eta]$. The intrinsic viscosity corresponds to the hydrodynamic volume of the polymer chain in a given solvent. Classically, it can be obtained by measuring the viscosity at different low concentrations in the Newtonian domain, where the overlap between hydrodynamic volume of individual polymer chains is negligible and applying the equations of Huggins (Eq. 2) and Kraemer (Eq. 3) (Launay et al., 1986):

$$\eta_{sp}/C = [\eta] + \lambda_H[\eta]^2C \tag{2}$$

$$\ln(\eta_r)/C = [\eta] + \lambda_k[\eta]^2C \text{ with} \tag{3}$$

$$\eta_r = \eta/\eta_0 \tag{4}$$

$$\eta_{sp} = (\eta - \eta_0)/\eta_0 \tag{5}$$

where η is the solution viscosity, η_0 the solvent viscosity, η_r the relative viscosity, η_{sp} the specific viscosity, $[\eta]$ the intrinsic viscosity and λ_H and λ_k are constants which are functions of the hydrodynamic interaction between molecules. In dynamic measurement, the sample is submitted to a deformation $\gamma^*(t)$ or a stress $\sigma^*(t)$ which are sinusoidal function of time. When the system is viscoelastic linear, the stress $\sigma^*(t)$ is a sinusoidal function of time with the same frequency ω as $\gamma^*(t)$ and a phase angle gap δ.

$$\gamma^*(t) = \gamma_0(\cos \omega t + i \cdot \sin \omega t) \quad (6)$$

$$\sigma^*(t) = \sigma_0(\cos (\omega t + \delta) + i \cdot \sin (\omega t + \delta)) \quad (7)$$

If $\delta = 0$, stress and deformation are proportional at every moment, which means that the behavior is elastic linear. If $\delta = \pi/2$, stress and deformation speed are proportional at every moment, the behavior is Newtonian. If $0 < \delta < \pi/2$ the behavior is viscoelastic. The complex modulus G^* can be defined as:

$$G^* = \sigma^*/\gamma^* = G' + i\, G'' \text{ with} \quad (8)$$

$$G' = (\sigma_{0\,/}\ \gamma_0) \cdot \cos \delta \quad (9)$$

$$G'' = (\sigma_{0\,/}\ \gamma_0) \cdot \sin \delta \quad (10)$$

$$\tan \delta = G''/G' \quad (11)$$

G' is the storage modulus which corresponds to the elastic component of the system, and G'' is the loss modulus which corresponds to the viscous component of the system. We can also define the complex viscosity:

$$\eta^* = \sigma^*/\acute{y}^* = \eta' + i\, \eta' \text{ with} \quad (12)$$

$$\eta' = G''/\omega \quad (13)$$

$$\eta'' = G'/\omega \quad (14)$$

The gel-like structure of xanthan solutions can be characterized by deformation measurements. Small deformation measurements characterize the intact network, whilst large deformation measurements destroy the network and therefore give lower values. It is possible to measure the static yield point of a xanthan solution by applying an increasing deformation at constant shear rate and measuring the shear stress. Initially, the stress generated in resistance to the deformation increases linearly with the deformation, as in an elastic solid. Ultimately, the resistance reaches a maximum corresponding to the breaking point of the network (i.e., the yield point) and then drops again, settling down at a constant value which defines the steady shear viscosity. Xanthan gum solutions can also be characterized using dynamic light scattering. This technique allows characterization of the boundary between dilute and semi-dilute solutions. The degree of dilution is important because in a truly diluted state, the xanthan coils occupy a defined hydrodynamic volume, whereas above a critical concentration the molecules interact. Dynamic light scattering experiments can demonstrate the onset of molecular interaction and the onset of anisotropic aggregation (Rodd et al., 2000).

7
Biosynthesis

The path of xanthan biosynthesis has been described and reviewed by several authors (Leigh and Coplin, 1992; Becker et al., 1998). Xanthan synthesis starts with the assembly of the pentasaccharide repeating units, and these are then polymerized to produce the macromolecule. The oligosaccharide repeating units of xanthan are formed by the sequential addition of monosaccharides from energy-rich sugar nucleotides, involving acetyl-CoA and phosphoenolpyruvate. A polyisoprenol phosphate from the inner membrane functions as an acceptor (Ielpi et al., 1993). The first step of the pentasaccharide assembly is the transfer of glycosyl-1-phosphate from UDP-glucose to polyisoprenol phosphate. This transfer is followed by sequential transfer of the other sugar residues, D-mannose and D-glucuronic acid from GDP-mannose and UDP-glucuronic acid, respectively, which gives the complete lipid-linked repeating pentasaccharide unit. Acetyl and pyruvyl residues are added at this lipid-linked pentasaccharide stage; these are donated by acetyl-CoA and phosphoenolpyruvate, respectively. Depending on the strain and on the fermentation conditions, O-acetyl groups are attached in varying quantities to the internal mannose residue, and pyruvate is added to the terminal mannose. The xanthan chains grow at the reducing end (Figure 2). The final steps of the biosynthesis, which means the secretion from the cytoplasmic membrane, the passage across the periplasm and outer membrane and the excretion into the extracellular environment has not yet been entirely elucidated. The process requires energy, and it is probably accomplished via a specific transport system, which ensures the release of the polymer from the lipid carrier and the transport across the outer membrane (Daniels and Leach, 1993).

Many genes which are involved in xanthan biosynthesis have been identified, isolated and characterized (Figure 3). In *Xanthomonas campestris* pv. *campestris*, the biosynthesis is directed by a cluster of 12 genes, *gumB* to *gumM* (Vanderslice et al., 1989; Vojnov et al., 1998). Seven gene products are required for the transfer of the monosaccharides and for the acylation at the lipid intermediate level to form the complete acylated repeating unit. This gene cluster is not linked to the genes which are required for the synthesis of the sugar nucleotide precursors. The 12 genes of the cluster are

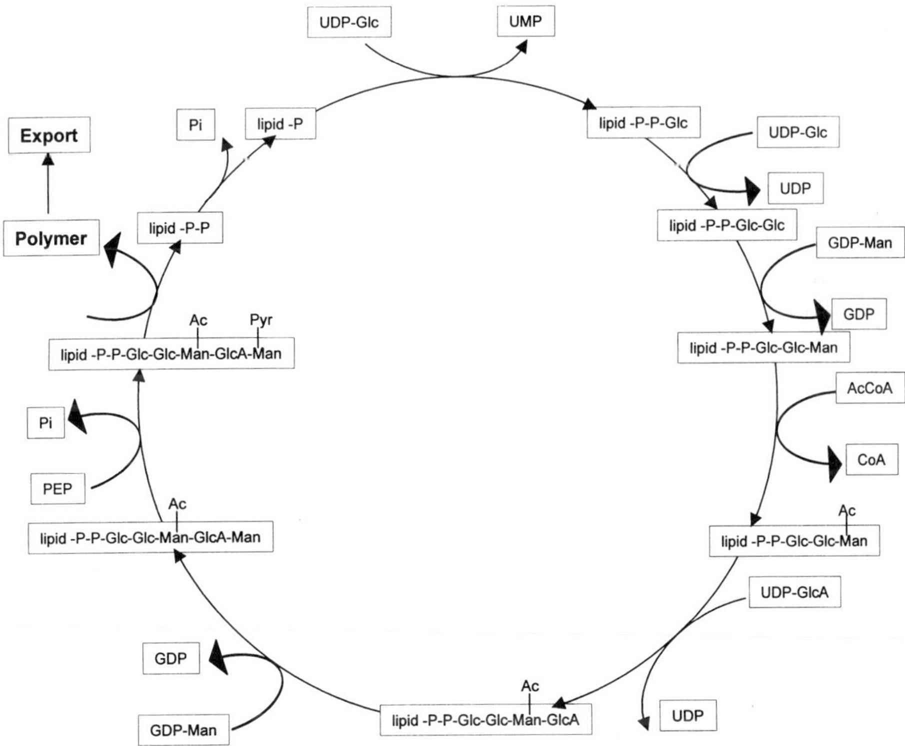

Fig. 2 Biosynthesis of xanthan. Ac: acetyl; Glc: glucose; Man: mannose; P: phosphate; PEP: phosphoenolpyruvate; Pyr: pyruvate (Adapted from Sutherland, 1993).

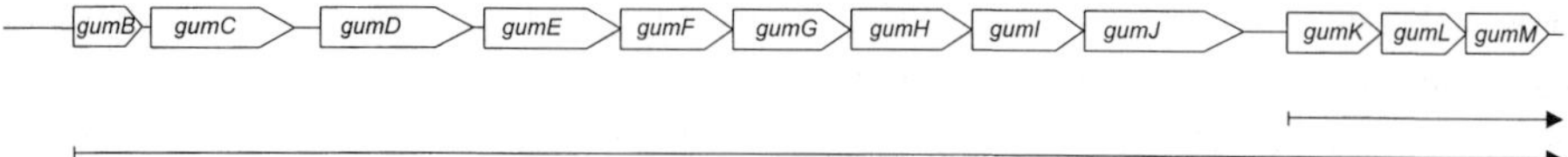

Fig. 3 Genetic map of the *Xanthomonas campestris* pv. *campestris gum* operon (adapted from Becker et al., 1998).

expressed as a single operon from a promoter which is located upstream of the first gene. Xanthan synthesis does not seem to be specifically controlled but is activated by the gene products of a cluster of at least five genes (Tang et al., 1990). The initial stages of xanthan biosynthesis appear to be regulated as part of a pathogenicity regulon (Coplin and Cook, 1990), and some promotors and transcription start sites have been characterized (Knoop et al., 1991; Kingsley et al., 1993; Katzen et al., 1998). It has been shown that a transposon insertion, located 15 bp upstream of the translational start site of the *gumB* gene leads to a defect in xanthan synthesis (Vojnov et al., 1998). Cells of the mutant strain were able to synthesize the lipid-linked pentasaccharide repeating unit of xanthan from the three nucleotide sugar donors (UDP-glucose, GDP-mannose, and UDP-glucuronic acid), but were unable to polymerize the pentasaccharide into mature xanthan. A subclone of the gum gene cluster carrying *gumB* and *gumC* restored xanthan production of the mutant strain to levels of almost 30% of the wild type. In contrast, subclones carrying *gumB* or *gumC* alone were not effective. These results indicate that *gumB* and *gumC* are both involved in the translocation of xanthan across the bacterial membranes. Tseng et al. (1999) have constructed a physical map of the *Xanthomonas campestris* pv. *campestris* chromosome, and the locations of eight loci involved in xanthan gum synthesis were determined. Furthermore, four loci were determined which may be involved in gum polymerization, secretion and regulation. Rodrigues and Aguilar (1997) have identified several mutants of *Xanthomonas campestris* showing increased viscosity and/or gum production after UV treatment. Xanthan solutions of the different mutants showed different intrinsic viscosity values, and no relationship was found between pyruvate or acetate contents and viscosifying ability of the xanthan. It is also possible to produce xanthan using genetically modified (GM) bacteria other than *Xanthomonas*. Pollock et al. (1997) have cloned 12 genes coding for assembly, acetylation, pyruvylation, polymerization and secretion of the polysaccharide xanthan gum in *Sphingomonas*, and these genes were sufficient for synthesis of xanthan gum. The polysaccharide from the recombinant microorganism was largely indistinguishable, both structurally and functionally, from the native xanthan gum.

8
Degradation

The xanthan molecule is very stable as long as it is in the ordered, double-stranded conformation. The double helix confers a good resistance to the molecule against degradation by free radicals, acid, enzymes or repeated freeze–thaw cycles. The stability of xanthan is significantly lowered if the molecule is in a disordered conformation, which is the case at temperatures above the transition temperature or at low ionic strength, for example in salt-free solutions at temperatures around 40–50 °C. Xanthan in solution will not degrade in a pH range from about 2 to 11, but at pH 9 xanthan gum starts to deacetylate. Temperatures up to

90 °C do not affect xanthan, and even after heat treatment at 120 °C for over 30 min the viscosity remained at about 90% of the original value. Chemical degradation of xanthan can be achieved through free radicals which are generated by strong oxidants such as hypochlorite and persulfate, at high temperature, or by a combination of H_2O_2 and Fe^{2+}. Oxidative–reductive depolymerization leads to a cleavage of the polymeric backbone (Herp, 1980). Acid hydrolysis leads preferentially to the hydrolysis of the terminal β-D-mannose in the side chain. Mechanical degradation is possible using high shear or ultrasound; however, the shear to be applied must be very high, in the order of $10^6\ s^{-1}$. Enzymatic degradation of xanthan is usually not very efficient, but if the molecules are in the disordered state they can be degraded by a class of enzymes called xanthanases. This group of enzymes comprises endo-1,4-β-glucanases and xanthan lyases. The glucanases can partially degrade the cellulosic backbone of xanthan (Rinaudo and Milas, 1980), and they are also able to degrade different celluloses. Xanthan lyases are able to cleave the β-D-mannosyl-D-glucuronic acid linkage of the trisaccharide side chain. The presence of acetate and pyruvate hampers the action of the cellulases, whereas the activity of xanthan lyase does not seem to be affected by either pyruvate or acetyl groups. It is likely that xanthan contains cellulosic regions that do not carry side chains and are thus preferred regions for an attack by cellulases. Christensen and Smidsrød (1996) have shown that acid hydrolysis prior to enzymatic degradation increases the depolymerization of xanthan by several orders of magnitude. Removal of the side chains from xanthan with a limited degree of backbone degradation can be obtained by partial hydrolysis. The terminal β-mannose residue linked to O-4 of the glucuronic acid is hydrolyzed quite rapidly (Christensen and Smidsrød, 1991; Christensen et al., 1993a,b). The α-mannose linked to the backbone is hydrolyzed about 10 times more slowly, the hydrolysis obeying pseudo first-order kinetics. Ruijssenaars et al. (1999) have identified several xanthan-degrading enzymes in a *Paenibacillus* strain; this strain is able to degrade about one-third of the xanthan molecule without attacking the backbone of the molecule. Nankai et al. (1999) have reported in detail an enzymatic route for the depolymerization of xanthan by a *Bacillus* strain. The enzymes which are necessary for the degradation of xanthan are xanthan lysase, glucanase, glucosidase, glucuronyl hydrolase and mannosidase. The degradation starts with the cleavage of the glycosidic bond between pyruvylated mannosyl and glucuronyl residues in xanthan side chains due to the action of an extracellular xanthan lyase. The modified xanthan can then be attacked by an extracellular β-D-glucanase, which produces a tetrasaccharide, representing the repeating unit of xanthan without the terminal mannosyl residue. The tetrasaccharide is then taken into the bacterial cell and is subsequently converted by β-D-glucosidase to yield the trisaccharide unsaturated glucuronyl-acetylated mannosyl-glucose. Afterwards, a glucuronyl hydrolase generates an unsaturated glucuronic acid and the disaccharide mannosyl-glucose. The last step in the degradation is the hydrolysis of this disaccharide to mannose and glucose by α-D-mannosidase. For more details, see Chapter 23.

9 Biotechnological Production

Xanthan gum is produced by fermentation; this is a very efficient and reliable process, but it does present some intrinsic problems.

9.1 General Description of the Process

Today, xanthan gum is the most successful industrial biopolymer produced by fermentation. Xanthan gum is produced by the aerobic fermentation of *Xanthomonas campestris*, and many different strains of this bacterium have been screened for their ability to produce xanthan gum (Gupte and Kamat, 1997; De Andrade Lima et al., 1997). Xanthan fermentation can be either batch-wise, semi-batch-wise, or continuous. Industrial production is usually carried out by a batch-wise, submersed fermentation in an aerated and agitated fermenter. The different steps of an industrial process are batch fermentation, alcoholic precipitation, first drying, rinsing with an alcohol/water mixture, final drying, grinding, quality control of the batch, and packaging (Figure 4), with the production strain usually preserved in a freeze-dried state. Galindo et al. (1994) and Salcedo et al. (1992) have reported the preservation of the production strain on agar slopes as well as preservation in sterile seeds. The strain is activated by inoculation into a nutrient medium containing a carbohydrate source, a nitrogen source and mineral salts. After growth, the culture can be used to inoculate successive fermenters through to the industrial scale. Throughout the fermentation process, pH, aeration, temperature and agitation are monitored and controlled. The optimal temperature for cell growth is between 24 and 27 °C, and the best temperature for xanthan production is 30–33 °C. A pH of 6–7.5 is most suitable for growth, while for xanthan production the pH optimum range is 7–8. During the fermentation, pH decreases; however, in an industrial fermentation process the pH is maintained close to neutral in order to allow the process to continue until complete exhaustion of the carbohydrate substrate. Xanthan is produced during the bacterial growth phase as well as during the stationary phase, though maximum production is seen during the exponential growth phase. The pyruvate and acetate contents of xanthan are highest in polymers synthesized immediately after the end of exponential growth (Sutherland, 1993). The quantity of polymer increases until about 30 h after inoculation, after which time a steady state is reached with termination of bacterial growth and polymer production, at which point the fermentation is stopped. The achievable productivity and concentrations reported in the literature range from 15 to 30 g L^{-1} (sometimes higher) and up to 0.7 g $L^{-1} \cdot h$) (mostly lower). Productivity in industrial fermentations may be significantly higher, however. At the end of the fermentation, the fermenter is emptied, cleaned and sterilized before the next fermentation takes place. The post-fermentation steps include a pasteurization by thermal treatment, and sometimes also

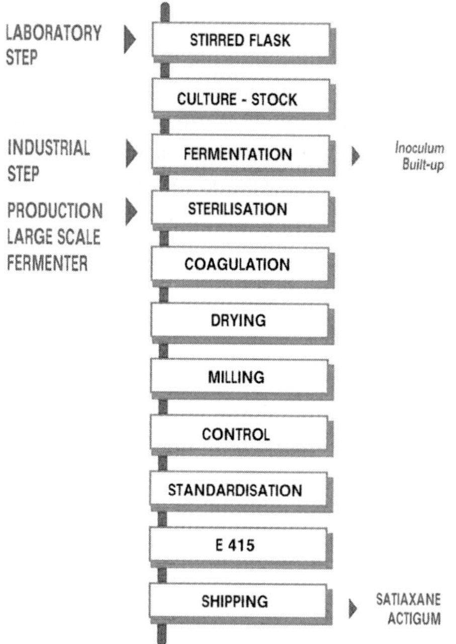

Fig. 4 Industrial production of xanthan.

purification. The heat treatment serves to kill the bacteria. Since *Xanthomonas* do not form spores, the fermentation can be stopped by heat treatment. Pasteurization also improves the properties of the gum, as it leads to a (partial) transformation of the native to the renatured state. Xanthan gum is recovered by precipitation in isopropyl alcohol or ethanol. Usually, the whole broth including biomass is precipitated, but for specific applications, purification may be necessary (see Section 8.4). In the different growth phases, xanthan molecules with different molecular weight and different degree of substitution are produced; hence, the xanthan recovered at the end of a batch fermentation is a mixture of structurally different molecules. The coagulum obtained is isolated by solid–liquid separation, rinsed with alcohol, pressed and dried. The dried xanthan is ground to obtain a white or cream-colored powder with the desired particle size (usually 80–250 μm). Xanthan powder may also be granulated in order to obtain products with specific dispersion and dissolution properties. The final steps in the industrial production chain are quality control, including chemical, physical, rheological and microbiological analysis, packaging and shipping.

9.2 Process Improvement

Although the process for xanthan production is well established, numerous and also recent publications deal with the improvement of xanthan production. Due to the rheological properties of the xanthan molecule, the production of xanthan by fermentation is technically challenging. During the course of the fermentation, the excretion of the polysaccharide results in a highly viscous and shear-thinning broth. As the fermentation broth becomes non-Newtonian, the rheology of the broth causes serious problems of mixing, oxygen supply and heat transfer, thereby limiting the maximum gum concentration achievable, as well as the product quality. Changes in viscosity during culture exceed four orders of magnitude. At the beginning of the fermentation the broth is more or less like water, whereas at the end of the process it is highly viscous and pseudoplastic, exhibiting yield stress. Besides the technical difficulties which always demand improvement, economic aspects of the fermentation process are also important, and improvement of the fermentation costs has a high priority for industrial xanthan producers.

9.2.1 General Improvement

Instead of batch fermentation, other approaches such as continuous fermentation, fed-batch or emulsion fermentation are possible. Continuous fermentation allows high productivity to be obtained, but the process is not easy to set up, the equipment needed is not standard, and stirring in the continuous process is often insufficient. Furthermore, the sterility of the process is difficult to maintain (Silman and Rogovin, 1972). Hence, continuous processes are usually not suitable for industrial production.

Fed-batch fermentation is more successful, and towards the end of the production nutrients are recharged and very good productivity and process efficiency can be obtained. Fermentation in emulsions or in water/oil dispersions can result in high yield in the water phase, but low yields in the total volume.

Emulsion fermentation can reduce the viscosity of the broth by dispersing it into an organic liquid, such as isoparaffin, with the help of an emulsifier, for example fatty acid alkanolamide. The water-in-oil emulsions obtained develop less viscosity and show less pseudoplastic behavior than normal fermentation broth. High oxygen transfer rates can be achieved, and mixing is efficient even at low power input (Schumpe et al., 1991). However, downstream processing is much more difficult than in one-phase submerged cultivation, and the gum may be contaminated with the emulsifier. The separation of cell growth and biopolymer production in a two-stage process can improve productivity (Banks and Browning, 1983; DeVuyst et al., 1987). Xanthan production by immobilized cells has also been described (Lebrun et al., 1994, Yang et al., 1996).

9.2.2 Oxygen Supply

Xanthomonas campestris is a strictly aerobic bacterium, and so oxygen is required for growth and for xanthan production. Sufficient oxygen supply is a prerequisite for efficient polymer production and good productivity. Oxygen is provided by agitation of the culture broth, but this becomes increasingly difficult during the process as the viscosity of the broth increases. Agitation is much affected by vessel geometry, liquid volume, the method of agitation and impeller properties. The productivity of *Xanthomonas* can be improved through the use of impeller systems and suitable reactor configurations which lead to increased oxygen supply for the bacteria. The various reactor configurations which have been proposed to increase oxygen transfer into the medium include stirred tank reactors, external circulation loop reactors, bubble column or air lift reactors. The impeller types which have been proposed are helical stirrers, which cause axial motion, and anchor and multiple rod stirrers which have the tangential action of flat-blade turbines (Galindo, 1994). Amanullah et al. (1998b) have investigated the agitator speed and dissolved oxygen effects in xanthan fermentation. Changes of shear stress in the vicinity of the microorganism which are caused by changes of agitation speed did not influence xanthan production up to a xanthan concentration of 20 g L^{-1}. At higher gum concentrations, xanthan production was enhanced at the higher agitation speed, due to higher microbial oxygen uptake rates.

9.2.3 Nutrients

A standard medium for xanthan production contains a nitrogen source, a carbon source, phosphate and magnesium ions, and trace elements. A typical composition of a medium is 0.06% ammonium nitrate, 0.5% potassium dihydrogen phosphate, 0.01% magnesium sulfate heptahydrate, 2.25% glucose and 97.18% water (Daniel et al., 1994). As carbon substrates, *Xanthomonas* can use different sources such as carbohydrates, amino acids and intermediates of the citric acid cycle. The most commonly used carbon source is glucose, with concentrations around 40 g L^{-1}. In fact, glucose concentrations above 50 g L^{-1} usually lead to poorer conversion into the polysaccharide. Nitrogen sources which allow growth and efficient xanthan production are ammonium salts or amino acids. The degree of substitution with pyruvate increases if $(NH_4)_2HPO_4$ is used as nitrogen source (US Secretary of Agriculture, 1978). The polymer production is promoted by a high ratio of carbon to nitrogen in the culture medium. The mode of glucose feeding can influence the productivity of the fermentation (Amanullah et al., 1998a). An improved performance cannot be achieved by increasing the initial glucose concentration above 50 g L^{-1}, nor by a single 10 g L^{-1} pulse addition with an initial glucose concentration of 40 g L^{-1}, while significant nitrogen is still present. However, a simple pulse and continuous feeding strategy, after nitrogen has been essentially exhausted and under conditions of nonlimiting dissolved oxygen, can result in a greatly enhanced performance compared with batch fermentations. A xanthan gum concentration > 60 g L^{-1} and a productivity of 0.72 g $L^{-1} \cdot$ h) were obtained, which is higher than the values usually reported in the literature for conventional stirred bioreactors.

Between 20–30% of the total process cost can be attributed to the culture medium. In order to reduce the costs of xanthan production, several cheap substrates – many of which are agricultural waste or byproducts of the agronomic industry – have been proposed for xanthan fermentation. Cost reduction is only one aspect in the selection of a nutrient, however; another important aspect is that cheap nutrients must allow both satisfactory productivity and good product quality to be obtained. Among the proposed alternative substrates are whey, corn-steep liquor, molasses, glucose syrup, and olive oil waste waters (De Vuyst and Vermeire, 1994). The cultivation of *Xanthomonas campestris* on solid substrates, such as spent malt grains, apple pomace, grape pomace and citrus peels is also possible. With most of these solid substrates, xanthan yields were comparable with those obtained in conventional submerged cultivation, the yield achieved ranging from 33 to 57 g L^{-1} (Stredansky and Conti, 1999; Stredansky et al., 1999). The biopolymers obtained were chemically and structurally similar to commercial xanthan. Moreno et al. (1998) have reported successful fermentations using different acid-hydrolyzed wastes – from melon, watermelon, cucumber, and tomato – in the culture medium for xanthan production. Melon acid-hydrolyzed waste was the best substrate for xanthan production, allowing an exopolysaccharide to be obtained the chemical composition of which was very similar to that of commercial xanthan. Sugar beet pulp added as a supplement to the xanthan culture medium can increase the yield of xanthan fermentation and allows the production of nonfood-grade xanthan gum (Yoo and Harcum, 1999). Chestnut flour can also be used as a nutrient in xanthan fermentation (Liakopoulou Kyriakides et al., 1999), with a maximum concentration of 33 g L^{-1} xanthan being obtained after 45 h of fermentation. Albergaria et al. (1999) have used LBG extracts as a growth medium in the fermentation; the maximum xanthan concentration obtained was 14 g L^{-1}, and the specific xanthan yield was 7 g g^{-1} cells. This substrate is more of academic interest however, since the yield is rather low and LBG extracts are not a cheap raw material source.

9.3 Modeling the Fermentation Process

Controlling a difficult fermentation process, especially of a highly viscous biopolymer such as xanthan gum, carries several intrinsic difficulties. The process is complex, varies over time, and is nonlinear; moreover, on-line measurements of many important variables is often not possible or is too difficult for industrial routine. A helpful approach for controlling the process can be mathematical modeling, and models for the control of a rheologically complex fermentation have been developed by several groups (Abraham et al., 1995; Garcia Ochoa et al., 1995). The proposed models are able to describe the process parameters including growth, polymer production, consumption of nutrients and evolution of dissolved oxygen, and can be used to improve significantly the production efficiency. Garcia Ochoa et al. (1998) have developed a kinetic model describing xanthan production by *Xanthomonas campestris* NRRL B-1459 in a batch stirred-tank. Parameters described include biomass production, carbon, nitrogen and oxygen consumption, xanthan production and temperature effects. The model can accurately describe fermentations over the temperature range 25–34 °C. Kuttuva et al. (1998) have simulated W/O xanthan fermentation by integrating the microbial kinetic behavior and the multiple-phase process characteristics. Two different models are proposed, one of which assumes a uniform redistribution of cells, substrates and product by frequent droplet breakup and coalescence, and another model which simulates the system of viscous aqueous phase with minimal droplet breakup and component redistribution. The evaluated parameters were the xanthan concentration in the aqueous phase and the volumetric productivity achieved at 200 h. According to the model, W/O fermentation can lead to very high xanthan concentrations in the aqueous phase of > 200 g L^{-1}, and a much increased volumetric productivity of > 0.8 g $L^{-1} \cdot$ h compared with conventional fermentations where the xanthan concentration reaches ~50 g L^{-1} and a volumetric productivity of ~0.5 g $L^{-1} \cdot$h). Serrano Carreon et al. (1998) have modeled xanthan gum fermentation through the interactions between the kinetics of growth and product formation and mixing, which is related to the rheological behavior of the broth. The mixing was linked to the power drawn, as a function of the agitation rate and xanthan content.

9.4 Post-fermentation Treatment

Isolation and purification of xanthan are costly procedures, with post-fermentation treatments accounting for up to 50% of the total production costs. Industrial xanthan is usually not purified. Following pasteurization, which takes place at the end of the fermentation process, the entire fermentation broth is precipitated with isopropanol or a similar alcohol – which means that the bacterial cells remain in the product after processing. The final product contains about 85% pure xanthan gum, 5% biomass, and 10% moisture. The alcohol is recovered by distillation. The biomass imparts turbidity of the final product, and so removal of the bacterial cells is required when high transparency or clarity of the product is required. The fermentation broth can be clarified by filtration, though Yang et al. (1998) removed the cells present in xanthan fermentation broth by adsorption of the bacterial cells onto fibers. These authors were able to show that rough-surfaced cotton fibers were better than smooth-surfaced polyester fibers in the cell adsorption process. Cell adsorption was facilitated by the anionic xanthan gum present in the solution, as in the absence of xanthan gum the cell adsorption to fibers was poor. Alternatively, proteolytic enzymes which degrade the cellular debris from *Xanthomonas* can be used for clarification; this is possible for example using extracellular enzymes secreted by *Trichoderma koningii* (Triveni and Shamala, 1999).

The thermal denaturation and renaturation of a fermentation broth of xanthan can affect the rheological properties of xanthan (Capron et al., 1998b). Depending on the concentration and degree of purification, thermal treatment will have different effects. Changes in both the viscoelastic properties and molecular weight are observed after heating above the melting order–disorder temperature, these being related to the order–disorder conformational transition of the xanthan molecules. At concentrations below 10 g L^{-1}, heat denaturation occurs with dissociation of the native double-stranded structure into two single strands. Upon cooling, the single strands will fold back on themselves to form renatured helices, but at higher concentrations no complete dissociation occurs. Xanthan, which is renatured in concentrated conditions (> 10 g L^{-1}) has a higher viscosity than that of the native sample, and displays more gel-like properties. In order to optimize the economy of the post-fermentation treatment of the fermentation broth, Lo et al. (1997) proposed that ultrafiltration be carried out before the alcoholic precipitation. In this way, the xanthan solution could be concentrated to up to 15% (w/v). Mixing modes can also influence the recovery of xanthan gum by precipitation. Zhao et al. (1997) measured the rheological properties of xanthan solution at varying alcohol concentration, thereby introducing the concepts of mixing alcohol concentration and precipitation alcohol concentration. The best design for the precipitation was a continuous precipitation system, as this gave the highest yield and the least amount of impurities. Parlin (1997) proposed a scalable vibrating membrane filter system designed for the separation of solid–liquid foods. Instead of recovering the polymer by alcohol precipitation it is also possible to recover the gum by direct drying with drum dryers or spray dryers (Harrison et al., 1999). Alternative agents for precipitation which have been studied include quaternary ammonium salts or calcium salts. Commercial xanthan is usually dried until it reaches a moisture content of 8–12%, after which it is ground to a particle size usually between 80 and 250 μm. Easily-dispersible commercial products may also be granulated.

10 Properties

Xanthan has very interesting rheological properties. Generally speaking, xanthan gum is a thickener, and produces extremely viscous solutions in water, even at very low concentrations, though under certain conditions xanthan may also gellify.

The properties of pure xanthan gum and xanthan gum blends will be described in this section. As detailed above, xanthan molecules can exist in either an ordered or a disordered conformation. The most interesting properties of xanthan are due to the ordered form, in which the macromolecules adopt a helical conformation whereby the secondary structure resembles rigid rods that do not have any tendency to associate. The form and the rigidity of the xanthan macromolecules determine the rheology of xanthan solutions. The organized state is stabilized in the presence of electrolytes. Parker et al. (1999) determined the dissolution kinetics of xanthan powder, and showed the solubilization of xanthan powder to depend on the size and distribution of the particles; for example, a polydisperse powder will have tendency to form lumps. The dissolution behavior is different in pure water and salt solutions, however. In pure water, dissolution begins with a burst of aggregates, whereas the addition of salt diminishes the aggregate population. The presence of anionic side chains on xanthan gum molecules enhances hydration and renders it soluble in cold water.

10.1 Viscosity

At rest, xanthan solutions develop high viscosity values at concentrations as low as 1%, this being due to the high M_w of the molecules and to their secondary structure. The rigid double-stranded helix confers stability to the molecule; this is because the rigid rods cannot gyrate freely and, at rest, the macromolecules adopt an equilibrium position, stabilized by low interactions. Xanthan solutions are quite resistant to hydrolysis, temperature, and electrolytes.

Salts usually do not reduce the viscosity, though xanthan might precipitate in the presence of polyvalent cations under alkaline conditions. The nature of the counterion in single counterion xanthan influences xanthan viscosity. Solutions of xanthan gum are generally not affected by changes in pH value, and xanthan gum dissolves in most acids or bases. The viscosity of xanthan gum is stable at pH values ranging from 2 to 11 for long periods of time, whereas other hydrocolloids lose their viscosity under the same conditions. The viscosity varies only slightly with temperature, and xanthan solutions remain viscous at temperatures even exceeding 100 °C as long as the ordered conformation is maintained. Stability is enhanced by high gum concentrations.

10.2 Flow Behavior

At very low shear rate, xanthan solutions may present a Newtonian behavior. Milas et al. (1990) showed that transition from the Newtonian regime to viscoelastic behavior was characterized by a critical value of the shear rate, $\acute{y}_r$, i.e., a relaxation time $\acute{y}_r^{-1}$. Above a certain low polymer concentration, this critical shear rate depends on the concentration: the higher the concentration, the lower the shear rate, which can be explained by the change in the entanglement degree of the system. For $\acute{y} > \acute{y}_r$, the xanthan solutions present a shear-thinning behavior which, together with a high yield value, give xanthan solutions exceptional suspension properties. The shear-thinning effect is due to the orientation of the polymer. At low shear, the molecules are randomly oriented, and the flowing properties are bad. This state is relatively stable until the shear-stress exceeds a certain value, called the yield-value. At high shear, the molecules align in the direction of the applied force and the xanthan solution flows easily. Above the yield value, the shear-thinning behavior is more pronounced than with other gums with a random coil conformation. It has been reported (Stokke et al., 1998) that the dynamic viscosity $\eta^*(\omega)$ of a xanthan solution is higher than the steady flow viscosity $\eta(\acute{y})$ at a shear rate equal to ω, which means that xanthan does not follow the empirical Cox–Mertz rule, except at low concentration ($<1\ g\ L^{-1}$), in the absence of salt and in low shear rate and frequency regions (Milas et al., 1990).

10.3 Weak Network Formation

Xanthan chains in solution form a continuous three-dimensional network, with weak cross-links. Therefore, xanthan solutions may also be characterized as weak gels. The chains separate easily under shearing, which allows the xanthan solution to flow and also accentuates the shear-thinning behavior. The mechanical spectra of xanthan solutions resembles the spectra of gels with $G' > G''$, with little frequency dependence in either modulus. Association of the negatively charged xanthan is promoted by metal ions; the order of effectiveness in inducing weak gels is $Ca^{2+} > K^+ > Na^+$.

10.4 Gelation

Under certain conditions, xanthan may gellify. This may occur in the presence of certain metal ions, or as a synergistic effect with other polymers. Heavy metal ions, such as Cr^{3+}, Al^{3+} or Fe^{3+} cross-link xanthan; in the case of Cr^{3+}, the cross-link occurs through a ligand exchange reaction, where the water molecule which is bound to the Cr^{3+} is exchanged with a carboxylic group on the xanthan molecule.

10.5 Interaction of Xanthan with Other Macromolecules

Xanthan gum can develop additive, synergistic and/or antagonistic effects with other molecules, such as galactomannans and glucomannans. Xanthan molecules in their rigid rod helical form can be cross-linked via the smooth zones of galactomannans. Xanthan shows a synergistic viscosity increase with the galactomannan guar, and it can form a strong thermoreversible gel in the presence of the galactomannan LBG, the glucomannans konjac mannan, the galactomannan tara, or modified guar (Morris, 1990). With LBG, the maximum gel strength is obtained at a xanthan:LBG ratio of 1:1. The formation of thermoreversible gels by synergistic mixtures of xanthan with certain galacto- and glucomannans has been ascribed to intermolecular binding through cocrystallization of denatured xanthan chains within the mannan crystallite (Morris, 1990). The cellulosic backbone of xanthan and the stereochemically similar mannan backbone both form ribbon-like structures. The mixed crystallites probably act as strong junction zones to consolidate the weak xanthan network. The xanthan–galactomannan system is interesting because each polysaccharide alone does not form a gel, but mixing results in a synergistic gelation. The mannose on galactose (M:G) ratio in galactomannans controls the mechanism and the temperature at which gelation occurs.

11 Applications

Xanthan is soluble in both hot and cold water, it develops high viscosity at low concentrations, is compatible with many salts even at elevated concentrations, and is stable at both acid and alkaline pH. Xanthan solutions are also quite resistant to high temperatures, even above 100 °C. These properties, in addition to the concept of a yield stress, and along with the shear-thinning behavior and water-binding capacity of xanthan gum, make it a highly valuable texturizing agent, and consequently it is used in a wide variety of applications. In the food industry, xanthan can act as thickener, as a suspending agent, as an emulsion stabilizer, and as a foam enhancer. Technical applications of xanthan make use of the same properties which are also important for the food industry, and include oil drilling emulsions, paints and glues, as well as fire-fighting formulations. Xanthan applications have been reviewed recently by Kang and Pettitt (1993) and Nussinovitch (1997). From the total volume of xanthan produced worldwide, about 65% is used in food applications, 15% in the petroleum industry, and about 20% in technical applications other than oil drilling.

11.1 Food Applications

Some examples of the multitude of potential food applications include:

- Use in dry mixes for products such as sauces, dressing, gravies, or desserts. Xanthan is used in these applications because it dissolves quite easily in hot or cold water, and provides a rapid build-up of viscosity. In dressings, xanthan is particularly suitable for the suspension of herbs, spices and vegetables, of which it assures an even distribution throughout the bottle over a long period of time (Figure 5). Xanthan provides the necessary high yield value and strong pseudoplasticity, and its properties are not hampered by acid pH, high salt concentra-

Fig. 5 Application of xanthan for the stabilization of herbs in salad dressing. Left: dressing without xanthan. Right: dressing containing 0.125% (w/v) xanthan.

tions, or heat treatment. Due to the shear-thinning behavior, the dressing flows easily from the bottle; however, once poured onto the salad, the sauce clings to the food, which is important for visual presentation.

- In syrup or chocolate toppings, xanthan confers a good consistency and flow properties due to the high yield value and high at-rest viscosity.
- In beverages, xanthan gives body and a good mouthfeel to the liquid and can also stabilize pulps, especially in combination with other polysaccharides such as carboxymethylcellulose.
- Applications in dairy products are often in combination with LBG or guar gum. These combinations stabilize cream emulsions, prevent whey-off and improve the physical and organoleptic properties of pasteurized products. Xanthan stabilizes air cells as well as particles, which makes it useful for mousses, whipped creams and pourable aerated desserts (Sanofi, 1993).
- In frozen desserts, xanthan gum confers heat-shock resistance, and it can also protect foods from freeze–thaw instability (MacNeely and Kang, 1973; Vafiadis, 1999). Xanthan gum is able to increase perceived creaminess and minimize the effects of temperature variation during storage. Freeze–thaw cycles often destroy the delicate texture of frozen desserts, as under unfavorable storing conditions small ice crystals will migrate towards large ice crystals during the thaw stage. During refreezing they will attach to large ice crystals, causing them to grow. Xanthan gum, together with other ingredients such as starch, slows down the recrystallization of smaller ice crystals and prevents the formation of large ice crystals.
- The heat stability and the stabilizing and suspending properties of xanthan are also used widely in canned foods.
- In baking, xanthan is helpful in dough preparation by preventing lump formation during kneading and improving dough homogeneity and volume (Collar et al., 1999). Xanthan also facilitates pumping of the dough during production. In baking, xanthan is also used to suspend larger solid particles such as fruits or nuts.

Other applications, as described by Tilly (1991) include:

- the stabilization of chocolate milk and yogurt-based beverages;
- in combination with other polysaccharides for low-fat spreads;
- the production of fat-reduced biscuits (Conforti et al., 1997); and
- reduced-calorie grape juice jellies which will show similar texture characteristics, for example gel hardness, cohesiveness and springiness, as a reduced-calorie grape juice jelly texturized with low methoxyl pectin (Gaspar et al., 1998).

The application of xanthan in combination with LBG the production of "melt-in-the-mouth" polysaccharide gelling systems for foods has also been described (Marrs, 1997).

11.2 Non-food Applications

The most important technical application for xanthan gum is its use oil drilling. In this application, xanthan is particularly useful due to its pseudoplastic behavior, temperature stability, and salt tolerance. The use of xanthan in petroleum production has been reviewed by Kang and Pettitt (1993). During oil drilling, a low viscosity is required at the drill bit, whereas a high viscosity is required in the annulus. The pseudoplasticity of xanthan solutions meets these requirements as xanthan develops low viscosity at the drill bit where the shear is high, and high viscosity in the annulus where shear is low. Therefore, drilling fluids containing xanthan allow a rapid penetration at the bit, and a suspension of cuttings in the annulus.

In textile printing, xanthan is used to provide the specific rheological properties needed in the production of sharp and clean patterns by preventing migration of the dye. Xanthan is compatible with most components of printing pastes, and it is also removed easily by washing. Xanthan is also used in ceramic glazes where it prevents agglomeration of the different components. In cleaning liquids, xanthan can be useful by providing a high viscosity to the solution at low shear, thereby allowing the cleaner to cling to inclined surfaces. Other technical applications described for xanthan gum include paint and ink, where it can stabilize and suspend pigments, wallpaper adhesives, formulation of pesticides and toothpastes, and industrial emulsions. The use of xanthan gum as a support for enzyme and cell immobilization has been described by Dumitriu and Chornet (1998). Xanthan gum can also be useful as controlled-release agent for pharmaceuticals (Philipon, 1997), with microspheres of gellified xanthan encapsulating active ingredients. The spheres are swallowed in a dry state and swell in the stomach, thereby gradually releasing the active ingredient. The active molecule can also be linked covalently to the polysaccharide and then released in the body by enzymatic hydrolysis.

12 Relevant Patents

As long ago as 1959, Esso Research patented the use of xanthan gum for the displacement of oil from partially depleted reservoirs. Later, several patents followed concerning processes for fermentation (1963, 1966), recovery (1963) and application (1975) of xanthan. In 1960, several patents of Jersey Production Research Co. appeared, covering a thickening agent and the process for its production, a substituted heteropolysaccharide, and a process for synthesizing polysaccharides. Kelco Biospecialities Ltd. has patented processes for xanthan production and recovery as well as xanthan application, since the early 1960s. This includes several patents concerning processes for producing the polysaccharides deposited in 1966, a process for xanthan gum production (1981), an inoculation procedure for xanthan gum production (1981), or the production of xanthan gum by emulsion fermentation (1981). Patents concerning the post-fermentation treatment include patents regarding the treatment of a *Xanthomonas* hydrophilic colloid and the resulting product (1964, 1968) or the precipitation of xanthan gum (1981). Special xanthans such as etherified *Xanthomonas* hydrophilic colloids (1965), polymeric derivatives of a cationic *Xanthomonas* colloid (1964), cationic ethers of *Xanthomonas* hydrophilic colloids (1963) or xanthan having a low pyruvate content (1981) have also been patented by Kelco. Applications patented by Kelco include edible compositions comprising oil-in-water emulsions (1960), joint-filling compositions (1963), and a dehydrated food product (1971). Patents of Rhone-Poulenc appeared mainly in the 1970s and 1980s; those of the 1970s covered improvements of the fermentation process (1973, 1976, 1978) and applications as a thickening agent (1975) and in oil drilling (1977). During the 1980s, other patents relating to the fermentation process were deposited (1984, 1987, 1988). Merck began working on xanthan in the 1970s, and this resulted in several patents concerning improvements in the process to obtain xanthan solutions with increased viscosity (1975), the preparation of a xanthan copolymer (1975), a deacetylated borate-biosynthetic gum composition (1979), a process for producing low-calcium xanthan gum by fermentation (1978), a process to prepare a xanthan gum which does not contain any cellulases (1979), the production of low-calcium, smooth-flow xanthan gum (1979), and a process for improved recovery of xanthan gum with high viscosity (1979). In the 1980s, the Merck patents covered a dispersible xanthan gum composite (1980) and a heteropolysaccharide and its production and use (1985). In 1986, Merck patented a recombinant DNA plasmid for xanthan gum synthesis containing some essential genetic material for xanthan gum synthesis and in 1991, a patent about low-ash xanthan gum (< 2%) appeared which is claimed to be particularly useful in the preparation of ceramics. Patents of Standard Oil Co. about xanthan gum appeared mainly in 1980/1981, and referred to different strains and the corresponding processes, such as *Xanthomonas campestris* ATCC 31600 (1980), *Xanthomonas campestris* NRRL B-12075 and NRRL B-12074 (1980), *Xanthomonas campestris* ATCC 31602 (1980) and *Xanthomonas campestris* ATCC 31601 (1980). Patents relating to the production process appeared in 1980; these included the production of xanthan gum from a chemically defined medium, a method for improving xanthan yield, and also for improving specific xanthan productivity during continuous fermentation. Standard Oil Co. has also patented a semicontinuous method for the production of xanthan gum using different *Xanthomonas* strains (1981). A method of producing a low-viscosity xanthan gum was patented in 1981. Sanofi has patented production processes (1986, 1987) and the application of xanthan (1993), as well as a mutant strain which produces a xanthan that does not develop any viscosity (1990). Oil companies other than Esso which have worked on xanthan development include Shell, which has patented the treatment of polysaccharide solutions (1980), a process for preparing *Xanthomonas* heteropolysaccharide, the heteropolysaccharide as prepared by this process and its use (1983), as well as a method for improving the filterability of a microbial broth and its use (1984). Mobil Oil Corporation, with patents relating to a waterflood process employing thickened water (1966) and a method for clarifying polysaccharide solutions (1973), and Phillips Petroleum Co., with a patent about recovery of a microbial polysaccharide (1971). Other companies which have patented xanthan-related processes or applications include: Archer Daniels Midland Co., who claim the biochemical synthesis of industrial gums (1962); Pfizer, who have patented a batch process (1979) and a continuous production of *Xanthomonas* biopolymer (1982); Stauffer Chemical Co., who have patented the fermentation of whey to produce a thickening polymer (1981); and Hoechst AG, who claim the lowering of viscosity of fermentation broths (1983). In addition, Celanese Corp. patented concentrated xanthan gum solutions obtained by ultrafiltration (1980), and Jungbunzlauer AG claimed an improved fermentation process (1982) and a process for obtaining polysaccharides as grains (1989). Several xanthan-related patents have been deposited by General Mills Chemicals Inc. about *Xanthomonas* gum amine salts (1974), about flash-drying of xanthan gum (1976), or about a cationic polysaccharide obtained by reacting a xanthan with isopropanol and NaOH (1986). Hercules Inc. has patented a xanthan recovery process (1976), and Henkel KGaA has patented some biopolymers obtained from *Xanthomonas* (1981) as well as a process for the preparation of exocellular biopolymers (1986). The Akademie der Wissenschaften der DDR of the former East Germany has worked on a method for microbial production of xanthan (1986). Getty Scientific Development Co. has patented a polysaccharide polymer made by *Xanthomonas* in 1985, and the recombinant-DNA mediated production of xanthan gum in 1986. The latter patent claims a gene cluster encoding enzymes necessary for the biosynthesis of xanthan gum which was isolated from *Xanthomonas campestris*. The U.S. Secretary of Agriculture has patented a method for producing an atypically salt-responsive alkali-deacetylated polysaccharide (1959), a method of recovering microbial polysaccharides from their fermentation broths (1962), a continuous process for producing *Xanthomonas* heteropolysaccharide (1967) as well as a nitrogen source for improved production of microbial polysaccharides (1967). The Institut Français du Petrole has patented the clarification of xanthan in 1980, and a process for the production of an improved xanthan for the oil drilling application in 1993. Cerestar Holding B.V. patented a fermentation feedstock containing simple sugars including mannose and glucose (1993). One of the most recent patents has been from Rhodia Inc. (1997), and claims the use of a liquid carbohydrate fermentation product in food. The claimed product can be either food grade or pharmaceutical grade, and it is delivered in a carbohydrate medium other than dairy whey, without any drying steps prior to use (Table 1).

Tab. 1 Relevant patents relating to xanthan products

Publication date	*Patent holder*	*Inventors*	*Patent No.*	*Patent title*
1961	U.S. Secretary of Agriculture	Jeanes, A. R., Sloneker, J. H.	US 3 000 790	Method of producing an atypically salt-responsive alkali-deacetylated polysaccharide.
1962	Jersey Production Research Co.	Patton, J. T., Lindblom, G. P.	US 3 020 206	Process for synthesizing polysaccharides.
1962	Jersey Production Research Co.	Patton, J. T.	US 3 020 207	Thickening agent and process for producing same.
1962	Kelco Co.	O'Connell, J. J.	US 3 067 038	Edible compositions comprising oil in water emulsions.
1964	Jersey Production Research Co.	Lindblom, G. P., Patton, J. T.	US 3 163 602	Substituted heteropolysaccharide.
1964	U.S. Secretary of Agriculture	Rogovin, S. P., Albrecht, W. J.	US 3 119 812	Method of recovering microbial polysaccharides from their fermentation broths.

Tab. 1 (cont.)

Publication date	Patent holder	Inventors	Patent No.	Patent title
1966	Archer Daniels Midland Co.	Weber, R. O., Horan, F. E.	US 3 271 267	Biochemical synthesis of industrial gums.
1966	Esso Production Research Co.	Lipps, B. J.	US 3 251 749	Fermentation process for preparing polysaccharides.
1966	Esso Production Research Co.	Lipps, B. J.	US 3 281 329	Fermentation process for producing a heteropolysaccharide.
1966	Kelco Co.	Schweiger, R. G.	US 3 244 695	Cationic ethers of *Xanthomonas* hydrophilic colloids.
1966	Kelco Co.	Schuppner, H. R.	US 3 279 934	Joint filling composition.
1967	Esso Production Research Co.	Lindblom, G. P., Ortloff, G. D., Patton, J. T.	US 3 305 016	Displacement of oil from partially depleted reservoirs.
1967	Esso Production Research Co.	Lindblom, G. P., Patton, J. T.	US 3 328 262	Heteropolysaccharide fermentation process.
1967	Kelco Co.	O'Connell, J. J.	US 3 355 447	Treatment of *Xanthomonas* hydrophilic colloid and resulting product.
1967	Kelco Co.	Schweiger, R. G.	US 3 349 077	Etherified *Xanthomonas* hydrophilic colloids and process of preparation.
1968	Esso Production Research Co.	Patton, J. T., Holman, W. E.	US 3 382 229	Polysaccharide recovery process.
1968	Kelco Co.	Schweiger, R. G.	US 3 376 282	Polymeric derivatives of cationic *Xanthomonas* colloid derivatives.
1968	Kelco Co.	MacNeely, W. H.	US 3 391 060	Process for producing polysaccharides.
1968	Kelco Co.	MacNeely, W. H.	US 3 391 061	Process for producing polysaccharides.
1968	Mobil Oil Corp.	Sherrod, A. W.	US 3 373 810	Waterflood process employing thickened water.
1969	General Mills Inc.	Nordgren, R., Wittcoff, H. A.	DE 1 919 790	Polysaccharide B1459 cationique.
1969	Kelco Co.	Macneely, W. H.	US 3 427 226	Process for preparing polysaccharide.
1969	U.S. Secretary of Agriculture	Rogovin, S. P.	US 3 485 719	Continuous process for producing *Xanthomonas* heteropolysaccharide.
1970	Kelco Co.	Colegrove, G. T.	US 3 516 983	Treatment of *Xanthomonas* hydrophilic colloid and resulting product.
1970	Mobil Oil Corp.	Abdo, M. K.	US 3 711 462	Method of clarifying polysaccharide solutions.
1972	Kelco Co.	Edlin, R. L.	US 3 694 236	Method of producing a dehydrated food product.
1973	Philips Petroleum Co.	Buchanan, B. B., Cottle, J. E.	US 3 773 752	Recovery of microbial polysaccharide.
1975	General Mills Chemicals Inc.	Jordan, W. A., Carter, W. H.	US 3 928 316	*Xanthomonas* gum amine salts.
1975	Société des Usines Chimiques Rhone-Poulenc		FR 2 251 620	Perfectionnement à la production de polysaccharides par fermentation.
1976	Rhone-Poulenc Industries	Falcoz, P., Celle, P., Campagne, J. C.	FR 2 299 366	Nouvelles compositions épaississantes à base d'hétéropolysaccharides.

Tab. 1 (cont.)

Publication date	*Patent holder*	*Inventors*	*Patent No.*	*Patent title*
1977	Exxon Research & Engineering Co.	Naslund, L. A., Laskin, A. I.	FR 2 330 697	Procédé de traitement d'un hétéropolysaccharide et son application.
1977	General Mills Chemicals Inc.	Cahalan, P. T., Peterson, J. A., Arndt, D. A.	US 4 053 699	Flash drying of xanthan gum and product produced thereby.
1977	Hercules Inc.	Towle, G. A.	US 4 051 317	Xanthan recovery process.
1977	Merck & Co. Inc.	Kang, K. S., Burnett, D. B.	FR 2 318 926	Perfectionnement aux procédés pour accroitre la viscosité de solutions aqueuses de gomme xanthane et aux compositions obtenues.
1977	Merck & Co. Inc.	Cottrell, I. W.	FR 2 325 666	Copolymère greffe d'un colloide hydrophile de *Xanthomonas* et procédé pour sa préparation.
1977	Rhone-Poulenc Industries	Campagne, J. C.	FR 2 342 339	Procédé de production de polysaccharides par fermentation.
1979	Institut Français du Pétrole des Carburants & Lubrifiants - Rhone-Poulenc Industries	Ballerini, D., Claude, O., Chauveteau, G., Kohler, N., Vandecasteele, J. P.	FR 2 398 874	Utilisation de mouts de fermentation pour la récupération assistée du pétrole.
1979	Rhone-Poulenc Industries	Contat, F., Lartigau, G., Nocolas, O.	FR 2 414 555	Procédé de production de polysaccharides par fermentation.
1980	Merck & Co. Inc.	Empey, R., Dominik, J. G.	EP 20 097	Production of low-calcium smooth-flow xanthan gum.
1980	Merck & Co. Inc.	Roche, R. E.	FR 2 457 322	Procédé de récupération pour améliorer la viscosité de la gomme xanthane.
1980	Merck & Co. Inc.	Racciato, J. S., Cottrell, I. W.	US 4 214 912	Deacetylated borate-biosynthetic gum composition.
1981	Celanese Corp.	Lee, H. L.	US 4 299 825	Concentrated xanthan gum solutions.
1981	Merck & Co. Inc.	Kang, K. S.	FR 2 458 589	Procédé de préparation de gomme xanthane exempte de cellulase.
1981	Pfizer Inc.	Wernau, W. C.	US 4 282 321	Fermentation process for production of xanthan.
1981	Standard Oil Co.	Weisrock, W. P.	US 4 301 247	Method for improving xanthan yield.
1982	Kelco Biospecialities Ltd.	Jarman, T. R.	EP 66 957	Inoculation procedure for xanthan gum production.
1982	Henkel KGaA	Bahn, M., Engelskirchen, K., Schieferstein, L., Schindler, J., Schmid, R.	EP 58 364	Biopolymères à partir de *Xanthomonas.*

Tab. 1 (cont.)

Publication date	*Patent holder*	*Inventors*	*Patent No.*	*Patent title*
1982	Institut Français du Pétrole	Rinaudo, M., Milas, M., Kohler, N.	FR 2 491 494	Procédé enzymatique de clarification de gommes de xanthane permettant d'améliorer leur injectivité et leur filtrabilité.
1982	Kelco Biospecialities Ltd.	Jarman, T. R., Pace, G. W.	EP 66 961	Production of xanthan having a low pyruvate content.
1982	Kelco Biospecialties Ltd.	Jarman, T. R.	EP 66 377	Process for xanthan gum production.
1982	Kelco Biospecialties Ltd.	Maury, L. G.	US 4 352 882	Production of xanthan gum by emulsion fermentation.
1982	Merck & Co. Inc.	Sandford, P. A., Baird, J. K.	US 4 357 260	Dispersible xanthan gum composite.
1982	Shell Internationale Research Maatschappij B.V.	Van Lookeren Campagne, C. J., Roest, J. B.	EP 49 012	Treatment of polysaccharide solutions.
1982	Standard Oil Co.	Weisrock, W. P., MacCarthy, E. F.	EP 46 007	*Xanthomonas campestris* ATCC 31601 and process for use.
1982	Standard Oil Co.	Weisrock, W. P.	US 4 311 796	Method for improving specific xanthan productivity during continuous fermentation.
1982	Standard Oil Co.	Weisrock, W. P.	US 4 328 310	Semi-continuous method for production of xanthan gum using *Xanthomonas campestris* ATCC 31601.
1983	Kelco Biospecialties Ltd.	Smith, I. H.	EP 68 706	Precipitation of xanthan gum.
1983	Merck & Co. Inc.	Richmon, J. B.	US 4 375 512	Process for producing low calcium xanthan gum by fermentation.
1983	Standard Oil Co.	Weisrock, W. P., Klein, H. S.	US 4 374 929	Production of xanthan gum from a chemically defined medium introduction.
1983	Standard Oil Co.	Bauer, K. A., Khosrovi, B.	US 4 400 467	Process for using *Xanthomonas campestris* NRRL B-12075 and NRRL B-12074 for making heteropolysaccharide.
1983	Standard Oil Co.	Weisrock, W. P., MacCarthy, E. F.	US 4 407 950	*Xanthomonas campestris* ATCC 31602 and process for use.
1983	Standard Oil Co.	Weisrock, W. P., MacCarthy, E. F.	US 4 407 951	*Xanthomonas campestris* ATCC 31600 and process for use.
1983	Standard Oil Co.	Weisrock, W. P.	US 4 377 637	Method of producing a low viscosity xanthan gum.
1983	U.S. Secretary of Agriculture	Cadmus, M. C., Knutson, C. A.	US 4 394 447	Production of high-pyruvate xanthan gum on synthetic medium.
1984	Jungbunzlauer A.G.	Kirkovits, A. E., Waltenberger, I.	AT 373 916	Xanthan.

Tab. 1 (cont.)

Publication date	Patent holder	Inventors	Patent No.	Patent title
1984	Pfizer Inc.	Young, T. B.	EP 115 154	Continuous production of *Xanthomonas* biopolymer.
1984	Stauffer Chemical Co.	Schwartz, R. D., Bodie, E. A.	US 4 444 792	Fermentation of whey to produce a thickening polymer.
1985	Hoechst A.G.	Voelskow, H., Keller, R., Schlingmann, M.	DE 3 330 328	Lowering the viscosity of fermentation broths.
1985	Shell International Research Maatschappij B.V.	Downs, J. D.	EP 130 647	Process for preparing *Xanthomonas* heteropolysaccharide; heteropolysaccharide as prepared by the latter process and its use.
1986	Rhone-Poulenc Specialités Chimiques	Leproux, V., Peignier, M., Cros, P., Beucherie, J., Kennel, Y.	EP 187 092	Procédé de production de polysaccharides de type xanthane.
1986	Shell International Research Maatschappij B.V.	Drozd, J. W., Rye, A. J.	EP 184 882	Method for improving the filtrability broth and the use.
1987	Akademie der Wissenschaften der DDR	Behrens, U., Stottmeister, U.	DD 250720	Method for microbial synthesis of polysaccharides.
1987	Getty Scientific Development Co.	Vanderslice, R. W., Shanon, P.	EP 211 288	A polysaccharide polymer made by *Xanthomonas*.
1987	Getty Scientific Development Co.	Capage, M. A., Doherty, D. H., Betlach, M. R., Vanderslice, R. W.	WO 87/05 938	Recombinant-DNA mediated production of xanthan gum.
1987	Merck & Co. Inc.	Peik, J. A., Steenbergen, S. M., Veeder, G. T.	EP 209 277	Heteropolysaccharide and its production and use.
1987	Merck & Co. Inc.	Cleary, J. M., Rosen, I. G., Harding, N. E., Cabanas, D. K.	EP 233 019	Recombinant DNA plasmid for xanthan gum synthesis.
1988	Sanofi - Méro Rousselot Satia S.A.	Eyssautier, B.	EP 296 965	Procédé de fermentation pour l'obtention d'une polysaccharide de type xanthane.
1988	Sanofi Elf Bio-Industries	Eyssautier, B.	FR 2 606 423	Procédé d'obtention d'un xanthane à fort pourvoir épaississant et application de ce xanthane.
1989	Henkel KGaA	Viehweg, H.	US 4 871 665	Process for the preparation of exocellular biopolymers.
1989	Rhone-Poulenc Chimie	Tavernier, C.	FR 2 624 135	Procédé de production de polysaccharides.

Tab. 1 (cont.)

Publication date	*Patent holder*	*Inventors*	*Patent No.*	*Patent title*
1990	Jungbunzlauer A.G.	Westermayer, R., Stojan, O., Eder, J.	FR 2 646 857	Procédé pour obtenir sous forme grenue des polysaccharides formés par les bactéries du genre *Xanthomonas* ou *Arthrobacter.*
1990	Rhone-Poulenc chimie	Nicolas, O.	EP 365 390	Procédé de production de polysaccharides par fermentation d'une source carbonée à l'aide de microorganismes.
1992	Merck & Co. Inc.	Talashek, T., Cleary, J. M.	EP 511 784	Low-ash xanthan gum.
1992	Sanofi - Société Nationale Elf Aquitaine	Salome, M.	FR 2 671 097	Souche mutante de *Xanthomonas campestris.* Procédé d'obtention de xanthane et xanthane non visqueux.
1994	Cerestar Holding B.V.	De Troostemberghm J. C., Beck, R. H. F., De Wannemaeker, B. L. T.	EP 609 995	Fermentation feedstock.
1994	Institut Français du Pétrole	Monot, F., Noik, C., Ballerini, D.	FR 2 701 490	Procédé de production d'un mout de xanthane ayant une propriété améliorée. Composition obtenue et application de la composition dans une boue de forage de puits.
1995	Sanofi	Tilly, G.	EP 649 599	Stabilizer composition enabling the production of a pourable aerated dairy dessert.
1999	Rhodia Inc.	Hoppe, C. A., Lawrence, J., Shaheed, A.	WO 99/25 208	Use of liquid carbohydrate fermentation product in foods.

13 Current Problems and Limitations

As a food additive which is produced by fermentation, xanthan is affected by the current debate regarding the use and danger of GM organisms. Even though the xanthan production strains used today are not GM, the consumer is skeptical about a product which is obtained via a biotechnological process. Indeed, the consumer demands that the whole process from the very beginning is accomplished without using any substrate that may have a GM source. The culture medium used for xanthan production must contain carbon and nitrogen sources; in order for the production to be cost-competitive, these sources must be cheap and efficient, easy to handle during the fermentation, and lead to a product of good quality. These requirements are fulfilled by sources such as corn-steep water as a carbon source and soy protein as a nitrogen source. However, a significant part of the world production of soy and corn today is obtained from GM plants. Although it seems very unlikely that GM carbon and nitrogen sources, after having been metabolized by the bacterium to yield the biopolymer, will confer any risk to the final xanthan product, the customer – especially the European customer – is not accepting GM sources in the culture medium. One of the reasons is certainly that standard commercial xanthans are not 100% pure, but contain about 5% of biomass. The components of the culture medium – especially glucose and nitrogen – should be used completely at the end of the production and no GM material should be left in the final product, though no guarantee can be given for this. Hence, the culture medium must be adapted to contain only non-GM sources, yet costs, productivity and product quality should remain unchanged. This is a major challenge for today's xanthan producers.

The current refusal of genetic modification for food additives by the consumer also limits innovation concerning xanthan. GM *Xanthomonas* strains could be developed in order to increase productivity, to improve the use of nutrients, to enable the metabolism of cheap substrates and their conversion into xanthan gum. It has been proved in the past that this is possible. Fu and Tseng (1990) have constructed a *Xanthomonas campestris* strain which carries a recombinant β-galactosidase-encoding gene and which is able to convert whey into xanthan. GM strains could also be modified in order to produce biopolymers with different substitution patterns, different properties in terms of rheology, dissolution and dispersion behavior, or to provide new or different synergetic interactions with other molecules. For the oil drilling industry, xanthan with increased temperature stability would be valuable. However, the current GM debate makes such development highly unlikely in the near future.

14
Outlook and Perspectives

Today, xanthan gum is a very successful biopolymer, and this success is likely to continue as no other biopolymer with similar properties is available commercially at a similar price. The world market for xanthan gum is growing and new markets are emerging for xanthan consumption as well as for xanthan production; for example, China is currently estimated to produce 5–10% of the world's xanthan. As discussed above, it appears highly unlikely that any future development of xanthan gum will be based on genetic modification, and developments will rather focus on specialty xanthans with improved handling properties. However, some perspectives based on genetic modification and improvement will clearly be required, and these are discussed in the following section.

Common genetic methods such as conjugation, electroporation, chemical and transposon mutagenesis and site-directed mutagenesis can be used with *Xanthomonas campestris*. By modifying the biosynthetic pathway for xanthan production, the carbohydrate structure and substitution pattern of the polymer can be genetically controlled to produce polysaccharides with quite different properties (Betlach et al., 1987). Mutants that lack glucuronic acid and pyruvate residues have been constructed, as well as strains producing xanthan with an increased pyruvate content. Tait and Sutherland (1989) have constructed a strain producing truncated xanthan. Until today, strains for xanthan production have been selected and improved by conventional methods. Attempts to construct strains with improved xanthan yield have been unsuccessful in the past, and are not very likely in the future since the conversion rate of carbon to xanthan is very high (Linton, 1990). Improvement of yield by genetic methods seems less promising than improvement by a more efficient fermenter design and better culture media. Another perspective for the development of xanthan is the improvement of the molecule by chemical modification. Potential modifications might include oxidation, reduction, changing of the substitution pattern, altering the side chains, or grafting other molecules onto the xanthan molecule. However, any such modification would lead to products that today are not food-approved. Controlled degradation of xanthan might lead to oligosaccharides with bioactive properties.

Acknowledgements
The authors thank Annick Bourdais and Patricia Poutrel, without whom this chapter would never have been accomplished.

15
References

Abraham, N. H., Kent, C. A., Satti, S. M. (1995) Modeling for control of poorly-mixed bioreactors, *I. Chem E. Research Event* **2**, 1049–1051.

Akademie der Wissenschaften der DDR (1986) Method for microbial synthesis of polysaccharides, DD 250720.

Albergaria, H., Roseiro, J. C., Amaral Collaco, M. T. (1999) Technological aspects and kinetics analysis of microbial gum production in carob, *Agro-Food-Ind. Hi-Tech* **10**, 24–26.

Amanullah, A., Satti, S., Nienow, A. W. (1998a) Enhancing xanthan fermentations by different modes of glucose feeding, *Biotechnology* **14**, 265–269.

Amanullah, A., Tuttiet, B., Nienow, A. W. (1998b) Agitator speed and dissolved oxygen effects in xanthan fermentation, *Biotechnol. Bioeng.* **57**, 198–210.

Archer-Daniels-Midland Co. (1962) Biochemical synthesis of industrial gums, U.S. Patent No. 3 271 267.

Banks, G., Browning, F. (1983) The development of a two stage xanthan gum fermentation, *Process Biochem. Suppl. Proc. Conf. Adv. Ferment.* 163–170.

Becker, A., Katzen, F., Puhler, A., Ielpi, L. (1998) Xanthan gum biosynthesis and application: a biochemical/genetic perspective, *Appl. Microbiol. Biotechnol.* **50**, 145–152.

Betlach, M. R., Capage, M. A., Doherty, D. H., Hassler, R. A., Henderson, N. M., Vanderslice, R. W., Marreli, J. D., Ward, M. B. (1987) Genetically engineered polymers: manipulation of xanthan biosynthesis, in: *Industrial Polysaccharides: Genetic Engineering, Structure/Property Relations and Applications* (Yalpani, M., Ed.), Amsterdam: Elsevier, 35–50.

Bih-Ying, Y., Tseng, Y. H. (1988) Production of exopolysaccharide and levels of protease and pectinase activity in pathogenic and non-pathogenic strains of *Xanthomonas campestris* pv. *campestris, Bot. Bull. Acad. Sinica* **29**, 93–99.

Capron, I., Brigand, G., Muller, G. (1997) About the native and renatured conformation of xanthan exopolysaccharide, *Polymer* **38**, 5289–5295.

Capron, I., Alexandre, S., Muller, G. (1998a) An atomic force microscopy study of the molecular organisation of xanthan, *Polymer* **39**, 5725–5730.

Capron, I., Brigand, G., Muller, G. (1998b) Thermal denaturation and renaturation of a fermentation broth of xanthan: rheological consequences, *Int. J. Biol. Macromol.* **23**, 215–225.

Celanese Corp. (1980) Concentrated xanthan gum solutions, U.S. Patent No. 4 299 825.

Cerestar Holding B.V. (1993) Fermentation feedstock, EP 609 995.

Chan, J. W. Y. F., Goodwin, P. H. (1999) The molecular genetics of virulence of *Xanthomonas campestris, Biotechnol. Adv.* **17**, 489–508.

Cheetham, N. W. H., Punruckvong, A. (1985) An HPLC method for the determination of acetyl and pyruvyl groups in polysaccharides, *Carbohydr. Polym.* **5**, 399–406.

Christensen, B. E., Smidsrød, O. (1991) Hydrolysis of xanthan in dilute acid: effects on chemical composition, conformation, and intrinsic viscosity, *Carbohydr. Res.* **214**, 55–69.

Christensen, B. E., Smidsrød, O. (1996) Dependence of the content of unsubstituted (cellulosic) regions in prehydrolysed xanthans on the rate of hydrolysis by *Trichoderma reesei* endoglucanase, *Int. J. Biol. Macromol.* **18**, 93–99.

Christensen, B. E., Smidsrød, O., Elgsaeter, A., Stokke, B. T. (1993a) Depolymerization of double-stranded xanthan by acid hydrolysis: characterization of partially degraded double strands and single-stranded oligomers released from the ordered structures, *Macromolecules* **26**, 6111–6120.

Christensen, B. E., Smidsrød, O., Stokke, B. T. (1993b) Xanthans with partially hydrolysed side chains: conformation and transitions, in: *Carbohydrates and Carbohydrate Polymers, Analysis, Biotechnology, Modification, Antiviral, Biomedical and Other Applications* (Yalpani, M., Ed.), ATL Press, 166–173.

Collar, C., Andreu, P., Martinez, J. C., Armero, E. (1999) Optimisation of hydrocolloid addition to improve wheat bread dough functionality: a response surface methodology study, *Food Hydrocolloids* **13**, 467–475.

Conforti, F. D., Charles, S. A., Duncan, S. E. (1997) Evaluation of a carbohydrate-based fat replacer in a fat-reduced baking powder biscuit, *J. Food Qual.* **20**, 247–256.

Coplin, D. L., Cook, D. (1990) Molecular genetics of extracellular polysaccharide biosynthesis in vascular phytopathogenic bacteria, *Mol. Plant-Microbe Interact.* **3**, 271–279.

Daniel, J. R., Whistler, R. L., Voragen, A. C. J., Pilnik W. (1994) Starch and other polysaccharides, in: *Ullmann's Encyclopedia of Industrial Chemistry* (Elvers, B., Hawkins, S., Russey, W., Eds.), Weinheim: VCH, 1–62.

Daniels, M. J., Leach, J. E. (1993) Genetics of *Xanthomonas* in: *Xanthomonas* (Swings, J. G., Civerolo, E. L., Eds.), London: Chapman & Hall, 301–339.

De Andrade Lima, M. A. G., De Araujo, J. M., De Franca, F. P. (1997) The evaluation of different parameters to characterize xanthan gum-producing strains of *Xanthomonas* pv. *campestris*, *Arq. Biol. Tecnol.* **40**, 179–187.

De Crecy Lagard, V., Glaser, P., Lejeune, P., Sismeiro, O., Barber, C. E., Daniels, M. J., Danchin, A. (1990) A *Xanthomonas campestris* pv. *campestris* protein similar to catabolite activation factor is involved in regulation of phytopathogenicity, *J. Bacteriol.* **172**, 5877–5883.

De Vuyst, L., Vermeire, A. (1994) Use of industrial medium components for xanthan production by *Xanthomonas campestris* NRRL-B-1459, *Appl. Microbiol. Biotechnol.* **42**, 187–191.

De Vuyst, L., Van Loo, J., Vandamme, E. J. (1987) Two stage fermentation process for improved xanthan production by *Xanthomonas campestris* NRRL B-1459, *J. Chem. Technol. Biotechnol.* **39**, 263–273.

Duckworth, M., Yaphe, W. (1970) Definitive assay for pyruvic acid in agar and other algal polysaccharide, *Chem. Ind.* **23**, 747–748.

Dumitriu, S., Chornet, E. (1998) Polysaccharides as support for enzyme and cell immobilisation, in: *Polysaccharides. Structural Diversity and Functional Versatility* (Dumitriu, S., Ed.), New York: Marcel Dekker, 629–748.

Esso Production Research Co. (1959) Displacement of oil from partially depleted reservoirs, U.S. Patent No. 3 305 016.

Esso Production Research Co. (1963) Fermentation process for preparing polysaccharides, U.S. Patent No. 3 251 749.

Esso Production Research Co. (1963) Fermentation process for producing a heteropolysaccharide, U.S. Patent No. 3 281 329.

Esso Production Research Co. (1963) Polysaccharide recovery process, U.S. Patent No. 3 382 229.

Esso Production Research Co. (1966) Heteropolysaccharide fermentation process, U.S. Patent No. 3 328 262.

Exxon Research & Engineering Co. (1975) Procédé de traitement d'un hétéropolysaccharide et son application, FR 2 330 697.

Fu, J. F., Tseng, Y. H. (1990) Construction of lactose-utilizing *Xanthomonas campestris* and production of xanthan gum, *Appl. Environ. Microbiol.* **56**, 919–923.

Galindo, E. (1994) Aspects of the process for xanthan production, *Trans. I. Chem. E* **72**, 227–237.

Galindo, E., Salcedo, G., Ramirez, M. E. (1994) Preservation of *Xanthomonas campestris* on agar slopes: effects on xanthan production, *Appl. Microbiol. Biotechnol.* **40**, 634–637.

Garcia Ochoa, F., Santos, V. E., Alcon, A. (1995) Xanthan gum production: an unstructured kinetic model, *Enzyme Microb. Technol.* **17**, 206–217.

Garcia Ochoa, F., Santos, V. E., Alcon, A. (1998) Metabolic structured kinetic model for xanthan production, *Enzyme Microb. Technol.* **23**, 75–82.

Gaspar, C., Laureano, O., Sousa, I. (1998) Production of reduced-calorie grape juice jelly with gellan, xanthan and locust bean gums: sensory and objective analysis of texture, *Z. Lebensm. Unters. Forsch.* **206**, 169–174.

General Mills Chemicals Inc. (1974) *Xanthomonas* gum amine salts, U.S. Patent No. 3 928 316.

General Mills Chemicals Inc. (1976) Flash drying of xanthan gum and product produced thereby, U.S. Patent No. 4 053 699.

General Mills Inc. (1968) Polysaccharide B1459 cationique, DE 1 919 790.

Getty Scientific Development Co. (1985) A polysaccharide polymer made by *Xanthomonas*, EP 211 288.

Getty Scientific Development Co. (1986) Recombinant-DNA mediated production of xanthan gum, WO 87/05 938.

Gupte, M. D., Kamat, M. Y. (1997) Isolation of wild *Xanthomonas* strains from agricultural produce, their characterisation and potential related to polysaccharide production, *Folia Microbiol.* **42**, 621–628.

Harris, P. J., Fergusson, L. R., (1999) Dietary fibres may protect or enhance carcinogenesis, *Mutat. Res.* **443**, 95–110.

Harrison, G. M., Mun, R., Cooper, G., Boger, D. V. (1999) A note on the effect of polymer rigidity and concentration on spray atomisation, *J. Non-Newtonian Fluid Mech.* **85**, 93–104.

Henkel KGaA (1981) Biopolymères à partir de *Xanthomonas*, EP 58 364.

Henkel KGaA (1986) Process for the preparation of exocellular biopolymers, U.S. Patent No. 4 871 665.

Hercules Inc. (1976) Xanthan recovery process, U.S. Patent No. 4 051 317.

Herp, A. (1980) Oxidative-reductive depolymerization of polysaccharides, in: *The Carbohydrates,* Vol. Ib (Pigman, W., Horton, D., Eds.), New York: Academic Press, 1276–1297.

Hestrin, S. (1949) Reaction of acetylcholine and other carboxylic acid derivatives with hydroxylamine, and its analytical application, *J. Biol. Chem.* **180**, 249–261.

Hoechst A.G. (1983) Lowering the viscosity of fermentation broths, DE 3 330 328.

Ielpi, L., Couso, R. O., Dankert, M. A. (1993) Sequential assembly and polymerization of the polyprenol linked pentasaccharide repeating unit of the xanthan polysaccharide in *Xanthomonas campestris, J. Bacteriol.* **175**, 2490–2500.

Institut Français du Pétrole des Carburants & Lubrifiants - Rhone-Poulenc Industries (1977) Utilisation de mouts de fermentation pour la récupération assistée du pétrole, FR 2 398 874.

Institut Français du Pétrole (1980) Procédé enzymatique de clarification de gommes de xanthane permettant d'améliorer leur injectivité et leur filtrabilité, FR 2 491 494.

Institut Français du Pétrole (1993) Procédé de production d'un mout de xanthane ayant une propriété améliorée. Composition obtenue et application de la composition dans une boue de forage de puits, FR 2 701 490.

Janca, J. (Ed.) (1988) Field-Flow Fractionation: analysis of macromolecules and particles, New York: Marcel Dekker.

Jarman, T. R., Pace, G. W. (1984) Energy requirement for microbial expolysaccharide synthesis, *Arch. Microbiol.* **137**, 231–235.

Jersey Production Research Co. (1960) Process for synthesizing polysaccharides, U.S. Patent No. 3 020 206.

Jersey Production Research Co. (1960) Thickening agent and process for producing same, U.S. Patent No. 3 020 207.

Jersey Production Research Co. (1960) Substituted heteropolysaccharide, U.S. Patent No. 3 163 602.

Jungbunzlauer A.G. (1982) Xanthan, AT 373 916.

Jungbunzlauer A.G. (1989) Procédé pour obtenir sous forme grenue des polysaccharides formés par les bactéries du genre *Xanthomonas* ou *Arthrobacter,* FR 2 646 857.

Kang, K. S., Pettitt, D. L. (1993) Xanthan, gellan, welan and rhamsan, in: *Industrial Gums,* 3rd edn (Whistler, R. L., BeMiller, J. N., Eds.), San Diego, CA: Academic Press, 341–397.

Katzen, F., Ferreiro, D. U., Oddo, C. G., Ielmini, M. V., Becker, A., Puhler, A., Ielpi, L. (1998) *Xanthomonas campestris* pv. *campestris* gum mutants: effects on xanthan biosynthesis and plant virulence, *J. Bacteriol.* **180**, 1607–1617.

Kelco Biospecialities Ltd. (1981) Inoculation procedure for xanthan gum production, EP 66 957.

Kelco Biospecialities Ltd. (1981) Production of xanthan having a low pyruvate content, EP 66 961.

Kelco Biospecialties Ltd. (1981) Process for xanthan gum production, EP 66 377.

Kelco Biospecialties Ltd. (1981) Precipitation of xanthan gum, EP 68706.

Kelco Biospecialties Ltd. (1981) Production of xanthan gum by emulsion fermentation, U.S. Patent No. 4 352 882.

Kelco Co. (1960) Edible compositions comprising oil in water emulsions, U.S. Patent No. 3 067 038.

Kelco Co. (1963) Cationic ethers of *Xanthomonas* hydrophilic colloids, U.S. Patent No. 3 244 695.

Kelco Co. (1963) Joint filling composition, U.S. Patent No. 3 279 934.

Kelco Co. (1964) Treatment of *Xanthomonas* hydrophilic colloid and resulting product, U.S. Patent No. 3 355 447.

Kelco Co. (1964) Polymeric derivatives of cationic *Xanthomonas* colloid derivatives, U.S. Patent No. 3 376 282.

Kelco Co. (1965) Etherified *Xanthomonas* hydrophilic colloids and process of preparation, U.S. Patent No. 3 349 077.

Kelco Co. (1966) Process for producing polysaccharides, U.S. Patent No. 3 391 060.

Kelco Co. (1966) Process for producing polysaccharides, U.S. Patent No. 3 391 061.

Kelco Co. (1966) Process for preparing polysaccharide, U.S. Patent No. 3 427 226.

Kelco Co. (1968) Treatment of *Xanthomonas* hydrophilic colloid and resulting product, U.S. Patent No. 3 516 983.

Kelco Co. (1971) Method of producing a dehydrated food product, U.S. Patent No. 3 694 236.

Kingsley, M., Gabriel, D., Marlow, G., Roberts, P. (1993) The *opsX* locus of *Xanthomonas campestris* affects host range and biosynthesis of lipopolysaccharide and extracellular polysaccharide, *J. Bacteriol.* **175**, 5839–5850.

Knoop, V., Staskawicz, B., Bonas, U. (1991) Expression of the avirulence gene *avr*Bs3 from *Xanthomonas campestris* pv. *vesicatoria* is not under the control of *hrp* genes and is independent of plant factors, *J. Bacteriol.* **173**, 7142–7150.

Kuo, T. T., Lin, B. C., Li, C. C. (1970) Bacterial leaf blight of rice plant. III – Phytotoxic polysaccharides produced by *Xanthomonas oryzae*, *Bot. Bull. Acad. Sinica* **11**, 46–54.

Kuttuva, S. G., Sundararajan, A., Ju, L. K. (1998) Model simulation of water-in-oil xanthan fermentation, *J. Dispersion Sci. Technol.* **19**, 1003–1029.

Launay, B., Doublier, J.L., Cuvelier, G. (1986) Flow properties of aqueous solutions and dispersion of polysaccharides, in: *Functional Properties of Food Macromolecules* (Mitchell, J. R., Ledward, D. A., Eds.), London, New York: Elsevier Applied Science Publisher, 1–78.

Leach, J. G., Lilly, V. G., Wilson, H. A., Purvis, M. R. (1957) Bacterial polysaccharides: the nature and function of the exudate produced by *Xanthomonas phaseoli*, *Phytopathology* **47**, 113–120.

Lebrun, L., Junter, G. A., Jouenne, T., Mignot, L. (1994) Exopolysaccharide production by free and immobilized microbial cultures, *Enzyme Microb. Technol.* **16**, 1048–1054.

Leigh, J. A., Coplin, D. L. (1992) Exopolysaccharides in plant–bacterial interactions, *Annu. Rev. Microbiol.* **46**, 307–346, 1048–1054.

Li, H., Rief, M., Oesterhelt, F., Gaub, H. E. (1999) Force spectroscopy on single xanthan molecules, *Appl. Physics A* **68**, 407–410.

Liakopoulou Kyriakides, M., Psomas, S. K., Kyriakidis, D. A. (1999) Xanthan gum production by *Xanthomonas campestris* w.t. fermentation from chestnut extract, *Appl. Biochem. Biotechnol.* **82**, 175–183.

Linton, J. D. (1990) The relationship between metabolite production and the growth efficiency of the producing organism, *FEMS Microbiol. Rev.* **75**, 1–18.

Liu, W., Sato, T., Norisuye, T., Fujita, H. (1987) Thermally induced conformational change of xanthan in 0.01M aqueous sodium chloride, *Carbohydr. Res.* **160**, 267–281

Lo, Y. M., Yang, S. T., Min, D. B. (1997) Ultrafiltration of xanthan gum fermentation broth: process and economic analyses, *J. Food Eng.* **31**, 219–236.

Marrs, M. (1997) Melt-in-mouth gels, *World Ingr.* June, 39–40.

MacNeely, W. H., Kang, K. S. (1973) Xanthan and some other biosynthetic gums, in: *Industrial Gums*, 2nd edn (Whistler, R. L., BeMiller, J. N., Eds.), New York: Academic Press, 473–497.

Merck & Co. Inc. (1975) Copolymère greffe d'un colloide hydrophile de *Xanthomonas* et procédé pour sa préparation, FR 2 325 666.

Merck & Co. Inc. (1975) Perfectionnement aux procédés pour accroitre la viscosité de solutions aqueuses de gomme xanthane et aux compositions obtenues, FR 2 318 926.

Merck & Co. Inc. (1978) Process for producing low calcium xanthan gum by fermentation, U.S. Patent No. 4 375 512.

Merck & Co. Inc. (1979) Production of low-calcium smooth-flow xanthan gum, EP 20 097.

Merck & Co. Inc. (1979) Procédé de préparation de gomme xanthane exempte de cellulase, FR 2 458 589.

Merck & Co. Inc. (1979) Deacetylated borate-biosynthetic gum composition, U.S. Patent No. 4 214 912.

Merck & Co. Inc. (1980) Dispersible xanthan gum composite, U.S. Patent No. 4 357 260.

Merck & Co. Inc. (1985) heteropolysaccharide and its production and use, EP 209 277.

Merck & Co. Inc. (1986) Recombinant DNA plasmid for xanthan gum synthesis, EP 233 019.

Merck & Co. Inc. (1991) Low-ash xanthan gum, EP 511 784.

Milas, M., Rinaudo, M., Knipper, M., Schuppiser, J.L. (1990) Flow and viscoelastic properties of xanthan gum solutions, *Macromolecules* **23**, 2506–2511.

Mobil Oil Corp. (1966) Waterflood process employing thickened water, U.S. Patent No. 3 373 810.

Mobil Oil Corp. (1973) Method of clarifying polysaccharide solutions, U.S. Patent No. 3 711 462.

Moorhouse, R., Walkinshaw, M. D., Arnott, S. (1977) Xanthan gum. Molecular conformation and interactions, in: *ACS Symposium Series 45, Extracellular Microbial Polysaccharides* (Sandford, P. A., Laskin, A., Eds.), Washington, DC: American Chemical Society, 90–102.

Moreno, J., Lopez, M. J., Vargas Garcia, C., Vasquez, R. (1998) Use of agricultural wastes for xanthan production by *Xanthomonas campestris*, *J. Ind. Microbiol. Biotechnol.* **21**, 242–246.

Morris, V. J. (1990) Science, structure and applications of microbial polysaccharides, in: *Gums and Stabilisers for the Food Industry* 5 (Phillips, G. O., Wedlock, D. J., Williams, P. A., Eds.), New York: IRL Press, 315–328.

Morris, V. J., Kirby, A. R., Gunning, A. P. (1999) Using atomic force microscopy (AFM) to probe food biopolymer functionality, *Scanning* **21**, 287–292.

Nankai, H., Hashimoto, W., Miki, H., Kawai, S., Murata, K. (1999) Microbial system for polysaccharide depolymerisation: enzymatic route for xanthan depolymerisation by *Bacillus* sp. strain GL1, *Appl. Environ. Microbiol.* **65**, 2520–2526.

Nussinovitch, A. (Ed.) (1997) Xanthan gum, in: *Hydrocolloids Applications: Gum Technology in the Food and Other Industries*, London: Blackie Academic & Professional, 154–168.

Parker, A., Michel, R., Vigouroux, F., Reed, W. F. (1999) Dissolution kinetics of polymer powders, *Polym. Prep. Amer. Chem. Soc. Div. Polym. Chem.* **40**, 685–686.

Parlin, S. (1997) Good vibrations. New scalable vibrating membrane filter system separates liquids, solids, *Food Process* **58**, 106–107.

Pfizer Inc. (1979) Fermentation process for production of xanthan, U.S. Patent No. 4 282 321.

Pfizer Inc. (1982) Continuous production of *Xanthomonas* biopolymer, EP 115 154.

Philipon, P. (1997) Des médicaments libérés sur commande, *Biofutur* **171**, 25–27.

Philips Petroleum Co. (1971) Recovery of microbial polysaccharide, U.S. Patent No. 3 773 752.

Pollock, T. J., Mikolajczak, M., Yamazaki, M., Thorme, L., Armentrout, R. W. (1997) Production of xanthan gum by *Sphingomonas* bacteria carrying genes from *Xanthomonas campestris*, *J. Ind. Microbiol. Biotechnol.* **19**, 92–97.

Ramirez, M. E., Fucikovsky, L., Garcia-Jimenez, F., Quintero, R., Galindo, E. (1988) Xanthan gum production by altered pathogenicity variants of *Xanthomonas campestris*, *Appl. Microbiol. Biotechnol.* **29**, 5–10.

Rhodia Inc. (1997) Use of liquid carbohydrate fermentation product in foods, WO 99/25 208.

Rhone-Poulenc Chimie (1987) Procédé de production de polysaccharides, FR 2 624 135.

Rhone-Poulenc Chimie (1988) Procédé de production de polysaccharides par fermentation d'une source carbonée à l'aide de microorganismes, EP 365 390.

Rhone-Poulenc Industries (1975) Nouvelles compositions épaississantes à base d'hétéropolysaccharides, FR 2 299 366.

Rhone-Poulenc Industries (1976) Procédé de production de polysaccharides par fermentation, FR 2 342 339.

Rhone-Poulenc Industries (1978) Procédé de production de polysaccharides par fermentation, FR 2 414 555.

Rhone-Poulenc Specialités Chimiques (1984) Procédé de production de polysaccharides de type xanthane, EP 187 092.

Rinaudo, M., Milas, M. (1980) Enzymic hydrolysis of the bacterial polysaccharide xanthan by cellulase, *Int. J. Biol. Macromol.* **2**, 45–48.

Rodd, A. B., Dunstan, D. E., Boger, D. V. (2000) Characterisation of xanthan gum solutions using dynamic light scattering and rheology, *Carbohydr. Polym.* **42**, 159–174.

Rodriguez, H., Aguilar, L. (1997) Detection of *Xanthomonas campestris* mutants with increased xanthan production, *J. Ind. Microbiol. Biotechnol.* **18**, 232–234.

Ruijssenaars, H. J., De Bont, J. A. M., Hartmans, S. (1999) A pyruvated mannose-specific xanthan lyase involved in xanthan degradation by *Paenibacillus alginolyticus* XL-1, *Appl. Environ. Microbiol.* **65**, 2446–2452.

Salcedo, G., Ramirez, M. E., Flores, C., Galindo, E. (1992) Preservation of *Xanthomonas campestris* in *Brassica oleracea* seeds, *Appl. Microbiol. Biotechnol.* **37**, 723–727

Sanofi Elf Bio-Industries (1986) Procédé d'obtention d'un xanthane à fort pourvoir épaississant et application de ce xanthane, FR 2 606 423.

Sanofi (1993) Stabilizer composition enabling the production of a pourable aerated dairy dessert, EP 649 599.

Sanofi, Méro Rousselot Satia S.A. (1987) Procédé de fermentation pour l'obtention d'une polysaccharide de type xanthane, EP 296 965.

Sanofi, Société Nationale Elf Aquitaine (1990) Souche mutante de *Xanthomonas campestris*, procédé d'obtention de xanthane et xanthane non visqueux, FR 2 671 097.

Schumpe, A., Diedrichs, S., Hesselink, P. G. M., Nene, S., Deckwer, W. D. (1991) Xanthan production in emulsions, *Proceedings of the Second*

International Symposium on Biochemical Engineering, Stuttgart, 196–199.
Serrano Carreon, L., Corona, R. M., Sanchez, A., Galindo, E. (1998) Prediction of xanthan fermentation development by a model linking kinetics, power drawn and mixing, *Proc. Biochem.* **33**, 133–146.
Shell International Research Maatschappij B.V. (1983) Process for preparing *Xanthomonas* heteropolysaccharide, heteropolysaccharide as prepared by the latter process and its use, EP 130 647.
Shell International Research Maatschappij B.V. (1984) Method for improving the filterability broth and the use, EP 184 882.
Shell International Research Maatschappij B.V. (1980) Treatment of polysaccharide solutions, EP 49 012.
Silman, R. W., Rogovin, P. (1972) Continuous fermentation to produce xanthan biopolymer: effect of dilution rate, *Biotechnol. Bioeng.* **14**, 23–31.
Sloneker, J. H., Orentas, D. G. (1962) Exocellular bacterial polysaccharide from *Xanthomonas campestris* NRRL B61459. II – Linkage of the pyruvic acid, *Can. J. Chem.* **40**, 2188–2189.
Société des Usines Chimiques Rhone-Poulenc (1973) Perfectionnement à la production de polysaccharides par fermentation, FR 2 251 620.
Standard Oil Co. (1980) *Xanthomonas campestris* ATCC 31601 and process for use, EP 46 007.
Standard Oil Co. (1980) Method for improving xanthan yield, U.S. Patent No. 4 301 247.
Standard Oil Co. (1980) Method for improving specific xanthan productivity during continuous fermentation, U.S. Patent No. 4 311 796.
Standard Oil Co. (1980) Production of xanthan gum from a chemically defined medium introduction, U.S. Patent No. 4 374 929.
Standard Oil Co. (1980) Process for using *Xanthomonas campestris* NRRL B-12075 and NRRL B-12074 for making heteropolysaccharide, U.S. Patent No. 4 400 467.
Standard Oil Co. (1980) *Xanthomonas campestris* ATCC 31602 and process for use, U.S. Patent No. 4 407 950.
Standard Oil Co. (1980) *Xanthomonas campestris* ATCC 31600 and process for use, U.S. Patent No. 4 407 951.
Standard Oil Co. (1981) Semi-continuous method for production of xanthan gum using *Xanthomonas campestris* ATCC 31601, U.S. Patent No. 4 328 310.
Standard Oil Co. (1981) Method of producing a low viscosity xanthan gum, U.S. Patent No. 4 377 637.
Stauffer Chemical Co. (1981) Fermentation of whey to produce a thickening polymer, U.S. Patent No. 4 444 792.
Stokke, B. J., Christensen, B. E., Smidsrød, O. (1998) Macromolecular properties of xanthan, in: *Polysaccharides. Structural, Diversity and Functional Versatility* (Dumitriu, S., Ed.), New York: Marcel Dekker, 433–472.
Stredansky, M., Conti, E. (1999) Xanthan production by solid state fermentation, *Process Biochem.* **34**, 581–587.
Stredansky, M., Conti, E., Navarini, L., Bertocchi, C. (1999) Production of bacterial exopolysaccharides by solid substrate fermentation, *Process Biochem.* **34**, 11–16.
Sutherland, I. W. (1993) Xanthan, in: *Xanthomonas* (Swings, J. G., Civerolo, E. L., Eds.), London: Chapman & Hall, 363–388.
Tait, M. I., Sutherland, I. W. (1989) Synthesis and properties of a mutant type of xanthan, *J. Appl. Bacteriol.* **66**, 457–460.
Tait, M. I., Sutherland, I. W., Clarke-Sturman, A. J. (1990) Acid hydrolysis and high-performance liquid chromatography of xanthan, *Carbohydr. Polym.* **13**, 133–148.
Tang, J. L., Gough, C. L., Daniels, M. J. (1990) Cloning of genes involved in negative regulation of production of extracellular enzymes and polysaccharide of *Xanthomonas campestris* pathovar *campestris*, *Mol. Gen. Genet.* **222**, 157–160.
Tilly, G. (1991) Stabilization of dairy products by hydrocolloids, in: *Food Ingredients Europe: Conference Proceedings* (Van Zeijst, R., Ed.) Maarsen: Expoconsult Publishers, 105–121.
Triveni, R., Shamala, T. R., (1999) Clarification of xanthan gum with extracellular enzymes secreted by *Trichoderma koningii*, *Process Biochem.* **34**, 49–53.
Tseng, Y. H., Choy, K. T., Hung, C. H., Lin, N. T., Liu, J. Y., Lou, C. H., Yang, B. Y., Wen, F. S., Wu, J. R. (1999) Chromosome map of *Xanthomonas campestris* pv. *campestris 17* with locations of genes involved in xanthan gum synthesis and yellow pigmentation, *J. Bacteriol.* **181**, 117–125.
U. S. Secretary Agriculture (1959) Method of producing an atypically salt-responsive alkali-deacetylated polysaccharide, U.S. Patent No. 3 000 790.
U. S. Secretary Agriculture (1962) Method of recovering microbial polysaccharides from their fermentation broths, U.S. Patent No. 3 119 812.
U. S. Secretary Agriculture (1967) Continuous process for producing *Xanthomonas* heteropolysaccharide, U.S. Patent No. 3 485 719.

U. S. Secretary Agriculture (1978) Production of high-pyruvate xanthan gum on synthetic medium, U.S. Patent No. 4 394 447.

Vafiadis, D. (1999) Anti-shock treatment (frozen dairy products formulation), *Dairy Field* **182**, 85–88.

Vanderslice, R. W., Doherty, D. H., Capage, M. A., Betlach, M. R., Hassler, R. A., Henderson, N. M., Ryan-Graniero, J., Tecklenburg, M. (1989) Genetic engineering of polysaccharide structure in *Xanthomonas campestris*, in: *Biomedical and Biotechnological Advances in Industrial Polysaccharides* (Crescenzi, V., Dea, I. C. M., Paoletti, S., Stivala, S. S., Sutherland, I. W., Eds.), New York: Gordon and Breach, 145–156.

Vidhyasekaran, P., Alvenda, M. E., Mew, T. W. (1989) Physiological changes in rice seedlings induced by extracellular polysaccharide produced by *Xanthomonas campestris* pv. *oryzae*, *Physiol. Mol. Plant Pathol.* **35**, 391–402.

Vojnov, A. A., Zorreguieta, A., Dow, J. M., Daniels, M. J., Dankert, M. A. (1998) Evidence for a role for the gumB and gumC gene products in the formation of xanthan from its pentasaccharide repeating unit by *Xanthomonas campestris*, *Microbiology* **144**, 1487–1493.

Watabe, M., Yamaguchi, M., Kitamura, S., Horino, O. (1993) Immunohistochemical studies on localization of the extracellular polysaccharide produced by *Xanthomonas oryzae* pv. *oryzae* in infected rice leaves, *Can. J. Microbiol.* **39**, 1120–1126.

Yang, B. Y., Tseng, Y. H. (1988) Production of exopolysaccharide and levels of protease and pectinase activity in pathogenic and non-pathogenic strains of *Xanthomonas campestris* pv. *campestris*, *Bot. Bull. Academia Sinica* **29**, 93–99.

Yang, S. T., Lo, Y. M., Min, D. B. (1996) Xanthan gum fermentation by *Xanthomonas campestris* immobilized in a novel centrifugal fibrous-bed bioreactor, *Biotechnol. Prog.* **12**, 630–637.

Yang, S. T., Lo, Y. M., Chattopadhyay, D. (1998) Production of cell-free xanthan fermentation broth by cell adsorption on fibers, *Biotechnol. Prog.* **14**, 259–264.

Yoo, S. D., Harcum, S. W. (1999) Xanthan gum production from paste waste sugar beet pulp, *Bioresource Technol.* **70**, 105–109.

Zhao, X. M., Li, X. H., Ban, R., Zhu, Y. (1997) The effects of mixing modes on recovery of xanthan gum by precipitation, *BHR Group Conf. Ser. Publ.* **25**, 3–8.

12
Dextran

Dr. Timothy D. Leathers
Fermentation Biochemistry Research Unit, National Center for Agricultural Utilization Research, Agricultural Research Service, United States Department of Agriculture;1815 N. University St.; Peoria, IL, 61604, USA; Tel. +1-309-681-6377; Fax: +1-309-681-6427; E-mail: leathetd@ncaur.usda.gov

ATP	adenosine 5′-triphosphate
CM	carboxymethyl
DEAE	diethylaminoethyl
EC	enzyme commission
FTIR	fourier-transform infrared
GRAS	Generally Regarded as Safe
HIV	human immunodeficiency virus
IU	international units
K_m	Michaelis-Menten constant
MRI	magnetic resonance imaging
NMR	nuclear magnetic resonance
QAE	diethyl(2-hydroxypropyl)aminoethyl
SP	sulfopropyl

Names are necessary to report factually on available data; however, the USDA neither guarantees nor warrants the standard of the product, and the use of the name by USDA implies no approval of the product to the exclusion of others that also may be suitable.

1 Introduction

Dextrans are defined as homopolysaccharides of glucose that feature a substantial number of consecutive α-(1→6) linkages in their major chains, usually more than 50% of total linkages. These α-D-glucans also possess side chains stemming from α-(1→2), α-(1→3), or α-(1→4) branch linkages. The exact structure of each type of dextran depends on its specific microbial strain of origin. Dextrans are produced by certain lactic acid bacteria, particularly strains of *Streptococcus* species and *Leuconostoc mesenteroides*. Dextrans from oral *Streptococcus* species are of clinical interest as components of dental plaque. Dextrans from *L. mesenteroides* are of commercial interest, primarily as specialty chemicals for clinical, pharmaceutical, research, and industrial uses. Numerous reviews have appeared on dextrans, including those by Evans and Hibbert (1946), Neely (1960), Jeanes (1966, 1978), Murphy and Whistler (1973), Sidebotham (1974), Walker (1978), Alsop (1983), Robyt (1986, 1992, 1995), de Belder (1990, 1993), and Cote and Ahlgren (1995).

2 Historical Outline

Because dextrans are formed from sucrose, they have long been known as troublesome contaminants of food products and sugar refineries. Pasteur made an early applied study of dextran formation in wine, proving that this phenomenon was caused by microbial activity (Pasteur, 1861). Scheibler (1874) determined that dextran was a carbohydrate of the empirical formula $(C_6H_{10}O_6)_n$ having a positive optical rotation and thus coined the term “dextran”. Van Tieghem (1878) identified a dextran-forming bacterium and named it *Leuconostoc mesenteroides*. Beijerinck (1912) investigated the phenomenon, and Hehre (1941) demonstrated dextran synthesis by a cell-free culture filtrate. Scientific interest in dextrans was stimulated by studies suggesting its value as a blood-plasma volume expander (Gronwall and Ingelman, 1945, 1948). In the late 1940s, an extensive research program on dextrans was initiated at the Northern Regional Research Laboratory (now the National Center for Agricultural Utilization Research) of the Agricultural Research Service, U.S. Department of Agriculture, in Peoria, Illinois. Numerous dextran-producing strains were characterized, including *L. mesenteroides* strain NRRL B-512F used today for commercial dextran production in North America and Western Europe.

3 Chemical Structure

Within the general definition of dextran as a glucan in which α-(1 → 6) linkages predominate, chemical structures vary considerably as a function of the specific microbial strain of origin. The article of commerce is the product of a single strain of *L. mesenteroides*, NRRL B-512F. As shown in Figure 1, dextran from this strain features α-(1 → 6) linkages in the main chains with a relatively low level (about 5%) of α-(1 → 3) branch linkages (Van Cleve et al., 1956; Jeanes et al., 1954; Slodki et al., 1986). Larm et al. (1971) estimated that 40% of these side chains are one subunit long and 45% are two subunits long. The remaining side chains are probably greater than 30 subunits long, and branches appear to be distributed randomly (Bovery, 1959; Covacevich and Richards, 1977; Taylor et al., 1985; Kuge et al., 1987).

Dextrans are produced by numerous additional strains of bacteria, and the structures of these dextrans are diverse. In the classic study of Jeanes et al. (1954), 96 strains of *Leuconostoc* and *Streptococcus* were surveyed for the formation of polysaccharides from sucrose. In this study it was reported that α-(1 → 6) linkages in dextran varied from 50% to 97% of total linkages. The balance represented α-(1 → 2), α-(1 → 3), or α-(1 → 4) linkages, usually at branch points. Several isolates produced more than one type of polysaccharide, which were named on the basis of their greater (fraction *S*, for soluble) or lesser (fraction *L*) degree of solubility in water–ethanol mixtures.

Leuconostoc citreum strain NRRL B-742 (formerly *L. mesenteroides*; Takahashi et al., 1992) produces a fraction *L* dextran that contains approximately 15% α-(1 → 4) branch linkages and a fraction *S* dextran that contains a variable (30% to 45%) percentage of α-(1 → 3) branches (Jeanes et al., 1954; Seymour et al., 1979a; Côté and Robyt, 1983; Slodki et al., 1986). Dextran from *L. mesenteroides* strain NRRL B-1299 has a high percentage (27% to 35%) of α-(1 → 2) linked single-glucose branches (Jeanes et al., 1954; Kobayashi and Matsuda, 1977; Slodki et al., 1986). Soluble dextran produced by *Streptococcus sobrinus* strain 6715 (formerly *Streptococcus mutans* strain 6715) appears to be structurally similar to the *S* dextran from *L. citreum* strain NRRL B-742 in that it is an α-(1 → 6) glucan with a relatively high percentage of α-(1 → 3) branch linkages (Shimamura et al., 1982). Many strains of oral *Streptococcus* species, and some strains of *L. mesenteroides*, also make α-D-glucans containing linear sequences of consecutive α-(1 → 3) linkages. These low solubility glucans were once proposed to be "Class 3″ dextrans (Seymour and Knapp, 1980) but now are considered to be forms of mutan, an important component of dental plaque.

L. mesenteroides strains NRRL B-1355, NRLL B-1498, and NRRL B-1501 produce an *S* fraction glucan with a unique backbone structure of regularly alternating α-(1 → 3) and α-(1 → 6) linkages (Côté and Robyt, 1982; Misaki et al., 1980; Seymour and

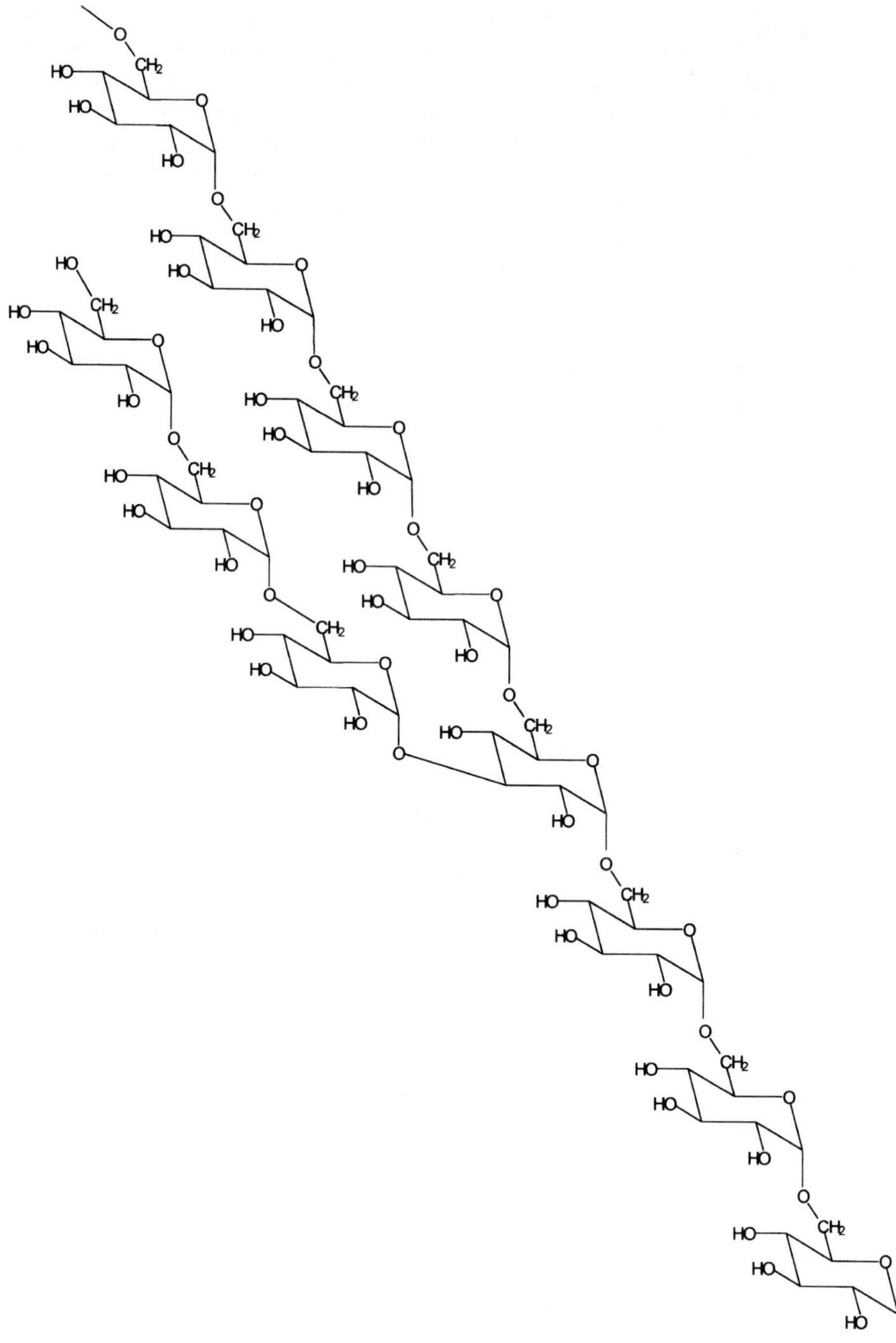

Fig. 1 Chemical structure of a representative portion of dextran from *Leuconostoc mesenteroides* strain NRRL B-512F. Figure courtesy of Dr. Gregory L. Côté.

Knapp, 1980). Because this polysaccharide does not contain significant regions of contiguous α-(1→6) linkages, it is not considered to be a true dextran and has been named alternan (Côté and Robyt, 1982). Alternan has distinctive properties of significant basic and applied interest, and it is the subject of Chapter 13, this volume. A mutant derivative of strain NRRL B-1355 recently was reported to produce a third polysaccharide, an insoluble α-D-glucan containing linear (1→3) and (1→6) linkages with (1→2) and (1→3) branch points (Smith et al., 1998; Côté et al., 1999).

4 Physiological Function

Oral *Streptococcus* species produce both dextran and mutan, an insoluble glucan in which α-(1→3) linkages predominate. These glucans comprise the matrix of dental plaque, which provides an environment for the proliferation of these bacteria (Hamada and Slade, 1980; Shimamura et al., 1982; Loesche, 1986). The physiological function of dextrans produced by *L. mesenteroides* is unknown. Since dextran-producing bacteria do not break down the polymers, dextrans presumably do not serve as storage materials. It is possible that these polysaccharides serve to protect cells from dessication or help them adhere to environmental substrates.

5 Chemical Analyses

General methods applicable to polysaccharides and polyglucans may be used to detect and quantitate dextrans. More specifically, dextrans show high positive specific optical rotation and exhibit characteristic infrared absorption bands at about 917, 840, and 768 cm^{-1} (Barker et al., 1956; Jeanes, 1966). Since dextrans are homopolysaccharides of glucose subunits, many studies have focused on the analysis of specific linkage patterns. Methods have included periodate analysis (Rankin and Jeanes, 1954; Dimler et al., 1955), methylation analysis (Van Cleve et al., 1956; Lindberg and Svensson, 1968; Seymour et al., 1977; Jeanes and Seymour, 1979; Seymour et al., 1979a; Slodki et al., 1986), nuclear magnetic resonance (NMR) spectroscopy (Seymour et al., 1976; Seymour et al., 1979b; Cheetham et al., 1991), and Fourier-transform infrared (FTIR) spectroscopy (Seymour and Julian, 1979). Enzymes that attack dextran in a specific fashion also have been exploited to reveal useful structural information (Covacevich and Richards, 1977; Sawai et al., 1978; Taylor et al., 1985; Pearce et al., 1990).

6 Occurrence

Several lactic acid bacteria have been reported to produce dextrans, principally including *Streptococcus* species and *L. mesenteroides* (Jeanes, 1966; Sidebotham, 1974; Cerning, 1990). *Leuconostoc* and *Streptococcus* are related genera, both composed of Gram-positive, facultatively anaerobic cocci. *L. mesenteroides* (incorporating as subspecies the former species *L. dextranicum* and *L. cremoris*) generally is found on plant materials, particularly on mature or harvested crops, and often plays a role in spoilage (Holzapfel and Schillinger, 1992; Stiles and Holzapfel, 1997). Because sucrose is the natural substrate for dextran synthesis, contamination problems are most evident in food products containing this sugar. *L. mesenteroides* can cause significant problems in sugar (particularly cane) refineries, where dextrans can clog filters and inhibit sugar crystallization (Jeanes, 1977). *Leuconostoc* species are considered Generally Regarded as Safe (GRAS) organisms because of their common appearance in natural fermented foods. In fact, certain strains are valued as starter cultures for buttermilk, cheese, and other dairy products (Holzapfel and Schillinger, 1992). *L. mesenteroides* also plays a role in the production of sauerkraut and other fermented vegetables. Recently, a strain of *L. mesenteroides* that produces both dextran and the related glucan alternan was isolated from the traditional fermented food Kim-Chi (Jung et al., 1999). Oral *Streptococcus* species produce dextran and the insoluble glucan mutan as components of dental plaque (Hamada and Slade, 1980; Shimamura et al., 1982). In addition, certain strains of the ubiquitous Gram-negative bacterium *Gluconobacter oxydans* (formerly *Acetobacter capsulatus*) produce dextran from starch-derived dextrins (Hehre and Hamilton, 1951; Kooi, 1958; Yamamoto et al., 1993a,b; Mountzouris et al., 1999).

7 Biosynthesis

Dextrans are produced extracellularly by secreted enzymes commonly referred to as glucansucrases or, more specifically, dextransucrases. These enzymes are glycosyltransferases (EC 2.4.1.5) that catalyze the transfer of D-glucopyranosyl subunits from sucrose to dextrans. Fructose is released and consumed by growing cells, if they are present. No adenosine 5′-triphosphate (ATP) or cofactors are required for these reactions, as the enzymes utilize energy available in the glycosidic bond between glucose and fructose. Glucansucrase synthesis in wild-type strains of *L. mesenteroides* is induced by growth on sucrose, while *Streptococcus* species produce this enzyme constitutively. Dextransucrase production by *L. mesenteroides* strain NRRL B-512F appears to be regulated at the transcriptional level (Quirasco et al., 1999).

Characterizations of dextransucrases have been complicated by the appearance of multiple enzyme species in culture fluids. Many of these species appear to be proteolytic-processing or degradation products, although some retain activity (Sanchez-Gonzalez et al., 1999). However, strains that produce more than one type of glucan appear to produce a separate glucansucrase for each. Furthermore, enzymes may exist as aggregates and typically are associated with their polysaccharide products, making purifications difficult. Despite these problems, a number of glucansucrases have been studied. Dextransucrase from *L. mesenteroides* strain NRRL B-512F has been purified and characterized (Robyt and Walseth, 1979; Kobayashi and Matsuda, 1980; Paul et al., 1984; Miller et al., 1986; Fu and Robyt, 1990; Kitaoka and Robyt, 1998a; Kim and Kim, 1999). The enzyme appears to have an initial molecular mass of 170 kDa, a pI value of 4.1, and a Michaelis-Menten constant (K_m) for sucrose of approximately 12 to 16 mM. Optimal reaction conditions are pH 5.0 to 5.5 and 30 °C. Kim and Kim (1999) reported a specific activity of up to 250 IU mg^{-1} protein for highly purified enzyme. Low levels of calcium are necessary for optimal enzyme production and activity. Various forms of dextransucrase have been described from *L. mesenteroides* strain NRRL B-1299, with molecular masses of 48 to 79 kDa, temperature optima of 35 °C to 45 °C, pH optima of 5.0 to 6.5, and K_m values for sucrose of 13 to 30 mM (Kobayashi and Matsuda, 1975, 1976; Dols et al., 1997). Multiple forms of dextransucrase have been described from *Streptococcus* species, having molecular masses of 94 to 170 kDa, pI values of approximately 4.0, temperature optima of

34 °C to 42 °C, pH optima of 5.0 to 5.7, and K_m values for sucrose of 2 to 9 mM (Chludzinski et al., 1974; Fukui et al., 1974; Shimamura et al., 1982; Furuta et al., 1985; McCabe, 1985). In addition, dextran-dextrinase (also called dextrin dextranase) has been purified and characterized from *G. oxydans* strain ATTC 11894 (Yamamoto et al., 1992; Suzuki et al., 1999).

Robyt and colleagues have developed a model for the reaction mechanism of dextransucrase from *L. mesenteroides* strain NRRL B-512F (Robyt et al., 1974; Robyt, 1992, 1995; Su and Robyt, 1994). According to this model, two nucleophilic reaction sites exist in the catalytic domain of the enzyme. Sucrose is hydrolyzed at one or both sites, and the glucosyl residues are bound covalently to the enzyme in high-energy bonds conserved from sucrose. A dextran chain grows by successive glucosyl insertions between the enzyme and the reducing end of the chain, which remains bound to the enzyme. Branches are formed when glucosyl units or dextran chains are transferred to secondary hydroxyl positions on the dextran chains (Robyt and Taniguchi, 1976). Termination of chain extension occurs by transfer to an acceptor molecule.

Sucrose is strongly preferred as the glucosyl donor, although other natural and synthetic donors have been identified (Hehre and Suzuki, 1966; Binder and Robyt, 1983). However, a number of sugars and derivatives may function as alternative acceptors, including maltose, isomaltose, nigerose, α-methyl glucoside, and others (Robyt and Taniguchi, 1976; Robyt and Walseth, 1978; Robyt and Eklund, 1983; Fu et al., 1990). These acceptor reactions can be utilized to produce dextrans of lower average molecular weights, including clinical dextrans (Koepsell et al., 1955; Tsuchiya et al., 1955; Remaud et al., 1991; Robyt, 1992) and oligosaccharides of interest (Pelenc et al., 1991; Remaud et al., 1992; Remaud-Simeon et al., 1994; Dols et al., 1999). Maltose, the most effective alternative acceptor, accepts a glucosyl residue to form the trisaccharide panose (Killey et al., 1955; Heincke et al., 1999). The acceptor reaction with fructose, the natural co-product of dextran synthesis, has been studied for production of the disaccharide leucrose, a potential alternative sweetener and substrate for industrial conversions (Stodola et al., 1956; Swengers, 1991; Reh et al., 1996; Heincke et al., 1999). A minor product of this reaction is isomaltulose, also known as palatinose, likewise of interest as an alternative sweetener (Sharpe et al., 1960; Takazoe, 1989).

Robyt and Martin (1983) found evidence that a similar reaction mechanism exists for glucansucrases from *S. sobrinus* strain 6715. Alternative models for the glucansucrase reaction mechanism have been reviewed (Monchois et al., 1999). Because the glucansucrases from *Leuconostoc* and *Streptococcus* species appear to be closely related on a molecular level, it seems likely that they share a common reaction mechanism. If so, differences among the polymer structures might be determined by subtle differences in the stereochemistry of the reaction sites.

8 Genetics and Molecular Biology

L. mesenteroides strain NRRL B-512F, used for commercial production of dextran, has been described as a laboratory "substrain" that supplanted natural isolate NRRL B-512 in 1950 (Van Cleve et al., 1956). A dextransucrase hyperproducer mutant of NRRL B-512F was isolated as NRRL B-512FM (Miller and Robyt, 1984). Wild-type strains of *L. mesenteroides* form glucansucrases only when cultured on sucrose, and further mutations were obtained that allowed strain

B-512FMC to produce dextransucrase constitutively (Kim and Robyt, 1994). Further improvements in enzyme productivity were obtained through additional rounds of mutagenesis (Kim et al., 1997; Kitaoka and Robyt, 1998b). Although commercial dextran currently is produced by a fermentative process, such strains would be particularly valuable for dextran production by an enzymatic process. Similar glucansucrase mutants have been obtained for other strains of *Leuconostoc,* including NRRL B-742, NRRL B-1142, NRRL B-1299, and NRRL B-1355 (Kim and Robyt, 1994, 1995a,b, 1996; Kitaoka and Robyt, 1998b). Dextransucrase-deficient mutants of strain NRRL B-1355 also have been isolated for improved production of alternan (Smith et al., 1994; Leathers et al., 1995, 1997, 1998). Alternan is the subject of Chapter 13, this volume.

Because of clinical interest in developing anti-caries vaccines, a number of glucansucrase genes have been cloned from oral *Streptococcus* species (Shiroza et al., 1987; Ueda et al., 1988; Honda et al., 1990). Glucansucrase genes have been cloned and sequenced from *L. mesenteroides* NRRL B-512F (Wilke-Douglas et al., 1989; Bhatnagar and Singh, 1999; Arguello-Morales et al., 2000a; Funane et al., 2000; Ryu et al., 2000), *L. mesenteroides* strain NRRL B-1299 (Monchois et al., 1996, 1998), *L. citreum* strain NRRL B-742 (Kim et al., 2000), and *L. mesenteroides* strain NRRL B-1355 (Arguello-Morales et al., 2000b; Kossman et al., 2000). Interestingly, some of these genes apparently do not specify enzymes normally secreted *in vivo*. Glucansucrase genes appear to be closely related and exhibit a common organizational structure, with a conserved N-terminal catalytic domain and a C-terminal glucan-binding domain that contains a series of direct tandem repeat sequences (Monchois et al., 1999; Remaud-Simeon et al., 2000). Based on site-directed mutageneses and consensus sequences, potentially important catalytic sites have been proposed (Monchois et al., 1997; Arguello-Morales, 2000b; Monchois et al., 2000; Remaud-Simeon et al., 2000). On a broader scale, glucansucrases resemble enzymes in glycosyl hydrolase family 13, which includes α-amylases (Fujiwara et al., 1998; Janecek et al., 2000; Remaud-Simeon et al., 2000).

9
Biodegradation

A variety of fungi produce dextranases, including *Aspergillus* species (Carlson and Carlson, 1955b; Hiraoka et al., 1972), *Chaetomium gracile* (Hattori et al., 1981), *Fusarium* species (Simonson and Liberta, 1975; Shimizu et al., 1998), *Lipomyces starkeyi* (Webb and Spencer-Martins, 1983; Koenig and Day, 1988), *Paecilomyces lilacinus* (Lee and Fox, 1985; Sun et al., 1988; Galvez-Mariscal and Lopez-Munguia, 1991), and *Penicillium* species (Tsuchiya et al., 1956; Chaiet et al., 1970). These enzymes are endodextranases with specificity for internal α-(1 → 6) linkages, and they produce mainly isomaltose or isomaltotriose from dextran. Dextranases from *C. gracile* and *Penicillium* sp. are produced commercially and used for treatment of dextran contamination problems in sugar processing (Godfrey, 1983). Endodextranases have shown potential for the enzymatic production of specific molecular weight fractions of dextran (Carlson and Carlson, 1955a; Corman and Tsuchiya, 1957; Novak and Stoycos, 1958; Day and Kim, 1992; Kim and Day, 1995; Kim and Robyt, 1996). These enzymes also have been tested for the treatment of dental plaque (Fitzgerald et al., 1968; Caldwell et al., 1971), although the more highly branched dextrans are far less susceptible to endodextranase digestion. Limit endodextranase digestion of the branched dextran from *L. citreum* strain NRRL B-742 produces an interesting branched fraction with rheological characteristics similar to those of polydextrose (Cote et al., 1997).

Dextranases also have been reported from a number of bacteria. *Arthrobacter globiformis* produces isomaltodextranase, an exodextranase that successively releases isomaltose from the non-reducing ends of dextrans and oligosaccharides (Torii et al., 1976; Okada et al., 1988). This enzyme recognizes not only α-(1$\rightarrow$6) linkages but also α-(1$\rightarrow$2), α-(1$\rightarrow$3), and α-(1$\rightarrow$4) linkages. Unlike endodextranases, isomaltodextranase is able to partially hydrolyze alternan, producing an interesting limit alternan (Sawai et al., 1978; Cote, 1992). An isomaltodextranase from *Actinomadura* sp. exhibits slightly different specificities (Sawai et al., 1981). A dextran α-(1$\rightarrow$2) debranching enzyme also has been described from a *Flavobacterium* sp. (Mitsuishi et al., 1979). Dextran from *L. mesenteroides* strain NRRL B-512F is degraded by intestinal bacteria and enzymes in mammalian tissues other than blood (Sery and Hehre, 1956; Fischer and Stein, 1960). Intravenously administered clinical dextrans are metabolized slowly and completely in the body.

10 Production

To date, commercial production of dextran has employed primarily simple batch fermentation methods, using live cultures grown on sucrose. Methods and conditions for dextran fermentation have been detailed (Tarr and Hibbert, 1931; Hehre et al., 1959; Jeanes, 1965b, 1966; Alsop, 1983; de Belder, 1993). *L. mesenteroides* is a fastidious organism, and its special nutritional requirements include glutamic acid, valine, biotin, nicotinic acid, thiamine, and pantothenic acid (Holzapfel and Schillinger, 1992). In dextran production, these needs are met by combinations of complex medium components, such as yeast extract, corn steep liquor, casamino acids, malt extract, peptone, and tryptone. Sucrose serves as a carbon source, inducer of dextransucrase, and substrate for dextran production. Low levels of calcium (e.g., 0.005%) are necessary for optimal enzyme and dextran yields, and other basal salts, including a source of phosphate, complete the medium. Operative production factors include initial pH (typically pH 6.7 to 7.2), temperature (about 25 °C), initial sucrose concentration (usually 2%), and time (usually 24 to 48 h). Dextran branching appears to increase at elevated temperatures (Sabatie et al., 1988). High levels of sucrose (10% to 50%) reduce the yield of high-molecular-weight dextran, and this observation has been exploited for the production of intermediate sized dextran (Tsuchiya et al., 1955; Alsop, 1983). The organism is facultatively anaerobic or microaerophilic, and fermentations are not aerated. During the first 20 h of fermentation, culture pH falls to approximately 5.0 because of the formation of organic acids, favorably near the optimal pH of dextran sucrase. Dextran may be recovered by precipitation with solvents, particularly alcohols (Hehre et al., 1959; Jeanes, 1965b).

It has long been recognized that dextran also can be produced enzymatically, using cell-free culture supernatants that contain dextransucrase (Hehre, 1941; Tsuchiya and Koepsell, 1954; Hellman et al., 1955; Behrens and Ringpfeil, 1962; Jeanes, 1965a). Accordingly, improved dextransucrase production and purification methods have been developed (Lawford et al., 1979; Paul et al., 1984; Miller et al., 1986; Fu and Robyt, 1990; Kim and Kim, 1999). Glucansucrases also have been immobilized, although this approach may be most useful for production of oligosaccharides (Kaboli and Reilly, 1980; Monsan et al., 1987; Cote and Ahlgren, 1994; Reh et al., 1996; Alcalde et al., 1999). Enzymatic synthesis offers advantages of product molecular weight and quality control, as well as the benefit of obtaining fructose as a valuable co-product. However, this approach has been largely ignored for commercial production, presumably for economic reasons. Dextran production from maltodextrins, using dextran-dextrinase from *Gluconobacter oxydans*, also has attracted interest (Hehre and Hamilton, 1951; Kooi, 1958; Yamamoto et al., 1993a,b; Mountzouris et al., 1999).

Clinical dextran fractions are produced primarily by simple methods of partial acid hydrolysis followed by differential fractionation in solvents (Wolff et al., 1955; Gronwall, 1957; de Belder, 1990). Attractive alternative methods to produce these fractions include the use of dextranases (Carlson and Carlson, 1955a; Corman and Tsuchiya, 1957; Novak and Stoycos, 1958; Day and Kim, 1992; Kim and Day, 1995; Kim and Robyt, 1996) and chain-terminating acceptor reactions (Koepsell et al., 1955; Tsuchiya et al., 1955; Remaud et al., 1991; Robyt, 1992).

Dextran has been produced commercially for many years and by a number of companies, including Dextran Products, Ltd., Toronto, Canada; Pfeifer und Langen, Dormagen, Germany; Pharmachem Corp., Bethlehem, Pennsylvania, USA; and Pharmacia, Uppsala, Sweden. Annual world production was recently estimated at 2000 tons per year (Vandamme et al., 1996). The wholesale price of dextran varies, but recently it has been near 3 USD per pound.

11 Properties and Applications

Purified dextrans are white, tasteless solids. Other physical and chemical properties vary depending on the specific chemical structure, which is determined by the microbial strain of origin and method of production. Dextrans with the highest percentages of α-(1→6) linkages are generally the most soluble in water. Dextran from *L. mesenteroides* strain NRRL B-512F is freely soluble in water and other solvents, including 6 M urea, 2 M glycine, formamide, glycerol, etc. (Jeanes, 1966; de Belder, 1990). Dextran solutions behave as Newtonian fluids, and their viscosity is a function of concentration, temperature, and average molecular weight (Granath, 1958; Gekko and Noguchi, 1971; Carrasco et al., 1989). Native dextran is polydisperse and typically of high average molecular weight (generally between 10^6 and 10^9 daltons). However, many of the current applications for dextran depend on the convenience with which it can be broken down to fractions of specific weight ranges. The relative linearity of dextran from strain NRRL B-512F is crucial for the production of such fractions. Dextrans exhibit characteristic serological reactions, apparently related to their molecular weight and degree of branching (Gronwall, 1957; Kabat and Bezer, 1958; Jeanes, 1986). However, intravenously administered clinical dextrans are of relatively low antigenicity, although individuals can exhibit hypersensitivity. The pharmacological properties of clinical dextrans have been reviewed recently (de Belder, 1996). Free hydroxyl groups in dextran are potential targets for chemical derivatizations, and dextran from strain NRRL B-512F is particularly suitable for these reactions because of its low level of branch linkages.

A number of bulk chemical applications have been demonstrated for dextran, including uses in oil-drilling operations, agriculture, food products, and the manufacture of photographic films and other products (Murphy and Whistler, 1973; Alsop, 1983; Glicksman, 1983). It should be noted that dextrans are not explicitly approved as food additives in the United States or Europe, although *L. mesenteroides* is a GRAS organism commonly found in fermented foods. Currently, dextran and dextran derivatives are used primarily as specialty chemicals in clinical, pharmaceutical, research, and industrial applications (Yalpani, 1986; de Belder, 1996; Vandamme et al., 1996). Early work by Gronwall and Ingelman (1945, 1948) established the potential of using a hydrolyzed dextran fraction as a blood-plasma volume expander (Figure 2). Clinical dextrans used today are Dextran 40 and Dextran 70, which are 40,000 and 70,000 dalton average molecular weight fractions, respectively. Dextrans are less expensive than the albumins and starch derivatives also used in plasma therapies (Lilley and Aucker, 1999). These colloids essentially replace normal blood proteins in providing osmotic pressure to pull fluid from the interstitial space into the plasma. This treatment is useful to prevent shock from hemorrhage, burns, surgery, or trauma and to reduce the risk of thrombosis and embolisms. Dextran 40 also improves blood flow and inhibits the aggregation of erythrocytes (de Belder, 1996).

Iron dextran is a colloidal preparation used especially in veterinary medicine for the treatment of anemia, particularly in newborn piglets. Special iron dextran preparations also have been developed to enhance magnetic resonance imaging (MRI) techniques (de Belder, 1996). Dextran sulfate has been used as a substitute for heparin in anticoagulant therapy, and, more recently, it is being studied as an antiviral agent, particularly in the treatment of human immunodeficiency virus (HIV) (Mitsuya et al., 1988; Piret et al., 2000). Dextran can be crosslinked by epichlorohydrin (Flodin and Porath, 1961) to form beads (Sephadex®) that have become widely used in

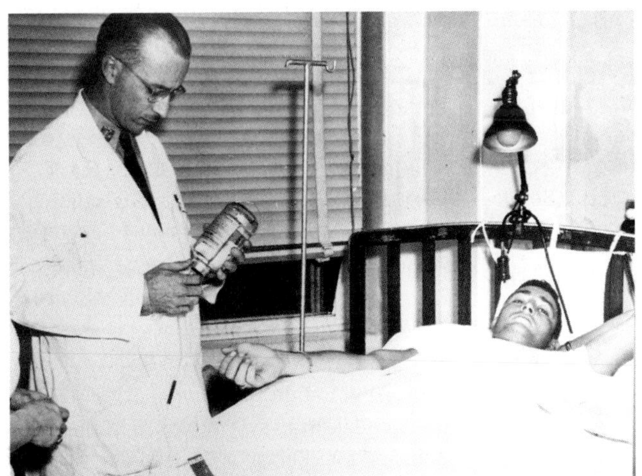

Fig. 2 Administration of dextran to a soldier at Walter Reed General Hospital, 1952. U.S. Dept. of Agriculture photograph.

research and industry for separations based on gel filtration. Anion and cation exchange resins based on Sephadex derivatives are widely used, including carboxymethyl (CM) Sephadex, diethylaminoethyl (DEAE) Sephadex, diethyl(2-hydroxypropyl)aminoethyl (QAE) Sephadex, and sulfopropyl (SP) Sephadex (Figure 3). Dextran is also an important component of many aqueous two-phase extraction systems, usually used in conjunction with polyethylene glycol (Tjerneld, 1992; Sinha et al., 2000).

Recently, oligosaccharides have received a great deal of attention as potential prebiotic compounds in food products, animal feeds, and cosmetics (Hidaka and Hirayama, 1991; Kohmoto et al., 1991; Monsan and Paul, 1995; Lamothe et al., 1996; Monsan et al., 2000). In contrast to clinical dextran preparations, prebiotic oligosaccharides must be resistant to digestion and preferentially utilized by beneficial bifidobacteria and lactic acid bacteria in the intestinal or skin microflora. Accordingly, dextran oligosaccharides of interest as prebiotics include the more branched varieties, containing α-(1 → 3) linkages from *L. citreum* strain NRRL B-742 (Remaud et al., 1992), α-(1 → 2) linkages from *L. mesenteroides* strain NRRL B-1299 (Remaud-Simeon et al., 1994; Dols et al., 1999), or the alternating α-(1 → 3) and α-(1 → 6) linkages from *L. mesenteroides* strain NRRL B-1355 (Pelenc et al., 1991).

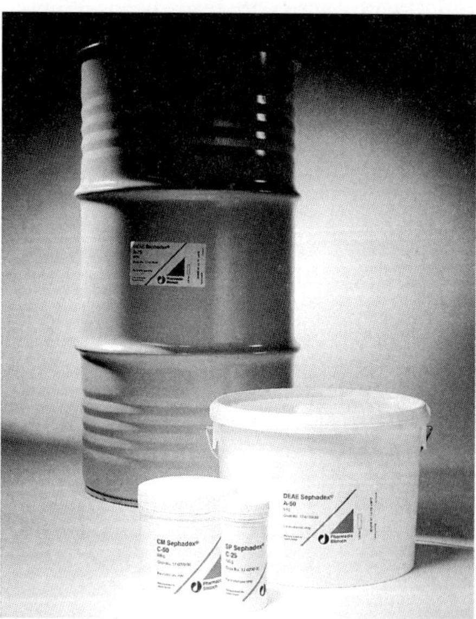

Fig. 3 Sephadex ion-exchange media are used widely for process scale applications. Photograph courtesy of Amersham Pharmacia Biotech, Inc.

12 Patents

Numerous patents claim methods for the production of dextran and dextran derivatives. The following examples, also summarized in Table 1, are illustrative. An early patent by Gronwall and Ingelman (1948) suggested that dextran might be useful as a blood-plasma volume expander. Hehre et al. (1959) described methods for dextran production by fermentation. Enzymatic production of dextran using dextransucrase also has been claimed (Tsuchiya and Koepsell, 1954; Hellman et al., 1955; Behrens and Ringpfeil, 1962). Wolff et al. (1955) described partial acid hydrolysis and fractionation of dextran for clinical applications. Alternative methods for the production of clinical dextrans include the use of dextranases (Carlson and Carlson, 1955a; Corman and Tsuchiya, 1957; Novak and Stoycos, 1958; Day and Kim, 1992) and chain-terminating acceptor reactions (Koepsell et al., 1955). Flodin and Porath (1961) described the cross-linking of dextran to form beads (Sephadex®) useful for gel filtration. Methods have been patented for the production of therapeutic iron dextrans (London and Twigg, 1958; Herb, 1979) and dextran sulfate (Morii et al., 1964; Usher, 1989). Recently, dextran oligosaccharides have garnered interest as potential prebiotic compounds (Lamothe et al., 1996).

Tab. 1 Selected patents related to dextran

Patent number	*Holder*	*Inventors*	*Title*	*Date*
U.S. Patent 2,437, 518	Pharmacia AB, Sweden	A. Gronwall, B. Ingelman	Manufacture of infusion and injection fluids	1948
U.S. Patent 2,686,147	U.S. Dept. Agriculture	H. M. Tsuchiya, H. J. Koepsell	Production of dextransucrase	1954
U.S. Patent 2,709,150	Enzmatic Chemicals, Inc., Delaware, USA	V. W. Carlson, W. W. Carlson	Method of producing dextran material by bacteriological and enzymatic action	1955
U.S. Patent 2,712,007	U.S. Dept. Agriculture	I. A. Wolff, R. L. Mellies, C. E. Rist	Fractionation of dextran products	1955
U.S. Patent 2,726,190	U.S. Dept. Agriculture	H. J. Koepsell, N. N. Hellman, H. M. Tsuchiya	Modification of dextran synthesis by means of alternate glucosyl acceptors	1955
U.S. Patent 2,726,985	U.S. Dept. Agriculture	N. N Hellman, H. M. Tsuchiya, S. P. Rogovin, R. W. Jackson, F. R. Senti	Controlled enzymatic synthesis of dextran	1955
U.S. Patent 2,841,578	The Commonwealth Engineering Co. of Ohio	L. J. Novak, G. S. Stoycos	Method for producing clinical dextran	1958
U.S. Patent 2,776,925	U.S. Dept. Agriculture	J. Corman, H. M. Tsuchiya.	Enzymic production of dextran of intermediate molecular weights	1957
U.S. Patent 2,820, 740	Benger Laboratories Ltd., England	E. London, G. D. Twigg	Therapeutic preparation of iron	1958
U.S. Patent 2,906,669	U.S. Dept. Agriculture	E. J. Hehre, H. M. Tsuchiya, N. N. Hellman, F. R. Senti	Production of dextran	1959
U.S. Patent 3,002,823	Pharmacia AB, Sweden	P. G. M. Flodin, J. O. Porath	Process of separating materials having different molecular weights and dimensions	1961
U.S. Patent 3,044,940	VEB Serum-Werk Bernburg, Germany	U. Behrens, M. Ringpfeil	Process for enzymatic synthesis of dextran	1962
U.S. Patent 3,141,014	Meito Sangyo Kabushiki Kaishu, Japan	E. Morii, K. Iwata, H. Kokkoku	Sodium and potassium salts of the dextran sulfate acid ester having substantially no anticoagulant activity but having lipolytic activity and the method of preparation thereof	1964
U.S. Patent 4,180,567	Pharmachem Corp., USA	J. R. Herb	Iron preparations and methods of making and administering the same	1979

Tab. 1 (cont.)

Patent number	Holder	Inventors	Title	Date
U.S. Patent 4,855,416	Polydex Pharmaceuticals, Ltd., The Bahamas	T. C. Usher	Method for the manufacture of dextran sulfate and salts thereof	1989
U.S. Patent 5,229,277	Louisiana State Univ.	D. F. Day, D. Kim	Process for the production of dextran polymers of controlled molecular size and molecular size distributions	1992
U.S. Patent 5,518,733.	Bioeurope, France	J.-P. Lamothe, Y. G. Marchenay, P. F. Monsan, F. M. B. Paul, V. Pelenc	Cosmetic compositions containing oligosaccharides	1996

13 Outlook and Perspectives

Advances in the molecular biology of glucansucrases promise not only to resolve fundamental questions concerning enzyme structure, function, and regulation but also to open new avenues for dextran applications. Recombinant organisms that overproduce dextransucrases may reduce the cost of dextran production by enzymatic synthesis, making dextrans more competitive for bulk chemical applications. Alternatively, dextrans might be produced in transgenic crops, as has been demonstrated recently for fructans (Caimi et al., 1996; Pilon-Smits et al., 1996; Sevenier et al., 1998) and *Streptococcus* glucans (Nichols, 2000a,b,c). Novel dextransucrases might be created by site-directed mutagenesis, chimeric recombination, or shuffling of dextransucrase genes. At the same time, dextran oligosaccharides appear to have considerable potential to find new and expanded markets as prebiotic supplements in foods, cosmetics, and animal feeds.

14 References

Alcalde, M., Plou, F. J., Gomez de Segura, A., Remaud-Simeon, M. Willemot, R. M., Monsan, P., Ballesteros, A. (1999) Immobilization of native and dextran-free dextransucrases from *Leuconostoc mesenteroides* NRRL B-512F for the synthesis of glucooligosaccharides, *Biotechnol. Tech.* **13**, 749–755.

Alsop, R. M. (1983) Industrial production of dextrans, in: *Progress in Industrial Microbiology*, (Bushell, M. E., Ed.), London: Elsevier, 1–44, Vol. 18.

Arguello-Morales, M. A., Remaud-Simeon, M., Pizzut, S., Sarcabal, P., Willemot, R.-M., Monsan, P. (2000a) *Leuconostoc mesenteroides* NRRL B-1355 *dsrC* gene for dextransucrase. GenBank Accession No. AJ250172.

Arguello-Morales, M. A., Remaud-Simeon, M., Pizzut, S., Sarcabal, P., Willemot, R.-M., Monsan, P. (2000b) Sequence analysis of the gene encoding alternansucrase, a sucrose glucosyltransferase from *Leuconostoc mesenteroides* NRRL B-1355. *FEMS Microbiol. Lett.* **182**, 81–85.

Barker, S. A., Bourne, E. J., Whiffen, D. H. (1956) Use of infrared analysis in the determination of carbohydrate structure, in: *Methods of Biochemical Analysis*, (Glick, D., Ed.), New York: Interscience Publishers, Inc., 213–245, Vol. 3.

Behrens, U., Ringpfeil, M. (1962) Process for enzymatic synthesis of dextran. U.S. Patent 3,044,940.

Beijerinck, M. W. K. (1912) Mucilaginous substances of the cell wall produced from cane sugar by bacteria. *Folia Microbiol.* **1**, 377.

Bhatnagar, R., Singh, D. K. S. (1999) Cloning and characterization of dextransucrase gene from *Leuconostoc mesenteroides* NRRL B-512F. GenBank Accession No. U81374.

Binder, T. P., Robyt, J. F. (1983) *p*-nitrophenyl α-D-glucopyranoside, a new substrate for glucansucrases, *Carbohydr. Res.* **124**, 287–299.

Bovery, F. A. (1959) Enzymatic polymerization. I. Molecular weight and branching during the formation of dextran, *J. Polymer Sci.* **35**, 167–182.

Caimi, P. G., McCole, L. M., Klein, T. M., Kerr, P. S. (1996) Fructan accumulation and sucrose metabolism in transgenic maize endosperm expressing a *Bacillus amyloliquefaciens SacB* gene, *Plant Physiol.* **110**, 355–363.

Caldwell, R. C., Sandham, H. J., Mann, W. V., Finn, S. B., Formicola, A. J. (1971) The effect of a dextranase mouthwash on dental plaque in young adults and children, *J. Amer. Dent. Assoc.* **82**, 124–131.

Carlson, V. W., Carlson, W. W. (1955a) Method of producing dextran material by bacteriological and enzymatic action. U.S. Patent 2,709,150.

Carlson, V. W., Carlson, W. W. (1955b) Production of endodextranase by *Aspergillus wentii*. U.S. Patent 2,716,084.

Carrasco, F., Chornet, E., Overend, R. P., Costa, J. (1989) A generalized correlation for the viscosity of dextrans in aqueous solutions as a function of temperature, concentration, and molecular weight at low shear rates, *J. Appl. Polymer Sci.* **37**, 2087–2098.

Cerning, J. (1990) Exocellular polysaccharides produced by lactic acid bacteria, *FEMS Microbiol. Rev.* **87**,113–130.

Chaiet, L., Kempf, A. J., Harman, R., Kaczka, E., Weston, R., Nollstadt, K., Wolf, F. J. (1970) Isolation of a pure dextranase from *Penicillium funiculosum*, *Appl. Microbiol.* **20**, 421–426.

Cheetham, N. W. H., Fiala-Beer, E., Walker, G. J. (1991) Dextran structural details from high-field proton NMR spectroscopy, *Carbohydr. Polymers* **14**, 149–158.

Chludzinski, A. M., Germaine, G. R., Schachtele, C. F. (1974) Purification and properties of dextransucrase from *Streptococcus mutans, J. Bacteriol.* **118**, 1–7.

Corman, J., Tsuchiya, H. M. (1957) Enzymic production of dextran of intermediate molecular weights. U.S. Patent 2,776,925.

Côté, G. L. (1992) Low-viscosity α-D-glucan fractions derived from sucrose which are resistant to enzymatic digestion, *Carbohydr. Polym.* **19**, 249–252.

Côté, G. L., Ahlgren, J. A. (1994) Production, isolation, and immobilization of alternansucrase. Amer. Chem. Soc. 207 Natl. Meeting, Abstract CARB#12.

Côté, G. L., Ahlgren, J. A. (1995) Microbial polysaccharides, in: *Kirk-Othmer Encyclopedia of Chemical Technology* (Kroschvitz, J. I., Howe-Grant, M., Eds), New York: John Wiley & Sons, Inc., 578–612, 4th Ed., Vol. 16.

Côté, G. L., Robyt, J. F. (1982) Isolation and partial characterization of an extracellular glucansucrase from *L. mesenteroides* NRRL B-1355 that synthesizes an alternating (1 → 6), (1 → 3)-α-D-glucan, *Carbohyd. Res.* **101**, 57–74.

Côté, G. L., Robyt, J. F. (1983) The formation of α-D-(1 → 3) branch linkages by an exocellular glucansucrase from *Leuconostoc mesenteroides* NRRL B-742, *Carbohydr. Res.* **119**, 141–156.

Côté, G. L., Leathers, T. D., Ahlgren, J. A., Wyckoff, H. A., Hayman, G. T., Biely, P. (1997) Alternan and highly branched limit dextrans: Low-viscosity polysaccharides as potential new food ingredients, in: *Chemistry of Novel Foods* (Spanier, A. M., Tamura, M., Okai, H., Mills, O., Eds.), Carol Stream, IL: Allured Publishing Corp., 95–110.

Côté, G. L., Ahlgren, J. A., Smith, M. R. (1999) Some structural features of an insolube α-D-glucan from a mutant strain of *Leuconostoc mesenteroides* NRRL B-1355, *J. Ind. Microbiol. Biotechnol.* **23**, 656–660.

Covacevich, M. T., Richards, G. N. (1977) Frequency and distribution of branching in a dextran: an enzymic method, *Carbohydr. Res.* **54**, 311–315.

Day, D. F., Kim, D. (1992) Process for the production of dextran polymers of controlled molecular size and molecular size distributions. U.S. Patent 5,229,277.

de Belder, A. N. (1990) *Dextran.* Uppsala, Sweden: Pharmacia.

de Belder, A. N. (1993) Dextran, in: *Industrial Gums. Polysaccharides and Their Derivatives,* Third Edition (Whistler, R. L., BeMiller, J. N., Eds.), San Diego, CA: Academic Press, 399–425.

de Belder, A. N. (1996) Medical applications of dextran and its derivatives, in: *Polysaccharides. Medical Applications* (Dumitriu, S., Ed.), New York: Marcel Dekker, 505–523.

Dimler, R. J., Wolff, I. A., Sloan, J. W., Rist, C. E. (1955) Interpretation of periodate oxidation data on degraded dextran, *J. Am. Chem. Soc.* **77**, 6568–6573.

Dols, M., Remaud-Simeon, M., Willemot, R.-M., Vignon, M., Monsan, P. F. (1997) Characterization of dextransucrases from *Leuconostoc mesenteroides* NRRL B-1299, *Appl. Biochem. Biotechnol.* **62**, 47–59.

Dols, M., Remaud-Simeon, M., Willemot, R.-M., Demuth, B., Joerdening, H.-J., Buchholz, K., Monsan, P. (1999) Kinetic modeling of oligosaccharide synthesis catalyzed by *Leuconostoc mesenteroides* NRRL B-1299 dextransucrase, *Biotechnol. Bioeng.* **63**, 308–315.

Evans, T. H., Hibbert, H. (1946) Bacterial polysaccharides, in: *Adv. Carbohydr. Chem.* (Pigman, W. W., Wolfrom, M. L., Eds.). New York: Academic Press, 203–233, Vol. 2.

Fischer, E. H., Stein, E. A. (1960) Cleavage of O- and S-glycosidic bonds (survey), in: *The Enzymes* (Boyer, P. D., Lardy, H., Myrback, K., Eds.), New York: Academic Press, 301–312, Vol. 4.

Fitzgerald, R. J., Spinell, D. M., Stoudt, T. H. (1968) Enzymatic removal of artificial plaques, *Arch. Oral Biol.***13**, 125–128.

Flodin, P. G. M., Porath, J. O. (1961) Process of separating materials having different molecular weights and dimensions. U.S. Patent 3,002,823.

Fu, D., Robyt, J. F. (1990) A facile purification of *Leuconostoc mesenteroides* B- 512FM dextransucrase, *Prep. Biochem.* **20**, 93–106.

Fu, D., Slodki, M. E., Robyt, J. F. (1990) Specificity of acceptor binding to *Leuconostoc mesenteroides* B512F dextransucrase: binding and acceptor-product structure of α-methyl-D-glucopyranoside analogs modified at C-2, C-3, and C-4 by inversion of the hydroxyl and by replacement of the hydroxyl with hydrogen, *Arch. Biochem. Biophys.* **276**, 460–465.

Fujiwara, T. Terao, Y., Hoshino, T., Kawabata, S., Ooshima, T., Sobue, S., Kimura, S., Hamada, S. (1998) Molecular analyses of glucosyltransferase genes among strains of *Streptococcus mutans, FEMS Microbiol. Lett.* **161**, 331–336.

Fukui, K., Fukui, Y., Moriyama, T. (1974) Purification and properties of dextransucrase and invertase from *Streptococcus mutans, J. Bacteriol.* **118**, 796–804.

Funane, K., Mizuno, K., Takahara, H., Kobayashi, M. (2000) Gene encoding a dextransucrase-like protein in *Leuconostoc mesenteroides* NRRL B-512F, *Biosci. Biotechnol. Biochem.* **64**, 29–38.

Furuta, T., Koga, T., Nisizawa, T., Okahashi, N., Hamada, S. (1985) Purification and characterization of glucosyltransferases from *Streptococcus mutans* 6715, *J. Gen. Microbiol.* **131**, 285–293.

Galvez-Mariscal, A., Lopez-Munguia, A. (1991) Production and characterization of a dextranase from an isolated *Paecilomyces lilacinus* strain, *Appl. Microbiol. Biotechnol.* **36**, 327–331.

Gekko, K., Noguchi, H. (1971) Physicochemical studies of oligodextran. I. Molecular weight dependence of intrinsic viscosity, partial specific compressibility and hydrated water, *Biopolymers* **10**, 1513–1524.

Glicksman, M. (1983) Dextran, in: *Food Hydrocolloids* (Glicksman, M., Ed.), Boca Raton: CRC Press 157–166.

Godfrey, T. (1983) Dextranase and sugar processing, in: *Industrial Enzymology. The Application of Enzymes in Industry* (Godfrey, T., Reichelt, T., Eds.), New York: Nature Press, 422–424.

Granath, K. A. (1958) Solution properties of branched dextrans, *J. Colloid Sci.* **13**, 308–328.

Gronwall, A. (1957) *Dextran and Its Use in Colloidal Infusion Solutions.* Stockholm: Almqvist & Wiksell.

Gronwall, A., Ingelman, B. (1945) Dextran as a substitute for plasma, *Nature* **155**, 45.

Gronwall, A., Ingelman, B. (1948) Manufacture of infusion and injection fluids. U.S. Patent 2,437, 518.

Hamada, S., Slade, H. D. (1980) Biology, immunology, and cariogenicity of *Streptococcus mutans*, *Microbiol. Rev.* **44**, 331–384.

Hattori, A., Ishibashi, K., Minato, S. (1981) The purification and characterization of the dextranase from *Chaetomium gracile*, *Agric. Biol. Chem.* **45**, 2409–2416.

Hehre, E. J. (1941) Production from sucrose of a serologically reactive polysaccharide by a sterile bacterial extract, *Science* **93**, 237–238.

Hehre, E. J., Hamilton, D. M. (1951) The biological synthesis of dextran from dextrins, *J. Biol. Chem.* **192**, 161–174.

Hehre, E. J., Tsuchiya, H. M., Hellman, N. N., Senti, F. R. (1959) Production of dextran. U.S. Patent 2,906,669.

Hehre, E. J., Suzuki, H. (1966) New reactions of dextransucrase: α-D-glucosyl transfers to and from the anomeric sites of lactulose and fructose, *Arch. Biochem. Biophys.* **113**, 675–683.

Heincke, K., Demuth, B., Jordening, H.-J., Buchholz, K. (1999) Kinetics of the dextransucrase acceptor reaction with maltose - experimental results and modeling, *Enzyme Microbial Technol.* **24**, 523–534.

Herb, J. R. (1979) Iron preparations and methods of making and administering the same. U.S. Patent 4,180,567.

Hellman, N. N., Tsuchiya, H. M., Rogovin, S. P., Jackson, R. W., Senti, F. R. (1955) Controlled enzymatic synthesis of dextran. U.S. Patent 2,726,985.

Hidaka, H., Hirayama, M. (1991) Useful characteristics and commercial applications of fructooligosaccharides, *Biochem. Soc. Trans.* **19**, 561–565.

Hiraoka, N., Fukumoto, J., Tsuru, D. (1972) Studies on mold dextranases: III. Purification and some enzymatic properties of *Aspergillus carneus* dextranase, *J. Biochem.* **71**, 57–64.

Holzapfel, W. H., Schillinger, U. (1992) The genus *Leuconostoc*, in: *The Procaryotes*, 2nd Ed., (Ballows, A., Truper, H. G., Dworkin, M., Harder, W., Schleifer, K.-H., Eds.), New York: Springer-Verlag, 1508–1534, Vol. 2.

Honda, O., Kato, C., Kuramitsu, H. K. (1990) Nucleotide sequence of the *Streptococcus mutans gtfD* gene encoding the glucosyltransferase-S enzyme, *J. Gen. Microbiol.* **136**, 2099–2105.

Janecek, S., Svensson, B, Russell, R. R. B. (2000) Location of repeat elements in glucansucrases of *Leuconostoc* and *Streptococcus* species, *FEMS Microbiol. Lett.* **192**, 53–57.

Jeanes, A. (1965a) Dextrans. Preparation of a water soluble dextran by enzymic synthesis, in: *Methods in Carbohydrate Chemistry* (Whistler, R. L., BeMiller, J. N., Eds.), New York: Academic Press, 127–132, Vol. 5.

Jeanes, A. (1965b) Dextrans. Preparation of dextrans from growing *Leuconostoc* cultures, in: *Methods in Carbohydrate Chemistry* (Whistler, R. L., BeMiller, J. N., Eds.), New York: Academic Press, 118–126, Vol. 5.

Jeanes, A. (1966) Dextran, in: *Encyclopedia of Polymer Science and Engineering*, (Mark, H. F.; Bikales, N. M.; Overberger, C. G.; Menges, G.; Kroschwitz, J. I., Eds.), New York: John Wiley & Sons, 752–767, Vol. 4.

Jeanes, A. (1977) Dextrans and pullulans: industrially significant α-D-glucans, in: *ACS Symp. Series No. 45, Extracellular Microbial Polysaccharides* (Sandford, P. A., Laskin, A., Eds.), Washington, D. C.: American Chemical Society, 284–298.

Jeanes, A. (1978) *Dextran Bibliography.* Washington, D. C.: U. S. Dept. Agriculture.

Jeanes, A. (1986) Immunochemical and related interactions with dextrans reviewed in terms of improved structural information. *Mol. Immun.* **23**, 999–1028.

Jeanes, A., Seymour, F. R. (1979) The α-D-glucopyranosidic linkages of dextrans: comparison of percentages from structural analysis by periodate oxidation and by methylation, *Carbohydr. Res.* **74**, 31–40.

Jeanes, A., Haynes, W. C., Wilham, C. A., Rankin, J. C., Melvin, E. H., Austin, M. J., Cluskey, J. E., Fisher, B. E., Tsuchiya, H. M., Rist, C. E. (1954) Characterization and classification of dextrans from ninety-six strains of bacteria, *J. Amer. Chem. Soc.* **76**, 5041–5052.

Jung, H-K., Kim, K-N., Lee, H-S., Jung, S-H. (1999) Production of alternan by *Leuconostoc mesenteroides* CBI-110, *Kor. J. Appl. Microbiol. Biotechnol.* **27**, 35–40.

Kabat, E. A., Bezer, A. E. (1958) The effect of variation in molecular weight on the antigenicity of dextran in man, *Arch. Biochem. Biophys.* **78**, 306–318.

Kaboli, H., Reilly, P. J. (1980) Immobilization and properties of *Leuconostoc mesenteroides* dextransucrase, *Biotechnol. Bioeng.* **22**, 1055–1069.

Killey, M., Dimler, R. J., Cluskey, J. E. (1955) Preparation of panose by the action of NRRL B-512 dextransucrase on a sucrose-maltose mixture, *J. Amer. Chem. Soc.* **77**, 3315–3318.

Kim, D., Day, D. F. (1995) Isolation of a dextranase constitutive mutant of *Lipomyces starkeyi* and its use for the production of clinical size dextran, *Lett. Appl. Microbiol.* **20**, 268–270.

Kim, D., Kim, D-W. (1999) Facile purification and characterization of dextransucrase from *Leuconostoc mesenteroides* B-512FMCM, *J. Microbiol. Biotechnol.* **9**, 219–222.

Kim, D., Robyt, J. F. (1994) Production and selection of mutants of *Leuconostoc mesenteroides* constitutive for glucansucrases, *Enzyme Microb. Technol.* **16**, 659–664.

Kim, D., Robyt, J. F. (1995a) Dextransucrase constitutive mutants of *Leuconostoc mesenteroides* B-1299, *Enzyme Microb. Technol.* **17**, 1050–1056.

Kim, D., Robyt, J. F. (1995b) Production, selection, and characteristics of mutants of *Leuconostoc mesenteroides* B-742 constitutive for dextransucrases, *Enzyme Microb. Technol.* **17**, 689–695.

Kim, D., Robyt, J. F. (1996) Properties and uses of dextransucrases elaborated by a new class of *Leuconostoc mesenteroides* mutants, *Prog. Biotechnol.* **12**, 125–144.

Kim, D., Kim, D-W., Lee, J-H., Park, K-H., Day, L. M., Day, D. F. (1997) Development of constitutive dextransucrase hyper-producing mutants of *Leuconostoc mesenteroides* using the synchrotron radiation in the 70–1000 eV region, *Biotechnol. Tech.* **11**, 319–321.

Kim, H., Kim, D., Ryu, W-H., Robyt, J. F. (2000) Cloning and sequencing of the α-1 → 6 dextransucrase gene from *Leuconostoc mesenteroides* B-742CB, *J. Microbiol. Biotechnol.* **10**, 559–563.

Kitaoka, M., Robyt, J. F. (1998a) Large-scale preparation of highly purified dextransucrase from a high-producing constitutive mutant of *Leuconostoc mesenteroides* B-512FMC, *Enzyme Microb. Technol.* **23**, 386–391.

Kitaoka, M., Robyt, J. F. (1998b) Use of a microtiter plate screening method for obtaining *Leuconostoc mesenteroides* mutants constitutive for glucansucrase, *Enzyme Microb. Technol.* **22**, 527–531.

Kobayashi, M., Matsuda, K. (1975) Purification and characterization of two activities of the intracellular dextransucrase from *Leuconostoc mesenteroides* NRRL B-1299, *Biochim. Biophys. Acta* **397**, 69–79.

Kobayashi, M., Matsuda, K. (1976) Purification and properties of the extracellular dextransucrase from *Leuconostoc mesenteroides* NRRL B-1299, *J. Biochem.* **79**, 1301–1308.

Kobayashi, M., Matsuda, K. (1977) Structural characteristics of dextrans synthesized by dextransucrases from *Leuconostoc mesenteroides* NRRL B-1299, *Agric. Biol. Chem.* **41**, 1931–1937.

Kobayashi, M., Matsuda, K. (1980) Characterization of the multiple forms and main component of dextransucrase from *Leuconostoc mesenteroides* NRRL B-512F, *Biochim. Biophys. Acta* **614**, 46–62.

Koenig, D. W., Day, D. F. (1988) Production of dextranase by *Lipomyces starkeyi*, *Biotechnol. Lett.* **10**, 117–122.

Koepsell, H. J., Hellman, N. N., Tsuchiya, H. M. (1955) Modification of dextran synthesis by means of alternate glucosyl acceptors. U.S. Patent 2,726,190.

Kohmoto, T., Fukui, F., Takaku, H., Mitsuoka, T. (1991) Dose-response test of isomaltooligosaccharides for increasing fecal bifidobacteria, *Agric. Biol. Chem.* **55**, 2157–2159.

Kooi, E. R. (1958) Production of dextran-dextrinase. U.S. Patent 2,833,695.

Kossman, J., Welsh, T., Quanz, M., Knuth, K. (2000) Nucleic acid molecules encoding alternansucrase. PCT Patent WO00/47727.

Kuge, T., Kobayashi, K., Kitamura, S., Tanahashi, H. (1987) Degrees of long-chain branching in dextrans, *Carbohydr. Res.* **160**, 205–214.

Lamothe, J.-P., Marchenay, Y. G., Monsan, P. F., Paul, F. M. B., Pelenc, V. (1996) Cosmetic compositions containing oligosaccharides. U.S. Patent 5,518,733.

Larm, O., Lindberg, B., Svensson, S. (1971) Studies on the length of the side chains of the dextran elaborated by *Leuconostoc mesenteroides* NRRL B-512, *Carbohydr. Res.* **20**, 39–48.

Lawford, G. R., Kligerman, A., Williams, T. (1979) Dextran biosynthesis and dextransucrase production by continuous culture of *Leuconostoc mesenteroides, Biotechnol. Bioeng.* **21**, 1121–1131.

Leathers, T. D., Hayman, G. T., Cote, G. L. (1995) Rapid screening of *Leuconostoc mesenteroides* mutants for elevated proportions of alternan to dextran, *Curr. Microbiol.* **31**, 19–22.

Leathers, T. D., Hayman, G. T., Cote, G. L. (1997) Microorganism strains that produce a high proportion of alternan to dextran. U.S. Patent 5,702,942.

Leathers, T. D., Hayman, G. T., Cote, G. L. (1998) Rapid screening method to select microorganism strains that produce a high proportion of alternan to dextran. U.S. Patent 5,789,209.

Lee, J. M., Fox, P. F. (1985) Purification and characterization of *Paecilomyces lilacinus* dextranase, *Enzyme Microb. Technol.* **7**, 573–577.

Lilley, L. L., Aucker, R. S. (1999) Fluids and electrolytes, in: *Pharmacology and the Nursing Process.* St. Louis, MO: Mosby, Inc., 335–348.

Lindberg, B., Svensson, S. (1968) Structural studies on dextran from *Leuconostoc mesenteroides* NRRL B-512, *Acta Chem. Scand.* **22**, 1907–1912.

Loesche, W. J. (1986) Role of *Streptococcus mutans* in human dental decay, *Microbiol. Rev.* **50**, 353–380.

London, E., Twigg, G. D. (1958) Therapeutic preparation of iron. U.S. Patent 2,820, 740.

McCabe, M. M. (1985) Purification and characterization of a primer-independent glucosyltransferase from *Streptococcus mutans* 6715-13 mutant 27, *Infect. Immun.* **50**, 771–777.

Miller, A. W., Robyt, J. F. (1984) Stabilization of dextransucrase from *Leuconostoc mesenteroides* NRRL B-512F by nonionic detergents, poly(ethylene glycol) and high-molecular-weight dextran, *Biochim. Biophys. Acta* **785**, 89–96.

Miller, A. W., Eklund, S. H., Robyt, J. F. (1986) Milligram to gram scale purification and characterization of dextransucrase from *Leuconostoc mesenteroides* NRRL B-512F, *Carbohydr. Res.* **147**, 119–133.

Misaki, A., Torii, M., Sawai, T., Goldstein, I. J. (1980) Structure of the dextran of *Leuconostoc mesenteroides* B-1355, *Carbohydr. Res.* **84**, 273–285.

Mitsuishi, Y., Kobayashi, M., Matsuda, K. (1979) Dextran α-1, 2 debranching enzyme from *Flavobacterium* sp. M-73: its production and purification, *Agric. Biol. Chem.* **43**, 2283–2290.

Mitsuya, H., Looney, D. J., Kuno, S., Ueno, R., Wong-Staal, F., Broder, S. (1988) Dextran sulfate suppression of viruses in the HIV family: inhibition of virion binding to $CD4^+$ cells, *Science* **240**, 646–649.

Monchois, V., Willemot, R.-M., Remaud-Simeon, M., Croux, C., Monsan, P. (1996) Cloning and sequencing of a gene coding for a novel dextransucrase from *Leuconostoc mesenteroides* NRRL B-1299 synthesizing only a α(1-6) and α(1-3) linkages, *Gene* **182**, 23–32.

Monchois, V., Remaud-Simeon, M., Russell, R. R. B., Monsan, P., Willemot, R.-M. (1997) Characterization of *Leuconostoc mesenteroides* NRRL B512F dextransucrase (DSRS) and identification of amino-acid residues playing a key role in enzyme activity, *Appl. Microbiol. Biotechnol.* **48**, 465–472.

Monchois, V., Remaud-Simeon, M., Monsan, P., Willemot, R.-M. (1998) Cloning and sequencing of a gene coding for an extracellular dextransucrase (DSRB) from *Leuconostoc mesenteroides* NRRL B-1299 synthesizing only α(1-6) glucan, *FEMS Microbiol. Lett.* **159**, 307–315.

Monchois, V., Willemot, R.-M., Monsan, P. (1999) Glucansucrases: mechanism of action and structure-function relationships, *FEMS Microbiol. Rev.* **23**, 131–151.

Monchois, V., Vignon, M., Russell, R. R. B. (2000) Mutagenesis of asp-569 of glucosyltransferase I glucansucrase modulates glucan and oligosaccharide synthesis, *Appl. Environ. Microbiol.* **66**, 1923–1927.

Monsan, P., Paul, F. (1995) Enzymatic synthesis of oligosaccharides, *FEMS Microbiol. Rev.* **16**, 187–192.

Monsan, P., Paul, F., Auriol, D., Lopez, A. (1987) Dextran synthesis using immobilized *Leuconostoc mesenteroides* dextransucrase, in: *Methods in Enzymology: Immobilized Enzymes and Cells* (Mosbach, K., Ed.), Orlando, FL: Academic Press, Inc, 239–254, Vol. 136.

Monsan, P., Potocki de Montalk, G., Sarcabal, P., Remaud-Simeon, M., Willemont, R.-M. (2000) Glucansucrases: efficient tools for the synthesis of oligosaccharides of nutritional interest, in: *Food Biotechnology* (Bielecki, S., Tramper, J., Polak, J., Eds.), Amsterdam: Elsevier Science B. V., 115–122.

Morii, E., Iwata, K., Kokkoku, H. (1964) Sodium and potassium salts of the dextran sulphate acid ester having substantially no anticoagulant activity but having lipolytic activity and the method of preparation thereof. U.S. Patent 3,141,014.

Mountzouris, K. C., Gilmour, S. G., Jay, A. J., Rastall, R. A. (1999) A study of dextran production from maltodextrin by cell suspensions of *Gluconobacter oxydans* NCIB 4943, *J. Appl. Microbiol.* **87**, 546–556.

Murphy, P.T., Whistler, R. L. (1973) Dextrans, in *Industrial Gums*, Second Ed. (Whistler, R. L., BeMiller, J. N., Eds.), New York: Academic Press, 513–542.

Neely, W. B. (1960) Dextran: structure and synthesis, in: *Adv. Carbohydr. Chem.* (Wolfrom, M. L., Tipson, R. S., Eds.), New York: Academic Press, 341–369, Vol. 15.

Nichols, S. E. (2000a) Plant cells and plants transformed with *Streptococcus mutans* gene encoding glucosyltransferase C enzyme. U.S. Patent 6,127,603.

Nichols, S. E. (2000b) Plant cells and plants transformed with *Streptococcus mutans* genes encoding wild-type or mutant glucosyltransferase B enzymes. U.S. Patent 6,087,559.

Nichols, S. E. (2000c) Plant cells and plants transformed with *Streptococcus mutans* genes encoding wild-type or mutant glucosyltransferase D enzymes. U.S. Patent 6,127,602.

Novak, L. J., Stoycos, G. S. (1958) Method for producing clinical dextran. U.S. Patent 2,841,578.

Okada, G., Takayanagi, T., Sawai, T. (1988) Improved purification and further characterization of an isomaltodextranase from *Arthrobacter globiformis* T6, *Agric. Biol. Chem.* **52**, 495–501.

Pasteur, L. (1861) On the viscous fermentation and the butyrous fermentation, *Bull. Soc. Chim. Paris*, 30–31.

Paul, F., Auriol, D., Oriol, E. Monsan, P. (1984) Production and purification of dextransucrase from *Leuconostoc mesenteroides* NRRL B-512(F). *Ann. N. Y. Acad. Sci.* **434**, 267–270.

Pearce, B. J., Walker, G. J., Slodki, M. E., Schuerch, C. (1990) Enzymic and methylation analysis of dextrans and (1-3)-α-D-glucans, *Carbohydr. Res.* **203**, 229–246.

Pelenc, V., Lopez-Munguia, A., Remaud, M., Biton, J., Michel, J. M., Paul, F., Monsan, P. (1991) Enzymatic synthesis of oligoalternans, *Sci. Aliments* **11**, 465–476.

Pilon-Smits, E. A. H., Ebskamp, M. J. M., Jeuken, M. J. W., van der Meer, I. M., Visser, R. G. F., Weisbeek, P. J., Smeekens, S. C. M. (1996) Microbial fructan production in transgenic potato plants and tubers, *Ind. Crops Products* **5**, 35–46.

Piret, J., Lamontagne, J., Bestman-Smith, J., Roy, S., Gourde, P., Desormeaux, A., Omar, R. F., Juhasz, J., Bergeron, M. G. (2000) *In vitro* and *in vivo* evaluations of sodium lauryl sulfate and dextran sulfate as microbicides against herpes simplex and human immunodeficiency viruses, *J. Clin. Microbiol.* **38**, 110–119.

Quirasco, M., Lopez-Munguia, A., Remaud-Simeon, M., Monsan, P., Farres, A. (1999) Induction and transcriptional studies of the dextransucrase gene in *Leuconostoc mesenteroides* NRRL B-512F, *Appl. Environ. Microbiol.* **65**, 5504–5509.

Rankin, J. C., Jeanes, A. (1954) Evaluation of periodate oxidation method for structural analysis of dextrans, *J. Am. Chem. Soc.* **76**, 4435–4441.

Reh, K.-D., Noll-Borchers, M., Buchholz, K. (1996) Productivity of immobilized dextransucrase for leucrose formation, *Enzyme Microb. Technol.* **19**, 518–524.

Remaud, M., Paul, F., Monsan, P., Heyraud, A., Rinaudo, M. (1991) Molecular weight characterization and structural properties of controlled molecular weight dextrans synthesized by acceptor reaction using highly purified dextransucrase, *J. Carbohydr. Chem.* **10**, 861–876.

Remaud, M., Paul, F., Monsan, P. (1992) Characterization of α-(1 → 3) branched oligosaccharides synthesized by acceptor reaction with the extracellular glucosyltransferases from *L. mesenteroides* NRRL B-742, *J. Carbohydr. Chem.* **11**, 359–378.

Remaud-Simeon, M., Lopez-Munguia, A., Pelenc, V., Paul, F., Monsan, P. (1994) Production and use of glucosyltransferases from *Leuconostoc mesenteroides* NRRL B-1299 for the synthesis of oligosaccharides containing α-(1 → 2) linkages, *Appl. Biochem. Biotechnol.* **44**, 101–117.

Remaud-Simeon, M., Willemot, R.-M., Sarcabal, P., Potocki de Montalk, G., Monsan, P. (2000) Glucansucrases: molecular engineering and oligosaccharide synthesis, *J. Mol. Catalysis B: Enzymatic* **10**, 117–128.

Robyt, J. F. (1986) Dextran, in: *Encyclopedia of Polymer Science and Engineering*, (Mark, H. F., Gaylord, N. G., Bikales, N. M., Eds.), New York: John Wiley & Sons, 752–767, Vol. 4.

Robyt, J. F. (1992) Structure, biosynthesis, and uses of nonstarch polysaccharides: dextran, alternan, pullulan, and algin, in: *Developments in Biochemistry and Biophysics* (Alexander, R. J., Zobel, H. F., Eds.), St. Paul: Amer. Assoc. Cereal Chemists, 261–292.

Robyt, J. F. (1995) Mechanisms in the glucansucrase synthesis of polysaccharides and oligosaccharides from sucrose, in: *Adv. Carbohydr. Chem. Biochem.* (Horton, D., Ed.), San Diego: Academic Press, 133–168, Vol. 51.

Robyt, J. F., Eklund, S. H. (1983) Relative, quantitative effects of acceptors in the reaction of *Leuconostoc mesenteroides* B-512F dextransucrase, *Carbohydr. Res.* **121**, 279–286.

Robyt, J. F., Martin, P. J. (1983) Mechanism of synthesis of D-glucans by D-glucosyltransferases from *Streptococcus mutans* 6715, *Carbohydr. Res.* **113**, 301–315.

Robyt, J. F., Taniguchi, H. (1976) The mechanism of dextransucrase action. Biosynthesis of branch linkages by acceptor reactions with dextran, *Arch. Biochem. Biophys.* **174**, 129–135.

Robyt, J. F., Walseth, T. F. (1978) The mechanism of acceptor reactions of *Leuconostoc mesenteroides* B-512F dextransucrase, *Carbohydr. Res.* **61**, 433–445.

Robyt, J. F., Walseth, T. F. (1979) Production, purification and properties of dextransucrase from *Leuconostoc mesenteroides* NRRL B-512F, *Carbohydr. Res.* **68**, 95–111.

Robyt, J. F., Kimble, B. K., Walseth, T. F. (1974) The mechanism of dextransucrase action. Direction of dextran biosynthesis, *Arch. Biochem. Biophys.* **165**, 634–640.

Ryu, H-J., Kim, D., Kim, D-W., Moon, Y Y., Robyt, J. F. (2000) Cloning of a dextransucrase gene (*fmcmds*) from a constitutive dextransucrase hyper-producing *Leuconostoc mesenteroides* B-512FMCM developed using VUV, *Biotechnol. Lett.* **22**, 421–425.

Sabatie, J., Choplin, L., Moan, M., Doublier, J. L., Paul, F., Monsan, P. (1988) The effect of synthesis temperature on the structure of dextran NRRL B 512F, *Carbohydr. Polymers* **9**, 87–101.

Sanchez-Gonzalez, M., Alagon, A., Rodriguez-Sotres, R., Lopez-Munguia, A. (1999) Proteolytic processing of dextransucrase of *Leuconostoc mesenteroides, FEMS Microbiol. Lett.* **181**, 25–30.

Sawai, T., Tohyama, T., Natsume, T. (1978) Hydrolysis of fourteen native dextrans by *Arthrobacter* isomaltodextranase and correlation with dextran structure, *Carbohydr. Res.* **66**, 195–205.

Sawai, T., Ohara, S., Ichimi, Y., Okaji, S., Hisada, K., Fukaya, N. (1981) Purification and some properties of the isomaltodextranase of *Actinomadura* strain R10 and comparison with that of *Arthrobacter globiformis* T6, *Carbohydr. Res.* **89**, 289–299.

Scheibler, C. (1874) Investigation on the nature of the gelatinous excretion (so-called frog's spawn) which is observed in production of beet-sugar juices, *Z. Ver. Dtsch. Zucker-Ind.* **24**, 309–335.

Sery, T. W., Hehre, E. J. (1956) Degradation of dextrans by enzymes of intestinal bacteria, *J. Bacteriol.* **71**, 373–380.

Sevenier, R., Hall, R. D., van der Meer, I. M., Hakkert, H. J. C., van Tunen, A. J., Koops, A. J. (1998) High level fructan accumulation in a transgenic sugar beet, *Nature Biotechnol.* **16**, 843–846.

Seymour, F. R., Julian, R. L. (1979) Fourier-transform, infrared difference-spectrometry for structural analysis of dextrans, *Carbohydr. Res.* **74**, 63–75.

Seymour, F. R., Knapp, R. D. (1980) Unusual dextrans: 13. Structural analysis of dextrans from strains of *Leuconostoc* and related genera, that contain 3-O-α-D-glucosylated α-D-glucopyranosyl residues at the branch points, or in consecutive linear position, *Carbohyd. Res.* **81**, 105–129.

Seymour, F. R., Knapp, R. D., Bishop, S. H. (1976) Determination of the structure of dextran by ^{13}C-nuclear magnetic resonance spectroscopy, *Carbohydr. Res.* **51**, 179–194.

Seymour, F. R., Slodki, M. E., Plattner, R. D., Jeanes, A. (1977) Six unusual dextrans: methylation structural analysis by combined G. L. C.-M. S. of per-O-acetylaldononitriles, *Carbohydr. Res.* **53**, 153–166.

Seymour, F. R., Chen, E. C. M., Bishop, S. H. (1979a) Methylation structural analysis of unusual dextrans by combined gas-liquid chromatography-mass spectrometry, *Carbohydr. Res.* **68**, 113–121.

Seymour, F. R., Knapp, R. D., Bishop, S. H. (1979b) Correlation of the structure of dextrans to their ^{1}H-NMR spectra, *Carbohydr. Res.* **74**, 77–92.

Sharpe, E. S., Stodola, F. H., Koepsell, H. J. (1960) Formation of isomaltulose in enzymatic dextran synthesis, *J. Org. Chem.* **25**, 1062–1063.

Shimamura, A., Tsumori, H., Mukasa, H. (1982) Purification and properties of *Streptococcus mutans* extracellular glucosyltransferase, *Biochim. Biophys. Acta* **702**, 72–80.

Shimizu, E., Unno, T., Ohba, M., Okada, G. (1998) Purification and characterization of an isomaltotriose-producing *endo*-dextranase from a *Fusarium* sp., *Biosci. Biotechnol. Biochem.* **62**, 117–122.

Shiroza, T., Ueda, S., Kuramitsu, H. K. (1987) Sequence analysis of the *gftB* gene from *Streptococcus mutans, J. Bacteriol.* **169**, 4263–4270.

Sidebotham, R. L. (1974) Dextrans, in: *Adv. Carbohydr. Chem. Biochem.* (Tipson, R. S., Horton, D., Eds.), New York: Academic Press, 371–444, Vol. 30.

Simonson, L. G., Liberta, A. E. (1975) New sources of fungal dextranase. *Mycologia* **4**, 845–851.

Sinha, J., Dey, P. K., Panda, T. (2000) Aqueous two-phase: the system of choice for extractive fermentations, *Appl. Microbiol. Biotechnol.* **54**, 476–486.

Slodki, M. E., England, R. E., Plattner, R. D., Dick Jr., W. E. (1986) Methylation analyses of NRRL dextrans by capillary gas-liquid chromatography, *Carbohyd. Res.* **156**,199–206.

Smith, M. R., Zahnley, J., Goodman, N. (1994) Glucosyltransferase mutants of *Leuconostoc mesenteroides* NRRL B-1355, *Appl. Environ. Microbiol.* **60**, 2723–2731.

Smith, M. R., Zahnley, J. C., Wong, R. Y., Lundin, R. E., Ahlgren, J. A. (1998) A mutant strain of *Leuconostoc mesenteroides* B-1355 producing a glucosyltransferase synthesizing $\alpha(1\rightarrow 2)$ glucosidic linkages, *J. Ind. Microbiol. Biotechnol.* **21**, 37–45.

Stiles, M. E., Holzapfel, W. H. (1997) Lactic acid bacteria of foods and their current taxonomy, *Int. J. Food Microbiol.* **36**, 1–29.

Stodola, F. H., Sharpe, E. S., Koepsell, H. J. (1956) The preparation, properties and structure of the disaccharide leucrose, *J. Amer. Chem. Soc.* **78**, 2514–2518.

Su, D., Robyt, J. F. (1994) Determination of the number of sucrose and acceptor binding sites for *Leuconostoc mesenteroides* B-512FM dextransucrase, and the confirmation of the two-site mechanism for dextran synthesis, *Arch. Biochem. Biophys.* **308**, 471–476.

Sun, J., Cheng, X., Zhang, Y., Yan, Z., Zhang, S. (1988) A strain of *Paecilomyces lilacinus* producing high quality dextranase, *Ann. N. Y. Acad. Sci.* **542**, 192–194.

Suzuki, M., Unno, T., Okada, G. (1999) Simple purification and characterization of an extracellular dextrin dextranase from *Acetobacter capsulatum* ATTC 11894, *J. Appl. Glycosci.* **46**, 469–473.

Swengers, D. (1991) Leucrose, a ketodisaccharide of industrial design, in: *Carbohydrates as Organic Raw Materials* (Lichtenthaler, F. W., Ed.), Weinheim, Germany: VCH, 183–195.

Takahashi, M., Okada, S., Uchimura, T., Kozaki, M. (1992) *Leuconostoc amelibiosum* Schillinger, Holzapfel, and Kandler 1989 is a later subjective synonym of *Leuconostoc citreum* Farrow, Facklam, and Collins 1989, *Int. J. Syst. Bacteriol.* **42**, 649–651.

Takazoe, I. (1989) Palatinose - an isomeric alternative to sucrose, in: *Progress in Sweeteners* (Grenby, T. H., Ed.), New York: Elsevier Science, 143–167.

Tarr, H. L. A., Hibbert, H. (1931) Studies on reactions relating to carbohydrates and polysaccharides. XXXVII. The formation of dextran by *Leuconostoc mesenteroides, Can. J. Res.* **5**, 414–427.

Taylor, C., Cheetham, N. W. H., Walker, G. J. (1985) Application of high-performance liquid chromatography to a study of branching dextrans, *Carbohydr. Res.* **137**, 1–12.

Tjerneld, F. (1992) Aqueous two-phase partitioning on an industrial scale, in: *Poly(Ethylene Glycol) Chemistry: Biotechnical and Biomedical Applications* (Harris, J. M., Ed.), New York: Plenum Press, 85–102.

Torii, M., Sakakibara, K., Misaki, A., Sawai, T. (1976) Degradation of alpha-linked D-gluco-oligosaccharides and dextrans by an isomaltodextranase preparation from *Arthrobacter globiformis* T6, *Biochem. Biophys. Res. Comm.* **70**, 459–464.

Tsuchiya, H. M., Koepsell, H. J. (1954) Production of dextransucrase. U.S. Patent 2,686,147.

Tsuchiya, H. M., Hellman, N. N., Koepsell, H. J., Corman, J., Stringer, C. S., Rogovin, S. P., Bogard, M. O., Bryant, G., Feger, V. H., Hoffman, C. A., Senti, F. R., Jackson, R. W. (1955) Factors affecting molecular weight of enzymatically synthesized dextran, *J. Amer. Chem. Soc.* **77**, 2412–2419.

Tsuchiya, H. M., Jeanes, A., Bricker, H. M., Wilham, C. A. (1956) Production of dextranase. U.S. Patent 2,742,399.

Ueda, S., Shiroza, R., Kuramitsu, H. K. (1988) Sequence analysis of the *gtfC* gene from *Streptococcus mutans* GS-5, *Gene* **69**, 101–109.

Usher, T. C. (1989) Method for the manufacture of dextran sulfate and salts thereof. U.S. Patent 4,855,416.

Van Cleve, J. W., Schaefer, W. C., Rist, C. E. (1956) The structure of NRRL B-512 dextran. Methylation studies, *J. Amer. Chem. Soc.* **78**, 4435–4438.

Vandamme, E. J., Bruggeman, G., De Baets, S., Vanhooren, P. T. (1996) Useful polymers of microbial origin, *Agro-Food-Industry Hi-Tech* **Sept./Oct.**, 21–25.

Van Tieghem, P. (1878) On sugar-mill gum, *Ann. Sci. Nat. Bot. Biol. Veg.* **7**, 180–203.

Walker, G. J. (1978) Dextrans, in: *International Review of Biochemistry. Biochemistry of Carbohydrates II,* (Manners, D. J., Ed.), Baltimore, MD: University Park Press, 75–125, Vol. 16.

Webb, E., Spencer-Martins, I. (1983) Extracellular endodextranase from the yeast *Lipomyces starkeyi*, *Can. J. Microbiol.* **29**, 1092–1095.

Wilke-Douglas, M., Perchorowicz, J. T., Houck C. M., Thomas, B. R. (1989) Methods and compositions for altering physical characteristics of fruit and fruit products. PCT Patent WO89/12386.

Wolff, I. A., Mellies, R. L., Rist, C. E. (1955) Fractionation of dextran products. U.S. Patent 2,712,007.

Yalpani, M. (1986) Preparation and applications of dextran-derived products in biotechnology and related areas, *CRC Crit. Rev. Biotechnol.* **3**, 375–421.

Yamamoto, K., Yoshikawa, K., Kitahata, S., Okada, S. (1992) Purification and some properties of dextrin dextranase from *Acetobacter capsulatus* ATTC 11894, *Biosci. Biotechnol. Biochem.* **56**, 169–173.

Yamamoto, K., Yoshikawa, K., Okada, S. (1993a) Dextran synthesis from reduced maltooligosaccharides by dextrin dextranase from *Acetobacter capsulatus* ATTC 11894, *Biosci. Biotechnol. Biochem.* **57**, 136–137.

Yamamoto, K., Yoshikawa, K., Okada, S. (1993b) Effective dextran production from starch by dextrin dextranase with debranching enzyme, *J. Ferment. Bioeng.* **76**, 411–413.

13
Alternan

Dr. Gregory L. Côté
Fermentation Biotechnology Research Unit, National Center for Agricultural Utilization Research, Agricultural Research Service, United States Department of Agriculture, 1815 North University Street Peoria, Illinois 61604, USA;
Tel.: +1-309-681-6319; Fax: +1-309-681-6427; E-mail: cotegl@ncaur.usda.gov

DP	degree of polymerization (number of monomer units in a polymer)
EC	Enzyme Commission
Glc	glucose
Glc*p*	glucopyranose
kDa	kilodaltons (1000 atomic mass units)
kHz	kilohertz (10^3 cycles s^{-1})
MHz	megahertz (10^6 cycles s^{-1})
NRRL	Northern Regional Research Laboratory (in reference to USDA microbial culture collection strain numbers)
TEMPO	2,2,6,6-tetramethyl-1-piperidine-*N*-oxide
USDA	United States Department of Agriculture

The use of brand or trade names may be necessary to report factually on available data. The USDA neither guarantees nor warrants the standard of the product, and the use of the name by USDA implies no approval of the product to the exclusion of others that may also be suitable. All programs and services of the USDA are offered on a nondiscriminatory basis without regard to race, color, national origin, religion, sex, age, marital status, or handicap.

1 Introduction

Alternan is the name given to an extracellular polysaccharide produced from sucrose by the lactic acid bacterium *Leuconostoc mesenteroides*. Specifically, it refers to the α-D-glucan in which the D-glucopyranosyl units are linked by alternating α(1→6) and α(1→3) linkages. This name was coined by Côté and Robyt (1982a) to differentiate it from the well-known dextran. Although not currently produced for commercial applications, alternan is under investigation for a number of uses, including low-viscosity bulking agents, gum arabic substitutes, and paper additives (Hardin, 1999). In addition, the enzyme alternansucrase, which converts sucrose to alternan may be useful for the synthesis of prebiotic oligosaccharides (Côté and Robyt, 1982b; Argüello-Morales, 2001).

2 Historical Outline

In 1954, researchers at the USDA research laboratory in Peoria, Illinois, published a landmark paper describing the structure and properties of dextrans from 96 strains of bacteria (Jeanes et al., 1954). The work was part of a directed effort to find dextrans suitable for use as blood plasma extenders. Most of the bacteria studied were strains of *L. mesenteroides*, and all secreted dextrans into the fermentation broth. At that time, a dextran was defined as a bacterial exopolysaccharide consisting of D-glucose units linked primarily through α(1→6) linkages. Six of the strains described by Jeanes et al. (1954) produced two distinct dextran fractions. These were referred to as fractions "L" and "S", for "less-soluble" and "soluble" respectively in the water-ethanol mixtures used selectively to precipitate them. Three of these six strains produced a typical dextran, containing over 90% α(1→6) linkages, as well as an atypical fraction consisting of only 57–65% of α(1→6) linkages. These three fractions, synthesized *by L. mesenteroides* NRRL strains B-1355, B-1498, and B-1501, were unusually low in viscosity. Since they did not possess the requisite properties for blood plasma extenders, no attempt was made to develop them for commercial use. In fact, of the 100 or so dextrans described, only NRRL B-512F was commercialized.

Two decades later, methods for the structural analysis of polysaccharides had evolved to a much greater degree of sophistication. Seymour et al. (1977) first proposed that the fraction "S" glucans from *L. mesenteroides* strains B-1355, B-1498, and B-1501 consisted of an unusual structure in which α(1→6) linkages alternated with α(1→3) linkages in the main polysaccharide chain. They later confirmed this view by use of ^{13}C-NMR spectrometry (Seymour et al., 1979b). Based on this and other work, Côté and Robyt (1982a) concluded that the term "dextran" should be reserved for polymers in which the main chain contained only α(1→6) linked D-glucose units. To differentiate the alternating type of polysaccharides from "true dextrans", they coined the trivial name *alternan* to refer to the former.

During the 1990s, interest in the use of polysaccharides and oligosaccharides as low-calorie food ingredients and prebiotics led several research groups to reconsider the potential uses of alternan. Investigators at the United States Department of Agriculture (USDA) laboratory in Peoria, Illinois studied the physical properties of alternan and some low molecular-weight fractions derived therefrom (Côté, 1992a; Côté et al., 1997). At the same time, scientists in France were investigating the enzymatic synthesis of oligosaccharides by alternansucrase, the enzyme responsible for the biosynthesis of alternan (Pelenc et al., 1991; Castillo et al., 1992). A recent resurgence of interest in alternan and other unusual bacterial glucans has led to the cloning of enzymes responsible for their biosynthesis (Argüello-Morales et al., 2000; Kossmann et al., 2000).

3
Chemical Structure

In their survey, Jeanes et al. (1954) used periodate oxidation to determine the proportions of different linkages in the various dextrans (Jeanes and Wilham, 1950; Rankin and Jeanes, 1954; Sloan et al., 1954). There are numerous potential sources of error or ambiguity inherent in this method, but it did reveal that B-1355 dextran fraction S (i.e., alternan) contained 57% 1,6-like linkages, 35% 1,3-like linkages, and 8% 1,4-like linkages. It should be pointed out that "1,4-like linkages" include 1,2-linkages, since they are indistinguishable by periodate oxidation (Sloan et al., 1954). According to unpublished USDA in-house reports, the 8% of 1,4-like linkages is very close to the limit of error for the protocol used at that time, so the only conclusion that could be drawn from these data was that B-1355 fraction S dextran (alternan) contained a very high proportion of 1,3-linkages. The S-fractions from NRRL strains B-1498 and B-1501 yielded similar results, although the proportions of 1,3-linkages were somewhat less.

Periodate oxidation studies did not differentiate between 1,3-linkages at branch points and those in linear, unbranched residues. Early results from partial acid hydrolysis and methylation analysis showed that the α(1→3) linkages in alternan occur at branch points and in linear, nonbranching residues, and that the branches tend to be at least four glucose units long (Dimler et al., 1956). No α(1→2) or α(1→4) linkages were detected. Optical rotation studies on the cuprammonium complexes of alternan also indicated that nearly all of the α(1 → 3) linkages were in linear, rather than branching positions (Scott et al., 1957). This was supported by Taylor et al. (1959), who used X-ray diffraction to study the water sorption and degree of order of various dextrans. These authors found that B-1355 fraction S was anomalous in that it gave a high degree of order and low water sorption, despite the high percentage of α(1 → 3) linkages.

Goldstein and Whelan (1962) subjected alternan to acetolysis, which preferentially cleaves α(1 → 6) linkages, and identified some of the oligomeric products. The major products were D-glucose (21%), isomaltose (α-D-Glc*p*(1 → 6)-D-Glc, 2%), nigerose (α-D-Glc*p*(1 → 3)-D-Glc, 20%), and the trisaccharide α-D-Glc*p*(1 → 6)-D-Glc*p*(1 → 3)-D-Glc (2%). No nigerotriose was detected in the acetolysis mixture. These results revealed two important features of the alternan structure: first, that the α(1 → 3) linkages and α(1 → 6) linkages occur together in the same regions of the same molecules; and second, that very few (if any) of the α(1 → 3) linkages were contiguous. Torii and Sakakibara (1974) carried out a controlled acetolysis of alternan, followed by borate ion-exchange chromatography of the products, and reached the same conclusion as previous workers – namely that a high percentage (> 30%) of the D-glucose units in alternan were linked by nonbranching α(1 → 3) linkages. They also confirmed previous evidence showing an absence of α(1 → 2) and α(1 → 4) linkages.

By the mid-1970s, more precise methods were available for the study of polysaccharide structures. Chemists at the Peoria laboratory examined several dextrans by methylation analysis and combined gas-liquid chromatography and mass spectrometry (Seymour et al., 1977). The fraction S glucan from *L. mesenteroides* strain NRRL B-1355 yielded 6.9% of the 2,3,4,6-tetra-*O*-methyl derivative of D-glucose, 46.9% of the 2,3,4-tri-*O*-methyl, 35.0% of the 2,4,6-tri-*O*-methyl, and 11.2% of the 2,4-di-*O*-methyl derivative. The difference between the percentages of the di-*O*-methyl and the tetra-*O*-methyl derivatives suggests a certain degree of error, probably due to degradation of acetylated residues released during prolonged acetolysis and hydrolysis. Later methylation analysis by Slodki et al. (1986) yielded more precise data, indicating 45% of 1,6-disubstituted residues, 34% of 1,3-disubstituted residues, 11% 1,3,6-trisubstituted (branched) residues, and 11% nonreducing end residues. These results were in good agreement with those reported by Hare et al. (1978). The methylation studies conclusively demonstrated the presence of linear α(1 → 3) linkages in the alternan chain. They also showed the molecule to be branched through 3,6-di-substituted D-glucosyl residues, with approximately 10% branching. Based on their methylation data and a review of the evidence previously cited, Seymour et al. (1977) proposed two possible structural variants to account for their observations. Both structures were characterized by the presence of alternating sequences of α(1 → 3) and α(1 → 6) linked D-glucospyranosyl residues, the two differing mainly in the interpretation of the branching data. Seymour et al. (1979c) later extended their studies to include the fraction S dextrans from strains NRRL B-1498 and B-1501. Results of methylation analyses of these fractions were similar to B-1355 fraction S, although the percentages of α(1 → 3) linkages were somewhat lower. These values were 35% for B-1355 fraction S, 29% for B-1498 fraction S, and 24% for B-1501 fraction S. In the same study, they correlated methylation data with ^{13}C-NMR spectroscopic results and proposed an alternating structure for all three glucans. In their proposed structure, the excess α(1 → 6) linkages are located predominantly in short branches. These authors also discussed the immunological evidence for the alternating structure in great detail, and later used ^{1}H-NMR spectrometry to provide additional evidence for the alternating structure with α(1 → 6) linkages occurring in short (two to three residues) branches (Seymour et al., 1979a). In that paper, it was stated that "dextran B-1355 fraction S and dextran B-1501 fraction S are apparently not true dextrans". This distinction was based on a definition of dextran as a D-glucan in which the main chains are linked through α(1 → 6) linkages, with branching through α(1 → 2), α(1 → 3), or α(1 → 4) linkages at di-*O*-substituted residues. Since the alternating structures of B-1355 fraction S, B-1498 fraction S, and B-1501 fraction S do not meet these criteria, it made sense to classify them as nondextrans.

At the same time that Seymour and his colleagues were using methylation analysis and NMR spectrometry to study dextran structures, a group of Japanese workers were using methylation analysis combined with selective chemical and enzymatic degradation to achieve the same goal. In 1980, these authors published a comprehensive study on the structure of alternan from strain B-1355 (Misaki et al., 1980). They found the polysaccharide to be practically inert to enzymatic hydrolysis by (1 → 3)-α-D-endoglucanase, and only slightly hydrolyzed by (1 → 6)-α-D-endoglucanase (endodextranase) and (1 → 6)-α-D-exoglucanase (glucodextranase), although it was hydrolyzed extensively by isomaltodextranase. Based on their results, these authors proposed a structure in which the alternating sequences of (1 → 3) and (1 → 6)-α-D-glucopyranosyl residues are branched in a highly ramified manner. The structure proposed by Misaki et al. (1980) differs in some key aspects from that proposed by Seymour et al. (1979a). Whereas Seymour's model places all of the excess α(1 → 6) linkages in the short external branches, Misaki places them uniformly throughout the molecule. Also, Misaki's structure contains a small proportion of α(1 → 3) linkages adjacent to one another at some of the branch points, whereas Seymour's structure contains no adjacent α(1 → 3) linkages. Acetolysis of Misaki's structure would be expected to yield a small amount of nigerotriose. Seymour et al., on the one hand, based their structure partly on the work of Goldstein and Whelan (1962), in which acetolysis yielded no nigerotriose. On the other hand, Misaki et al. performed their own acetolysis and claimed to have isolated a small proportion of nigerotriose.

Both Seymour and Misaki considered the immunological evidence for certain structural features of alternan. One problem with this approach is that the binding specificity of many of the antidextrans has been elucidated based on structural evidence obtained through chemical and spectroscopic means. It appears somewhat tautological then to use the binding specificities thus derived to study further the chemical structures of those same polysaccharides. Nonetheless, some structural features of alternan have been confirmed, if not initially discovered, by studying the inhibition of antibody binding by oligosaccharides of known structures. In such a manner did Torii et al. (1981) show evidence for nigerosyl units at the nonreducing termini of alternan.

Ogawa and Kaburagi (1982) synthesized an α-D-glucoheptaose that contained alternating (1 → 3) and (1 → 6)-α-D-glucopyranosyl units with a single branch point. Spectroscopic and chemical evidence indicate that this is very close, if not identical, to the repeating structural unit of alternan from *L. mesenteroides* NRRL B-1355. Figure 1 represents an idealized model of the structure of alternan.

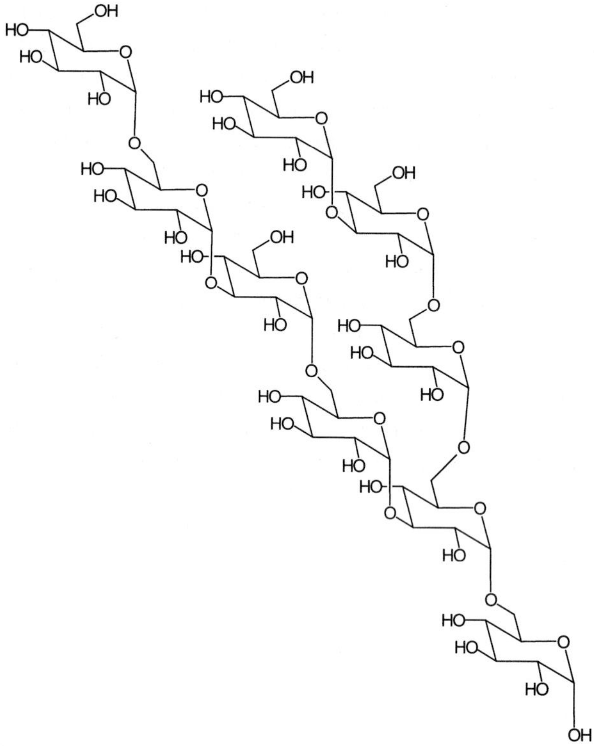

Fig. 1 Chemical structure of a representative portion of the alternan molecule.

Although the primary chemical structure of alternan has been extensively studied, it is noteworthy that relatively little is known about its molecular weight. Polysaccharides, being characteristically polydisperse, cannot be said to have a definite molecular weight. Instead, it is customary to determine a weight-average or number average molecular mass. A search of the literature reveals only two papers that report any values at all for the molecular weight of alternan. During the 1950s, R. Tobin at the Peoria USDA laboratory used light scattering to determine the molecular weight of fermentatively produced B-1355 fraction S to be 4×10^7 daltons. This value was not reported in the literature until it was cited as a footnote by Seymour et al. (1979b). More recently, Côté (1992a) used multi-angle laser light scattering to obtain a weight-average molecular weight of 1.0×10^7 for enzymatically synthesized alternan. Due to its high molecular weight, a reliable value could not be obtained using size-exclusion chromatography.

4
Occurrence and Physiological Function

Relatively few alternan-producing bacteria have been isolated, but of the three strains described by Jeanes et al. (1954), *L. mesenteroides* NRRL B-1355 has been the most intensively studied. This strain was first isolated by R. Patrick at the USDA citrus laboratory in Winter Haven, Florida. The other two strains described at that time were B-1498 and B-1501; both were obtained from

C. S. McCleskey, who isolated them from cane juice being processed at an experimental sugar plant in Louisiana. McCleskey et al. (1947) classified a number of strains according to their colony types. Both B-1498 and B-1501 belonged to group F, which was characterized by smooth, opaque colonies. At least two other type F strains were also studied by Jeanes et al. (1956). Designated NRRL B-1500 and NRRL B-1502, these produced typically $\alpha(1 \rightarrow 6)$ linked dextrans, but no alternan.

Preobrazhenskaya et al. (1978) described a glucan from *L. mesenteroides* strain 61-2 which appeared remarkably similar to alternan from NRRL strain B-1355. Methylation analysis yielded 9.8% of the 2,3,4,6-tetra-*O*-methyl D-glucose derivative, 46.9% of the 2,3,4-tri-*O*-methyl, 32.8% of the 2,4,6-tri-*O*-methyl, and 10.5% of the 2,4-di-*O*-methyl D-glucose derivative. Quantitative precipitation of the strain 61-2 D-glucan with concanavalin A followed a curve nearly identical to B-1355 fraction S alternan, and its anti-complementary activity against guinea pig serum was likewise nearly identical. Acetolysis and periodate oxidation data were also similar; thus, it is reasonable to conclude that the D-glucan from the strain 61-2 is alternan. More recently, Jung et al. (1999) isolated a strain of *L. mesenteroides* from kimchi, a traditional Korean dish made from fermented cabbage and other vegetables. According to their data, strain CBI-110 produces both dextran and alternan. This evidence that alternan occurs naturally in some fermented food products is an important point in demonstrating the safety of alternan in foods. Claims that *Streptococcus mutans* strains also produce alternan (Kossmann et al. 2000) appear to be in error, as the original data indicate the glucans to be either highly branched dextran (Tsumori et al., 1985), mutan, or graft copolymers thereof (Mukasa et al. 1989).

A number of mutants have been created from *L. mesenteroides* NRRL B-1355, which are altered in their ability to synthesize dextran and alternan. These will be discussed under the heading of biosynthesis, as these mutations affect the production and properties of the biosynthetic enzymes.

All that can be said about the physiological role of alternan is that it probably plays the same role as dextrans. Microbial exopolysaccharides in general may aid the survival of an organism by preventing desiccation, maintaining an environment of favorable pH and ionic strength, and by acting as a resistant barrier to phages and other microbes. They are also believed to be important in the adhesion of cells to surfaces.

5 Biosynthesis

Much of what can be said about dextran biosynthesis is also true of alternan biosynthesis. As discussed in the preceding sections, alternan is very similar to dextran in many ways. In fact, most, if not all of the bacteria that produce alternan also produce dextran. For a general discussion of the biosynthesis of dextran, the reader is referred to Chapter 12, this volume.

5.1 Alternansucrase: Purification and Properties

Alternan is synthesized from sucrose by the extracellular enzyme alternansucrase (EC 2.4.1.140). Alternansucrase is a glycosyltransferase that polymerizes the α-D-glucopyranosyl moiety of sucrose and releases the D-fructose portion. Like dextransucrases from *L. mesenteroides*, alternansucrase is induced only when the bacteria are grown in the presence of sucrose. As a result, the enzymes are associated with copious amounts of their polysaccharide products. Côté and Robyt (1982a) took advantage of this fact when they first showed alternansucrase to be a separate and distinct enzyme from the dextransucrase complex produced by *L. mesenteroides* NRRL B-1355. They treated a crude glucansucrase mixture from the sucrose-grown culture fluid with endodextranase. The dextranase hydrolyzed the dextran present in the mixture, allowing the dextransucrase to disaggregate into a lower molecular-weight form. The high molecular-weight alternan–alternansucrase complex was separated from the dissociated dextransucrase by gel filtration chromatography. The alternansucrase thus isolated was reversibly inhibited by the substrate analogue 3-deoxy-3-fluoro-α-D-glucopyranosyl fluoride, as well as by the general glycosidase inhibitor *tris*-hydroxymethyl aminomethane. Some inhibition was also observed with 1-*O*-octyl β-D-glucopyranoside. The enzyme was active in the pH range of 3 to 7.5, with optimal activity occurring between pH 5 and 6.

Alternansucrase has also been prepared from *L. mesenteroides* NRRL B-1355 by aqueous two-phase liquid extraction (López-Munguía et al., 1991). Dextransucrase was selectively heat-denatured, leaving alternansucrase as the only active glucansucrase in the preparation. This was possible because dextransucrase from this strain is completely inactivated at 45 °C, whereas alternansucrase retains a substantial proportion of its activity at that temperature. It should be noted that this procedure is not a purification *per se*, but it does serve the purpose of isolating active alternansucrase free of contaminating dextransucrase activity. The enzyme thus prepared was found to exhibit optimum activity at pH 5.6, with a temperature optimum of 40 °C (López-Munguía et al., 1993). Much of the glucansucrase activity was found to be in an insoluble fraction of the cultures. Some of this activity was associated with cell membranes and could be released into solution with the nonionic detergent polysorbate 80. However, a substantial fraction was associated with insoluble aggregates containing high molecular-weight glucan–enzyme complexes. Zahnley and Smith (2000) reported similar results, concluding that alternansucrase is associated primarily with cell walls or membranes.

Until recently, the molecular mass of alternansucrase was a matter of some debate. Unpublished results from the author's laboratory indicated more than one band of activity when analyzed by SDS–polyacrylamide gel (SDS–PAGE) electrophoresis. At least two alternansucrase bands were observed, at 135 kDa and 145 kDa, and both exhibited pI values of approximately 4.9. A constitutive mutant of B-1355 was reported by Kim and Robyt (1994) to produce a single activity band at 173 kDa when grown on glucose, but at least three when grown on sucrose, the additional bands appearing at 184 kDa and 125 kDa. Smith et al. (1994, 1998) have described mutants of B-1355 that produce several glucansucrase bands on SDS–PAGE. It was later determined that the 200 kDa band in one of their mutants was alternansucrase, whereas other bands represented dextransucrase and a novel, previously undescribed glucansucrase (Smith et al., 1998; Smith and Zahnley, 1999). One likely explanation for the reported variability in molecular weights is proteolytic cleavage of the enzyme. It has been shown that various molecular weight fragments of dextransucrase are due to proteolysis (Sanchez-Gonzalez et al., 1999).

L. mesenteroides NRRL B-1355 alternansucrase has recently been cloned (Arguello-Morales et al., 2000; Kossmann et al., 2000), whilst the B-1355 alternansucrase gene (*asr*; GenBank accession no. AJ250173) and a B-

1355 dextransucrase gene (*dsrC*; GenBank accession no. AJ250172) have both been sequenced. Based on its deduced amino acid sequence, alternansucrase has a molecular mass of 228,971 daltons and a pI of 5.1. It possesses an N-terminal signal sequence of 39 amino acids, a large variable region, a catalytic domain, and a C-terminal glucan-binding domain. Alternansucrase is the largest glucansucrase cloned to date. Based on analysis of the gene sequence, Janeček et al. (2000) have noted that alternansucrase bears a number of unique features not observed in other glucansucrases, including the presence of multiple long N-terminal repeat sequences and novel short repeats.

5.2
Alternansucrase: Reactions and Mechanism of Action

By analogy with similar glucansucrases, alternansucrase probably synthesizes alternan from sucrose via an insertion-type mechanism similar to that described for dextransucrase. The only major difference in the mechanism for alternansucrase is likely to be the orientation of one of the glucosyl residues bound at the active site (Robyt, 1995). In Robyt's model, D-glucopyranosyl units are added to the reducing end of the growing D-glucan chain by a stepwise transfer process. In the first step, D-glucopyranosyl units are transferred from sucrose to form a covalent intermediate with the enzyme, thereby releasing fructose into the medium. A second D-glucopyranosyl unit is then bound to the enzyme in an orientation that allows its 6-OH group to displace the first D-glucopyranosyl unit from the enzyme, forming an $\alpha(1 \rightarrow 6)$ bond. This results in the formation of an isomaltosyl–enzyme covalent complex. The isomaltosyl unit is then displaced by the 3-OH group of a newly bound D-glucopyranosyl unit, forming the basic alternating $\alpha(1 \rightarrow 3)$, $\alpha(1 \rightarrow 6)$ sequence of alternan. This mechanism suggests the existence of two D-glucopyranosyl binding sites, such that one is oriented to form $\alpha(1 \rightarrow 3)$ linkages and the other to form $\alpha(1 \rightarrow 6)$ linkages.

Some degree of branching is present in essentially all dextrans, as well as in alternan. Robyt and Taniguchi (1976) have explained the formation of these branches in B-512F dextran by the insertion mechanism. According to their model, branches are formed by acceptor reactions, whereby single D-glucopyranosyl units or growing dextran chains are transferred to 3-OH groups on another dextran chain. Côté and Robyt (1983, 1984) extended this mechanism to include highly branched comb polymers. It is likely that branch formation by alternansucrase is by the same mechanism.

The so-called acceptor reactions of glycansucrases are a very interesting and potentially useful aspect of these enzymes. Although the usual reaction of these enzymes is considered to be polysaccharide synthesis via glycosyl transfer to growing polymer chains, in the presence of certain hydroxyl-bearing acceptor molecules, glycosyl units are transferred to the acceptor compounds instead. In many of these acceptor reactions, the initial acceptor product often acts as an acceptor for additional transfer reactions. Thus, a single acceptor often gives rise to a homologous series of acceptor products. In this manner, D-glucose acceptor reactions with dextransucrase in the presence of sucrose yield a series of isomaltooligosaccharides (Koepsell et al., 1953).

Acceptor reactions of alternansucrase were first described by Côté and Robyt (1982b), who tested a number of sugars as acceptors. They found that D-glucose added as an acceptor first gave rise to isomaltose, an $\alpha(1 \rightarrow 6)$ linked glucose disaccharide, but no nigerose (the $\alpha(1 \rightarrow 3)$ linked glucose disaccharide) was formed. However, isomaltose also acted as an acceptor, giving rise to two different trisaccharides. One of these was isomaltotriose, arising from the formation of a second α(1 → 6) linkage; the other isomaltose acceptor product was found to be α-D-Glc*p*(1 → 3)-α-D-Glc*p*(1 → 6)-D-Glc. Alternansucrase catalyzed similar acceptor reactions with 1-*O*-methyl α- and β-D-glucopyranosides. When nigerose was tested as an acceptor, the initial product was found to be α-D-Glc*p*(1 → 6)-α-D-Glc*p*(1 → 3)-D-Glc. Tetrasaccharide products arising from subsequent transfers of D-glucose to the trisaccharide product were α-D-Glc*p*(1 → 6)-α-D-Glc*p*(1 → 6)-α-D-Glc*p*(1 → 3)-D-Glc and α-D-Glc*p*(1 → 3)-α-D-Glc*p*(1 → 6)-D-Glc*p*(1 → 3)-D-Glc. Likewise, maltose first gave rise to panose, which then gave rise to both α-D-Glc*p*(1 → 6)-α-D-Glc*p*(1 → 6)-α-D-Glc*p*(1 → 4)-D-Glc and α-D-Glc*p*(1 → 3)-α-D-Glc*p*(1 → 6)-D-Glc*p*(1 → 4)-D-Glc.

The pentasaccharide alternansucrase products from maltose were subsequently shown to include at least three different oligosaccharides (Castillo et al., 1992), as follows:

α-D-Glc*p*(1 → 6)-α-D-Glc*p*(1 → 6)-
α-D-Glc*p*(1 → 6)-α-D-Glc*p*(1 → 4)-D-Glc

α-D-Glc*p*(1 → 3)-α-D-Glc*p*(1 → 6)-
α-D-Glc*p*(1 → 6)-α-D-Glc*p*(1 → 4)-D-Glc

α-D-Glc*p*(1 → 6)-α-D-Glc*p*(1 → 3)-
α-D-Glc*p*(1 → 6)-α-D-Glc*p*(1 → 4)-D-Glc

Pelenc et al. (1991) treated a mixture of maltose acceptor products with endodextranase and glucoamylase, which eliminated the α(1 → 6)-linked oligomers. This allowed them to isolate some of the higher degree of polymerization (DP) oligoalternans. Figure 2 shows the acceptor reactions catalyzed by alternansucrase in the presence of sucrose and maltose.

It has been concluded that alternansucrase is capable of synthesizing both α(1 → 6) and α(1 → 3) linkages by acceptor reactions, but that an α(1 → 3) linkage can only be formed through attachment to a D-glucopyranosyl residue linked to another residue through an α(1 → 6) linkage (Côté and Robyt, 1982b). This means that consecutive α(1 → 3) linkages cannot be formed by acceptor reactions. By contrast, there seems to be no such limitation on the formation of consecutive α(1 → 6) linkages. This specificity may explain in part some of the observations regarding the relative occurrence and distribution of these linkage types in the alternan structure, particularly around branch points which may arise via acceptor reactions.

Since D-fructose is released whenever sucrose is present, substantial concentrations of D-fructose can accumulate in the reaction medium. Like dextransucrase, alternansucrase can utilize this D-fructose as an acceptor. Transfer of a single D-glucopyranosyl residue to OH-5 of D-fructose is the major reaction, resulting in the formation of leucrose (Stodola et al., 1952, 1956; Côté and Robyt, 1982b). However, leucrose appears not to be an acceptor, since no additional transfers take place and no homologous series of products is produced.

One effect of acceptors on the reaction of alternansucrase is to divert the transfer of glucosyl residues away from alternan synthesis and toward the production of oligosaccharide acceptor products. Better acceptors are more effective at diverting glucosyl units away from polysaccharide synthesis. For alternansucrase, the best known acceptor is maltose, but other acceptors (in order of their effectiveness) are nigerose > 1-*O*-methyl α-D-glucopyranoside > isomaltose > D-glucose > 1-*O*-methyl β-D-glucopyranoside. These were measured by analyzing the relative proportion of radiolabeled glu-

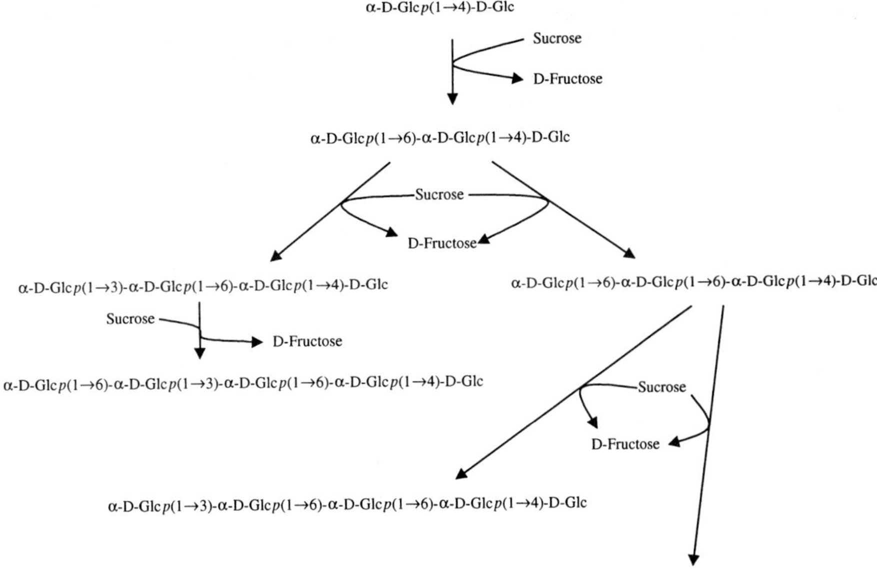

Fig. 2 Acceptor reactions of alternansucrase in the presence of sucrose and maltose. Glucosyl transfer from sucrose results in an acceptor product that can subsequently accept additional glucosyl transfers.

cose incorporated into each product from ^{14}C-sucrose (Côté and Robyt, 1982b). Other acceptors include D-fructose, D-allose, D-galactose, cellobiose, melibiose, raffinose, 3-deoxy-3-fluoro D-glucose, D-xylose, and L-sorbose. Yields and distributions of acceptor products depend on the relative and absolute concentrations of sucrose and acceptors (Pelenc et al., 1991).

The catalytic action of alternansucrase may be viewed either as the synthesis of alternan or as the cleavage of sucrose. It can thus be measured in three different ways: (1) by monitoring the disappearance of sucrose; (2) by the release of D-fructose; or (3) by the formation of polymeric alternan. Each of these approaches has certain advantages and drawbacks, as discussed in the section on dextran and dextransucrase. If one is interested in the cleavage of sucrose, the most convenient measurement is to analyze the amount of reducing sugar (i.e., D-fructose) released. However, if one is interested in the formation of polymeric product, then the radiometric analysis of alternan synthesis is most convenient (Côté and Robyt, 1982a). This procedure measures the amount of ^{14}C-glucose from ^{14}C-sucrose incorporated into alcohol-insoluble polysaccharide within a given time period (Germaine et al., 1974). In nature, dextransucrase and alternansucrase typically occur in the same organisms; therefore, in order to analyze for one or the other, some sort of selective assay is required. One convenient method for accomplishing this involves performing a radiometric assay in the usual manner. After measuring the total incorporation of ^{14}C-glucose into polysaccharide, the ^{14}C-labeled polysaccharide can then be digested with dextranase. After washing away the ^{14}C-dextran hydrolysis products, the ^{14}C-label remaining undigested is assumed to be alternan. In this manner, the relative proportion of total glucansucrase activity due to alternansucrase may be estimated with a reasonable degree of accuracy (Leathers et al., 1995).

For most crude or partially purified alternansucrase preparations, the incorporation of labeled glucose into alternan proceeds in a linear fashion with time. However, this is not always the case with purified enzyme preparations. Figure 3 shows a typical assay curve for purified alternansucrase, obtained by measuring the amount of ^{14}C-glucose incorporated into methanol-insoluble polysaccharide over time. The reaction rate clearly increases exponentially during the earlier stages of the reaction. Addition of exogenous alternan results in a slight enhancement of the reaction rate, but does not markedly affect the shape of the curve. Crude alternansucrase preparations that contain high concentrations of associated alternan do not usually exhibit this type of activity curve.

The recent finding that alternansucrase possesses a glucan-binding site (Arguelo-Morales, 2000) offers one possible explanation for this phenomenon. It may be that binding of endogenously formed alternan to this site results in an allosteric activation of the enzyme. This would be analogous to the allosteric activation of B-512F dextransucrase by dextran (Robyt et al., 1995).

5.3 Alternansucrase: Genetics and Regulation

In practical terms, *L. mesenteroides* NRRL B-1355 alternansucrase is an inducible enzyme, and significant levels are not produced unless sucrose is present in the growth medium. Dextransucrase is simultaneously produced at approximately equal levels, and this results in a mixture containing an alternan–alternansucrase complex and a dextran–dextransucrase complex. Several groups have successfully created mutants which lack the ability to produce one of the two glucansucrases, or in which one of the enzymes is produced constitutively.

Kim and Robyt (1994) described several chemically mutated strains of *L. mesenteroides* which were constitutive for glucansucrases. One of these, B-1355C, was derived

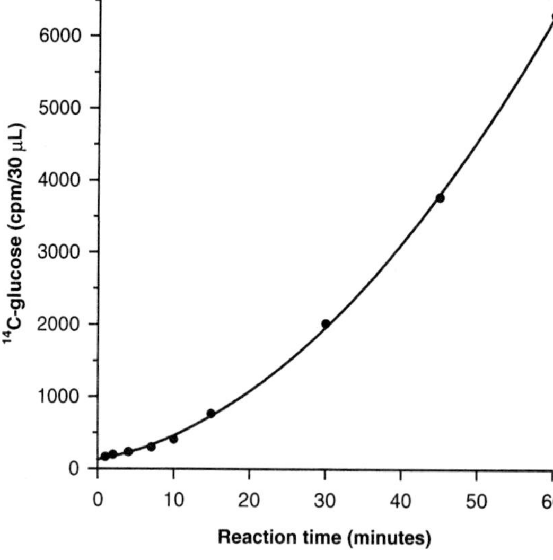

Fig. 3 Alternansucrase-catalyzed incorporation of ^{14}C-glucose into labeled alternan as a function of time. The substrate was uniformly labeled ^{14}C-sucrose; alternan was determined as methanol-insoluble ^{14}C-polysaccharide. The exponential increase is typical of purified alternansucrase preparations.

from NRRL B-1355 and found to produce alternansucrase when grown on glucose media, although at lower levels than on sucrose media. It could not be determined if the lowered activity was due to the production of less enzyme or to decreased stability of the enzyme in the absence of alternan. Using an improved screening procedure, Kitaoka and Robyt (1998) later isolated a mutant that produced enhanced levels of alternansucrase when grown on glucose media.

Researchers at the USDA Western Regional Research Laboratory have described a number of chemically induced mutants that are altered in their ability to produce alternansucrase and dextransucrase (Smith et al., 1994). Some of these produce alternansucrases of lowered molecular mass (Smith and Zahnley, 1997), while others are constitutive for alternansucrase production (Smith and Zahnley, 1999). A new type of glucan was discovered in at least one of these strains which differs from both alternan and the previously known B-1355 fraction L dextran (Smith et al., 1998; Côté et al., 1999a). The glucansucrase responsible for its synthesis, GTF-1, was previously undetected in the wild-type parent strain, but has since been observed in minute amounts (Zahnley and Smith, 1995). Its relationship to alternansucrase and dextransucrase from this strain is uncertain.

Leathers et al. (1995, 1997a) have described the isolation and properties of some UV-induced mutants of *L. mesenteroides* NRRL B-1355 in which the ability to produce dextransucrase is diminished or lacking, and the production of alternansucrase is enhanced. One of these mutants, NRRL B-21138, produced alternansucrase at the same level as the parent strain but secreted little or no dextransucrase. Strain B-21297, which was derived from B-21138, produces the same ratio of alternansucrase to dextransucrase as B-21138, but the total glucansucrase activity is enhanced.

6 Biodegradation

There is no evidence that the *L. mesenteroides* strains that make alternan are capable of degrading the polymer, which apparently does not serve any role in the storage of energy or carbon. In fact, alternan is remarkably resistant to biodegradation by other microorganisms, suggesting that it might play some sort of protective role for the bacterium. Whereas many fungi and bacteria have been described which secrete endodextranases, until recently, none was known that secreted an endoalternanase, and only two exo-glucanases were known to degrade alternan. Several enzymes that have been tested have no action on alternan, including α-amylase, yeast isomaltase, fungal glucoamylase, and α,α-trehalase (Côté, 1992a). Glucodextranase, an exo-$\alpha(1 \rightarrow 6)$ glucosidase, does release some D-glucose from alternan, but hydrolysis is extremely limited (Misaki et al., 1980). A report that *Penicillium* endodextranase hydrolyzes alternan to the extent of about 7% (Misaki et al., 1980) has not been reproducible in our hands, which suggests that the alternan preparation used in that experiment may have been contaminated with traces of B-1355 fraction L dextran. This scarcity of alternanase-degrading organisms may simply be a reflection of the relatively limited distribution of alternan-producing bacteria.

The structural studies of alternan were aided in part by the use of an enzyme known as isomaltodextranase (Torii et al., 1976; Misaki et al., 1980). This exolytic enzyme was first described as a dextranase, but was also found to degrade alternan extensively

(Sawai et al., 1978). The main product from NRRL B-512F dextran is isomaltose. From alternan, the major product is also isomaltose, but significant amounts of the trisaccharide α-D-Glc*p*(1 → 3)-α-D-Glc*p*(1 → 6)-D-Glc are also formed. Torii et al. (1976) showed that *Arthrobacter globiformis* T6 (NRRL strain B-4425) isomaltodextranase is capable of bypassing the α-(1 → 3) linked residues at the nonreducing ends of alternan to give rise to this trisaccharide. The remaining biopolymer is then hydrolyzed to isomaltose units, except where branch points cannot be bypassed. A low molecular-weight "limit alternan" remains after extensive digestion with isomaltodextranase (Misaki et al. 1980); this limit alternan represents a more highly branched core of the alternan molecule. It may be considered analogous to the β-limit dextrin remaining after hydrolysis of amylopectin by β-amylase. A similar isomaltodextranase has been isolated from an *Actinomadura* sp. (NRRL strain B-11411) (Sawai et al., 1981), its action on alternan being similar to that of the *Arthrobacter* enzyme, though it is apparently capable of more completely degrading alternan. The rate of hydrolysis of α(1 → 3) linkages is also more rapid.

Walker and Hare (1979) subjected B-1355 alternan to hydrolysis by an exo-(1 → 3)-α-D-glucanase. The enzyme released only about 1% of D-glucose, but subsequent hydrolysis of the treated alternan by *A. globiformis* isomaltodextranase yielded far less of the trisaccharide α-D-Glc*p*(1 → 3)-α-D-Glc*p*(1 → 6)-D-Glc than observed on previously untreated alternan. This was interpreted as evidence for the presence of a significant proportion of α-(1 → 3) linkages at the nonreducing ends of alternan.

Wyckoff et al. (1996) have described several bacterial isolates capable of growth on alternan. These bacteria – all of which are members of a unique species of *Bacillus* – secrete an enzyme that endohydrolytically cleaves alternan into oligosaccharide fragments. Enzyme from the most productive strain was purified (Biely et al., 1994) and its products characterized (Côté and Biely, 1994; Bradbrook et al., 2000). Alternanase (as the enzyme is known) converts alternan into some unusual products. Instead of simply cleaving the alternan chain into linear fragments, it produces a mixture of linear and cyclic oligosaccharides. In its pure form, alternanase does not produce any D-glucose from alternan, its main products being isomaltose, the trisaccharide α-D-Glc*p*(1 → 3)-α-D-Glc*p*(1 → 6)-D-Glc, and the cyclic tetrasaccharide *cyclo*{→ 6)α-D-Glc*p*(1 → 3)-α-D-Glc*p*(1 → 6)-α-D-Glc*p*(1 → 3)-α-D-Glc*p*(1 →} (Figure 4). A branched pentamer is also formed which contains a D-glucopyranosyl residue linked to the cyclic tetramer through an α-(1 → 6) linkage. This branched compound may arise from branch points in the alternan molecule.

This cyclic alternan tetramer represents the smallest naturally produced cyclic oligosaccharide known. It is noteworthy that alternanase produces no nigerose from alternan, nor has any linear tetrasaccharide

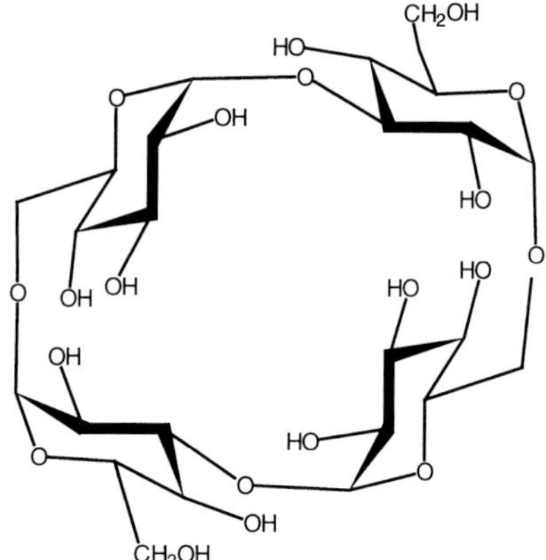

Fig. 4 Chemical structure of cyclic tetrasaccharide formed by alternanase-mediated degradation of alternan.

been isolated from alternanase hydrolyzates. Recent studies indicate that alternanase is a hydrolase/transglycosylase that recognizes the isomaltosyl functional group and either transfers it to water (hydrolysis) or to another isomaltosyl unit to form the cyclic tetramer (Côté et al., 2001).

7
Biotechnological Production

At present, alternan is not produced on a commercial scale. The traditional method for producing the polysaccharide for laboratory use has been to grow *L. mesenteroides* NRRL B-1355 on a suitable medium containing approximately 2% sucrose. Cells and other insoluble matter are removed by centrifugation, and ethanol is slowly mixed into the supernatant solution to a concentration of 38% by volume. This causes the fraction L dextran to precipitate. After removal of the dextran by centrifugation, the ethanol concentration is then adjusted to 42%, which precipitates the alternan. Alternan is collected by centrifugation and subsequently dried (Wilham et al., 1955). Alternan thus prepared may be redissolved in water for additional purification steps, such as dialysis to eliminate traces of sucrose, fructose and other sugars, or refractionation by ethanol precipitation to remove coprecipitated dextran. Careful control of the ethanol precipitation is necessary to ensure the purity of alternan prepared in this way. The ethanol concentration at which the two D-glucans precipitate depends to some degree on the polysaccharide concentration and temperature, as well as the molecular weight distribution of the product.

Much of what has been published on the optimization of fermentation parameters for dextran production (see Chapter 12) can also be applied to the fermentative production of alternan. Some data have been published pertaining specifically to the production of alternansucrase, but the purpose of those researchers was to optimize enzyme production, and not polysaccharide production (López-Munguía et al., 1991, 1993; Raemaekers and Vandamme, 1997).

Because of the need to separate alternan from dextran in cultures of the wild-type organism, strains have been developed which produce alternan but no dextran (Section 4.3). Screening large numbers of mutants for strains that produce only alternan was the challenge facing Leathers et al. (1995), who solved the problem by treating colonies with endodextranase, followed by a reagent that would reveal the presence of reducing sugars. This method allowed them to determine which mutated strains were still capable of producing dextran. Strains that produced extracellular polysaccharide from sucrose but did not produce dextran were presumed to be alternan producers. Confirmation was provided by isolating the polysaccharides produced by the dextran-negative strains and subjecting them to structural analysis. One stable strain thus isolated, NRRL B-21297, produces enhanced levels of alternansucrase and negligible levels of dextransucrase, and is currently used in the author's laboratory for routine production of alternan. This strain and others derived by the same screening protocol have been patented (Leathers et al., 1997b, 1998).

Kitaoka and Robyt (1998) approached the problem of developing mutant strains in a different manner. They were not interested in finding a strain that produced only alternansucrase, but instead sought strains that would produce glucansucrase constitutively. Part of their goal was to obtain polysaccharide-free enzyme for research purposes. In their method, cultures of chemically mutated bacteria were first grown on glucose media, after which microplate assays were performed to seek cultures that contained measurable levels of glucansucrase activity. Although the method did not specifically screen for dextransucrase-lacking mutants, they were successful in isolating a constitutive mutant that produced only alternansucrase (strain B-1355C/CF10G3).

Although alternan can be produced fermentatively using the aforementioned mutant strains, most researchers find it convenient use cell-free enzyme preparations to produce alternan on a laboratory scale. In a typical example, *L. mesenteroides* NRRL B-21297 is grown on a suitable medium in the presence of both sucrose and maltose as carbon and energy sources. Maltose plays two roles in the growth medium: (1) it serves as an additional carbon and energy source; and (2) more importantly, it acts as an acceptor for alternansucrase. By including a good acceptor in the medium, alternan production is greatly curtailed, as glucosyl units are instead transferred to maltose to form oligosaccharides. This results in a much less viscous medium that can be filtered and concentrated much more easily. After cells are removed by centrifugation, the cell-free culture medium is concentrated by ultrafiltration. The concentrate can be dialyzed against pH 5.4 buffer to remove low molecular-weight compounds and to bring the pH into the optimal range for enzyme activity. At this point, the concentrated culture fluid is suitable for most applications. It can be stored frozen for many months with minimal loss of activity, and is also stable to freeze-drying, provided that sufficient alternan is associated with the enzyme to maintain stability.

For laboratory production of alternan by use of soluble enzyme, sucrose is dissolved in either water or pH 5.4 acetate buffer. A suitable amount of enzyme solution is added, the amount depending on the specific activity and the calculated time required to convert all of the sucrose. A sucrose concentration around 10% (w/v) is typically used, though higher concentrations are feasible. The molecular weight of alternan depends somewhat on the sucrose concentration used for its synthesis. At high sucrose concentrations (20–50%) a much more polydisperse product has been noted, with a lower average molecular weight. Temperature also plays a role, with lower temperatures requiring longer reaction times but yielding slightly higher average molecular weights. A temperature range of 25 to 35 °C represents a suitable compromise between greater enzyme activity at higher temperatures and greater enzyme stability at lower temperatures.

Once all of the sucrose is consumed (calculated according to the enzyme activity, or measured by thin-layer chromatography), the alternan can be harvested. The customary method involves adding ethanol to a final concentration of at least 42% by volume. The alternan that precipitates is collected by centrifugation and can be dried or further purified. In our laboratory, we have found it to be convenient first to concentrate the alternan solution by ultrafiltration using 300,000 dalton molecular weight cut-off membranes in a tangential flow apparatus. The highly viscous solution is then mixed with an equal volume of ethanol, and the precipitated alternan is collected by suction filtration on sintered glass. Washing the precipitate with 50% ethanol removes most of the coprecipitated fructose and buffer salts, whereupon the product can be oven-dried. After grinding or milling, alternan prepared in this way is a white powder that is suitable for most applications.

Like dextran, alternan is produced by a single enzyme, and consequently such a system would seem ideal for immobilization. However, certain problems are inherent in the production of a high molecular-weight polysaccharide by use of an immobilized enzyme. If the enzyme is immobilized on a porous support such as polyacrylamide gel, the newly synthesized polysaccharide quickly fills up the gel matrix and clogs the pores. The problem is somewhat mitigated if a nonporous matrix is used, but surface area then becomes a limiting factor (Monsan et al., 1987). We found that alternansucrase could be immobilized very easily by adsorption to a hydrophobic matrix such as phenyl agarose beads, but that enzyme immobilized in this way is practically useless for the production of alternan, due to rapid pore clogging by newly synthesized alternan. We were, however, able to immobilize alternansucrase by covalent coupling to glutaraldehyde-activated polyvinyl chloride (PVC)–silica sheets in a spiral-wound cartridge; the immobilized-enzyme cartridges showed good retention of activity and could be used repeatedly and continuously for the production of alternan (Côté and Ahlgren, 1994). Chemical and spectroscopic analysis of the product showed it to be identical to alternan produced by soluble enzyme. Alternansucrase immobilized in this way is extremely stable and suitable for the production of oligosaccharides via acceptor reactions.

Kossmann et al. (2000) have filed a patent application describing the cloning of alternansucrase and methods for the creation of transgenic plants containing the gene. If alternansucrase can be successfully produced in crop plants, this will provide a cheap source of the enzyme. Furthermore, if the enzyme is expressed in plants which accumulate sufficient levels of sucrose, it may be possible to simply extract alternan from the plants themselves.

8 Properties of Alternan

As normally produced (either fermentatively or enzymatically), alternan is a high molecular-weight, water-soluble gum. When dissolved in water, the native material produces an opalescent, bluish-white solution. Solutions of at least 10% (w/v) are readily obtained by simply adding dry alternan to water and stirring on a magnetic stirrer. Higher concentrations can be attained by blending or mechanical mixing and tend to be increasingly pasty or gel-like at very high concentrations. Heating may be used to hasten hydration and dissolution. These solutions are stable to heat and prolonged standing. Since few microorganisms are capable of degrading alternan, microbial growth is usually not observed unless the sample contains traces of sugars or protein.

8.1 Physical Properties

Compared with other microbial gums such as xanthan or dextran, alternan exhibits a remarkably low viscosity for such a high molecular-weight polymer. According to Jeanes et al. (1954), the intrinsic viscosity of B-1355 alternan was 0.193 dL g^{-1}, compared with values of 0.953 for B-512F dextran and 1.115 for B-1355 fraction L dextran.

As Figure 5 clearly illustrates, native alternan is much less viscous than native B-512F dextran. For example, a 5% (w/v) solution of alternan is a thin liquid not much different from water in viscosity, whereas a 5% (w/v) solution of commercial dextran is so thick as to be very difficult to pipette or pump at room temperature. The viscosity of alternan does increase exponentially with concentration (Figure 5).

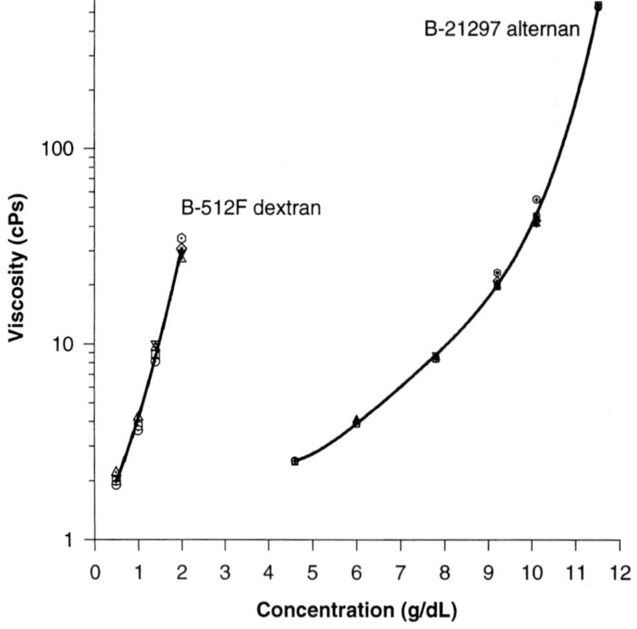

Fig. 5 Comparison of viscosities of commercial *L. mesenteroides* NRRL B-512F dextran and native, high molecular-weight *L. mesenteroides* NRRL B-21297 alternan. Points at each concentration value represent readings taken at various shear rates, according to the methods described in Côté et al. (1997).

In order to obtain fractions with even lower viscosity, it is possible to depolymerize alternan either physically, chemically, or enzymatically to lower molecular-weight forms. Sonication at 20 kHz was used to depolymerize alternan from its native weight-average molecular weight of 1×10^7 to approximately 9×10^5 (Côté, 1992a). Sonication, unlike most chemical methods of depolymerization, breaks the polymer chains near the middle and yields a fairly uniform distribution of products. Figure 6 depicts the viscosity of alternan sonicated for several hours at 1.5 MHz relative to the native material. At low concentrations, the viscosities of both are so low that they are essentially indistinguishable. At higher concentrations, the divergence of the two curves shows that the viscosity of native alternan increases at a much greater rate than that of the sonicated material. In fact, the viscosity of sonicated alternan is very similar to that of gum arabic (Côté, 1992a). It should be noted that the frequency and time of sonication play important roles in determining the degree of depolymerization of any polymer (Côté and Willett, 1999).

The action of isomaltodextranase on alternan was discussed earlier, where it was noted that extensive hydrolysis with *A. globiformis* NRRL B-4425 isomaltodextranase resulted in a limit alternan. The molecular weight of this limit alternan is in the range of 3000–5000, with a number-average molecular weight of 3500. This corresponds to an average DP of approximately 22. This fraction is extremely soluble in water, and is rheologically similar to maltodextrins with an average DP of 10 (Côté, 1992a).

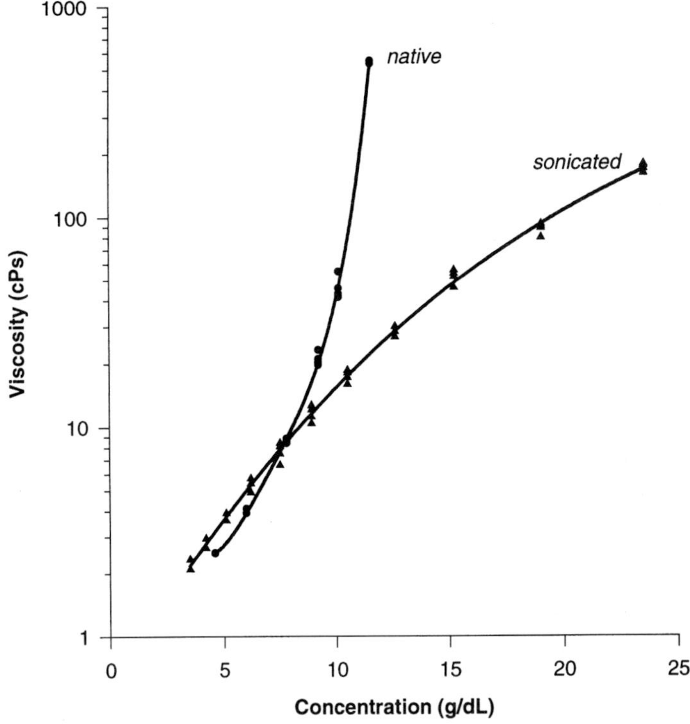

Fig. 6 Comparison of viscosities of native high molecular-weight alternan and alternan extensively sonicated at 1.5 MHz. Methods described in Côté et al. (1997) and Côté and Willet (1999).

8.2
Biological Properties

Until recently, most scientific interest in alternan has focused on its immunochemical properties. Because of its use as a blood plasma extender, the antigenicity of dextran has been studied in great detail. Dextrans are also known to react with other carbohydrate-binding proteins, including lectins and antibodies, and have been used to study the specificities of these proteins. The strong interaction with the lectin concanavalin A was exploited in one procedure for the separation of alternansucrase from dextransucrase (Côté, 1992b).

Due to the high value reported for $\alpha(1 \rightarrow 3)$ linkage content in B-1355 alternan (Jeanes et al., 1954), this glucan has been extensively used as the prototypical "high $\alpha(1 \rightarrow 3)$ content" dextran in immunochemical studies. Unfortunately, some of the early workers tended not to differentiate between $\alpha(1 \rightarrow 3)$ linkages at branch points and those in linear, nonconsecutive sequences. This led to conflicting results among the various reports – a problem that was compounded by the somewhat erroneous values for various linkage types obtained by periodate oxidation (Jeanes et al., 1954; Jeanes and Seymour, 1979). As the structures of alternan and other branched dextrans became more clearly understood, the immunochemistry of dextrans became somewhat less ambiguous. As synthetic oligosaccharides become more readily available, as the use of monoclonal antibodies has become standard practice, and as the field of glycobiology has developed, the use of dextrans in immunological studies has fallen off drastically. In her last scientific publication, Jeanes (1986) extensively reviewed the immunochemistry of dextrans, including alternan, in light of improved structural information on the various dextrans.

9 Production and Applications

Although at the time of writing alternan is not used for any commercial application, and is not produced outside research laboratories, it does show some promise for a number of uses. The initial renewal of interest in alternan during the late 1980s arose from its potential as a noncaloric filler and binder in low-calorie foods. Currently, there are two major obstacles preventing its use in such applications. The first is cost, and the second is the uncertain legal status of alternan and alternan oligosaccharides as food ingredients. The recent finding that alternan occurs naturally in some fermented Asian foods (Jung et al., 1999) may be of help in removing the latter obstacle.

Côté et al. (1997) have reviewed the commercial potential for alternan and related fractions as food additives, with several different fractions being considered. Native alternan is potentially the least expensive of these fractions, as it would require no additional processing to alter its molecular weight. Although its applicability as a food additive is probably limited to use as a binder and filler, other nonfood uses have also been investigated.

Alternan can be used as a wet-end additive in the papermaking process, where it acts as a filter aid and also enhances the dry strength of paper. It is effective at lower concentrations than those at which modified starch is typically used. If the issue of cost can be addressed, this may represent a large potential market for alternan and, as previously mentioned, the use of immobilized enzymes may help to mitigate the cost factor in this respect. Growth of *Leuconostoc* on inexpensive media is also expected to bring down the cost of producing the enzyme (Leathers, 1998). Similar glucans from various species of *Streptococcus* are also being investigated for this purpose. Indeed, the streptococcal enzymes have been cloned and expressed in maize, with the intention of producing a grain containing mixtures of starch and dextran for use in papermaking (Nichols, 2000a–c).

Although few chemical derivatives of alternan have been described, one interesting derivative is an oxidized alternan in which the $\alpha(1 \rightarrow 3)$ substituted residues have been converted to D-glucuronic acid by catalytic oxidation in the presence of 2,2,6,6-tetramethyl-1-piperidine-*N*-oxide (TEMPO) (Chang and Robyt, 1996). Oxidation in this manner reportedly enhanced solubility while not significantly altering the intrinsic viscosity, although the viscosity value reported for native alternan differed from that reported by Jeanes et al. (1954). The oxidized alternan did not appear to form a gel in the presence of calcium ions. The molecular weight distribution was not reported.

One modified alternan product that has been investigated for its commercial potential is sonicated alternan. This product is rheologically similar to gum arabic (Côté, 1992a) and has been considered as a potential substitute for this plant gum (Hardin, 1999). Unfortunately, the major use of gum arabic is as an emulsifier, its ability to form and stabilize emulsions being due largely to its proteoglycan nature, with the protein portion of the molecule playing a major role in its amphiphilic properties (Qi et al., 1991). Alternan, being a pure polysaccharide, does not exhibit the same capacity for emulsification, and thus lacks the key properties that would allow it to compete with gum arabic for this market. Nevertheless, it may be possible to combine alternan or sonicated alternan with a protein to produce a complex with suitable emulsifying capabilities (Kato et al., 1990; Ho et al., 2000).

Isomaltodextranase-limit alternan, as previously mentioned, is similar to DP 10 maltodextrin in many respects. It is also very similar to the commercial bulking agent polydextrose (Côté, 1992a). As a more natural material, it may possess certain market advantages over polydextrose if the issue of cost could be overcome. Côté et al. (1997) have suggested possible approaches for addressing some of the cost-related issues, including enhanced production of alternansucrase and isomaltodextranase, improved processing, and utilization of the coproducts fructose and isomaltose.

The oligosaccharides derived from alternan may also find uses in foods. There is much interest in the use of oligosaccharides as prebiotics (VanLoo et al., 1999; Flickinger et al., 2000; LaMont, 2000). It is encouraging to note that related oligosaccharides from *L. mesenteroides* NRRL B-1299 are being produced on a commercial scale as prebiotics for use in cosmetics and personal healthcare products (Valette et al., 1993; Djouzi et al., 1995; Lamothe et al., 1996). Oligosaccharides arising from alternansucrase acceptor reactions (Côté and Robyt, 1982b; Pelenc et al.,1991; Argüello-Morales et al., 2001) or from alternanase hydrolyzates (Côté and Biely, 1994) may also find similar uses. Although the oligoalternan acceptor products are in the public domain, alternanase and the cyclic oligosaccharides arising therefrom are covered by several recent patents (Côté et al., 1998, 1999b; Côté and Biely, 1999).

Alternan may be useful in many applications in which low molecular-weight dextran fractions have been used. One application we have investigated is in two-phase aqueous extraction systems with poly(ethylene glycol) for the purification of proteins (Kula, 1990), though more research is still needed in this area. Others possible uses include ink formulations (Kondo and Matsubara, 1995), ceramic casting (Schilling et al., 1999), concrete blends (DeWacker, 1996), and aluminum ore processing (Connelly et al., 1995).

10 Outlook and Perspectives

It is difficult to predict exactly where or how alternan will first enter the market as a commercial product. Native, high molecular-weight alternan may be most readily produced using the same fermentation technologies that are used to produce dextran, and this would require little in the way of new infrastructure, as existing plants could be adapted to use alternan-producing bacteria rather than dextran-producing strains. For specialty applications, where a price of US$1–1.5 per kilogram is acceptable, this would be suitable. However, in commodity applications, where the price must be significantly less than US$0.5 per kilogram, other production technologies will be necessary – this is a factor, for instance, in paper processing and production. One possible solution might be to express the alternansucrase gene in a plant host, where endogenous sucrose can be converted to alternan. Recent patents describe the expression of various glucansucrases in maize (Nichols, 2000a–c). Since alternansucrase has been recently cloned, the expression of alternansucrase in maize or other crop plants may not be very far off (Kossmann et al., 2000).

As discussed in Section 9, alternan oligomers and oligosaccharides arising from alternansucrase acceptor reactions may be suitable as prebiotics in animal feeds, cosmetics, and human food and nutritional supplements. In the case of acceptor products, it appears likely that immobilized enzymes will be very useful in converting sugar to oligosaccharides on a commercially viable scale. Insoluble enzymes are currently used in Europe to produce branched oligodextrans on a scale of approximately 50 tonnes per year (P. Monsan, personal communication) and, since the prebiotics market is relatively new, this could be expected to grow substantially in the coming years. The range of acceptors for alternansucrase appears to be significantly greater than that for dextransucrase, and its utility in the production of heterooligosaccharides has yet to be fully explored.

11 Patents

Few patents exist that pertain specifically to alternan and alternansucrase, and none of these dates earlier than 1996. It should be noted that prior to 1982, alternan was considered to be a dextran, so some patents issued prior to that date may or may not include alternan as a member of that broad class. The known patents that relate directly to alternan are listed in Table 1.

Tab. 1 Patents and patent applications pertaining to alternan

Patent No.	*Date*	*Inventors*	*Assignee*	*Title*
WO 96/04365*	15 February 1996	Leathers, T.D., Hayman, G.T., Cote, G.L.	USDA	Screening method and strains to produce more alternan than dextran
US 5702942	30 December 1997	Leathers, T.D., Hayman, G.T., Cote, G.L.	USDA	Microorganisms that produce a high proportion of alternan to dextran
US 5786196	28 July 1998	Cote, G.L., Wyckoff, H., Biely, P.	USDA	Bacteria and enzymes for production of alternan fragments
US 5789209	4 August 1998	Leathers, T.D., Hayman, G.T., Cote, G.L.	USDA	Rapid screening method to identify microorganism strains that produce a high proportion of alternan to dextran
US 5888776	30 March 1999	Cote, G.L., Wyckoff, H., Biely, P.	USDA	Bacteria and enzymes for production of alternan fragments
US 5889179	30 March 1999	Cote, G.L., Biely, P.	USDA	Bacteria and enzymes for production of alternan fragments
KR260849	1 July 2000	Kwon, C.H.	Charmzone	*Leuconostoc mesenteroides* CBI-110 and method for preparing alternan using this
WO 00/47727 (DE19905069)	17 August 2000	Kossmann, J., Welsh, T., Quanz, M., Knuth, K.	Plantec Biotechnologie	Nucleic acid molecules encoding alternansucrase

*The following patents are essentially identical: AU3235295, CA2196596, EP0804545.

12
References

Argüello-Morales, M. A., Remaud-Simeon, M., Pizzut, S., Sarcabal, P., Willemot, R. M., Monsan, P. (2000) Sequence analysis of the gene encoding alternansucrase, a sucrose glucosyltransferase from *Leuconostoc mesenteroides* NRRL B-1355. *FEMS Microbiol. Lett.* **182**, 81–85.

Argüello-Morales, M. A., Remaud-Simeon, M., Willemot, R. M., Vignon, M. R., Monsan, P. (2001) Novel oligosaccharides synthesized from sucrose donor and cellobiose acceptor by alternansucrase. *Carbohydr. Res.* **331**, 403–411.

Biely, P., Côté, G. L., Burgess-Cassler, A. (1994) Purification and properties of alternanase, a novel endo-α-1,3-α-1,6-D-glucanase. *Eur. J. Biochem.* **226**, 633–639.

Bradbrook, G. M., Gessler, K., Côté, G. L., Momany, F., Biely, P., Bordet, P., Pérez, S., Imberty, A. (2000) X-ray structure determination and modeling of the cyclic tetrasaccharide *cyclo*{→6)-α-D-Glc*p*(1→3)-α-D-Glc*p*(1→6)-α-D-Glc*p*(1→3)-α-D-Glc*p*(1→}. *Carbohydr. Res.* **329**, 655–665.

Castillo, E., Iturbe, F., López-Munguía, A., Pelenc, V., Paul, F., Monsan, P. (1992) Dextran and oligosaccharide production with glucosyltransferases from different strains of *Leuconostoc mesenteroides*. *Ann. NY Acad. Sci.* **672**, 425–430.

Chang, P. S., Robyt, J. F. (1996) Oxidation of primary alcohol groups of naturally occurring polysaccharides with 2,2,6,6-tetramethyl-1-piperidine oxoammonium ion. *J. Carbohydr. Chem.* **15**, 819–830.

Connelly, L. J., Mahoney, R. P., Kaesler, R. W., Wetegrove, R. L. (1995) Bayer liquor polishing. U.S. Patent 5387405.

Côté, G. L. (1992a) Low-viscosity α-D-glucan fractions derived from sucrose which are resistant to enzymatic digestion. *Carbohydr. Polym.* **19**, 249–252.

Côté, G. L. (1992b) The use of immobilized concanavalin A for the separation of alternansucrase from dextransucrase in culture broth of *Leuconostoc mesenteroides* NRRL B-1355. *Biotechnol. Techn.* **6**, 45–48.

Côté, G. L., Ahlgren, J. A. (1994) Production, isolation, and immobilization of alternansucrase. *Am. Chem. Soc. (Abstracts of meeting)* **207**, CARB 12 (Abstract).

Côté, G. L., Ahlgren, J. A. (2001) The hydrolytic and transferase action of alternanase on oligosaccharides. *Carbohydr. Res.* **332**, 373–379.

Côté, G. L., Biely, P. (1994) Enzymically produced cyclic α-1,3-linked and α-1,6-linked oligosaccharides of D-glucose. *Eur. J. Biochem.* **226**, 641–648.

Côté, G. L., Biely, P. (1999) Bacteria and enzymes for production of alternan fragments. U.S. Patent 5889179.

Côté, G. L., Robyt, J. F. (1982a) Isolation and partial characterization of an extracellular glucansucrase from *Leuconostoc mesenteroides* NRRL B-1355 that synthesizes an alternating (1→6),(1→3)-α-D-glucan. *Carbohydr. Res.* **101**, 57–74.

Côté, G. L., Robyt, J. F. (1982b) Acceptor reactions of alternansucrase from *Leuconostoc mesenteroides* NRRL B-1355. *Carbohydr. Res.* **111**, 127–142.

Côté, G. L., Robyt, J. F. (1983) The formation of α-D-(1→3) branch linkages by an exocellular glucansucrase from *Leuconostoc mesenteroides* NRRL B-742. *Carbohydr. Res.* **119**, 141–156.

Côté, G. L., Robyt, J. F. (1984) The formation of α-D-(1→3) branch linkages by a D-glucansucrase from *Streptococcus mutans* 6715 producing a soluble D-glucan. *Carbohydr. Res.* **127**, 95–107.

Côté, G. L., Willet, J. L. (1999) Thermomechanical depolymerization of dextran. *Carbohydr. Polym.* **39**, 119–126.

Côté, G. L., Wyckoff, H. A., Biely, P. (1998) Bacteria and enzymes for production of alternan fragments. U.S. Patent 5786196.

Côté, G. L., Leathers, T. D., Ahlgren, J. A., Wyckoff, H. A., Hayman, G. T., Biely, P. (1997) Alternan and highly branched limit dextrans: low-viscosity polysaccharides as potential new food ingredients. In: Okai, H., Mills, S., Spanier, A. M., Tamura, M. (Eds.), *Chemistry of Novel Foods*, Carol Stream, IL, USA: Allured Publishing, 95–109.

Côté, G. L., Ahlgren, J. A., Smith, M. R. (1999a) Some structural features of an insoluble α-D-glucan from a mutant strain of *Leuconostoc mesenteroides* NRRL B-1355. *J. Ind. Microbiol. Biotechnol.* **23**, 656–660.

Côté, G. L., Wyckoff, H. A., Biely, P. (1999b) Bacteria and enzymes for production of alternan fragments. U.S. Patent 5888776.

DeWacker, D. R. (1996) Cement mortar systems using blends of cold-water-soluble, unmodified starches. U.S. Patent 5575840.

Dimler, R. J., Jones, R. W., Schaefer, W. C., VanCleve, J. W. (1956) Variations in structure among selected dextrans as indicated by hydrolysis and methylation techniques. *Am. Chem. Soc. (Abstracts of meeting)* **129**, 2D:#4 (Abstract).

Djouzi, Z., Andrieux, C., Pelenc, V., Somarriba, S., Popot, F., Paul, F., Monsan, P., Szylit, O. (1995) Degradation and fermentation of α-gluco-oligosaccharides by bacterial strains from human colon: *in vitro* and *in vivo* studies in gnotobiotic rats. *J. Appl. Bacteriol.* **79**, 117–127.

Flickinger, E. A., Wolf, B. W., Garleb, K. A., Chow, J., Leyer, G. J., Johns, P. W., Fahey, G. C. (2000) Glucose-based oligosaccharides exhibit different *in vitro* fermentation patterns and affect in vivo apparent nutrient digestibility and microbial populations in dogs. *J. Nutr.* **130**, 1267–1273.

Germaine, G. R., Schachtele, C. F., Chludzinski, A. M. (1974) Rapid filter paper assay for the dextransucrase activity from *Streptococcus mutans*. *J. Dent. Res.* **53**, 1355–1360.

Goldstein, I. J., Whelan, W. J. (1962) Structural studies of dextrans. Part I. A dextran containing α-1,3-glucosidic linkages. *J. Chem. Soc.* 170–175.

Hardin, B. (1999) Nonfattening food additives – from sugar? *Agric. Res.* **47**, 10–11.

Hare, M. D., Svensson, S., Walker, G. J. (1978) Characterization of the extracellular, water-insoluble α-D-glucans of oral streptococci by methylation analysis and by enzymic synthesis and degradation. *Carbohydr. Res.* **66**, 245–264.

Ho, Y. T., Ishizaki, S., Tanaka, M. (2000) Improving emulsifying activity of ε-polylysine by conjugation with dextran through the Maillard reaction. *Food Chem.* **68**, 449–455.

Janeček, Š., Svensson, B., Russell, R. R. B. (2000) Location of repeat elements in glucansucrases of *Leuconostoc* and *Streptococcus* species. *FEMS Microbiol. Lett.* **192**, 53–57.

Jeanes, A. (1986) Immunochemical and related interactions with dextrans reviewed in terms of improved structural information. *Mol. Immunol.* **23**, 999–1028.

Jeanes, A., Seymour, F. R. (1979) The α-D-glucopyranosidic linkages of dextrans: comparison of percentages from structural analysis by periodate oxidation and by methylation. *Carbohydr. Res.* **74**, 31–40.

Jeanes, A., Wilham, C. A. (1950) Periodate oxidation of dextran. *J. Am. Chem. Soc.* **72**, 2655–2657.

Jeanes, A., Haynes, W. C., Wilham, C. A., Rankin, J. C., Melvin, E. H., Austin, M. J., Cluskey, J. E., Fisher, B. E., Tsuchiya, H. M., Rist, C. E. (1954) Characterization and classification of dextrans from ninety-six strains of bacteria. *J. Am. Chem. Soc.* **76**, 5041–5052.

Jeanes, A., Haynes, W. C., Wilham, C. A. (1956) Characterization of dextrans from four types of *Leuconostoc mesenteroides*. *J. Bacteriol.* **71**, 167–173.

Jung, H.-K., Kim, K.-N., Lee, H.-S., Jung, S.-H. (1999) Production of alternan by *Leuconostoc mesenteroides* CBI-110. *Kor. J. Appl. Microbiol. Biotechnol.* **27**, 35–40.

Kato, A., Sasaki, Y., Furuta, R., Kobayashi, K. (1990) Functional protein-polysaccharide conjugate prepared by controlled dry heating of ovalbumin-dextran mixtures. *Agric. Biol. Chem.* **54**, 107–112.

Kim, D., Robyt, J. F. (1994) Production and selection of mutants of *Leuconostoc mesenteroides* constitutive for glucansucrases. *Enzyme Microb. Technol.* **16**, 659–664.

Kitaoka, M., Robyt, J. F. (1998) Use of a microtiter plate screening method for obtaining *Leuconostoc mesenteroides* mutants constitutive for glucansucrase. *Enzyme Microb. Technol.* **22**, 527–531.

Kondo, M., Matsubara, N. (1995) Aqueous ink composition for writing instrument. U.S. Patent 5466283.

Koepsell, H. J., Tsuchiya, H. M., Hellman, N. N., Kazenko, A., Hoffman, C. A., Sharpe, E. S., Jackson, R. W. (1953) Enzymatic synthesis of dextran. Acceptor specificity and chain initiation. *J. Biol. Chem.* **200**, 793–801.

Kossmann, J., Welsh, T., Quanz, M., Knuth, K. (2000) Nucleic acid molecules encoding alternansucrase. WIPO Pat. Appl. WO 00/47727.

Kula, M.-R. (1990) Trends and future prospects of aqueous two-phase extraction. *Bioseparation* **1**, 181–189.

Kwon, C. H. (2000) *Leuconostoc mesenteroides* CBI-110 and method for preparing alternan using this. Korean Patent 260849.

LaMont, J. T. (2000) The renaissance of probiotics and prebiotics. *Gastroenterology* **119**, 291.

Lamothe, J.-P. H. G., Marchenay, Y. G., Monsan, P. F., Paul, F. M. B., Pelenc, V. (1996) Cosmetic compositions containing oligosaccharides. U.S. Patent 5518733.

Leathers, T. D. (1998) Utilization of fuel ethanol residues in the production of the biopolymer alternan. *Process Biochem.* **33**, 15–19.

Leathers, T. D., Hayman, G. T., Côté, G. L. (1995) Rapid screening of *Leuconostoc mesenteroides* mutants for elevated proportions of alternan to dextran. *Curr. Microbiol.* **31**, 19–22.

Leathers, T. D., Ahlgren, J. A., Côté, G. L. (1997a) Alternansucrase mutants of *Leuconostoc mesenteroides* strain NRRL B-21138. *J. Ind. Microbiol. Biotechnol.* **18**, 278–283.

Leathers, T. D., Hayman, G. T., Côté, G. L. (1997b) Microorganism strains that produce a high proportion of alternan to dextran and rapid screening method to select same. U.S. Patent 5702942.

Leathers, T. D., Hayman, G. T., Côté, G. L. (1998) Method to identify microorganism strains that produce a high proportion of alternan to dextran. U.S. Patent 5789209.

López-Munguía, A., Pelenc, V., Remaud, M., Paul, F., Monsan, P., Biton, J., Michel, J. M., Lang, C. (1991) Production and purification of *Leuconostoc mesenteroides* NRRL B-1355 alternansucrase. *Ann. N. Y. Acad. Sci.* **613**, 717–722.

López-Munguía, A., Pelenc, V., Remaud, M., Biton, J., Michel, J. M., Lang, C., Paul, F., Monsan, P. (1993) Production and purification of alternansucrase, a glucosyltransferase from *Leuconostoc mesenteroides* NRRL B-1355, for the synthesis of oligoalternans. *Enzyme Microb. Technol.* **15**, 77–85.

McCleskey, C. S., Faville, L. W., Barnett, R. O. (1947) Characteristics of *Leuconostoc mesenteroides* from cane juice. *J. Bacteriol.* **54**, 697–708.

Misaki, A., Torii, M., Sawai, T., Goldstein, I. (1980) Structure of the dextran of *Leuconostoc mesenteroides* B-1355. *Carbohydr. Res.* **84**, 273–285.

Monsan, P., Paul, F., Auriol, D., Lopez, A. (1987) Dextran synthesis using immobilized *Leuconostoc mesenteroides* dextransucrase. In: Mosbach, K., (Ed.), *Methods in Enzymology*, Orlando, FL: Academic Press, 239–254, vol. 136.

Mukasa, H., Shimamura, A., Tsumori, H. (1989) Purification and characterization of cell-associated glucosyltransferase synthesizing insoluble glucan from *Streptococcus mutans* serotype c. *J. Gen. Microbiol.* **135**, 2055–2063.

Nichols, S. E. (2000a) Plant cells and plants transformed with *Streptococcus mutans* genes encoding wild-type or mutant glucosyltransferase B enzymes. U.S. Patent 6087559.

Nichols, S. E. (2000b) Plant cells and plants transformed with *Streptococcus mutans* genes encoding wild-type or mutant glucosyltransferase D enzymes. U.S. Patent 6127602.

Nichols, S. E. (2000c) Plant cells and plants transformed with *Streptococcus mutans* gene encoding glucosyltransferase C enzyme. U.S. Patent 6127603.

Ogawa, T., Kaburagi, T. (1982) Synthesis of a branched D-glucoheptaose: the repeating unit of extracellular α-D-glucan 1355-S of *Leuconostoc mesenteroides* NRRL B-1355. *Carbohydr. Res.* **110**, C12–C15.

Pelenc, V., López-Munguía, A., Remaud, M., Biton, J., Michel, J. M., Paul, F., Monsan, P. (1991) Enzymatic synthesis of oligoalternans. *Sciences des Aliments* **11**, 465–476.

Preobrazhenskaya, M. E., Rosenfeld, E. L., Kandra, L. (1978) Studies on some biologically active dextrans. *Carbohydr. Res.* **66**, 213–223.

Qi, W. U., Fong, C., Lamport, D. T. A. (1991) Gum arabic glycoprotein is a twisted hairy rope. *Plant Physiol.* **96**, 848–855.

Raemaekers, M. H. M., Vandamme, E. J. (1997) Production of alternansucrase by *Leuconostoc mesenteroides* NRRL B-1355 in batch fermentation with controlled pH and dissolved oxygen. *J. Chem. Technol. Biotechnol.* **69**, 470–478.

Rankin, J. C., Jeanes, A. (1954) Evaluation of the periodate oxidation method for structural analysis of dextrans. *J. Am. Chem. Soc.* **76**, 4435–4441.

Robyt, J. F. (1995) Mechanisms in the glucansucrase synthesis of polysaccharides and oligosaccharides from sucrose. In: Horton, D. (Ed.), *Advances in Carbohydrate Chemistry and Biochemistry.* San Diego: Academic Press, Inc., 133–168, vol. 51.

Robyt, J. F., Taniguchi, H. (1976) The mechanism of dextransucrase action. Biosynthesis of branch linkages by acceptor reactions with dextran. *Arch. Biochem. Biophys.* **174**, 129–135.

Robyt, J. F., Kim, D., Yu, L. (1995) Mechanism of dextran activation of dextransucrase. *Carbohydr. Res.* **266**, 293–299.

Sanchez-Gonzalez, M., Alagon, A., Rodriguez-Sotres, R., López-Munguía, A. (1999) Proteolytic processing of dextransucrase of *Leuconostoc mesenteroides. FEMS Microbiol. Lett.* **181**, 25–30.

Sawai, T., Tohyama, T., Natsume, T. (1978) Hydrolysis of fourteen native dextrans by *Arthrobacter* isomaltodextranase, and correlation with dextran structure. *Carbohydr. Res.* **66**, 195–205.

Sawai, T., Ohara, S., Ichimi, Y., Okaji, S., Hisada, K., Fukaya, N. (1981) Purification and some properties of the isomaltodextranase of *Actinomadura* strain R10 and comparison with that of *Arthrobacter globiformis* T6. *Carbohydr. Res.* **89**, 289–299.

Schilling, C. H., Tomasik, P., Kim, J. C. (1999) Processing technical ceramics with maltodextrins: crosslinking by acetalation. *Starch-Stärke* **51**, 397–405.

Scott, T. A., Hellman, N. N., Senti, F. R. (1957) Characterization of dextrans by the optical rotation of their cuprammonium complexes. *J. Am. Chem. Soc.* **79**, 1178–1182.

Seymour, F. R., Slodki, M. E., Plattner, R. D., Jeanes, A. (1977) Six unusual dextrans: Methylation structural analysis by combined GLC-MS of per-*O*-acetyl aldononitriles. *Carbohydr. Res.* **53**, 153–166.

Seymour, F. R., Knapp, R. D., Bishop, S. H. (1979a) Correlation of the structure of dextrans to their ^{1}H-NMR spectra. *Carbohydr. Res.* **74**, 77–92.

Seymour, F. R., Knapp, R. D., Bishop, S. H., Jeanes, A. (1979b) High-temperature enhancement of ^{13}C-NMR chemical shifts of unusual dextrans and correlation with methylation structural analysis. *Carbohydr. Res.* **68**, 123–140.

Seymour, F. R., Knapp, R. D., Chen, E. C. M., Bishop, S. H., Jeanes, A. (1979c) Structural analysis of *Leuconostoc* dextrans containing 3-*O*-α-D-glucosylated residues in both linear-chain and branch-point positions, or only in branch-point positions, by methylation and by ^{13}C-NMR spectroscopy. *Carbohydr. Res.* **74**, 41–62.

Sloan, J. W., Alexander, B. H., Lohmar, R. L., Wolff, I. A., Rist, C. E. (1954) Determination of dextran structure by periodate oxidation techniques. *J. Am. Chem. Soc.* **76**, 4429–4434.

Slodki, M. E., England, R. E., Plattner, R. D., Dick, W. E. (1986) Methylation analyses of NRRL dextrans by capillary gas-liquid chromatography. *Carbohydr. Res.* **156**, 199–206.

Smith, M. R., Zahnley, J. C. (1997) *Leuconostoc mesenteroides* B-1355 mutants producing alternansucrases exhibiting decreases in apparent molecular mass. *Appl. Environ. Microbiol.* **63**, 581–586.

Smith, M. R., Zahnley, J. C. (1999) Production of glucosyltransferases by wild-*type Leuconostoc mesenteroides* in media containing sugars other than sucrose. *J. Ind. Microbiol. Biotechnol.* **22**, 139–146.

Smith, M. R., Zahnley, J., Goodman, N. (1994) Glucosyltransferase mutants of *Leuconostoc mesenteroides* NRRL B-1355. *Appl. Environ. Microbiol.* **60**, 2723–2731.

Smith, M. R., Zahnley, J. C., Wong, R. Y., Lundin, R. E., Ahlgren, J. A. (1998) A mutant strain of *Leuconostoc mesenteroides* B-1355 producing a glucosyltransferase synthesizing α(1 → 2) glucosidic linkages. *J. Ind. Microbiol. Biotechnol.* **21**, 37–45.

Stodola, F. H., Koepsell, H. J., Sharpe, E. S. (1952) A new disaccharide produced by *Leuconostoc mesenteroides. J. Am. Chem. Soc.* **74**, 3202–3203.

Stodola, F. H., Sharpe, E. S., Koepsell, H. J. (1956) The preparation, properties, and structure of the disaccharide leucrose. *J. Am. Chem. Soc.* **78**, 2514–2518.

Taylor, N. W., Zobel, H. F., Hellman, N. N., Senti, F. R. (1959) Effect of structure and crystallinity on water sorption of dextrans. *J. Phys. Chem.* **63**, 599–603.

Torii, M., Sakakibara, K. (1974) Column chromatographic separation and quantitation of α-linked glucose oligosaccharides. *J. Chromatogr.* **96**, 255–257.

Torii, M., Sakakibara, K., Misaki, A., Sawai, T. (1976) Degradation of α-linked D-glucooligosaccharides and dextrans by an isomaltodextranase preparation from *Arthrobacter globiformis* T6. *Biochem. Biophys. Res. Commun.* **70**, 459–464.

Torii, M., Tanaka, S., Sawai, T. (1981) An epitopic structure of dextran B-1355. *Microbiol. Immunol.* **25**, 969–973.

Tsumori, H., Shimamura, A., Mukasa, H. (1985) Purification and properties of extracellular glucosyltransferase synthesizing 1,6-,1,3-α-D-glucan from *Streptococcus mutans* serotype a. *J. Gen. Microbiol.* **131**, 3347–3353.

Valette, P., Pelenc, V., Djouzi, Z., Andrieux, C., Paul, F., Monsan, P., Szylit, O. (1993) Bioavailability of new synthesised glucooligosaccharides in the intestinal tract of gnotobiotic rats. *J. Sci. Food Agric.* **62**, 121–127.

VanLoo, J., Cummings, J., Delzenne, N., Englyst, H., Franck, A., Hopkins, M., Kok, N., Macfarlane, G., Newton, D., Quigley, M., Roberfroid, M., vanVliet, T., vandenHeuvel, E. (1999) Functional food properties of non-digestible oligosaccharides: a consensus report from the ENDO project

(DGXII AIRII-CT94-1095). *Br. J. Nutr.* **81**, 121–132.

Walker, G. J., Hare, M. D. (1979) Hydrolysis of (1 → 3)-α-D-glucosidic linkages in oligosaccharides and polysaccharides by *Cladosporium resinae* exo-(1 → 3)-α-D-glucanase. *Carbohydr. Res.* **77**, 289–292.

Wilham, C. A., Alexander, B. H., Jeanes, A. (1955) Heterogeneity in dextran preparations. *Arch. Biochem. Biophys.* **59**, 61–75.

Wyckoff, H.A., Côté, G.L., Biely, P. (1996) Isolation and characterization of microorganisms with alternan hydrolytic activity. *Curr. Microbiol* **32**, 343–348.

Zahnley, J. C., Smith, M. R. (1995) Insoluble glucan formation by *Leuconostoc mesenteroides* B-1355. *Appl. Environ. Microbiol.* **61**, 1120–1123.

Zahnley, J. C., Smith, M. R. (2000) Cellular association of glucosyltransferases in *Leuconostoc mesenteroides* and effects of detergent on cell association. *Appl. Biochem. Biotechnol.* **87**, 57–70.

14
Levan

Dr. Sang-Ki Rhee[1], Dr. Ki-Bang Song[2], Dr. Chul-Ho Kim[3], Dr. Buem-Seek Park[4], Ms. Eun-Kyung Jang[5], Dr. Ki-Hyo Jang[6]

[1] Biomolecular Engineering Laboratory, Korea Research Institute of Bioscience and Biotechnology (KRIBB), 52 Eoeun-dong, Yuseong, Daejeon 305-333, Korea; Tel.: +82-42-860-4450; Fax: +82-42-860-4594; E-mail: rheesk@mail.kribb.re.kr

[2] Biomolecular Engineering Laboratory, Korea Research Institute of Bioscience and Biotechnology (KRIBB), 52 Eoeun-dong, Yuseong, Daejeon 305-333, Korea; Tel.: +82-42-860-4457; Fax: +82-42-860-4594; E-mail: songkb@mail.kribb.re.kr

[3] Biomolecular Engineering Laboratory, Korea Research Institute of Bioscience and Biotechnology (KRIBB), RealBioTech Co., Ltd., #202 Bioventure Center, KRIBB, 52 Eoeun-dong, Yuseong, Daejeon 305-333, Korea; Tel.: +82-42-860-4452; Fax: +82-42-860-4594; E-mail: kim3641@mail.kribb.re.kr

[4] Biomolecular Engineering Laboratory, Korea Research Institute of Bioscience and Biotechnology (KRIBB), 52 Eoeun-dong, Yuseong, Daejeon 305-333, Korea; Tel.: +82-42-860-4454; Fax: +82-42-860-4594; E-mail: buemseekpk@mail.kribb.re.kr

[5] RealBioTech Co., Ltd., #202 Bioventure Center, KRIBB, 52 Eoeun-dong, Yuseong, Daejeon 305-333, Korea; Tel.: +82-42-863-4381; Fax: +82-42-863-4382; E-mail: levanis@realbio.com

[6] Department of Medical Nutrition, Graduate School of East-West Medical Science, Kyung Hee University, Suwon 449-701, Korea; Tel.: +82-2-961-0506; Fax: +82-2-961-9215; E-mail: kihyojang@hotmail.com

1-kestose	*O*-β-D-fructofuranosyl-(2 → 1)-β-D-fructofuranosyl-(2 → 1)-β-D-glucopyranoside
1-SST	sucrose:sucrose 1-fructosyltransferase
6-SFT	sucrose:fructan 6-fructosyltransferase
DFA	di-β-D-fructofuranose dianhydride
DP	degree of polymerization
EPS	exopolysaccharides
FFT	fructan:fructan fructosyltransferase
FOS	fructo-oligosaccharides
HPr	histidine-containing phosphocarrier protein
LBT	levanbiosyl transfer
LFT	levanfructosyl transfer
LFTase	levan fructotransferase
PEG	polyethylene glycol
PTS	phosphoenolpyruvate-dependent carbohydrate
RBT	RealBioTech Co., Ltd.
TLC	thin-layer chromatography

1 Introduction

Fructan, one of the most highly distributed biopolymers in nature, is a homopolysaccharide composed of D-fructofuranosyl residues joined by β-(2,6) and β-(2,1) linkages. Two types of fructan, distinguishable by the type of linkage present, are inulin and levan. The term levan is used to describe the microbial polyfructan which consists of D-fructofuranosyl residues linked predominantly by β-(2,6) linkage as a main chain, but with some β-(2,1) branching points. The other polyfructan, inulin, is mainly isolated from natural vegetable sources and serves as a reserve carbohydrate in the Compositate and Gramineae (Vandamme and Derycke, 1983), although inulins from the microbial origin have also been reported in *Streptococcus mutans* and *Streptococcus sanguis*, the human pathogens involved in dental caries (Birkhed et al., 1979). The fructose homopolymer, levan, is found in plants and especially in bioproducts of microorganisms. Plant levans, graminans, and phleins have shorter residues (varying from 10 to ~200 fructose residues) than microbial levans, of which molecular weights are up to several million daltons, with multiple branches. Microbial levans are produced extracellularly from sucrose- and raffinose-based substrates by levansucrase (sucrose 6-fructosyltransferase, EC 2.4.1.10) from a wide range of taxa such as bacteria, yeasts, and fungi (Han, 1990; Hendry and Wallace, 1993). Microbial levans are produced mainly by bacteria such as *Bacillus subtilis, Zymomonas mobilis, Bacillus polymyxa, Aerobacter levanicum, Erwinia amylovora, Rhanella aquatilis* and *Pseudomonas*. The production and utilization of levan in the industrial field have been strictly limited until very recently, and very few reports have been made on the production of levan using fermentation techniques (Elisashvili, 1984; Beker et al., 1990; Han, 1990; Keith et al., 1991; Ohtsuka et al., 1992; Uchiyama, 1993). Recently, great interest in this fructan has been renewed to discover novel applications for levan as a new industrial gum in the fields of cosmetics, foods (e.g., as dietary fiber), and pharmaceuticals. In this chapter, we describe the production and degradation of levan by use of enzymatic reactions, the genetic regulation and control of such reactions, and outline the properties of levan and the current status of its industrial applications.

2 Historical Outline

Historically, levan was generally considered to be an undesirable byproduct of sugar and juice processing because it increases the viscosity of the processing liquor (Fuchs, 1959; Avigad, 1965). Levan was first described by Lippmann in Germany in 1881, when the name "laevulan" was proposed. Greig-Smith (1901) later showed that a strain of *Bacillus*, when grown on sucrose, produced fructans, and the name "levan" was then introduced as being analogous to dextran. The term laevulan now denotes partially degraded levan fractions. However, early reports on levan were confusing because the microbial nomenclature was unsystematic and the materials were inadequately described.

The biosynthesis of levan was elucidated some years later. The mechanism was shown to involve two enzymes, sucrose fructosyltransferase and fructan fructosyltransferase, and was proposed by Edelman and Jefford in 1968. The enzyme kinetics of the transfructosylation reaction was revealed by Chambert and Gonzy-Treboul in 1976. The enzyme which is now generally recognized as levansucrase was named by Hestrin et al. in 1943, and is responsible for the synthesis of levan from sucrose. The most extensive studies of levansucrase were performed in *B. subtilis*, and focused on the localization of the enzyme as well as its properties, expression regulation, genetic organization, and kinetics (Suzuki and Chatterton, 1993). As levan began to receive more attention based on its potential applications, many levan-producing microorganisms were identified (Han, 1990; Hendry and Wallace, 1993). The mass production of levan from *Z. mobilis*, and the secretion of levansucrase were reported relatively recently in detail by Song and coworkers (1996) and Ananthalakshmy and Gunasekaran (1999).

Although extensive research and searches for industrial applications have been conducted with dextran (which is also known as a bacterial biopolymer), much less attention has been focused on levan, mainly because of the very poor yields obtained in its industrial production. Nonetheless, great interest has been expressed in the diverse aspects of this fructose homopolymer over the past decade, despite the applications of levan having remained relatively few in number because of the limited supplies. Levans are now available commercially in reagent grade from microbial sources (e.g., from Sigma Chemical Co., IGI Biotechnology), but these are used only for research purposes. Since the time when levan was first produced on a large scale by using levansucrase from genetically engineered *Escherichia coli* (Song et al., 1996), attention has been renewed on the potential industrial application of levan and its derivatives in the fields of agriculture, cosmetics, food ingredients, animal feed and pharmaceuticals (Clarke et al., 1997; Kim et al., 1998; Vijn and Smeekens, 1999; Rhee et al., 2000d), as well as being a good source of pure fructose and di-β-D-fructofuranose dianhydride (DFA) (Saito and Tomita, 2000).

3 Chemical Structures of Levan

Fructans are chemically versatile molecules, and consist of a single glucose unit attached to two or more fructose units. Three fructan trisaccharides are known, each being produced through a glycosidic linkage of fructose to one of the three primary hydroxyl groups of sucrose. Fructose linked to the primary carbon of the fructose moiety of sucrose forms 1-kestose (also called isokestose), while fructose linked to the sixth carbon of the fructose moiety of sucrose forms 6-kestose (also called kestose). Both of these trisaccharides have a terminal glucose and a terminal fructose. Linkage of a fructose moiety to the sixth carbon of glucose moiety of sucrose forms neokestose, with both end groups being fructose (Nelson and Spollen, 1987).

Chemically, levan consists of β-D-fructofuranosyl residues linked predominantly through β-(2,6) as 6-kestose of the basic trisaccharide, with extensive branching through β-(2,1) linkages (Figure 1). In contrast, inulin is composed of β-D-fructofuranose attached by β-(2,1) linkages. The first monomer of the chain is either a β-D-glucopyranosyl or a β-D-fructofuranosyl residue. Although they are similar fructose homopolymers, it is evident that levan is different from inulin-type fructan since microbial inulin contains 5–7% of β-(2,6)-linked branches (Wolff et al., 2000).

The molecular shape of levan, as visualized by electron microscopy (Newbrun et al., 1971), is spheroidal, indicating that the constituent chains are extended radially at the same synthetic rate. The molecular

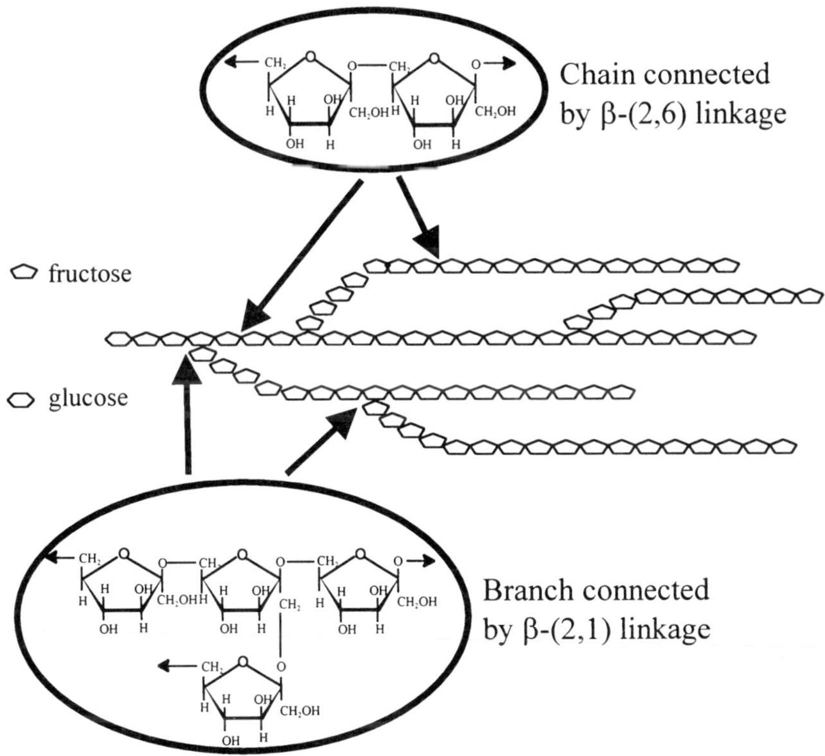

Fig. 1 Structure of levan. The main chain is connected by β-(2,6) linkages and the branch is connected to the main chain by a β-(2,1) linkage; the branch then continues with β-(2,6) linkages.

weight of bacterial levans is typically in the range of 2×10^6 to 10^8 (Keith et al., 1991), with the final molecular size being influenced by the synthesizing conditions such as ionic strength, temperature, and co-solutes. Although microbial levans have the similar structure, several types of (IX) levan are produced by different microorganisms, and this may be attributed to a varying degree of polymerization (DP) and branching of the repeating unit.

A cell-free enzyme system could be used to synthesize levans which have both β-(2,6) and β-(2,1) linked fructosyl units, and with similar structure to those of a whole-cell enzyme system. However, the structure of levans synthesized in the cell-free enzyme system is also known to differ in length compared with that synthesized by whole-cell systems (Han, 1990).

4
Occurrence

Various polysaccharides are produced as structural components in living organisms, and levan is one of the most diversely distributed components in plants, yeasts, fungi, and bacteria in particular. Levan is produced by grass (*Dactylis glomerata, Poa secunda* and *Agropyron cristatum*), wheat and barley (*Hordeum vulgare*), fungi (*Aspergillus sydawi* and *Aspergillus versicolor*) and yeasts (Han, 1990). Levans produced by microorganisms have been reported by Han (1990), and Hendry and Wallace (1993), and are listed in Table 1.

Previously, oral bacteria such as *Streptococcus, Rothis* and *Odontomyces* had received much attention due to their presence in human dental caries, together with soil microorganisms, especially *Bacillus*. Subsequently, focus was centered on the biological and functional aspects of levan rather than on its oral accumulation. The most extensive studies of levan were performed using *B. subtilis* (Suzuki and Chatterton, 1993). Furthermore, levans from *Bacillus polymyxa* (Aymerich, 1990) and *Pseudomonas* sp. (Hettwer et al., 1995, 1998) were identified as playing a role in the plant defense response. Levan from *B. subtilis* was shown to be tolerant against salt stress (Kunst and Rapoport, 1995), while that obtained from *Z. mobilis* exhibited antitumor activity (Calazans et al., 2000). The synthesis of levan using the genus *Lactobacillus* was also recently reported (Van Geel-Schutten et al., 1999).

Tab. 1 Levan-producing microorganisms

Microorganism	***Reference***
Acetobacter xylinum	Tajima et al. (1998)
Actinomyces naeslundii	Bergeron et al. (2000)
Bacillus circulans	Perez Oseguera et al. (1996)
Bacillus stearothermophilus	Li et al. (1997)
Gluconacetobacter (formerly *Acetobacter*) *diazotrophicus*	Arrieta et al. (1996)
Lactobacillus reuteri	Van Geel-Schutten et al. (1999)
Pseudomonas syringae pv. phaseolicola	Hettwer et al. (1995)
Pseudomonas syringae pv. glycinea	Hettwer et al. (1998)
Rahnella aquatilis	Ohtsuka et al. (1992); Song et al. (1998)
Serratia levanicum	Kojima et al. (1993)
Zymomonas mobilis	Song et al. (1993)

5 Physiological Functions of Levan

Bacterial polysaccharides are found either as a dense layer of more or less regularly arranged polymer structures attached to the bacterial cell walls (capsules) or as loosely associated exopolysaccharides (EPS) (Beveridge and Graham, 1991). Levans produced microbiologically have a number of interesting features. The levan which is synthesized extracellularly by bacteria may be visualized in a sucrose-containing medium, giving rise to a typical mucoid morphology. This type of mucoid feature provides a role in the symbiosis, phytopathogenesis, or participation in the defense mechanism against cold and dry conditions (Kunst and Rapoport, 1995). Extracellular levan produced by bacterial plant pathogens increases bacterial fitness and also acts as a detoxifying barrier against plant defence compounds (Hettwer et al., 1998). Among natural polysaccharides, glucans and fructans possess antitumor activity, and levan is included in this list. The antitumor activity of levan against sarcoma 180 depends on the molecular weight of the polysaccharides (Calazans et al., 2000), and this may indicate the polydiversity of levan.

Levans produced in plants are present as storage carbohydrates in the stem and leaf sheaths, and are degraded in a later stage of the growing season to provide plants with carbohydrates for grain filling (Pollock and Cairns, 1991). The biological role of polysaccharides in protection is less clearly understood, but a hypothesis has recently emerged. Levan penetrates into lipid membranes composed of monomolecular lipid layers, after which interactions occur which are orders of magnitude greater than the interaction between disaccharides and lipids. An extended layer of levan adheres to the lipids and partially protrudes into the aqueous phase. It is also possible that the membranes present are coated with levan; this coating of membranes imparts a reduction in accessibility of the membrane surface to proteins. In this way, in a biological system levan is able to protect membranes by interacting with the membrane lipid fraction (Vereyken et al., 2001).

6 Chemical Analysis and Detection

Several methods can be used in the qualitative and quantitative analysis of levan and the estimation of its concentration in solutions, with spectrophotometry and chromatography being the major techniques.

6.1 Spectrophotometry

Low concentrations of levan are measured by monitoring the optical density at 450–550 nm, as the presence of levan creates turbidity within the enzyme reaction mixture.

6.2 High-Performance Liquid Chromatography (HPLC)

The HPLC method is employed for both qualitative and quantitative determination of levan and other components (oligosaccharides, sucrose, fructose, and glucose). Details of the method are described here. The enzyme reaction mixture or levan solution is filtered using a 0.45 μm pore size membrane filter, and the filtrate is analyzed by HPLC equipped with a gel filtration column and refractive index detector (Shodex Ionpack KS-802, 300×8 mm; Showa Denko Co., Japan) (Jang et al., 2000). Deionized water is used as a mobile phase at a flow rate of

$0.4\,mL\,min^{-1}$. The DP of levan is also determined by HPLC equipped with successive columns, GPC 4000–GPC 1000 (Polymer Laboratories, USA), and a refractive index detector (Jang et al., 2001). The analyses of sugar components and linkage type of levan are determined by using acid hydrolysis, methylation and nuclear magnetic resonance (NMR) shift experiments (Suzuki and Chatterton, 1993). In ^{13}C-NMR, signals of carbons of levan obtained from *Z. mobilis* and *Aerobacter levanicum* are identical, and show six main resonances at 104.9, 81.0, 76.9, 64.1, and 60.6 p.p.m. (Song and Rhee, 1994), these signals differing from those obtained with inulin.

6.3 Other Methods

Levan (nonmobiles) can be distinguished from oligosaccharides, sucrose, and other byproducts, either qualitatively or quantitatively, by the use of thin-layer chromatography (TLC). The sucrose-hydrolyzing activity of bacterial levansucrase is also used in the determination of levan concentration. The methods established are based on the fact that glucose is formed stoichiometrically in relation to the amount of fructose incorporated into levan (major product) and oligosaccharides (minor products). The amounts of glucose generated by the enzymatic reaction can be determined quantitatively by commercially available kits from the suppliers (Song et al., 1993).

7 Biosynthesis of Levan

The biosynthesis of levan requires the involvement of an extracellular enzyme levansucrase, which shows specificity for sucrose. Genetic characterization of the enzyme and the regulation of levan synthesis have been extensively studied, mostly using levansucrase genes from *B. subtilis* and *Z. mobilis*.

7.1 Enzymology of Levan Synthesis

Levansucrase (sucrose:2,6-β-D fructan:6-D-fructosyltransferase, sucrose 6-fructosyltransferase, EC 2.4.1.10.) was first named by Hestrin et al. (1943), and is responsible for the synthesis of levan from sucrose. Levansucrase exists as constituent intracellular and inducible extracellular forms in microorganisms (Han, 1990). The function of the levan-producing enzyme located intracellularly in some bacteria is not yet understood. The most abundant substrate for levansucrase in nature is sucrose, but raffinose also serves as a substrate.

Levansucrase is a type of transferase which catalyzes a fructosyl transfer from sucrose to various acceptor molecules. The enzyme catalyzes the following reactions:

1. Polymerization:

$(\text{Sucrose})_n \rightarrow (\text{Glucose})_n + \text{Levan} + \text{Oligosaccharides}$

2. Hydrolysis:

$\text{Sucrose} + H_2O \rightarrow \text{Fructose} + \text{Glucose}$

$(\text{Levan})_n + H_2O \rightarrow (\text{Levan})_{n-1} + \text{Fructose}$

3. Acceptor:

$\text{Sucrose} + \text{Acceptor molecules} \rightarrow \text{Fructosyl-acceptor} + \text{Glucose}$

4. Exchange:

$\text{Sucrose} + [^{14}C]\text{Glucose} \rightarrow \text{Fructose-}[^{14}C]\text{Glucose} + \text{Glucose}$

5. Disproportionation:

$$[\text{Levan}]_m + [\text{Levan}]_n \rightarrow [\text{Levan}]_{m-1} + [\text{Levan}]_{n+1}$$

The enzyme catalyzes hydrolysis and polymerization reactions concomitantly (Reaction 1), resulting in a fructose homopolymer (levan) and free glucose. This reaction occurs when sucrose exists as the sole fructosyl donor and acceptor, and involves three steps: initiation, propagation, and termination (Chambert et al., 1974). The chains of levan grow step-wise by repeated transfer of a hexosyl group from the donor to growing acceptor molecules. The enzyme primarily catalyzes a coupled reaction by a ping-pong mechanism, i.e., sucrose hydrolysis followed by transfructosylation involving a fructosyl-enzyme intermediate (Chambert et al., 1976).

When water acts as an acceptor, a free fructose is generated from both sucrose and levan (Reaction 2). This reaction occurs in all the levansucrase-catalyzed reactions mentioned above, but the rate is much slower when compared with a sugar acceptor. Reaction 3 occurs in the presence of an acceptor in the environment. The enzyme transfers the fructosyl residue of sucrose specifically to the C-1 hydroxyl group of aldose in the acceptor. Compounds containing hydroxyl groups, such as methanol, glycerol and oligosaccharides, can act as fructosyl acceptors.

The reaction mechanism yields a non-reducing sugar compound and a series of oligosaccharides, in which the sugar molecule with one more fructose moiety remains as a major reaction product. The reaction occurs predominantly in the presence of a high concentration of fructosyl donors, such as sucrose or raffinose. Reaction 4 might be considered analogous to Reactions 2 and 3, but differs in the regeneration of sucrose, which has a high-energy bond. The enzyme also catalyzes Reaction 5, a disproportionation reaction, in which the degree of polydispersity of levan or oligomers is modified. The above five reactions compete with one another, yielding a specific major product with some minor products but they are predominantly controlled by environmental factors.

At present, little is known of plant levans, and their biosynthesis is not fully understood (Heyer et al., 1999). One plant levan, known as "graminan", is synthesized by sucrose:fructan 6-fructosyltransferase (6-SFT) which catalyzes the formation and extension of β-(2,6)-linked fructans. The 6-SFT is closely related to vacuolar invertase and transfers the fructosyl residues from sucrose preferentially to 1-kestose or larger fructans (Sprenger et al., 1995). However, most fructan synthesis in plants occurs in two steps (Edelman and Jefford, 1968). Initially, sucrose:sucrose 1-fructosyltransferase (1-SST; EC 2.4.1.99.) catalyzes the formation of the trisaccharide 1-kestose and glucose from two molecules of sucrose. Later, fructan:fructan 1-fructosyltransferase (1-FFT; EC 2.4.1.100.) reversibly transfers fructosyl residues from one fructan with a DP of ≥ 3 to another DP of ≥ 2, producing a mixture of fructans with different chain lengths.

7.2 Genetic Basis of Levan Synthesis

As yet, levansucrase genes have been cloned and biochemically characterized in seven Gram-negative strains; namely, *Acetobacter diazotrophicus* (Arrieta et al., 1996), *Acetobacter xylinum* (Tajima et al., 2000), *Erwinia amylovora* (Geier and Geider, 1993), *P. syringae* pv. glycinea, *P. syringae* pv. phaseolicola (Hettwer et al., 1998), *Rahnella aquatilis* (Song et al., 1998) and *Z. mobilis* (Song et al., 1993). Several levansucrase genes have also been cloned in Gram-positive strains, such as *Bacillus* (Gay et al., 1983; Li et al., 1997) and *Streptococcus* species (Sato et al., 1984). All levansucrase genes share several conserved regions, which are thought to be important for the enzyme activity. Although conservation is observed, dissimilarity exists depending on the source of the enzyme. Levansucrase genes from a Gram-negative origin show relatively high similarity (>50%) when compared with the genes from Gram-positive bacterial enzymes. However, very little similarity (<30%) exists among the genes from two different sources (Song and Rhee, 1994). The deduced amino acid sequences are aligned in Figure 2.

Although the amino acid sequences of levansucrases do not show any considerable homology to those of sucrose-related enzymes, the third (-EWS/AGT/SP/A-) and the

```
                     I                                                II
Psp-Lsc    22 YEPTVWSRAD ALKVNENDPT TT-Q-PLVSA DFPVM--SDT VF-IWDTMPL RELDGTVVSV NGWSVILTLT ADRHPNDPQY
Psg-Lsc     6 YAPTIWSRAD ALKVNENDPT TT-Q-PLVSP DFPVM--SDT VF-IWDTMPL RELDGTVVSV NGWSVIVTLT ADRHPDDPQY
Ra-LsrA     6 YTPTIWTRAD ALKVNENDPT TT-Q-PIVDA DFPVM--SDE VF-IWDTMPL RSLDGTVVSV DGWSVIFTLT AQRNNNNSEY
Ea-Lsc      6 YKPTLWTRAD ALKVHEDDPT TT-Q-PVIDI AFPVM--SEE VF-IWDTMPL RDFDGEIISV NGWCIIFTLT ADRNTDNPQF
Zm-LevU     8 AEPSLWTRAD AMKVHTDDPT AT-M-PTIDY DFPVM--TDK YW-VWDTWPL RDINGQVVSG QGWSVIFALV ADRTKY----
Bs-SacB    41 YGISHITRHD MLQIPEQQKN EKYQVPEFDS STIKNISSAK GLDVWDSWPL QNADGTVANY HGYHIVFALA GDPKNADDTS
Bst-SurB   41 YGISHITRHD MLQIPEQQKN EKYQVPEFDS STIKNISSAK GLDVWDSWPL QNADGTVANY HGYHIVFALA GDPKNADDTS

                                                                  III
Psp-Lsc    97 LDANGRYDIK RDWEDRHGRA RMSYWYSRTG KDWIFGGRVM AEGVSPTTRE WAGTPILLND KGDIDLYYTC VTPGAAIAKV
Psg-Lsc    81 VGANGRYDIK RDWEDRHGRA RMCYWYSRTG KDWIFGGRVM AEGVSPTTRE WAGTPVLLND KGDIDLYYTC VTPGAAIAKV
Ra-LsrA    81 LDAEGNYDIT SDWNNRHGRA RICYWYSRTG KDWIFGGRVM AEGVSPTSRE WAGTPILLNE DGDIDLYYTF VTPGATIAKV
Ea-Lsc     81 QDENGNYDIT RDWEKRHGRA RICYWYSRTG KDWIFGGRVM AEGVAPTTRE WAGTPILLND RGDIDLYYTG VTPGATIAKV
Zm-LevU    79 ----G----- --WHNRNDGA RIGYFYSRGG SNWIFGGHLL KDGANPRSWE WSGCTIMAPG TANSVEVFFT SVNDTPSESV
Bs-SacB   121 IYMFYQKVGE TSIDSWKNAG RVFKDSDKFD AN-------- DSILKDQTQE WSGSATFTSD -GKIRLFYTD FSGKHYGKQT
Bst-SurB  121 IYMFYQKVGE TSIDSWKTPG RVFKDSDKFD AN-------- DSILKDQTQE WSGSATFTSD -GKIRLFYTD FSGKHYGKQT

                                                                     IV
Psp-Lsc   177 ----RGRIVT SDQGVELKDF TQVKKLFEAD GTYYQTEAQ- --------NS SWNFRDPSPF IDPNDGKL-- ---------Y
Psg-Lsc   161 ----RGRIVT SDKGVELKDF TEVKTLFEAD GKYYQTEAQ- --------NS TWNFRDPSPF IDPNDGKL-- ---------Y
Ra-LsrA   161 ----RGKVLT SEEGVTLAGF NEVKSLFSAD GVYYQTESQ- --------NP YWNFRDPSPF IDPHDGKL-- ---------Y
Ea-Lsc    161 ----RGKIVT SDQSVSLEGF QQVTSLFSAD GTIYQTEEQ- --------NA FWNFRDPSPF IDRNDGKL-- ---------Y
Zm-LevU   148 PAQCKGYIYA DDKSVWFDGF DKVTDLFQAD GLYYADYAE- --------NN FWDFRDPHVF ITPKDGKL-- ---------Y
Bs-SacB   192 LTTAQVNVSA SDSSLNINGV EDYKSIFDGD GKTYQNVQQF IDEGNYSSGD NHTLRDPHYV EDKGHKYLVF EANTGTEDGY
Bst-SurB  192 LTTAQVNVSA SDSSLNINGV EDYKSIFDGD SKTYQNVQQF IDEGNYSSGD NHTLRDPHYV EDKGHKYLVF EANTGTEDGY

                                                                               V
Psp-Lsc   233 MVFEGNVAGE RGSHTVGAAE LGPVPPGHED VGGARFQVGC IGLAVAKDLS GEEWEILPPL VTAVGVNDQT ERPHYVFQDG
Psg-Lsc   212 MVFEGNVAGE RGTHTVGAAE LGPVPPGHEE TGGARFQVGC IGLAVAKDLS GDEWEILPPL VTAVGVNDQT ERPHYVFQDG
Ra-LsrA   217 MVFEGNVAGE RGSHVIFKQE MGTLPPGHRD VGNARYQAGC IGMAVAKDLS GDEWEILPPL VTAVGVNDQT ERPHEVFQDG
Ea-Lsc    217 MLFEGNVAGP RGSHEITQAE MGNVPPGYED VGGAKYQAGC VGLAVAKDLS GSEWQILPPL ITAVGVNDQT ERPHFVFQDG
Zm-LevU   208 ALFEGNVAME RGTVAVGEEE IGPVPPKTET PDGARYCAAA IGIAQALNEA RTEWKLLPPL VTAFGVNDQT ERPHVVFQNG
Bs-SacB   272 QGEESLFNKA YYGKSTSFFR QESQKLLQSD KKRTAELANG ALGMIELNDD YTLKKVMKPL IASNTVTDEI ERANVFKMNG
Bst-SurB  272 QGEESLFNKA YYGKSTSFFR QESQKLLQSD KNRTAELANG ALGMIELNDD YTLKKVMKPL IASNTVTDEI ERANVFKMNG

                VI                               VII
Psp-Lsc   313 KYYLFTISHK -FTYAEGLTG PDGVY--GFV GEH-LFGPYR PMNASGLVLG NPPEQPFQTY SHCVMPNGLV TSFIDSVPTE
Psg-Lsc   297 KYYLFTISHK -FTYADGVTG PDGVY--GFV GEH-LFGPYR PMNASGLVLG NPPAQPFQTY SHCVMPNGLV TSFIDSVPTS
Ra-LsrA   297 KYYLFTISHK -ETYADGLTG PDGVY--GFL SDN-LTGPYS PMNGSGLVLG NPPSQPFQTY SHCVMPNGLV TSFIDNVPTS
Ea-Lsc    297 KYYLFTISHK -YTFADNLTG PDGVU--GFV SDK-LTGPYT PMNSSGLVLG NPSSQPFQTY SHYVMPNGLV TSFIDSVPWK
Zm-LevU   288 LTYLFTISHH S-TYADGLSG PDGVY--GFV SENGIFGPYE PLNGSGLVLG NPSSQPYQAY SHYVMINGLV TSFIDTIPSS
Bs-SacB   352 KWYLFTDSRG SKMTIDGITS NDI-YMLGYV S-NSLTGPYK PLNKTGLVLK MDLDPNDVTF TYSHFAVPQA KGNNVVITSY
Bst-SurB  352 KWYLSTDSRG SQMTIDGITS NDI-YMLGYV S-NSLTGPYK PLNKTGLVLK MDLDPNDVTF TYSHFAVPQA TGNNVVITSY

Psp-Lsc   389 GED-YRIGGT EAPTVRILLK GDRSFVQEEY DYGYIPAMKD VTLS 431  P.syringae pv.phaseolicola
Psg-Lsc   373 GED-YRIGGT EAPTVRILLE GDRSFVQEVY DYGYIPAMKN VVLS 415  P.syringae pv.glycinea
Ra-LsrA   373 DGN-YRIGGT EAPTVKIVLK GNRSFVERVF DYGYIPPMKN IILN 415  R.aquatilis
Ea-Lsc    373 GKD-YRIGGT EAPTVKILLK GDRSFIVDSF DYGYIPAMKD ITLK 415  E.amylovora
Zm-LevU   365 DPNVYRYGGT LAPTIKLELV GHRSFVTEVK GYGYIPPQIE WLAE 408  Z.mobilis
Bs-SacB   430 MTNRGFYADK QSTFAPSFLL NIKGKKTSVV KDSILEQGQL TVNK 473  B.subtilis
Bst-SurB  430 MTNRGFYADK QSTFAPSFLL NIQGKKTSVV KASILDQGQL TVNQ 473  B.stearothermophilus
```

Fig. 2 Multiple alignment of deduced amino acid sequences of bacterial levansucrases. Origins of levansucrase are indicated in ends of the sequences. Asterisks indicate identical- and similar-residues in all levansucrases. Regions considered as important for activity are boxed (I–VII). Amino acid residues that are different between Gram-negative and Gram-positive origin are indicated by dots.

fourth (-FRDP-) conserved regions are found in all fructosyl- and glucosyltransferases, sucrase, sucrose-phosphate hydrolase and even in fructan-hydrolyzing enzymes. The fact that the regions are preserved in all of the sucrose-related enzymes implies that they may be catalytically important regions for the hydrolysis of sucrose. The serine residue in the sixth region (-YLFTI/DS-) has been proposed as the putative residue of the catalytic site (Chambert and Petit-Glatron, 1991).

7.3 Regulation of Levan Synthesis

The genes encoding sucrose-hydrolyzing enzymes may not be linked to each other on the chromosome, but linked only with accessory genes coding for proteins belonging to the phosphoenolpyruvate-dependent carbohydrate phosphotransferase (PTS) system. The expression of these genes is regulated by many regulatory protein systems such as the *glk* operon of *Z. mobilis* and the pleiotropic system of *B. subtilis*, etc.

7.3.1 Regulation at the Protein Level

In microorganisms, two types of sucrose utilization were found: intra- and extracellular. Commonly, these sucrose-uptake utilization systems exist within the cell; sucrose is transported by the PTS system. The sucrase system of this type is well known in *B. subtilis* (Klier and Rapoport, 1988). In contrast, some bacteria such as *Z. mobilis* that lack the PTS system first hydrolyze sucrose extracellularly to monomeric sugars, after which these sugars are transported inside the cell (Di Marco and Romano, 1985).

The co-contribution of both saccharolytic enzymes for the sucrose utilization of *Z. mobilis* has been well characterized. The glucose uptake and utilization system (*glk* operon), which is located very close to the *levU* operon and is also linked metabolically with the sucrose utilization system of *Z. mobilis*, is also regulated by the mechanism of tightly linked gene expression (Liu et al., 1992). In the intervening sequence of the *levU* and *glf* operon, two putative ORFs, encoding Lrp-like regulatory protein and aspartate racemase respectively, were found (Song et al., 1999).

At the molecular level, the genes encoding sucrose-hydrolyzing enzymes reported to date are not linked to each other on the chromosome, but are linked only with accessory genes coding for proteins belonging to the PTS system (Bruckner et al., 1993). The expression of these genes is modulated by regulatory mechanisms, such as anti-termination or repression, which is controlled by the complex regulatory network system including many regulatory proteins (Klier and Rapoport, 1988).

7.3.2 Regulation at Transcriptional and Translational Levels

The genes encoding the extracellular levansucrase and sucrase have been isolated and characterized. The nucleotide sequences of the DNA segment containing the genes encoding extracellular levansucrase and sucrase of *Z. mobilis* and *B. subtilis* were reported recently (Kyono et al., 1995). The two genes are located together in an operon on the chromosome, whereas almost all other genes coding for saccharolytic enzymes in other bacteria and yeasts are dispersed on the chromosomes (Carson and Botstein, 1983). The levansucrase gene of *B. subtilis* is activated in the presence of an inducer (sucrose or fructose), and is under a pleiotropic regulatory system controlling the expression of the sucrose operon (Lepesant et al., 1976; Shimotsu and Henner, 1986).

The pleiotropic system involving the *degS/degU*, *degQ* (formerly *sacU* and *sacQ*) and *degR* genes affects the expression of *sacB* (Débarbouillé et al., 1991). Levansucrase is encoded by the *sacB* gene and expressed from a constitutive promoter in the closely linked *sacR* locus. The *sacR* locus contains a palindromic structure acting as a transcription terminator. In the presence of sucrose, an anti-terminator, the *sacY* gene product that belongs to the *sacS* operon allows transcription of the *sacB* gene. The expression of this gene is also controlled by other regulatory genes such as two- component system *degS/degU* and also by *degQ* (Rapoport and Klier, 1990) (Figure 3).

8
Biodegradation of Levan

Although the biodegradation of levan involves several enzymes including levanase, levansucrase, and levan fructotransferase, the genetic characterization of these is limited to levanase.

8.1
Enzymology of Levan Degradation

Levan is degraded to D-fructose, levanbiose, sucrose, levan oligomers or low molecular-weight levan by the hydrolytic activity of levanase, levansucrase or levan fructotransferase from some plants and microorganisms. The mode and degree of hydrolysis depend on the enzyme sources and the reaction conditions.

8.1.1
Levanase

Many levan-forming microorganisms also produce hydrolytic enzymes–levanases–that degrade levan (Hestrin and Goldblum, 1953; Avigad, 1965). Certain strains of *Bacillus, Pseudomonas, Actinomyces, Aerobacter, Clostridium* and *Streptococcus* produce exocellular levanase (2,6-β-fructan 6-levanbiohydrolase, EC 3.2.1.64.) (Fuchs, 1959; Uchiyama, 1993). The enzyme hydrolyzes only levan, and the resulting product is usually levanbiose, indicating that a terminal fructosyl unit is removed. An exo-hydrolytic enzyme

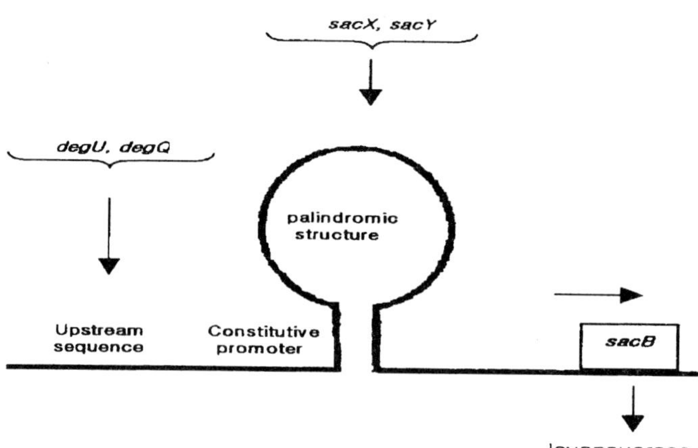

Fig. 3 Specific and pleiotropic control mechanisms affecting the *sacB* gene in *Bacillus subtilis*. *degU*: transcriptional regulator of degradation enzymes; *degQ*: pleiotropic regulatory gene; *sacX*: negative regulatory protein of *sacY*; *sacY*: positive levansucrase synthesis regulatory protein; *sacB*: gene encoding levansucrase. (Reprinted from Débarbouillé et al., 1991, p. 758, with permission from Elsevier Science.)

which has a 2,6-β-linkage-specific fructan-β-fructosidase activity (Marx et al., 1997) was reported from the grass *Lolium perenne*, and a 2,1-β-linkage-specific exohydrolase from Jerusalem artichoke. The other exo-levanase (fructan-β-fructosidase, EC 3.2.1.80; beta-D-fructofuranosidase, EC 3.2.1.26.) hydrolyzes levan to produce D-fructose. Endo-levanases (2,6-β-D-fructan fructanohydrolase, EC 3.2.1.65.) hydrolyze levan and levan oligomers consisting of more than three fructosyl units.

8.1.2
Levansucrase

Levan may be degraded not only by levanases, but also by levansucrase itself, which may catalyze the hydrolysis under certain conditions (Rapoport and Dedonder, 1963). The degree of levan hydrolysis depends on the enzyme sources and reaction conditions. For example, levansucrase from *R. aquatilis* showed a higher degradation activity than did that from *Z. mobilis* (Song et al., 1998), though for both enzymes a higher degradation activity of levan was seen as the reaction temperature was increased from 4 °C to 30 °C.

Although indirect evidence of the reversal of enzymatic synthesis of levan has been observed, little is known regarding the nature of such enzymatic degradation. Smith (1976) showed that beta-fructofuranosidase present in tall fescue degraded levan by removing one fructose residue at a time until a molecule of sucrose remained. Levansucrase of *B. subtilis* has a hydrolytic effect on small levans (Dedonder, 1966), the hydrolytic action stopping at branch points. Neither inulin, inulobiose, inulintriose, nor methyl D-fructofuranoside is hydrolyzed, despite these substrates being hydrolyzed by inulinase and yeast invertase. This hydrolytic activity may be responsible for the appearance of heterogeneous short-chain polysaccharides, rather than uniform high molecular-weight polymers, in the final product of many levan preparations.

8.1.3
Levan Fructotransferase

Microbial levan is an interesting starting material for the production of valuable oligosaccharides such as DFA IV (Yun, 1996; Saito and Tomita, 2000). DFA IV (di-D-fructose-2,6′:6,2′-dianhydride) is an oligosaccharide which is produced from levan by microbial enzymes, i.e., levan fructotransferase (LFTase) and a type of levanase (Tanaka et al., 1981, 1983; Saito et al., 1997). Currently, two LFTases have been isolated and cloned from *Arthrobacter nicotinovorans* GS-9 (Saito et al., 1997) and *A. ureafaciens* (Tanaka et al., 1981; Song et al., 2000). The enzymes have also been shown to degrade levan molecules from the nonreducing fructose end of the outer chains, and to catalyze intermolecular levanbiosyl and levanfructosyl transfer (LBT and LFT, respectively) reactions (Tanaka et al., 1983).

8.2
Genetic Basis of Levan Degradation

In *B. subtilis*, the expression of the levanase operon is inducible by fructose and is subjected to catabolite repression. A fructose-inducible promoter has been characterized 2.7 kb upstream from the gene *sacC*, which encodes levanase. The *sacC* gene is the distal gene of an operon containing five genes: *levD*, *levE*, *levF*, *levG*, and *sacC* (Martin et al., 1989; Débarbouillé et al., 1991) and is expressed under the regulated control of *sacR*, the inducible levansucrase leader region. The first four gene products are involved in a fructose-PTS system. In *Pseudomonas*, levanase is an exohydrolase of levan and produces levanbiose as a sole product; the limits of hydrolysis of levan from *Z. mobilis* and *Serratia* sp. were 65% and 80%, respectively (Jung et al., 1999).

8.3
Regulation of Levan Degradation

There are two levels on which the expression of the levanase operon in *B. subtilis* is controlled: (1) an induction by fructose, which involves a positive regulator, LevR, and the fructose phosphotransferase system encoded by this operon (lev-PTS); and (2) a global regulation of catabolite repression (Débarbouillé et al., 1991) (Figure 4).

The LevR protein is an activator for the expression of the levanase operon from *B. subtilis*. RNA polymerase containing the sigma 54-like factor sigma L recognizes the promoter of this operon. One domain of the LevR protein is homologous to activators of the NtrC family, and another resembles anti-terminator proteins of the BglG family (Débarbouillé et al., 1991). It has been proposed that the domain, which is similar to anti-terminators, is a target of phosphoenolpyruvate:sugar phosphotransferase system (PTS)-dependent regulation of LevR activity. The LevR protein is not only negatively regulated by the fructose-specific enzyme IIA/B of the phosphotransferase system encoded by the levanase operon (lev-PTS), but is also positively controlled by the histidine-containing phosphocarrier protein (HPr) of PTS (Martin et al., 1990; Stülke et al., 1995). This second type of control of LevR activity depends on phosphoenolpyruvate-dependent phosphorylation of HPr histidine 15, as demonstrated with point mutations in the ptsH gene encoding HPr. *In vitro* phosphorylation of partially purified LevR was obtained in the presence of phosphoenolpyruvate, enzyme I, and HPr. The dependence of truncated LevR polypeptides on stimulation by HPr indicates that the domain homologous to anti-terminators is the target of HPr-dependent regulation of LevR activity. This domain appears to be duplicated in the LevR protein. The first anti-terminator-like domain seems to be the target of enzyme I and HPr-dependent phosphorylation and the site of LevR activation, whereas the carboxy-terminal anti-terminator-like domain could be the target for negative regulation by lev-PTS (Débarbouillé et al., 1990).

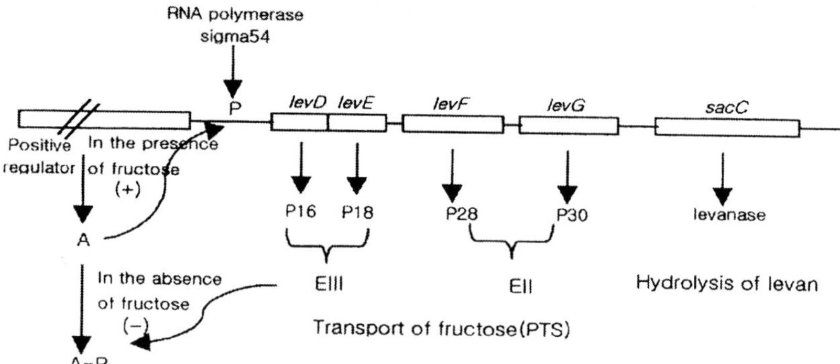

Fig. 4 Regulation model of levanase operon from *Bacillus subtilis*. P represents the fructose-inducible promoter. The *levD*, *levE*, *levF*, and *levG* gene products correspond to a fructose-specific phosphoenolpyruvate-dependent carbohydrate (PTS). *LevR* encodes a positive regulator. The activator may exist in two forms: (A-P), an inactive phosphorylated form; or (A), an active non-phosphorylated form. (Reprinted from Débarbouillé et al., 1991, p. 759, with permission from Elsevier Science.)

9
Biotechnological Production of Levan

Levan can be produced by either microbial fermentation or enzymatic synthesis, but the conversion yield of sucrose to levan is higher in the latter process than in the former.

9.1
Isolation and Screening for Levan-producing Strains

In the screening of levan-producing microbial strains, the levan formation activity can be determined using solid agar plates containing sucrose; the strains producing levan are then isolated by following the analytical procedures. The presence of levansucrase is positively selected by inducing mucoid morphology to the microorganisms. Subsequently, levan is collected from the agar plates by precipitation with alcohol (methanol, ethanol, or isopropanol). The identity of levan is then determined after acid hydrolysis of the polymers, followed by TLC analysis; fructose is identified as a single spot on the TLC plates.

9.2
Fermentative Production of Levan

The microbial production of levan requires fermentation and handling of highly viscous solutions. The conditions for producing levan by growing cultures of bacteria vary according to the microorganisms used, but yields of levan production are fairly low (Table 2); this is due to the utilization of sucrose as energy source, the formation of byproducts, the low level of levansucrase production, and the presence of levanase activity in bacteria. In addition, the recovery process of levan from the fermentation broth is often very difficult due to the high viscosity of levan. In theory, the yield of levan production by levansucrase is 50% (w/w) when sucrose is used as a substrate. Routinely, the yields of levan production based on the amount of sucrose consumed are no higher than 58% of the theoretical yield by fermentation (Table 2).

9.3
In vitro Biosynthesis of Levan

In the *in vitro* formation of levan by bacterial levansucrase, sucrose serves as fructosyl donor while the released glucose inhibits levan formation. The inhibitory action is influenced by competition with the glucose moiety of sucrose for the enzyme activity. The glucose moiety of sucrose can be replaced by D-xylose, L-arabinose, lactose, etc. Although the catalytic properties of levansucrase vary, the substrate specificity for acceptors is relatively broad where alco-

Tab. 2 Levan production by fermentation processes.

Strains	*Type of production*[a]	*Substrate conc. [%]*	*Levan Yield [%,w/w]*[b]	*Reference*
Bacillus spp.	BF	12	23.5	Elisashvili (1984)
Bacillus polymyxa	BF	15	26.6	Han (1990)
Erwinia herbicola	CF	5	19.2	Keith et al. (1991)
Gluconobacter oxydans	BF	6.2	23.3	Uchiyama (1993)
Rahnella aquatilis	BF	10	29	Ohtsuka et al. (1992)
Zymomonas mobilis	CF	12	23	Beker et al. (1990)

[a] BF, batch fermentation; CF, continuous fermentation. [b] Based on sucrose consumed.

hol, monosaccharides, disaccharides, sugar alcohols and levan are available, but not for levanbiose, levantriose, and levantetraose.

The optimal temperature range for the *in vitro* synthesis of levan is from 0 °C to 40 °C. Levansucrase prepared from *Z. mobilis* displayed an optimum levan formation activity at 0 °C (Song et al., 1996), from *B. subtilis* at >10 °C (Elisashvili, 1984), from *Pseudomonas* at 18 °C (Hettwer et al., 1995), and from *Rahnella* at 40 °C (Ohtsuka et al., 1992). Most levansucrase activities are inactivated at temperatures higher than 45 °C, with the exception of *R. aquatilis* ATCC 33071, which shows the maximum velocity of levan formation at 50 °C within 3 h, after which the rate declines slightly. Interestingly, at lower temperatures, transfructosylation rather than hydrolysis of sucrose is preferentially catalyzed; however, at higher temperatures hydrolysis is preferentially catalyzed, and this thermolabile feature may have advantages for large-scale levan production. In particular, levan production by *Z. mobilis* levansucrase was most active at the lowest temperature (0 °C), so that it could provide a stable operating condition with minimized contamination opportunities (Song et al., 1996). Plant fructosyltransferases lose 50% of their activity at 5 °C compared with that obtained at the optimum temperature of 20–25 °C (Koops and Jonker, 1996). The *Z. mobilis* levansucrase is stable at pH 4–7, and no activity is observed below pH 3 and above pH 9, similar to the enzymes from *Bacillus, Pseudomonas,* and *Rahnella*, the optimum pH of which was 6.0. However, the enzyme from *B. licheniformis* NRRL B-18962 retains 50% of its maximal activity at 55 °C and pH 4.

When *Z. mobilis* levansucrase was immobilized onto hydroxyapatite, the enzymatic and biochemical properties were similar to those of native enzyme towards salt and detergent effects (Jang et al., 2000). However, immobilization of the enzyme on the surface of a matrix shifts the optimum pH to acidic conditions (pH 4.0). The cell-free system synthesized two types of levan which differ in molecular weight. Levans produced by the immobilized system consisted of a higher proportion of low molecular-weight levan to total levan generated than those obtained by the native enzyme. Toluene-permeabilized whole-cell systems produced levan similarly to immobilized systems (Jang et al., 2001).

9.4 Recovery and Purification of Levan

In the microbial production of levan, the yield is low and, as a consequence, costly processes are required to extract the levan from the fermentation broth. The separation process of levan with high purity from the reaction mixture containing sucrose, glucose, fructose, and fructo-oligosaccharides is both laborious and inefficient. Likewise, the separation of levan by using solvents requires huge amounts of ethanol, methanol, isopropanol, or acetone to be used (Rhee et al., 1998). Subsequently, this solvent is lost as waste or recovered by distillation. Recently, membrane processes have been developed to separate polysaccharides from the fermentation broth or enzymatic reaction mixture, without organic waste. However, the resulting solution contains a low concentration ($<5\%$) of levan, and this must be recovered using various types of drier.

9.5 Commercial Production of Levan

Currently, a Korean start-up company, RealBioTech Co., Ltd. (RBT), is the first and only company worldwide to produce levan on a commercial basis. RBT produces levan in large-scale quantities for supply to companies as a moisturizing ingredient in cosmetics, as a dietary fiber, as food and feed additives, and as a fertilizer (http://www.realbio.com). High-purity levan (>99%) is used in cosmetics and functional foods, while low-grade levan (<15%) containing glucose and oligosaccharides is used as a supplement for feeds and fertilizers.

9.6 Market Analysis and Cost of Levan Production

As levan became available commercially only recently, it is difficult to estimate the size of its current market. It is clear, however, that the expected world market is huge, since levan has a variety of functions and applications within the bioindustries. The production cost of levan depends mainly on the cost of the raw materials, including sucrose and levansucrase, for which the depreciation costs account for 30% and 18%, respectively (RealBioTech Co., unpublished data).

9.7 Levan Competitors

Many types of oligosaccharides and polysaccharides may represent potential competitors of levan in industrial applications. The strongest competitors in the food industry are the fructo-oligosaccharides (FOS), and especially inulin, which belongs to the same fructan category as levan. The use of inulin is limited as a dietary fiber by its insolubility in water at room temperature, whereas levan is a water-soluble and viscous fructan. Other potential competitors are dextran in the food and pharmaceutical industries, β-glucans in the feed industry, and hyaluronic acid in the cosmetic industry. In addition, levan could replace other potential competitors such as xanthan gum, pullulan, and mannan in the food industry. While most commercially available polysaccharides are produced either by microbial fermentation or direct extraction from natural sources, it is possible to produce levan from sucrose by a simple one-step enzyme reaction, which in turn makes it more competitive in terms of production costs.

10 Properties of Levan

While levan is highly soluble in water at room temperature, inulin is almost insoluble (<0.5%), this difference being most likely due to the presence of β-(2,6) linkages in levan. Despite their highly branched molecular structures, microbial levans have several common interesting features in soluble form (Kasapis et al., 1994), including an exceptionally low intrinsic viscosity for a polymer of high molecular weight, an unusual sensitivity of viscosity to increasing concentration between the beginning and end of the intermediate zone, and an extreme concentration-dependence of viscosity at the intermediate zone (Figure 5). These properties may be derived from intermolecular interaction by physical entanglement rather than from any form of specific noncovalent association.

The viscosity of a solution of bacterial levan, originating from *P. syringae* pv. phaseolicola, exhibited Newtonian characteristics up to a concentration of 20%. The concentration-dependence of the 'zero-shear' specific viscosity for the levan solution was unusually high, as would be expected from the molecule's branched structure (Kasapis et al., 1994). In Figure 5, there are three linear regions with changes of slope at levan concentrations of about 4% and 20%. The linear region, including the intermediate region, was also observed in xanthan gum (Milas et al., 1990). However, the concentration-dependence of viscosity may

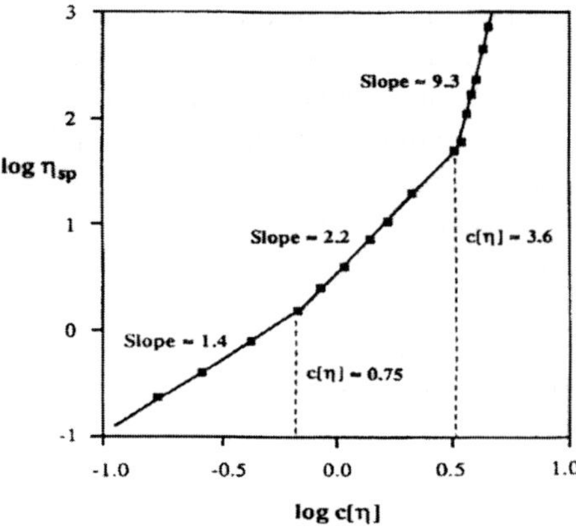

Fig. 5 Concentration dependence of 'zero-shearapos; specific viscosity for levan at 20 °C. The parameter η is used for solution viscosity and η_{sp} for specific viscosity which defines the fractional increase in viscosity due to the presence of the polymer ($\eta_{sp} = (\eta - \eta_s)/\eta_s = \eta_{rel} - 1$). η_{rel} indicates relative viscosity, which is the ratio of η to η_s (η_s for solvent viscosity). (Reprinted from Kasapis et al., 1994, p. 59, with permission from Elsevier Science.)

vary with DP, pH, temperature, and salt concentrations (Vina et al., 1998).

11 Applications of Levan

Levan has a wide range of industrial applications, for example in medicine, pharmacy, agriculture, and food. However, it is likely that the low production cost of levan will permit a much increased use of levan in the near future.

11.1 Medical Applications

Microbially produced levan has a direct effect on tumor cells that is related to a modification in the cell membrane, including changes in cell permeability (Leibovici and Stark, 1985; Calazans et al., 2000), as well as radioprotective and antibacterial activities (Vina et al., 1998). Levan derivatives have also been suggested as inhibitors of smooth muscle cell proliferation, as excipients in making tablets, and as agents to transit water into gels. Sulfated, phosphated and acetylated levans have also been suggested as anti-AIDS agents (Clarke et al., 1997).

11.2 Pharmaceutical Applications

Water-soluble polymers, including levan, can be used in a wide variety of applications in the pharmaceutical industry. Water-soluble polymers such as cellulose derivatives, pectin and carrageenan play key roles in the formulation of solid, liquid, semisolid and even controlled release dosage forms (Guo et al., 1998). The viscosity of levan varies with its DP and degree of branching, which relates to the number of side fructose chains attached to one fructose unit in the main fructose chain, and in this respect levan can be used in pharmaceutical formulations in various ways. Low molecular-weight, less branched levan usually provides a low viscosity, and can be used as a tablet binder in immediate-release dosage forms, while levans of medium- and high-viscosity grade are used in controlled-release matrix formulations. Levan has also been suggested as a possible substitute for blood expanders (Imam and Abd-Allah, 1974).

11.3 Agricultural Applications

Microbial levans were introduced in plants to promote their agronomic performance in temperature zones, as well as their natural storage capacities (Vijn and Smeekens, 1999). Transgenic tobacco plants expressing levansucrase genes from *B. subtilis* (Pilon-

Smits et al., 1995) or *Z. mobilis* (Park et al., 1999) showed an increased tolerance to drought and cold stresses. Transgenic plants accumulating fructan have been suggested as novel nutritional feed for ruminants (Biggs and Hancock, 1998, 2001). Recently, microbial levan produced enzymatically was developed as an animal feed (Rhee et al., 2000d) and also as a soil conditioner to improve the germination of various seeds (Imam and Abd-Allah, 1974).

11.4 Food Applications

A number of novel applications of levan have been suggested, particularly in food (Han, 1990; Suzuki and Chatterton, 1993). Levan may act as a prebiotic to change the intestinal microflora, thereby offering beneficial effects when present in the human diet. Levan and its partially hydrolyzed products are fermented by intestinal bacteria including bifidobacteria and *Lactobacillus* sp. (Müller and Scyfarth, 1997; Yamamoto et al., 1999; Marx et al., 2000). Levanheptaose was also suggested as a carbon source for selective intestinal microflora, including *Bifidobacterium adolescentis, Lactobacillus acidophilus,* and *Eubacterium limosum,* whereas *Clostridium perfringens, E. coli* and *Staphylococcus aureus* did not utilize levan (Kang et al., 2000). Cholesterol- and triacylglycerol-lowering effects of levan have also been reported (Yamamoto et al., 1999) and may be applied to develop levans as health foods or nutraceuticals.

Today, it also seems possible that levan might be used in the dairy industry, as *Lactobacillus reuteri* produces levan-type EPS. EPS-producing lactic acid bacteria, including the genera of *Streptococcus, Lactobacillus,* and *Lactococcus,* are used *in situ* to improve the texture of fermented dairy products such as yogurt and cheese. This group of food-grade bacteria produces a wide variety of structurally different polymers, including levan with potential use for new applications. A number of Japanese companies use microbial levans as additives in their milk products which contain *Lactobacillus* species. In addition, the replacement with levan of thickeners or stabilizers that are produced by nonfood-grade bacteria has recently emerged (Van Kranenburg et al., 1999).

11.5 Other Applications

One of the striking consequences of the densely branched structure of levan is its effectiveness in resisting interpenetration by other polymers, leading to macroscopic phase-separation (Kasapis et al., 1994; Chung et al., 1997). Dextran is often used to create two-phase liquid systems (e.g., polyethylene glycol (PEG)/dextran), and to purify biological materials of interest by selectively partitioning them into one phase (Albertsson and Tjerneld, 1994). Microbial levans also display phase-separation phenomena with pectin, locust bean gum, and PEG. Solutions of levan and locust bean gum showed a substantial reduction in viscosity, similar to the mixture levan/pectin. Levan/locust bean gum phases can be separated into discrete phases in mixed solutions in which the lower one consisted predominantly of the denser polymer, the levan phase (Kasapis et al., 1994). The PEG/levan two-phase system was prepared by combining PEG (60%, w/w) and levan (6.77%, w/w) (Chung et al., 1997). This aqueous two-phase system showed a good partitioning with six model proteins including horse heart cytochrome c, horse hemoglobin, horse heart myoglobin, hen egg albumin, bovine serum albumin, and hen egg lysozyme.

12 Patents

Many patents have been filed for the application of levan as functional food additives (Table 3). It was claimed that levan from *S. salivarius* can be used as a food additive with a hypocholesterolemic effect (Kazuoki, 1996). New applications of levan as food and feed additives (Rhee et al., 2000d) as well as a raw material for the production of difructose dianhydride IV (Rhee et al., 2000c) were also developed. Levan derivatives such as sulfated, phosphated or acetylated levans, were claimed to be anti-AIDS agents (Robert and Garegg, 1998), while a fructan *N*-alkylurethane which has excellent surface-active properties in combination with good biodegradability was patented as a surfactant for household use and industrial applications by means of replacing a hydroxyl group of fructose with an alkylaminocarbonyloxyl group (Stevens et al., 1999). A glycol/levan aqueous two-phase system which can substitute the glycerol/dextran system was developed for the partitioning of proteins (Rhee et al., 2000b). Besides levan, the fructosyl transferase activity of levansucrase has been used in the production of alkyl β-D-fructoside (Rhee et al., 2000a).

In spite of many studies on the production and application of levan, few of the levansucrase genes isolated from the microorganisms *Z. mobilis* (Rhee and Song, 1998), *R. aquatilis* (Rhee et al., 1999) and *A. diazotrophicus* (Juan et al., 1998) have been patented. A method for the production of levan using a recombinant levansucrase from *Z. mobilis* was claimed (Rhee et al., 1998), and a creative patent preparing transgenic plants harboring levansucrase gene was applied for by German researchers (Roeber et al., 1994). In the case of levansucrase, a process was claimed for the production of acid-stable

Tab. 3 Relevant patents for levan production and applications

Publication number	*Applicants*	*Inventors*	*Title of invention*	*Date*
US 4,769,254	IBI	Mays, T.D., Dally, E.L.	Microbial production of polyfructose.	September 6, 1988
US 4,927,757	IRFI	Hatcher, H.J. et al.	Production of substantially pure fructose.	May 22, 1990
WO 94/04692	IGF; Roeber, M.; Geier, G.; Geider, K.; Willmitzer, L.	Roeber, M. et al.	DNA sequences which lead to the formation of polyfructans (levans), plasmids containing these sequences as well as a process for preparing transgenic plants.	March 3, 1994
US 5,334,524	SOLVAY ENZYMES INC.	Robert, L.C.	Process for producing levansucrase using *Bacillus*.	August 2, 1994
US 5,527,784	–	Kazuoki, I.	Antihyperlipidemic and antiobesity agent comprising levan or hydrolysis products thereof obtained from *Streptococcus salivarius*.	June 17, 1996
US 5,547,863	USASA	Han, Y.W., Clarke, M.A.	Production of fructan (levan) polyfructose polymers using *Bacillus polymyxa*.	August 20, 1996

Tab. 3 (cont.)

Publication number	*Applicants*	*Inventors*	*Title of invention*	*Date*
WO 98/03184	Clarke, G; Margaret, A; SPRI -	Robert, E.J. et al.	Levan derivatives, their preparation, composition and applications including medical and food applications.	January 29, 1998
US 5,731,173		Juan, G.A.S. et al.	Fructosyltransferase enzyme, method for its production and DNA encoding the enzyme.	March 24, 1998
Korean patent 145946	KRIBB	Rhee, S. K. et al.	Method for production of levan using levansucrase.	May 6, 1998
Korean patent 176410	KRIBB	Rhee, S. K. et al.	A novel levansucrase.	November 13, 1998
Korean patent 0207960	KRIBB	Rhee, S. K. et al.	Base and amino acid sequence of levansucrase derived from *Rahnella aquatilis*.	April 14, 1999
EP 0964054 A1	TS N.V.	Stevens, C.V. et al.	Surface-active alkylurethanes of fructans.	December 15, 1999
Korean patent 0257118	KRIBB	Rhee, S. K. et al.	A process for preparation of alkyl β-D-fructoside using levansucrase.	Febraury 28, 2000
Korean patent 262769	KRIBB	Rhee, S. K. et al.	Novel polyethylene glycol/levan aqueous two-phase system and protein partitioning method using thereof.	May 6, 2000
WO 01/29185	KRIBB and RBT	Rhee, S. K. et al.	Enzymatic production of difructose dianhydride IV from sucrose and elevant enzymes and genes coding for them.	October 19, 2000
WO 01/49127	KRIBB and RBT	Rhee, S. K. et al.	Animal feed containing simple polysaccharides.	December 29, 2000

EP: European Patent; IBI: Igene Biotechnology Institute; IGF: INST GENBIOLOGISCHE FORSCHUNG (DE); IRFI: Idaho Research Foundation Institute; KRIBB: Korea Research Institute of Bioscience and Biotechnology; PCT: World Intellectual Property Organization; RBT: RealBioTech Co., Ltd.; SPRI: Sugar Processing Research Institute; TS N.V: Tiense Suikerraffinaderij N.V.; US: United States Patent; USASA: The United States of America as represented by the Secretary of the Agriculture; WO: World IPO(PCT).

levansucrase from *Bacillus* which is not induced by sucrose (Robert, 1994).

13 Current Problems and Limits

Although levan is a water-soluble and low-viscosity polysaccharide, its applications might be limited due to its turbidity and low fluidity. In order to expand the application areas of levan, the development of biosynthetic methods, including the involvement of novel enzymes, will be essential in order to control the DP (molecular weight) and the degree of branching. More important factors from a commercial aspect include process development for the large-scale production which is comparable with that used for other competitive polysaccharides, especially inulin from chicory. In order to produce levan more economically, a cost-effective purification process of levan from the reaction mixture containing sucrose, glucose, fructose and fructo-oligosaccharides must be developed. The membrane processes will likely serve as one of these solutions.

14 Outlook and Perspectives

Levan has a great potential as a functional biomaterial in the food, cosmetic, pharmaceutical and other industries. However, use of this biopolymer has yet not been practicable due to a paucity of information on its polymeric properties highlighting its industrial applicability, as well as a lack of feasible processes for large-scale production.

For technical applications, fructans with a high molecular mass and a low degree of branching would be desirable. However, microbial levans and their oligomers have been less well characterized with regard to carbohydrate structure analysis. In order to utilize the versatility of water-soluble levans to a maximum, a broader understanding of the behavior of levan is required. The characterization of polysaccharides has advanced considerably during the past two decades due to the introduction of powerful methods such as mass spectrometry, nuclear magnetic resonance, atomic force microscopy, scanning probe microscopy, small-angle X-ray scattering, small-angle neutron scattering, and molecular-mechanic-based carbohydrate modeling (Brant, 1999). As a consequence, the complete characterization of levan should be attained in the near future. Furthermore, the fundamental rheological properties of levan in solution, for example viscosity, thixotropy, dilatancy, elasticity, pseudoelasticity, and viscoelasticity will become increasingly important for new applications of this compound.

15 References

Albertsson, P. A., Tjerneld, F. (1994) Phase diagrams, *Methods Enzymol.* **228**, 3–13.

Ananthalakshmy, V. K., Gunasekaran, P. (1999) Overproduction of levan in *Zymomonas mobilis* by using cloned *sacB* gene, *Enzyme Microb. Technol.* **25**, 109–115.

Arrieta, J., Hernández, L., Coego, A., Suárez, V., Balmori, E., Menéndez, C., Petit-Glatron, M. F., Chambert, R., Selman-Housein, G. (1996) Molecular characterization of the levansucrase gene from the endophytic sugarcane bacterium *Acetobacter diazotrophicus* SRT4, *Microbiology* **142**, 1077–1085.

Avigad, G. (1965) In *Methods in Carbohydrate Chemistry*, New York, London: Academic Press, vol. V, 161–165

Aymerich, S. (1990) What is the role of levansucrase in *Bacillus subtilis*? *Symbiosis* **9**, 179–184.

Beker, M. J., Shvinka, J. E., Pankova, L. M., Laivenienks, M. G., Mezhbarde, I. N. (1990) A simultaneous sucrose bioconversion into ethanol and levan by *Zymomonas mobilis*, *Appl. Biochem. Biotechnol.* **24/25**, 265–274.

Bergeron, L. J., Morou-Bermudez, E., Burne, R. A. (2000) Characterization of the fructosyltransferase gene of *Actinomyces naeslundii* WVU45, *J. Bacteriol.* **182**, 3649–3654.

Beveridge, T. J., Graham, L. L. (1991) Surface layers of bacteria, *Microbiol. Rev.* **55**, 684–705.

Biggs, D. R., Hancock, K. R. (1998) *In vitro* digestion of bacterial and plant fructans and effects on ammonia accumulation in cow and sheep rumen fluids, *J. Gen. Appl. Microbiol.* **44**, 167–171.

Biggs, D. R., Hancock, K. R. (2001) Fructan 2000, *Trends Plant Sci.* **6**, 8–9.

Birkhed, D., Rosell, K.-G, Granath, K. (1979) Structure of extracellular water-soluble polysaccharides synthesized from sucrose by oral strains of *Streptococcus mutans*, *Streptococcus salivarius*, *Streptococcus sanguis*, and *Actinomyces viscosus*, *Arch. Oral Biol.* **24**, 53–61.

Brant, D. A. (1999) Novel approaches to the analysis of polysaccharide structures, *Curr. Opin. Struct. Biol.* **9**, 556–562.

Bruckner, R., Wagner, E., Gotz, F. (1993) Cloning and characterization of the scrA gene encoding the sucrose-specific Enzyme II of the phosphotransferase system from *Staphylococcus xylosus*, *Mol. Gen. Genet.* **241**, 33–41.

Calazans, G. M. T., Lima, R. C., de França, F. P., Lopes, C. E. (2000) Molecular weight and antitumour activity of *Zymomonas mobilis* levans, *Int. J. Biol. Macromol.* **27**, 245–247.

Carson, M., Botstein, D. (1983) Organization of the SUC gene family in *Saccharomyces*, *Mol. Cell. Biol.* **3**, 351–359.

Chambert, R. G., Gonzy-Treboul, G. (1976) Levansucrase of *Bacillus subtilis*. Characterization of a stabilized fructosyl-enzyme complex and identification of an aspartyl residue as the binding site of the fructosyl group, *Eur. J. Biochem.* **71**, 493–508.

Chambert, R., Petit-Glatron, M. F. (1991) Polymerase and hydrolase activities of *Bacillus subtilis* levansucrase can be separately modulated by site-directed mutagenesis, *Biochem. J.* **279**, 35–41.

Chambert, R. G., Gonzy-Treboul, G., Dedonder, R. (1974) Kinetic studies of levansucrase of *Bacillus subtilis*, *Eur. J. Biochem.* **41**, 285

Chung, B. H., Kim, W. K., Song, K. B., Kim, C. H., Rhee, S. K. (1997) Novel polyethylene glycol/levan aqueous two-phase system for protein partitioning, *Biotech. Techn.* **11**, 327–329.

Clarke, M. A., Roberts, E. J., Garegg, P. J. (1997) New compounds from microbiological products of sucrose, *Carbohydr. Polym.* **34**, 425.

Débarbouillé, M., Arnaud, M., Fouet, A., Klier, A., Rapoport, G. (1990) The *sacT* genes regulating the *sacPA* operon in *Bacillus subtilis* shares strong homology with transcriptional antiterminators, *J. Bacteriol.* **172**, 3966–3973.

Débarbouillé, M., Martin. V, Arnaud, M., Klier, A., Rapoport, G. (1991) Positive and negative regulation controlling expression of the sac genes in *Bacillus subtilis, Res. Microbiol.* **142**, 757–764.

Dedonder, R. (1966) Levansucrase from *Bacillus subtilis, Methods Enzymol.* **8**, 500–505.

Di Marco, A. A., Romano, A. H. (1985) D-Glucose transport system of *Zymomonas mobilis, Appl. Environ. Microbiol.* **49**, 151–157.

Edelman, J., Jefford, T. G. (1968) The mechanism of fructosan metabolism in higher plants as exemplified in *Helianthus tuberosus, New Phytol.* **67**, 517–531.

Elisashvili, V. I. (1984) Levan synthesis by *Bacillus* sp., *Appl. Biochem. Microbiol.* **20**, 82–87.

Fuchs, A. (1959) *On the synthesis and breakdown of levan by bacteria,* Thesis, Uitgeverij Waltman, Delft.

Gay, P., Le Coq, D., Steinmetz, M., Ferrari, E., Hoch, J. A. (1983) Cloning structural gene *sacB*, which codes for exoenzyme levansucrase of *Bacillus subtilis*: expression of the gene in *Escherichia coli, J. Bacteriol.* **153**, 1424–1431.

Geier, G., Geider, K. (1993) Characterization and influence on virulence of levansucrase gene from the fireblight pathogen *Erwinia amylovora, Physiol. Mol. Plant Pathol.* **42**, 387–404.

Greig-Smith, R. (1901) The gum fermentation of sugar cane juice, *Proc. Linn. Soc. N.S.W.* **26**, 589.

Guo, J.-H, Skinner, G. W., Harcum, W. W., Barnum, P. E. (1998) Pharmaceutical applications of naturally occurring water-soluble polymers, *Pharmaceutical Science & Technology Today* **1**, 254–261.

Han, Y. W. (1989) Levan production by *Bacillus polymyxa, J. Ind. Microbiol.* **4**, 447–452.

Han, Y. W. (1990) Microbial levan, *Adv. Appl. Microbiol.* **35**, 171–194.

Han, Y. W., Clarke, M. A. (1996) Production of fructan(levan) polyfructose polymers using *Bacillus polymyxa*, U. S. Patent 5,547,863.

Hatcher, H. J., Gallian, J. J., Leeper, S. A. (1990) Production of substantially pure fructose, U. S. Patent 4,927,757.

Hendry, G. A. F., Wallace, R. K. (1993) The origin, distribution, and evolutionary significance of fructans, in: *Science and Technology of Fructans* (Suzuki, M., Chatterton, N. J., Eds.), Boca Raton: CRC Press, 119–139.

Hestrin, S., Goldblum, J. (1953) Levanpolyase, *Nature* **172**, 1047–1064.

Hestrin, S., Avineri-Shapiro, S., Aschner, M. (1943) The enzymatic production of levan, *Biochem. J.* **37**, 450.

Hettwer, U., Gross, M., Rudolph, K. (1995) Purification and characterization of an extracellular levansucrase from *Pseudomonas syringae* pv. phaseolicola, *J. Bacteriol.* **177**, 2834–2839.

Hettwer, U., Jaeckel, F. R., Boch, J., Meyer, M., Rudolph, K., Ullrich, M. S. (1998) Cloning, nucleotide sequence, and expression in *Escherichia coli* of levansucrase genes from the plant pathogens *Pseudomonas syringae* pv. glycinea and *P. syringae* pv. phaseolicola, *Appl. Environ. Microbiol.* **64**, 3180–3187.

Heyer, A. G., Lloyd, J. R., Kossmann, L. (1999) Production of polymeric carbohydrates, *Curr. Opin. Biotechnol.* **10**, 169–174.

Imam, G. M., Abd-Allah, N. M. (1974) Fructosan, a new soil conditioning polysaccharide isolated from the metabolites of *Bacillus polymyxa* AS-1 and its clinical applications, *Egypt. J. Bot.* **17**, 19–26.

Jang, K. H., Kim, J. S., Song, K. B., Kim, C. H., Chung, B. H., Rhee, S. K. (2000) Production of levan using recombinant levansucrase immobilized on hydroxyapatite, *Bioprocess Eng.* **23**, 89–93.

Jang, K. H, Song, K. B., Kim, C. H., Chung, B. H., Kang, S. A., Chun, U. H., Choue, R. W., Rhee, S. K. (2001) Comparison of characteristics of levan produced by different preparations of levansucrase from *Zymomonas mobilis, Biotechnol. Lett.* **23**, 339–344.

Juan, G. A. S., Lazaro, H. G., Alberto, C. G., Guillermo, S. H. S. (1998) Fructosyltransferase enzyme, method for its production and DNA encoding the enzyme, U. S. Patent 5,731,173.

Jung, K. E., Lee, S. O., Lee, J. D., Lee, T. H. (1999) Purification and characterization of a levanbiose-producing levanase from *Pseudomonas* sp. No. 43, *Biotechnol. Appl. Biochem.* **28**, 263–268.

Kang, S. K., Park, S. J., Lee, J. D., Lee, T. H. (2000) Physiological effects of levanoligosaccharide on growth of intestinal microflora, *J. Korean Soc. Food Sci. Nutr.* **29**, 35–40.

Kasapis, S., Morris, E. R., Gross, M., Rudolph, K. (1994) Solution properties of levan polysaccharide from *Pseudomonas syringae* pv. phaseolicola, and its possible primary role as a blocker of recognition during pathogenesis, *Carbohydr. Polym.* **23**, 55–64.

Kazuoki, I. (1996) Antihyperlipidemic and anti-obesity agent comprising levan or hydrolysis products thereof obtained from *Streptococcus salivarius*, U. S. Patent 5,527,784.

Keith, K., Wiley, B., Ball, D., Arcidiacono, S., Zorfass, D., Mayer, J., Kaplan, D. (1991) Continuous culture system for production of biopolymer levan using *Erwinia herbicola, Biotechnol. Bioeng.* **38**, 557–560.

Kim, M. G., Seo, J. W., Song, K.-B., Kim, C. H., Chung, B. H, Rhee, S. K. (1998) Levan and fructosyl derivatives formation by a recombinant levansucrase from *Rahnella aquatilis, Biotechnol. Lett.* **20**, 333–336.

Klier, A. F., Rapoport, G. (1988) Genetics and regulation of carbohydrate catabolism in *Bacillus, Annu. Rev. Microbiol.* **42**, 65–95.

Kojima, I., Saito, T., Iizuka, M., Minamiura, N., Ono, S. (1993) Characterization of levan produced by *Serratia* sp., *J. Ferment. Bioeng.* **75**, 9–12.

Koops, A. J., Jonker, H. H. (1996) Purification and characterization of the enzymes of fructan biosynthesis in tubers of *Helianthus tuberosus* Colombia. II. Purification of sucrose:sucrose 1-fructosyltransferase and reconstitution of fructan synthesis *in vitro* with purified sucrose:sucrose 1-fructosyltransferase and fructan:fructan 1-fructosyltransferase, *Plant Physiol.* **110**, 1167–1175.

Kunst, F., Rapoport, G. (1995) Salt stress is an environmental signal affecting degradative enzyme synthesis in *Bacillus subtilis, J. Bacteriol.* **177**, 2403–2407.

Kyono, K., Yanase, H., Tonomura, K., Kawasaki, H., Sakai, T. (1995) Cloning and characterization of *Zymomonas mobilis* genes encoding extracellular levansucrase and invertase, *Biosci. Biotechnol. Biochem.* **59**, 289–293.

Leibovici, J., Stark, Y. (1985) Increase in cell permeability to a cytotoxic agent by the polysaccharide levan, *Cell. Mol. Biol.* **31**, 337–341.

Lepesant, J. A., Kunst, F., Pascal, M., Steinmetz, M., Dedonder, R. (1976) Specific and pleiotropic regulatory mechanisms in the sucrose system of *Bacillus subtilis, Microbiology* **168**, 58–69.

Li, Y., Triccas, J. A., Ferenci, T. (1997) A novel levansucrase-levanase gene cluster in *Bacillus stearothermophilus* ATCC 12980, *Biochim. Biophys. Acta* **1353**, 203–208.

Liu, J., Barnell, W. O., Conway, T. (1992) The polycistronic mRNA of the *Zymomonas mobilis glf-zwf-edd-glk* operon is subject to complex transcript processing, *J. Bacteriol.* **174**, 2824–2833.

Martin, I., Débarbouillé, M., Klier, A., Rapoport, G. (1989) Induction and metabolic regulation of levanase synthesis in *Bacillus subtilis, J. Bacteriol.* **171**, 1885–1892.

Martin, I., Débarbouillé, M., Klier, A., Rapoport, G. (1990) Levanase operon of *Bacillus subtilis* includes a fructose-specific phosphotransferase system regulating the expression of the operon, *J. Mol. Biol.* **214**, 657–671.

Marx, S. P., Nosberger, J., Frehner, M. (1997) Hydrolysis of fructan in grasses: 2,6-β-linkage-specific fructan-β-fructosidase from stubble of *Lolium perenne, New Phytol.* **135**, 279–290.

Marx, S. P., Winkler, S., Hartmeier, W. (2000) Metabolization of β-(2,6)-linked fructose-oligosaccharides by different bifidobacteria, *FEMS Microbiol. Lett.* **182**, 163–169.

Mays, T. D., Dally, E. L. (1988) Microbial production of polyfructose, U.S. Patent 4,769,254.

Milas, M., Rinaudo, M., Knipper, M., Schuppise, J. L. (1990) Flow and viscoelastic properties of xanthan gum solutions, *Macromolecules* **23**, 2506–2511.

Müller, M., Seyfarth, W. (1997) Purification and substrate specificity of an extracellular fructan-hydrolase from *Lactobacillus paracasei* ssp. *paracasei* P 4134, *New Phytol.* **136**, 89–96.

Nelson, C. J., Spollen, W. G. (1987) Fructans, *Physiol. Plant.* **71**, 512–516.

Newbrun, E., Lacy, R., Christie, T. M. (1971) The morphology and size of the extracellular polysaccharide from oral streptococci, *Arch. Oral Biol.* **16**, 863–872.

Ohtsuka, K., Hino, S., Fukushima, T., Ozawa, O., Kanematsu, T., Uchida, T. (1992) Characterization of levansucrase from *Rahnella aquatilis* JCM-1638, *Biosci. Biotechnol. Biochem.* **56**, 1373–1377.

Park, J. M., Kwon, S. Y., Song, K. B., Kwak, J. W., Lee, S. B., Nam, Y. W., Shin, J. S., Park, Y. I., Rhee, S. K., Paek, K. H. (1999) Transgenic tobacco plants expressing the bacterial levansucrase gene show enhanced tolerance to osmotic stress, *J. Microbiol. Biotechnol.* **9**, 213–218.

Perez Oseguera, M. A., Guereca, L., Lopez-Munguia, A. (1996) Properties of levansucrase from *Bacillus circulans, Appl. Microbiol. Biotechnol.* **45**, 465–471.

Pilon-Smits, E. A. H., Ebskamp, M. J. M., Paul, M. J., Jeuken, M. J. W., Weisbeek, P. J., Smeekens, J. C. M. (1995) Improved performance of transgenic fructan-accumulating tobacco under drought stress, *Plant Physiol.* **107**, 125–130.

Pollock, C. J., Cairns, A. J. (1991) Fructan metabolism in grasses and cereals, *Annu. Rev. Plant Physiol. Plant Mol. Biol.* **42**, 77–101.

Rapoport, G., Dedonder, R. (1963) Le lévane-sucrase de *B. subtilis* III. Reaction d'hydrolyse, de transfert et d'échange avec des analogues du saccharose, *Bull. Soc. Chim. Biol.* (France) **45**, 515–535.

Rapoport, G., Klier, A. (1990) Gene expression using *Bacillus*, *Curr. Opin. Biotechnol.* **1**, 21–27.

Rhee, S. K., Song, K. B. (1998) A novel levansucrase, Korean Patent 176410.

Rhee, S. K., Song, K. B., Kim, C. H. (1998) Method for production of levan using levansucrase, Korean Patent 145946.

Rhee, S. K., Song, K. B., Seo, J. W., Kim, C. H., Chung, B. H. (1999) Base and amino acid sequence of levansucrase derived from *Rahnella aquatilis*, Korean Patent 0207960.

Rhee, S. K., Kim, C. H., Song, K. B., Kim, M. G., Seo, J. W., Chung, B. H. (2000a) A process for preparation of alkyl β-D-fructoside using levansucrase, Korean Patent 0257118.

Rhee, S. K, Chung, B. H., Kim, W. K., Song, K. B., Kim, C. H. (2000b) Novel polyethylene glycol/levan aqueous two-phase system and protein partitioning method using thereof, Korean Patent 262769.

Rhee, S. K., Song, K. B., Kim, C. H., Ryu, E. J., Lee, Y. B. (2000c) Enzymatic production of difructose dianhydride IV from sucrose and elevant enzymes and genes coding for them, PCT-KR00-01183.

Rhee, S. K., Song, K. B., Yoon, B. D., Kim, C. H. (2000d) Animal feed containing simple polysaccharides, PCT-KR00-01556.

Robert, E. J., Garegg, P. J. (1998) Levan derivatives, their preparation, composition and applications including medical and food applications, WO 98/03184.

Robert, L. C. (1994) Process for producing levan sucrase using *Bacillus*, U. S. Patent 5,334,524.

Roeber, M., Geier, G., Geider, K., Willmitzer, L. (1994) DNA sequences which lead to the formation of polyfructans (levans), plasmids containing these sequences as well as a process for preparing transgenic plants, WO-94/004692.

Saito, K., Tomita, F. (2000) Difructose anhydrides: their mass-production and physiological functions, *Biosci. Biotechnol. Biochem.* **64**, 1321–1327.

Saito, K., Goto, H., Yokoda, A., Tomita, F. (1997) Purification of levan fructotransferase from *Arthrobacter nicotinovorans* GS-9 and production of DFA IV from levan by the enzyme, *Biosci. Biotechnol. Biochem.* **61**, 1705–1709.

Sato, S., Koga, T., Inoue, M. (1984) Isolation and some properties of extracellular D-glucosyltransferases and D-fructosyltransferase from *Streptococcus mutans* serotypes c, e, and f, *Carbohydr. Res.* **134**, 293–304.

Shimotsu, H., Henner, D. J. (1986) Modulation of *Bacillus subtilis* levansucrase gene expression by sucrose and regulation of the steady-state mRNA level by *sacU* and *sacQ* genes, *J. Bacteriol.* **168**, 380–388.

Smith, A. E. (1976) Beta-fructofuranosidase and invertase activity in tall fescue culm bases, *J. Agric. Food Chem.* **24**, 476–478.

Song, K. B., Rhee, S. K. (1994) Enzymatic synthesis of levan by *Zymomonas mobilis* levansucrase overexpressed in *Escherichia coli*, *Biotechnol. Lett.* **16**, 1305–1310.

Song, K. B., Joo, H. K., Rhee, S. K. (1993) Nucleotide sequence of levansucrase gene (*levU*) of *Zymomonas mobilis* ZM1 (ATCC10988), *Biochim. Biophys. Acta* **1173**, 320–324.

Song, K. B., Belghith, H., Rhee, S. K. (1996) Production of levan, a fructose polymer, using an overexpressed recombinant levansucrase, *Ann. N. Y. Acad. Sci.* **799**, 601–607.

Song, K. B., Seo, J. W., Kim, M. K., Rhee, S. K. (1998) Levansucrase from *Rahnella aquatilis*: gene cloning, expression and levan formation, *Ann. N. Y. Acad. Sci.* **864**, 506–511.

Song, K. B., Seo, J. W., Rhee, S. K. (1999) Transcriptional analysis of *levU* operon encoding saccharolytic enzymes and two apparent genes involved in amino acid biosynthesis in *Zymomonas mobilis*, *Gene* **232**, 107–114.

Song, K. B., Bae, K. S., Lee, Y. B., Lee, K. Y., Rhee, S. K. (2000) Characteristics of levan fructotransferase from *Arthrobacter ureafaciens* K2032 and difructose anhydride IV formation from levan, *Enzyme Microb. Technol.* **27**, 212–218.

Sprenger, N., Bortlik, K., Brandt, A., Boller, T., Wiemken, A. (1995) Purification, cloning, and functional expression of sucrose:fructan 6-fructosyltransferase, a key enzyme of fructan synthesis in barley, *Proc. Natl. Acad. Sci. USA* **92**, 11652–11656.

Stevens, C. V., Karl, B., Isabelle, M.-A., Lucien, D. (1999) Surface-active alkylurethanes of fructans, EP 0964054 A1.

Stülke, J., Martin, V. I., Charrier, V., Klier, A., Deutscher, J., Rapoport, G. (1995) The HPr protein of the phosphotransferase system links induction and catabolite repression of the *Bacillus subtilis* levanase operon, *J. Bacteriol.* **177**, 6928–6936.

Suzuki, M., Chatterton, N. J. (1993) *Science and Technology of Fructans*. Boca Raton: CRC Press.

Tajima, K., Uenishi, N., Fujiwara, M., Erata, T., Munekata, M., Takai, M. (1998) The production of a new water-soluble polysaccharide by *Acetobacter xylinum* NCI 1005 and its structural analysis by NMR spectroscopy, *Carbohydr. Res.* **305**, 117–122.

Tajima, K., Tanio, T., Kobayashi, Y., Kohno, H., Fujiwara, M., Shiba, T., Erata, T., Munekata, M., Takai, M. (2000) Cloning and sequencing of the levansucrase gene from *Acetobacter xylinum* NCI 1005, *DNA Res.* **7**, 237–242.

Tanaka, K., Kawaguchi, H., Ohno, K., Shohji, K. (1981) Enzymatic formation of difructose IV from bacterial levan, *J. Biochem.* **90**, 1545–1548.

Tanaka, K., Karigane, T., Yamaguchi, F., Nishikawa, S., Yoshida, N. (1983) Action of levan fructotransferase of *Arthrobacter ureafaciens* on levanoligosaccharides, *J. Biochem.* **94**, 1569–1578.

Uchiyama, T. (1993) Metabolism in microorganisms. Part II. Biosynthesis and degradation of fructans by microbial enzymes other than levansucrase. in: *Science and Technology of Fructans* (Suzuki, M., Chatterton, N. J., Eds.), Boca Raton: CRC Press, 169–190.

Vandamme, E. J., Derycke, D. G. (1983) Microbial inulinase: fermentation process, properties, and applications, *Adv. Appl. Microbiol.* **29**, 139–176.

Van Geel-Schutten, G. H., Faber, E. J., Smit, E., Bonting, K., Smith, M. R., Ten Brink, B., Kamerling, J. P., Vliegenthart, J. F. G., Dijkhuizen, L. (1999) Biochemical and structural characterization of the glucan and fructan exopolysaccharides synthesized by the *Lactobacillus reuteri* wild-type strain and mutant strains, *Appl. Environ. Microbiol.* **65**, 3008–3014.

Van Kranenburg, R., Boels, I. C., Kleerebezem, M., de Vos, W. M. (1999) Genetics and engineering of microbial exopolysaccharides for food: approaches for the production of existing and novel polysaccharides, *Curr. Opin. Biotechnol.* **10**, 498–504.

Vereyken, I. J., Chupin, V., Demel, R. A., Smeekens, S. C. M., Kruijff, B. (2001) Fructans insert between the headgroups of phospholipids, *Biochim. Biophys. Acta* **1510**, 307–320.

Vijn, I., Smeekens, S. (1999) Fructan: more than reserve carbohydrate? *Plant Physiol.* **120**, 351–359.

Vina, I., Karsakevich, A., Gonta, S., Linde, R., Bekers, M. (1998) Influence of some physicochemical factors on the viscosity of aqueous levan solutions of *Zymomonas mobilis, Acta Biotechnol.* **18**, 167–174.

Wolff, D., Czapla, S., Heyer, A. G., Radosta, S., Mischnick, P., Springer, J. (2000) Globular shape of high molar mass inulin revealed by static light scattering and viscometry, *Polymer* **41**, 8009–8016.

Yamamoto, Y., Takahashi, Y., Kawano, M., Iizuka, M., Matsumoto, T., Saeki, S., Yamaguchi, H. (1999) *In vitro* digestibility and fermentability of levan and its hypocholesterolemic effects in rats, *J. Nutr. Biochem.* **10**, 13–18.

Yun, J. W. (1996) Fructooligosaccharides – occurrence, preparation, and application, *Enzyme Microb. Technol.* **19**, 107–117.

15
Hyaluronan

Prof. Dr. Peter Prehm
Institut für Physiologische Chemie und Pathobiochemie, Waldeyerstr. 15,
D-48129 Münster, Germany; Tel.: +49-251-8355579; Fax: +49-251-8355596;
E-mail: prehm@uni-muenster.de

CHO Chinese hamster ovary
PMA phorbol-12-myristate-13-acetate
RHAMM receptor for hyaluronan-mediated motility

1 Introduction

Although hyaluronan has a very simple structure, almost everything else concerning the molecule is unusual. Sometimes its role is mechanical and structural (as in the synovial fluid, the vitreous humor, or the umbilical cord), whereas sometimes it interacts in tiny concentrations in cells to trigger important responses. Hyaluronan has an unusual mechanism of biosynthesis and exceptional physical properties; consequently, research on this compound was cumbersome, with progress often impeded by failures – often because established procedures from other fields were not applicable and new techniques needed to be developed before any progress could be made.

During the decades of hyaluronan research, several books and reviews have marked such progress including Balazs (1970), Laurent (1989, 1998), Laurent and Fraser (1992), Goa and Benfield (1994), Lapcik et al. (1998), and Abatangelo and Weigel (2000), whilst reviews are published continually on a new web-site: http://www.glycoforum.gr.jp/science/hyaluronan.

2 Historical Outline

In 1934, Karl Meyer described a procedure for isolating a novel glycosaminoglycan from the vitreous humor of bovine eyes, and named it hyaluronic acid (from the Greek, hyalos = glassy, vitreous) (Meyer and Palmer, 1934). These authors showed that this substance contained a uronic acid and an amino sugar, but no sulfoesters. At physiological pH all carboxyl groups are dissociated, and hence the polysaccharide should be called hyaluronate. Today, this macromolecule is most frequently referred to as hyaluronan, in order to emphasize its polysaccharide nature. During the 1930s and 1940s, hyaluronan was isolated from many sources such as the vitreous body, synovial fluids, umbilical cord, skin, and rooster comb (Meyer, 1947) and also from streptococci (Kendall et al., 1937).

The physico-chemical characterization of hyaluranon was carried out during the 1950s and 1960s. The molecular weight is in the order of several millions, whilst in solution the chain behaves as an extended random coil, with a diameter of ~500 nm. At concentrations as low as 0.1%, the chains are entangled, and this results in extremely high, shear-dependent viscosity (Laurent, 1970). These properties enable hyaluronan to regulate water balance, osmotic pressure and flow resistance, to interact with proteins, and also to act as a sieve, as a lubricant, and to stabilize structures by virtue of electrostatic interactions (Comper and Laurent, 1978).

In 1972, Hardingham and Muir discovered that hyaluronan interacts with cartilage proteoglycans and serves as the central structural backbone of cartilage. This was the first example of a specific interaction between hyaluronan and a protein, and many more such interactions were discovered during the 1990s.

After 1980, the research spread in many directions, mainly because until that time it had been assumed that hyaluronan belonged to the proteoglycans, and that its biosynthesis proceeded in a similar manner. In fact, many studies were conducted to detect the protein moiety, but this assumption was disproved when a plausible mechanism of biosynthesis was proposed (Prehm, 1983a,b). It had also been assumed that the synthesis of hyaluronan occurred in the Golgi body – as was the case for all other secretory eucaryotic polysaccharides – until it was shown that hyaluronan was in fact synthesized at plasma membranes and the chains were extruded directly into the extracellular matrix (Prehm, 1984). The catabolic pathways of hyaluranon were also elucidated at about this time (Fraser et al., 1981).

Subsequently, it became possible to measure hyaluronan specifically in body fluids with high sensitivity (Tengblad et al., 1980), and also to visualize it histochemically. These advances opened the way to assess the role of hyaluronan in many pathological disturbances.

Balazs pioneered the application of hyaluronan for medical purposes, and produced highly viscous and noninflammatory preparations on a commercial scale both as an aid for ophthalmic surgery and as viscosupplementation for synovial fluids in patients with osteoarthritis (Balazs, 1982; Balazs and Denlinger, 1989).

Although the importance of hyaluronan in cellular behavior had been recognized for decades, it was not until 1986 that its requirement for detachment in mitotic cell division was proven (Brecht et al., 1986). Underhill and Toole (1979, 1982) reported that hyaluronan was an adhesive cell surface component that formed large coats around untransformed fibroblasts, and smaller coats around transformed cells.

Cell surface hyaluronan-binding proteins were discovered during the late 1980s (Turley et al., 1987; Aruffo et al., 1990), and studied intensively during the 1990s. Although hyaluranon was already known to be involved in both metastasis (Toole et al., 1979) and cell differentiation (Toole et al., 1977), it was investigations into the molecular biology of the receptors which led to a fundamental understanding of these processes. In particular, the receptors CD44 and RHAMM (Receptor for Hyaluronan-Mediated Motility) have attracted much enthusiasm, mainly because they are believed to be involved in cancer metastasis (Arch et al., 1992; Hall et al., 1995). However, a sobering shock reached the scientific community, when CD44-deficient mice were bred that had only marginal physiological impairments (Schmits et al., 1997). In addition, the receptor RHAMM became a matter of bitter scientific debate when it was found to be located mainly intracellularly (Hofmann et al., 1998a; Turley et al., 1998). Subsequently, a number of other intracellular hyaluronan-binding proteins have been found (Huang et al., 2000), though their function remains somewhat of a mystery. During the 1990s, hyaluronan synthases were cloned from different sources (Weigel et al., 1997), each capable of producing hyaluronan of different chain lengths and quantities (Itano et al., 1999).

The actions of hyaluronan as an adhesive component and also as a detachment factor appeared paradoxical. This paradox has recently been solved however, when it was realized that the cellular functions are mediated through cell-surface receptors that are susceptible to proteases (Dube et al., 2001). It appears that hyaluronan acts as an amplifier for active proteases, but as an attachment factor when proteases are inactive.

It has long been known that hyaluronan is very sensitive to breakdown by oxygen radicals (Wong et al., 1981), and it has become clear that it is the breakdown products which mediate important biological functions. Oligosaccharides of hyaluronan induce angiogenesis (West et al., 1985) and also activate lymphocytes (McKee et al., 1997; Termeer et al., 2000). Radical degradation generates reactive aldehydes (Uchiyama et al., 1990) which modify proteins into the main antigenic structures of rheumatoid arthritis (Prehm, 2000). This discovery finally terminated a long and oppressive period of ignorance in a medically important problem, and may eventually lead to a curative treatment of these diseases that currently are treated only symptomatically.

3 Chemical Structure

The complete structure of hyaluranon was elucidated by the group of Karl Meyer, who characterized the oligosaccharides obtained by the action of testicular hyaluronidases (Weissman and Meyer, 1954). Hyaluronan consists of basic disaccharide units of D-glucuronic acid and D-*N*-acetylglucosamine, these being linked together through alternating β-1,4 and β-1,3 glycosidic bonds (Figure 1).

The number of repeat disaccharides in a completed hyaluronan molecule can reach 10,000 or more, with a molecular mass of $\sim 4 \times 10^6$ daltons (each disaccharide is ~400 daltons). In a physiological solution, the backbone of a hyaluronan molecule is stiffened by a combination of the chemical structure of the disaccharide, internal hydrogen bonds, and interactions with solvent. In addition, the preferred shape in water features hydrophobic patches on alternating sides of the flat, tape-like secondary structure. The two sides are identical, so that hyaluronan molecules are ambidextrous, enabling them to aggregate via specific interactions in water to form meshworks, even at low concentrations (Scott et al., 1991). In physiological solutions a hyaluronan molecule assumes an expanded random coil structure which occupies a very large domain. The actual mass of hyaluronan within this domain is very low, and ~0.1% molecules would overlap each other at a hyaluronan concentration of 1 mg mL^{-1}, or higher. This domain structure of hyaluronan has interesting and important consequences. Small molecules such as water, electrolytes and nutrients can diffuse freely through the solvent, within the domain; however, large molecules such as proteins will be partially excluded from the domain because of their hydrodynamic sizes in solution. This leads both to slower diffusion of macromolecules through the network and to their lower concentration in the network compared with the surrounding hyaluronan-free compartments. At pH 7, the carboxyl groups are predominantly ionized, and the hyaluronan molecule is a polyanion that has associated exchangeable cation counterions to maintain charge neutrality.

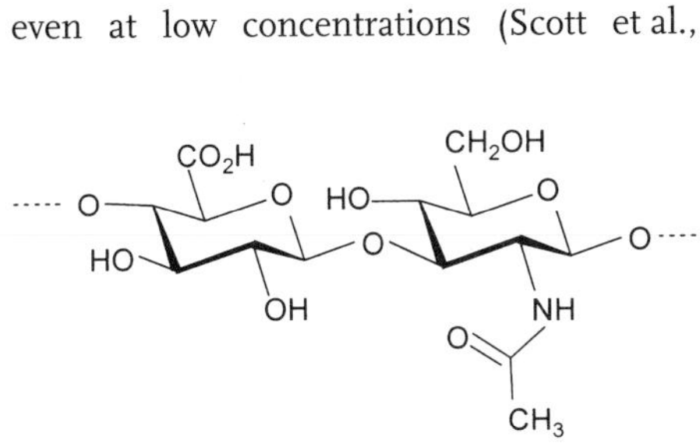

Fig. 1 Repeating unit of hyaluranon.

4 Occurrence of Hyaluronan

Hyaluronan is present in all vertebrates, and also in the capsule of some pathogenic bacteria such as *Streptococcus* sp. and *Pasteurella*. It is a component of extracellular matrices in most tissues, and in some tissues it is a major constituent. The concentration of hyaluronan is particularly high in rooster comb (7.5 mg mL^{-1}), in the synovial fluid (3–4 mg mL^{-1}), in umbilical cord (3 mg mL^{-1}), in the vitreous humor of the eye (0.2 mg mL^{-1}), and in skin (0.5 mg mL^{-1}). In other tissues that contain less hyaluronan, it forms an essential structural component of the matrix. In cartilage it forms the aggregation center for aggrecan, the large chondroitin sulfate proteoglycan, and re-

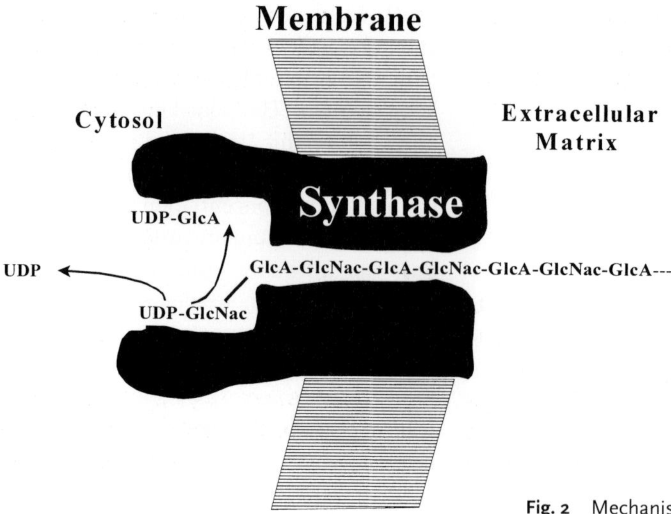

Fig. 2 Mechanism of hyaluranon synthesis.

tains this macromolecular assembly in the matrix by specific hyaluronan–protein interactions. It also forms a scaffold for binding of other matrix components around smooth muscle cells on the aorta, and on fibroblasts in the dermis of skin. The largest deposit of hyaluronan resides in the skin; in an adult human this totals ~8 g. Hyaluronan has also been detected intracellularly in proliferating cells (Evanko and Wight, 1999)

5 Mechanism of Hyaluronan Synthesis

The unusual mechanism of hyaluranon synthesis has impeded progress for a long time – a situation which has also occurred with other important polysaccharides such as cellulose and chitin. However, it now appears that two mechanisms have evolved independently – for mammalian cells and for streptococci on the one hand, and for *Pasteurella* on the other hand.

5.1 Chain Elongation

Hyaluronan synthesis in mammalian cells differs from that of other polysaccharides in many aspects. The molecule is elongated at the reducing end by alternate transfer of UDP-hyaluronan to the substrates UDP-GlcNac and UDP-GlcA, thereby liberating the UDP-moiety (Figure 2) (Prehm, 1983a,b).

Other glucosaminoglycans grow at the non-reducing end and hence require a protein backbone. Hyaluronan is synthesized at plasma membranes, the nascent chains being extruded directly into the extracellular matrix (Prehm, 1984). In contrast, other glucosaminoglycans are made in the Golgi body. Chain initiation does not require either a protein backbone (as for proteoglycans) or preformed oligosaccharides as starters; only the presence of the nucleotide sugar precursors is needed to initiate new chains. During elongation the chain is retained on the membrane-integrated synthase, this mechanism of synthesis being in operation for hyaluranon synthesis in both vertebrates and in Gram-positive streptococci. However, a different mechanism seems to exist for hyaluranon synthesis in Gram-negative *Pasteurella* in which the chains are elongated at the non-reducing end (DeAngelis, 1999).

5.2 Chain Size

One point for discussion is what determines the size of the synthesized hyaluronan, and this aspect of polymerization also applies to other macromolecular syntheses such as for proteins, DNA, or RNA. An answer was provided from experiments on isolated membranes from fibroblasts or streptococci, whereby the removal of nascent hyaluronan from the hyaluronan synthase enzyme stimulated its production. This was demonstrated in isolated streptococcal membranes (Nickel et al., 1998) and also in intact fibroblasts (Philipson et al., 1985). It thus appeared that high molecular-weight hyaluronan inhibited its own chain elongation, when it was retained on the synthase. This phenomenon may occur for solely thermodynamic reasons, because the decrease in entropy during the synthesis of a macromolecule must be compensated by free energy from cleavage of the nucleotide sugars and the subsequent formation of ordered structures. In fact, this explains why macromolecules such as proteins, RNA or DNA do not exceed a certain chain length (Peller, 1980).

5.3 Chain Export

The growing chain must be exported through a membrane pore, and consequently the proposal was made by Weigel that this pore is formed by the synthase itself, because the inactivation rate of the synthase by irradiation did not permit the participation of other proteins (Tlapak-Simmons et al., 1998). However, these authors did not show that transport of hyaluronan through the vesicle membrane was inactivated, and methods should be developed to confirm this finding.

5.4 Swelling

Hydration and swelling of nascent hyaluronan occurs at the site of synthesis on the cell surface. While swelling to enormous volumes (diameters up to 500 nm), one molecule of hyaluronan will displace many other cell-surface components by virtue of exclusion. Hence, it is conceivable that this swelling provides a mechanism whereby adhesive components are disrupted from the anchored cell.

5.5 Macromolecular Assembly

Macromolecular assembly with other matrix molecules such as proteoglycans also occurs at the cell surface. The compartmentation of hyaluronan and proteoglycan syntheses to the Golgi complex and the plasma membranes thus ensures that, during synthesis, very large aggregates are formed at the site of final deposition, and not intracellularly.

6 Hyaluronan Synthases

Hyaluronan is synthesized at the protoplast membrane of group A and group C streptococci (Markovitz and Dorfman, 1962), the enzymatic activity being solubilized by very mild detergents such as digitonin (Triscott and van de Rijn, 1986). Conventional purification procedures such as ion-exchange chromatography of detergent-solubilized membrane proteins yielded inhomogeneous protein mixtures that could not be separated into individual constituents without loss of enzymatic activity. Therefore, a new method, based on the phase separation of a detergent solution, was developed to allow purification of the synthase in its active form (Prehm et al., 1996). It was known that membrane proteins can be separated from soluble proteins by phase separation of a Triton X-114 extract. Phase separation can be induced in 1% Triton X-114 solutions by a temperature shift from 0 °C to 37 °C, with soluble proteins remaining in the aqueous phase and membrane proteins in the detergent phase. However, Triton X-114 was shown to inactivate the hyaluronan synthase. It was found that digitonin can undergo phase separation by the addition of polyethylene glycol 6000 at 0 °C, and that the synthase will remain in the aqueous phase, where it is associated with hyaluronan. Final purification of the hyaluronan synthase was achieved by ion-exchange chromatography and yielded an electrophoretically homogeneous protein of 42 kDa. This study proved that a single protein was sufficient to direct hyaluronan synthesis, and that the method may be generally applicable to other membrane proteins that are associated with polysaccharides, because it combines the advantages of the mild detergent digitonin with phase separation of all membrane proteins from polysaccharide-binding proteins.

Molecular cloning of the streptococcal hyaluronan synthase was reported independently by DeAngelis and van de Rijn (DeAngelis et al., 1993; Dougherty and van de Rijn, 1994). The gene was designated *HasA*. The *Streptococcus pyogenes* operon encodes two other proteins: HasB is a UDP-glucose-dehydrogenase, which is required to convert UDP-glucose to UDP-GlcA (Dougherty and van de Rijn, 1993), while HasC is a UDP-glucose-pyrophosphorylase, which is required to convert glucose-1-phosphate and UTP to UDP-glucose (Crater et al., 1995). Mammalian synthases were cloned simultaneously from a mutant mouse mammary carcinoma (Itano and Kimata, 1996) and *Xenopus laevis* (Meyer and Kreil, 1996). Now, three mammalian synthases are known and have been designated Has1, Has2, and Has3 (reviewed by Weigel et al., 1997). Because these proteins have 30% identity in terms of amino acid sequence with the streptococcal synthase, the genes may have a common ancestor. The synthase from *Pasteurella* has been cloned by DeAngelis et al. (1998), and is structurally unrelated to the other synthases.

7
Hyaluronan-binding Proteins and Receptors

Hyaluronan-binding proteins are constituents of the extracellular matrix, and stabilize its integrity. Hyaluronan receptors are involved in cellular signal transduction; one receptor family includes the binding proteins aggrecan, link protein, versican and neurocan and the receptors CD44, TSG6 (Lee et al., 1992), hyaluronectin (Delpech and Halavent, 1981), GHAP (Perides et al., 1990), and Lyve-1 (Banerji et al., 1999). The RHAMM receptor is an unrelated hyaluronan-binding protein, and the hyaluronan-binding sites contain a motif of a minimal site of interaction with hyaluronan. This is represented by B(X7)B, where B is any basic amino acid except histidine, and X is at least one basic amino acid and any other moiety except acidic residues. CD44 and RHAMM have attracted much attention, because they were believed to be involved in metastasis (Arch et al., 1992; Hall et al., 1995).

7.1 CD44

CD44 is a pleiomorphic extracellular matrix receptor that also binds to fibronectin and laminin, and also interacts with hyaluronan with relatively low affinity ($K_D = 10^{-8}$ M). CD44 exists as many isoforms, though some do not bind to hyaluronan. The hyaluronan-binding properties of CD44 have been implicated in promoting cell–cell aggregation and migration upon hyaluronan and collagen substrates. CD44 binds to oligosaccharides of up to 18 residues in a monovalent manner, and above 20 residues in a divalent manner, and this results in increased avidity (Lesley et al., 2000). When murine mammary carcinoma cells were transfected to produce high levels of soluble CD44, growth and metastasis *in vivo* was inhibited. In contrast, transfection with a CD44 mutant that does not bind CD44 has no effect (Peterson et al., 2000). The role of CD44 in hyaluronan synthesis shedding was analyzed in detail in metastatic and nonmetastatic melanoma cell lines that differed in degradation of CD44 and hyaluronan production (Lüke and Prehm, 1999). The nonmetastatic melanoma cell line IF6 did not significantly degrade CD44, while the metastatic cell line MV3 produced a soluble fragment. Increased hyaluronan synthesis and shedding correlated with proteolysis of CD44 on the melanoma cell lines. Intact cell surface CD44 retains hyaluronan to the vicinity of the synthase to inhibit shedding and the initiation of new chains.

The binding of hyaluronan to cells is also influenced by the level of CD44 expression (Takahashi et al., 1995; Miyake et al., 1998), by CD44 modifications such as expression of variants in different cell types (Van der Voort et al., 1995), by glycosylation (Skelton et al., 1998), by intracellular binding to ankyrin (Zhu and Bourguignon, 1998), or by phosphorylation (Puré et al., 1995). Treatment of cells with phorbol-12-myristate-13-acetate (PMA), calcium and forskolin stimulated phosphorylation of CD44, reduced the hyaluronan-binding activity (Liu et al., 1998), and stimulated hyaluronan synthesis (Klewes and Prehm, 1994). All these factors may be involved in the regulation of hyaluronan synthesis and exert profound effects on migration, growth and metastasis.

Reduced hyaluronan binding to CD44 might have two consequences:

- Loss of signal transduction from the extracellular hyaluronan to intracellular phosphorylation by the CD44-associated kinase.
- Overproduction of hyaluronan that swells to enormous volumes on the cell surface to displace cellular adhesion molecules.

Proteolysis of CD44 plays a key role in hyaluronan-mediated effects on cell growth, migration, metastasis, and adhesion (Dube et al., 2001) (Figure 3). Hyaluronan serves as an additional adhesive component that inhibits migration and proliferation, when it is retained on the cell surface by hyaluronan-binding proteins. In contrast, most transformed cells produce higher levels of surface proteases that also degrade CD44, resulting in the release of surface-bound hyaluronan and the stimulation of synthesis. In these cells, hyaluronan serves as an additional detachment factor, as the hyaluronan molecules which swell on the cell surface also displace undegraded adhesive components. Thus, hyaluronan acts as an amplifier of both active and inactive cell-surface proteases. It will therefore be difficult to identify those proteases which cleave CD44 at the surface of different cell lines. Recently, a chymotrypsin-like sheddase was thought responsible for shedding CD44 from a myoepithelial cell line (Lee et al., 2000).

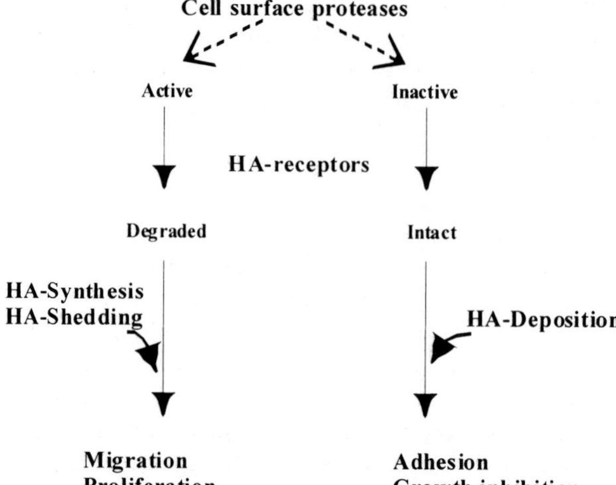

Fig. 3 Dual functions of hyaluronan.

The belief in a central role for CD44 in migration and metastasis met with severe disillusionment when CD44-deficient mice showed only marginal health disturbances (Schmits et al., 1997). This is an apparent contradiction to many studies which showed severe defects in the tissues of originally healthy animals when the function of CD44 was inhibited. A possible explanation might be that an unknown substitute of CD44 is expressed during development in CD44-defective mice (Ponta et al., 1998).

7.2 RHAMM

The hyaluronan-binding protein RHAMM was originally identified as a cell-surface protein that was also thought to be involved in hyaluronan-mediated migration and metastasis (Turley, 1989; Hall et al., 1995). Recently, a bitter scientific debate questioned its location and its function (Hofmann et al., 1998a; Turley et al., 1998), but it now appears that RHAMM is expressed either at the cell surface at low amounts under specific growth conditions, or is not expressed at all and is mainly localized intracellularly (Assmann et al., 1998; Hofmann et al., 1998b).

7.3 Other Hyaluronan-binding Proteins

Other intracellular hyaluronan-binding proteins have also been detected (Huang et al., 2000), but their functions remain elusive. Proliferating cells contain also detectable amounts of intracellular hyaluronan (Evanko and Wight, 1999) though again, the function of this is unknown.

8 Mechanisms of Hyaluronan Release from the Cell Surface

Hyaluronan could be released from cells by either enzymatic or radical degradation, or by dissociation as the intact macromolecule. Hyaluronidases were present in some transformed cells (Orkin et al., 1982), but their pH optima favor an intralysosomal function (Bernanke and Orkin, 1984); moreover, hyaluronidase-deficient cells also shed hyaluronan (Klein and von Figura, 1980). The mechanism of hyaluronan shedding from eucaryotic cells lines was analyzed by Prehm (1990). All cell lines shed identical sizes of hyaluronan as retained on the surface, but differed in the amount of hyaluronan synthesized and the ratio of released and retained hyaluronan. A method was developed which discriminated between intramolecular degradation and dissociation as the intact macromolecule. The cells were pulse-labeled to form hyaluronan chains with labeled and unlabeled segments, after which the sizes of labeled hyaluronan released into the media during the pulse extension period were determined by gel filtration. B6 cells released the same sizes as were retained on the cell surface, indicating that no intramolecular degradation occurred, and hyaluronan dissociated as the intact macromolecule. In contrast, SV3T3 cells released hyaluronan of varying molecular weight distribution during extension of the labeled segment. Shedding of smaller fragments could be prevented by radical scavengers such as superoxide dismutase and tocopherol. Therefore, SV3T3 cells released hyaluronan not only by dissociation, but also by radical degradation.

9 Regulation of Hyaluronan Synthesis

Almost any disturbance of cellular homeostasis results in a stimulation of hyaluronan production, though very few reports have investigated the inhibition of hyaluronan synthesis. Three levels of regulation have been recognized:

1) Expression of the synthase.
2) Stimulation and inhibition of the synthase by growth or differentiation factors acting on a specific target cell.
3) Disturbances in the integrity of the extracellular matrix, particularly the degradation of cell surface-bound hyaluronan.

9.1 Expression of the Synthase

The synthase has a high turnover rate and a half-life of 82 min, indicating a strict control of its activity by the transcription rate (Bansal and Mason, 1986). The three mammalian synthases are expressed in different tissues, thereby producing hyaluronan in different amounts and sizes (Itano et al., 1999).

9.2 Stimulation and Inhibition of the Synthase

Many growth or differentiation factors have been shown to stimulate hyaluronan synthesis (Tomida et al., 1975, 1977a,b; Lembach, 1976; Prehm, 1980; D'Arville and Mason, 1983; Hamerman and Wood, 1984; Hamerman et al., 1986; Pulkki, 1986; Tammi and Tammi, 1986; Wu and Wu, 1986; Heldin et al., 1989). During the early part of the 20th century, Kabat (1939) had already recognized enhanced hyaluronan production in tumors of sarcosis and leukosis. Viral transformation by Rous sarcoma virus (RSV) stimulates hyaluronan production in chondrocytes (Mikuni Takagaki and Toole, 1981), in chicken fibroblasts (Hamerman et al., 1965; Ishimoto et al., 1966), and in myoblasts (Yoshimura, 1985), whilst SV40 transformation induces a similar increase in 3T3 fibroblasts (Hopwood and Dorfman, 1977; Goldberg et al., 1984).

It is surprising that only two inhibitors of hyaluronan synthases have been discovered. Periodate-oxidized UDP-GlcA or UDP-GlcNac have been used as suicide inhibitors of the synthase (Prehm, 1985). These inhibitors do not surmount cell membranes, and must be imported into the cytoplasm by osmotic lysis of pinocytotic vesicles. Vesnarinone suppresses the synthase activity in intact myofibroblasts (Ueki et al., 2000). Hence, it may be worthwhile developing or isolating inhibitors of hyaluronan synthesis that might be used as curative drugs in pathological disturbances related to hyaluranon production, for example edema.

9.2.1 Signal Transduction at Membranes

The hyaluronan receptor CD44 is responsible for outside-in signaling (Perschl et al., 1995) and, intracellularly, this is associated with the tyrosine kinase $p56^{lck}$. Binding of hyaluronan stimulates intracellular phosphorylation (Taher et al., 1996). Similar phosphorylation was observed with bivalent antibodies to CD44, indicating that receptor cross-linking on the cell surface instigates the signal transduction cascade. Thus, the mechanism resembles growth factor-dependent signal transduction pathways. It is also possible that the angiogenic effects of hyaluronan oligosaccharides are transduced through CD44, though this mechanism does not apply to all cell types because CD44-deficient lymphocytes might also be activated by hyaluronan oligosaccharides (Termeer et al., 2000). Hence, other unknown receptors or signal transduction pathways must exist.

9.2.2 Intracellular Signal Transduction

Intracellular signal transduction pathways are dependent on protein synthesis and activation of protein kinase C (Heldin et al., 1992), and have been studied in detail in both B6 cells and 3T3 fibroblasts (Klewes and Prehm, 1994). Activation by fetal calf serum was inhibited by cycloheximide or by the protein kinase inhibitors H-7 or H-8, indicating that transcription as well as phosphorylation were required for activation. The activation by serum was markedly prolonged, when serum was added together with cholera toxin or theophylline. Without serum stimulation the hyaluronan synthase could also be activated by PMA, by dibutyryl-c-AMP, or forskolin. Increasing the intracellular Ca^{2+} concentration with a Ca-ionophore also led to an activation. In isolated plasma membranes the synthase activity could be decreased by phosphatase treatment, and enhanced by ATP in B6 cells and by ATP in the presence of PMA in 3T3 fibroblasts. In conclusion, hyaluronan synthase is induced by transcription and activated by phosphorylation by protein kinase C, c-AMP-dependent protein kinases or Ca^{2+}-dependent protein kinases.

9.3 Influence of Chain Length on Further Elongation

Another factor is the amount of cell surface hyaluronan which influenced its own production. Notably, hyaluronidase treatment of intact cells stimulated synthesis (Philipson et al., 1985). Interesting features on the regulation of hyaluronan synthesis by the growing hyaluronan chain itself have been obtained from studies with streptococci (Nickel et al., 1998). Group A and C streptococci differ in their capacity to retain hyaluronan as a coat on their cell surface. In group C streptococci, a 56-kDa hyaluronan receptor was closely associated with the synthase; this protein had an intrinsic kinase activity that performed autophosphorylation in response to extracellular ATP. Autophosphorylation of the 56-kDa protein led in turn to a reduction in hyaluronan binding and increased shedding of the hyaluronan capsule. Simultaneously, the synthase increased its activity to replace the lost hyaluronan chains. It thus appears that a large hyaluronan chain inhibited its own elongation, when it was retained in the vicinity of the synthase.

A similar mechanism for the regulation of hyaluronan synthesis operates on eucaryotic cells that express the CD44 receptor (Lüke and Prehm, 1999). Proteolytic degradation of CD44 in melanoma cells correlated with the metastatic potential and activation of the hyaluronan synthesis. This activation was the result of facilitated dissociation of growing hyaluronan chains from plasma membranes. These results add a new function to the CD44 receptor as a regulator of hyaluronan synthesis.

10 Turnover and Catabolism

Within the circulation of an adult human, a total of 34 mg of hyaluronan is turned over each day (Fraser and Laurent, 1989; Lebel et al., 1994). The major origins of hyaluronan are the joints, skin, eyes, and intestine. In skin and joints the half-life is about 12 h (Reed et al., 1990; Coleman et al., 2000), whilst in the anterior chamber of the eye it is 1–1.5 h (Laurent et al., 1993), and in the vitreous humor it is 70 days. The rapid turnover is somewhat surprising because hyaluronan has been regarded as a structural component of connective tissue.

Hyaluronan is drained through the lymph and reaches the circulating blood, where the serum concentration may reach ~31 ng L^{-1}. Hyaluranon is effectively endocytosed by the liver, the liver endothelial cells expressing a membrane receptor that endocytically clears hyaluronan from the circulation (Zhou et al., 1999). It is then transported into lysosomes that contain hyaluronidase, β-glucuronidase and β-*N*-acetyl-glucosaminidase (Roden et al., 1989).

11 Functions of Hyaluronan

Many different functions have been assigned to hyaluronan, but these can be grouped into cellular, physiological, and pathological functions. Most functions are determined either by the physical properties or by interactions with hyaluronan-binding proteins.

11.1 Cellular Functions

Fibroblasts are surrounded by a coat of hyaluronan that is lost upon transformation (Underhill and Toole, 1982) but which modifies cell–cell aggregation (Underhill, 1982) and cell–substratum adhesion (Erickson and Turley, 1983). Hyaluronan synthesis is coordinated with cell growth, with proliferating cells producing more hyaluronan than resting cells (Tomida et al., 1975; Mian, 1986). Synchronized fibroblasts show the highest hyaluronan production during mitosis, because hyaluronan synthesis is required for the detachment and mitosis of fibroblasts (Brecht et al., 1986).

The function of hyaluronan on cellular behavior appeared paradoxical, as hyaluronan synthesis was seen to be increased during cell migration (Toole, 1991), mitosis (Brecht et al., 1986), and tumor invasion (Toole et al., 1979). Overproduction of hyaluronan in the human tumor cell line HT1080 enhanced anchorage-independent growth and tumorigenicity (Kosaki et al., 1999). In contrast, hyaluronan promotes cell adhesion (Miyake et al., 1990) and cell–cell aggregation (Underhill, 1982), but inhibits cell proliferation (Goldberg and Toole, 1987; West and Kumar, 1989).

A similar paradoxical situation applies for hyaluronidases in tumor progression. In some tumors, hyaluronidase treatment blocks lymph node invasion by tumor cells for T-cell lymphoma (Zahalka et al., 1995), but overexpression of hyaluronidase correlates with disease progression in bladder, breast and prostate cancer (Lokeshwar et al., 1996, 1997; Bertrand et al., 1997).

A solution to this paradox was obtained from a comparison of the cellular behavior of hyaluronan-deficient Chinese hamster ovary (CHO) cells and CHO cells transfected with the hyaluronan synthase. Surprisingly, hyaluronan synthesis reduced initial cell adhesion, migration, growth and the density at contact inhibition. All these apparent contradictions were combined into a model for the cellular functions of hyaluronan (Dube et al., 2001). Thus, hyaluronan serves as an adhesive component when it is retained on the cell surface by intact CD44, and as a detachment factor when it is released. Migration- and growth- inhibitory effects of hyaluronan are mediated by the proteolytic cleavage of CD44. The cellular function of hyaluronan might also be an amplifier of cell-surface proteases: when proteases are inactive, hyaluranon amplifies the action of cellular adhesion factors by deposition and binding to receptors. However, when proteases are active, it amplifies cell detachment from the environment by activation of synthesis and shedding (see Figure 3).

11.2
Physiological Functions

A prominent physiological function of hyaluronan is the creation of hydrated pathways that allow the cells to penetrate cellular and fibrous barriers. Such hydrated pericellular matrices are required not only for cell rounding in mitosis, but also for cell migration during morphogenesis and wound healing.

11.2.1
Differentiation and Morphogenesis

Studies carried out on embryonic limb development have illustrated how hyaluronan modulates differentiation *in vivo* (Toole, 1991). Early limb mesodermal cells are surrounded and separated by an extensive, hyaluronan-enriched matrix. At this stage, pericellular hyaluronan appears to be tethered to the membrane-integrated synthase. This hydrated matrix facilitates the proliferation and migration of early mesenchymal precursors. Subsequently, the mesoderm condenses, i.e., the intercellular matrix decreases in volume at the sites of future cartilage and muscle differentiation. Mesodermal cells lose the ability to form hydrated matrices, the level of lysosomal hyaluronidases increases, and much of the pericellular hyaluronan is removed. The remaining hyaluronan is now retained at the cell surface by receptors. Similar events occur in the early mesodermal development of skin and teeth. Further differentiation of condensed limb mesoderm into cartilage is accompanied by recovery of matrix formation and by extensive production of proteoglycans that are tethered to the cell surface via hyaluronan and CD44.

$Has2^{-/-}$ mouse mutants contain virtually no hyaluranon at the E9.5 (embryo day 9.5) stage of development (Camenisch et al., 2000), and at this stage they exhibit multiple defects, including yolk sac, vasculature and heart abnormalities. The cardiac jelly and cardiac cushions fail to form valves and other structures.

11.2.2
Wound Healing

The following scenario has been proposed as a model for wound healing by (Weigel et al., 1986). After a wound has been sealed by the platelet plug and fibrin clot, platelet activators stimulate the inflammation and surrounding cells to synthesize hyaluronan into the matrix of the fibrin clot. The clot swells and becomes more porous in order to facilitate the migration of cells into the fibrin matrix. As more cells migrate into the wound, the hyaluronan–fibrin matrix is degraded and replaced by collagens and proteoglycans. The hyaluronan degradation products stimulate angiogenesis and the formation of new blood vessels.

Wound healing in fetal tissues occurs without scarring and is correlated with the prolonged presence of hyaluronan (Longaker et al., 1991). Hyaluronidases that are present in adult wounds were considered responsible for fibrotic healing (West et al., 1997).

11.2.3
Synovia

Hyaluronan is produced by the fibroblast-like type A cells in the upper synovial lining (Pitsillides et al., 1993), and not only serves as a lubricant but also enhances the resistance of the synovial linings to fluid outflow (Coleman et al., 1997). The concentration of hyaluronan in the synovial fluid ranges from 1.4 to 3.6 mg mL^{-1}. Its molecular mass is variable, and in normal synovial fluid has been estimated as 7.0×10^6 daltons, though this is decreased to $3-5 \times 10^6$ daltons in rheumatoid synovial fluids (Balazs et al., 1967; Dahl et al., 1985). The turnover rate depends on the size, indicating a partial reflection of hyaluronan by the lining (Coleman et al., 2000). Hyaluronan is drained through the lymphatic vessels and catabolized in lymph nodes (Fraser et al., 1996).

11.3
Pathological Functions

Aberrant hyaluronan synthesis will lead to disturbances of cell behavior and tissue integrity. For *Streptococcus* sp. and *Pasteurella*, this serves as a non-antigenic disguising capsule in the infected host.

11.3.1
Metastasis

Most malignant solid tumors contain elevated levels of hyaluronan; such enrichment of hyaluronan in tumors may be caused by increased production by the tumors themselves, or by induction in the surrounding stromal cells (Toole et al., 1979; Knudson et al., 1984). The mechanisms whereby hyaluronan–receptor interactions influence tumor cell behavior are not clearly understood, and this is currently a highly active area of investigation.

11.3.2
Edema

Inflammation of various organs is often accompanied with an accumulation of hyaluronan. This can cause edema due to the osmotic activity, and can in turn lead to dysfunction of the organs. For example, hyaluronan accumulation in the rheumatoid joint impedes flexibility of the joint (Engström-Laurent, 1997), whilst accumulation in rejected transplanted kidneys can cause edema and increased intracapsular pressure (Hällgren et al., 1990). Hyaluranon also accumulates in pulmonary edema (Nettelbladt et al., 1989) and during myocardial infarction (Waldenstrom et al., 1991).

11.3.3
Streptococci

Hyaluronan is produced by group A and C streptococci and deposited into a capsule which serves as a major virulence factor of pathogenic streptococci (Kass and Seastone, 1944; Wessels et al., 1991, 1994) because it protects the bacteria against phagocytosis (Whitnack et al., 1981) and oxygen damage (Cleary and Larkin, 1979). Another virulence factor is the hyaluronidase, which is a spreading factor and facilitates penetration of the bacteria through infected tissue (McClean, 1941; MacLennan, 1956). When streptococci enter the stationary phase, they lose the capsule (van de Rijn, 1983), but remain virulent. A constant rate of hyaluronan shedding from the capsule might be advantageous for bacteria, because they will eliminate host components such as attacking antibodies.

A 56-kDa protein was identified as the first example of a prokaryotic extracellular protein kinase. This is expressed on the surface of group C streptococci, and can bind and retain hyaluronan in the absence of ATP. The capsule is shed in the presence of extracellular ATP (Nickel et al., 1998). An equivalent hyaluronan-binding protein was not detected on group A streptococci that do not retain their capsule to the same extent.

Group A and C streptococci differ both in their hosts and their infection routes. Group A streptococci are virulent human pathogens and cause tonsilitis, scarlet fever and rheumatic fever, the infection route being via the throat. In contrast, group C streptococci are mainly animal pathogens, but can also infect humans; their infection route is mainly opportunistic through wounds in the skin.

Group C streptococci have clearly evolved a mechanism to retain their hyaluronan capsule that protects them against desiccation when they are localized on the skin surface. When they enter a wound, they encounter high concentrations of ATP from necrotic tissue and also need to defend themselves against attacking antibodies and macrophages. This they do by shedding the capsule, thereby preventing any attachment.

The cell surface-bound kinase on group C streptococci also elicits antibodies in the infected host. These antibodies show an immunological cross-reaction with cell surfaces proteins from fibroblasts, and are cytotoxic in the presence of complement (Prehm et al., 1995). This protein has recently been used as a vaccine to protect mice against fatal group C streptococci (Chanter et al., 1999).

12 Hyaluronan Degradation

As a very large molecule, hyaluronan is prone to mechanical degradation either by ultrasonic treatment or by thermal degradation at elevated temperatures. Physiologically, hyaluronan can be degraded by oxygen free radicals or by hyaluronidases.

12.1 Degradation by Free Radicals

Oxygen free radical degradation is mostly a side reaction of activated neutrophil granulocytes or monocytes in an inflammation. Radical degradation of hyaluronan results in a dramatic drop of the viscosity and function of the synovial fluid in the inflamed joints of patients with rheumatoid arthritis. This mechanism was hypothesized by Greenwald and Moak (1986). Although a hyaluronan structure modified by radical damage has been detected in the synovial fluid of the inflamed rheumatoid joint (Grootveld et al., 1991), direct biochemical proof for hyaluronan degradation was obtained from organ cultures of healthy and rheumatoid synovial tissue (Schenck et al., 1995). Healthy tissues and some arthritic tissues did not contain significant amounts of granulocytes and produced high molecular-weight hyaluronan. In contrast, arthritic tissue infiltrated with granulocytes released low molecular-weight hyaluronan. Hyaluronan degradation was accompanied by massive oxygen radical production. Radical scavengers protected hyaluronan from degradation in synovial tissue, in particular by the iron chelators DETAPAC or Desferal that block the formation of hydroxyl radicals. Hydroxyl radicals degrade hyaluronan with an efficiency of 100%, with 65% being cleaved within 15 min and the remaining radicals giving rise to thermally labile products that eventually lead to chain scission (Al Assaf et al., 2000). A variety of free aldehydes and oligosaccharides are produced in the rheumatoid synovial fluid (Chapman et al., 1989; Grootveld et al., 1991). The structures formed by oxygen radical damage of hyaluronan have been analyzed in detail by Uchiyama et al. (1990), the main degradation product being L-threotetradialdose.

12.2 Degradation by Hyaluronidases

Hyaluronidases are classified into three groups (Kreil, 1995).

1) Mammalian-type hyaluronidases (EC 3.2.1.35) are endo-β-*N*-acetylhexosaminidases and produce tetrasaccharides and hexasaccharides as the major end-products. They have both hydrolytic and transglycosidase activities, and can degrade hyaluronan and chondroitin sulfates.
2) Bacterial hyaluronidases (EC 4.2.99.1) are endo-β-*N*-acetylhexosaminidases and yield primarily disaccharides. They operate by a β-elimination reaction.
3) Hyaluronidases (EC 3.2.1.36) from leeches, other parasites, and crustaceans are endo-β-glucuronidases that generate tetrasaccharide and hexasaccharide end-products.

Hyaluronidase from testis has long been known as a spreading factor (Chain and Duthie, 1940). Other hyaluronidases from vertebrate tissues are more difficult to purify and analyze, because they occur in low concentrations and are often unstable (Csóka et al., 1997).

13 Production

Hyaluronan is produced either from rooster combs (Pharmacia AB, Uppsala, Sweden, have developed a special strain of roosters with highly luxuriant combs) or from streptococci. Commercially available hyaluranon is produced in molecular weights ranging from less than 10^6 daltons to as high as 8×10^6 daltons. There are many hyaluronan producers worldwide. For example, Genzyme Corp. (Framingham, MA, USA) merged with Biomatrix and operates a plant producing hyaluronan from mammalian sources in Canada. Anika (Woburn, MA, USA); Lifecore Biomedical (Chaska, MN, USA) are other suppliers. Pharmacia produces hyaluronan in Sweden, Fidia Advanced Biopolymers (Brindisi) in Italy, Bio-Technology General Corp. (Iselin, NJ, USA) in Israel, and a number of companies, including Kibun Food Chemifa Co. and Seikagaku Corp. (both Tokyo), in Japan.

Today there are still many products on the market that contain hyaluronan isolated from rooster combs, because this has set the standards for high molecular weight, purity and noninflammatory properties. Even if hyaluronan from streptococci meets these criteria, its market penetration is hampered by the reluctance of customers to change to cheaper medical or cosmetic products. In addition many of the streptococcal preparations are not characterized as thoroughly, or are less pure (Manna et al., 1999). The development of large-scale production from *Streptococcus zooepidemicus* cultures had to overcome many obstacles, including: growth in chemically defined media (Kjems and Lebech, 1976; van de Rijn and Kessler, 1980; Armstrong et al., 1997; Cooney et al., 1999); the production of high molecular-weight material (Kim et al., 1996); the elimination of toxic impurity such as streptolysin; and increasing the yield to about 7 g L^{-1} (Lowther and Rogers, 1956; Thonard et al., 1964; Gerlach and Kohler, 1970).

13.1 Patents

The patent literature concerning hyaluranon is extensive, the number of annual patent applications having increased from fewer than 10 before 1985 to about 200 in 2000. The major inventors in this field and their corporate affiliations are summarized in Table 1. Many different processing techniques and uses for hyaluronan have been developed and patented by Balazs and his coworkers. For example, the important Balazs patent (which was issued in 1979 but now has expired) on hyaluronan isolated from animal tissue that does not cause an inflammatory response when tested in the eye of the owl monkey (Balazs, 1979). This is marketed by Upjohn-Pharmacia as Healon®, a sterile, pyrogen-free, nonantigenic and noninflammatory, high molecular-weight fraction of hyaluronan.

13.2 Market

The world market of hyaluronan is difficult to estimate, because many companies produce it for medical and cosmetic applications. The current US market of $157 million for viscosupplementation in osteoarthritis is driven by two products: Synvisc® from Biomatrix (now Genzyme) and Hyalgan® from Fidia Pharmaceutical. In Europe, Fidia's Hyalgan® is the leading hyaluronan-based viscosupplement product. Details of the major pharmaceutical products are listed in Table 2.

Tab. 1 Some hyaluronan-related patents

Patent no.	*Patent holder*	*Inventors*	*Title*	*Date*
U.S. 4,141,973	Biomatrix	Balazs	Ultrapure hyaluronic acid and the use thereof	1979
U.S. 4,713,448	Biomatrix	Balazs	Chemically modified hyaluronic acid preparation and method of recovery thereof from animal tissues	1987
U.S. 4,957,744	Fidia	della Valle	Esters of hyaluronic acid	1990
U.S. 4,636,524	Biomatrix	Balazs	Cross-linked gels of hyaluronic acid and products containing such gels	1987
U.S. 4,937,270	Genzyme	Hamilton	Water-insoluble derivatives of hyaluronic acid	1990
U.S. 5,644,049	Murst Italian	Giusti	Biomaterial comprising hyaluronic acid and derivatives thereof in interpenetrating polymer networks	1997
U.S. 4,663,233	Biocaot	Beavers	Lens with hydrophilic coating	1987
U.S. 5,585,361	Burns	Genzyme	Method for the inhibition of platelet adherence and aggregation	1996

Tab. 2 Some commercially available pharmaceuticals containing hyaluronan

	Concentration [mg mL^{-1}]	*Size [mL]*	*Manufacturer*	*Application*	*Price (US $)*
AMO Vitrax Syringe	30	65	Allergen	Ophthalmology	138
Amvisc Plus Syringe	16	5	Chiron	Ophthalmology	145
Amvisc Plus Syringe	16	8	Chiron	Ophthalmology	190
Amvisc Syringe	12	5	Chiron	Ophthalmology	112
Healon GV Syringe	14	55	Kabi	Ophthalmology	101
Healon GV Syringe	14	85	Kabi	Ophthalmology	131
Healon Syringe	10	55	Kabi	Ophthalmology	94
Hyalgan SDV	10	2	Sanofi	Osteoarthritis	166
Hyalgan Syringe	10	2	Sanofi	Osteoarthritis	166
Provisc Syringe	10	4	Alcon	Ophthalmology	117
Provisc Syringe	10	55	Alcon	Ophthalmology	142
Provisc Syringe	10	85	Alcon	Ophthalmology	178
Synvisc Syringe	8	3×2.25	Wyeth	Osteoarthritis	705
Viscoat Syringe	40	5	Alcon	Ophthalmology	151

14 Medical Applications

Hyaluronan preparations or higher cross-linked products are increasingly used for many medical applications, such as ophthalmic viscosurgery, supplementation of the synovial fluid in patients with osteoarthritis (Balazs and Denlinger, 1989), as membranes for postsurgical separation of tissues (Burns et al., 1997), and as drug delivery systems (Vercruysse and Prestwich, 1998). Many cosmetics contain hyaluronan as an ingredient, because it is believed to keep skin young and fresh by preventing dryness as a result of its water-binding capacity, though scientific evidence on this subject is lacking.

14.1 Ophthalmics

Hyaluronan was first marketed in the early 1980s as Healon® in the ophthalmic field as a viscous gel which could be injected into the anterior chamber of the eye to protect tissues such as the corneal endothelium (Miller and Stegmann, 1983). A number of other hyaluronan products are now marketed by competing firms in the ophthalmic market.

14.2 Arthritis

Intra-articular administration of hyaluronan has been used in animals and man with reported clinical efficacy. In man, hyaluronan is being used to relieve pain and improve joint mobility in the treatment of osteoarthritis with intra-articular injections of Hyalgan® (Sanofi Pharmaceuticals), Orthovisc® (Anika Therapeutics), and SynVisc® (Biomatrix, now Genzyme). It has also been proposed for several degenerative joint diseases as an alternative to the traditional steroid therapy (Altman and Moskowitz, 1998; Wobig et al., 1999; Adams et al., 2000).

14.3 Wound Healing and Scarring

Hyaluronan products have been developed to foster the healing process. For burn and chronic ulcer patients, a line of modified hyaluronan products based on a HYAFF™

polymer is being marketed by Convatec in Europe, and is currently undergoing clinical trials in the US (Goa and Benfield, 1994).

14.4 Adhesion Prevention

Most surgical procedures are accompanied by undesired tissue damage caused by cutting, desiccation, lack of adequate blood supply and manipulative abrasion, and undesired connective tissue bridges (adhesions) are often formed on the damaged surfaces of organs. Hyaluronan preparations such as Seprafilm® from Genzyme can reduce such adhesions and improve the surgical outcome (Beck, 1997).

14.5 Drug Delivery

Hyaluronan is an ideal molecule for use as a carrier of drugs, particularly for local administration. As the polysaccharide is a ubiquitous component of tissues and fluids, it is immunologically inert. It can be metabolized in the lysosomes of certain cells, and the backbone of the molecule provides different chemical groups for drug attachment. Investigations are on-going for topical and intravenous drug delivery systems using modified hyaluronans (Vercruysse and Prestwich, 1998).

15 Effects of Hyaluronan Oligosaccharides

Hyaluronan oligosaccharides augment fibroblast proliferation, migration, stimulate the formation of new blood vessels, and also activate macrophages. These effects have been studied in detail in cell culture, where they stimulate proliferation of synovial fibroblasts *in vitro* (Goldberg and Toole, 1987), induce angiogenesis (Montesano et al., 1996), stimulate chemokine production on macrophages (McKee et al., 1996, 1997; Horton et al., 1998), and activate lymphocytes (Termeer et al., 2000).

16 Outlook and Perspectives

In times characterized by an enthusiasm for the achievements of molecular biology that are often announced with great fanfare for the welfare of mankind, it may be wise to issue a reminder that the understanding and management of some diseases require a horizon beyond the application of gene technology. Indeed, research on hyaluronan may be a typical illustration that pertinent yet silent progress contributes significantly to the solution of medical problems.

Nonetheless, many cellular and physiological functions of hyaluronan remain elusive. There is a notion amongst the scientific community that hyaluronan participates in the pathogenesis of metastasis and rheumatoid autoimmune diseases, and research into hyaluronan–cell interactions will clearly contribute to therapeutic interventions in these diseases. A second promising area is the development of hyaluronan-based biomaterials and hyaluronan-modified surfaces. Indeed, a number of important products have already reached the market, and the introduction of many hyaluronan-derived devices and drugs are eagerly anticipated during the next decade.

17 References

Abatangelo, G. and Weigel, P.H. (2000) *Redefining Hyaluronan*. Amsterdam: Wheley.

Adams, M. E., Lussier, A. J., Peyron, J. G. (2000) A risk-benefit assessment of injections of hyaluronan and its derivatives in the treatment of osteoarthritis of the knee. *Drug Safety* **23**, 115–130.

Al Assaf, S., Meadows, J., Phillips, G. O., Williams, P. A., Parsons, B. J. (2000) The effect of hydroxyl radicals on the rheological performance of hylan and hyaluronan. *Int. J. Biol. Macromol.* **27**, 337–348.

Altman, R. D., Moskowitz, R. (1998) Intraarticular sodium hyaluronate (Hyalgan®) in the treatment of patients with osteoarthritis of the knee: a randomized clinical trial. *J. Rheumatol.* **25**, 2203–2212.

Arch, R., Wirth, K., Hofmann, M., Ponta, H., Matzku, S., Herrlich, P., Zoller, M. (1992) Participation in normal immune responses of a metastasis-inducing splice variant of CD44. *Science* **257**, 682–685.

Armstrong, D. C., Cooney, M. J., Johns, M. R. (1997) Growth and amino acid requirements of hyaluronic-acid-producing *Streptococcus zooepidemicus*. *Appl. Microbiol. Biotechnol.* **47**, 309–312.

Aruffo, A., Stamenkovic, I., Melnick, M., Underhill, C. B., Seed, B. (1990) CD44 is the principal cell surface receptor for hyaluronate. *Cell* **61**, 1303–1313.

Assmann, V., Marshall, J. F., Fieber, C., Hofmann, M., Hart, I. R. (1998) The human hyaluronan receptor RHAMM is expressed as an intracellular protein in breast cancer cells. *J. Cell Sci.* **111**, 1685–1694.

Balazs, E. A. (1970) Chemistry and Molecular Biology of the Intercellular Matrix. London: Academic Press.

Balazs, E. A. (1979) Ultrapure hyaluronic acid and the use thereof. U.S. Patent 4,141,973.

Balazs, E. A. (1982) Use of hyaluronic acid in eye surgery. *Ann. Ther. Clin. Ophthalmol.* **33**, 95–110.

Balazs, E. A. (1987a) Chemically modified hyaluronic acid preparation and method of recovery thereof from animal tissues. U.S. Patent 4,713,448.

Balazs, E. A. (1987b) Cross-linked gels of hyaluronic acid and products containing such gels. U.S. Patent 4,636,524.

Balazs, E. A., Denlinger, J. L. (1989) Clinical uses of hyaluronan. *Ciba Found. Symp.* **143**, 265–275.

Balazs, E. A., Watson, D., Duff, I. F., Roseman, S. (1967) Hyaluronic acid in synovial fluid. I. Molecular parameters of hyaluronic acid in normal and arthritis human fluids. *Arthritis Rheum.* **10**, 357–376.

Banerji, S., Ni, J., Wang, S.X., Clasper, S., Su, J., Tammi, R., Jones, M., Jackson, D. G. (1999) LYVE-1, a new homologue of the CD44 glycoprotein, is a lymph-specific receptor for hyaluronan. *J. Cell Biol.* **144**, 789–801.

Bansal, M. K., Mason, R. M. (1986) Evidence for rapid metabolic turnover of hyaluronate synthetase in Swarm rat chondrosarcoma chondrocytes. *Biochem. J.* **236**, 515–519.

Beavers, E. M. (1987) Lens with hydrophilic coating. U.S. Patent 4,663,233.

Beck, D. E. (1997) The role of Seprafilm™ bioresorbable membrane in adhesion prevention. *Eur. J. Surg.*, **163** (Suppl. 577), 49–55.

Bernanke, D. H., Orkin, R. W. (1984) Hyaluronate binding and degradation by cultured embryonic chick cardiac cushion and myocardial cells. *Dev. Biol.*, **106**, 360–367.

Bertrand, P., Girard, N., Duval, C., D'Anjou, J., Chauzy, C., Ménard, J. F., Delpech, B. (1997)

Increased hyaluronidase levels in breast tumor metastases. *Int. J. Cancer* **73**, 327–331.
Brecht, M., Mayer, U., Schlosser, E., Prehm, P. (1986) Increased hyaluronate synthesis is required for fibroblast detachment and mitosis. *Biochem. J.* **239**, 445–450.
Burns, J. W., Valeri, C. R. (1996) Method for the inhibition of platelet adherence and aggregation. U.S. Patent 5,585,361.
Burns, J. W., Colt, M. J., Burgess, L. S., Skinner, K. C. (1997) Preclinical evaluation of Seprafilm™ bioresorbable membrane. *Eur. J. Surg.* **163** (Suppl. 577), 40–48.
Callegaro, L., Giusti, P. (1997) Biomaterial comprising hyaluronic acid and derivatives thereof in interpenetrating polymer networks. U.S. Patent 5,644,049.
Camenisch, T. D., Spicer, A. P., Brehm-Gibson, T., Biesterfeldt, J., Augustine, M. L., Calabro, A., Jr., Kubalak, S., Klewer, S. E., McDonald, J. A. (2000) Disruption of hyaluronan synthase-2 abrogates normal cardiac morphogenesis and hyaluronan-mediated transformation of epithelium to mesenchyme. *J. Clin. Invest.* **106**, 349–360.
Chain, E., Duthie, E. S. (1940) Identity of hyaluronidase and spreading factor. *Br. J. Exp. Pathol.* **21**, 324–338.
Chanter, N., Ward, C. L., Talbot, N. C., Flanagan, J. A., Binns, M., Houghton, S. B., Smith, K. C., Mumford, J. A. (1999) Recombinant hyaluronate associated protein as a protective immunogen against *Streptococcus equi* and *Streptococcus zooepidemicus* challenge in mice. *Microbiol. Pathog.* **27**, 133–143.
Chapman, M. L., Rubin, B. R., Gracy, R. W. (1989) Increased carbonyl content of proteins in synovial fluid from patients with rheumatoid arthritis. *J. Rheumatol.* **16**, 15–18.
Cleary, P. P., Larkin, A. (1979) Hyaluronic acid capsule: strategy for oxygen resistance in group A streptococci. *J. Bacteriol.* **140**, 1090–1097.
Coleman, P. J., Scott, D., Ray, J., Mason, R. M., Levick, J. R. (1997) Hyaluronan secretion into the synovial cavity of rabbit knees and comparison with albumin turnover. *J. Physiol.* **503** , 645–656.
Coleman, P. J., Scott, D., Mason, R. M., Levick, J. R. (2000) Role of hyaluronan chain length in buffering interstitial flow across synovium in rabbits. *J. Physiol.* **526**, 425–434.
Comper, W. D., Laurent, T. C. (1978) Physiological function of connective tissue polysaccharides. *Physiol. Rev.* **58**, 255–315.
Cooney, M. J., Goh, L. T., Lee, P. L., Johns, M. R. (1999) Structured model-based analysis and control of the hyaluronic acid fermentation by *Streptococcus zooepidemicus*: physiological implications of glucose and complex nitrogen-limited growth. *Biotechnol. Prog.* **15**, 898–910.
Crater, D. L., Dougherty, B. A., van de Rijn, I. (1995) Molecular characterization of *has*C from an operon required for hyaluronic acid synthesis in group A streptococci – Demonstration of UDP-glucose pyrophosphorylase activity. *J. Biol. Chem.* **270**, 28676–28680.
Csóka, T. B., Frost, G. I., Wong, T., Stern, R. (1997) Purification and microsequencing of hyaluronidase isozymes from human urine. *FEBS Lett.* **417**, 307–310.
D'Arville, C., Mason, R. M. (1983) Effects of serum and insulin on hyaluronate synthesis by cultures of chondrocytes from the Swarm rat chondrosarcoma. *Biochim. Biophys. Acta* **760**, 53–60.
Dahl, L. B., Dahl, I. M., Engstrom Laurent, A., Granath, K. (1985) Concentration and molecular weight of sodium hyaluronate in synovial fluid from patients with rheumatoid arthritis and other arthropathies. *Ann. Rheum. Dis.* **44**, 817–822.
DeAngelis, P. L. (1999) Molecular directionality of polysaccharide polymerization by the *Pasteurella multocida* hyaluronan synthase. *J. Biol. Chem.* **274**, 26557–26562.
DeAngelis, P. L., Papaconstantinou, J., Weigel, P. H. (1993) Molecular cloning, identification, and sequence of the hyaluronan synthase gene from group A *Streptococcus pyogenes*. *J. Biol. Chem.* **268**, 19181–19184.
DeAngelis, P. L., Jing, W., Drake, R. R., Achyuthan, A. M. (1998) Identification and molecular cloning of a unique hyaluronan synthase from *Pasteurella multocida*. *J. Biol. Chem.* **273**, 8454–8458.
della Valle, F. (2001) Esters of hyaluronic acid. U.S. Patent 4,957,744.
Delpech, B., Halavent, C. (1981) Characterization and purification from human brain of a hyaluronic acid-binding glycoprotein, hyaluronectin. *J. Neurochem.* **36**, 855–859.
Dougherty, B. A., van de Rijn, I. (1993) Molecular characterization of hasB from an operon required for hyaluronic acid synthesis in group A streptococci. Demonstration of UDP-glucose dehydrogenase activity. *J. Biol. Chem.* **268**, 7118–7124.
Dougherty, B. A., van de Rijn, I. (1994) Molecular characterization of hasA from an operon required for hyaluronic acid synthesis in group A streptococci. *J. Biol. Chem.* **269**, 169–175.
Dube, B., Luke, H. J., Aumailley, M., Prehm, P. (2001) Hyaluronan reduces migration and pro-

liferation in CHO cells. *Biochim. Biophys. Acta* **1538**, 283–289.

Engström-Laurent, A. (1997) Hyaluronan in joint disease. *J. Intern. Med.* **242**, 57–60.

Erickson, C. A., Turley, E. A. (1983) Substrata formed by combination of extracellular matrix components alter neural crest cell migration. *J. Cell Sci.* **61**, 299–323.

Evanko, S. P., Wight, T. N. (1999) Intracellular localization of hyaluronan in proliferating cells. *J. Histochem. Cytochem.* **47**, 1331–1341.

Fox, E. M., Walts, A. E., Acharya, R. A., Hamilton, R. (1990) Water insoluble derivatives of hyaluronic acid. U.S. Patent 4,937,270.

Fraser, J. R., Laurent, T. C. (1989) Turnover and metabolism of hyaluronan. *Ciba Found. Symp.* **143**, 41–53.

Fraser, J. R., Laurent, T. C., Pertoft, H., Baxter, E. (1981) Plasma clearance, tissue distribution and metabolism of hyaluronic acid injected intravenously in the rabbit. *Biochem. J.* **200**, 415–424.

Fraser, J. R. E., Cahill, R. N., Kimpton, W. G., Laurent, T. C. (1996) Lymphatic system, in: *Extracellular Matrix. 1. Tissue Function* (Comper, W. D., Ed.), Amsterdam: Harwood Academic Publications, 110–131.

Gerlach, D., Kohler, W. (1970) [Production and isolation of streptococcal hyaluronic acid]. *Zentralbl. Bakteriol. Orig.* **215**, 187–195.

Goa, K. L., Benfield, P. (1994) Hyaluronic acid. A review of its pharmacology and use as a surgical aid in ophthalmology, and its therapeutic potential in joint disease and wound healing. *Drugs* **47**, 536–566.

Goldberg, R. L., Toole, B. P. (1987) Hyaluronate inhibition of cell proliferation. *Arthritis Rheum.* **30**, 769–778.

Goldberg, R. L., Seidman, J. D., Chi Rosso, G., Toole, B. P. (1984) Endogenous hyaluronate-cell surface interactions in 3T3 and simian virus-transformed 3T3 cells. *J. Biol. Chem.* **259**, 9440–9446.

Greenwald, R. A., Moak, S. A. (1986) Degradation of hyaluronic acid by polymorphonuclear leukocytes. *Inflammation* **10**, 15–30.

Grootveld, M., Henderson, E. B., Farrell, A., Blake, D. R., Parkes, H. G. H.-P. (1991) Oxidative damage to hyaluronate and glucose in synovial fluid during exercise of the inflamed rheumatoid joint. Detection of abnormal low-molecular-mass metabolites by proton-n.m.r. spectroscopy. *Biochem. J.* **273**, 459–467.

Hall, C. L., Yang, B., Yang, X., Zhang, S., Turley, M., Samuel, S., Lange, L. A., Wang, C., Curpen, G. D., Savani, R. C., Greenberg, A. H., Turley, E. A. (1995) Overexpression of the hyaluronan receptor RHAMM is transforming and is also required for H-*ras* transformation. *Cell* **82**, 19–28.

Hamerman, D., Wood, D. D. (1984) Interleukin 1 enhances synovial cell hyaluronate synthesis. *Proc. Soc. Exp. Biol. Med.* **177**, 205–210.

Hamerman, D., Todaro, G. J., Green, H. (1965) The production of hyaluronate by spontaneously established cell lines and viral transformed lines of fibroblastic origin. *Biochim. Biophys. Acta* **101**, 343–351.

Hamerman, D., Sasse, J., Klagsbrun, M. (1986) A cartilage-derived growth factor enhances hyaluronate synthesis and diminishes sulfated glycosaminoglycan synthesis in chondrocytes. *J. Cell Physiol.* **127**, 317–322.

Hardingham, T. E., Muir, H. (1972) The specific interaction of hyaluronic acid with cartilage proteoglycans. *Biochim. Biophys. Acta* **279**, 401–405.

Hällgren, R., Gerdin, B., Tufveson, G. (1990) Hyaluronic acid accumulation and redistribution in rejecting rat kidney graft. Relationship to the transplantation edema. *J. Exp. Med.* **171**, 2063–2076.

Heldin, P., Laurent, T. C., Heldin, C. H. (1989) Effect of growth factors on hyaluronan synthesis in cultured human fibroblasts. *Biochem. J.* **258**, 919–922.

Heldin, P., Asplund, T., Ytterberg, D., Thelin, S., Laurent, T. C. (1992) Characterization of the molecular mechanism involved in the activation of hyaluronan synthetase by platelet-derived growth factor in human mesothelial cells. *Biochem. J.* **283**, 165–170.

Hofmann, M., Assmann, V., Fieber, C., Sleeman, J. P., Moll, J., Ponta, H., Hart, I. R., Herrlich, P. (1998a) Problems with RHAMM: a new link between surface adhesion and oncogenesis? *Cell* **95**, 591–592.

Hofmann, M., Fieber, C., Assmann, V., Göttlicher, M., Sleeman, J., Plug, R., Howells, N., Von Stein, O., Ponta, H., Herrlich, P. (1998b) Identification of IHABP, a 95kDa intracellular hyaluronate binding protein. *J. Cell Sci.* **111**, 1673–1684.

Hopwood, J. J., Dorfman, A. (1977) Glycosaminoglycan synthesis by cultured human skin fibroblasts after transformation with Simian virus 40. *J. Biol. Chem.* **252**, 4777–4785.

Horton, M. R., McKee, C. M., Bao, G., Liao, F., Farber, J. M., Hodge DuFour, J., Pure, E., Oliver, B. L., Wright, T. M., Noble, P. W. (1998) Hyaluronan fragments synergize with interferon-gamma to induce the C-X-C chemokines Mig and interferon-inducible protein-10 in mouse macrophages. *J. Biol. Chem.* **273**, 35088–35094.

Huang, L., Grammatikakis, N., Yoneda, M., Banerjee, S. D., Toole, B. P. (2000) Molecular characterization of a novel intracellular hyaluronan-binding protein. *J. Biol. Chem.* 275, 29829–29839.

Ishimoto, N., Temin, H. M., Strominger, J. L. (1966) Studies of carcinogenesis by avian sarcoma viruses. II. Virus- induced increase in hyaluronic acid synthetase in chicken fibroblasts. *J. Biol. Chem.* **241**, 2052–2057.

Itano, N., Kimata, K. (1996) Expression cloning and molecular characterization of HAS protein, a eukaryotic hyaluronan synthase. *J. Biol. Chem.* **271**, 9875–9878.

Itano, N., Sawai, T., Yoshida, M., Lenas, P., Yamada, Y., Imagawa, M., Shinomura, T., Hamaguchi, M., Yoshida, Y., Ohnuki, Y., Miyauchi, S., Spicer, A. P., McDonald, J. A., Kimata, K. (1999) Three isoforms of mammalian hyaluronan synthases have distinct enzymatic properties. *J. Biol. Chem.* **274**, 25085–25092.

Kabat, E. A. (1939) A polysaccharide in tumors due to a virus of leukosis and sarcoma in fowls. *J. Biol. Chem.* **130**, 143–147.

Kass, E. H., Seastone, C. V. (1944) The role of the mucoid polysaccharide hyaluronic acid in the virulence of group A hemolytic streptococci. *J. Exp. Med.* **70**, 319–330.

Kendall, F. E., Heidelberger, M., Dawson, M. H. (1937) A serologically inactive polysaccharide elaborated by mucoid strains of group A hemolytic streptococci. *J. Biol. Chem.* **118**, 61–69.

Kim, J. H., Yoo, S. J., Oh, D. K., Kweon, Y. G., Park, D. W., Lee, C. H., Gil, G. H. (1996) Selection of a *Streptococcus equi* mutant and optimization of culture conditions for the production of high molecular weight hyaluronic acid. *Enzyme Microb. Technol.* **19**, 440–445.

Kjems, E., Lebech, K. (1976) Isolation of hyaluronic acid from cultures of streptococci in a chemically defined medium. *Acta Pathol. Microbiol. Scand. B.* **84**, 162–164.

Klein, U., von Figura, K. (1980) Characterization of dermatan sulfate in mucopolysaccharidosis VI. Evidence for the absence of hyaluronidase-like enzymes in human skin fibroblasts. *Biochim. Biophys. Acta* **630**, 10–14.

Klewes, L., Prehm, P. (1994) Intracellular signal transduction for serum activation of the hyaluronan synthase in eukaryotic cell lines. *J. Cell Physiol.* **160**, 539–544.

Knudson, W., Biswas, C., Toole, B. P. (1984) Interactions between human tumor cells and fibroblasts stimulate hyaluronate synthesis. *Proc. Natl. Acad. Sci. USA* **81**, 6767–6771.

Kosaki, R., Watanabe, K., Yamaguchi, Y. (1999) Overproduction of hyaluronan by expression of the hyaluronan synthase Has2 enhances anchorage-independent growth and tumorigenicity. *Cancer Res.* **59**, 1141–1145.

Kreil, G. (1995) Hyaluronidases – A group of neglected enzymes. *Protein Sci.* **4**, 1666–1669.

Lapcik, L., De Smedt, S., Demeester, J., Chabrecek, P. (1998) Hyaluronan: preparation, structure, properties, and applications. *Chem. Rev.* **98**, 2663–2684.

Laurent, T. C. (1970) Structure of hyaluronic acid, in: *Chemistry and Molecular Biology of the Intercellular Matrix* (Balazs, E. A., Ed.), New York: Academic Press, 703–732.

Laurent, T. C. (1989) *The Biology of Hyaluronan.* Ciba Foundation Symposium **143**. Chichester: Wiley.

Laurent, T. C. (1998) *The Chemistry, Biology and Medical Applications of Hyaluronan and its Derivatives.* Wenner-Gren International Series **72**. London: Portland Press.

Laurent, T. C., Fraser, J. R. (1992) Hyaluronan. *FASEB J.* **6**, 2397–2404.

Laurent, T. C., Dahl, L. B., Lilja, K. (1993) Hyaluronan injected in the anterior chamber of the eye is catabolized in the liver. *Exp. Eye Res.* **57**, 435–440.

Lebel, L., Gabrielsson, J., Laurent, T. C., Gerdin, B. (1994) Kinetics of circulating hyaluronan in humans. *Eur. J. Clin. Invest.* **24**, 621–626.

Lee, M. C., Alpaugh, M. L., Nguyen, M., Deato, M., Dishakjian, L., Barsky, S. H. (2000) Myoepithelial-specific CD44 shedding is mediated by a putative chymotrypsin-like sheddase. *Biochem. Biophys. Res. Commun.* **279**, 116–123.

Lee, T. H., Wisniewski, H. G., Vilcek, J. (1992) A novel secretory tumor necrosis factor-inducible protein(TSG-6) is a member of the family of hyaluronate binding proteins, closely related to the adhesion receptor CD44. *J. Cell Biol.* **116**, 545–557.

Lembach, K. J. (1976) Enhanced synthesis and extracellular accumulation of hyaluronic acid during stimulation of quiescent human fibroblasts by mouse epidermal growth factor. *J. Cell Physiol.* **89**, 277–288.

Lesley, J., Hascall, V. C., Tammi, M., Hyman, R. (2000) Hyaluronan binding by cell surface CD44. *J. Biol. Chem.* **275**, 26967–26975.

Liu, D. C., Liu, T., Sy, M. S. (1998) Identification of two regions in the cytoplasmic domain of CD44 through which PMA, calcium, and forskolin differentially regulate the binding of CD44 to hyaluronic acid. *Cell. Immunol.* **190**, 132–140.

Lokeshwar, V. B., Lokeshwar, B. L., Pham, H. T., Block, N. L. (1996) Association of elevated levels of hyaluronidase, a matrix-degrading enzyme, with prostate cancer progression. *Cancer Res.* **56**, 651–657.

Lokeshwar, V. B., Öbek, C., Soloway, M. S., Block, N. L. (1997) Tumor-associated hyaluronic acid: a new sensitive and specific urine marker for bladder cancer. *Cancer Res.* **57**, 773–777.

Longaker, M. T., Chiu, E. S., Adzick, N. S., Stern, M., Harrison, M. R., Stern, R. (1991) Studies in fetal wound healing. V. A prolonged presence of hyaluronic acid characterizes fetal wound fluid. *Ann. Surg.* **213**, 292–296.

Lowther, D. A., Rogers, H. J. (1956) The role of glutamine in the biosynthesis of hyaluronate by streptococcal suspensions. *Biochem. J.* **62**, 304–314.

Lüke, H. J., Prehm, P. (1999) Synthesis and shedding of hyaluronan from plasma membranes of human fibroblasts and metastatic and non-metastatic melanoma cells. *Biochem. J.* **343**, 71–75.

MacLennan, A. P. (1956) The production of capsules, hyaluronic acid and hyaluronidase by group A and group C streptococci. *J. Gen. Microbiol.* **14**, 134–142.

Manna, F., Dentini, M., Desideri, P., De Pita, O., Mortilla, E., Maras, B. (1999) Comparative chemical evaluation of two commercially available derivatives of hyaluronic acid (hylaform from rooster combs and restylane from *Streptococcus*) used for soft tissue augmentation. *J. Eur. Acad. Dermatol. Venereol.* **13**, 183–192.

Markovitz, A., Dorfman, A. (1962) Synthesis of capsular polysaccharide hyaluronic acid by protoplast membrane preparations of group A streptococci. *J. Biol. Chem.* **237**, 273–279.

McClean, D. (1941) The capsulation of streptococci and its relation to diffusion factor (hyaluronidase). *J. Pathol. Bacteriol.* **53**, 13.

McKee, C. M., Penno, M. B., Cowman, M., Burdick, M. D., Strieter, R. M., Bao, C., Noble, P. W. (1996) Hyaluronan (HA) fragments induce chemokine gene expression in alveolar macrophages – The role of HA size and CD44. *J. Clin. Invest.* **98**, 2403–2413.

McKee, C. M., Lowenstein, C. J., Horton, M. R., Wu, J., Bao, C., Chin, B. Y., Choi, A. M. K., Noble, P. W. (1997) Hyaluronan fragments induce nitric-oxide synthase in murine macrophages through a nuclear factor kappaB-dependent mechanism. *J. Biol. Chem.* **272**, 8013–8018.

Meyer, K. (1947) The biological significance of hyaluronic acid and hyaluronidase. *Physiol. Rev.* **27**, 335–359X.

Meyer, K., Palmer, J. W. (1934) The polysaccharide of the vitreous humor. *J. Biol. Chem.* **107**, 629–634.

Meyer, M. F., Kreil, G. (1996) Cells expressing the DG42 gene from early *Xenopus* embryos synthesize hyaluronan. *Proc. Natl. Acad. Sci. USA* **93**, 4543–4547.

Mian, N. (1986) Analysis of cell-growth-phase-related variations in hyaluronate synthase activity of isolated plasma-membrane fractions of cultured human skin fibroblasts. *Biochem. J.* **237**, 333–342.

Mikuni Takagaki, Y., Toole, B. P. (1981) Hyaluronate-protein complex of Rous sarcoma virus-transformed chick embryo fibroblasts. *J. Biol. Chem.* **256**, 8463–8469.

Miller, D., Stegmann, R. (1983) Healon. A Guide to its use in Ophthalmic Surgery. New York: Wiley.

Miyake, H., Hara, I., Okamoto, I., Gohji, K., Yamanaka, K., Arakawa, S., Saya, H., Kamidono, S. (1998) Interaction between CD44 and hyaluronic acid regulates human prostate cancer development. *J. Urol.* **160**, 1562–1566.

Miyake, K., Underhill, C. B., Lesley, J., Kincade, P. W. (1990) Hyaluronate can function as a cell adhesion molecule and CD44 participates in hyaluronate recognition. *J. Exp. Med.* **172**, 69–75.

Montesano, R., Kumar, S., Orci, L., Pepper, M. S. (1996) Synergistic effect of hyaluronan oligosaccharides and vascular endothelial growth factor on angiogenesis in vitro. *Lab. Invest.* **75**, 249–262.

Nettelbladt, O., Tengblad, A., Hällgren, R. (1989) Lung accumulation of hyaluronan parallels pulmonary edema in experimental alveolitis. *Am. J. Physiol.* **257**, L379–L384.

Nickel, V., Prehm, S., Lansing, M., Mausolf, A., Podbielski, A., Deutscher, J., Prehm, P. (1998) An ectoprotein kinase of group C streptococci binds hyaluronan and regulates capsule formation. *J. Biol. Chem.* **273**, 23668–23673.

Orkin, R. W., Underhill, C. B., Toole, B. P. (1982) Hyaluronate degradation in 3T3 and simian virus-transformed3T3 cells. *J. Biol. Chem.* **257**, 5821–5826.

Peller, L. (1980) Thermodynamic considerations in the synthesis and assembly of biological macromolecules. *Macromolecules* **13**, 609–615.

Perides, G., Asher, R., Dahl, D., Bignami, A. (1990) Glial hyaluronate-binding protein (GHAP) in optic nerve and retina. *Brain Res.* **512**, 309–316.

Perschl, A., Lesley, J., English, N., Trowbridge, I., Hyman, R. (1995) Role of CD44 cytoplasmic domain in hyaluronan binding. *Eur. J. Immunol.* **25**, 495–501.

Peterson, R. M., Yu, Q., Stamenkovic, I., Toole, B. P. (2000) Perturbation of hyaluronan interactions by soluble CD44 inhibits growth of murine mammary carcinoma cells in ascites. *Am. J. Pathol.* **156**, 2159–2167.

Philipson, L. H., Westley, J., Schwartz, N. B. (1985) Effect of hyaluronidase treatment of intact cells on hyaluronate synthetase activity. *Biochemistry* **24**, 7899–7906.

Pitsillides, A. A., Wilkinson, L. S., Mehdizadeh, S., Bayliss, M. T., Edwards, J. C. (1993) Uridine diphosphoglucose dehydrogenase activity in normal and rheumatoid synovium: the description of a specialized synovial lining cell. *Int. J. Exp. Pathol.* **74**, 27–34.

Ponta, H., Wainwright, D., Herrlich, P. (1998) The CD44 protein family. *Int. J. Biochem. Cell Biol.*, **30**, 299–305.

Prehm, P. (1980) Induction of hyaluronic acid synthesis in teratocarcinoma stem cells by retinoic acid. *FEBS Lett.* **111**, 295–298.

Prehm, P. (1983a) Synthesis of hyaluronate in differentiated teratocarcinoma cells. Characterization of the synthase. *Biochem. J.* **211**, 181–189.

Prehm, P. (1983b) Synthesis of hyaluronate in differentiated teratocarcinoma cells. Mechanism of chain growth. *Biochem. J.* **211**, 191–198.

Prehm, P. (1984) Hyaluronate is synthesized at plasma membranes. *Biochem. J.* **220**, 597–600.

Prehm, P. (1985) Inhibition of hyaluronate synthesis. *Biochem. J.* **225**, 699–705.

Prehm, P. (1990) Release of hyaluronate from eukaryotic cells. *Biochem. J.* **267**, 185–189.

Prehm, P. (2000) Antigene von rheumatischen Autoimmunerkrankungen. Patent PCT/EP00/05279.

Prehm, S., Herrington, C., Nickel, V., Völker, W., Briko, N. I., Blinnikova, E. A., Schmiedel, A., Prehm, P. (1995) Antibodies against proteins of streptococcal hyaluronate synthase bind to human fibroblasts and are present in patients with rheumatic fever. *J. Anat.* **187**, 271–277.

Prehm, S., Nickel, V., Prehm, P. (1996) A mild purification method for polysaccharide binding membrane proteins: phase separation of digitonin extracts to isolate the hyaluronate synthase from *Streptococcus* sp. in active form. *Protein Expression and Purification* **7**, 343–346.

Pulkki, K. (1986) The effects of synovial fluid macrophages and interleukin-1 on hyaluronic acid synthesis by normal synovial fibroblasts. *Rheumatol. Int.* **6**, 121–125.

Puré, E., Camp, R. L., Peritt, D., Panettieri, R. A., Jr., Lazaar, A. L., Nayak, S. (1995) Defective phosphorylation and hyaluronate binding of CD44 with point mutations in the cytoplasmic domain. *J. Exp. Med.* **181**, 55–62.

Reed, R. K., Laurent, T. C., Taylor, A. E. (1990) Hyaluronan in prenodal lymph from skin: changes with lymph flow. *Am. J. Physiol.* **259**, H1097–H1100.

Roden, L., Campbell, P., Fraser, J. R., Laurent, T. C., Pertoft, H. T.-J. (1989) Enzymic pathways of hyaluronan catabolism. *Ciba Foundation Symposium* **143**, 60–76.

Schenck, P., Schneider, S., Miehlke, R., Prehm, P. (1995) Synthesis and degradation of hyaluronate by synovia from patients with rheumatoid arthritis. *J. Rheumatol.* **22**, 400–405.

Schmits, R., Filmus, J., Gerwin, N., Senaldi, G., Kiefer, F., Kundig, T., Wakeham, A., Shahinian, A., Catzavelos, C., Rak, J., Furlonger, C., Zakarian, A., Simard, J. J., Ohashi, P. S., Paige, C. J., Gutierrez, R. J., Mak, T. W. (1997) CD44 regulates hematopoietic progenitor distribution, granuloma formation, and tumorigenicity. *Blood* **90**, 2217–2233.

Scott, J. E., Cummings, C., Brass, A., Chen, Y. (1991) Secondary and tertiary structures of hyaluronan in aqueous solution, investigated by rotary shadowing-electron microscopy and computer simulation. Hyaluronan is a very efficient network-forming polymer. *Biochem. J.* **274**, 699–705.

Skelton, T. P., Zeng, C. X., Nocks, A., Stamenkovic, I. (1998) Glycosylation provides both stimulatory and inhibitory effects on cell surface and soluble CD44 binding to hyaluronan. *J. Cell Biol.* **140**, 431–446.

Taher, T. E. I., Smit, L., Griffioen, A. W., Schilder-Tol, E. J. M., Borst, J., Pals, S. T. (1996) Signaling through CD44 is mediated by tyrosine kinases – Association with $p56^{lck}$ in T lymphocytes. *J. Biol. Chem.* **271**, 2863–2867.

Takahashi, K., Stamenkovic, I., Cutler, M., Saya, H., Tanabe, K. K. (1995) CD44 hyaluronate binding influences growth kinetics and tumorigenicity of human colon carcinomas. *Oncogene* **11**, 2223–2232.

Tammi, R., Tammi, M. (1986) Influence of retinoic acid on the ultrastructure and hyaluronic acid synthesis of adult human epidermis in whole skin organ culture. *J. Cell Physiol.* **126**, 389–398.

Tengblad, A., Caputo, C. B., Raisz, L. G. (1980) Quantitative analysis of hyaluronate in nanogram amounts. *Biochem. J.* **185**, 101–105.

Termeer, C. C., Hennies, J., Voith, U., Ahrens, T., Weiss, M., Prehm, P., Simon, J. C. (2000) Oligosaccharides of hyaluronan are potent activators of dendritic cells. *J. Immunol.* **165**, 1863–1870.

Thonard, J. C., Migliore, S. A., Blustein, R. (1964) Isolation of hyaluronic acid from broth cultures of streptococci. *J. Biol. Chem.* **239**, 726–728.

Tlapak-Simmons, V. L., Kempner, E. S., Baggenstoss, B. A., Weigel, P. H. (1998) The active streptococcal hyaluronan synthases (HASs) contain a single HAS monomer and multiple cardiolipin molecules. *J. Biol. Chem.* **273**, 26100–26109.

Tomida, M., Koyama, H., Ono, T. (1975) Induction of hyaluronic acid synthetase activity in rat fibroblasts by medium change of confluent cultures. *J. Cell Physiol.* **86**, 121–130.

Tomida, M., Koyama, H., Ono, T. (1977a) A serum factor capable of stimulating hyaluronic acid synthesis in cultured rat fibroblasts. *J. Cell Physiol.* **91**, 323–328.

Tomida, M., Koyama, H., Ono, T. (1977b) Effects of adenosine 3′:5′-cyclic monophosphate and serum on synthesis of hyaluronic acid in confluent rat fibroblasts. *Biochem. J.* **162**, 539–543.

Toole, B. P. (1991) Proteoglycans and hyaluronan in morphogenesis and differentiation, in: *Cell Biology of the Extracellular Matrix* (Hay, E. D., Ed.), New York: Plenum Press, 305–341.

Toole, B. P., Okayama, M., Orkin, R. W., Yoshimura, M., Muto, M., Kaji, A. (1977) Developmental roles of hyaluronate and chondroitin sulfate proteoglycans. *Soc. Gen. Physiol. Ser.* **32**, 139–154.

Toole, B. P., Biswas, C., Gross, J. (1979) Hyaluronate and invasiveness of the rabbit V2 carcinoma. *Proc. Natl. Acad. Sci. USA* **76**, 6299–6303.

Triscott, M. X., van de Rijn, I. (1986) Solubilization of hyaluronic acid synthetic activity from streptococci and its activation with phospholipids. *J. Biol. Chem.* **261**, 6004–6009.

Turley, E. A. (1989) The role of a cell-associated hyaluronan-binding protein in fibroblast behaviour. *Ciba Foundation Symposium* **143**, 121–133.

Turley, E. A., Moore, D., Hayden, L. J. (1987) Characterization of hyaluronate binding proteins isolated from 3T3 and murine sarcoma virus transformed 3T3 cells. *Biochemistry* **26**, 2997–3005.

Turley, E. A., Pilarski, L., Nagy, J. I. (1998) Problems with RHAMM: a new link between surface adhesion and oncogenesis? Response. *Cell* **95**, 592–593.

Uchiyama, H., Dobashi, Y., Ohkouchi, K., Nagasawa, K. (1990) Chemical change involved in the oxidative reductive depolymerization of hyaluronic acid. *J. Biol. Chem.* **265**, 7753–7759.

Ueki, N., Taguchi, T., Takahashi, M., Adachi, M., Ohkawa, T., Amuro, Y., Hada, T., Higashino, K. (2000) Inhibition of hyaluronan synthesis by vesnarinone in cultured human myofibroblasts. *Biochim. Biophys. Acta* **1495**, 160–167.

Underhill, C. B. (1982) Interaction of hyaluronate with the surface of simian virus40- transformed 3T3 cells: aggregation and binding studies. *J. Cell Sci.* **56**, 177–189.

Underhill, C. B., Toole, B. P. (1979) Binding of hyaluronate to the surface of cultured cells. *J. Cell Biol.* **82**, 475–484.

Underhill, C. B., Toole, B. P. (1982) Transformation-dependent loss of the hyaluronate-containing coats of cultured cells. *J. Cell Physiol.* **110**, 123–128.

van de Rijn, I. (1983) Streptococcal hyaluronic acid: proposed mechanisms of degradation and loss of synthesis during stationary phase. *J. Bacteriol.* **156**, 1059–1065.

van de Rijn, I., Kessler, R. E. (1980) Growth characteristics of group A streptococci in a new chemically defined medium. *Infect. Immun.* **27**, 444–448.

Van der Voort, R., Manten-Horst, E., Smit, L., Ostermann, E., Van den Berg, F., Pals, S. T. (1995) Binding of cell-surface expressed CD44 to hyaluronate is dependent on splicing and cell type. *Biochem. Biophys. Res. Commun.* **214**, 137–144.

Vercruysse, K. P., Prestwich, G. D. (1998) Hyaluronate derivatives in drug delivery. *Crit. Rev. Therap. Drug Carrier Systems* **15**, 513–555.

Waldenstrom, A., Martinussen, H. J., Gerdin, B., Hällgren, R. (1991) Accumulation of hyaluronan and tissue edema in experimental myocardial infarction. *J. Clin. Invest.* **88**, 1622–1628.

Weigel, P. H., Fuller, G. M., LeBoeuf, R. D. (1986) A model for the role of hyaluronic acid and fibrin in the early events during the inflammatory response and wound healing. *J. Theor. Biol.* **119**, 219–234.

Weigel, P. H., Hascall, V. C., Tammi, M. (1997) Hyaluronan synthases. *J. Biol. Chem.* **272**, 13997–14000.

Weissman, B., Meyer, K. (1954) The structure of hyalobiuronic acid and of hyaluronic acid from umbilical cord. *J. Am. Chem. Soc.* **76**, 1753–1757.

Wessels, M. R., Moses, A. E., Goldberg, J. B., DiCesare, T. J. (1991) Hyaluronic acid capsule is a virulence factor for mucoid group A streptococci. *Proc. Natl. Acad. Sci. USA* **88**, 8317–8321.

Wessels, M. R., Goldberg, J. B., Moses, A. E., DiCesare, T. J. (1994) Effects on virulence of mutations in a locus essential for hyaluronic acid capsule expression in group A streptococci. *Infect. Immun.* **62**, 433–441.

West, D. C., Kumar, S. (1989) The effect of hyaluronate and its oligosaccharides on endothelial cell proliferation and monolayer integrity. *Exp. Cell Res.* **183**, 179–196.

West, D. C., Hampson, I. N., Arnold, F., Kumar, S. (1985) Angiogenesis induced by degradation products of hyaluronic acid. *Science* **228**, 1324–1326.

West, D. C., Shaw, D. M., Lorenz, P., Adzick, N. S., Longaker, M. T. (1997) Fibrotic healing of adult and late gestation fetal wounds correlates with increased hyaluronidase activity and removal of hyaluronan. *Int. J. Biochem. Cell Biol.* **29**, 201–210.

Whitnack, E., Bisno, A. L., Beachey, E. H. (1981) Hyaluronate capsule prevents attachment of group A streptococci to mouse peritoneal macrophages. *Infect. Immun.* **31**, 985–991.

Wobig, M., Bach, G., Beks, P., Dickhut, A., Runzheimer, J., Schwieger, G., Vetter, G., Balazs, E. A. (1999) The role of elastoviscosity in the efficacy of viscosupplementation for osteoarthritis of the knee: a comparison of hylan G-F 20 and a lower-molecular-weight hyaluronan. *Clin. Ther.* **21**, 1549–1562.

Wong, S. F., Halliwell, B., Richmond, R., Skowroneck, W. R. (1981) The role of superoxide and hydroxyl radicals in the degradation of hyaluronic acid induced by metal ions and by ascorbic acid. *J. Inorg. Biochem.* **14**, 127–134.

Wu, R., Wu, M. M. (1986) Effects of retinoids on human bronchial epithelial cells: differential regulation of hyaluronate synthesis and keratin protein synthesis. *J. Cell Physiol.* **127**, 73–82.

Yoshimura, M. (1985) Change of hyaluronic acid synthesis during differentiation of myogenic cells and its relation to transformation of myoblasts by Rous sarcoma virus. *Cell Differ.* **16**, 175–185.

Zahalka, M. A., Okon, E., Gosslar, U., Holzmann, B., Naor, D. (1995) Lymph node (but not spleen) invasion by murine lymphoma is both CD44- and hyaluronate-dependent. *J. Immunol.* **154**, 5345–5355.

Zhou, B., Oka, J. A., Singh, A., Weigel, P. H. (1999) Purification and subunit characterization of the rat liver endocytic hyaluronan receptor. *J. Biol. Chem.* **274**, 33831–33834.

Zhu, D., Bourguignon, L. Y. W. (1998) The ankyrin-binding domain of CD44s is involved in regulating hyaluronic acid-mediated functions and prostate tumor cell transformation. *Cell Motil. Cytoskeleton* **39**, 209–222.

16
Exopolysaccharides of Lactic Acid Bacteria

Ir. Isabel Hallemeersch[1], Ir. Sophie De Baets[2], Prof. Dr. Ir. Erick J. Vandamme[3]
[1] Laboratory of Industrial Microbiology and Biocatalysis, Department of Biochemical and Microbial Technology, Faculty of Agricultural and Applied Biological Sciences, Ghent University, Coupure links 653, B-9000 Gent, Belgium; Tel.: +32-9-264-6029; Fax: +32-9-264-6231; E-mail: Isabel.Hallemeersch@rug.ac.be
[2] Laboratory of Industrial Microbiology and Biocatalysis, Department of Biochemical and Microbial Technology, Faculty of Agricultural and Applied Biological Sciences, Ghent University, Coupure links 653, B-9000 Gent, Belgium; Tel.: +32-9-264-6028; Fax: +32-9-264-6231; E-mail: Sophie.DeBaets@rug.ac.be
[3] Laboratory of Industrial Microbiology and Biocatalysis, Department of Biochemical and Microbial Technology, Faculty of Agricultural and Applied Biological Sciences, Ghent University, Coupure links 653, B-9000 Gent, Belgium; Tel.: +32-9-264-6027; Fax: +32-9-264-6231; E-mail: Erick.Vandamme@rug.ac.be

EPS	exopolysaccharides
GRAS	generally recognized as safe
LAB	lactic acid bacteria
NMR	nuclear magnetic resonance

1 Introduction

The exopolysaccharides (EPS) produced by lactic acid bacteria (LAB) can be divided into three major groups, based on their composition: (1) glucans, namely dextrans, alternans, and mutans; (2) fructans such as levan; and (3) heteropolysaccharides produced by mesophilic and thermophilic LAB (Cerning, 1990). As details of the glucans and fructans are described in Chapters 12–14 of this Volume, only the latter group will be discussed here. These EPS are composed of linear and branched repeating units, and vary in size from disaccharides to heptasaccharides (Petry et al., 2000).

Over the past few decades, there has been a growing interest in ropy LAB, such as *Lactoccocus lactis* subsp. *cremoris, L. lactis* subsp. *lactis, Lactobacillus delbreuckii* subsp. *bulgaricus, Lb. helveticus, Streptococcus salivarius* subsp. *thermophilus*, etc. Based on the "generally recognized as safe" (GRAS) status of these bacteria, their EPS preparations are widely applied as thickening, gelling and stabilizing agents in the food industry (Sutherland, 1994). Their most important application area in this respect is undoubtedly in the dairy industry, where they are used in the production of various fermented milk products and contribute to both texture and mouthfeel (Cerning, 1995). Moreover, it was suggested that they also possess advantageous biological functions and are beneficial for human health (Nakajima et al., 1992; Kitazawa et al., 1993; Gibson and Roberfroid, 1995; Hosono et al., 1997).

2 Historical Outline

Although ropy LAB such as *S. thermophilus* and *Lb. delbreuckii* subsp. *bulgaricus* have been available commercially as dairy starter cultures since the early 1900s, the chemical composition and structures of their EPS were reported in detail only as recently as the 1980s (Marshall and Rawson, 1999). Little information existed regarding the level of polysaccharide produced, the culture conditions and the rheological properties, in contrast to the situation with glucans such as dextran, on which most attention was focused and which had already undergone intensive study (Forsén and Häivä, 1981).

Nonetheless, during the 1950s and 1970s, a number of articles were published on the composition of heteropolysaccharides produced by LAB. These early investigations did not refer to any precise culture conditions, nor analytical details, and should therefore be treated with some caution (Sundman, 1953; Nilsson and Nilsson, 1958; Groux, 1973; Tamime, 1978).

It was not until the 1980s that EPS from LAB received the attention of numerous investigators, particularly in France and the Netherlands, where the use of stabilizing agents from plant or animal origin was prohibited. Among the early publications on the chemical composition of EPS no clear picture emerged (Cerning et al., 1986), and EPS were seen either as proteinaceous material, a carbohydrate–protein complex, or simply as a complex carbohydrate. These differences were due to different isolation and purification methods, each with their variable efficiencies. This was especially so because EPS purification was hampered by the use of complex culture media, such as milk or whey; moreover, some bacterial strains were able to synthesize more than one type of EPS (Marshall et al., 1995).

Subsequently, a great deal of research was carried out on various aspects of EPS synthesis, such as the influence of nutritional and physico-chemical fermentation parameters, the isolation and characterization of the EPS, and the rheological properties. Detailed structural and rheological studies were performed in order to provide a better insight into the mechanisms by which the LAB and their EPS influence the consistency of dairy products (Staaf et al., 2000; Faber et al., 2001).

During the late 1980s, investigations into the instability of the mucoid character of LAB revealed the involvement of plasmids in EPS synthesis (Vedamuthu and Neville, 1986; Neve et al., 1988). This was the beginning of a series of studies on the molecular biology and genetics of EPS produced by various LAB.

During the past decade, a large number of genes involved in the synthesis of repeating units, polymerization, chain length determination and export have been characterized and functionally analyzed. In addition, the entire genome of *L. lactis* has been reported recently (Bolotin et al., 1998; Kleerebezem et al., 2000; Ricciardi and Clementi, 2000). Models were developed in order to predict the behavior and the effect of EPS addition to food products, and this has led to the identification of several important properties, though these models require further validation and fine-tuning (Kleerebezem et al., 1999).

Nowadays, the challenge is to increase production levels and to modify the structure and hence the properties of EPS, by using genetic approaches and enzyme technology. The purpose of this type of polysaccharide engineering is the synthesis of tailor-made oligo- and polysaccharides, for a variety of specific applications (De Vuyst and Degeest, 1999a; Kleerebezem et al., 2000).

3 Chemical Structure

Heteropolysaccharides produced by LAB have been investigated to a lesser extent when compared with other polymers. The overall level of EPS production is low and is often characteristically variable (Cerning, 1990). Most strains produce a limited quantity of polymer, perhaps up to 200 mg L^{-1}, although some strains have been found to produce up to 4 g L^{-1}.

Many LAB synthesize heteropolysaccharides with a molecular weight in excess of 10^6 daltons. In addition, their composition, structure and physico-chemical properties are highly variable, these are being influenced by the composition of the culture medium (Ricciardi and Clementi, 2000).

Today, it is generally recognized that EPS from LAB are composed of linear and branched repeating units that vary in size from disaccharides to heptasaccharides, and which contain α- and β- linkages. Most EPS contain D-glucose, D-galactose and/or L-rhamnose in different ratios, but other hexoses such as D-mannose, D-fructose and pentoses such as L-fucose, D-xylose and D-arabinose also appear. Hexosamines and uronic acids are also found in minor quantities (Andaloussi et al., 1995; Marshall et al., 1995; Petry et al., 2000). The chemical composition of a few EPS is summarized in Table 1, while Table 2 shows the structure of typical repeating units.

The EPS from mesophilic LAB have a more varied composition than those from thermophilic LAB, and contain sometimes acetylated and phosphorylated residues (Ricciardi and Clementi, 2000). Different authors have described the chemical composition of EPS produced by *Lactobacillus* strains (Toba et al., 1991; Kojic et al., 1992; Cerning, 1994; Van den Berg et al., 1995; Robijn et al., 1996). Several lactococci have also been described that produce EPS; these usually contain glucose and galactose, but rhamnose and charged residues are also found quite commonly (Cerning, 1994). Most *Pediococcus* strains, which occur most often as spoilers in beer and wine, produce β-glucans of high molecular weight (Ricciardi and Clementi, 2000).

Galactose is seen as the major monosaccharide in the EPS produced by thermophilic

Tab. 1 Chemical composition of several exopolysaccharides (EPS) synthesized by lactic acid bacteria (LAB)

Strain	***Monosaccharides***				***Reference***
	Gal	***Glc***	***Rha***	***Other***	
Lb. delbreukii subsp	+	+	+	–	
bulgaricus NCFB 2772	+	+	–	–	Grobben et al. (1995, 1996)
Lb. delbreukii subsp. *bulgaricus* CNRZ 1187	+	+	–	–	Petry et al. (2000)
Lb. delbreukii subsp. *bulgaricus* CNRZ 416	+	+	+	–	Petry et al. (2000)
Lb. casei CG11	–	+	+	–	Kojic et al. (1992)
Lb. casei CG11	+	+	+	–	Cerning (1994)
Lb. helveticus var. *jugurti*	+	+	–	–	Oda et al. (1983)
Lb. helveticus LB161	+	+	–	Ac, P	Staaf et al. (2000)
Lb. kefiranofaciens	+	+	–	–	Toba et al. (1991)
Lb. paracasei	+	+	+	–	Van Calsteren (2001)
Lb. paracasei 34-1	+	–	–	GP	Robijn et al. (1996)
Lb. rhamnosus	+	+	+	–	Van Calsteren (2001)
Lb. sake 0-1	–	+	+	Ac, GP	Robijn et al. (1996)
Lb. sake 0-1	–	+	+	–	Van den Berg et al. (1995)
L. lactis subsp. *cremoris* LC330	+	+	–	GlcNAc	Marshall et al. (1995)
L. lactis subsp. *cremoris* LC330	+	+	+	GlcNAc P	Marshall et al. (1995)
L. lactis subsp. *cremoris* SBT 0495	+	+	+	P	Nakajima and Toyoda (1990)
Pediococcus	–	+	–	–	Cerning (1994)
S. thermophilus EU20	+	+	+	–	Marshall et al. (2001)
S. thermophilus CNCMI 733	+	+	–	GalNAc	Doco et al. (1991)
S. thermophilus OR 901	+	–	+	–	Ariga et al. (1992)
S. thermophilus S3	+	–	+	Ac	Faber et al. (2001)

Gal, galactose ; Glc, glucose ; Rha, rhamnose ; Ac, acetate ; P, phosphate ; GP, glycerol-3-phosphate ; Nac, *N*-acetyl.

Tab. 2 Primary structure of several EPS synthesized by LAB

Strain	***Structure***	***Reference***
Lb. delbreukii subsp. *bulgaricus* NCFB 2772	β-D-Gal*p* β-D-Gal*p* α-L-Rha*p* ↓ ↓ ↓ →2)-α-D-Gal*p*-(1→3)-β-D-Glc*p*-(1→3)-β-D-Gal*p*-(1→4)-α-D-Gal*p*-(1→	Grobben et al. (1995, 1996)
Lb. helveticus LB161	β-D-Glc*p* β-D-Glc*p* ↓ ↓ →4)-α-D-Glc*p*-(1→4)-β-D-Gal*p*-(1→3)-α-D-Gal*p*-(1→2)-α-D-Glc*p*-(1→3)- β-D-Glc*p*-(1→	Staaf et al. (2000)
Lb. paracasei 34-1	*sn*-Glycerol-3-phosphate-3 ↓ →3)-β-D-Gal*p*NAc-(1→4)-β-D-Gal*p*-(1→6)-β-D-Gal*p*-(1→6)-β-D-Gal*p*-(1→	Robijn et al. (1996)
Lb. sake 0-1	β-D-Glc*p*-(1→6) (Ac) $_{0.85}$ ↓ ↓ →4)-β-D-Glc*p*-(1→4)-α-D-Glc*p*-(1→3)-β-L-Rha*p*-(1→ ↑ -*sn*-Glycerol-3-phosphate-3→4)-α-L-Rha*p*-(1→3)	Robijn et al. (1996)
L. lactis **subsp.** *cremoris* **SBT 0495**	α-L-Rha*p*-(1→2) ↓ →4)-β-D-Glc*p*-(1→4)-β-D-Gal*p*-(1→4)-β-D-Glc*p*-(1→ ↑ α-D-Gal*p*-1-phosphate	Nakajima and Toyoda, 1990

Tab. 2 (cont.)

Strain	Structure	Reference
S. thermophilus CNCMI 733	α-D-Gal*p*-(1→6) ↓ →3)-β-D-Gal*p*-(1→3)-β-D-Glc*p*-(1→3)-α-D-Gal*p*NAc-(1→	Doco et al., 1991
S. thermophilus OR 901	β-D-Gal*p*-(1→6)-β-D-Gal*p*-(1→4) ↓ →2)-α-D-Gal*p*-(1→3)-α-D-Gal*p*-(1→3)-α-L-Rha*p*-(1→2)-α-L-Rha*p*-(1→	Ariga et al., 1992
S. thermophilus S3	β-D-Gal*f*2Ac ↓ →3)-β-D-Gal*p*-(1→3)-α-D-Gal*p*-(1→3)-α-L-Rha*p*-(1→2)-α-L-Rha*p*-(1→2)-α-D-Gal*p*-(1→	Faber et al., 2001

Gal = Galactose, Glc = Glucose, Rha = Rhamnose, Ac = Acetate

LAB, most likely because it is metabolized less quickly than glucose and hence is available for polymer synthesis (Cerning, 1990).

Many EPS types from *S. thermophilus* strains have been investigated as to their composition and characteristics. Most such heteropolysaccharides contain D-glucose and D-galactose and, on occasion, also L-rhamnose.

Heteropolysaccharide production was also observed in other thermophilic LAB such as *Lactobacillus helveticus* and *Lactobacillus delbreuckii* subsp. *bulgaricus*. Grobben et al. (1995, 1996) demonstrated with a *Lb. delbreuckii* subsp. *bulgaricus* strain, that the EPS composition varied with different sugars present in the growth medium. When using glucose or lactose as a carbon source, the EPS consisted of glucose, galactose, and rhamnose in a ratio of 1:6.8:0.7, but in the presence of fructose, the EPS formed contained only glucose and galactose in a ratio of 1:2.4. However, when glucose and fructose are used together, the composition of the EPS is similar to that obtained when either glucose or lactose was the carbon source. Marshall et al. (1995) isolated two EPS types from the culture broth of the same strain–a high and a low molecular fraction. Finally, De Vuyst and Degeest (1999b) showed that both fractions were similar in their monomeric composition and that the growth medium was in fact influencing EPS composition.

Petry et al. (2000) also investigated the EPS of two *Lb. delbreuckii* subsp. *bulgaricus* strains and concluded that the monomeric sugar composition remained the same, but that the relative proportions of the individual monosaccharides varied under different fermentation conditions.

In order to obtain insight into the structure–function relationship, it is necessary to study the conformation of the polymer in solution as well as its dynamic behavior. The three-dimensional structure of a polysaccharide is dependent on the time-averaged ring conformation of the monosaccharides and the relative orientations of adjacent monosaccharides. Besides the conformations around the glycosidic linkages, the intramolecular hydrogen bonds also contribute to the conformation, and this can be important for a complete understanding of the solution behavior. NMR spectroscopy does not provide sufficient information to elucidate the complete conformation (Vliegenthart et al., 2001). As the secondary and tertiary conformation of a polysaccharide is mainly dependent on its primary (chemical) structure, relatively small changes in the chemical structure might have a major effect on the conformation and the physical and chemical properties of the polysaccharide. However, despite all information already currently available, prediction of the properties on basis of polymer structure is not yet possible.

4 Occurrence

The LAB form a diverse group of bacteria, including the genera *Lactobacillus, Lactococcus, Streptococcus, Enterococcus, Leuconostoc,* and *Pediococcus.* They are Gram positive, have a low DNA GC-content (32–53 mol.%), are catalase-, reductase-, and oxidase-negative, and are also nonmotile and nonsporulating. *Bifidobacteria* are often associated with these genera as they are also added to traditional yogurt cultures (Dellaglio, 1994).

EPS-producing LAB can be isolated from dairy products such as the well-known Scandinavian fermented milk products "viili" and "longfil", yogurts, fermented milk drinks, and kefir (Kandler and Kunath, 1983; Toba et al., 1991; Ariga et al., 1992). Cheese and fermented meat and vegetables are also an important source of ropy LAB (Kojic et al., 1992; Van den Berg et al., 1993). Recently, Smitinont et al. (1999) isolated two EPS producing *Pediococcus* strains from traditional Thai fermented foods; although grown on sucrose, the EPS differed from dextran.

EPS-producing LAB can also be disadvantageous however, for example when causing spoilage of beer, wine, and vacuum-packed cooked meat products (Korkeala et al., 1988; Morin, 1998).

More severe problems arise when biofilm formation occurs on heat exchanger plates in cheese and milk factories, resulting in either excessive openness in cheese texture, or taste changes in milk. In particular, thermophilic LAB can cause significant problems by contaminating heat exchanger plates on the downstream side of the pasteurizers, whereby already pasteurized milk may become re-contaminated (Bouwman et al., 1982; Neu and Marshall, 1990).

5 Physiological Function

Although EPS are not essential to ensure the viability of bacterial cells, their exact function in nature has not yet been clearly defined and is most likely highly complex (Cerning, 1994).

In the past, it has generally been accepted that EPS do not serve as storage materials, since most EPS-producing strains are not able to catabolize EPS, nor use them as a carbon source. In the case of LAB this remains doubtful however, as the amount of EPS produced often decreases upon prolonged fermentation. One possible explanation is that the synthesized EPS are degraded enzymatically by the LAB themselves. Recently, Pham et al. (2000) reported on the synthesis of several glycohydrolases by a *Lb. rhamnosus* strain which was able to hydrolyze the EPS to a limited extent.

The role of EPS in pathogenesis has been investigated, since their synthesis–which results mainly in capsule formation–is widespread among pathogenic bacteria. The presence of EPS alone does not appear to be sufficient to turn bacterial cells virulent. It has been suggested that microorganisms producing EPS are more hydrophilic, and thus are less susceptible to phagocytosis (Jann and Jann, 1977). EPS also seem to play a role in the protection against phage attack, antibiotics and other toxic compounds.

Another proposed function is the protection against desiccation and other extreme physical conditions; this is due to the fact that EPS are present as a highly hydrated layer surrounding the bacterial cells. EPS also play an important role in the adhesion of microorganisms onto solid surfaces, although the adsorption process is complex and the exact impact of EPS in the process has not yet been completely elucidated (Cerning, 1994).

6 Chemical Analysis and Detection

6.1 Separation of EPS and Microbial Cells

The first step in the purification of EPS is the separation of microbial cells from the EPS-containing culture broth. This is normally done by centrifugation, and when working at laboratory scale, ultracentrifugation can be used. It is difficult to define an optimal g-value as this depends on the specific viscosity of the culture broth (Sutherland, 1972). Sometimes a too-high viscosity may prevent easy sedimentation of bacterial cells. The addition of electrolytes (NaCl) may facilitate the separation by neutralizing the charges on the polysaccharides (Cerning, 1994). Because EPS are thermostable, heat treatment can also be used to reduce the viscosity.

When present as a capsule, the EPS must first be dissociated from the cells and, depending on the nature of the association between the cells and the capsule, extreme conditions such as alkaline treatment, sonication or heating may be required (Morin, 1998).

6.2 Isolation and Purification

EPS are mainly recovered by precipitation with organic solvents, such as ethanol, acetone, or isopropanol. Organic solvents permit separation by lowering the solubility of the EPS. Although the EPS:solvent ratio is variable, about one up to three volumes of solvents are normally used. The precipitate is finally recovered by centrifugation, filtration, and pressing or settling, after which dialysis and freeze-drying of the final product is carried out (Garcia-Garibay and Marshall, 1991; Marshall et al., 1995).

EPS from LAB are often synthesized in complex media such as milk, milk ultrafiltrate and whey; the proteins present in these media, such as casein, seem to coprecipitate with the EPS and should be removed prior to the isolation step. This can be achieved enzymatically by the addition of proteases such as pronase, trypsin, or proteinase K (Cerning, 1994). Residual peptides can also be removed by repeated trichloroacetic acid precipitation, followed by centrifugation and precipitation of the EPS. Gel filtration is another possibility, although it is much slower and problems often arise due to the high viscosity of the media. Further purification of the EPS can be achieved by anion-exchange chromatography (Nakajima and Toyoda, 1990; Doco et al., 1991; Andaloussi et al., 1995).

Although EPS yields are higher in milk or whey-based media, the EPS isolation from such complex media is often tedious and time-consuming. More recently, several semi-defined and chemically defined media have been developed; these are more suitable to investigate the influence of nutrients on the growth and on the EPS production, the metabolic pathways involved, the composition of the EPS, and the rheological properties (Petry et al., 2000).

6.3 Structural Analysis

Several structures of repeating units of branched EPS from LAB have been elucidated. General methods used to study the composition and specific linkages present in polysaccharides are normally applied, including acid hydrolysis, periodate oxidation, methylation analysis, enzymatic degradation, Smith-degradation and NMR-spectroscopy. With these methods it is possible to determine not only the neutral sugars but also the anomeric configuration, the specific linkages present, and the sequence of the repeating units.

The molecular weights of polysaccharides are determined by gel filtration and high-performance liquid chromatography.

On occasion, a strain produces more than one type of EPS, and these can usually be separated using chromatography based on the differences in either molecular weight or charge (Toba et al., 1991; Marshall et al., 1995; Robijn et al., 1996; Grobben et al., 1997).

7 Biosynthesis

Unlike the biochemical pathways involved in the biosynthesis of exocellular homopolysaccharides, those of exocellular heteropolysaccharides are much more complex (Cerning, 1994).

Heteropolysaccharides are synthesized at the cytoplasmic membrane through polymerization of precursors, and are formed in the cytoplasm. These precursors are mainly UDP-nucleotide diphosphate sugars. As with the large number of enzymes involved, the sugar nucleotide precursors are not all unique to EPS synthesis; some are also involved in the synthesis of cell-wall polymers such as peptidoglycans, lipopolysaccharides, and teichoic acids. However, as they are freely soluble in the cytoplasm, they can be readily channeled to the appropriate biosynthetic process (Sutherland, 1990).

The sugar nucleotides serve different functions. First, they play an important role in sugar activation, supplying energy for the assembly of glycosyl units on appropriate carrier molecules, with the release of a diphosphate nucleotide. Second, they play a role in sugar interconversions, which involve several mechanisms such as epimerization (UDP-D-glucose → UDP-D-galactose), oxidation (UDP-D-glucose → UDP-D-glucuronic acid), decarboxylation (UDP-D-glucose → UDP-D-xylose), reduction and rearrangement (GDP-D-mannose → GDP-L-fucose) (Sutherland, 1972, 1990).

LAB can metabolize a variety of mono- and disaccharides, but as EPS are mainly produced in milk or whey-based media, the main carbon source present is lactose. Based on studies investigating the influence of sugars on EPS synthesis and the enzymes involved in their anabolism, it might be concluded that glucose or the glucose moiety of lactose is the most important sugar for

EPS production in LAB (De Vos and Vaughan, 1994; Grobben et al., 1996). Hence, glucose-1-phosphate serves as an important precursor for EPS synthesis, and phosphoglucomutase might be seen as a key enzyme linking energy generation and sugar nucleotide formation (Kleerebezem et al., 2000) (Figure 1).

De Vos (1996) suggested that if the flux via phosphoglucomutase was high enough, then EPS overproduction could be achieved. As such, the galactose moiety of lactose would be catabolized via glycolysis, while the glucose moiety would serve for EPS production. A major problem is the inability of *S. thermophilus* and *Lb. delbreuckii* subsp. *bulgaricus* to catabolize galactose, this being due to the absence of the enzyme galactokinase; hence galactose is excreted via the lactose/galactose antiport transport system. However, the gene coding for this enzyme is present but is not transcribed. Galactose-fermenting mutants were constructed and isolated by Kleerebezem et al. (1999) through repair of the promotor mutations, and as a result galactose can be fully metabolized.

The composition and amount of EPS produced are not only dependent on the type of carbon source and on the sugar nucleotide level, but also on the assembly process of the repeating units.

The involvement of an isoprenoid glycosyl lipid carrier was reported in 1971 by Troy et al. This lipid is a C55-isoprenyl phosphate (bactoprenyl phosphate, undecaprenyl phosphate), and is the same acceptor lipid that functions in the formation of several cell-wall polymers (Figure 2). As such, the availability of the isoprenoid carrier is one of the most important factors affecting EPS production. Because of competition for the same acceptor lipid, it was postulated that EPS production is stimulated under conditions which lead to a reduction in growth (e.g., a lower temperature), and so cell-wall polymer biosynthesis is reduced (Sutherland, 1972).

Few authors have reported the exact role of the lipid carrier, and details of the mechanism remain unknown. However, possibilities include facilitation of the formation of the repeating units, solubilization of hydrophilic oligosaccharides in a hydrophobic membrane environment, and transport across the membrane (Troy, 1979; Cerning, 1990).

Once transported through the cell membrane, the polysaccharide can be excreted into the environment or will remain attached to the cell as a capsule. The transport of polysaccharides is an energy-demanding process and has not been fully clarified (Van den Berg et al., 1995).

Recently, *in vitro* experiments have been conducted to elucidate the biosynthesis of the polysaccharide backbone of *L. lactis* NIZO B40. As shown in Figure 3, EPS synthesis occurs through sequential addition of sugar residues by specific glycosyl transferases from sugar nucleotides to a growing repeating unit anchored to the lipid acceptor, thereby yielding the EPS (Kleerebezem et al., 1999, 2000).

8
Genetics and Regulation

A well-known problem of LAB in the dairy industry is the instability of EPS production at the genetic level, as well as the instability of the ropy texture itself (Cerning 1990, 1994). Loss of the ropy character may occur after numerous transfers and prolonged incubation periods, even at optimum growth temperatures. Therefore, strains have to be reselected regularly from the master culture to conserve the ropy character in industrially applied strains.

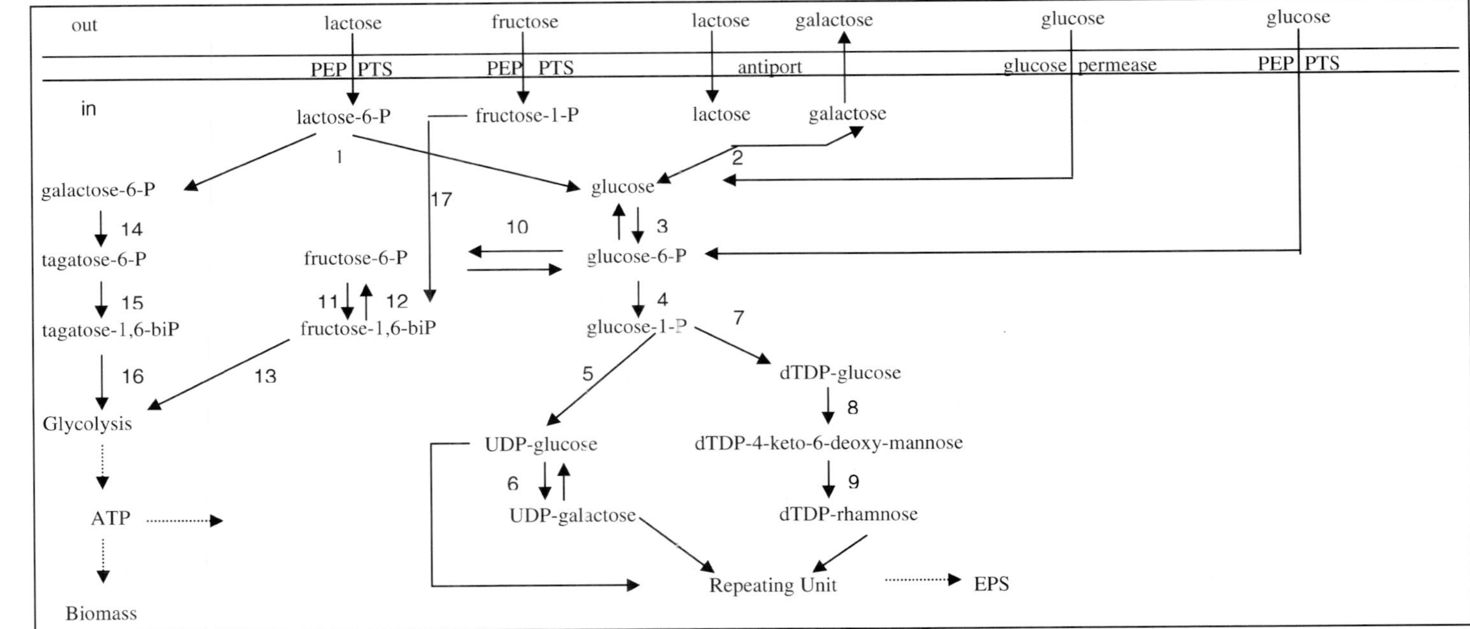

Fig. 1 Schematic representation of the lactose-, glucose-, and fructose-ut lizing pathways and exopolysaccharide biosynthesis. *Lactococcus lactis* transports lactose via the lactose phosphoenolpyruvate (PEP) -dependent phosphotransferase system (PTS), while the galactose-negative *Streptococcus thermophilus* and *Lactobacillus delbreuckii* subsp. *bulgaricus* transport lactose via a lactose/galactose antiport transport system. Enzymes involved are: 1, phospho-β-galactosidase; 2, β-galactosidase; 3, glucokinase; 4, phosphoglucomutase; 5, UDP-glucose pyrophosphorylase; 6, UDP-galactose-4-epimerase; 7, dTDP-glucose pyrophosphorylase; 8, dehydratase; 9, epimerase reductase; 10, phosphoglucoisomerase; 11, 6-phosphofructokinase; 12, fructose-1,6-biphosphatase; 13, fructose-1,6-biphosphate aldolase; 14, galactose-6-phosphate isomerase; 15, tagatose-6-phosphate kinase; 16, tagatose-1,6-ɔiphosphate aldolase; 17, fructokinase (Kleerebezem et al., 1999).

a

$$CH_3-\overset{CH_3}{\overset{|}{C}}=CH-CH_2-\left[CH_2-\overset{CH_3}{\overset{|}{C}}=CH-CH_2\right]_9-CH_2-\overset{CH_3}{\overset{|}{C}}=CH-CH_2OP$$

b

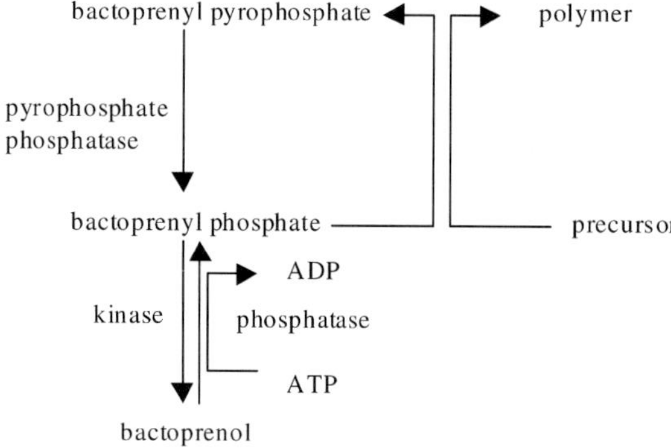

Fig. 2 Structure of the active form of the isoprenoid lipid carrier. (b) Activation and deactivation of the isoprenoid lipid; the polymer can be EPS, lipopolysaccharide, or peptidoglycan (Cerning, 1994).

Numerous plasmids encoding for EPS production have already been identified in mesophilic LAB. Moreover, it has been proven that the loss of the EPS-producing capacity in these bacteria is associated with the loss of plasmids (Vedamuthu and Neville, 1986; Neve et al., 1988). In thermophilic LAB, the EPS production is not encoded by plasmids, but the required genes are located on the chromosome. It is likely that the genetic instability is due to mobile genetic elements or to a generalized genomic instability, including deletions and rearrangements. Both phenomena were observed for *Lb. delbreuckii* subsp. *bulgaricus* and *S. thermophilus* (Stingele et al., 1996).

During the past few years, specific gene clusters have been characterized for some EPS-producing lactococci and streptococci; in addition, the entire genome sequence of *L. lactis* has recently been completed (Kleerebezem et al., 2000). These gene clusters contain *eps* genes, which are involved in the synthesis of the repeating units, export, polymerization and chain length determination. In addition to these *eps* genes, a number of housekeeping genes are also required for EPS synthesis. These genes are involved in the metabolic pathways leading to the EPS building blocks, namely the sugar nucleotides. Elucidation of the function of the glycosyltransferase genes might create opportunities for EPS engineering, while the identification and characterization of the housekeeping genes allows the design of metabolic engineering strategies, leading to increased EPS production levels (Kleerebezem et al., 1999).

The chromosomally located gene cluster of *S. thermophilus* Sfi6 consists of a 15.25 kb region coding for 15 genes; only 13 genes, *eps*A to *eps*M, were involved in EPS synthesis (Stingele et al., 1996). Low et al. (1998) partially identified and characterized the *eps* gene cluster of *S. thermophilus* MR-1C and found an organization similar to that of *S. thermophilus* Sfi6.

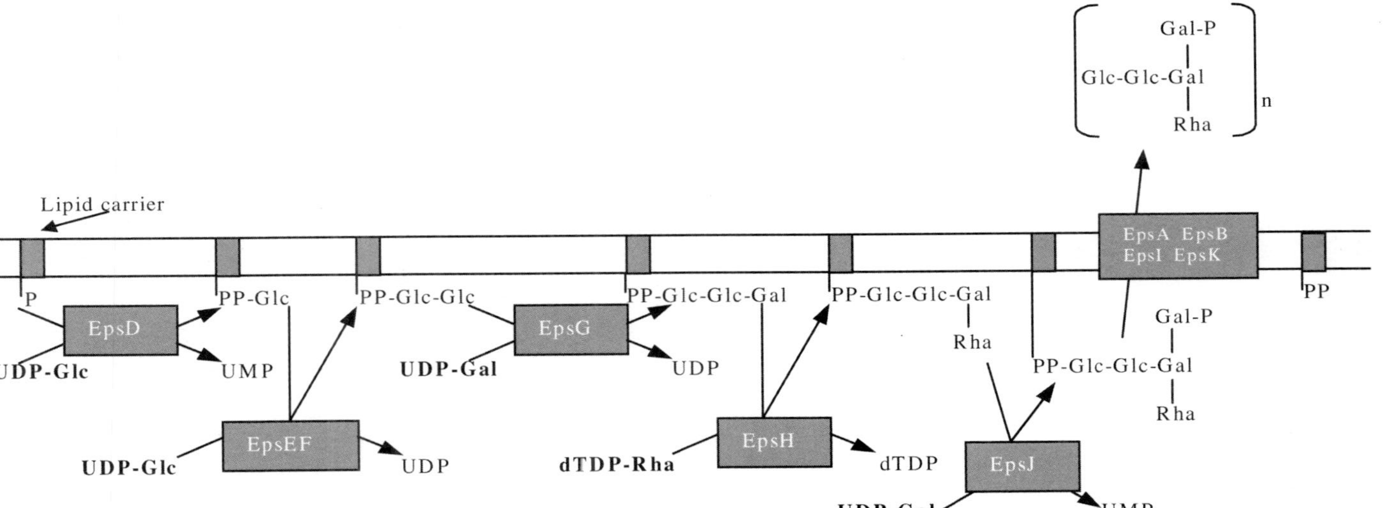

Fig. 3 Schematic representation of EPS biosynthesis by *L. lactis* NIZO B-40. Nucleotide sugars are indicated in bold; enzymes are indicated in the gray boxes by their encoding genes using established genetic nomenclature. Gal, galactose; Glc, glucose; Rha, rhamnose (Kleerebezem et al., 1999).

For *L. lactis*, the *eps* genes are encoded on large plasmids (> 20 kb) and can be transferred from one strain to another by conjugation. The best-characterized lactococcal strain is *L. lactis* NIZO B40, which produces a phosphopolysaccharide. A 12-kb gene cluster specific for EPS biosynthesis, contains 14 coordinately expressed genes and is inserted between IS sequences on a 40-kb plasmid.

Major similarities exist between the organization of the genes in the different gene clusters involved in EPS biosynthesis; four functional regions can be distinguished: a region involved in regulation is located at the beginning of the gene cluster; a central region which is involved in the synthesis of the EPS repeating units; and two flanking regions that are involved in polymerization and export (Stingele et al., 1996).

Further detailed studies can offer perspectives for EPS engineering and evaluation of the possibilities and restrictions towards the production of tailor-made EPS. This concerns not only changing the chemical structure of the polymers but also other relevant features, such as its chain length.

It was shown recently that a *L. lactis* NIZO B40 EpsD deletion mutant could functionally be complemented by the homologous *Eps*D gene itself, but more important also by heterologous glucosyltransferase encoding genes, yielding an EPS with a similar molecular weight but with a different structure (Van Kranenburg et al., 1999). Other examples, which indicate possibilities for the synthesis of heterologous or new EPS structures, have also been reported (Gilbert et al., 1998; Stingele, 1998).

Another approach would be to engineer the metabolic pathways in order to increase EPS production levels, perhaps by increasing the flux towards sugar nucleotides. The so-called "housekeeping" genes are involved in the synthesis of sugar nucleotides; in this respect, the physiology of sugar nucleotide synthesis should be investigated in relation to EPS production levels (Kleerebezem et al., 1999).

At present, all genes of *L. lactis* MG1363 encoding for enzymes involved in the biosynthesis of sugar nucleotides from glucose-1-phosphate have been cloned, and specific enzyme activities, which form bottlenecks in EPS synthesis, can now be investigated. The potential of these metabolic engineering strategies to improve production levels can also be evaluated (Boels et al., 1998). The role of the genes which encode for important enzymes in these metabolic pathways is now also being investigated in other LAB.

It is expected that engineering of EPS and their biosynthetic pathways in LAB will be an interesting challenge in future metabolic research. However, the use of genetically modified microorganisms or their products will require not only legal approval, but also acceptance by the consumer (De Vuyst and Degeest, 1999b; Kleerebezem et al., 2000).

9 Factors Influencing Growth and EPS Production

9.1 Effect of Physico-chemical Parameters

Generally, EPS are produced at temperatures below the optimal growth temperature. For mesophilic LAB, EPS production increased by 50–60% if the strains are cultured at 25°C instead of at 30°C (Cerning et al., 1992). *L. lactis* subsp. *lactis* exhibited the highest EPS production at 22°C; at temperatures higher than 37°C, no further EPS production occurred. The highest EPS production for *L. lactis* subsp. *cremoris* was noted at 20°C (Marshall et al., 1995), while for *Lb. casei* EPS are produced at 25°C as well as at 30°C (Cerning, 1994; Mozzi et al., 1996).

Mozzi et al. (1995) investigated the influence of the temperature on EPS production for the thermophilic species *Lb. delbreuckii* subsp. *bulgaricus, Lb. acidophilus* and *S. salivarius* subsp. *thermophilus*. In milk-based media without pH control, the optimum temperature for EPS production was 42°C for *Lb. acidophilus* and 37°C for *Lb. delbreuckii* subsp. *bulgaricus*, which coincides with optimal growth temperatures. Garcia-Garibay and Marshall (1991) found a maximal EPS production for *Lb. delbreuckii* subsp. *bulgaricus* at 45°C. Grobben et al. (1995) noted an increased EPS production with an increased temperature during batch fermentation, using a defined medium with controlled pH; the EPS production was maximal at 40°C.

For *S. salivarius* subsp. *thermophilus*, the highest EPS production was noted at 30°C. These results were confirmed by Gancel and Novel (1994) and Gassem et al. (1995).

As mentioned previously, EPS synthesis is expected to be stimulated by reduced growth temperatures; as such, cell growth is reduced as well as the synthesis of cell-wall polymers. This would result in a better availability of the lipid isoprenoid carrier which is necessary for export of EPS (Sutherland, 1982).

For some LAB, the effect of the initial pH of the growth medium on EPS biosynthesis was investigated; during this type of experiment the pH was no longer controlled during fermentations and was allowed to vary freely. On occasion, no clear effect was noted, but for *S. thermophilus* S22 an optimal initial pH of 7.0 for lactose as a carbon source and an optimal initial pH of 5.5 for sucrose was determined (Gancel and Novel, 1994).

A maximal EPS yield was obtained for *Lb. casei* CRL 87 when the pH was controlled at 6.0 during the fermentation; moreover, no decrease of EPS level occurred upon prolonged incubation (Mozzi et al., 1994).

In a continuous fermentation with *Lb. delbreuckii* subsp. *bulgaricus*, and using a whey-based medium, maximal EPS production occurred at pH 6.5, while the optimal pH for EPS production in batch fermentations was 5.8 (Van den Berg et al., 1995).

With *Lb. delbreuckii* subsp. *bulgaricus* CNRZ 1187 and CNRZ 416, about three-fold and four-fold respectively more EPS was produced when the pH was controlled at 6.0 (Petry et al., 2000).

In general, higher EPS yields are obtained when fermentations are carried out under controlled pH conditions. The optimal pH control range for EPS production by LAB is often close to pH 6–6.5 (Mozzi et al., 1994), and this may be a limiting factor when considering industrial use of these bacteria in fermented milk drinks, as the pH is allowed to vary freely during such fermentations (De Vuyst and Degeest, 1999a).

With regard to temperature, EPS production appears to increase under less than optimal pH conditions for growth, though this might also be explained by a better availability of the lipid carriers (Ricciardi and Clementi, 2000).

According to Cerning (1990), EPS biosynthesis by LAB is highest under aerobic growth conditions, and this explains why more EPS is excreted during growth on solid media rather than in liquid cultures. However, another reason might be that adhesion of the bacterial cells onto a solid surface stimulates EPS formation.

Mäkelä and Korkeala (1992) investigated the effect of aeration, temperature and medium composition for different LAB. Aerobic or anaerobic conditions resulted in better EPS production, depending on the strain.

9.2
Effect of Nutritional Parameters

In general, milk-based media or other complex media containing peptones or yeast extract are used for culturing EPS-producing

LAB. Higher EPS yields are indeed obtained, but isolation and purification of the polymers are hampered. Recently, some chemically defined media were developed which greatly facilitate the study of the influence of nutrients on EPS production, growth, chemical composition, and rheological properties.

Cerning et al. (1986) found that the addition of casein enhanced growth, but not EPS synthesis, although contrasting results were subsequently reported which indicated a positive effect on EPS production by *Lb. delbrueckii* subsp. *bulgaricus* following the addition of hydrolyzed casein (Garcia-Garibay and Marshall, 1991; Cerning, 1994).

Another point identified was that the carbon source influences not only the composition of the EPS produced by LAB but also the total amount produced. *Lb. casei* CG11 produced more EPS when glucose was added to milk or milk ultrafiltrate; moreover, the monomeric composition changed when glucose was used as the major monosaccharide rather than galactose (Cerning et al., 1992). In a later study, Mozzi et al. (1995) obtained the highest EPS yield for *Lb. casei* CRL 87 when using galactose as the carbon source.

Considerably more EPS was synthesized in a chemically defined medium by *Lb. delbreuckii* subsp. *bulgaricus* NCFB 2772 when grown on glucose, lactose, or glucose and fructose rather than on mannose or fructose. A continuous culture was used to study the effect of glucose and/or fructose on EPS production and composition. When grown on glucose or glucose and fructose, the EPS contained rhamnose, while in the presence of fructose alone, no rhamnose was present in the EPS and the overall EPS yield remained low. It was found that in this latter case, the enzymes involved in the synthesis of rhamnose were absent; moreover, when glucose was present in the culture medium, higher activities were determined for various enzymes involved in EPS synthesis (Grobben et al., 1995, 1996).

Later, it was shown that this strain produces two types of EPS: a high-molecular fraction, which appears to be influenced by the carbon source; and a low-molecular fraction, which is produced more consistently (Grobben et al., 1997).

For another *Lb. delbreuckii* subsp. *bulgaricus* strain, the composition of the EPS was found to be independent of the nature of the carbon source (Manca de Nadra et al., 1985). No influence of the carbon source on the amount of EPS produced by *Lb. sake* 01 was noted by Van den Berg et al. (1995).

Petry et al. (2000) designed a chemically defined medium that allowed good growth and EPS production for several *Lb. delbreuckii* subsp. *bulgaricus* strains. The presence of excess sugar in the growth medium (10–20 g L^{-1}) did not stimulate EPS production, as was proven for *Lb. casei* and *Lb. rhamnosus*. Independently of the carbon source, the monomeric composition remained similar.

In the case of *S. thermophilus* LY03, a clear influence of the type of carbon source on the amount of EPS produced was noted, while the composition remained constant. This strain produces two polysaccharides with identical composition, but different molecular weights. The proportion in which these polysaccharides are synthesized is strongly dependent on the C/N ratio of the culture medium (De Vuyst and Degeest, 1999b).

Other nutrients such as vitamins, amino acids and minerals might also affect the composition and quantity of EPS produced. *Lb. delbreuckii* subsp. *bulgaricus* NCFB 2722 required only riboflavin, nicotinic acid and calcium pantothenate for EPS synthesis. The absence of glutamine, asparagine and threonine reduced growth considerably (Grobben et al., 1997).

Clearly, contrasting results have often been obtained, and so it is advisable to evaluate separately the influence of physicochemical and nutritional parameters on EPS synthesis for each strain. Moreover, additional studies on the effect of medium composition and of culture conditions on the amount and composition of EPS might result in a better understanding of the biosynthesis regulatory mechanisms.

9.3 Kinetics of EPS Biosynthesis

In general, mesophilic LAB produce greater amounts of EPS under conditions not optimal for growth, and so EPS production appears to be nongrowth-associated. In contrast, in thermophilic LAB, EPS production appears to be growth-associated (Andaloussi et al., 1995; Grobben et al., 1995; Manca de Nadra et al., 1985).

Growth-associated EPS production was noted for *S. thermophilus* LY03 (De Vuyst and Degeest, 1999a) and for *Lb. delbreuckii* subsp. *bulgaricus* NCFB 2772 (Grobben et al., 1995). However, Gancel and Novel (1994) reported that the EPS production of a *S. thermophilus* strain was not growth-associated, and Petry et al. (2000) noted that EPS production occurred mainly during the stationary growth phase for a *Lb. delbreuckii* subsp. *bulgaricus* strain.

For some mesophilic LAB, a growth-associated EPS production was observed, this being the case for *Lb. sake* 01 and a *Lb. rhamnosus* strain (Van den Berg et al., 1995). Other investigations observed a continued EPS synthesis during the stationary growth phase (Manca de Nadra et al., 1985; Gancel and Novel 1994).

Often, the amount of EPS produced declines upon prolonged fermentation. Several explanations were proposed, such as an enzymatic degradation, or a change in the physical parameters of the culture. Moreover, this trend is not rare with regard to some other microbial EPS types such as gellan (Kennedy and Sutherland, 1994) or sphingan (Hashimoto and Murata, 1998).

Pham et al. (2000) investigated this phenomenon for a *Lb. rhamnosus* strain and identified the presence of a large spectrum of glycohydrolases, three of them extracellular and the others either cell-wall bound or intracellular. When incubated with EPS, the enzymes were capable of liberating reducing sugars and of lowering the viscosity; indeed, a viscosity loss of 33% was noted after 27 h.

Gassem et al. (1995) observed a reduction in viscosity of the culture liquid after 18 h of fermentation with *S. salivarius* subsp. *thermophilus* ST3. Cerning (1994) noted similar results for *S. salivarius* subsp. *thermophilus,* although EPS degradation was also influenced by pH and temperature.

Various authors have reported this phenomenon, but few have investigated it further; as such, the complete elucidation of this degradation mechanism requires further study.

10 Applications

One way to improve the rheological properties and functional characteristics of dairy products, such as yogurt, is the addition of stabilizers. Normally, hydrocolloids of animal or plant origin are applied, but their use is prohibited in several countries, including France and the Netherlands. EPS from LAB, which are GRAS bacteria, may serve as alternative biothickeners (Grobben et al., 1995). In order to be suitable, the polymers should be compatible with other food components and with applied processing conditions, and should exhibit thixotropic or pseudoplastic behavior, notably when considering aspects such as processing costs, mouthfeel and texture (Van den Berg et al., 1995).

Since the amount of EPS produced by LAB is rather limited, these bacteria are mostly used to produce EPS *in situ*, to improve consistency and viscosity as well as to avoid syneresis (whey separation) in yogurts. Also, for the production of other fermented milk products and cheeses, EPS-producing LAB are used in starter cultures (Cerning, 1995; Marshall and Rawson, 1999; Ricciardi and Clementi, 2000).

Over the past 40 years, yogurt production has evolved substantially. One variation, which is now dominating the market, is stirred yogurt in various forms, obtained by further processing of set yogurt. Due to different mechanical processing steps, stirred yogurt tends to lose some of its initial viscosity. This might be solved by using ropy LAB as starter cultures, because yogurt containing viscosifying EPS is less susceptible to damage by mechanical pumping, blending and filling (Cerning, 1990; Rawson and Marshall, 1997).

Based on rheological studies, it was found that no clear correlation existed between the EPS concentration and apparent viscosity. The effect of EPS appeared to be much more complex, however. The contribution of EPS to specific food properties depends not only on the properties of the EPS itself, such as molar mass and structural composition, but also on the interaction of the EPS with various milk constituents, and in particular caseins (Kleerebezem et al., 1999; Marshall and Rawson, 1999).

Yogurt consists of a network of casein micelles, associated to form a gel-like structure within which small spaces are filled with liquid whey phase, while larger spaces contain the EPS and the LAB. The typical viscoelastic property of yogurt is obtained by retention of the whey. By using scanning electron microscopy, EPS strands were observed between the cells and the protein network; hence, besides the protein strands and protein–protein bonds, protein–polysaccharide bonds also occur (Wacher-Rodarte et al., 1993; Ricciardi and Clementi, 2000).

The use of ropy strains results in a yogurt which is more stable towards shear forces. Due to EPS synthesis, protein strand formation and protein–protein bond formation is partly prevented, resulting in less rigid gels with an increased viscosity, an ability to recover lost viscosity, and a certain adhesiveness. Although these characteristics are strain-dependent, it appears likely that EPS synthesis contributes to adhesiveness while the protein network has a greater influence on firmness and elasticity (Rawson and Marshall, 1997).

The thermophilic *S. thermophilus* and *Lb. delbreuckii* subsp. *bulgaricus* are traditionally used in yogurt and are both able to form EPS. Ropy, mesophilic LAB are also widely used in starter cultures, but they produce less EPS and their ropiness is often an unstable characteristic (Wacher-Rodarte et al., 1993).

In Scandinavian countries, slime-producing mesophilic LAB are used in the manufacture of different fermented milks. In Finnish "viili" and Nordic "longfil", a ropy strain of *L. lactis* subsp. *cremoris* is essential for correct consistency (Forsén and Häivä, 1981; Kontusaari and Forsén, 1988; Toba et al., 1990).

Kefir is an acidic, alcoholic beverage that is popular in Eastern Europe. The EPS produced by *Lb. kefiranofaciens* comprises the mass of the kefir grain, and was named kefiran. Other microorganisms are also associated with kefir, such as *Lb. acidophilus, Lb. kefir granum, Lb. kefir, Lb. parkefir* and the yeast *Candida kefir* (Yokoi et al., 1990).

The capacity of EPS to increase water retention was used to improve the functional properties of cheese. Perry et al. (1988) produced low-fat mozzarella with a significantly higher moisture content and a better stretching ability, using capsulated nonropy cultures of *S. thermophilus* and *Lb. delbreuckii* subsp. *bulgaricus.* Eventually, Low et al. (1998) showed that *S. thermophilus* was mainly responsible for the increased moisture content. Scanning electron microscopy was used to elucidate the positive effect on the microstructure of the cheese, and revealed that bridges were formed between the EPS filaments and the casein fibers, thus preventing collapse of the whey channels.

EPS produced by LAB may also contribute to human health because of their prebiotic (Gibson and Robertfroid, 1995), cholesterol-lowering (Nakajima et al., 1992), immunomodulating (Kitazawa et al., 1993; Hosono et al., 1997) or anti-tumor (Oda et al., 1983; Toba et al., 1991) activities.

Other health-promoting activities can be attributed to the LAB themselves, for example the removal of lactose in lactose-intolerant persons. While the pH permits metabolic activity of the LAB, lactose is utilized; however, by re-routing the sugar metabolism in *L. lactis* towards L-alanine rather than towards lactic acid, complete conversion of lactose was obtained. Moreover, the sweetness and general taste of the yogurt were enhanced (Kleerebezem et al., 2000).

Although LAB are known to be vitamin-auxotrophic bacteria, *S. thermophilus* is able to produce folic acid, an essential compound in human nutrition, and which is further metabolized by another yogurt bacterium, *Lb. delbreuckii* subsp. *bulgaricus.* Applying high folic acid-producing strains or relatively high numbers of *S. thermophilus* should lead to the production of yogurt with an increased folic acid content (Kleerebezem et al., 2000).

During the fermentation of dairy products, LAB are responsible for the synthesis of typical flavor compounds such as acetaldehyde in yogurt and diacetyl in buttermilk (Kleerebezem et al., 2000). Moreover, LAB also produce antimicrobial peptides ("bacteriocins") which have an antagonistic effect against genetically closely related strains; they are mostly thermostable, digestible and active in low concentrations. The antimicrobial spectrum of most bacteriocins is restricted to Gram-positive bacteria, but bacteria responsible for food spoilage, such as *Bacillus cereus, Clostridium perfringens, Listeria monocytogenes* and *Staphylococcus aureus* are often included. Consequently, bacteriocins may serve as a natural food preservative (De Vuyst and Vandamme, 1994).

It is clear that the metabolic activity of LAB results in "totally natural" products with a clear added value, and which have not only the desired nutritional properties but also contribute to human health in general. However, most bacterial activities may still be improved as they do not reach maximal functionality during *in situ* milk fermentation.

11
References

Andaloussi, S. A., Talbaoui, H., Marczak, R., Bonaly, R. (1995) Isolation and characterization of exocellular polysaccharides produced by *Bifidobacterium longum*, *Appl. Microbiol. Biotechnol.* **43**, 995.

Ariga, H., Urashima, T., Michihata, E., Ito, M., Morizono, N., Kimura, T., Takahashi, S. (1992) Extracellular polysaccharide from encapsulated *Streptococcus salivarius* subsp. *thermophilus* OR 901 isolated from commercial yogurt, *J. Food Sci.* **57**, 625–628.

Boels, I. C., Kleerebezem, M., Hugenholtz, J., de Vos, W. M. (1998) Metabolic engineering of exopolysaccharide production in *Lactococcus lactis*, in: Proceedings, 5th ASM on the genetics and molecular biology of streptococci, enterococci and lactococci, 66.

Bolotin, A., Mauger, S., Malarme, K., Sorokin, A., Ehrlich, D. S. (1998) *Lactococcus lactis* IL1403 diagnostic genomics, in: Proceedings, 5th ASM on the genetics and molecular biology of streptococci, enterococci and lactococci, 10–11.

Bouwman, S., Lund, D., Driessen, F. M., Schmidt, D. G. (1982) Growth of thermoresistant streptococci and deposition of milk constituents on plates of heat exchangers during long operating times, *Food Prot.* **45**, 806–812.

Cerning, J. (1990) Exocellular polysaccharides produced by lactic acid bacteria, *FEMS Microbiol. Rev.* **87**, 113–130.

Cerning, J. (1994) Polysaccharides exocellulaires produits par les bactéries lactiques, in: *Bactéries Lactiques: Aspects fondamentaux et technologiques* (de Roissart, H., Luquet, F. M., Eds), Uriage: Lorica, 309–329.

Cerning, J. (1995) Production of exopolysaccharides by lactic acid bacteria and dairy propionibacteria, *Le Lait* **75**, 463–472.

Cerning, J., Bouillanne, C., Desmazeaud, M. J. (1986) Isolation and characterization of exocellular polysaccharide produced by *Lactobacillus bulgaricus*, *Biotechnol. Lett.* **8**, 625–628.

Cerning, J., Bouillanne, C., Landon, M., Desmazeaud, M. J. (1990) Comparison of exocellular polysaccharide production by thermophilic lactic acid bacteria, *Sciences des Aliments* **10**, 443–451.

Dellaglio, F. (1994) Caractéristiques générales des bactéries lactiques, in: *Bactéries Lactiques: Aspects fondamentaux et technologiques* (de Roissart, H., Luquet, F. M., Eds), Uriage: Lorica, 10–58.

De Vos, W. M. (1996) Metabolic engineering of sugar catabolism in lactic acid bacteria, *Antonie van Leeuwenhoek* **70**, 223–242.

De Vos, W. M., Vaughan, E. E. (1994) Genetics of lactose utilization in lactic acid bacteria, *FEMS Microbiol. Rev.* **15**, 216–237.

De Vuyst, L., Degeest, B. (1999a) Exopolysaccharides from lactic acid bacteria: technological bottlenecks and practical solutions, *Macromol. Symp.* **140**, 31–41.

De Vuyst, L., Degeest, B. (1999b) Heteropolysaccharides from lactic acid bacteria, *FEMS Microbiol. Rev.* **23**, 153–177.

De Vuyst, L., Vandamme, E. J. (1994) Lactic acid bacteria and bacteriocins: their practical importance, in: *Bacteriocins of lactic acid bacteria, microbiology, genetics and applications* (De Vuyst, L., Vandamme, E. J., Eds), London: Blackie Academic and Professional, 1–11.

Doco, T., Carcano, D., Ramos, P., Loones, A., Fournet, B. (1991) Rapid isolation and estimation of polysaccharide from fermented skim milk with *Streptococcus salivarius* subsp. *thermophilus* by coupled anion exchange and gel permeation high-performance liquid chromatography, *J. Dairy Res.* **58**, 147–150.

Faber, E. J., van den Haak, M. J., Kamerling, J. P., Vliegenthart, J. F. G. (2001) Structure of the exopolysaccharide produced by *Streptococcus thermophilus* S3, *Carbohydr. Res.* **331**, 173–182.

Forsén, R., Häivä, V. (1981) Induction of stable slime-forming and mucoid states by p-fluorophenylalanine in lactic streptococci, *FEMS Microbiol. Lett.* **12**, 409–413.

Gancel, F., Novel, G. (1994) Exopolysaccharide production by *Streptococcus salivarius* subsp. *thermophilus* cultures: two distinct modes of polymer production and degradation among clonal variants, *J. Appl. Bacteriol.* **77**, 689–695.

Garcia-Garibay, M., Marshall, V. M. (1991) Polymer production by *Lactobacillus delbreuckii* ssp. *bulgaricus*, *J. Appl. Bacteriol.* **70**, 325–328.

Gassem, M. A., Schmidt, K. A., Frank, J. F. (1995) Exopolysaccharide production in different media by lactic acid bacteria, *Cult. Dairy Prod. J.* **30**, 18–21.

Gibson, G. R., Roberfroid, M. D. (1995) Dietary modulation of the human colonic microbiota: introducing the concept of prebiotics, *J. Nutr.* **124**, 1401–1412.

Gilbert, C., Robinson, K., LePage, R. W. F., Wells, J. M. (1998) Heterologous biosynthesis of pneumococcal type 3 capsule in *Lactococcus lactis* and immunogenicity studies, in: Proceedings, 5th ASM on the genetics and molecular biology of streptococci, enterococci and lactococci, 67.

Grobben, G. J., Sikkema, J., Smith, M. R., De Bont, J. A. M. (1995) Production of extracellular polysaccharides by *Lactobacillus delbreuckii* ssp. *bulgaricus* NCFB 2772 grown in a chemically defined medium, *J. Appl. Bacteriol.* **79**, 103–107.

Grobben, G. J., Smith, M. R., Sikkema, J., De Bont, J. A. M. (1996) Influence of fructose and glucose on the production of exopolysaccharides and the activities of enzymes involved in the sugar metabolism and the synthesis of sugar nucleotides in *Lactobacillus delbreuckii* subsp. *bulgaricus* NCFB 2772, *Appl. Microbiol. Biotechnol.* **46**, 279–284.

Grobben, G. J., Van Casteren, W. H. M., Schols, H. A., Oosterveld, A., Sala, G., Smith, M. R., Sikkema, J., De Bont, J. A. M. (1997) Analysis of the exopolysaccharides produced by *Lactobacillus delbreuckii* subsp. *bulgaricus* NCFB 2772 grown in continuous culture on glucose and fructose, *Appl. Microbiol. Biotechnol.* **48**, 516–521.

Groux, M. (1973) Kritische betrachtungen der heutigen joghurt-herstellung mit berücksichtigung des proteinabbaues, *Schweiz. Milchz.* **4**, 2–8.

Hashimoto, W., Murata, K. (1998) α-L-Rhamnosidase of *Sphingomonas* sp. R1 producing an unusual exopolysaccharide of sphingan, *Biosci. Biotechnol. Biochem.* **62**, 1068–1074.

Hosono, A., Lee, J., Ametani, A., Natsume, M., Adachi, T., Kaminogawa, S. (1997) Characterization of a water-soluble polysaccharide fraction with immunopotentiating activity from *Bifidobacterium adolescentis* M101-4, *Biosci. Biotechnol. Biochem.* **61**, 312–316.

Jann, K., Jann, B. (1977) Bacterial polysaccharide antigens, in: *Surface Carbohydrates of the Prokaryotic Cell* (Sutherland, I. W., Ed.), New York: Academic Press, 247–287.

Kandler, O., Kunath, P. (1983) *Lactobacillus kefir* sp. nov., a component of the microflora of kefir, *Syst. Appl. Microbiol.* **4**, 286–294.

Kennedy, L., Sutherland, I. W. (1994) Gellan lyases – novel polysaccharide lyase, *Microbiology* **140**, 3007–3013.

Kitazawa, H., Yamaguchi, T., Miura, M., Saito, T., Itoh, H. (1993) B-cell mitogen produced by slime-forming, encapsulated *Lactococcus lactis* subspecies *cremoris* isolated from ropy sour milk, viili, *J. Dairy Sci.* **76**, 1514–1519.

Kleerebezem, M., van Kranenburg, R., Tuinier, R., Boels, I. C., Zoon, P., Looijesteijn, E., Hugenholtz, J., de Vos, W. M. (1999) Exopolysaccharides produced by *Lactococcus lactis*: from genetic engineering to improved rheological properties, *Antonie van Leeuwenhoek* **76**, 657–665.

Kleerebezem, M., Hols, P., Hugenholtz, J. (2000) Lactic acid bacteria as a cell factory: rerouting of carbon metabolism in *Lactococcus lactis* by metabolic engineering, *Enzyme Microb. Technol.* **26**, 840–848.

Kojic, M., Vujcic, M., Banina, A., Cocconcelli, P., Cerning, J., Topisirovic, L. (1992) Analysis of exopolysaccharide production by *Lactobacillus casei* CG11, isolated from cheese, *Appl. Environ. Microbiol.* **58**, 4086–4088.

Kontusaari, S., Forsèn, R. (1988) Finnish fermented milk "Viili": involvement of two cell surface proteins in production of slime by *Streptococcus lactis* spp. *cremoris*, *J. Dairy Sci.* **71**, 3197–3202.

Korkeala, H., Suortti, T., Mäkelä, P. (1988) Ropy slime formation in vacuum-packed cooked meat products caused by homofermentative *Lactobacilli* and a *Leuconostoc* species. *Int. J. Food Microbiol.* **7**, 339–347.

Low, D., Ahlgren, J. A., Horne, D., McMahon, D. J., Oberg, C., Broadbent, J. R. (1998) Role of *Streptococcus thermophilus* MR-1C capsular exopolysaccharide in cheese moisture retention, *Appl. Environ. Microbiol.* **64**, 2140–2147.

Mäkelä, P. M., Korkeala, H. J. (1992) The ability of the ropy slime-producing lactic acid bacteria to form ropy colonies on different culture media and at different incubation temperatures and atmospheres, *Int. J. Food Microbiol.* **16**, 161–166.

Manca de Nadra, M. C., Strasser de Saad, A. M., Pesce de Ruiz Holgado, A. A., Oliver, G. (1985) Extracellular polysaccharide production by *Lb. bulgaricus* CRL 420, *Milchwissenschaft* **40**, 409–411.

Marshall, V. M., Rawson, H. L. (1999) Effects of exopolysaccharide-producing strains of thermophilic lactic acid bacteria on the texture of stirred yoghurt, *Int. J. Food Sci. Technol.* **34**, 137–143.

Marshall, V. M., Cowie, E. N., Moreton, R. S. (1995) Analysis and production of two exopolysaccharides from *Lactobacillus casei*, *J. Dairy Res.* **62**, 621–628.

Marshall, V. M., Dunn, H., Elvin, M., McLay, N., Gu, Y., Laws, A. P. (2001) Structural characterisation of the exopolysaccharide produced by *Streptococcus thermophilus* EU20, in: Proceedings, 1st International Symposium on Exopolysaccharides from Lactic acid bacteria. Brussels, Belgium: VUB, 12.

Morin, A. (1998) Screening for polysaccharide-producing microorganisms, factors influencing the production, and recovery of microbial polysaccharides, in: *Polysaccharides: Structural Diversity and Functional Versatility* (Dumitriu, S., Ed.), New York: Marcel Dekker, Inc., vol. 8, 275–296.

Mozzi, F., de Giori, G. S., Oliver, G., de Valdez, G. F. (1994) Effect of culture pH on the growth characteristics and polysaccharide production by *Lactobacillus casei*, *Milchwissenschaft* **49**, 667–670.

Mozzi, F., Oliver, G., de Giori, G. S., de Valdez, G. F. (1995) Influence of temperature on the production of exopolysaccharides by thermophilic lactic acid bacteria, *Milchwissenschaft* **50**, 80–82.

Mozzi, F., de Giori, G. S., Oliver, G., de Valdez, G. F. (1996) Exopolysaccharide production by *Lactobacillus casei* under controlled pH, *Biotechnol. Lett.* **18**, 435–439.

Nakajima, H., Toyoda, S. (1990) A novel phosphopolysaccharide from slime-forming *Lactococcus lactis* subspecies *cremoris* SBT 0495, *J. Dairy Sci.* **73**, 1472–1477.

Nakajima, H., Suzuki, Y., Kaizu, H., Hirota, T. (1992) Cholesterol-lowering activity of ropy fermented milk, *J. Food Sci.* **57**, 1327–1329.

Neu, T. R., Marshall, K. C. (1990) Bacterial polymers: physicochemical aspects of their interactions at interfaces, *J. Biomater. Appl.* **5**, 107–133.

Neve, H., Geis, A., Teuber, M. (1988) Plasmid-encoded functions of ropy lactic acid streptococcal strains from Scandinavian fermented milk, *Biochimie* **70**, 437–442.

Nilsson, R., Nilsson, G. (1958) Studies concerning Swedish ropy milk, the antibiotic qualities of ropy milk, *Arch. Microbiol.* **31**, 191–197.

Oda, M., Hasegawa, H., Komatsu, S., Kambe, M., Tsuchiya, F. (1983) Anti-tumor polysaccharide from *Lactobacillus* sp., *Agr. Biol. Chem.* **47**, 1623–1625.

Perry, D. B., McMahon, D. J., Oberg, C. J. (1998) Manufacture of low fat mozzarella cheese using exopolysaccharide-producing starter cultures, *J. Dairy Sci.* **81**, 561–563.

Petry, S., Furlan, S., Crepeau, M. J., Cerning, J., Desmazeaud, M. (2000) Factors affecting exocellular polysaccharide production by *Lactobacillus delbreuckii* subsp. *bulgaricus* grown in a chemically defined medium, *Appl. Environ. Microbiol.* **66**, 3427–3431.

Pham, P. L., Dupont, I., Roy, D., Lapointe, G., Cerning, J. (2000) Production of exopolysaccharide by *Lactobacillus rhamnosus* R and analysis of its enzymatic degradation during prolonged fermentation, *Appl. Environ. Microbiol.* **66**, 2302–2310.

Rawson, H. L., Marshall, V. M. (1997) Effect of ropy strains of *Lactobacillus delbreuckii* ssp. *bulgaricus* and *Streptococcus thermophilus* on rheology of stirred yogurt, *Int. J. Food Sci. Technol.* **32**, 213–220.

Ricciardi, A., Clementi, F. (2000) Exopolysaccharides from lactic acid bacteria: structure, production and technological applications, *Ital. J. Food Sci.* **12**, 23–45.

Robijn, G. W., Wienk, H. L. J., Van den Berg, D. J. C., Haas, H., Kamerling, J. P., Vliegenthart, J. F. G. (1996) Structural studies of the exopolysaccharide produced by *Lactobacillus paracasei* 34-1, *Carbohydr. Res.* **285**, 129–139.

Smitinont, T., Tansakul, C., Tanasupawat, S., Keeratipibul, S., Navarini, I., Bosco, M., Cescutti, P. (1999) Exopolysaccharide-producing lactic acid bacteria strains from traditional Thai fermented foods: isolation, identification and exopolysaccharide characterization, *Int. J. Food Microbiol.* **51**, 105–111.

Staaf, M., Yang, Z., Huttunen, E., Widmalm, G. (2000) Structural elucidation of the viscous exopolysaccharide produced by *Lactobacillus helveticus* Lb161, *Carbohydr. Res.* **326**, 113–119

Stingele, F. (1998) Exopolysaccharide production and engineering in dairy *Streptococcus* and *Lactococcus*, in: Proceedings, 5th ASM on the genetics and molecular biology of streptococci, enterococci and lactococci, 31–32.

Stingele, F., Neeser, J. R., Mollet, B. (1996) Identification and characterization of the eps gene cluster from *Streptococcus thermophilus* Sfi6, *J. Bacteriol.* **178**, 1680–1690.

Sundman, V. (1953) On the protein character of a slime produced by *Streptococcus cremoris* in Finnish ropy sour milk, *Acta Chem. Scand.* **7**, 558–560.

Sutherland, I. W. (1972) Bacterial exopolysaccharides, *Adv. Microbiol. Physiol.* **8**, 143.

Sutherland, I. W. (1982) Biosynthesis of microbial exopolysaccharides, in: *Advances in Microbial Physiology* (Rose, A.H., Ed.), London: Academic Press, Vol. 33, 78–150.

Sutherland, I. W. (1990) *Biotechnology of Microbial Exopolysaccharides.* Cambridge, Sydney: Cambridge University Press, 163.

Sutherland, I. W. (1994) Structure–function relationships in microbial exopolysaccharides, *Biotech. Adv.* **12**, 393–448.

Tamime, A. Y. (1978) Some aspects of the production of a concentrated yoghurt (labneh) popular in the Middle East, *Milchwissenschaft* **33**, 209–212.

Toba, T., Nakajima, H., Tobitani, A., Adachi, S. (1990) Scanning electron microscopic and texture studies on characteristic consistency of Nordic ropy sour milk, *Int. J. Food Microbiol.* **11**, 313–320.

Toba, T., Kotani, T., Adachi, S. (1991) Capsular polysaccharide of a slime-forming *Lactococcus lactis* ssp. *cremoris* LAPT 3001 isolated from Swedish fermented milk longfil, *Int. J. Food Microbiol.* **12**, 167–172.

Troy, F. A. (1979) The chemistry and biosynthesis of selected bacterial polymers, *Annu. Rev. Microbiol.* **33**, 519–560.

Troy, F. A., Frerman, F. A., Heath, E. C. (1971) Biosynthesis of capsular polysaccharides in *Aerobacter aerogenes, J. Biol. Chem.* **243**, 118–133.

Van Calsteren, M. (2001) Structural characterisation of the exopolysaccharide produced by different strains of *Lactobacillus rhamnosus* and *Lactobacillus paracasei*, in: Proceedings, 1st International Symposium on Exopolysaccharides from Lactic acid bacteria, 13.

Van den Berg, D. J. C., Smits, A., Pot, B., Ledeboer, A. M., Kersters, K., Verbrakel, J. M. A., Verrips, C. T. (1993) Isolation, screening and identification of lactic acid bacteria from traditional food fermentation processes and culture collections, *Food Biotechnol.* **7**, 189–205.

Van den Berg, D. J. C., Robijn, G. W., Janssen, A. C., Giusepin, M. L. F., Vreeker, R., Kamerling, J. P., Vliegenthart, J. F. G., Ledeboer, A. M., Verrips, C. T. (1995) Production of a novel extracellular polysaccharide by *Lactobacillus sake* 0-1 and characterization of the polysaccharide, *Appl. Environ. Microbiol.* **61**, 2840–2844.

Van Kranenburg, R., Vos, H. R., Van Swam, I., Kleerebezem, M., De Vos, W. M. (1999) Functional analysis of glycosyltransferase genes from *Lactococcus* and other Gram-positive cocci: complementation, expression and diversity, *J. Bacteriol.* **181**, 6347–6353.

Vedamuthu, E. R., Neville, J. M. (1986) Involvement of a plasmid in production of ropiness in milk cultures by *Streptococcus cremoris, Appl. Environ. Microbiol.* **61**, 2840–2844.

Vliegenthart, J. F. G., Faber, E. J., Kamerling, J. P. (2001) Studies on structure of exopolysaccharides from lactic acid bacteria, in: Proceedings, 1st International Symposium on Exopolysaccharides from Lactic acid bacteria. Brussels, Belgium: VUB, 11.

Wacher-Rodarte, C., Galvan, M., Farres, A., Gallardo, F., Marshall, V. M. E., Garcia-Garibay, M. (1993) Yogurt production from reconstituted skim milk powders using different polymer and non-polymer forming starter cultures, *J. Dairy Res.* **60**, 247–254.

Yokoi, H., Watanabe, T., Fuji, Y., Toba, T., Adachi, S. (1990) Isolation and characterization of polysaccharide-producing bacteria from kefir grains, *J. Dairy Sci.* **73**, 1684–1689.

17
Murein (Peptidoglycan)

Dr. Christoph Heidrich[1], Dr. Waldemar Vollmer[2]
[1] Landesunfallkasse NRW, Ulenbergstrasse 1, Abteilung Prävention, D-40223 Düsseldorf, Germany; Tel.: +49-0211-9024 302; Fax: +49-0211-9024 480; E-mail: CHeidrich@luk-nrw.de
[2] Max-Planck-Institut für Entwicklungsbiologie, Abteilung Biochemie, Spemannstraße 35/II, D-72076 Tübingen; Phone: +49-7071-601 466; Fax: +49-7071-601 447; e-mail: waldemar.vollmer@tuebingen.mpg.de

m-A_2pm	m-2,6-diaminopimelic acid
L-Dab	L-diaminobutyric acid
DP	degree of polymerization
GlcNAc	N-acetylglucosamine
Gro	glycerol
HPLC	high-pressure lipid chromatography
L-Hsr	L-homoserine
m-HyA_2pm	m-2,6-diamino-3-hydroxypimelic acid
3-Hyg	threo-3-hydroxyglutamic acid
L-HyLys	L-hydroxylysine
IR	infrared
Mlt	membrane-bound lytic transglycosylase
MurNAc	*N*-acetylmuramic acid
MurNAcAnh	1,6-Anhydro-N-acetylmuramic acid
NADPH	nicotinamide adenine dinucleotide phosphate, reduced form
NMR	nuclear magnetic resonance
L-Orn	L-ornithine
PBP	Penicillin-binding protein
(p)ppGpp	guanosine 5′-di(tri)phosphate-3′-diphosphate
SDS	sodium dodecylsulfate
Slt	soluble lytic transglycosylase
UDP	uridine diphosphate

1 Introduction

The bacterial murein (peptidoglycan) sacculus is a unique biopolymer that completely encloses the cytoplasm, thereby forming a bag-shaped structure. As part of the bacterial envelope, this stress-bearing exoskeleton is essential for maintaining the shape of the bacterium. Murein is found in all eubacteria with the exception of L-forms, and of Mycoplasma. Archaebacteria contain the analogous pseudomurein.

The murein of Gram-negative bacteria such as *Escherichia coli* is extremely thin, and forms a single layer. It is localized in the periplasmic space between the cytoplasmic membrane and the outer membrane to which it is connected via Braun's lipoprotein. In contrast, Gram-positive bacteria such as *Staphylococcus aureus* have no outer membrane but instead have a thick (20–50 nm), multilayered murein shell.

Although there is variation in the chemical composition of the murein of different bacterial species, the principal architecture is conserved, and murein is composed of glycan strands that are cross-linked by short peptides. The enzymatically released murein building blocks (the muropeptides) of various bacteria have been separated by high-pressure liquid chromatography (HPLC) and analyzed by enzymatic and chemical methods and by mass spectrometry. Murein is a complex heteropolymer composed of a few major building blocks and many components present in lower amounts. The murein composition often varies with the growth conditions and physiological state of the bacterium.

The biochemical pathway of murein synthesis has been elucidated, and most of the enzymes involved in the formation of precursors, in the polymerization of the murein glycan strands and the cross-linking of the peptides (the murein synthases), and in the hydrolysis of murein (murein hydrolases or autolysins) have been characterized and their corresponding genes identified in *E. coli* and other bacteria. Enlargement of the murein sacculus during growth of the cell requires both murein synthases and murein hydrolases. The present research aims to understand the mechanisms of the incorporation of new murein into the existing stress-bearing sacculus that guarantees maintenance of the specific shape of the bacterium.

Because murein is an essential bacterial compound which is not found in eucaryotic cells, its unique structure as well as its biosynthetic enzymes represent excellent targets for components of the antibacterial host defense, and also for antibiotics. Higher organisms such as fungi, plants, animals and humans produce enzymes (e.g., lysozyme) that can hydrolyze the murein of invading bacteria. However, some pathogens modify their murein structure, making it nonsusceptible to certain murein hydrolases. Glycopeptides and β-lactams which inhibit the final steps of murein biosynthesis have been among the most widely used antibiotics during the past few decades. However, today an increasing number of isolates of pathogenic bacteria are found to be resistant to these antibiotics, and consequently efforts are being made to identify novel, alternative targets for antibacterial therapy in murein metabolism.

In this chapter, the current knowledge on murein structure and biosynthesis is reviewed, and a model is described for elongation of the sacculus during growth. Attention is also paid to the mechanism of antibiotic inhibition of murein synthesis, following which the occurrence of ubiquitous murein hydrolases and the biological activities of murein and its degradation products are described.

2 Historical Outline

The field of bacterial cell wall research may be traced back to about 1675, when bacteria were discovered by Antonie van Leeuwenhoek, and the first descriptions were made of differences in morphology between the various organisms. At this time, the question was raised as to what "held bacteria together?". Almost 300 years later, Weibull first isolated *Bacillus megaterium* protoplasts by removing the cell wall with lysozyme in an osmotically stabilizing medium, and showed that the cell wall enclosed the bacterial cell with a fairly rigid structure, thereby conferring a characteristic shape as well as providing protection against osmotic lysis (see Salton, 1994). In the same year, the surface structure of *Bacillus cereus* first became visible due to the superior resolution of the newly introduced electron microscope (Chapman and Hillier, 1953). Microscopy, together with new staining techniques, allowed visualization of the differences in cell wall organization of the two types of bacteria that could be distinguished by using Gram's staining procedure that was developed as early as 1884. Gram-positive bacteria were shown to have a thick murein (peptidoglycan) outside the cytoplasmic membrane, whilst the cell membrane of Gram-negative bacteria consisted of a thin murein layer localized in the periplasmic space and embedded between the inner and outer membranes (Murray et al., 1965). It also became clear that the murein sacculus is the stress-bearing structure that protected the cell against rupture by the internal turgor, and was also responsible for the specific shape of the bacterium, such as spheres, rods and spirals.

In 1951, Salton and Horne prepared chemically "pure" cell wall fractions from Gram-positive and Gram-negative bacteria, thereby making the cell wall available for precise analytical and structural characterization. This led to the discovery of unique bacterial compounds, such as the uridine nucleotides ("Park" nucleotides), diaminopimelic acid, muramic acid, D-amino acids and teichoic acids. In 1964, Weidel and Pelzer isolated and analyzed the thin murein sacculus of *E. coli*, which is a single, bag-shaped molecule with the size and shape of the cell (Figure 1). Paper chromatography of enzymatic digestion products of the murein indicated a rather simple structure with a few major building blocks in which polysaccharide chains are interlinked through peptide bridges. However, a careful re-examination of the murein structure employing a high-resolution HPLC method introduced by Glauner et al. (1988) showed a pattern of unexpected complexity. More than 80 different murein subunits were isolated, purified and structurally characterized.

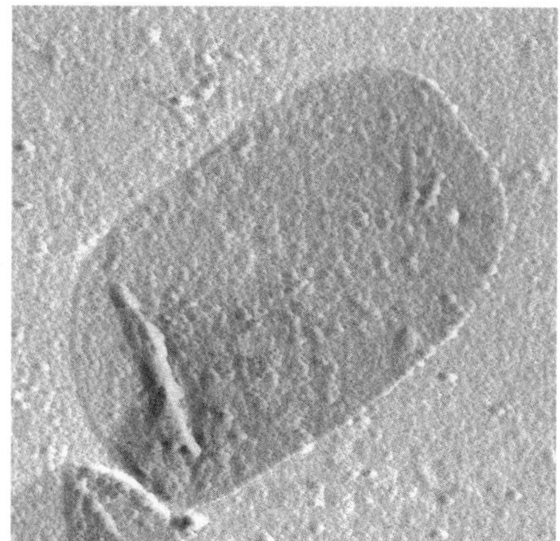

Fig. 1 An isolated murein sacculus of *Escherichia coli*.

Identification of the enzymes involved in the pathway of murein biosynthesis and in murein hydrolysis paralleled the work on murein structure. Alexander Fleming's powerful antibiotic, penicillin, was shown to bind covalently to a class of murein synthases – the penicillin-binding proteins (PBPs) – thus provoking bacteriolysis of the cells. Another antibacterial agent found by Fleming – lysozyme – is widely distributed in the tissues of both animals and plants, and later was shown to be an enzyme which hydrolyzed specific bonds in the murein (Jolles, 1996). Studies on bacterial culture supernatants and extracts revealed that bacteria are equipped with their own murein hydrolases (autolysins) that can eventually autolyze the cell (Shockman and Höltje, 1994). Being potentially dangerous, the autolysins were found to be involved in cellular processes such as cutting the septum during cell division and murein turnover during cell growth. Recent biochemical studies indicated that enlargement of the murein sacculus during bacterial growth is made possible by a coordinated action of the two opposing systems, murein synthases and hydrolases (Höltje, 1996). Moreover, today the interactions between different cell wall enzymes are analyzed to gain a better understanding of the molecular mechanisms that allow growth and division of the murein sacculus, as well as maintaining its specific shape and mechanical integrity.

3 Isolation, Analysis and Structure of Murein

In this section, we describe the procedures used to isolate murein and to analyze different solubilized products obtained after enzymatic digestions. The structural complexity of this polymer was only recognized after the introduction of high-resolution techniques which also allowed the detection of differences in the murein structure of different clonal types and changes in the structure depending on growth conditions.

3.1 Isolation of Murein

For the isolation of murein sacculi, the bacteria from an exponentially growing culture are pelleted, washed, and resuspended in ice-cold water. The suspension is dropped into an equal volume of boiling 8% sodium dodecylsulfate (SDS) solution, and boiled for 30 min. Ultracentrifugation of this suspension leads to a crude cell wall being obtained, which is washed free of SDS by repeated centrifugation/resuspension steps. Impurities such as glycogen, DNA and RNA are removed by sequential incubations with α-amylase, DNAse and RNAse before these enzymes, together with the proteins attached to the sacculi, are digested with pronase. The excess pronase is subsequently removed by the addition of SDS followed by several washing steps with centrifugation/resuspension. This procedure yields purified murein for Gram-negative bacteria and purified cell wall for Gram-positive bacteria which still contains covalently attached teichoic acid and capsular polysaccharides. By incubation of the cell walls of Gram-positive bacteria at low temperature with 48% hydrofluoric acid, these compounds are quantitatively removed and, after washing, the purified murein is obtained. Murein can be stored in aqueous suspension at 4 °C.

3.2 The Basic Chemical Structure of Murein

Chemically, murein is built of glycan strands which are cross-linked by short peptide bridges (Figure 2). Because the murein glycan strands are relatively uniform, the main goal of the early analytical work was to establish the primary structure (sequence) of the peptide side chains of the different murein types isolated from different bacterial species (Schleifer and Kandler, 1972). The following combination of chemical methods and hydrolytic enzyme reactions using autolysins was employed: (1) determination of the amino acid and amino sugar composition of isolated murein after total hydrolysis and determination of the configuration of the amino acids; (2) determination of the N- and C-terminal amino acids in intact murein; (3) isolation and analysis of

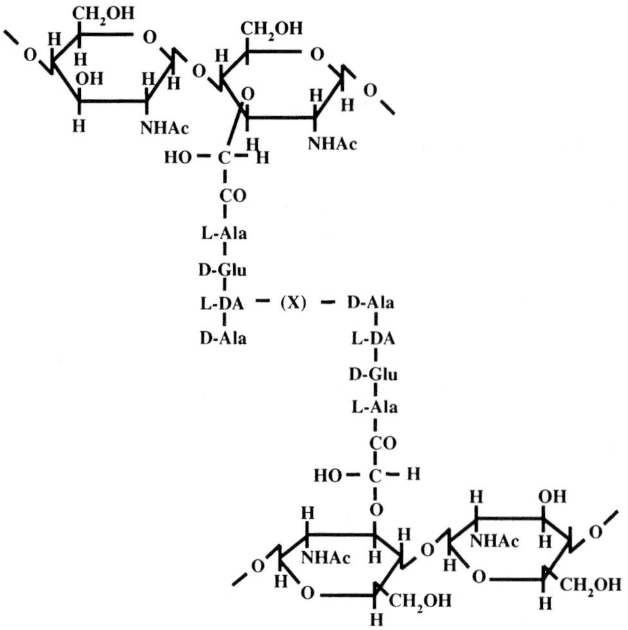

Fig. 2 Fragment of a typical murein structure. L-DA: L-diamino acid, X: interpeptide bridge.

peptides obtained by partial hydrolysis of murein; (4) analysis of murein precursors; and (5) analysis of muropeptides obtained by digestion of murein with a muramidase (lysozyme).

The murein glycan strands are formed from alternating *N*-acetylglucosamine (GlcNAc) and *N*-acetylmuramic acid (MurNAc) which is the 3-*O*-D-lactic acid ether of GlcNAc and is found only in bacteria and blue-green algae. The peptide moiety is bound through its N-terminal amino group to the carboxyl group of muramic acid, and contains alternating L and D amino acids. The presence of amino acids with D configuration is a typical feature of murein. In most cases, L-alanine is bound to MurNAc, followed by D-glutamic acid. The latter is linked by its γ-carboxyl group to the amino group at the C-atom with L configuration of a diamino acid in position 3. The D-alanine in position 4 is highly conserved, and may be followed by a second D-alanine in position 5. Mature murein often lacks the terminal alanine residues because of the action of hydrolytic enzymes. The free amino group of the diamino acid in position 3 is linked directly or via a peptide bridge to the carboxyl group of a D-alanine residue in position 4 of a peptide side chain of a neighboring glycan strand, thus cross-linking both strands. Investigation of the chemical structure from numerous bacterial species revealed that there is a broad variation, especially in the peptide moiety (Figure 3), whereas the structure of the glycan strands is remarkably uniform.

3.2.1 Species-specific Structural Variation (Murein Types)

Almost 100 different murein structures with variations in the amino acids of the peptides and in the mode of cross-linkage are

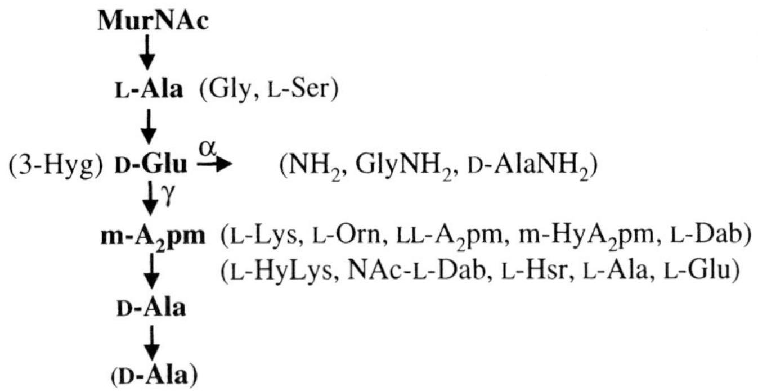

MurNAc
↓
L-Ala (Gly, L-Ser)
↓
(3-Hyg) **D-Glu** $\xrightarrow{\alpha}$ (NH_2, $GlyNH_2$, D-$AlaNH_2$)
↓γ
m-A_2pm (L-Lys, L-Orn, LL-A_2pm, m-HyA_2pm, L-Dab)
↓ (L-HyLys, NAc-L-Dab, L-Hsr, L-Ala, L-Glu)
D-Ala
↓
(D-Ala)

Fig. 3 Possible variations in the primary peptide structure in the murein of different species. 3-Hyg: threo-3-hydroxyglutamic acid; m-A_2pm: m-2,6-diaminopimelic acid; L-Orn: L-ornithine; m-HyA_2pm: m-2,6-diamino-3-hydroxypimelic acid; L-Dab: L-diaminobutyric acid; L-HyLys: L-hydroxylysine, L-Hsr: L-homoserine. (Adapted from Schleifer and Kandler, 1972.)

described in an excellent review by Schleifer and Kandler (1972). The authors introduced the following classification for the characterization and representation of murein types of different bacteria. In **type A** murein, the cross-linkage occurs between the diamino acid at position 3 of a peptide – directly or via an interpeptide bridge – to the D-alanine residue at position 4. The subtype **A1** with a direct linkage is further divided into **A1α** (L-lysine in position 3), **A1β** (L-ornithine) and **A1γ** (m-diaminopimelic acid). With a few exceptions, the Gram-negative bacteria have a murein of the **A1γ** subtype, which is shown in Figure 4A. The murein of subtype **A2** contains an interpeptide bridge which is built by a "head-to-tail" connection of one or up to four peptide subunits (Figure 4B). The murein of this type contains at least 50% of unsubstituted MurNAc residues, indicating that the peptide subunits forming the interpeptide bridges were released from this position. The **A3** subtype with an interpeptide bridge consisting of two to seven monocarboxylic L-amino acid residues or glycine (Figure 4C) is very common among Gram-positive bacteria. Branched amino acids, aromatic amino acids, methionine, cysteine, arginine, histidine and proline were not found in interpeptide bridges. Type **A4** murein contains cross-linking interpeptide bridges with dicarboxylic amino acids (Figure 4D) such as D-asparagine, D-glutamine or L-glutamine which are always linked in the peptide bridge by their distant carboxyl group (β or γ). The free α-carboxyl group is sometimes amidated. Murein of type **B** is much less frequent than that of type A, and is found mainly in coryneform bacteria. The amino acid in position 1 of the peptide is not L-alanine, but glycine or L-serine. Cross-linkage occurs between the α-carboxyl group of D-glutamine (position2) via an interpeptide bridge containing a diamino acid to the D-alanine (position 4) of an adjacent peptide subunit (Figure 4E). Subgroup **B1** contains an L-diamino acid in the peptide bridge; subtype **B2** a D-diamino acid.

3.2.2 Secondary Modifications in the Murein Structure

In many species, the structure of the murein is modified as a result of enzymatic reactions. The possible modifications occur both on the peptide and on the glycan part. The α-carboxylic group of the D-glutamine residue

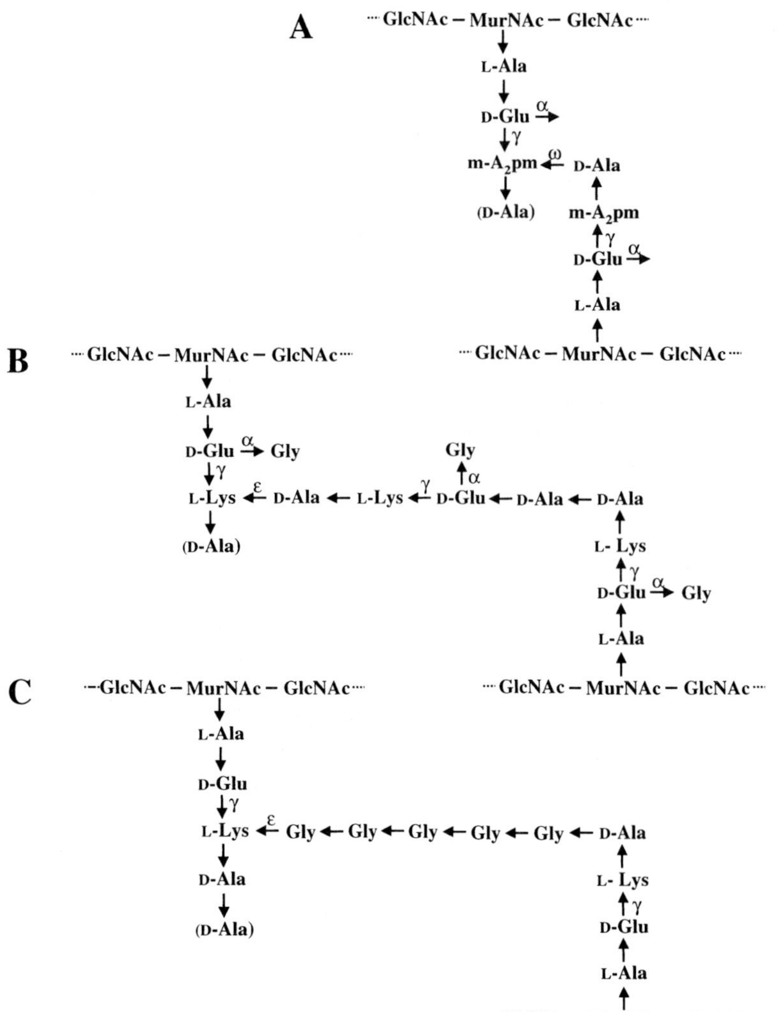

Fig. 4 Examples of different murein types. The A1γ type murein of *Escherichia coli* does not contain an interpeptide bridge (A). The peptide bridge of the A2 type murein of *Micrococcus luteus* (B) has the same structure as the peptide side chains. *Staphylococcus aureus* has a pentaglycine bridge in its A3α type murein (C). The hallmark of the A4 type murein, like the one in *Streptococcus faecium*, is the presence of a dicarboxylic acid in the peptide bridge (D). In the murein of type B, the cross-linkage is formed between positions 2 and 4 of the peptide side chains, as for example, in *Corynebacterium poinsettiae* (E), which belongs to the subtype B2.

in position 2 (*Staphylococcus aureus*) or the carboxylic group of the diamino acid in position 3 (*B. subtilis*) of the peptide is often amidated. The murein glycan strands of *E. coli* do not carry a reducing end because the terminal MurNAc residue carries an intramolecular 1,6-anhydro bond (Figure 5A) (Höltje et al., 1975). It is not known yet, if these nonreducing glycan ends are formed by the reaction of hydrolytic en-

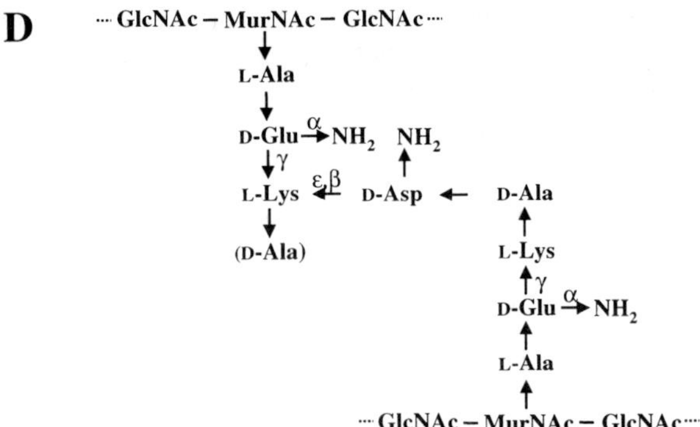

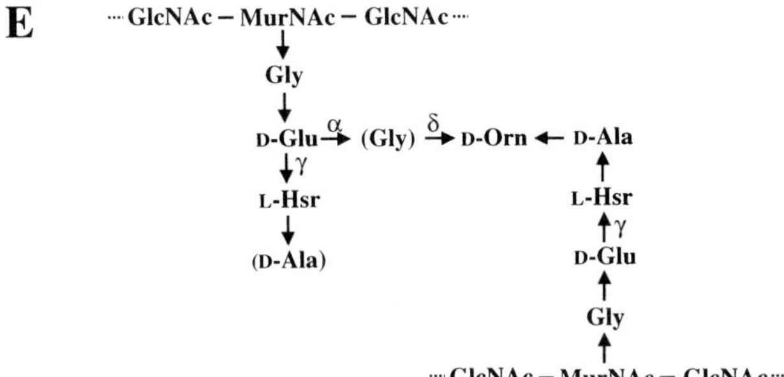

Fig. 4 (continued)

zymes (the lytic transglycosylases) or if they are generated upon termination of the glycan polymerizing reaction (synthetic transglycosylation), or by both reactions.

The murein glycan strands of several pathogenic species are modified in such a way that protective hydrolytic enzymes of the host organism, such as lysozyme, are less active or not active at all. This is the case for O-acetylated and for N-deacetylated murein. There are 11 known species of both Gram-positive and Gram-negative bacteria in which up to 70% of the MurNAc residues are O-acetylated at the hydroxyl group at C6 (Figure 5B) (Clarke and Dupont, 1992). These include *Streptococcus faecalis, Staphylococcus aureus, Neisseria gonorrhoea* and *Proteus mirabilis*. Glycan strands with N-deacetylated hexosamine residues are present in several *Bacillus* species (Hayashi et al., 1973). In *B. anthracis* and *B. cereus*, both glucosamine and muramic acid are present in nonacetylated form, whereas in *B. megaterium* and *B. subtilis* deacetylation is mainly a property of the GlcNAc residues (Figure 5C) (Zipperle et al., 1984). The same is true for *Streptococcus pneumoniae* in which up to 80% of the GlcNAc and 10% of the MurNAc residues are subject to enzymatic deacetylation by a recently identified peptidoglycan GlcNAc deacetylase PgdA (Vollmer and Tomasz, 2000). Both O-acetylation and N-deacetylation are assumed to be secondary modifications of the polymerized murein sacculus.

Upon starvation of *B. subtilis* cells, a complex genetic program is activated leading to asymmetric cell division and to the

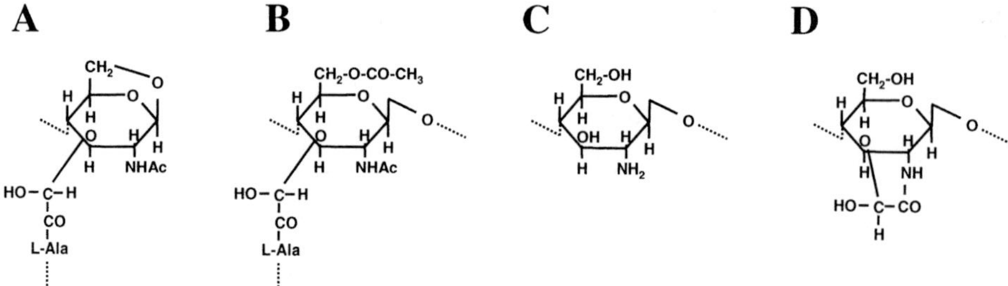

Fig. 5 Modifications in the murein glycan strands. The terminal MurNAc residues in the murein of *E. coli* carries a 1,6-anhydro-bond (A). In several pathogenic species the glycan strands are modified, making the murein a poorer substrate for defense enzymes of the host organism such as lysozyme. In *S. aureus*, some of the MurNAc residues are O-acetylated at C6 (B). N-Deacetylated GlcNAc residues (C) are found in the murein of *S. pneumoniae*. In the murein of the spore cortex of *B. subtilis* endospores, about 50% of the MurNAc residues are changed to δ-muramic lactam (D).

development of a dormant endospore which is dehydrated and extremely heat-resistant. The murein of the spore cortex is only poorly cross-linked (about 3% as compared with 33% in vegetative cells) and 25% of the MurNAc residues carry only a D-alanine side chain. Furthermore, approximately 50% of the muramic acid residues are converted to δ-muramic lactam (Figure 5D) (Atrih and Foster, 1999).

3.2.3
Covalent Attachment of Cell Wall Components to the Murein Sacculus

In Gram-positive bacteria, the multilayered murein sacculus serves as an anchor molecule for covalently attached oligosaccharides such as teichoic acid, teichuronic acid and capsular polysaccharides and for surface proteins. Because of their surface location, these compounds play important roles in the interaction of the bacterium with its environment. Teichoic acids and teichuronic acids are composed of simple or more complex repeating units and are attached via a linkage unit to the murein (Shockman and Barrett, 1983). The binding to murein occurs via a phosphodiester bond between the reducing end of the terminal sugar of the linkage unit and the hydroxyl group at C6 of MurNAc (Figure 6 A). Capsular polysaccharides provide protection for the bacterium against defense mechanisms of eucaryotic organisms (immune system). They are either covalently linked to the murein in the same way as teichoic acid or bound to the cell surface noncovalently.

Originally identified for protein A of *S. aureus*, it transpired that there is a widely distributed mechanism for the covalent anchoring of surface proteins to the murein of Gram-positive bacteria (Navarre and Schneewind, 1999). In addition to an N-terminal signal peptide typical for prokaryotic secretion, these proteins contain a C-terminal cell wall sorting signal consisting of the conserved LPXTG motif followed by a hydrophobic domain and a positively charged tail. It is proposed that the proteins are secreted by components of the general secretory pathway and retained at the cytoplasmic membrane by the C-terminal hydrophobic domain and the charged tail. An enzyme – the sortase – then cleaves the protein between the threonine and glycine residue of the LPXTG motif and forms an amide bond between the carboxylic group of the threonine and the amino group of the terminal amino acid of a noncross-linked murein interpeptide bridge (Figure 6 B).

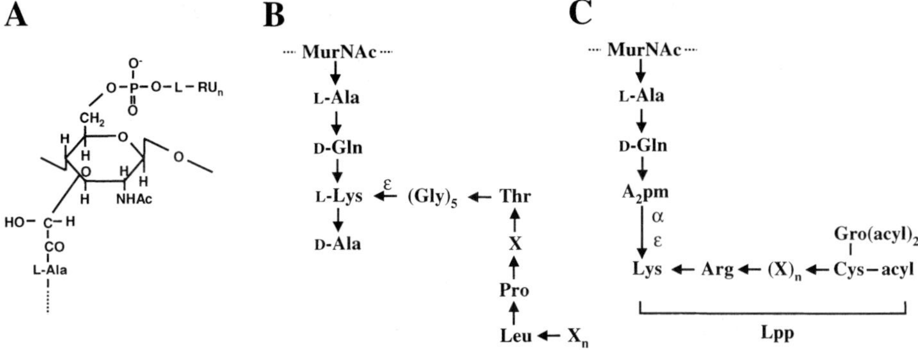

Fig. 6 Covalent attachment of cell wall polymers to the murein. In Gram-positive bacteria, surface oligosaccharides (teichoic acid, teichuronic acid, capsular polysaccharide) made up of repeating units (RU) are covalently linked with their linkage unit (L) to the C6 of MurNAc residues via a phosphodiester bond (A). Surface proteins of Gram-positive bacteria are attached by a sortase at their LPXTG motif to free amino groups of the noncross-linked interpeptide bridges, as shown for *S. aureus* (B). In the Gram-negative *E. coli*, Braun's lipoprotein (Lpp) is attached with its C-terminal lysine residue to the α-carboxylic group of m-A_2pm (C); Gro, glycerol; acyl, acyl residue.

There are numerous examples of surface proteins of Gram-positive bacteria attached to the murein in this way.

In the Gram-negative *E. coli*, approximately one-third of the major lipoprotein (Braun's lipoprotein), which is localized by its lipid anchor in the outer membrane, is covalently attached to the murein and thereby connects the outer membrane with the sacculus (Braun, 1975). An as-yet unknown enzyme forms an amide bond between the carboxylic group at the L center of m-A_2pm and the ε-amino group at the C-terminal lysine residue of the lipoprotein (Figure 6C).

3.3 The Fine Structure of Murein

The analytical methods for the analysis of the fine structure of murein were established by Harz, Burgdorf and Höltje (glycan strands) and by Glauner and Höltje (muropeptides) for *E. coli*. The latter HPLC method was adapted for the analysis of the mureins of many other bacteria, and is now used in combination with mass spectrometry (Allmaier and Schmid, 1993).

3.3.1 Analysis of the Length Distribution of the Murein Glycan Strands

As described in Harz et al. (1990), the oligo-(GlcNAc-MurNAc) glycan strands of *E. coli* are released from the murein with purified human serum amidase, an enzyme which cleaves the amide bond between the lactyl group of MurNAc and the first amino acid of the peptide side chain, the L-alanine residue. The glycan material is then separated from the peptides on a MonoS cation exchange column and the glycan strands are fractionated according to their degree of polymerization (DP) (from 1 to 30 disaccharides) on a reversed-phase C_{18} Nucleosil HPLC column (Figure 7). The predominant length of this glycan material is between 5 and 10 disaccharide units. About 30% of the glycan material consists of glycan strands, with more than 30 disaccharide units which cannot be separated by this method and which has an average DP of about 80

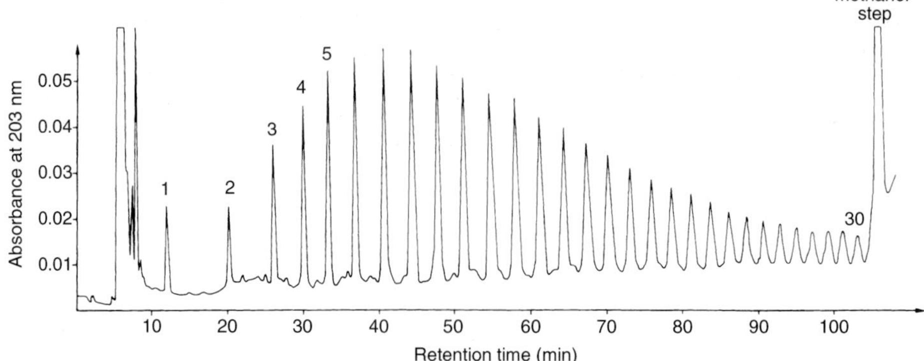

Fig. 7 Analysis of the length distribution of the purified murein glycan strands of *E. coli* by reversed-phase HPLC. The oligomers with 1 to 30 disaccharide units are separated by their size. Longer glycan strands with more than 30 disaccharide units are eluted in the methanol step.

disaccharide units. Therefore, the glycan strands of *E. coli* show a broad variation in their length distribution. The method was also applied in combination with mass spectrometry for the analysis of glycan strands from the murein of *S. aureus*, which show a similar length distribution to those from *E. coli*. However, in *S. aureus* the glycan strands contain reducing ends and strands with odd numbers of amino sugars were observed which are the products of an *N*-acetylglucosaminidase activity (Boneca et al., 2000). In contrast, the glycan material isolated from the murein of *S. pneumoniae* consists quantitatively of strands with more than 30 disaccharide units which were not suitable for separation by reversed-phase HPLC (W. Vollmer and A. Tomasz, unpublished results).

3.3.2
Analysis of the Muropeptide Composition

Muropeptides are generated by the treatment of murein with a muramidase such as lysozyme, mutanolysin, or cellosyl. After reduction with sodium borohydride, they are separated by HPLC on a C_{18} ODS-Hypersil column and the structures are analyzed by chemical methods and with mass spectrometry. The pioneering analytical studies by Glauner and Höltje (Glauner, 1988; Glauner et al., 1988) revealed the high diversity of about 80 different muropeptide structures in the murein of *E. coli*. This complexity is explained by the occurrence of different monomeric structures with peptides that consist of two to five amino acids and with peptides with Lys-Arg (in positions 4 and 5) which indicate attachment sites for Braun's lipoprotein. These monomers can be combined to dimeric, trimeric and tetrameric structures. Next to compounds with the major DD-type of cross-linkage between the D-center of A_2pm and a D-alanine residue of another peptide side chain, there are a few percent of oligomeric muropeptides having an LD-cross-bridge between the D-center of A_2pm and the L-center of a second A_2pm residue. The muropeptides released from the glycan strand end(s) contain one (or more) 1,6-anhydro-MurNAc residue(s). The muropeptide composition of *E. coli* KN126 is shown in Table 1.

Today, there is structural information and data available about the muropeptide or peptide (released by an amidase) composi-

Tab. 1 The muropeptide composition of *E. coli* KN 126. Di, GlcNAcMurNAc(dipeptide); Tri, GlcNAcMurNAc(tripeptide); Tetra, GlcNAcMurNAc(tetrapeptide); Penta, GlcNAcMurNAc(pentapeptide); Gly^4/Gly^5, glycine at position 4 or 5 of peptide side chain; Lys-Arg, Lys-Arg residue from lipoprotein; Anh, 1,6-anhydro-MurNAc; A_2pm (in short formula), A_2pm- A_2pm cross-bridge. (Adapted from Höltje, 1998)

Muropeptide	Relative amount [%]	Standard deviation [%]
Tri	7.41	±5
Tetra(Gly^4)	1.71	±8
Tetra	35.90	±4
Penta(Gly^5)	0.26	±18
Di	2.13	±11
Penta	0.07	±23
Tri-Lys-Arg	4.86	±6
Tri(Anh)	0.36	±8
Tetra-Tri(A_2pm,Gly^4)	0.07	±22
Tri-Tri(A_2pm)	0.28	±13
Tetra-Tri(A_2pm)	1.55	±11
Tetra-Tetra(Gly^4)	1.47	±10
Tetra-Tri	2.99	±9
Tetra-Penta(Gly^5)	0.32	±18
Tetra-Tetra	27.30	±3
Tetra(Anh)	0.60	±11
Tetra-Penta	0.17	±14
Tri-Tri-Lys-Arg(A_2pm)	0.43	±20
Tetra-Tetra-Tri(A_2pm)	0.24	±14
Tetra-Tetra-Tri	0.28	±10
Tetra-Tri-Lys-Arg	2.92	±7
Tetra-Tetra-Tetra	2.33	±5
Tetra-Tri(A_2pm,Anh)I	0.26	±15
Tetra-Tri(A_2pm,Anh)II	0.27	±16
Tetra-Tri(Anh)I	0.42	±12
Tetra-Tri(Anh)II	0.14	±15
Tetra-Tetra-Tetra-Tetra	0.08	±17
Tetra-Tetra-Tri-Lys-Arg	0.27	±10
Tetra-Tetra(Anh)I	0.67	±7
Tetra-Tetra(Anh)II	0.67	±5
Tetra-Tetra-Tri(A_2pm,Anh)	0.08	±13
Tetra-Tetra-Tri(Anh)	0.18	±12
Tetra-Tetra-Tetra(Anh)	0.55	±10
Tetra-Tetra-Tetra-Tetra(Anh)	0.05	±12
Tetra-Tri-Lys-Arg(Anh)I	0.13	±13
Tetra-Tri-Lys-Arg(Anh)II	0.16	±21
Tetra-Tetra-Tri-Lys-Arg(Anh)I	0.03	±20
Tetra-Tetra-Tri-Lys-Arg(Anh)II	0.05	±24
Tetra-Tetra-Tri-Lys-Arg(Anh)III	0.05	±17

tion of different strains of many bacterial species, including *S. aureus, S. pneumoniae, B. subtilis, Enterococcus faecalis, Neisseria meningitidis, Enterobacter cloacae, Pseudomonas aeruginosa, Yersinia enterocolitica, A. tumefaciens* and *Helicobacter pylori.*

3.4 Variability of the Species-specific Murein Composition

As in the case of *E. coli*, the detailed analysis of the murein structure of several other species revealed a more complex composition than indicated only by the murein type. Also, variations in the murein structure depending on growth conditions, growth phase, clonal type or the presence of antibiotics were described. This is illustrated in the following examples.

1) The glycine content of the murein of *S. aureus* is higher if the cells are grown in glycine-rich medium. The same is true for *E. coli* in which glycine can partially substitute D-alanine in positions 4 and 5 of the peptide. This bacterium also incorporates unusual D-amino acids such as D-cysteine into the murein if they are present in the growth medium (Caparros et al., 1992).
2) Besides the major A1γ type of DD-cross-linkage between m-A_2pm and D-Ala, the murein of *E. coli* contains about 2% of LD-cross-linkages formed between two m-A_2pm residues of neighboring peptide subunits. The relative amount of this unusual linkage can double in the stationary phase (Glauner and Höltje, 1990).
3) Similarly, the murein of *Enterococcus faecium* contains a low percentage of an unusual L-Lys- D-Asn- L-Lys LD-cross-bridge besides the major L-Lys- D-Asn-D-Ala DD-linkage, the biosynthesis of which is inhibited by β-lactam antibiotics such as penicillin. A laboratory mutant which was selected for penicillin resistance, exclusively contained the unusual cross-links, indicating that the biosynthesis of LD cross-links is not sensitive to β-lactam antibiotics (Mainardi et al., 2000), and also illustrating that some bacteria drastically change their murein structure upon antibiotic pressure.
4) Major differences in the murein composition were also observed in some penicillin-resistant *S. pneumoniae* isolates in which normally minor substructures with indirect Ala-Ala or Ala-Ser cross-bridges became dominant. Interestingly, penicillin resistance was completely lost in all mutant pneumococci which were unable to synthesize indirect cross-bridges, even if these structures were only minor compounds in the murein of the resistant parental strain (Filipe and Tomasz, 2000).
5) The glycopeptide antibiotic vancomycin inhibits cell wall biosynthesis by formation of a complex with the terminal D-Ala-D-Ala moiety of the murein precursors. In vancomycin-resistant strains of enterococci, the precursors contain a D-Ala-D-lactate which eliminates the high affinity of vancomycin to this structure (Arthur et al., 1996).
6) Another strategy to acquire high-level resistance to glycopeptide antibiotics is used by both laboratory mutants and some clinical isolates of *S. aureus*. The thicker murein of the resistant strains is not as highly cross-linked as usual, but has an increased number of noncross-linked, D-Ala-D-Ala-containing peptide side chains which are able to capture vancomycin from the medium. It is assumed that the binding of vancomycin to those residues forms a barrier which prevents the critical binding of the antibiotic to precursor molecules at the site of murein biosynthesis at the cytoplasmic membrane. In addition, the cell walls loaded with vancomycin are more protected from the action of intrinsic autolytic enzymes (Sieradzki and Tomasz, 1997).

3.5 Three-dimensional Structure of Murein

A number of physical-chemical methods have been employed to acquire insights into the three-dimensional architecture of murein. However, until now there has been no analytical method available which would provide detailed structural information of native murein sacculi. Data from X-ray-, infra-red (IR)- and nuclear magnetic resonance (NMR) spectroscopy and electron microscopy indicated that high molecular-weight murein has a noncrystalline structure (Labischinski et al., 1983). The three-dimensional network of glycan strands cross-linked by peptides is not rigid, but has a high elasticity. Osmotic swelling/shrinking experiments showed that isolated, relaxed murein sacculi of *E. coli* can reversibly increase in surface up to three-fold. This flexibility is mainly mediated by the peptides, whereas the glycan strands behave as rigid, straight rods. Conformational energy calculations led to a model in which the glycan strands are twisted because of the bulky lactyl groups at the C3 atom of MurNAc (Barnikel et al., 1983). This results in a helical glycan structure in which the peptide substituents of two successive GlcNAc-MurNAc disaccharides are shifted, relative to one another, by about 90 °. Every second peptide side chain on a glycan strand would then be in a plane, and a single layer of murein could be formed by cross-linking of those peptides. Small-angle neutron scattering experiments indicated that 75–80% of the sacculus of *E. coli* consists of a single layer and 20–25% of a triple layer of murein (Labischinski et al., 1991). Peptides which point out of the plane of the layer might be cross-linked to connect the multiple layers in the murein of Gram-positive bacteria. The net-like murein structure has pores allowing transport of molecules across the bacterial cell wall. The permeability of isolated murein sacculi was determined by penetration experiments using fluorescent probes (Demchick and Koch, 1996). The mean radius of the holes in the relaxed murein of *E. coli* was 2.06 nm, and 2.12 nm in that of *B. subtilis*. It was estimated that this would allow the free diffusion of a globular hydrophilic molecule (e.g., a protein) with a maximum size of about 25 kDa, and that this value might double for stretched murein (during growth).

4 Biosynthesis of Murein

The biosynthesis of murein can be divided into several steps with respect to the location of the biosynthetic reactions (Höltje and Schwarz, 1985). The last monomeric murein precursor is the membrane-bound lipid II, which is polymerized at the periplasmic side of the cytoplasmic membrane. The insertion of the newly synthesized material into the existing murein sacculus is accompanied by a turnover of pre-existing, old material which is subsequently digested and recycled by the cell. In this section, we focus mainly on the biosynthetic pathway of the murein synthesis in *E. coli*.

4.1 Biosynthesis of the Murein Precursor Lipid II

The soluble UDP-MurNAc(pentapeptide) is synthesized in the cytoplasm and serves as an intermediate for the formation of lipid II. The lipid modification facilitates the translocation of the precursor molecules across the cytoplasmic membrane to the site of murein synthesis.

4.1.1
Formation of UDP-MurNAc(pentapeptide) in the Cytoplasm

The pathway of the biosynthesis of lipid II (van Heijenoort, 1998) is shown in Figure 8. Fructose-6-phosphate is first converted to glucosamine-6-phosphate by an L-glutamine:D-fructose-6-P amidotransferase encoded by the *glmS* gene. Next, glucosamine-6-phosphate is converted by the GlmM mutase to glucosamine-1-phosphate, which is then N-acetylated and uridinylated by the bifunctional GlmU enzyme, to yield UDP-*N*-acetylglucosamine (UDP-GlcNAc). UDP-MurNAc is formed in two steps from UDP-GlcNAc. First, a transfer of enolpyruvate from phosphoenolpyruvate to the hydroxyl group at C3 of GlcNAc is catalyzed by the transferase MurA, followed by the reduction of the formed UDP-GlcNAc-enolpyruvate in a NADPH-dependent reaction by the MurB. Next, the first three amino acids of the peptide side chain are subsequently added by the highly specific ligases MurC (L-alanine), MurD (D-glutamine) and MurE (m-A_2pm) yielding UDP-MurNAc(tripeptide). The addition of the amino acids by these enzymes is dependent on the hydrolysis of ATP. The D-amino acids present in the peptide subunits are formed from the corresponding L-amino acids by the racemases MurI (D-glutamine) and DadX (D-alanine). The two D-alanine residues (positions 4 and 5 of the peptide) are first ligated by the D-alanine:D-alanine ligases DdlA/B before the resulting dipeptide is attached to UDP-MurNAc(tripeptide) by the MurF ligase. The UDP-MurNAc(pentapeptide) molecule is the last soluble murein precursor.

4.1.2
Formation of the Lipid Intermediates

Two membrane-bound enzymes are involved in the next steps. The MraY transferase catalyzes the reaction of UDP-MurNAc(pentapeptide) with the lipophilic carrier undecaprenyl phosphate (C_{55}-isoprenylphosphate or bactoprenol) to yield undecaprenyl-pyrophosphoryl-MurNAc(pentapeptide) (lipid I). The UDP-sugar transferase MurG then catalyzes the transglycosylation of GlcNAc from UDP-GlcNAc to lipid I to form undecaprenyl-pyrophosphoryl-MurNAc(pentapeptide)-GlcNAc (lipid II). After it is translocated to the outer leaflet of the cytoplasmic membrane, the lipid II molecule is the substrate for the synthetic murein-polymerizing enzymes, the transglycosylases and transpeptidases. The undecaprenyl pyrophosphate which is released in the transglycosylation reaction is dephosphorylated to undecaprenyl phosphate which is re-used as carrier for the precursors. In many bacterial species, there are interpeptide bridges attached to the diamino acid in position 3 of the peptide side chain (see above). It was shown for *S. aureus*, that the glycine residues forming the cross-bridge are transferred to the lipids I and II from the corresponding glycyl-tRNA by the gene products of *fmhB*, *femA*, and *femB*.

4.2
Murein Synthases and Hydrolases

The final steps of the murein biosynthesis take place at the outer leaflet of the cytoplasmic membrane, where the precursor lipid II is inserted into the pre-existing murein sacculus by two enzymatic reactions: the precursor molecules are polymerized by transglycosylation reactions and the peptide side chains of the nascent glycan strands are cross-linked by transpeptidation reactions (Figure 9). Interestingly, not only monofunctional murein synthases with transglycosylase (MtgA) or transpeptidase (PBP2, PBP3) activity, but also bifunctional enzymes exist in which both activities are

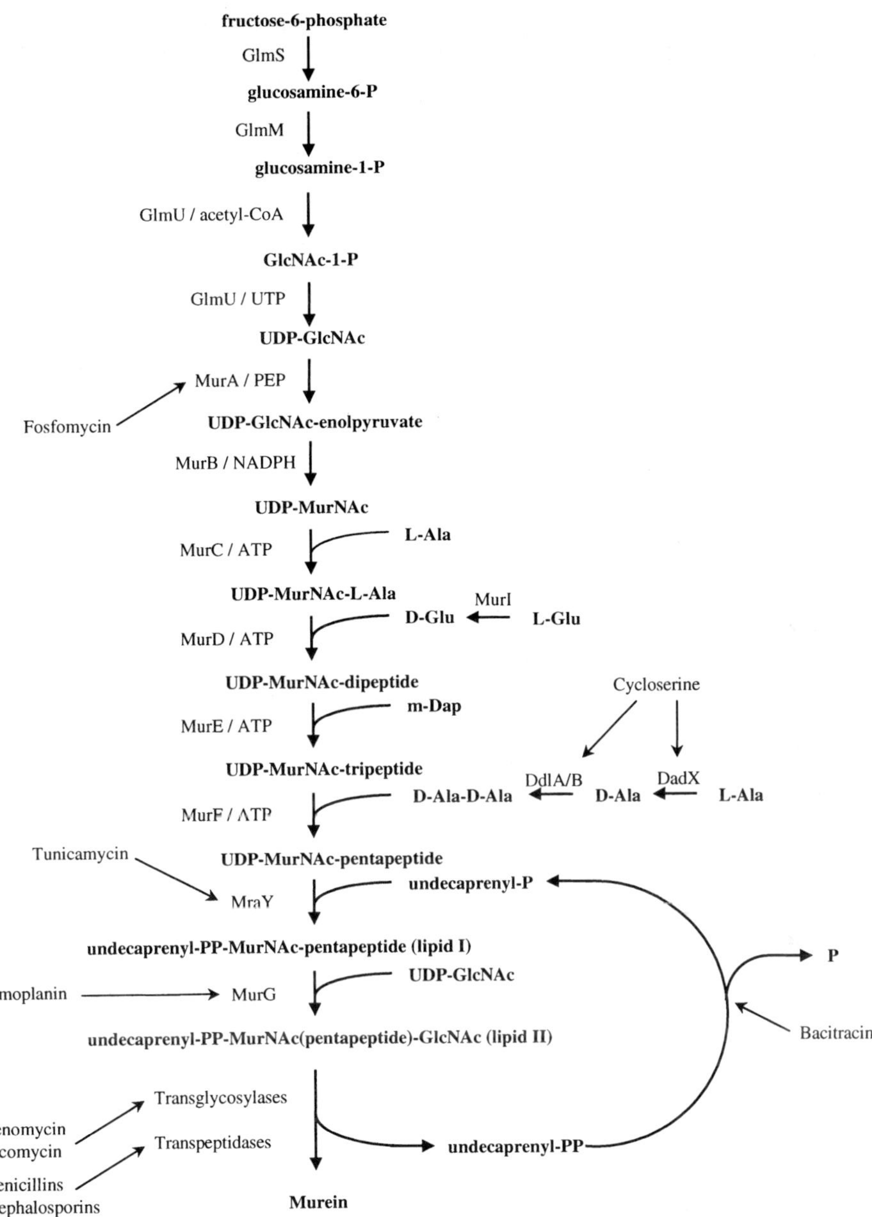

Fig. 8 Biosynthetic reactions for the synthesis of murein in *E. coli* from fructose-6-phosphate. Some antibiotics and their target reactions in murein metabolism are also indicated.

combined (PBP1A, PBP1B, PBP1C) (Table 2). The DD-transpeptidation creates dimeric, trimeric and tetrameric muropeptide structures. This reaction is sensitive towards β-lactam antibiotics such as penicillin (see below; Ghuysen, 1991). The murein synthases are anchored by a short C- or N-terminal stretch in the cytoplasmic mem-

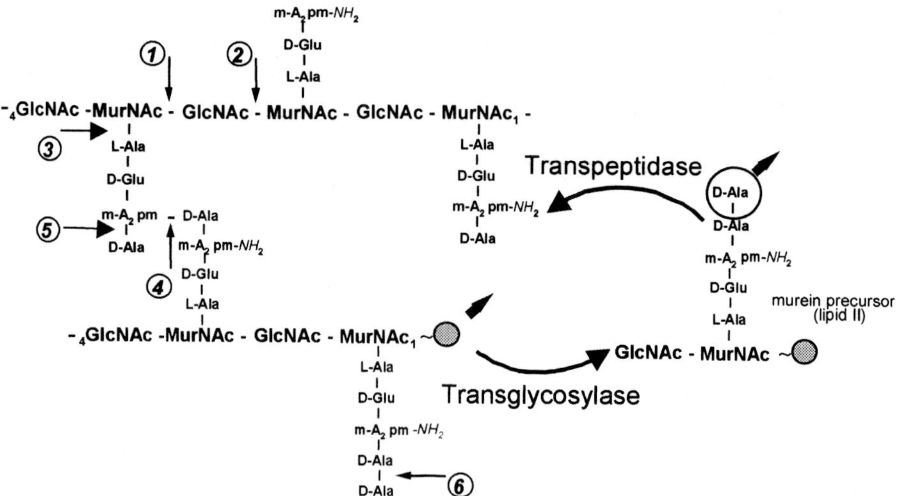

Fig. 9 Formation and cleavage of covalent bonds in the murein net. Polymerization of the murein precursor is achieved by the activity of transpeptidases and transglycosylases, respectively. During transglycosylation, a glycosidic bond between the MurNAc of the nascent glycan strand and the C4 at the GlcNAc of the precursor is formed. The transpeptidase creates a DD-peptide bond between the carboxyl group of D-Ala at position four of the murein precursor and the ε-amino group of m-A_2pm of an adjacent peptide side chain. The energy for this reaction is generated by the simultaneous cleavage of the terminal D-Ala residue (position 5). The numbers indicate the covalent bonds of the murein net that are cleaved by different classes of murein hydrolases: 1: lytic transglycosylase; 2: *N*-Acetylglucosaminidase; 3: *N*-Acetylmuramyl-L-alanine amidase; 4: DD-endopeptidase; 5: LD-carboxypeptidase; 6: DD-carboxypeptidase. (Adapted from Höltje, 1998.)

Tab. 2 Murein synthases in *E. coli*

Enzyme	*Gene*	*Map position*	*Size [kDa]*	*Localization*
Transpeptidase/transglycosylase PBP1A	*pon*A (*mrc*A)	75.9 min	94.5	inner membrane
Transpeptidase/transglycosylase PBP1B	*pon*B (*mrc*B)	3.6 min	94.3	inner membrane
Transpeptidase/transglycosylase PBP1C	*pbp*C	57.0 min	85.1	inner membrane
Transpeptidase PBP2	*pbp*A (*mrd*A)	14.4 min	70.8	inner membrane
Transpeptidase PBP3	*fts*I (*pbp*B)	2.0 min	63.9	inner membrane
Monofunctional glycosyltransferase MtgA	*mtgA*	72.1 min	27.3	inner membrane

For references, see Höltje (1998).

brane, whereas their bulky, multi-modular part points to the murein net.

The meshwork character of the murein sacculus implies that its enlargement and division is only possible if covalent bonds within the polymer are cleaved to allow insertion of the new material. Hence, murein hydrolases are believed to act as spacemakers and pacemakers during murein enlargement and splitting of the septum. Obviously, the activity of these murein hydrolases can also lead to autolysis if covalent bonds are opened in an uncontrolled fashion (Höltje, 1998). About 20 different murein hydrolases are known to be expressed in *E. coli* (Table 3). These participate in cell separation, in murein turnover, in the recycling of the turnover products or in bacterial autolysis (Höltje, 1995). Muramidases (lysozymes) and gluco-

Tab. 3 Murein hydrolases in *E. coli*

Enzyme	*Gene*	*Map position*	*Size [kDa]*	*Localization*
Lytic transglycosylases				
Slt70	*slt*Y	99.7 min	70	Periplasm
MltA	*mlt*A	63.4 min	38	Outer membrane
MltB	*mlt*B	60.9 min	39	Outer membrane
MltC	*mlt*C	66.9 min	40	Outer membrane
MltD	*mlt*D	5.0 min	49	Outer membrane
EmtA**	*emt*A	26.8 min	22	Outer membrane
β-*N*-Acetyl-glucosaminidase				
NagZ	*nag*Z	25.1 min	37	Cytoplasm
N-Acetylmuramyl-L-alanine amidase				
AmiA	*ami*A	55.0 min	28	Periplasm
AmiB	*ami*B	94.7 min	48	Periplasm
AmiC	*ami*C	63.5 min	43	Periplasm
AmpD	*amp*D	2.6 min	20.5	Cytoplasm
DD-Endopeptidases				
PBP4	*dac*B	71.7 min	48	Membrane*
PBP7	*pbp*G	48.1 min	31	Membrane*
MepA	*mep*A	52.7 min	28	Periplasm
DD-Carboxypeptidase				
PBP5	*dac*A	14.3 min	40	Inner membrane
PBP6	*dac*C	19.0 min	40	Inner membrane
PBP6B	*dac*D	44.8 min	43.5	Inner membrane
LD-Carboxypeptidase				
LdcA	*ldc*A	26.8 min	33.4	Cytoplasm

For references, see Höltje (1998); concerning LdcA, see Templin et al. (1999); for MltD, see Pellegrini et al. (1999).**Endospecific lytic transglycosylase; *membrane-associated.

saminidases cleave the glycosidic bonds in the murein glycan strands. In *E. coli*, a special class of muramidases – the lytic transglycosylases Slt70, MltA, MltB, MltC and MltD – cleave the β-1,4-glycosidic bond between MurNAc and GlcNAc and concomitantly catalyze the formation of an 1,6-anhydro bond on the muramic acid residue of the released fragment. Most lytic transglycosylases (Mlts) are lipoproteins anchored in the outer membrane. The only known β-*N*-acetylglucosaminidase (glucosaminidase) of *E. coli* (NagZ) is localized in the cytoplasm and is involved in the recycling of muropeptides, unlike many autolytic glucosaminidases of Gram-positive bacteria (Park, 1996; Vötsch and Templin, 2000). The DD-endopeptidases (PBP4, PBP7 and MepA) cleave peptide bonds in multimeric muropeptide structures to yield monomers. The *N*-acetylmuramyl-L-alanine amidases (amidases) AmiA, AmiB and AmiC cleave the amide bond between the lactyl-residue of MurNAc and the L-alanine of the peptide chain, and thereby separate peptide and glycan parts of the murein.

4.3 Growth Mechanism of the Murein Sacculus

A fascinating question is how the murein sacculus is enlarged during growth, and how it is divided into two sacculi during cell division. Any imprecision during the growth of this stress-bearing structure would lead to a loss of the specific shape of the bacterium, or even to bacteriolysis caused by endogenous autolysins.

4.3.1 Safe Strategies for the Enlargement of the Stress-bearing Murein

Koch introduced the concept of the "make-before-break" strategy for growth of the stress-bearing sacculus. It implies that in a safe mechanism covalent bonds are first formed before meshes are opened to allow the insertion of the new material. In Gram-positive bacteria, this concept is believed to be realized by an "inside-to-outside" growth mechanism which is depicted in Figure 10 (Koch and Doyle, 1985). New murein is synthesized and linked to the existing murein at the inside of the sacculus. As a result of the enzymatic digestion of the outermost layers, the new material subsequently becomes part of the stress-bearing layers.

Until now, no direct proof has been available for a role of murein hydrolases in enabling cell growth of Gram-positive bacteria by degradation of the outer murein layers, as depicted in Figure 10. On the other hand, murein hydrolases were shown to be involved in cutting the newly formed septum. The major staphylococcal autolysin (Atl) is bifunctional and after processing both enzymatically active parts – a *N*-acetylglucosaminidase and an amidase (Oshida et al., 1995) – are found to localize in an equatorial ring at the site of division on the surface of the bacterium (Yamada et al., 1996). In *S. pneumoniae*, the cell separation is ensured mainly by the LytB glucosaminidase as indicated by the "chaining" phenotype of a mutant strain lacking the enzyme (Garcia et al., 1999). In Gram-negative bacteria, the splitting of the new septum occurs simultaneously with the septum formation, leading to the progression of a constriction during cell division, and recent studies on deletion mutants indicate that in *E. coli* the periplasmic amidases are of major impor-

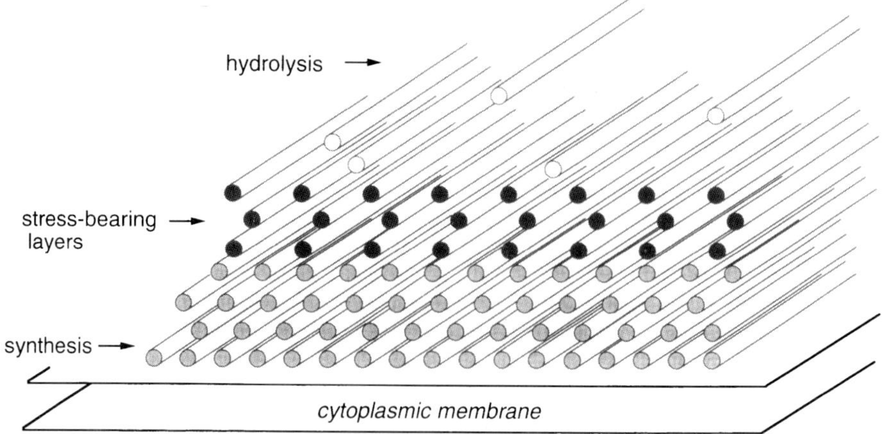

Fig. 10 "Inside-to-outside" growth mechanism in Gram-positive bacteria. The addition of new murein in a relaxed state is performed underneath the inner side of the multilayered, stress-bearing murein sacculus. Concomitantly, the outermost layers are specifically degraded by hydrolytic enzymes. The glycan strands are represented by rods; peptide cross-links are not indicated.

tance in this process (C. Heidrich and J.-V. Höltje, unpublished results).

Both, the formation and cutting of the septum and the enlargement of the sacculus during cell elongation, is a risky challenge in Gram-negative bacteria with their thin, mainly single-layered sacculus. It requires a strict regulation of the potentially autolytic murein hydrolases, which are present in the periplasm and the uncontrolled action of which causes immediate lysis. Different growth models for the murein sacculus of *E. coli* have been proposed. According to Park, single new glycan strands are inserted between the pre-existing ones by the action of a bifunctional transglycosylase/transpeptidase (Park and Burman, 1985). Insertion is possible because of the preceding cutting of the peptide bridges by endopeptidases. The hernia model of Norris implies that murein synthesis occurs preferentially at weak spots where the cytoplasmic membrane starts bulging out due to the action of murein hydrolases (Norris and Manners, 1993).

The "three-for-one" model introduced by Höltje is based on the "make-before-break" strategy, and on the control of the important but potentially dangerous murein hydrolase activities within multi-enzyme complexes.

4.3.2
"Three-for-one" Model (Höltje, 1996)

Results from pulse and pulse–chase experiments showed the dynamics of the maturation of new murein after its insertion into the sacculus (Glauner and Höltje, 1990). As

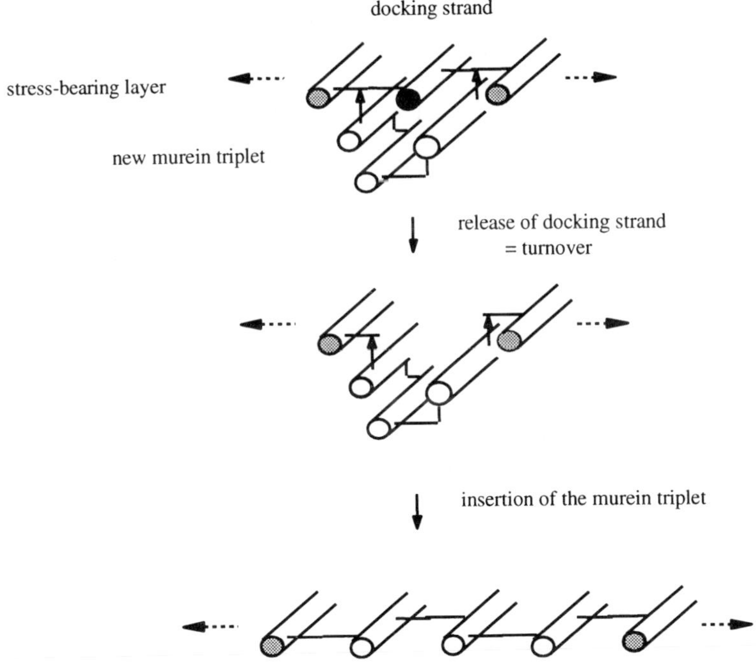

Fig. 11 The "three-for-one" growth mechanism. A triplet of newly synthesized murein is attached to the free amino groups present in the donor peptides on both sides of the central "docking strand". Removal of the docking strand allows for insertion of the new material in the stress-bearing murein layer. Glycan strands are indicated as rods; peptide links by black lines; the arrows point towards the direction of tension. (Adapted from Höltje, 1996.)

expected, peptides with pentapeptide side chains decrease rapidly because they are the substrates for transpeptidases and DD-carboxy/endopeptidases. At the same time, there is an increase in cross-linkage, but the label is shifting from being present in donor peptides (which donate the D-Ala for cross-linkage) to acceptor peptides (to which the cross-linkage occurs), and this shift is twice as fast in TetraTri dimers as compared with TetraTetra dimers. Trimeric cross-bridges have a short half-life; in addition, there is a decrease in the average length of the glycan strands and an increase in lipoprotein attachment sites.

These observations, and the finding that 40–50% of the material is released from the sacculus per generation (murein turnover) (Goodell, 1985; Goodell and Schwarz, 1985), led to the three-for-one growth model being proposed by Höltje (1998). According to this model, three new glycan strands are synthesized by murein synthases, cross-linked to each other, and cross-linked to dimeric peptide bridges on both sides of a docking strand in the existing murein. The removal of the docking strand by murein hydrolases (autolysins) provokes the insertion of the new cross-linked triplet into the sacculus (Figure 11).

Höltje also proposed that the murein synthases and hydrolases form multienzyme complexes in order to achieve a tight spatial and temporal cooperation of the different enzyme activities, and especially to control the autolysins which digest the docking strands. Indeed, specific interactions between several PBPs, lytic transglycosylases and endopeptidases were detected by affinity chromatography (Romeis and Höltje, 1994; von Rechenberg et al., 1996). In addition, a structural protein (MipA) was identified, which has a highly specific affinity to both a murein synthase (PBP1B) and to a murein hydrolase (MltA). The trimeric complex composed of these proteins was reconstituted on a BIAcore chip (Vollmer et al., 1999). The interaction studies indicated further that there is probably some variation in the composition of different complexes, but that the enzyme activities might be generally combined in the way shown in Figure 12. Two bifunctional transglycosylases/transpeptidases (PBP1A or PBP1B) and one monofunctional transglycosylase (MtgA) synthesize the new cross-linked murein triplet. The transpeptidases (PBP2 or PBP3) covalently attach the murein triplet to the free amino groups present in the donor peptides of the cross-links on both sides of the central docking strand in the pre-existing murein layer. Due to the specific action of the lytic members in the multi-enzyme complex (amidases, lytic transglycosylases, endopeptidases) the central docking strand is removed and digested. The mechanism can be applied for cell elongation as well as for cell division. Upon release of the central docking strand in the course of elongation the newly attached murein triplet is pulled into the layer under tension. This occurs at many locations on the cylindrical part of the cell wall. During cell division the attachment of the murein triplets is restricted to the site of septum formation. Multiple new triplets are attached to each other, followed by the release of the docking strands. An inward pull of the murein biosynthesis complex from the contracting FtsZ ring, the contractile component of the cytoplasmic cell division machinery (divisome), results in the formation of a septum. The monofunctional transpeptidases which attach the new triplet to the old murein are likely to be part of two distinct machines and involved in either elongation (PBP2) or cell division (PBP3).

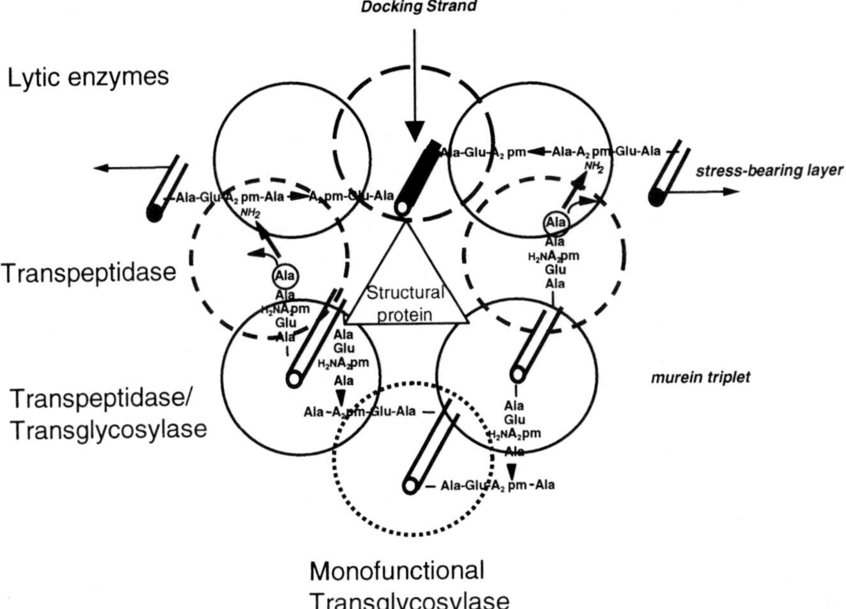

Fig. 12 Model for the proposed multienzyme complex of murein biosynthesis acting according to the "three-for-one" growth mechanism. The new murein is synthesized and attached to the stress-bearing murein layer due to the concerted action of different transpeptidases and transglycosylases. Concomitantly, the release of the central "docking strand" will be performed by the lytic enzymes in the complex. (Adapted from Höltje; 1998.)

4.4
Recycling of the Turnover Products

Turnover products are released during growth from the murein sacculus of both Gram-positive and Gram-negative bacteria. Released cell wall fragments are found in the culture supernatants of Gram-positive bacteria, but to a much lesser extent in Gram-negative bacteria in which they are effectively recycled. About 40–50% of the material of the sacculus of *E. coli* is released per generation, but only about 5% escapes recycling and is lost to the medium (Goodell, 1985; Goodell and Schwarz, 1985). The major turnover products are tripeptide, tetrapeptide, TetraAnh and TriAnh which are the products of amidases, lytic transglycosylases, DD-endopeptidases, and LD-carboxypeptidases, and which enter the recycling pathway (Figure 13).

Several uptake systems have been described to be involved in recycling. For the import of murein tripeptide into *E. coli*, the periplasmic murein tripeptide binding protein (MppA) is essential. Like the nonspecific peptide binding protein OppA, MppA is recognized by the membrane-bound components of the general oligopeptide uptake system (Opp) which transfers the oligopeptides, including murein tripeptide, into the cytoplasm (Goodell and Higgins, 1987; Park et al., 1998). The fate of the sugar moieties of the murein is still unclear. The 1,6-anhydromuropeptides are imported via the AmpG transporter. Inside the cytoplasm, further degradation and conversion takes place. The *N*-acetylmuramyl-L-alanine-amidase AmpD

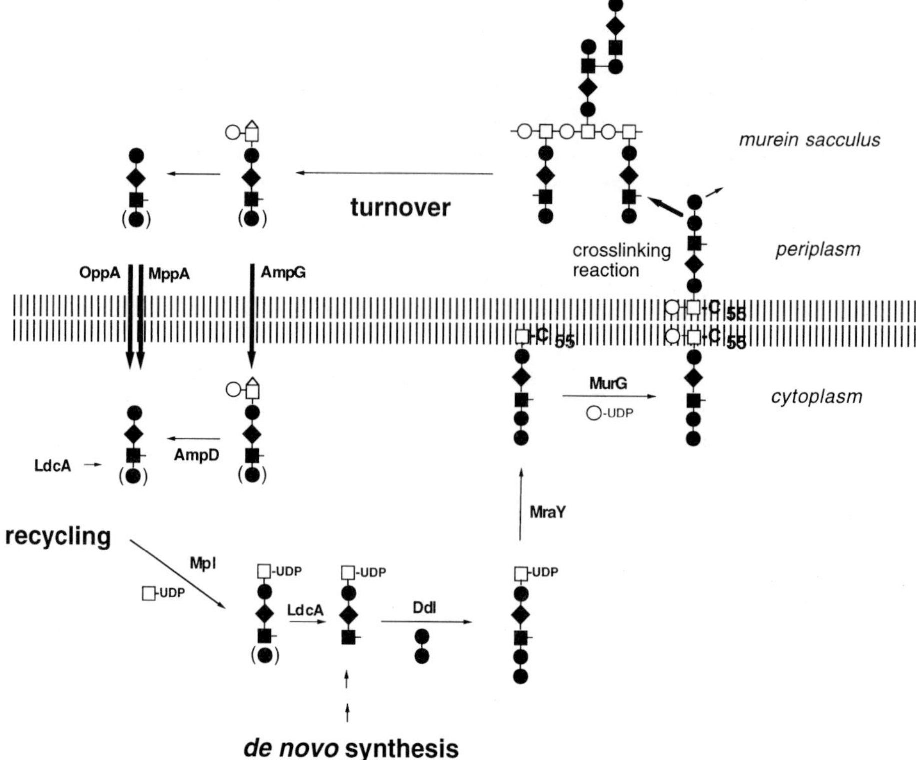

Fig. 13 Schematic drawing of the recycling of turnover products. Murein turnover products are internalized by the different import systems. They are degraded by the activity of AmpD into the peptides and saccharides. LdcA releases the D-Ala at position 4 of different intermediates. The muropeptide ligase Mpl attaches the peptides to UDP-MurNAc to form UDP-MurNAc(tripeptide) or UDP-MurNAc(tetrapeptide). Only UDP-MurNAc(tripeptide) enters the pathway of the synthesis of lipid II (see also Figure 8). Open square, MurNAc; open square plus triangle, MurNAcAnh; open circle, GlcNAc; closed circle, Ala; closed diamond, Glu; closed square, m-A_2pm; C_{55}, undecaprenyl pyrophosphate.

specifically releases the peptide moiety from the 1,6-anhydromuropeptides (Höltje et al., 1994), which is then trimmed to a tripeptide by the LD-carboxypeptidase LdcA (Templin et al., 1999). The tripeptide is ligated to UDP-MurNAc by Mpl, which results in the formation of the murein precursor UDP-MurNAc(tripeptide) that can be utilized for synthesis of new murein. Interestingly, the interference with the recycling pathway by inactivating the *ldcA* gene triggered autolysis in the stationary phase (Templin et al., 1999).

Previously, it was demonstrated that the murein turnover products containing 1,6-anhydroMurNAc induce the production of the chromosomal β-lactamase AmpC in the family of Enterobacteriaceae. They serve as signals for the conversion of the regulatory protein AmpR into an activator for the expression of the β-lactamase (Jacobs et al., 1997). In contrast, the murein precursor UDP-MurNAc(pentapeptide) keeps AmpR in the inactive state. Thus, the balance between turnover and synthesis can be measured by the cell. In the case of a disturbance of murein synthesis due to the presence of β-lactam antibiotics, the intracellular concentration of the murein turnover products is increased, which in turn induces the expression of the β-lactamase which is then secreted and capable of inactivating the β-lactam antibiotic (see below). Accordingly, the inducible expression of β-lactamase is blocked in deletion mutants of the lytic transglycosylases due to the reduction in inductor molecules (Kraft et al., 1999).

4.5 Inhibition of Murein Biosynthesis

The murein biosynthesis represents a metabolic pathway which is not only of vital importance for the bacteria but also is unique to bacteria, making it a fundamental target in antibacterial chemotherapy. Various agents have been found to target overall murein biosynthesis (see Figure 8). Among these are the still widely used β-lactam antibiotics, the targets of which are the PBPs (see above, and Ghuysen, 1991). Both, PBPs and the majority of β-lactamases belong to the family of penicilloyl serine transferases, the enzymes which catalyze transfer of the penicilloyl moiety of penicillin to the OH group of their active site serine. The acyl enzyme intermediate is relatively inert in the case of PBPs, and the murein synthases are inactivated with this suicide substrate. In the case of the β-lactamases, the penicilloyl residue is transferred to water, resulting in the hydrolysis of the β-lactam ring which makes the producer strain resistant to the antibiotic (Figure 14).

Glycopeptide antibiotics such as vancomycin and teicoplanin also inhibit murein biosynthesis, and are used to treat severe infections caused by several Gram-positive bacteria (Arthur et al., 1996). These compounds form stable complexes with the carboxy-terminal D-Ala-D-Ala residues of the murein precursors. Thereby they prevent the incorporation of the disaccharide pentapeptide precursors into the nascent murein by transglycosylation reaction. Binding to the nascent murein is also expected to inhibit D,D-transpeptidation and D,D-carboxypeptidation. Inducible resistance to high levels of vancomycin and teicoplanin

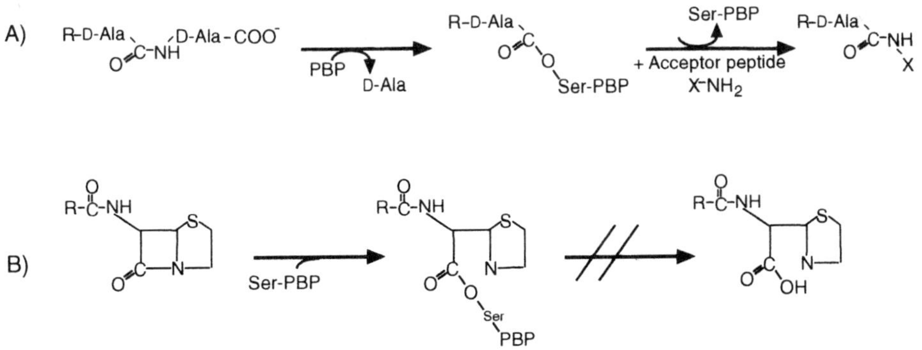

Fig. 14 Interaction of PBPs with the murein substrate and with penicillin, respectively. (A) The PBP cleaves the covalent bond of the terminal D-Ala to the penultimate D-Ala and creates a new bond between the carboxyl group of the latter D-Ala and the hydroxyl group of its active site serine. During the second step, the mureinyl moiety is transferred to the ε- amino group of a diaminopimelic acid ($X\text{-}NH_2$) in a neighboring peptide chain. (B) Interaction with penicillin results in cleavage of the β-lactam ring and formation of the penicilloyl enzyme intermediate, which is very stable. Thus, the PBP is blocked by a suicide substrate.

(VanA phenotype) is mediated by *Tn*1546-like transposons in enterococcci. The transposon encodes the dehydrogenase VanH that reduces pyruvate to D-lactate, and the ligase VanA that catalyzes the formation of an ester bond between D-Ala and D-Lac. The depsipeptide misses a hydrogen bond critical for antibiotic binding resulting in glycopeptide resistance.

Other antibiotics act at earlier stages of the murein biosynthesis (Lancini et al., 1995) (see Figure 8). Bacitracin inhibits recycling of the undecaprenylpyrophosphate to undecaprenylphosphate that is the acceptor of the muramyl-pentapeptide in the transfer reaction. Tunicamycin inhibits the formation of lipid I by MraY, whereas ramoplanin acts on the following step, the *N*-acetylglucosaminyl-transferase reaction that forms lipid II. Fosfomycin inhibits the addition of phosphoenolpyruvate to UDP-GlcNAc. Finally, cycloserine inhibits both the alanine racemase and the D-alanyl-D-alanine ligase because of its structural similarity to D-Ala.

The bacteriolysis following inhibition of murein biosynthesis is caused by autolysins, and is most probably the result of unbalancing the delicate equilibrium of the activities of murein synthetic and hydrolytic enzymes. Under some conditions, bacteria become tolerant against cell wall biosynthesis inhibitors, and they do not lose viability. This is true when there is no ongoing murein synthesis in stationary phase or in amino acid-deprived cells. In the latter case, the stringent response leads to the accumulation of (p)ppGpp in the cytoplasm, resulting in the down-regulation of the biosynthesis of phospholipids and of different cellular macromolecules, including murein.

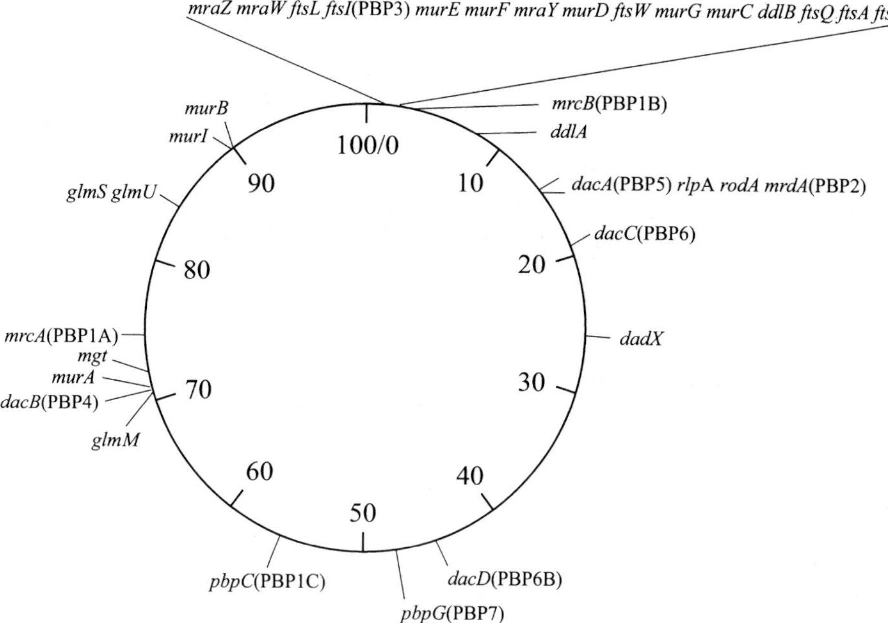

Fig. 15 The genetic map of *E. coli* with the location of the genes encoding murein biosynthesis enzymes and of the penicillin-sensitive murein hydrolases. The position of the genes encoding the other murein hydrolases are given in Table 3.

5 Molecular Genetics of Murein Biosynthesis

Most of the genes encoding murein biosynthesis enzymes and murein hydrolases have been identified in *E. coli*, and genes encoding homologous enzymes were found in many other bacteria. The localization of these genes on the *E. coli* chromosome is shown in Figure 15. There is the large dcw (mra) cluster at 2 min containing eight structural genes encoding the murein synthetic enzymes PBP3, MurE, MurF, MraY, MurD, MurG, MurC and DdlB and the associated protein of yet unproven function in murein biosynthesis FtsW, together with genes encoding cell division proteins FtsQ, FtsA and FtsZ (Matsuhashi, 1994). Most of the dcw genes are closely adjacent to each other and there are several overlaps. Transcription in the cluster proceeds in the same direction, and there is a complex regulation of gene expression which include promotors individually regulated by *cis*- and *trans*-acting signals (Vicente et al., 1998). Database analysis of completely sequenced genomes revealed that the dcw cluster is conserved in a variety of bacteria, including *B. subtilis* (Figure 16). A smaller gene cluster is found at 14.5 min in the map of *E. coli* containing the genes encoding PBP5, a putative lipoprotein of unknown function RlpA, the associated protein RodA and the murein synthase PBP2. Other genes encoding enzymes involved in murein precursor formation, murein synthases (PBP1A, PBP1B, PBP1C and MtgA) and the murein hydrolases are scattered at different locations on the chromosome, with the exceptions of the pairs of *glmS-glmU*, *mltA-amiC* and *emtA-ldcA* in which the genes are adjacent to each other (see also Table 3).

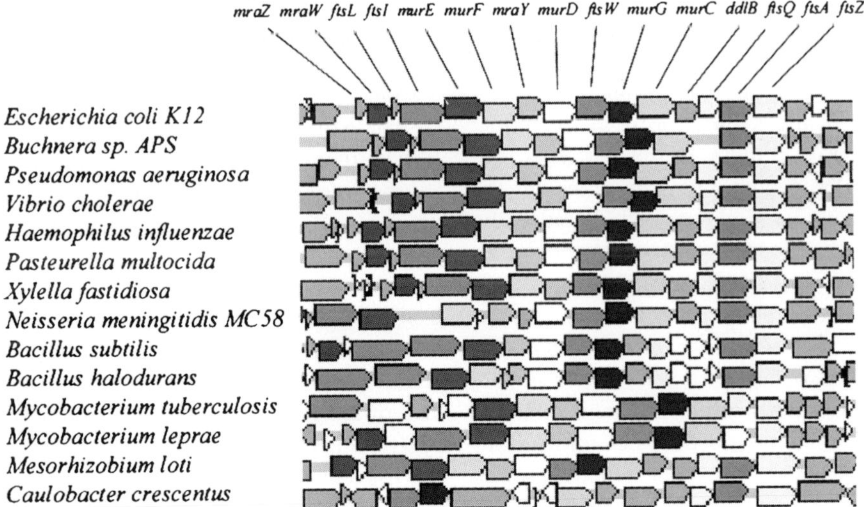

Fig. 16 The similarity of the chromosomal mra (dcw) region of *E. coli* with the corresponding chromosomal regions in other bacteria. The same colors indicate genes encoding ortholog or paralog proteins which were classified in the clusters of orthologous groups of proteins (COGs) at the NCBI (http://www.ncbi.nlm.-nih.gov). Note that only bacteria with a similar region are listed (14 out of 41 completely sequenced species, data obtained on June 1, 2001). The names of the genes of *E. coli* are shown in the first line.

6 Exogenous Murein Hydrolases

Murein hydrolases not only act on the murein of the producing bacterium, but are also ubiquitous in nature. They are involved in the release of bacteriophages from the infected bacterium, in the digestion of prey bacteria by predatory organisms, and in the host defense of higher organisms against bacterial invasion.

The bacteriophages utilize murein hydrolases for the release from the infected bacterium at the end of the lytic cycle. There are two main mechanisms by which phages are liberated from the bacterial cytoplasm where they were synthesized (Young, 1992):

1) Some small phages express a single protein which induces lysis of the host cell. For example, the lysis protein E of the coliphage ΦX174 blocks cell wall synthesis by inhibition of the formation of lipid I. The resulting lysis is caused by the autolysins of the host cell.
2) Most larger phages such as λ, T4 and T7 express two proteins to provoke lysis. A holin disrupts the cytoplasmic membrane and allows one or more phage-encoded murein hydrolases (lysozyme and/or amidase) to gain access to the murein of the host cell. Because the phage encoded hydrolases do not have an N-terminal secretion specific signal peptide, the membrane disruption by the holin is essential for their activation.

The Gram-negative soil bacterium *Myxococcus xanthus* feeds on other microorganisms. Next to antibacterial compounds, this predator produces hydrolytic enzymes including murein hydrolases, to digest the prey bacteria (Sudo and Dworkin, 1972). Murein hydrolases were also found in the culture supernatant of fungi such as the lysozyme (β-1,4-*N*-acetylmuramidase/ β-1,4-*N*,6-*O*-diacetylmuramidase) from *Chalaropsis* (Lyne et al., 1990) which proved to be a valuable tool for murein analysis. Similar lysozymes were found in many *Streptomyces* spp., e.g., the mutanolysin of *S. globisporus* (Lichenstein et al., 1990). The protozoan parasite *Entamoeba histolytica* colonizes the lower intestine of humans and feeds on bacteria in the colon. The amoeba engulfs the prey bacteria, and then kills and lyses them using a combination of antibacterial peptides and a lysozyme (Jacobs and Leippe, 1995).

Virtually all higher organisms, plants, insects and vertebrates including humans, express murein hydrolases in their tissues and fluids as a first line of defense against invading bacteria and to remove bacteria that have been killed by other factors. These enzymes are mostly lysozymes (reviewed in Jolles, 1996), amidases or endopeptidases which are capable of dissolving murein sacculi. In humans, lysozyme is present in tears, saliva, leukocytes, skin, human milk and serum, and it is secreted in high amounts by cells of the immune system at the sites of infection. In addition, there is a human *N*-acetylglucosaminidase and a *N*-acetylmuramic acid-L-alanine amidase present in granulocytes and in serum.

7 Biological Activities of Murein

Murein turnover products from growing bacteria and degradation products generated by lysis of invading bacteria are detectable in the tissues and serum of mammals, and have a variety of biological effects (reviewed in Seidl and Schleifer, 1986). Due to the complex structure of the bacterial cell wall, the fragments released by turnover and lysis may not represent pure murein structures, but may contain additional covalently attached compounds such as teichoic acid, capsular polysaccharides and/or surface proteins. In addition, mixtures of different bacterial cell wall components may potentiate the capacity to cause inflammatory reactions as in the case of lipopolysaccharide (LPS) and murein.

High molecular-weight cell wall fragments are usually cleared rapidly by the action of hydrolytic enzymes of macrophages or of serum. However, in some cases, large fragments of murein or murein–polysaccharide compounds may persist and circulate in the host organism because they are highly resistant to the action of host enzymes. In the case of the pathogens *S. aureus* and *N. gonorrhoeae*, the peptidoglycan persistence was attributed to the O-acetylated murein (Clarke and Dupont, 1992). The poorly lysozyme-digestable high molecular-weight fragments can induce chronic inflammatory processes in animal models, and they are believed to contribute to gonococcal arthritis in humans. *In vitro* digestion with a muramidase led to the arthropathic potential of the O-acetylated murein being lost. Similarly, it was found that murein or soluble high molecular-weight murein fragments obtained by an endopeptidase digestion activate the complement cascade, but not smaller fragments obtained from a digestion of murein with a muramidase.

In contrast, there are a variety of biological effects including induction of slow-wave sleep, fever, leukocytosis and septic shock that can be attributed to small murein fragments. The minimal structural requirement for the expression of these bioactivities is *N*-acetylmuramyl-L-Ala-D-Glu (muramyl dipeptide or MDP). The precise mechanisms of recognition of these molecules by cells of the immune system and the induced signal transduction cascade are only poorly understood. Soluble murein fragments most likely bind to the CD14 molecule on human peripheral monocytes (Rietschel et al., 1998), a receptor which also recognizes other bacterial activators of immune cells including LPS, polymannuronic acid, chitosan, and lipoarabinomannan. The binding of bacterial compounds to CD14 leads to the production and release of the endogenous mediators tumor necrosis factor α (TNFα), interleukin 1 (IL-1) and interleukin 6 (IL-6) which are ultimately responsible for the severe consequences of bacterial sepsis. Recently, a murein recognition protein (PGRP) was identified in the silkworm which turned out to have similarity in sequence to the *N*-acetylmuramyl-L-alanine amidase from bacteriophage T7 but did not have a hydrolytic activity against murein. PGRP was found to be conserved from insects to mammals, and is believed to be ubiquitously involved in innate immunity (Kang et al., 1998).

8 Production

At present, it is unclear whether murein is available commercially, although the murein degradation product MurNAc-L-Ala-D-(iso)Gln, its isomer MurNAc-L-Ala-L-(iso)Gln, and two acylated derivatives MurNAc(6-O-stearoyl)-L-Ala-D-(iso)Gln and MurNAc-L-Ala-D-(iso)Gln-*N*-ε-stearoyl-L-Lys are used as adjuvants to enhance the antigenicity of weak antigens (Ellouz et al., 1974). These compounds can be purchased from Sigma-Aldrich.

9 Current Problems and Outlook

Although much progress has been made in the understanding of the enzymology of murein biosynthesis and in the structural analysis of the sacculus over the past few decades, a number of important questions relating to bacterial cell wall polysaccharides remain unanswered. The overall structure of the sacculus can only be speculated on, because a powerful high-resolution method which would give a detailed molecular picture of the murein sacculus is not available. Therefore, the question remains as to whether the glycan strands in murein are arranged in parallel, or whether different glycan strands are carrying different sets of peptide side chains – as proposed by the three-for-one model. Also, the arrangement of the glycan strands and peptide bridges with respect to the geometry of the cell is not known, though experimental data exist which indicate that there are as yet unidentified (structural?) proteins participating in different murein-synthesizing multienzyme complexes (T. Kast, G. K. Schiffer, W. Vollmer and J.-V. Höltje, unpublished results). The interaction of those complexes involved in cell division with other components of the divisome remains to be elucidated. The questions of why some bacteria such as *E. coli* or *B. subtilis* produce an astonishing set of lytic enzymes while others such as *S. pneumoniae* have only a few, and whether the presence of murein hydrolases is essential for bacterial growth cannot yet be answered. Also, there is no convincing model for the controlled enlargement of the thick murein in Gram-positive bacteria, and the mechanism(s) for the regulation of the autolytic activities in these bacteria are mostly unknown. Moreover, there are various phenomena that are either directly or indirectly related to structural changes in the bacterial cell wall. The resistance of Gram-positive pathogens to cell wall-targeting antibiotics is often accompanied by changes in the biosynthetic enzymes or in the murein structure, and the exact mechanisms by which these changes lead to antibiotic resistance are often unclear. The world-wide spread of multi-resistant pathogenic strains of *S. aureus*, *S. pneumoniae* and other pathogens is of great clinical importance, and there is hope that a better understanding in the mechanisms of cell wall biosynthesis and of antibiotic resistance might lead to the discovery of new targets for future antibiotics. The interaction of the pathogenic bacteria with components of the host organism is mediated by its cell wall, and the present data indicate that the murein and murein degradation products are part of this interplay. However, the exact mechanisms by which bacterial cell wall components are recognized by different systems of the host, and how the response to bacterial challenge is regulated, remain to be elucidated. A better understanding of these processes may allow interventions with drugs to be made in order to avoid the severe septic shock syndrome caused by these cell wall compounds.

Acknowledgments

The authors are grateful to J.-V. Höltje for providing us with the opportunity to contribute to this series, for his continuous encouragement and for many stimulating discussions. We also thank U. Schwarz for his interest and support, and V. Kastner and D. Mauch for their critical reading of the manuscript.

10
References

Allmaier, G., Schmid, E. R. (1993) New mass spectrometric methods for peptidoglycan analysis, in: *Bacterial Growth and Lysis: Metabolism and Structure of the Bacterial Sacculus* (de Pedro, M. A., Höltje, J.-V., Löffelhardt, W., Eds.), London/New York: Plenum Press, 23–30.

Arthur, M., Reynolds, P., Courvalin, P. (1996) Glycopeptide resistance in enterococci, *Trends Microbiol.* **4**, 401–407.

Atrih, A., Foster, S. J. (1999) The role of peptidoglycan structure and structural dynamics during endospore dormancy and germination, *Antonie Van Leeuwenhoek* **75**, 299–307.

Barnikel, G., Naumann, D., Bradaczek, H., Labischinski, H., Giesbrecht, P. (1983) Computer aided molecular modelling of the three-dimensional structure of bacterial peptidoglycan, in: *The Target of Penicillin: The Murein Sacculus of Bacterial Cell Walls. Architecture and Growth* (Hakenbeck, R., Höltje, J.-V., Labischinski, H., Eds.), Berlin/New York: de Gruyter, 61–66.

Boneca, I. G., Huang, Z. H., Gage, D. A., Tomasz, A. (2000) Characterization of *Staphylococcus aureus* cell wall glycan strands, evidence for a new beta-N-acetylglucosaminidase activity, *J. Biol. Chem.* **275**, 9910–9918.

Braun, V. (1975) Covalent lipoprotein from the outer membrane of *Escherichia coli*, *Biochim. Biophys. Acta* **415**, 335–377.

Caparros, M., Pisabarro, A. G., de Pedro, M. A. (1992) Effect of D-amino acids on structure and synthesis of peptidoglycan in *Escherichia coli*, *J. Bacteriol.* **174**, 5549–5559.

Chapman, G. P., Hillier, J. (1953) Electron microscopy of ultra-thin sections of bacteria. I. Cellular division in *Bacillus cereus*, *J. Bacteriol.* **66**, 362–373.

Clarke, A. J., Dupont, C. (1992) O-acetylated peptidoglycan: its occurrence, pathobiological significance, and biosynthesis, *Can. J. Microbiol.* **38**, 85–91.

Demchick, P., Koch, A. L. (1996) The permeability of the wall fabric of *Escherichia coli* and *Bacillus subtilis*, *J. Bacteriol.* **178**, 768–773.

Ellouz, F., Adam, A., Ciorbaru, R., Lederer, E. (1974) Minimal structural requirement for adjuvant activity of bacterial peptidoglycan derivatives, *Biochem. Biophys. Res. Commun.* **59**, 1317–1325.

Filipe, S. R., Tomasz, A. (2000) Inhibition of the expression of penicillin resistance in *Streptococcus pneumoniae* by inactivation of cell wall muropeptide branching genes, *Proc. Natl. Acad. Sci. USA* **97**, 4891–4896.

Garcia, P., Gonzalez, M. P., Garcia, E., Lopez, R., Garcia, J. L. (1999) LytB, a novel pneumococcal murein hydrolase essential for cell separation, *Mol. Microbiol.* **31**, 1275–1277.

Ghuysen, J. M. (1991) Serine beta-lactamases and penicillin-binding proteins, *Annu. Rev. Microbiol.* **45**, 37–67.

Glauner, B. (1988) Separation and quantification of muropeptides with high-performance liquid chromatography, *Anal. Biochem.* **172**, 451–464.

Glauner, B., Höltje, J.-V. (1990) Growth pattern of the murein sacculus of *Escherichia coli*, *J. Biol. Chem.* **265**, 18988–18996.

Glauner, B., Höltje, J.-V., Schwarz, U. (1988) The composition of the murein of *Escherichia coli*, *J. Biol. Chem.* **263**, 10088–10095.

Goodell, E. W. (1985) Recycling of murein by *Escherichia coli*, *J. Bacteriol.* **163**, 305–310.

Goodell, E. W., Higgins, C. F. (1987) Uptake of cell wall peptides by *Salmonella typhimurium* and *Escherichia coli*, *J. Bacteriol.* **169**, 3861–3865.

Goodell, E. W., Schwarz, U. (1985) Release of cell wall peptides into culture medium by exponentially growing *Escherichia coli*, *J. Bacteriol.* **162**, 391–397.

Harz, H., Burgdorf, K., Höltje, J.-V. (1990) Isolation and separation of the glycan strands from murein of *Escherichia coli* by reversed-phase high-performance liquid chromatography, *Anal. Biochem.* **190**, 120–128.

Hayashi, H., Araki, Y., Ito, E. (1973) Occurrence of glucosamine residues with free amino groups in cell wall peptidoglycan from bacilli as a factor responsible for resistance to lysozyme, *J. Bacteriol.* **113**, 592–598.

Höltje, J.-V. (1995) From growth to autolysis: the murein hydrolases in *Escherichia coli, Arch. Microbiol.* **164**, 243–254.

Höltje, J.-V. (1996) Molecular interplay of murein synthases and murein hydrolases in *Escherichia coli, Microb. Drug Resist.* **2**, 99–103.

Höltje, J.-V. (1998) Growth of the stress-bearing and shape-maintaining murein sacculus of *Escherichia coli, Microbiol. Mol. Biol. Rev.* **62**, 181–203.

Höltje, J.-V., Schwarz, U. (1985) Biosynthesis and growth of the murein sacculus, in: *Molecular Cytology of Escherichia coli* (Nanninga, N., Ed.), London: Academic Press, 77–119.

Höltje, J.-V., Mirelman, D., Sharon, N., Schwarz, U. (1975) Novel type of murein transglycosylase in *Escherichia coli, J. Bacteriol.* **124**, 1067–1076.

Höltje, J.-V., Kopp, U., Ursinus, A., Wiedemann, B. (1994) The negative regulator of β-lactamase induction AmpD is a N-acetyl-anhydromuramyl-L-amidase, *FEMS Microbiol. Lett.* **122**, 159–164.

Jacobs, C., Frere, J. M., Normark, S. (1997) Cytosolic intermediates for cell wall biosynthesis and degradation control inducible beta-lactam resistance in Gram-negative bacteria, *Cell* **88**, 823–832.

Jacobs, T., Leippe, M. (1995) Purification and molecular cloning of a major antibacterial protein of the protozoan parasite *Entamoeba histolytica* with lysozyme-like properties, *Eur. J. Biochem.* **231**, 831–838.

Jolles, P. (Ed.) (1996) Lysozymes: Model enzymes in Biochemistry and Biology, Basel: Birkhäuser.

Kang, D., Liu, G., Lundstrom, A., Gelius, E., Steiner, H. (1998) A peptidoglycan recognition protein in innate immunity conserved from insects to humans, *Proc. Natl. Acad. Sci. USA* **95**, 10078–10082.

Koch, A. L., Doyle, R. J. (1985) Inside-to-outside growth and turnover of the wall of Gram-positive rods, *J. Theor. Biol.* **117**, 137–157.

Kraft, A. R., Prabhu, J., Ursinus, A., Höltje, J.-V. (1999) Interference with murein turnover has no effect on growth but reduces beta-lactamase induction in *Escherichia coli, J. Bacteriol.* **181**, 7192–7198.

Labischinski, H., Barnikel, G., Naumann, D. (1983) The state of order of bacterial peptidoglycan, in: *The Target of Penicillin: The Murein Sacculus of Bacterial Cell Walls. Architecture and Growth* (Hakenbeck, R., Höltje, J.-V., Labischinski, H., Eds.), Berlin/New York: de Gruyter, 49–54.

Labischinski, H., Goodell, E. W., Goodell, A., Hochberg, M. L. (1991) Direct proof of a 'more-than-single-layered' peptidoglycan architecture of *Escherichia coli* W7: a neutron small-angle scattering study, *J. Bacteriol.* **173**, 751–756.

Lancini, G., Parenti, F., Gallo, G. G. (1995) *Antibiotics: A Multidisciplinary Approach*, New York: Plenum Press.

Lichenstein, H. S., Hastings, A. E., Langley, K. E., Mendiaz, E. A., Rohde, M. F., Elmore, R., Zukowski, M. M. (1990) Cloning and nucleotide sequence of the N-acetylmuramidase M1-encoding gene from *Streptomyces globisporus, Gene* **88**, 81–86.

Lyne, J. E., Carter, D. C., He, X. M., Stubbs, G., Hash, J. H. (1990) Preliminary crystallographic examination of a novel fungal lysozyme from *Chalaropsis, J. Biol. Chem.* **265**, 6928–6930.

Mainardi, J. L., Legrand, R., Arthur, M., Schoot, B., van Heijenoort, J., Gutmann, L. (2000) Novel mechanism of beta-lactam resistance due to bypass of DD-transpeptidation in *Enterococcus faecium, J. Biol. Chem.* **275**, 16490–16496.

Matsuhashi, M. (1994) Utilization of lipid-linked precursors and the formation of peptidoglycan in the process of cell growth and division: membrane enzymes involved in the final steps of peptidoglycan synthesis and the mechanism of their regulation, in: *Bacterial cell wall, New Comprehensive Biochemistry Vol. 27* (Ghuysen, J.-M., Hakenbeck, R., Eds.), Amsterdam: Elsevier Science B. V., 55–71.

Murray, R. G. E., Steed, P., Elson, H. E. (1965) Location of mucopeptide in sections of cell wall of *Escherichia coli* and other Gram-negative bacteria, *Can. J. Microbiol.* **11**, 547–560.

Navarre, W. W., Schneewind, O. (1999) Surface proteins of Gram-positive bacteria and mechanisms of their targeting to the cell wall envelope, *Microbiol. Mol. Biol. Rev.* **63**, 174–229.

Norris, V., Manners, B. (1993) Deformations in the cytoplasmic membrane of *Escherichia coli* direct the synthesis of peptidoglycan. The hernia model, *Biophys. J.* **64**, 1691–1700.

Oshida, T., Sugai, M., Komatsuzawa, H., Hong, Y. M., Suginaka, H., Tomasz, A. (1995) A *Staphylococcus aureus* autolysin that has an *N*-acetylmuramoyl-L-alanine amidase domain and an endo-beta-*N*-acetylglucosaminidase domain: cloning, sequence analysis, and characterization, *Proc. Natl. Acad. Sci. USA* **92**, 285–289.

Pellegrini, M., Marcotte, E. M., Thompson, M. J., Eisenberg, D., Yeates, T. O. (1999) Assigning protein functions by comparative genome analysis: protein phylogenetic profiles, *Proc. Natl. Acad. Sci. USA* **96**, 4285–4288.

Park, J. T. (1996) The convergence of murein recycling research with β-lactamase research, *Microb. Drug Resist.* **2**, 105–112.

Park, J. T., Burman, L. G. (1985) Elongation of the murein sacculus of *Escherichia coli*, *Ann. Inst. Pasteur Microbiol.* **136A**, 51–58.

Park, J. T., Raychaudhuri, D., Li, H., Normark, S., Mengin-Lecreulx, D. (1998) MppA, a periplasmic binding protein essential for import of the bacterial cell wall peptide L-alanyl-gamma-D-glutamyl-meso-diaminopimelate, *J. Bacteriol.* **180**, 1215–1223.

Rietschel, E. T., Schletter, J., Weidemann, B., El-Samalouti, V., Mattern, T., Zahringer, U., Seydel, U., Brade, H., Flad, H. D., Kusumoto, S., Gupta, D., Dziarski, R., Ulmer, A. J. (1998) Lipopolysaccharide and peptidoglycan: CD14-dependent bacterial inducers of inflammation, *Microb. Drug Resist.*, **4**, 37–44.

Romeis, T., Höltje, J.-V. (1994) Specific interaction of penicillin-binding proteins 3 and 7/8 with soluble lytic transglycosylase in *Escherichia coli*, *J. Biol. Chem.* **269**, 21603–21607.

Salton, M. R. J. (1994) The bacterial cell envelope – a historical perspective, in: *Bacterial Cell Wall, New Comprehensive Biochemistry, Vol. 27* (Ghuysen, J.-M., Hakenbeck, R., Eds.), Amsterdam: Elsevier Science B. V., 1–22.

Salton, M. R. J., Horne, R. W. (1951) Studies of the bacterial cell wall. 2. Methods of preparation and some properties of cell walls, *Biochim. Biophys. Acta* **7**, 177–197.

Schleifer, K. H., Kandler, O. (1972) Peptidoglycan types of bacterial cell walls and their taxonomic implications, *Bacteriol. Rev.* **36**, 407–477.

Seidl, P. H., Schleifer, K. H. (Eds.) (1986) *Biological Properties of Peptidoglycan*. Berlin/New York: de Gruyter.

Shockman, G. D., Barrett, J. F. (1983) Structure, function, and assembly of cell walls of Gram-positive bacteria, *Annu. Rev. Microbiol.* **37**, 501–527.

Shockman, G. D., Höltje, J.-V. (1994) Bacterial peptidoglycan (murein) hydrolases, in: *Bacterial Cell Wall, New Comprehensive Biochemistry, Vol. 27* (Ghuysen, J.-M., Hakenbeck, R., Eds.), Amsterdam: Elsevier Science B. V., 131–166.

Sieradzki, K., Tomasz, A. (1997) Inhibition of cell wall turnover and autolysis by vancomycin in a highly vancomycin-resistant mutant of *Staphylococcus aureus*, *J. Bacteriol.* **179**, 2557–2566.

Sudo, S., Dworkin, M. (1972) Bacteriolytic enzymes produced by *Myxococcus xanthus*, *J. Bacteriol.* **110**, 236–245.

Templin, M., Ursinus A., Höltje, J.-V. (1999) A defect in cell wall recycling triggers autolysis during the stationary growth phase of *Escherichia coli*, *EMBO J.* **18**, 4108–4117.

van Heijenoort, J. (1998) Assembly of the monomer unit of bacterial peptidoglycan, *Cell. Mol. Life Sci.* **54**, 300–304.

Vicente, M., Gomez, M. J., Ayala, J. A. (1998) Regulation of transcription of cell division genes in the *Escherichia coli* dcw cluster, *Cell. Mol. Life Sci.* **54**, 317–324.

Vötsch, W., Templin, M. (2000) Characterization of a β-N-acetylglucosaminidase of *Escherichia coli* and elucidation of its role in muropeptide recycling and β-lactamase induction, *J. Biol. Chem.* **275**, 39032–39038.

Vollmer, W., Tomasz, A. (2000) The pgdA gene encodes for a peptidoglycan N-acetylglucosamine deacetylase in *Streptococcus pneumoniae*, *J. Biol. Chem.* **275**, 20496–20501.

Vollmer, W., von Rechenberg, M., Höltje, J.-V. (1999) Demonstration of molecular interactions between the murein polymerase PBP1B, the lytic transglycosylase MltA, and the scaffolding protein MipA of *Escherichia coli*, *J. Biol. Chem.* **274**, 6726–6734.

von Rechenberg, M., Ursinus, A., Höltje, J.-V. (1996) Affinity chromatography as a means to study multienzyme complexes involved in murein synthesis, *Microb. Drug Resist.* **2**, 155–157.

Weidel, W., Pelzer, H. (1964) Bagshaped macromolecules – a new outlook on bacterial cell walls, *Adv. Enzymol.* **26**, 193–232.

Yamada, S., Sugai, M., Komatsuzawa, H., Nakashima, S., Oshida, T., Matsumoto, A., Suginaka, H. (1996) An autolysin ring associated with cell separation of *Staphylococcus aureus*, *J. Bacteriol.* **178**, 1565–1571.

Young, R. (1992) Bacteriophage lysis: mechanism and regulation, *Microbiol. Rev.* **56**, 430–481.

Zipperle, G. F., Jr., Ezzell, J. W., Jr., Doyle, R. J. (1984) Glucosamine substitution and muramidase susceptibility in *Bacillus anthracis*, *Can. J. Microbiol.* **30**, 553–559.

18
Teichoic and Teichuronic Acids from Gram-Positive Bacteria

Dr. Vladimir Lazarevic[1], **Dr. Harold M. Pooley**[2], **Dr. Catherine Mauël**[3], **Prof. Dimitri Karamata**[4]

[1] Institut de génétique et de biologie microbiennes, University of Lausanne, Rue César-Roux 19, CH-1005 Lausanne, Switzerland; Tel. +41-21-320-6075; Fax +41-21-320-60-78; E-mail: Vladimir.Lazarevic@igbm.unil.ch

[2] Institut de génétique et de biologie microbiennes, University of Lausanne, Rue César-Roux 19, CH-1005 Lausanne, Switzerland; Tel. +41-21-320-6075; Fax +41-21-320-60-78; E-mail: Harold.Pooley@igbm.unil.ch

[3] Institut de génétique et de biologie microbiennes, University of Lausanne, Rue César-Roux 19, CH-1005 Lausanne, Switzerland; Tel. +41-21-320-6075; Fax +41-21-320-60-78; E-mail: Catherine.Mauel@igbm.unil.ch

[4] Institut de génétique et de biologie microbiennes, University of Lausanne, Rue César-Roux 19, CH-1005 Lausanne, Switzerland; Tel. +41-21-320-6075; Fax +41-21-320-60-78; E-mail: Dimitri.Karamata@igbm.unil.ch

^{13}C-NMR	^{13}C-Nuclear Magnetic Resonance
CDPGro	cytidine diphosphoglycerol
D-ala	D-alanine
GalNAc	N-acetylgalactosamine
Glc	glucose
GlcA	glucuronic acid
GlcNAc	N-acetylglucosamine
IPTG	isopropylthio-β-D-galactoside
LTA(s)	lipoteichoic acid(s)
LU	linkage unit
ManNAc	N-acetylmannosamine
MIC	minimal inhibitory concentration
PG	peptidoglycan, murein
poly(GlcGalNAcP)	poly(3-*O*-β-D-glucopyranosyl N-acetylgalactosamine 1-phosphate)
poly(GroP)	poly(glycerol phosphate)
poly(RboP)	poly(ribitol phosphate)
TUA(s)	teichuronic acid(s)
UDPGlc	uridine 5′-diphosphoglucose

UDPGlcPPase	UDPGlc pyrophosphorylase
WTA(s)	wall teichoic acid(s)

1 Introduction

Teichoic and teichuronic acids, two chemically diverse groups of linear anionic polymers, are distributed widely among Gram-positive (Firmicutes) eubacteria, a taxon characterized by a thick, 25 to 35 nm, cell wall. These polymers are linked covalently to the structurally defining cell-wall polymer, peptidoglycan or murein, present in practically all eubacteria. The repeating units of teichoic and teichuronic acids contain phosphodiester linkages and uronic acids, respectively.

The structure, biosynthesis and functions of anionic wall polymers have been reviewed regularly (Ward, 1981; Hancock and Baddiley, 1985; Archibald et al., 1993; Pooley and Karamata, 1994; Naumova and Shashkov, 1997).

This chapter will cover information obtained from structural, genetic, biochemical, and cell biology studies arising from a longstanding effort primarily aimed at understanding the biological role of the anionic cell wall polymers in *Bacillus subtilis*, the reference Gram-positive bacterium. It is centered, essentially, on the biosynthesis of anionic polymers in *B. subtilis* 168, the reference strain of this species, and describes our current knowledge of the nearly complete set of genes involved in their synthesis. A body of experimental findings relevant to the still unknown biological function(s) of these polymers also is provided.

Until now, no industrial applications of cell-wall anionic polymers as such have been reported. For this reason, large-scale preparative methods for these polymers have not been developed.

2 Historical Outline

The discovery of negatively charged polymers in the cell wall of several Gram-positive bacteria followed the identification of cytidine nucleotide-linked C3 and C5 sugar alcohols in cell extracts of *Lactobacillus arabinosus* (Baddiley and Mathias, 1954; Baddiley et al., 1956). Subsequently, two classes of these surface-associated polymers were identified (Critchley et al., 1962; Shockman and Slade, 1964), the so-called "membrane" or lipoteichoic acids (LTAs), associated with the cytoplasmic membrane, and the cell-wall-bound teichoic acids (WTAs). It is the latter that form the main object of the present review. Insight into the considerable diversity of chemical structures of these molecules was first obtained in studies from Baddiley's laboratory (Baddiley et al., 1962a,b; Sanderson et al., 1962; Baddiley, 1972; Hancock and Baddiley, 1976; Hancock et al., 1976; Heptinstall et al., 1978; McArthur et al., 1978). Further detailed structures of the WTA main chain and many aspects of its biosynthesis in a range of organisms followed (Liu and Gotschlich, 1963, 1967; Knox and Hall, 1965; Archibald and Stafford, 1972; Heckels et al., 1975; Naumova and Shashkov, 1997).

The presence of a specific intervening structure, the so-called linkage unit (LU), linking the first unit of the main chain of the WTA to the muramic C-6 hydroxyl group via a phosphodiester bond, also was recognized first in Baddiley's laboratory. The LU structure was identified fully in studies from Ito's laboratory (Sasaki et al., 1980, 1983; Murazumi et al., 1981; Yokoyama et al., 1986, 1987).

Several of the enzymes involved in the biosynthesis of WTA were revealed to be membrane associated (Glaser, 1964; Burger and Glaser, 1966; Bracha and Glaser, 1976a,b). Subsequently, the role of LTA as an intermediate acceptor for D-alanine substituants of the WTA (Mirelman et al., 1970) was shown by work of Neuhaus and collaborators (Reusch and Neuhaus, 1971), who also determined the number of glycerol phosphate repeating units making up the WTA chain of *B. subtilis* 168 (Pollack and Neuhaus, 1994). Teichuronic acid was described and structural details were obtained for the polymer isolated from the cell walls of *B. licheniformis* (previously *B. subilis*) NCTC 6346 (Janczura et al., 1961).

The effect on the overall polymer shape of the mutual repulsion of negative charges of these polyanions has been studied (Doyle et al., 1975) on isolated poly(glycerol phosphate) [poly(GroP)] molecules. They exhibited a rod-like rigidity or a random coil structure in solutions of low or high ionic strength, respectively, a behavior accounted for by the increased shielding of negative charges of the polymer through higher concentrations of positive ions.

The early identification of *B. subtilis* 168 mutants characterized by a conditionally expressed lethal phenotype and major disturbances of cell morphology (Boylan and Mendelson, 1969; Rogers et al., 1970), several of which were shown to be deficient in the synthesis of the major WTA (Rogers et al., 1970; Boylan et al., 1972), was not followed for nearly two decades by a thorough genetic analysis. The reason for this lack of interest, at least in part, was a failure to be open to the possibly essential role of the synthesis of these polymers for cell growth (Boylan et al., 1972; Rogers et al., 1974; Shiflett et al., 1977; Rogers and Taylor, 1978). An early mapping of mutations affecting the synthesis of these polymers (Karamata et al., 1972) was followed by the identification of a major cluster of WTA genes located at 314° on the *B. subtilis* chromosome (Briehl et al., 1989; Honeyman and Stewart, 1989; Mauël et al., 1989). Through the *B. subtilis* genome-sequencing project (Kunst et al., 1997) and the biochemical characterization of mutants (Pooley et al., 1991, 1992; Lazarevic and Karamata, 1995; Soldo, 1999; Da Silva, 2000; unpublished data), the roles of specific genes, as well as their transcriptional organisation, in *B. subtilis* 168 and W23 strains now have been determined almost entirely.

The question of the biological roles of the wall anionic polymers was raised long ago, and their importance for the cell wall's overall negative charge, immunogenicity, and capacity to adsorb bacteriophages has been well established (Young, 1967; Archibald, 1988). Nevertheless, the possibility of their synthesis being essential was not addressed until relatively recently (Karamata et al., 1987). Demonstration of this (Mauël et al., 1989; Pooley et al., 1991) is a key finding at the fundamental level. The attendant possibility of their synthesis playing a key role in cell surface extension (Pooley et al., 1993), as well as in the maintenance of the permeability properties of the cell wall, underlines how much remains to be understood. The essentiality of WTA synthesis offers a new potential antibiotic target (Pooley and Karamata, 1988; Pooley and Karamata, 2000), providing yet another incentive for identifying and understanding their role.

3
Chemical Structures

Following bacterial cell breakage, the cell wall fraction is purified by differential centrifugation and extensive washing. Isolation of the cell wall anionic polymers requires splitting the covalent linkages to the peptidoglycan macromolecule. This can be achieved either chemically or enzymatically. Two methods of chemical extraction have been widely used for extraction of WTA: a treatment with (1) 5% w/v trichloroacetic acid for 5 h at 60°C or (2) 0.1N NaOH for 5 h at 60°C. Either treatment leads to release of over 95% of the WTA (Pollack and Neuhaus, 1994). The first procedure leads to partial hydrolysis of the phosphodiester linkages, whereas with the second procedure (Pollack and Neuhaus, 1994), the WTA main chain remains apparently intact. Intact WTA and TUA molecules also can be prepared by a lysozyme degradation of the cell wall. This treatment yields wall teichoic acid-glycopeptide complexes containing peptide-linked oligosaccharide fragments of peptidoglycan (Ghuysen and Strominger, 1963; Araki and Ito, 1989).

TUA also can be released from the cell wall by mild acid treatment, which causes separation of the phosphodiester bond between TUA and the C-6 hydroxyl group of N acetyl muramic acid (Pavlik and Rogers, 1973). For cell walls containing both TUA and WTA, a selective extraction method, leaving WTA unreleased, consists of a treatment with a pH 3 buffer at 100°C for 20 min (Pavlik and Rogers, 1973). Purification of the extracted TUA by ionic exchange and Sephadex chromatography has been described (Lifely et al., 1980). A second procedure (Beveridge et al., 1982) for separating TUA and WTA is based on the alkaline extraction of WTA described above, since TUA is not released by this treatment.

Following WTA or TUA extraction from cell walls, the results of their chemical analysis have been reviewed previously (Archibald, 1974; Hancock and Poxton, 1988). The preceding decade has seen a considerable body of data on chemical shifts of carbon atoms in ^{13}C-Nuclear Magnetic Resonance (^{13}C-NMR) spectra of WTAs of different structural types (Naumova and Shashkov, 1997). New and significant features of WTAs have been revealed by the application of ^{13}C-NMR to WTA structure largely from the work of Naumova and Shashkov (1997), who claim that data on chemical shifts and splitting of carbon atom signals of a wide range of anionic polymers often can render unnecessary the time-consuming procedures of chemical analysis.

3.1
Cell Wall Teichoic Acid

The structures of cell wall anionic polymers from across a wide spectrum of genera belonging to the Gram-positive phylum were reported in a recent review (Naumova and Shashkov, 1997). Notwithstanding the extensive variation revealed, it was suggested that the WTA main chain structures can be grouped into a few classes. In the first group, the repeating units consist of polyol phosphate with some, or most, polyols carrying side-chain glycosyl substituents. The glucosylated poly(GroP) of *B. subtilis* 168 is an example of this group. The second group is made up of units of glycosyl polyol phosphate, the glycosyl group being in the main chain, an example of which is the poly(galactosyl glycerol phosphate) of *B. licheniformis* (Schipper, 1995).

Polymers belonging to the remaining group have a more complex structure, apparently consisting of an assemblage of elements found in both foregoing groups, i.e., with a repeating unit of alternating polyol phosphate (first group) and glycosyl polyol phosphate units (second group). An example of such a polymer is the glycerol phosphate N-acetylgalactosamine-β-D-glycerol phosphate of *Nocardiopsis dassonville* (Tul'skaya et al., 1993). A further kind of phosphodiester-linked sugar polymer, referred to as sugar 1-phosphate polymer, has long been recognized (Archibald and Stafford, 1972; Partridge et al., 1973). In some cases at least, it resembles a structural variant of the second group described above, with the polyol moiety replaced by an N-acetylated hexosamine (Shibaev et al., 1973).

3.1.1
Polyol Variation

Among the first group of organisms, *Bacilli, Staphylococci,* and *Lactobacilli,* to be studied, the commonest polyols are C3 (glycerol) and C5 (ribitol). However, WTAs with C4 (erythritol), C5 (arabitol), and C6 (mannitol) have since been identified (Naumova and Shashkov, 1997).

3.1.2
B. subtilis Cell Wall Teichoic Acids

Poly(GroP) and poly(ribitol phosphate) [poly-(RboP)], carrying side-chain glucosyl substituents on a free hydroxyl group of the sugar alcohol, identified in cell walls of *B. subtilis* 168 and W23, respectively, are widely distributed (Baddiley, 1972; Archibald, 1974). They belong to the first group of WTAs, i.e., consisting of linear polymers of phosphodiester-linked sugar alcohols (see above, Section 3.1). Structurally, they consist of two parts (Figure 1): the main chain and the so-called linkage unit (LU), a shorter intervening sequence between the main chain and the peptidoglycan (PG) (Araki and Ito, 1989). In *B. subtilis* 168, the former shows a certain "scattering" of length with an average of 53 glycerol 1,3 phosphate repeating units (Pollack and Neuhaus, 1994) whose 2-hydroxyl groups bear various substituents. The α or β C-1-linked glucose (Glc) and ester-linked D-alanine (D-ala) are found most frequently. The attachment points to PG via muramic 6-phosphate residues (Liu and Gotschlich, 1963, 1967), which represent about 6% of the total cell-wall muramate and are phosphodiester-linked to an N-acetylglucosamine (GlcNAc) C-1 of the intervening LU (Bracha and Glaser, 1976a,b; Coley et al., 1976; Sasaki et al., 1980; Harrington and Baddiley, 1985). The commonest form of the latter, present with minor variations, in several members of *Bacilli, Staphylococci,* and *Lactobacilli* (Kojima et al., 1985a,b), consists of the disaccharide N-acetylmannosamine (ManNAc), (β1-4) linked to GlcNAc. The distal, non-reducing extremity of this disaccharide is joined to the main polyol phosphate (see above) via one or several GroP residues. In a variety of organisms, modified forms of LU for anionic polymers also have been identified (Araki and Ito, 1989).

3.2
Teichuronic Acids

First identified in 1961, a teichuronic acid (TUA) polymer described (Janczura et al., 1961) in *B. licheniformis* was reported to contain α 1 – 3-linked glucuronic acid (GlcA) and N-acetylgalactosamine (GalNAc). Evidence for structurally similar molecules was found in several strains of *B. subtilis* (Ellwood and Tempest, 1969, 1972; De Boer, 1979). The TUA present in cell walls of *B. subtilis* W23 under phosphate-limited growth was reported (Hancock, 1983) to have a GlcA GalNAc repeating unit, a structure supported by biosynthetic data. Whereas the structure of the TUA of *B. subtilis* 168 has not been reported previously, evidence has been found for the presence, in this organism, of a more complex repeating unit or the presence of an LU (Soldo et al., 1999; unpublished). This suggests that a critical reexamination of the evidence on which such structures, containing disaccharide repeating units, are based may reveal a hitherto unsuspected complexity.

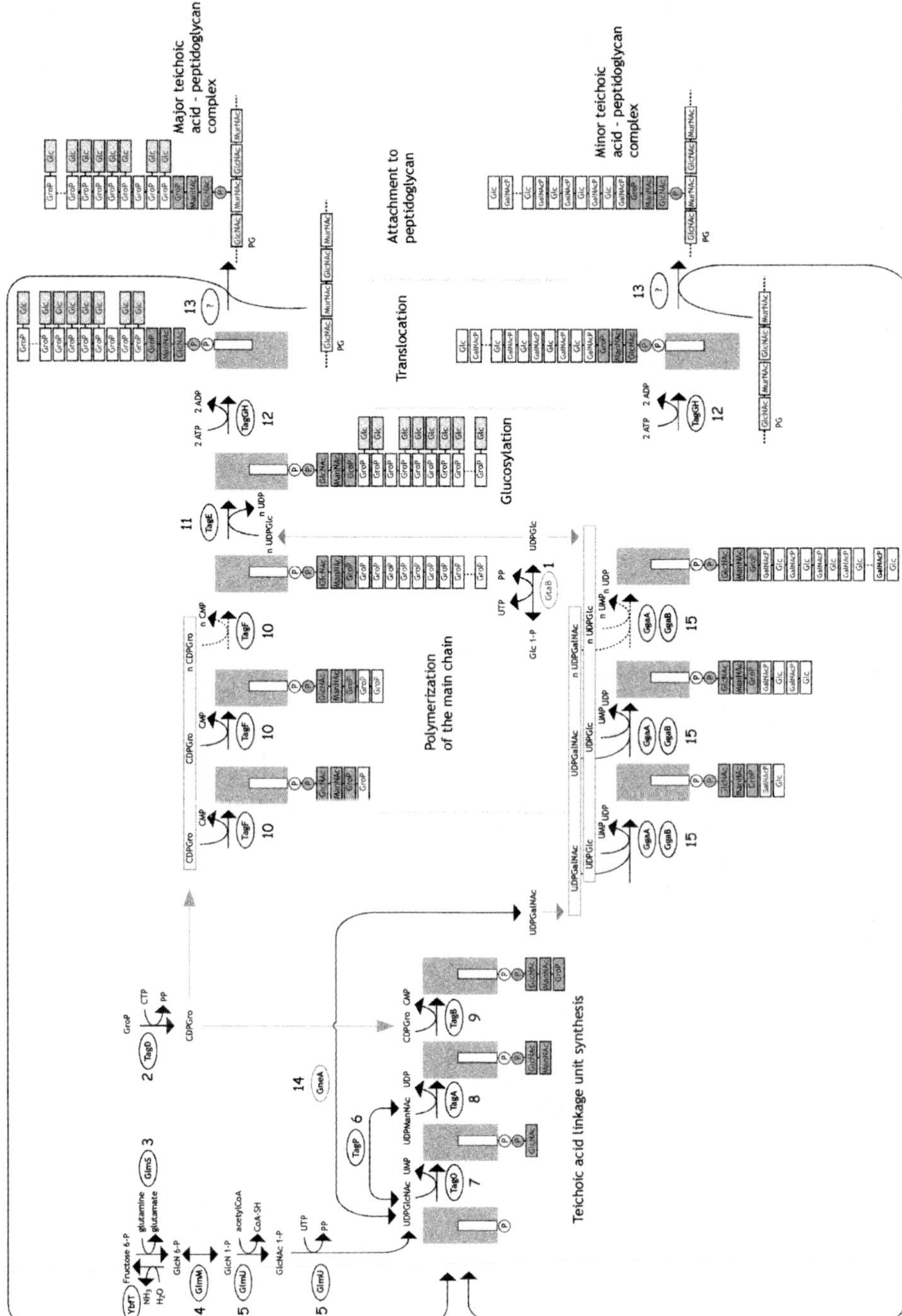

Fig. 1 Biosynthetic pathway for wall teichoic acids of *B. subtilis* 168, glucosylated poly(glycerol phosphate) and poly(3-*O*-β-D-glucopyranosyl N-acetylgalactosamine 1-phosphate), and for their corresponding soluble precursors.

TUAs are present in cell walls of the genera *Bacillus, Corynebacterium, Micrococcus, Propionobacterium, Streptococcus,* and *Staphylococcus.* In few cases have the chemical structures been fully identified.

Linkage of TUA to PG is believed to be direct (Archibald et al., 1993) via a phosphodiester to muramic acid C-6, as in the case of a TUA from *B. licheniformis* (Ward and Curtis, 1982). However, in one case, a TUA polymer of *M. luteus* ATCC 4698, the presence of an intervening sequence was reported (Hase and Matsushima, 1977; Rohr et al., 1977; Stark et al., 1977; Traxler et al., 1982; Johnson et al., 1984), while in many cases, evidence in favor of there being no LU was either absent or insufficient (Araki and Ito, 1989).

4 Biosynthesis and Genetics of Poly(GroP)

This section focuses on biosynthesis and genetics of poly(GroP), the major cell wall teichoic acid of strain 168, as well as on its linkage unit, an entity that appears to be common to the major and minor WTAs of this organism. The latter polymer is described in more detail in Section 5.

4.1 Polymerizing Enzyme System for Poly(Glycerol Phosphate) Teichoic Acid

Resting on a substantial body of mostly earlier work (Hancock, 1981; Kaya et al., 1984; Harrington and Baddiley, 1985; Yokoyama et al., 1986) and largely on the basis of their spatial and temporal organization, four aspects of the biosynthesis of this polymer are recognizable (Figure 1):

1) The cytoplasmically located pathways leading to the soluble nucleotide-linked polyol, hexose, and N-acetyl hexosamine diphosphates (including steps 1 to 6 and 14).
2) The membrane-located polymerization of the LU, pyrophosphate-linked to bactoprenol, by successive addition of GlcNAc, ManNAc, and, most likely, one (*B. subtilis* strain 168) or two (strain W23) GroP units by transfer from the soluble nucleotide-linked precursors (steps 7 to 9).
3) The polymerization, distal to bactoprenol, onto the preformed bactoprenol-linked LU, of the main polymer backbone by sequential addition to the terminal polyol phosphate residue of polyol phosphate units from the soluble cytidine diphosphoglycerol (CDPGro) or CDP ribitol monomers (step 10). Hexose linkage to the free hydroxyl group (step 11) may proceed in step with the polymerization of GroP units.
4) The translocation across the membrane (step 12) of the completed WTA molecule and its ligation (step 13) to a PG chain made at the same time takes place through formation of a phosphodiester link to muramic 6 hydroxyl. Free polyprenol phosphate formed in this step can take part in a new round of synthesis.

4.2 The Genetic Basis of the Biosynthesis of the Soluble Precursors

The soluble precursors required for LU assembly are UDPGlcNAc, UDPManNAc, and CDPGro, while those required for the synthesis of the main polymer chain are CDPGro (or CDP ribitol for poly(RboP)), uridine 5′-diphosphoglucose (UDPGlc), and D-ala. After its activation by linkage to a specific D-alanyl carrier protein (Heaton and Neuhaus, 1994; Perego et al., 1995), D-ala is first transferred to the membrane-associated LTA. Subsequently, it is transferred to the

WTA (Perego et al., 1995), possibly after its linkage to PG and incorporation into the cell-wall fabric.

4.2.1 UDPGlcNAc

This precursor, the major fate of which is to supply the N-acetylated hexosamines of the glycan chains of PG, is also essential for the synthesis of the disaccharide-forming part of the LU linking PG to the main polyol phosphate chain. The formation of UDPGlcNAc from fructose 6-phosphate and glutamine requires four enzyme activities encoded by genes *glmS*, *glmM*, and *glmU* (*gcaD* is the *B. subtilis* homolog of *glmU*), as shown in *E. coli* (Dobrogosz, 1968; White, 1968; Mengin-Lecreulx and van Heijenoort, 1996). Homologs of these genes are present in *B. subtilis* (Table 1), which contains an apparently identical pathway. Mutations associated with a conditional lethal phenotype have been mapped to the *B. subtilis* gene *glmM* (Da Silva, 2000). The complex phenotype developed by these *glmM*-deficient mutants under restrictive conditions for growth parallels in several respects that exhibited by *glmM*-deficient mutants in *E. coli* and is consistent with a defective synthesis of WTA (Da Silva, 2000).

4.2.2 UDPManNAc

UDPManNAc is formed from UDPGlcNAc through the activity of UDPGlcNAc 2-epimerase identified in cell extracts of *B. cereus* (Kawamura et al., 1978), *B. subtilis* (Harrington and Baddiley, 1985), and *E. coli* (Kawamura et al., 1975). The purified recombinant YvyH-His6 protein has UDPGlcNAc 2-epimerase activity (unpublished data). Mutants affected in *yvyH*, now renamed *tagP*, as predicted, show a specifically diminished rate of incorporation of [2C-^{3}H]Gro into WTA (Pooley and Karamata, 2000).

4.2.3 CDPGro

CDPGro, a precursor specific to WTA synthesis, is formed by glycerol 3-phosphate cytidylyltransferase, which is present in the cytosol fraction of disrupted *B. subtilis* 168 and W23 cells (Glaser and Loewy, 1979; Hancock, 1983). The structural gene for this enzyme was identified, and named *tagD* (see Figure 2), by mapping of a mutation, *tagD1* (Mauël et al., 1991), characterized at the non-permissive temperature by the absence of, or strongly reduced, CDPGro pool (Pooley et al., 1991).

4.2.4 UDPGlc

Absence of Glc in cell walls of *B. subtilis* 168 characterizes mutations in *gtaB* (Young, 1967). A large number of strains bearing *gtaB* mutations were found to be deficient in UDPGlc pyrophosphorylase (UDPGlcPPase) (Pooley et al., 1987). This finding and a marked overall homology of the deduced amino acid sequence of the *gtaB* locus with several prokaryotic homologs (Soldo et al., 1993) establish that *gtaB* is the structural gene for the UDPGlcPPase.

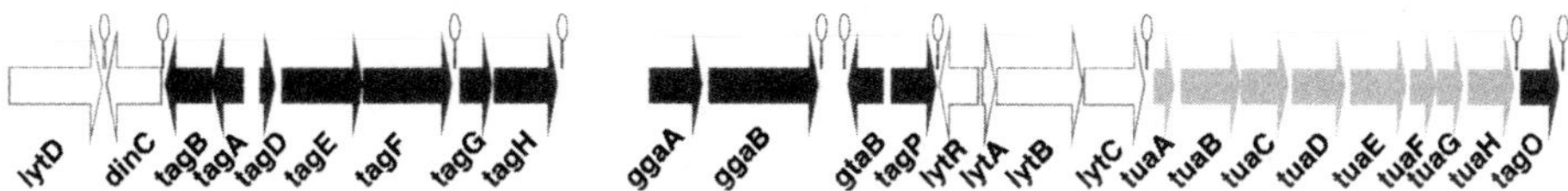

Fig. 2 Arrangement of teichoic (black) and teichuronic acid (gray) genes in the 310 ° region of the *B. subtilis* 168 chromosome. Intervening and flanking genes for other functions are also shown (unfilled).

Tab. 1 Genes and enzymes involved in synthesis of wall teichoic acids in *B. subtilis* 168

Biosynthetic step[a]	***Enzyme***[b]	***Genetic map position***[c]	***Gene***	***Operon***	***Polymer concerned***		***References***
					Glucosylated poly(GroP)	***Poly(Glc-GalNAcP)***	
Soluble precursors	1	313°	*gtaB*	gtaB	+	+	Pooley et al., 1987; Soldo et al., 1993
	2	314°	*tagD*	tagDEF	+	+[d]	Pooley et al., 1991
	6	313°	*tagP* (*yvyH*, *orfX*, *mnaA*)	*tagP*	+	+[d]	Soldo et al., 1993; unpublished data Pooley and Karamata, 2000
	14	335°[e]	*gneA* (= *galE*)[f]	*yxkAgalE*	–	+	Estrela et al., 1991; unpublished data
	3	17°	*glmS*	ybbPRglmMS	+	+	Mengin-Lecreulx and van Heijenoort, 1996
	4	17°	*glmM* (*ybbT*)	*ybbPRglmMS*	+	+	Mengin-Lecreulx and van Heijenoort, 1996; Da Silva, 2000
	5	5°	*gcaD* (*glmU*)	*gcaD...yabR*[g]	+	+	Ogasawara et al., 1994
Linkage unit	7	312°	*tagO* ?	*tagO*	+	+[d]	Soldo, 1999
	8	314°	*tagA* ?	*tagAB*	+	+[d]	Mauël et al., 1991
	9[h]	314°	*tagB* (*rodA*) ?	*tagAB*	+	+[d]	Mauël et al., 1991
Main-chain formation	10	314°	*tagF* (*rodC*)	*tagDEF*	+	–	Pooley et al., 1992
	11	314°	*tagE* (*gtaA*, *rodD*)	*tagDEF*	+	–	Young, 1967
	15	313°	*ggaA*	ggaAB	–	+	Estrela et al., 1991; this chapter
	15	313°	*ggaB*	ggaAB	–	+	Estrela et al., 1991; this chapter
Membrane translocation	12	314°	*tagG*	tagGH	+	+	Lazarevic et al., 1995
	12	314°	*tagH*	tagGH	+	+	Lazarevic et al., 1995
Linkage to peptidoglycan	13[i]	?	?				

[a] Soluble precursors synthesized in the cytoplasm. All other reactions are membrane associated. [b] Numbers refer to steps shown in Figure 1. [c] Chromosome location of relevant gene calculated from converting the entire genome sequence to 360°. [d] The existence of an LU or of covalent linkage to PG has received little attention. The possibility of linkage via the same LU as poly(GroP) remains open (see Text). [e] Map position estimated from transduction (Estrela et al., 1991). [f] *gneA* is the same gene as *galE* (see Text; Krispin and Allmansberger, 1998) which is located at 338° on the genetic and 341° on the calculated map. [g] Putative operon comprising genes *gcaD*, *prs*, *ctc*, *spoVC*, *yabK*, *mfd*, *spoVT*, *yabM*, *yabN*, *yabO*, *yabP*, *yabQ*, *divIC*, and *yabR*. [h] It is not known whether two separate enzymes are needed for this step. If the LU contains one molecule of glycerol phosphate, conceivably, one activity could be sufficient. For strain W23, the LU consists of two glycerol phosphate units. [i] Possibly, common enzyme responsible for this step of glucosylated poly(GroP) and of poly(GlcGalNAcP) biosynthesis (see Text).

4.2.5
The Linkage Unit Formation

The process is initiated by transfer of GlcNAc from UDPGlcNAc to bactoprenol (Figure 1, step 7). This step was shown, in membrane preparations from several Gram-positive organisms, to be highly sensitive to tunicamycin (Wyke and Ward, 1977; Ward et al., 1980), whereas all the following reactions in LU formation and main-chain polymerization were unaffected by this antibiotic (Wyke and Ward, 1977; Ward et al., 1980).

Inspection of the *B. subtilis* 168 genome sequence reveals that a gene located downstream of the TUA operon (Figure 2) encodes a protein similar to enzymes that transfer phospho N-acetyl glucosamine to bactoprenol phosphate (Soldo, 1999).The essentiality of this gene was established by inserting, in front of the start codon, an isopropylthio-β-D-galactoside (IPTG)-inducible Pspac promoter. During limited growth in the absence of the inducer, cells developed coccoid morphology, characteristic of mutants blocked in WTA synthesis (Soldo, 1999). In addition, a markedly diminished content of WTA led to this gene being designated *tagO*.

An earlier study (Bracha et al., 1978), based on incorporation of CDP-[^{3}H]Rbo by membrane preparations from *S. aureus* mutant 52A5, revealed an absence of ribitol from the mutant but comparable amounts of glycerol (0.05 to 0.1 μmole/mg) and muramate phosphate in cell walls of mutant and wild type (Mirelman et al., 1971). Contrary to the conclusion of the authors, these findings suggest the presence of an LU. The findings are inconsistent with a suggested block in the first step of LU formation, while militating in favor of a nonfunctional counterpart of TagF, the main-chain polymerase of *B. subtilis* 168.

Similarities between *B. subtilis* 168 TagF and TagB are consistent with the latter being involved in the addition of the GroP of the LU. It is likely that *tagA* encodes the enzyme adding ManNAc of the LU (Mauël et al., 1994). However, there is no direct experimental evidence to support this.

The proposed biosynthetic pathway for the LU and the main chain of poly(RboP) in *S. aureus* were strongly supported (Yokoyama et al., 1986) by the isolation and identification of the specific intermediates, together with demonstration of their stepwise formation through the activity of enzymes present in membranes (including steps 7 to 9 in Figure 1).

The observation that the cell morphological phenotype produced by addition of tunicamycin (Takatsuki et al., 1972) is comparable to that provoked by mutations in *tag* genes provided further support for this interpretation, specifically in view of the now-established fact that *in vitro* tunicamycin inhibits the first step of LU formation (see above). Recently, it has been shown (Pooley and Karamata, 2000) that, at the minimal inhibitory concentration (MIC) *in vivo*, this drug specifically inhibits the synthesis of both *B. subtilis* 168 WTAs, i.e., poly(GroP) and poly(3-*O*-β-D-glucopyranosyl N-acetylgalactosamine 1-phosphate) [poly(GlcGalNAcP)].

4.2.6
Genetic Basis of Synthesis of the Main Chain

Already 37 years ago, poly(GroP) polymerase activity was demonstrated *in vitro* with membrane preparations of *B. subtilis* and *B. licheniformis* (Burger and Glaser, 1964). The implication of *tagF*(*rodC*) as the gene that encodes this activity followed assays of this polymerase on membrane preparations obtained from cells grown at the permissive temperature of a series of thermosensitive mutants containing mutations mapping in the *tagAB-tagDEF* divergon (Mauël et al., 1991). Polymerase deficiency was invariably and exclusively associated with the 10 examined mutant alleles of *tagF* (Pooley et al., 1992). For several mutants, this activity was found to be thermosensitive *in vitro*, an observation that led to the conclusion that this protein was encoded by *tagF*. This result received strong support from the finding that, in the cell wall of the *tagF1* mutant at the restrictive temperature, the average chain length of the poly(GroP) chain was 8 compared with 53 for the parent *tagF* allele (Pollack and Neuhaus, 1994).

Staphylococcus epidermidis, which has a poly(GroP) WTA apparently identical to that of *B. subtilis* 168, was shown to contain a *tagF* homolog (Fitzgerald and Foster, 2000). Its product, *tagF* of *S. epidermidis*, was able to complement the conditionally lethal, temperature-sensitive mutation *tagF1* of *B. subtilis*, indicating an identical enzyme activity consistent with the presence of the major WTA. Interestingly, *tagF* of *S. epidermidis* is not found in a *tag* gene cluster. The failure to obtain viable *tagF* disruptants indicates that the gene product and WTA synthesis are essential in this organism also.

4.2.7
Genetic Basis of Glucosylation of Poly(Glycerol Phosphate)

Transfer from UDPGlc of glycosyl groups to the C-2 position of the 1,3 phosphodiester-linked glycerol units (step 11, Figure 1) was shown in isolated membrane preparations (Burger and Glaser, 1964; Brooks et al., 1971).

A specific deficiency in this activity was found in all strains examined bearing mutations in the *gtaA* locus (Young, 1967). This gene, sequenced under the name *rodD*, encodes a 78 kDa protein (Honeyman and Stewart, 1989). As the second gene in the *tagDEF* operon, it has been finally designated *tagE* (Mauël et al., 1991). Normal growth of mutants with inactivated *tagE* implies that the nonglucosylated poly(GroP) is able to fulfill all functions essential for growth in the laboratory.

4.2.8
Attachment to Newly Synthesized Peptidoglycan

Transfer of the completed molecule to a nascent PG chain through the formation of a phosphodiester linkage represents the final step in WTA biosynthesis. Attachment of WTA or TUA (see below) to simultaneously synthesized PG chains *in vivo* was demonstrated in *B. subtilis* 168 (Mauck and Glaser, 1972). The gene or genes involved in this attachment step (step 13) remain unidentified up to now.

4.3
Transcriptional Regulation of Wall Teichoic Acid Genes in *B. subtilis* 168

Transcription of WTA operons *gtaB*, *tagP*, *tagGH*, and *tagO* is initiated from at least one σ^A-controlled promoter (Soldo et al., 1993; Lazarevic and Karamata, 1995; Soldo, 1999). In the case of *gtaB*, an additional, σ^B-controlled promoter was identified (Soldo et al., 1993). Studies of the regulation of wall anionic polymer synthesis focused on a 400 bp region, located between *tagAB* and *tagDEF* (Figure 2), which plays a crucial role in the global regulation of the wall negative charge. In this region, two divergent mRNA starts, presumably resulting from σ^A-controlled promoters, were mapped by primer extension analysis (Mauël et al., 1995). Transcriptional fusions, located either at the *tag* locus or at an ectopic locus (*amyE*), allowed the analysis of *tag* gene expression under specific physiological conditions, known, or predicted, to entail changes in cell-wall synthesis (Mauël et al., 1994). As expected, phosphate limitation was accompanied by reduced *tag* gene expression. Following the onset of sporulation, transcription of *tag* genes diminished rapidly and was essentially abolished by stage II. During germination, the activity of *tag* genes was detectable before the rise in culture turbidity associated with spore outgrowth. Transcription was not affected by DNA damage that induces the repair machinery called the SOS response. Under all experimental conditions, *tagA* and *tagD* appeared coordinately expressed, the level of *tagD* transcription being always two to three times higher than that of *tagA*.

A more detailed analysis (Mauël et al., 1995) using *lacZ* and *gus* transcriptional fusions to the first genes of the operons *tagA* and *tagD*, revealed that, at the end of exponential growth, the rate of expression of both genes increased abruptly by a factor of about two, the inflection point coinciding exactly with the slowdown of mass increase. This acceleration was correlated with the morphological changes characterizing the transition from the exponential to the stationary growth phase, i.e., the progressive disappearance of long chains of large multinucleate cells and the segregation of chromosomes into individual cells. Therefore, the acceleration of *tagA* and *tagD* transcription was coupled with the increased septation frequency that occurs during the transition phase. This interpretation was supported by experiments in which transcription was monitored in media of different richness. It appeared that higher growth rates were paralleled with lower transcription rates, a behavior characteristic of genes involved in septation (Vicente et al., 1991). Although *tagD* and the neighboring downstream genes *tagEF* are transcribed from the same promoter, however, a *lacZ* fusion to *tagEF* revealed that these genes were expressed at a much lower level than was *tagD*. In addition, their transcription did not increase at the onset of the stationary phase. This finding indicates that additional regulatory signals, which may operate in the intergenic *tagD-tagE* region, govern *tagEF* transcription. These results, taken together with measurements of cell length in media of different richness, suggest that WTA synthesis at the cross-wall may mainly require the transcription of *tagAB* and *tagD*, i.e., the three genes most likely specifying the synthesis of the LU coupling the poly(GroP) chain to PG. Two conclusions were drawn. First, cross-wall WTA molecules may differ from those present in the cylindrical part of the wall. This is in agreement with several previous observations pointing to a difference between the septal and cylindrical wall, e.g., fluorescence and electron microscopy observations (Mobley et al., 1984; Clarke-Sturman et al., 1989), freeze-substitution (Graham and Beveridge, 1994), and endolysin susceptibility of purified cell wall (Fan et al., 1972). Secondly, it appears that *tag* genes synthesizing the septal or the cylindrical polymer may obey two different regulatory pathways, possibly responding to cell-cycle-specific signals coupling chromosome replication and cell division.

5 Biosynthesis and Genetics of Poly (3-O-β-D-Glucopyranosyl N-Acetylgalactosamine 1-Phosphate)

Walls of *B. subtilis* 168 contain a second, so-called minor WTA, poly(GlcGalNAcP) (Duckworth et al., 1972; Shibaev et al., 1973), which belongs to the group containing hexosamine 1-phosphate phosphodiester-linked polymers lacking sugar alcohols (polyols) but is structurally similar to the second group of polyol containing WTAs (see Section 3).

5.1
Genetic and Biochemical Analysis of Mutants

In cells grown in a rich medium at 30°C, about 20% of all wall hexosamine is present as galactosamine (Pooley et al., 1987). Twenty-five mutants specifically resistant to bacteriophage Ø3T, whose receptor contains poly(GlcGalNAcP), were analyzed genetically and biochemically. About one-quarter of the mutations were mapped to the *gneA* locus around 335° (Estrela et al., 1991), the remainder, named *gga*, were all localized to a second region very close to the *tagDEF* operon (Estrela et al., 1986, 1991; Mauël et al., 1991). No mutant deficient in poly(GlcGalNAcP) revealed growth deficiencies, in marked contrast to results obtained with mutants affected in poly(GroP). As expected (see below), mutations in genes responsible for the synthesis of the precursors of this polymer (i.e., *gtaC*, phosphoglucomutase-deficient, *gtaE*, phosphoglucomutase- and UDPGlcPPase-deficient, and *gtaB*, UDPGlcPPase-deficient) (Pooley et al., 1987), yielded mutants impaired in the synthesis of this polymer, as well as in the glucosylation of the major WTA.

5.2
Genes Involved in Synthesis of Soluble Precursors

Genes needed for the formation of the soluble sugar nucleotide-linked precursors for the main chain are those for UDPGlc, as described above, and UDPGalNAc. Evidence summarized here is strongly in favor of there being an LU, whose synthesis requires the same precursors as the LU (see Section 4.2) for glucosylated poly(GroP).

5.3
UDPGalNAc

UDPGalNAc is obtained by epimerization of UDPGlcNAc (step 14). The structural gene for this enzyme is presumed to be *gneA*, since cell extracts of six strains carrying mutations in this locus (Estrela et al., 1991) were deficient in UDPGlcNAc 4-epimerase activity. There is evidence (unpublished data) that this locus is identical to the previously described *galE* (Krispin and Allmansberger, 1998) located at 341° (see Table 1) on the calculated (DNA sequence) genetic map.

5.4
Genetics of Polymerizing Enzyme System

All *gga* mutations associated with a Ø3T-resistant phenotype were mapped to an approximately 4 kb region, located between the *tagAB-tagDEF* (Mauël and Karamata, 1990) and *gtaB-tagP* (Soldo et al., 1993) divergons (Estrela et al., 1991). Physical mapping and insertional mutagenesis revealed two genes named *ggaA* and *ggaB* (Freymond, 1995). Recently, another locus, homologous to *E. coli nagB* (Holmes and Russell, 1972), has been implicated in the synthesis of poly(GlcGalNAcP), since its insertional inactivation leads to cells devoid of this polymer and resistant to phages Ø3T and ρ11 (unpublished data). The predicted sequences of two *B. subtilis* genes, *nagA* and *nagB*, exhibit strong homologies to *E. coli nagA* and *nagB* genes, which were shown to be involved in the deacylation and the deamination, respectively, of GlcNAc 6-phosphate. However, another *B. subtilis* gene, *ybfT*, also has substantial homology with *nagB* of *E. coli* and of *B. subtilis*. Analysis of *B. subtilis* mutants deficient in either *nagB* or *ybfT* revealed that the latter gene is actually the true functional homolog of the *E. coli nagB* (Rivolta, 1999).

When membrane preparations of strains deficient in UDPGlc synthesis were supplemented by UDPGlc and UDPGalNAc only, synthesis of poly(GlcGalNAcP) took place (Hayes et al., 1977). The authors did not examine the possible involvement of an LU. Subsequently, experiments of Freymond (1995) established that conditional mutants in *tagA* or *tagB* genes, involved in the synthesis of the poly(GroP) LU, were unable to synthesize poly(GlcGalNAcP) at the non-permissive temperature, whereas *tagF*-deficient mutants, responsible for the polymerization of the main chain, were able to do so. These observations strongly suggest that both the major and the minor WTAs are likely to be attached to the same LU and that *tag*A and *tag*B are involved in its synthesis. In addition, more recent experiments (Pooley and Karamata, 2000) revealed that tunicamycin added at the MIC to growing cultures of *B. subtilis* 168 led to a rapid and complete block of [^{3}H]-labeled hexosamine incorporation into this polymer, an observation consistent with there being an LU joining this polymer to PG.

In contrast to the monosaccharidic phosphate repeating unit of poly(GroP), a linearly coupled disaccharide phosphate is the repeating unit of the secondary polymer, which shows some likeness to the WTA group 2 polymers (see above). As previously suggested (Pooley and Karamata, 1994), two alternative biosynthetic pathways appear possible. The first, consisting of alternate addition of GalNAc 1-phosphate and Glc, is illustrated by the polymerization of galactosylglycerol phosphate, a WTA of *B. coagulans* (Yokoyama et al., 1987) with a disaccharide repeating unit. The second pathway would involve the synthesis on a lipid carrier of disaccharides consisting of GalNAc 1-phosphate and Glc, followed by their polymerization. In either case, chain extension would be expected to proceed, as for the poly(GroP) polymer, by monomer addition to the extremity distal to the lipid carrier-linked LU. However, polymerization of disaccharide repeating units of PG or TUA of several organisms, as well as, apparently, of the WTA of *B. licheniformis*, the poly(glucosyl GroP), proceeds in the direction opposite to that of other WTAs that have been studied (Hancock and Baddiley, 1972; Archibald et al., 1993). This mechanism involves the transfer of the lipid carrier-linked growing polymer chain molecule to the repeating unit bound to a second lipid carrier. It has been suggested (Araki and Ito, 1989) that the lipid carrier distal addition, normal for monosaccharide phosphate units, may not apply for WTAs containing repeating units of tri- or oligo-saccharide phosphates. These polymers, synthesized in a lipid carrier proximal direction, would be directly attached to PG without any LU.

6 Biosynthesis and Genetics of the *B. subtilis* 168 Cell Wall Teichuronic Acids

In the conditions of phosphate excess, TUAs may be synthesized constitutively either together with WTAs, as in *B. licheniformis* (Hussey et al., 1978; Lifely et al., 1980), where each different anionic polymer is linked to different PG chains (Hughes et al., 1968), or as the only cell-wall polymer, as in *B. megaterium* and some *B. cereus* strains (White, 1977). In certain strains of the genus *Bacillus*, including *B. subtilis* 168 and W23, under phosphate-limiting conditions, WTAs are substituted by phosphate-free TUAs (Ellwood and Tempest, 1969). In the natural environment of *B. subtilis*, TUA plays an important role, since this soil bacterium often faces the problem of phosphate limitation.

The TUA of *B. subtilis* strain W23 was reported to be composed of alternating residues of GalNAc and D-GlcA (Wright and Heckels, 1975). The TUA of *B. subtilis* strain 168 may have the same structure.

Within the framework of the *B. subtilis* 168 genome-sequencing project (Kunst et al., 1997) an eight-gene operon, located next to the WTA determinants and predicted to encode enzymes for the synthesis of TUA, has been identified (Soldo et al., 1999). Indeed, inactivation of any of the *tuaABC-DEFGH* operon genes was accompanied by reduction in cell-wall TUA content of cells grown in phosphate-limited conditions (Soldo et al., 1999).

The hypothetical TUA synthesis pathway is given in Figure 3. *tuaD* is the structural gene of UDP-Glc dehydrogenase, the enzyme that converts UDP-Glc into UDP-GlcA (Pagni et al., 1999). The latter, as well as UDPGalNAc, are precursors of TUA (Archibald et al., 1993). Synthesis of the repeating disaccharide unit on a lipid carrier seems likely to require at least two of the four apparent sugar transferases, TuaA, TuaC, TuaG, and TuaH. Contrary to the situation with the WTAs, TUA is synthesized by addition of new blocks at the proximal end of the growing chain (Archibald et al., 1993). Such a mode of elongation would require a membrane exporter different from the ABC transporter involved in WTA translocation (Lazarevic and Karamata, 1995). After flipping to the outer membrane surface by TuaB, it was proposed that the disaccharide units are polymerized by TuaE. Both TuaB and TuaE contain periodically distributed hydrophobic domains, characteristic of integral membrane proteins. They may, together with TuaA and TuaF, be endowed with hydrophobic N-terminal domains, correspond to the core of the multienzyme complex involved in the polymer synthesis. TuaF, devoid of significant similarities to available protein sequences, might catalyze, for instance, TUA linkage to PG. Two of the four putative sugar transferases, TuaA, TuaC, TuaG, and TuaH, are surplus to those required of the disaccharide containing TUA structure represented in Figure 3. This suggests that there is additional structural complexity, either through a tetrasaccharide repeating unit or through the presence of an LU. It would be most interesting to have knowledge of the *tua* genes from other organisms.

The regulatory region upstream of the transcriptional start of the *tuaABCDEFGH* operon revealed the presence of the Pho box common to Pho regulon promoters (Soldo et al., 1999). The expression of the TUA and other Pho regulon operons is induced during phosphate starvation and depends on the two-component regulatory system PhoP/R (Seki et al., 1987; 1988; Soldo et al., 1999). This system is also responsible for the transcriptional repression of the *tagAB* and *tagDEF* operons during phosphate starvation (Liu et al., 1998). The synthesis of TUA additionally may contribute to the cessation of WTA incorporation in the cell wall. Indeed, a deficiency of TUA synthesis is, despite phosphate starvation, compensated by increased WTA incorporation into the cell wall (Müller et al., 1997; Soldo et al., 1999). Conversely, in phosphate-rich medium, expression of the TUA operon from an artificial phosphate-independent promoter leads to a lower WTA content. There appears to be a mechanism for coordinating the synthesis of the two types of anionic polymers so as to maintain the overall wall negative charge nearly constantly. Accordingly, at limiting phosphate concentrations, maintained for example by growth in a chemostat, the cell wall of *B. subtilis* may contain both WTAs and TUA (Lang et al., 1982). The constancy of negative charge is dependent on the pH of the cell's surround-

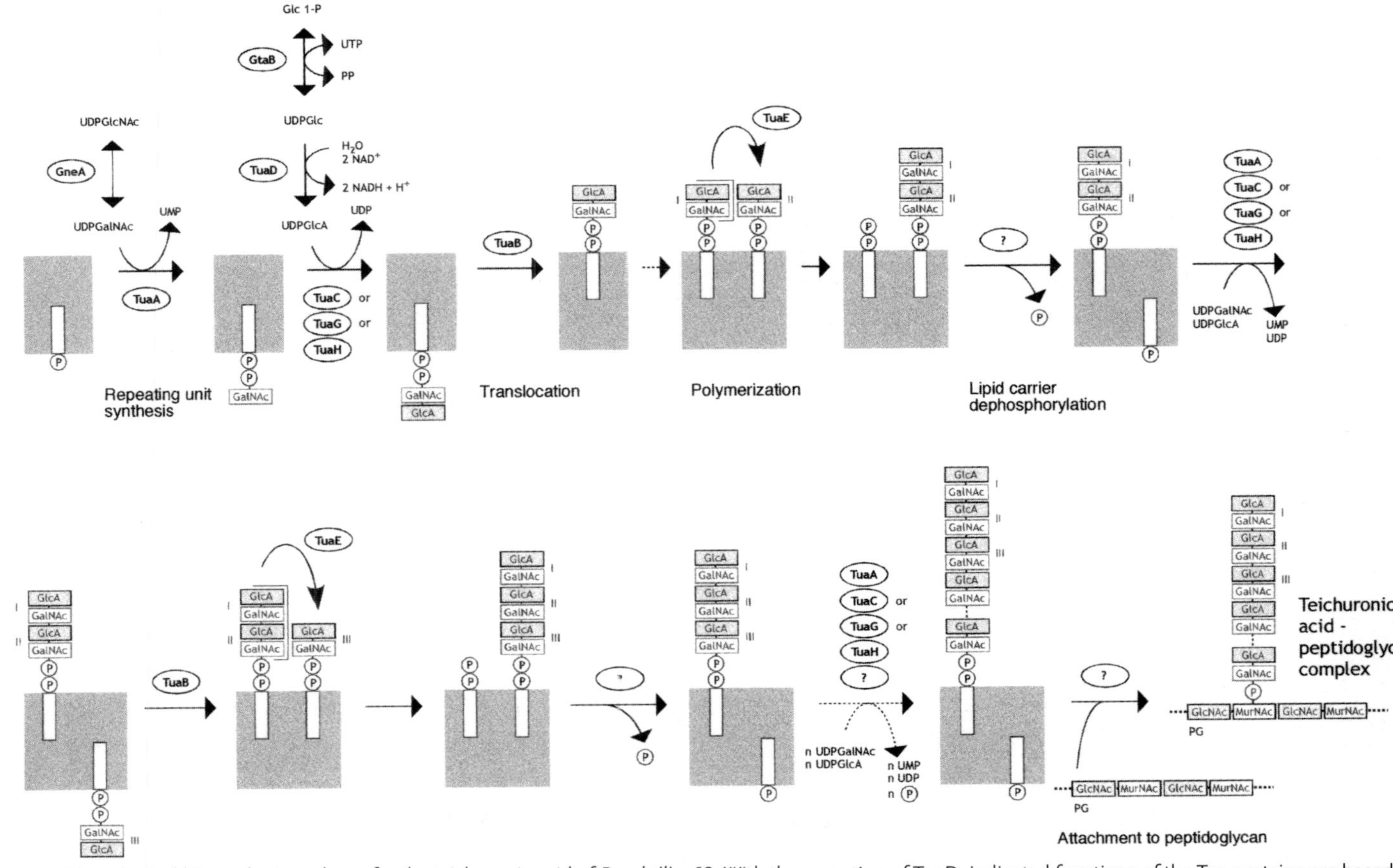

Fig. 3 Hypothetical biosynthetic pathway for the teichuronic acid of *B. subtilis* 168. With the exception of TuaD, indicated functions of the Tua proteins are based on homologies to other known proteins. A TUA containing a tetrasaccharice repeating unit would be more consistent with the number and putative functions of the TUA genes identified (see Section 6).

ings. The overall negative charge of the cell wall can be increased substantially by transferring alkalophilic bacteria from pH 7 to media of pH 10 (Aono et al., 1995; see Section 8.3).

7
Biodegradation

The little that is known of the processes of biodegradation of WTA is limited to *B. subtilis* 168. The use of biodegradation of WTA by growing cell populations to provide a source of phosphate was shown by growth in phosphate-limited media (Grant, 1979). Cultures of *B. subtilis* 168 in limited low concentrations of phosphate show a biphasic growth. The first phase, ending with a plateau that corresponds to phosphate depletion, is followed by the second phase, characterized by a slower rate of mass increase. During this latter phase, cells incorporate phosphate derived from extracellular WTA that had been released from the cell wall during previous growth in phosphate-replete media by the phenomenon of cell-wall turnover (Pooley, 1976a,b). At present, it is not known which WTA-degrading activity or activities are involved. However, phosphate starvation leads to the appearance in the periplasmic fraction of several proteins having phosphodiesterase or phosphatase activity, some of which are released to the medium (Merchante et al., 1995). A teichoicase has been purified from *B. subtilis* 168 (Kusser and Fiedler, 1982, 1983), but it was claimed that it was absent from other strains of *B. subtilis*.

8
Biological Roles Associated with Cell Wall Teichoic and Teichuronic Acids

Whereas these cell-wall anionic polymers have been known for more than 40 years, confirmation of their playing an essential role was obtained barely a decade ago. The nature of this role still remains to be determined. The following section deals with later results underlying the importance of WTAs and their substituents for processes active in the cell's periplasmic region, such as the secretion and stability of the proteins dependent upon the pH and negative charge of the zone encompassing the inner wall. The overriding importance for the cell of continued surface expansion may well represent the key role for these polymers. The importance of recent findings is raised for this fundamental and as yet unresolved question.

8.1
Horizontal Transfer of Genes for Wall Teichoic Acid Synthesis

In *B. subtilis* 168, the WTA divergon *tagAB*-*tagDEF* has an average G + C content of 33%, while the operon specific for the secondary polymer poly(GlcGalNAc) has an even lower one, i.e., 28%. These figures, well below the overall *B. subtilis* average of 43.5% G + C (Kunst et al., 1997), contrast with that characteristic of the TUA operon, i.e., 44.3%. This, it has been suggested (Soldo et al., 1999), indicates that the TUA operon is likely to encode enzymes synthesizing an ancestral *Bacillus* wall anionic polymer, strongly implying that the WTA operons were acquired by horizontal transfer from other Gram-positive bacteria characterized by a lower G + C content (Freymond, 1995; Lazarevic et al., 1995).

Possibly, resident ancestral WTA genes could have been exchanged for those encoding a different polymer (Karamata et al., 1987). A number of features suggest that the *tar* genes that encode the synthesis of poly(RboP), the major WTA of *B. subtilis* W23, represent such ancestral WTA genes (Lazarevic et al., 2002). They are organized in apparently coherent operons that are not interrupted, as in strain 168, by regions of noncoding DNA ("gray holes"). These observations are suggestive of a "crash landing" in strain 168 of WTA genes from another organism.

Horizontal transfer and substitution of WTA genes of *B. subtilis* 168 by *tar* genes from *B. subtilis* W23, encoding distinct polymers, have been achieved in the laboratory, albeit at frequencies orders of magnitude below those obtained with other markers (Karamata et al., 1987). Transfers in the reciprocal direction also have been possible, again at very low frequencies (unpublished data). The relatively unchanged phenotype (Karamata et al., 1987) accompanying the formation of certain of these hybrids provides further support for the view that, rather than the specific chemical components, it is the anionic nature and the overall structure that are actually important for the function of WTA.

Nonetheless, it would be surprising if such a substitution were not accompanied by a major loss of affinity by the endogenous-wall-associated proteins, present in *B. subtilis*, and, therefore, by altered cell-wall metabolism. For instance, the absence of affinity of the major cell-wall autolysin, LytC of *B. subtilis* 168, for the cell walls of *B. subtilis* W23 has been shown (Herbold and Glaser, 1975).

Replacement of the *lytC* gene of 168 by its homolog from W23 is not accompanied by any notable change in growth phenotype at pH 7.4 (unpublished data).

8.2
Role of the D-Alanine Substituent of Wall Teichoic Acid and of the Overall Cell Wall Negative Charge

Inactivation of any of the four *B. subtilis* genes leading to a block in D-alanylation of LTA and WTA, i.e., *dltA* to *dltD*, did not affect cell growth or morphology in growth media of pH 7.4 (Perego et al., 1995). However, grown under these conditions, *dlt* mutants of *B. subtilis* exhibited a more than two-fold increase in the cell-wall negative charge, as revealed by the quantity of cytochrome C bound by whole bacteria (Wecke et al., 1997). Blocking D-alanylation by *dlt* gene inactivation in *B. subtilis* was recently shown (Hyyryläinen et al., 2000) to allow increased secretion of several exoproteins in a strain bearing a defective PrsA protein, periplasmic foldase, or chaperone.

All polypeptides, having once crossed the cytoplasmic membrane, need to be folded efficiently, as for cytoplasmically located proteins. The importance of this process of PrsA, essential for cell viability, has been revealed by studies with a partially defective PrsA (Kontinen et al., 1991; Leskelä et al., 1999). Unfolded, or badly folded, mutant polypeptides, after transiting the cytoplasmic membrane, are degraded by periplasmic proteases. Recent evidence (Hyyryläinen et al., 2000; Tjalsma et al., 2000) for the influence on polypeptide folding of the overall negative charge of the cell-wall/cytoplasmic membrane underlines the key role of cell-surface anionic polymers, including probably both WTA and LTA, in this vital process, whether the effect is direct or indirect, e.g., by affecting the rate of PrsA folding.

8.3 The pH Gradient Across the Cell Wall

The role of polyanions in the cell wall and cytoplasmic membrane, in maintaining an appropriate ionic environment for membrane synthetic activities, has long been recognized (Hughes et al., 1973). Their presence is apparently essential for generating a pH gradient across the cell wall (Urrutia et al., 1992; Kemper and Doyle, 1993; Kemper et al., 1993) by sequestering protons released at the membrane level by oxidative or fermentative processes. It has been suggested that a low pH (Kemper and Doyle, 1993), close to the membrane, would inhibit PG hydrolases from degrading nascent PG. Generating and maintaining a sufficiently low periplasmic pH in cells growing in a high-pH medium would be more difficult. An increased pH in the vicinity of the membrane could lead to dysfunction capable of impeding growth. Support for this view (Pooley and Karamata, 1994) comes from changes of cell-wall composition shown by alkaliphilic bacteria (Aono and Horikoshi, 1983; Aono et al., 1995) when they are transferred from media of pH 7 to pH 10. For a majority of alkaliphilic organisms of the genus *Bacillus*, such a transfer is accompanied by a substantial increase of polyanionic wall components (Aono and Horikoshi, 1983; Aono et al., 1995), a response that should facilitate retention of protons near the membrane.

The capacity of organisms to grow at pH below 7 also is affected by the partial neutralization of the negatively charged LTA and WTA by the proton of D-ala. *dlt* mutants of *Streptococcus mutans* lost their acid-tolerant response, i.e., their ability to grow in a medium of pH 5 (Boyd et al., 2000). Cells of these mutants also were shown to be twice as permeable to the passive inflow of protons as were those with D-alanylated LTA.

8.4 Cell Wall Teichoic Acid and Upkeep of the Cell Shape

Taken as a whole, the processes determining cell-shape upkeep and surface growth remain mysterious. To ensure their orderly accomplishment, an important number of metabolic and biosynthetic pathways must be coordinated. Some of these pathways are known, but it is likely that many remain to be identified. After this, there remains the major task of understanding the mechanisms and structures involved and their articulation. Among the questions awaiting answers, one concerns the means (e.g., structures, signal molecules) by which transmembrane communication is maintained between the polymerization of cell-wall components taking place in the periplasm and the cytoplasmically located synthesis of the corresponding soluble precursors.

Mutations in *B. subtilis* WTA *tag* genes, including those encoding proteins involved in polymerization steps (*tagB, tagF*) as well as those (e.g., *tagD* and *tagP*) responsible for the synthesis of cytoplasmically located precursors, were shown to be associated with a conditional, lethal temperature-sensitive phenotype. The block in cylindrical surface expansion (Pooley et al., 1993) that occurs in precursor-deficient mutants at the restrictive temperature does not appear to be significantly different or slower from that associated with mutants deficient in membrane-associated polymerization steps. In one of these genes, *tagP*, the mutant phenotype is characterized by significant residual enzyme activity (Pooley and Karamata, 2000), strongly suggesting that the inhibition of cylindrical wall extension is not directly correlated with the degree of inhibition of the enzyme's activity. This latter phenotype may be due to an alteration of protein–protein interactions, as in a multienzyme complex. Such interactions may represent an essential part of the transmembrane signal transmission evoked here above.

Independent support linking cell extension to cell-wall anionic polymer synthesis comes from the study of mutants specifically deficient in TUA synthesis, growing in near phosphate-limited concentrations (Forsberg et al., 1973; Archibald, 1980). Decreasing phosphate concentration provokes a lowered cell-wall anionic polymer content that is accompanied by progressive shortening and increased cell width. This result implies a tight coupling of the rate of wall anionic polymer synthesis relative to that of PG and the rate of expansion of the cylindrical dimension of the rod-shaped cell.

Two *B. subtilis* proteins, MreB and Mbl, each essential for cell growth, are involved (Jones et al., 2001) in the synthesis of structures previously undescribed in bacteria, i.e., actin-like filaments. Each protein is necessary for the synthesis of a distinct helical structure lying close to the cell's surface. Both these structures, it was suggested, play a role in defining cell dimensions. The kind of interactions that must exist between these structures, or between MreB and Mbl and the protein complex responsible for synthesis of cell-wall components, remains to be defined. Nonetheless, proteins with significant homology to MreB and Mbl are absent from all coccal-shaped organisms, including *S. aureus,* for which genomic sequence data are currently available, implying that such actin-like structures may be limited to rod-shaped organisms. However, WTA synthesis is essential in organisms representative of both rod and coccal shape.

Evidence emerging from studies with Gram-positive bacteria besides *B. subtilis* suggests that WTA synthesis plays an essential part in the processes involved in the orderly surface expansion in organisms of both coccal and rod shape.

9 Applications

Neither WTA/TUA nor any of the genes involved in their synthesis have formed the subject of a patent application, to date, to the best of our knowledge. This notwithstanding, wall anionic polymers remain an important subject for applied research, not least in view of possible public health and medical applications, which include the following:

1) WTA of pathogenic bacteria is a potential new target for antibiotics (Pooley and Karamata, 1988; Pooley and Karamata, 2000).
2) *Micrococcus luteus* TUA addition stimulates the formation of tumor necrosis factor alpha and interleukins in human and murine blood cells (Monodane et al., 2001; Yang et al., 2001), which opens up the prospect of using TUA as an immunomostimulator.
3) WTA contributes to the cell wall's affinity for cations (Beveridge and Murray, 1976), a property that can be exploited for the detoxification of heavy-metal-polluted wastewaters and discharges.

10
References

Aono, R., Horikoshi, K. (1983) Chemical composition of cell walls of alkalophilic strains of *Bacillus*, *J. Gen. Microbiol.* **129**, 1083–1087.

Aono, R., Ito, M., Joblin, K. N., Horikoshi, K. (1995) A high cell wall negative charge is necessary for the growth of the alkaliphile *Bacillus lentus* C-125 at elevated pH, *Microbiology* **141**, 2955–2964.

Araki, Y., Ito, E. (1989) Linkage units in cell walls of gram-positive bacteria, *Crit. Rev. Microbiol.* **17**, 121–135.

Archibald, A. R., Stafford, G. H. (1972) A polymer of N-acetylglucosamine 1-phosphate in the wall of *Staphylococcus lactis* 2102, *Biochem. J.* **130**, 681–690.

Archibald, A. R. (1974) The structure, biosynthesis and function of teichoic acid, *Adv. Microb. Physiol.* **11**, 53–95.

Archibald, A. R. (1980) Phage receptors in Gram positive bacteria, in: *Receptors and Recognition*, Series B, Vol. 7, Virus receptors (Randall, L. L., Philipson, L., Eds.), London: Chapman and Hall, 5–26.

Archibald, A. R. (1988) Bacterial cell wall structure and the ionic environment, in: *Homeostatic Mechanisms in Microorganisms* (Whittenbury, R., Gould, G. W., Banks, J. G., Board, R. G., Eds.), Bath, England: Bath University Press, 159–173.

Archibald, A. R., Hancock, I. C., Harwood, C. R. (1993) Cell wall structure, synthesis and turnover, in: *Bacillus subtilis and Other Gram-positive Bacteria*, (Sonenshein, A. L., Hoch, J. A., Losick, R., Eds.), Washington, D.C.: American Society for Microbiology, 381–410.

Baddiley, J., Mathias, A. P. (1954) Cytidine nucleotides, Part I, Isolation from *Lactobacillus arabinosus*, *J. Chem. Soc.* 2723–2731.

Baddiley, J., Buchanan, J. G., Mathias, A. P., Sanderson, A. R. (1956) Cytidine diphosphate glycerol, *J. Chem. Soc.* 4186–4190.

Baddiley, J., Buchanan, J. G., Martin, R. O., RajBhandary, U. L. (1962a) Teichoic acids from the walls of *Staphylococcus aureus* H, *Biochem. J.* **85**, 49–56.

Baddiley, J., Buchanan, J. G., RajBhandary, U. L., Sanderson, A. R. (1962b) Teichoic acids from walls of *Staphylococcus aureus* H, *Biochem. J.* **82**, 439–448.

Baddiley, J. (1972) Teichoic acids in cell walls and membranes of bacteria, *Essays Biochem.* **8**, 35–77.

Beveridge, T. J., Murray, R. G. E. (1976) Uptake and retention of metals by cell walls of *Bacillus subtilis*, *J. Bacteriol.* **127**, 1502–1518.

Beveridge, T. J., Forsberg, C. W., Doyle, R. J. (1982) Major sites of metal binding in *Bacillus licheniformis* walls, *J. Bacteriol.* **150**, 1438–1448.

Boyd, D. A., Cvitkovitch, D. G., Bleiweis, A. S., Kiriukhin, M. Y., Debabov, D. V., Neuhaus, F. C., Hamilton, I. R. (2000) Defects in D-alanyl-lipoteichoic acid synthesis in *Streptococcus mutans* results in acid sensitivity, *J. Bacteriol.* **182**, 6055–6065.

Boylan, R. J., Mendelson, N. H. (1969) Initial characterization of a temperature-sensitive Rod⁻ mutant of *Bacillus subtilis*, *J. Bacteriol.* **100**, 1316–1321.

Boylan, R. J., Mendelson, N. H., Brooks, D., Young, F. E. (1972) Regulation of the bacterial cell wall: analysis of a mutant of *Bacillus subtilis* defective in biosynthesis of teichoic acid, *J. Bacteriol.* **110**, 281–290.

Bracha, R., Glaser, L. (1976a) An intermediate in teichoic acid biosynthesis, *Biochem. Biophys. Res. Commun.* **72**, 1091–1098.

Bracha, R., Glaser, L. (1976b) *In vitro* system for the synthesis of teichoic acid linked to peptidoglycan, *J. Bacteriol.* **125**, 872–879.

Bracha, R., Davidson, R., Mirelman, D. (1978) Defect in biosynthesis of the linkage unit between peptidoglycan and teichoic acid in a bacteriophage-resistant mutant of *Staphylococcus aureus*, *J. Bacteriol.* **134**, 412–417.

Briehl, M., Pooley, H. M., Karamata, D. (1989) Mutants of *Bacillus subtilis* 168 thermosensitive for growth and wall teichoic acid synthesis, *J. Gen. Microbiol.* **135**, 1325–1334.

Brooks, D., Mays, L. L., Hatefi, Y., Young, F. E. (1971) Glucosylation of teichoic acid : solubilization and partial characterization of the uridine diphosphoglucose:polyglycerolteichoic acid glucosyl transferase from membranes of *Bacillus subtilis*, *J. Bacteriol.* **107**, 223–229.

Burger, M. M., Glaser, L. (1964) The synthesis of teichoic acids. I. Polyglycerolphosphate, *J. Biol. Chem.* **239**, 3168–3177.

Burger, M. M., Glaser, L. (1966) CDPglycerol: acceptor phosphoglyceroltransferase from *Bacillus licheniformis* and *Bacillus subtilis*, *Methods Enzymol.* **8**, 430–435.

Clarke-Sturman, A. J., Archibald, A. R., Hancock, I. C., Harwood, C. R., Merad, T., Hobot, J. A. (1989) Cell wall assembly in *Bacillus subtilis*: partial conservation of polar wall material and the effect of growth conditions on the pattern of incorporation of new material at the polar caps, *J. Gen. Microbiol.* **135** , 657–665.

Coley, J., Archibald, A. R., Baddiley, J. (1976) A linkage unit joining peptidoglycan to teichoic acid in *Staphylococcus aureus* H, *FEBS Lett.* **61**, 240–242.

Critchley, P., Archibald, A. R., Baddiley, J. (1962) The intracellular teichoic acid from *Lactobacillus arabinosus* 17-5, *Biochem. J.* **85**, 420–431.

Da Silva, E. (2000) Characterization of the phosphoglucosamine mutase gene of *Bacillus subtilis* 168 and elaboration of a novel region of its defective prophage PBSX - Two approaches to antibiotic development, Ph.D. Thesis, University of Lausanne, Switzerland.

De Boer, W. (1979) Cell wall polymers of *Bacillus subtilis*, structure and metabolism. Thesis, Universiteit van Amsterdam, 59–65.

Dobrogosz, W. J. (1968) Effect of amino sugars on catabolite repression in *Escherichia coli*, *J. Bacteriol.* **95**, 578–584.

Doyle, R. J., McDannel, M. L., Helman, J. R., Streips, U. N. (1975) Distribution of teichoic acid in the cell wall of *Bacilus subtilis*, *J. Bacteriol.* **122**, 152–158.

Duckworth, M., Archibald, A. R., Baddiley, J. (1972) The location of N-acetylgalactosamine in the walls of *Bacillus subtilis* 168, *Biochem. J.* **130**, 691–696.

Ellwood, D. C., Tempest, D. W. (1969) Control of teichoic acid and teichuronic acid biosyntheses in chemostat cultures of *Bacillus subtilis* var. *niger*, *Biochem. J.* **111**, 1–5.

Ellwood, D. C., Tempest, D. W. (1972) Effects of environment on bacterial wall content and composition, *Adv. Microbiol. Physiol.* **7**, 83–117.

Estrela, A. I., de Lencastre, H., Archer, L. J. (1986) Resistance of a *Bacillus subtilis* mutant to a group of temperate bacteriophages, *J. Gen. Microbiol.* **132**, 411–415.

Estrela, A. I., Pooley, H. M., de Lencastre, H., Karamata, D. (1991) Genetic and biochemical characterization of *Bacillus subtilis* 168 mutants specifically blocked in the synthesis of the teichoic acid, poly(3-*O*-β-D-glucopyranosyl-*N*-acetylgalactosamine-1-phosphate); *gneA*, a new locus, is associated with UDP-N-acetylglucosamine 4-epimerase activity, *J. Gen. Microbiol.* **137**, 943–950.

Fan, D. P., Pelvit, M. C., Cunningham, W. P. (1972) Structural difference between walls from ends and sides of the rod-shaped bacterium *Bacillus subtilis*, *J. Bacteriol.* **109**, 1266–1272.

Fitzgerald, S. N., Foster, T. J. (2000) Molecular analysis of the *tagF* gene, encoding CDP-glycerol:poly(glycerophosphate) glycerophosphotransferase of *Staphylococcus epidermidis* ATCC 14990, *J. Bacteriol.* **182**, 1046–1052.

Forsberg, C. W., Wyrick, P. B., Ward, J. B., Rogers, H. J. (1973) Effect of phosphate limitation on the morphology and wall composition of *Bacillus licheniformis* and its phosphoglucomutase-deficient mutants, *J. Bacteriol.* **113**, 969–984.

Freymond, P.-P. (1995) Génétique et biochimie des acides téichoïques secondaires de *Bacillus subtilis* 168 et W23, Ph.D. Thesis, University of Lausanne, Switzerland.

Ghuysen, J. M., Strominger, J. L. (1963) Structure of the cell wall of *Staphylococcus aureus* strain Copenhagen I, preparation of fragments by enzymic hydrolysis, *Biochemistry*, New York **2**, 1110–1119.

Glaser, L. (1964) The synthesis of teichoic acids II polyribitol phosphate, *J. Biol. Chem.* **239**, 3178–3186.

Glaser, L., Loewy, A. (1979) Control of teichoic acid synthesis during phosphate limitation, *J. Bacteriol.* **137**, 327–331.

Graham, L. L., Beveridge, T. J. (1994) Structural differentiation of the *Bacillus subtilis* 168 cell wall, *J. Bacteriol.* **176**, 1413–1421.

Grant, W. D. (1979) Cell wall teichoic acid as a reserve phosphate source in *Bacillus subtilis*, *J. Bacteriol.* **137**, 35–43.

Hancock, I. C., Baddiley, J. (1972) Biosynthesis of the wall teichoic acid in *Bacillus licheniformis, Biochem. J.* **127**, 27–37.

Hancock, I. C., Baddiley J. (1976) *In vitro* synthesis of the unit that links teichoic acid to peptidoglycan, *J. Bacteriol.* **125**, 880–886.

Hancock, I. C., Wiseman, G., Baddiley, J. (1976) Biosynthesis of the unit that links teichoic acid to the bacterial wall: inhibition by tunicamycin, *FEBS Lett.* **69**, 75–80.

Hancock, I. C. (1981) The biosynthesis of ribitol teichoic acid by toluenised cells of *Bacillus subtilis* W23, *Eur. J. Biochem.* **119**, 85–90.

Hancock, I. C. (1983) Activation and inactivation of secondary wall polymers in *Bacillus subtilis* W23, *Arch. Microbiol.* **134**, 222–226.

Hancock, I. C., Baddiley, J. (1985) Biosynthesis of the bacterial envelope polymers teichoic acid and teichuronic acid, in: *The Enzymes of Biological Macromolecules*, (Martonosi, A. N., Ed.), New York: Plenum Press, Inc., 279–307, Vol. 2.

Hancock, I. C., Poxton, I. (1988) Teichoic acids and other accessory carbohydrates from gram-positive bacteria, in: *Bacterial Cell Surface Techniques* (Hancock I. C., Poxton I., Eds.), Chichester: John Wiley & Sons, 79–88.

Harrington, C. R., Baddiley, J. (1985) Biosynthesis of wall teichoic acids in *Staphylococcus aureus* H, *Micrococcus varians* and *Bacillus subtilis* W23. Involvement of lipid intermediates containing the disaccharide N-acetylmannosaminyl N-acetylglucosamine, *Eur. J. Biochem.* **153**, 639–645.

Hase, S., Matsushima, Y. (1977) The structure of the branching point between acidic polysaccharide and peptidoglycan in *Micrococcus lysodeikticus* cell wall, *J. Biochem.* (Tokyo) **81**, 1181–1186.

Hayes, M. V., Ward, J. B., Rogers, H. J. (1977) The synthesis of an N-acetylgalactosamine polymer by cell-free preparations of *Bacillus subtilis, Proc. Soc. Gen. Microbiol.* **4**, 85–86.

Heaton, M. P., Neuhaus, F. C. (1994) Role of the D-alanyl carrier protein in the biosynthesis of D-alanyl-lipoteichoic acid, *J. Bacteriol.* **176**, 681–690.

Heckels, J. E., Archibald, A. R., Baddiley, J. (1975) Studies on the linkage between teichoic acid and peptidoglycan in a bacteriophage-resistant mutant of *Staphylococcus aureus* H, *Biochem. J.* **149**, 637–647.

Heptinstall, J., Coley, J., Ward, P. J., Archibald, A, R., Baddiley, J. (1978) The linkage of sugar phosphate polymer to peptidoglycan in walls of *Micrococcus* sp. 2102, *Biochem. J.* **169**, 329–336.

Herbold, D. R., Glaser, L. (1975) Interaction of N-acetylmuramic acid L-alanine amidase with cell wall polymers, *J. Biol. Chem.* **250**, 7231–7238.

Holmes, R. P., Russell, R. R. B. (1972) Mutations affecting amino sugar metabolism in *Escherichia coli* K-12, *J. Bacteriol.* **111**, 290–291.

Honeyman, A. L., Stewart, G. C. (1989) The nucleotide sequence of the *rodC* operon of *Bacillus subtilis, Mol. Microbiol.* **3**, 1257–1268.

Hughes, A. H., Hancock, I. C., Baddiley, J. (1973) The function of teichoic acids in cation control in bacterial membranes, *Biochem. J.* **132**, 83–93.

Hughes, R. C., Pavlik, J. G., Rogers, H. J., Tanner, P. J. (1968) Organization of polymers in the cell walls of some bacilli, *Nature* **219**, 642–644.

Hussey, H., Sueda, S., Cheah, S. C., Baddiley, J. (1978) Control of teichoic acid synthesis in *Bacillus licheniformis* ATCC 9945, *Eur. J. Biochem.* **82**, 169–174.

Hyyryläinen, H.-L., Vitikainen, M., Thwaite, J., Wu, H., Sarvas, M., Harwood, C. R., Kontinen, V. P., Stephenson, K. (2000) D-alanine substitution of teichoic acids as a modulator of protein folding and stability at the cytoplasmic membrane/cell wall interface of *Bacillus subtilis, J. Biol. Chem.* **275**, 26696–26703.

Janczura, E., Perkins, H. R., Rogers, H. J. (1961) Teichuronic acid: a mucopolysaccharide present in wall preparations from vegetative cells of *Bacillus subtilis, Biochem. J.* **80**, 82–93.

Johnson, G. L., Hoger, J. H., Ratnayake, J. H., Anderson, J. S. (1984) Characterization of three intermediates in the biosynthesis of teichuronic acid of *Micrococcus luteus, Arch. Biochem. Biophys.* **235**, 679–691.

Jones, L. J. F., Carballido-Lopez, R., Errington, J. (2001) Control of cell shape in bacteria: helical, actin-like filaments in *Bacillus subtilis, Cell* **104**, 913–922.

Karamata, D., McConnel, M., Rogers, H. J. (1972) Mapping of *rod* mutants of *Bacillus subtilis, J. Bacteriol.* **111**, 73–79.

Karamata, D., Pooley, H. M., Monod, M. (1987) Expression of heterologous genes for wall teichoic acid in *Bacillus subtilis* 168, *Mol. Gen. Genet.* **207**, 73–81.

Kawamura, T., Ichihara, N., Ishimoto, N., Ito, E. (1975) Biosynthesis of uridine diphosphate N-acetyl-D-mannosaminuronic acid from uridine diphosphate N-acetyl-D-glucosamine in *Escherichia coli*: separation of enzymes responsible for epimerization and dehydrogenation, *Biochem. Biophys. Res. Commun.* **66**, 1506–1512.

Kawamura, T., Kimura, M., Yamamori, S., Ito, E. (1978) Enzymatic formation of uridine diphosphate N-acetyl-D-mannosamine, *J. Biol. Chem.* **253**, 3595–3601.

Kaya, S., Yokoyama, K., Araki, Y., Ito, E. (1984) N-acetylmannosaminyl(1-4)N-acetylglucosamine, a linkage unit between glycerol teichoic acid and peptidoglycan in cell walls, *J. Bacteriol.* **158**, 990–996.

Kemper, M., Doyle, R. J. (1993) The cell wall of *Bacillus subtilis* is protonated during growth, in: *Bacterial Growth and Lysis: Metabolism and Structure of the Bacterial Sacculus* (de Pedro, M. A., Höltje, J.-V., Löffelhardt, W., Eds.), New York, Plenum Publishing Corp, 245–252.

Kemper, M. A., Urrutia, M. M., Beveridge, T. J., Koch, A. L., Doyle, R. J. (1993) Proton motive force may regulate cell wall-associated enzymes of *Bacillus subtilis, J. Bacteriol.* **175**, 5690–5696.

Knox, K. W., Hall, K. J. (1965) The linkage between the polysaccharide and mucopeptide components of the cell wall of *Lactobacillus casei, Biochem. J.* **96**, 302–309.

Kojima, N., Araki, Y., Ito, E. (1985a) Structure of the linkage units between ribitol teichoic acids and peptidoglycan, *J. Bacteriol.* **161**, 299–306.

Kojima, N., Araki, Y., Ito, E. (1985b) Structural studies on the linkage unit of ribitol teichoic acid of *Lactobacillus plantarum, Eur. J. Biochem.* **148**, 29–34.

Kontinen, V. P., Saris, P., Sarvas, M. (1991) A gene (*prsA*) of *Bacillus subtilis* involved in a novel, late stage of protein export, *Mol. Microbiol.* **5**, 1273–1283.

Krispin, O., Allmansberger, R. (1998) The *Bacillus subtilis galE* gene is essential in the presence of glucose and galactose, *J. Bacteriol.* **180**, 2265–2270.

Kunst, F., Ogasawara, N., Moszer, I., Albertini, A. M., Alloni, G., et al. (1997) The complete genome sequence of the Gram positive bacterium *Bacillus subtilis, Nature* **390**, 249–256.

Kusser, W., Fiedler, F. (1982) Purification, Mr-value and subunit structure of a teichoic acid hydrolase from *Bacillus subtilis, FEBS Lett.* **149**, 67–70.

Kusser, W., Fiedler, F. (1983) Teichoicase from *Bacillus subtilis* Marburg, *J. Bacteriol.* **155**, 302–310.

Lang, W. K., Glassey, K., Archibald, A. R. (1982) Influence of phosphate supply on teichoic acid and teichuronic acid content of *Bacillus subtilis* cell walls, *J. Bacteriol.* **151**, 367–375.

Lazarevic, V., Karamata, D. (1995) The *tagGH* operon of *Bacillus subtilis* 168 encodes a two-component ABC transporter involved in the metabolism of two wall teichoic acids, *Mol. Microbiol.* **16**, 345–355.

Lazarevic, V., Mauël, C., Soldo, B., Freymond, P.-P., Margot, P., Karamata, D. (1995) Sequence analysis of the 308° to 311° segment of the *Bacillus subtilis* 168 chromosome, a region devoted to cell wall metabolism, containing non-coding grey holes which reveal chromosomal rearrangements, *Microbiology* **141**, 329–335.

Lazarevic, V., Abellan, F.-X., Beggah Möller, S., Karamata D., Mauël, C. (2002) Comparison of ribitol and glycerol teichoic acid genes in *Bacillus subtilis* W23 and 168: identical function, similar divergent organization, but different regulation, *Microbiology* **148**.

Leskelä, S., Wahlström, E., Kontinen, V. P., Sarvas, M. (1999) Lipid modification of prelipoproteins is dispensable for growth but essential for efficient protein secretion in *Bacillus subtilis*: characterization of the Lgt gene, *Mol. Microbiol.* **31**, 1075–1085.

Lifely, M. R., Tarelli, E., Baddiley, J. (1980) The teichuronic acid from the walls of *Bacillus licheniformis* ATCC 9945, *Biochem. J.* **191**, 305–318.

Liu, T. Y., Gotschlich, E. C. (1963) The chemical composition of pneumococcal C-polysaccharide, *J. Biol. Chem.* **238**, 1928–1934.

Liu, T. Y., Gotschlich, E. C. (1967) Muramic acid phosphate as a component of the mucopeptide of Gram-positive bacteria, *J. Biol. Chem.* **242**, 471–476.

Liu, W., Eder, S., Hulett, F. M. (1998) Analysis of *Bacillus subtilis tagAB* and *tagDEF* expression during phosphate starvation identifies a repressor role for PhoP-P, *J. Bacteriol.* **180**, 753–758.

Mauck, J., Glaser, L. (1972) On the mode of *in vivo* assembly of the cell wall of *Bacillus subtilis, J. Biol. Chem.* **247**, 1180–1187.

Mauël, C., Karamata, D. (1990) Identification of transcription units in the region encompassing teichoic acids genes of *Bacillus subtilis*, in: *Genetics and Biotechnology of Bacilli*, (Zukowski, M. M., Ganesan, A. T., Hoch, J. A., Eds.), New York: Academic Press Inc, 43–48, Vol. 3.

Mauël, C., Young, M., Margot, P., Karamata, D. (1989) The essential nature of teichoic acids in *Bacillus subtilis* as revealed by insertional mutagenesis, *Mol. Gen. Genet.* **215**, 388–394.

Mauël, C., Young, M., Karamata, D. (1991) Genes concerned with synthesis of poly(glycerol phosphate), the essential teichoic acid in *Bacillus subtilis* strain 168, are organized in two divergent transcription units, *J. Gen. Microbiol.* **137**, 929–941.

Mauël, C., Young, M., Monsutti-Grecescu, A., Marriot, S. A., Karamata, D. (1994) Analysis of *Bacillus subtilis tag* gene expression using transcriptional fusions, *Microbiology* **140**, 2279–2288.

Mauël, C., Bauduret, A., Chervet, C., Beggah, S., Karamata, D. (1995) In *Bacillus subtilis* 168, teichoic acid of the cross-wall may be different from that of the cylinder: a hypothesis based on transcription analysis of *tag* genes, *Microbiology* **141**, 2379–2389.

McArthur, H. A., Roberts, F. M., Hancock, I. C., Baddiley, J. (1978) Lipid intermediates in the biosynthesis of the linkage unit between teichoic acids and peptidoglycan, *FEBS Lett.* **86**, 193–200.

Mengin-Lecreulx, D., van Heijenoort, J. (1996) Characterization of the essential gene *glmM* encoding phosphoglucosamine mutase in *Escherichia coli*, *J. Biol. Chem.* **271**, 32–39.

Merchante, R., Pooley, H. M., Karamata, D. (1995) A periplasm in *Bacillus subtilis*, *J. Bacteriol.* **177**, 6176–6183.

Mirelman, D., Beck, B. D., Shaw, D. R. (1970) The location of the D-alanyl ester in the ribitol teichoic acid of *Staphylococcus aureus*, *Biochem. Biophys. Res. Commun.* **39**, 712–717.

Mirelman, D., Shaw, D. R., Park, J. T. (1971) Nature and origins of phosphorus compounds in isolated cell walls of *Staphylococcus aureus*, *J. Bacteriol.* **107**, 239–244.

Mobley, H. L., Koch, A. L., Doyle, R. J., Streips, U. N. (1984) Insertion and fate of the cell wall in *Bacillus subtilis*, *J. Bacteriol.* **158**, 169–179.

Monodane, T., Kawabata, Y., Yang, S., Hase, S., Takada, H. (2001) Induction of necrosis factor-alpha and interleukin-6 in mice *in vivo* and in murine peritoneal macrophages and human whole blood cells *in vitro* by *Micrococcus luteus* teichuronic acids, *J. Med. Microbiol.* **50**, 4–12.

Müller, J. P., An, Z., Merad, T., Hancock, I. C., Harwood, C. R. (1997) Influence of *Bacillus subtilis phoR* on cell wall anionic polymers, *Microbiology* **143**, 947–956.

Murazumi, N., Sasaki, Y., Okada, J., Araki, Y., Ito, E. (1981) Biosynthesis of glycerol teichoic acid in *Bacillus cereus*: formation of linkage unit disaccharide on a lipid intermediate, *Biochem. Biophys. Res. Commun.* **99**, 504–510.

Naumova, I. B., Shashkov, A. S. (1997) Anionic polymers in cell walls of gram-positive bacteria, *Biochemistry (Mosc)* **62**, 809–840.

Ogasawara, N., Nakai, S., Yoshikawa, H. (1994) Systematic sequencing of the 180 kilobase region of the *Bacillus subtilis* chromosome containing the replication origin, *DNA Res.* **1**, 1–14.

Pagni, M., Lazarevic, V., Soldo, B., Karamata, D. (1999) Assay for UDPglucose 6-dehydrogenase in phosphate-starved cells: gene *tuaD* of *Bacillus subtilis* 168 encodes the UDPglucose 6-dehydrogenase involved in teichuronic acid synthesis, *Microbiology* **145**, 1049–1053.

Partridge, M. D., Davison, A. L., Baddiley, J. (1973) The distribution of teichoic acids and sugar 1-phosphate polymers in walls of micrococci, *J. Gen. Microbiol.* **74**, 169–173.

Pavlik, J. G., Rogers, H. J. (1973) Selective extraction of polymers from cell walls of Gram-positive bacteria, *Biochem. J.* **131**, 619–621.

Perego, M., Glaser, P., Minutello, A., Strauch, M. A., Leopold, K., Fischer, W. (1995) Incorporation of D-alanine into lipoteichoic acid and wall teichoic acid in *Bacillus subtilis*, *J. Biol. Chem.* **270**, 15598–15606.

Pollack, J. H., Neuhaus, F. C. (1994) Changes in wall teichoic acid during the rod-sphere transition of *Bacillus subtilis* 168, *J. Bacteriol.* **176**, 7252–7259.

Pooley, H. M. (1976a) Turnover and spreading of old wall during surface growth of *Bacillus subtilis*, *J. Bacteriol.* **125**, 1127–1138.

Pooley, H. M. (1976b) Layered distribution, according to age, within the cell wall of *Bacillus subtilis*, *J. Bacteriol.* **125**, 1139–1147.

Pooley, H. M., Karamata, D. (1988) Can synthesis of cell wall anionic polymers in *Bacillus subtilis* be a target for antibiotics? in: *Antibiotic Inhibition of Bacterial Cell Surface Assembly and Function* (Actor, P., Daneo-Moore, L., Higgins, M. L., Salton, M. R. J., Shockman, G. D., Eds.), Washington, D.C.: Society for Microbiology, 591.

Pooley, H. M., Karamata, D. (1994) Teichoic acid synthesis in *Bacillus subtilis*: genetic organisation and biological roles, in: *Bacterial Cell Wall*, Book series: New comprehensive biochemistry, (Ghuysen, J.-M., Hakenbeck, R., Eds.), Amsterdam, The Netherlands: Elsevier Science Publishers B.V. 187–198.

Pooley, H. M., Karamata D. (2000) Incorporation of [2C-^{3}H]glycerol into cell surface components of *Bacillus subtilis* 168 and thermosensitive mutants affected in wall teichoic acid synthesis: effect of tunicamycin, *Microbiology* **146**, 797–805.

Pooley, H. M., Paschoud, D., Karamata, D. (1987) The *gtaB* marker in *Bacillus subtilis* 168 is associated with a deficiency in UDPglucose pyrophosphorylase, *J. Gen. Microbiol.* **133**, 3481–3493.

Pooley, H. M., Abellan, F.-X., Karamata, D. (1991) A conditional-lethal mutant of *Bacillus subtilis* 168 with a thermosensitive glycerol-3-phosphate cytidylyl transferase, an enzyme specific for the synthesis of the major wall teichoic acid, *J. Gen. Microbiol.* **137**, 921–928.

Pooley, H. M., Abellan, F.-X., Karamata, D. (1992) CDP-glycerol:poly(glycerophosphate) glycerophosphotransferase, involved in the synthesis of the major wall teichoic acid in *Bacillus subtilis* 168, is encoded by *tagF(rodC)*, *J. Bacteriol.* **174**, 646–649.

Pooley, H. M., Abellan, F.-X., Karamata, D. (1993) Wall teichoic acid, peptidoglycan synthesis and morphogenesis in *Bacillus subtilis*, in: *Bacterial Growth and Lysis: Metabolism and Structure of the Bacterial Sacculus* (de Pedro, M. A., Höltje, J.-V., Löffelhardt, W., Eds.), New York: Plenum Publishing Corp, 385–392.

Reusch, V. M. Jr., Neuhaus, F. C. (1971) D-Alanine: membrane acceptor ligase from *Lactobacillus casei*, *J. Biol. Chem.* **246**, 6136–6143.

Rivolta, C. (1999) Molecular genetics of *Bacillus subtilis*: from genome sequence to protein function, Ph.D. Thesis, University of Lausanne, Switzerland.

Rogers, H. J., Taylor, C. (1978) Autolysins and shape change in *rodA* mutants of *Bacillus subtilis*, *J. Bacteriol.* **135**, 1032–1042.

Rogers, H. J., McConnell, M., Burdett, I. D. J. (1970) The isolation and characterization of mutants of *Bacillus subtilis* and *Bacillus licheniformis* with disturbed morphology and cell division, *J. Gen. Microbiol.* **61**, 155–171.

Rogers, H. J., Thurman, P. F., Taylor, C., Reeve, J. N. (1974) Mucopeptide synthesis by *rod* mutants of *Bacillus subtilis*, *J. Gen. Microbiol.* **85**, 335–350.

Rohr, T. E., Levy, G. N., Stark, N. J., Anderson, J. S. (1977) Initial reactions in biosynthesis of teichuronic acid of *Micrococcus lysodeikticus* cell walls, *J. Biol. Chem.* **252**, 3460–3465.

Sanderson, A. R., Strominger, J. L., Nathenson, S. G. (1962) Chemical structure of teichoic acid from *Staphylococcus aureus* strain Copenhagen, *J. Biol. Chem.* **237**, 3603–3613.

Sasaki, Y., Araki, Y., Ito, E. (1980) Structure of the linkage region between glycerol teichoic acid and peptidoglycan in *Bacillus cereus* AHU cell walls, *Biochem. Biophys. Res. Commun.* **96**, 529–534.

Sasaki, Y., Araki, Y.. Ito, E. (1983) Structure of teichoic-acid-glycopeptide complexes from cell walls of *Bacillus cereus* AHU 1030, *Eur. J. Biochem.* **132**, 207–213.

Schipper, D. (1995) Structural studies of the teichoic acids from *Bacillus licheniformis*, *Carbohydr. Res.* **279**, 75–82.

Seki, T., Yoshikawa, H., Takahashi, H., Saito, H. (1987) Cloning and nucleotide sequence of *phoP*, the regulatory gene for alkaline phosphatase and phosphodiesterase in *Bacillus subtilis*, *J. Bacteriol.* **169**, 2913–2916.

Seki, T., Yoshikawa, H., Takahashi, H., Saito, H. (1988) Nucleotide sequence of the *Bacillus subtilis phoR* gene, *J. Bacteriol.* **170**, 5935–5938.

Shibaev, V. N., Duckworth, M., Archibald, A. R., Baddiley, J. (1973) The structure of a polymer containing galactosamine from walls of *Bacillus subtilis* 168, *Biochem. J.* **135**, 383–384.

Shiflett, M. A., Brooks, D., Young, F. E. (1977) Cell wall and morphological changes induced by temperature shift in *Bacillus subtilis* cell wall mutants, *J. Bacteriol.* **132**, 681–690.

Shockman, G. D., Slade, H. D. (1964) The cellular location of the streptococcal group D antigen, *J. Gen. Microbiol.* **37**, 297–305.

Soldo, B. (1999) Analyse génétique et physiologique de l'acide teichuronique de *Bacillus subtilis*, Ph.D. Thesis, University of Lausanne, Switzerland.

Soldo, B., Lazarevic, V., Margot, P., Karamata, D. (1993) Sequencing and analysis of the divergon comprising *gtaB*, the structural gene of UDP-glucose pyrophosphorylase of *Bacillus subtilis* 168, *J. Gen. Microbiol.* **139**, 3185–3195.

Soldo, B., Lazarevic, V., Pagni, M., Karamata, D. (1999) Teichuronic acid operon of *Bacillus subtilis* 168, *Mol. Microbiol.* **31**, 795–805.

Stark, N. J., Levy, G. N., Rohr, T. E., Anderson, J. S. (1977) Reactions of second stage of biosynthesis of teichuronic acid of *Micrococcus lysodeikticus* cell walls, *J. Biol. Chem.* **252**, 3466–3472.

Takatsuki, A., Shimizu, K.-I., Tamura, G. (1972) Effect of tunicamycin on microorganisms: morphological changes and degradation of RNA and DNA induced by tunicamycin, *J. Antibiot.* **25**, 75–85.

Tjalsma, H., Bolhuis, A., Jongbloed, J. D., Bron, S., van Dijl, J. M. (2000) Signal peptide-dependent protein transport in *Bacillus subtilis*: a genome-based survey of the secretome, *Microbiol. Mol. Biol. Rev.* **64**, 515–547.

Traxler, C. I., Goustin, A. S., Anderson, J. S. (1982) Elongation of teichuronic acid chains by a wall-membrane preparation from *Micrococcus luteus*, *J. Bacteriol.* **150**, 649–656.

Tul'skaya, E. M., Streshinskaya, G. M., Naumova, I. B., Shashkov, A. S., Terekhova, L. P. (1993) A new structural type of teichoic acid and some chemotaxonomic criteria of two species *Nocardiopsis dassonvillei* and *Nocardiopsis antarcticus*, *Arch. Microbiol.* **160**, 299–305.

Urrutia M. M., Kemper, M., Doyle, R. J., Beveridge, T. J. (1992) The membrane-induced proton motive force influences the metal binding ability of *Bacillus subtilis* cell walls, *Appl. Environ. Microbiol.* **58**, 3837–3844.

Vicente, M., Kushner, S. R., Garrido, T., Aldea, M. (1991) The role of the "gearbox" in the transcription of essential genes, *Mol. Microbiol.* **5**, 2085–2091.

Ward, J. B. (1981) Teichoic and teichuronic acids: biosynthesis, assembly and location, *Microbiol. Rev.* **45**, 211–243.

Ward, J. B., Curtis, C. A. M. (1982) The biosynthesis and linkage of teichuronic acid to peptidoglycan in *Bacillus licheniformis*, *Eur. J. Biochem.* **122**, 125–132.

Ward, J. B., Wyke, A. W., Curtis, C. A. M. (1980) The effect of tunicamycin on wall synthesis in *Bacilli*, *Biochem. Soc. Trans.* **8**, 164–166.

Wecke, J., Madela, K., Fischer, W. (1997) The absence of D-alanine from lipoteichoic acid and wall teichoic acid alters surface charge, enhances autolysis and increases susceptibility to methicillin in *Bacillus subtilis*, *Microbiology* **143**, 2953–2960.

White, P. J. (1977) A survey for the presence of teichuronic acid in walls of *Bacillus megaterium* and *Bacillus cereus*, *J. Gen. Microbiol.* **102**, 435–439.

White, R. J. (1968) Control of amino sugar metabolism in *Escherichia coli* and isolation of mutants unable to degrade amino sugars, *Biochem. J.* **106**, 847–858.

Wright, J., Heckels, J. E. (1975) The teichuronic acid of cell walls of *Bacillus subtilis* W23 grown in a chemostat under phosphate limitation, *Biochem. J.* **147**, 187–189.

Wyke, A. W., Ward, J. B. (1977) Biosynthesis of wall polymers in *Bacillus subtilis*, *J. Bacteriol.* **130**, 1055–1063.

Yang, S., Sugawara, S., Monodane, T., Nishijima, M., Adachi, Y., Akashi, S., Miyake, K., Hase, S., Takada, H. (2001) *Micrococcus luteus* teichuronic acids activate human and murine monocytic cells in a CD14- and toll-like receptor 4-dependent manner, *Infect. Immun.* **69**, 2025–2030.

Yokoyama, K., La Mar, G. N., Araki, Y., Ito, E. (1986) Structure and functions of linkage unit intermediates in biosynthesis of ribitol teichoic acids in *Staphylococcus aureus* H and *Bacillis subtilis* W23, *Eur. J. Biochem.* **161**, 479–489.

Yokoyama, K., Araki, Y., Ito, E. (1987) Biosynthesis of poly(galactosylglycerolphosphate) in *Bacillus coagulans*, *Eur. J. Biochem.* **165**, 47–53.

Young, F. E. (1967) Requirement of glucosylated teichoic acid for adsorption of phage in *Bacillus subtilis* 168, *Proc. Natl Acad. Sci. USA* **58**, 2377–2384.

19
Polysaccharide-containing Cell-wall Polymers of Archaea

Prof. Dr. Helmut König
Institut für Mikrobiologie und Weinforschung, Johannes Gutenberg-Universität, D-55099 Mainz, Germany; Tel.: +49-6131-39-24634; Fax: +49-6131-39-22695; E-mail: hkoenig@mail.uni-mainz.de

Ala	alanine
Asn	asparagine
Asp	aspartic acid
CM	cytoplasmic membrane
Gal	galactose
GG	glutaminylglycan
Glc	glucose
GalA	galacturonic acid
GlcA	glucuronic acid
GalN	galactosamine
GalNAc	*N*-acetylgalactosamine
GlcN	glucosamine
GlcNAc	*N*-acetylglucosamine
GC	glycocalyx
GG	glutaminylglycan
GlcA	glucuronic acid
Glu	glutamic acid
Gly	glycine
GulNA	gulosaminuronic acid
HP	heteropolysaccharide
IdA	iduronic acid
LP	lipoglycan
Lys	lysine
Man	mannose
Man*p*	mannopyranose
MC	methanochondroitin
3-O-MetGalA	3-*O*-methylgalacturonic acid
3-O-MetMan*p*	3-*O*-methylmannopyranose
NAcTalNUA	*N*-acetyltalosaminuronic acid
N^{α}-P-Glu	N^{α}-phosphoglutamic acid
N^{α}-UDP-Glu	N^{α}-uridine-5′-diphosphoglutamic acid
Orn	ornithine
PM	pseudomurein
Ser	serine
SL	S-layer
$-SO_4^{2-}$	sulfate residues
TalNUA	talosaminuronic acid

Thr	threonine
UA	uronic acid
UdP	undecaprenyl
UDP	uridine 5′-diphosphate
UDP-L-NAcAltNA	uridine-5′-diphospho-L-*N*-acetylaltrosaminuronic acid
UDP-D-GalNAc	uridine-5′-diphospho-D-*N*-acetylgalactosamine
UDP-D-GlcNAc	uridine-5′-diphospho-D-*N*-acetylglucosamine
UDP-L-NAcTalNA	uridine-5′-diphospho-L-*N*-acetyltalosaminuronic acid
X	amino acid

1 Introduction

The methanogens, extreme halophiles, thermoplasmas, sulfate- or sulfur- metabolizing extreme thermophiles belong to the Archaea. During the late 1970s, these prokaryotes were recognized as a separate third domain of life with the Bacteria and Eukarya (Woese, 1987). This domain was named Archaea, because many species live in habitats reminiscent of conditions encountered on the early Earth. The Archaea thrive in anaerobic niches, salt lakes, and marine hydrothermal systems and continental solfataras. The archaeal domain consists of three main phylogenetic lineages: the Crenarchaeota; the Euryarchaeota; and the Korarchaeota. These lineages include heterogeneous groups of prokaryotes, and it became obvious that the diversity of the Archaea is also reflected in a remarkable diversity of cell envelope types (Figure 1) and of chemical variations. A feature common to all Archaea is the lack of muramic acid and of a lipopolysaccharide-containing outer membrane. A universal cell-wall polymer is also missing, the cell walls of the Archaea being composed of different carbohydrate-containing polymers such as a glutaminylglycan, heterosaccharide, methanochondroitin, pseudomurein, glycoprotein, or a glycocalyx. The methanochondroitin of *Methanosarcina* is the only known cell-wall polymer which consists solely of carbohydrates. In the other polymers, the carbohydrates are conjugated with amino acids or peptides. The most common archaeal cell envelope is composed of a single protein or glycoprotein surface layer (S-layer), which is directly associated with the outside of the cytoplasmic membrane. The thermoplasmas lack a cell wall, but instead possess a glycocalyx. Some cell envelope types are restricted to the Archaea, but similar building blocks are also found in polymeric compounds (e.g., connective tissue) of the other domains.

2 Historical Outline

During the 1950s, significant differences from typical bacterial cell walls were established during cell envelope investigations of *Halobacterium* (Houwink, 1956). Later studies with cells of *Sulfolobus* (Weiss, 1974), *Halococcus* (Brown and Cho, 1970) and *Methanosarcina* (Kandler and Hippe, 1977) also showed unusual differences. The S-layer glycoprotein was the first to be discovered in prokaryotes (Mescher and Strominger, 1976), and initially these novel cell-wall structures were seen as curiosities. Their taxonomic significance was not realized before the publication of the new phylogenetic concept (Woese, 1987), but it then became obvious that no universal cell poly-

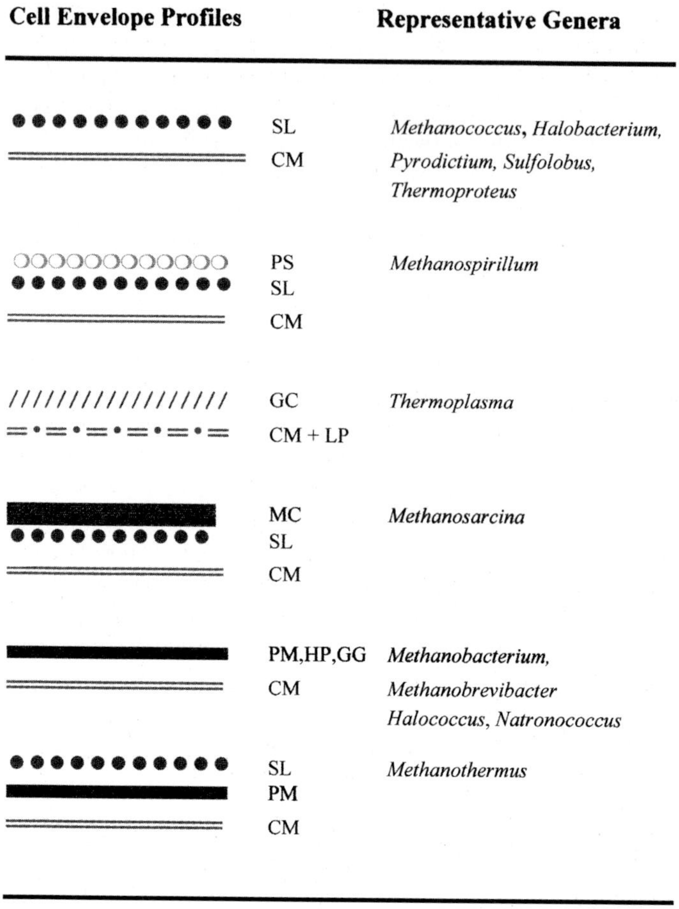

Cell Envelope Profiles	Representative Genera
SL CM	*Methanococcus, Halobacterium, Pyrodictium, Sulfolobus, Thermoproteus*
PS SL CM	*Methanospirillum*
GC CM + LP	*Thermoplasma*
MC SL CM	*Methanosarcina*
PM,HP,GG CM	*Methanobacterium, Methanobrevibacter Halococcus, Natronococcus*
SL PM CM	*Methanothermus*

Fig. 1 Cell wall profiles of Archaea.

mer exists and all Archaea lack murein (Kandler and König, 1985). The cell envelope structures were among the first biochemical characteristics studied in detail (Kandler and König, 1985, 1993), and the results obtained supported the Archaea concept of Woese (1987).

3
Chemical Structure

The majority of archaeal species produce different glycoconjugates such as glycoproteins, a peptidoglycan or heteropolysaccharides as cell-wall components. Cell-wall polymers lacking carbohydrates are found in only a few genera (e.g., *Methanococcus*) which form surface layers of crystalline protein subunits. This chapter provides an introduction to the chemical diversity of archaeal cell wall structures.

3.1
Glutaminylglycan

Beyond the major cell-wall components of Bacteria and Archaea, prokaryotes produce capsular exopolymers which differ widely in their chemical composition. The majority of the exopolymers are polysaccharides, but some composed of one type of amino acid, namely L- or D- glutamate, are also formed (Table 1). In general, it appears that in bacteria the α-linked polymers composed of glutamyl or glutaminyl residues possess the L-configuration, while in the case of the γ-linked exopeptides the glutamic acid residues have the D-configuration. The poly-γ-D-glutamyl polymers occur in the phylogenetically related genera *Bacillus, Sporosarcina*, and *Planococcus*.

Similar polymers have been found in extremely halophilic cocci of the genus *Natronococcus* (Niemetz et al., 1997), which survive optimally in extremely alkaline and saline biotopes. The cell polymer of *Natronococcus occultus* is formed by a polyglutamine glycoconjugate which, in contrast to the eubacterial polymer, is glycosylated. The isolated cell-wall polymer is composed of one amidated amino acid, namely L-glutamine, the two amino sugars glucosamine and galactosamine, the two uronic acids galacturonic acid and glucuronic acid, and the hexose glucose. In the intact polymer, the glutamine residues are linked via their γ-carboxylic group, thus forming a chain of about 60 monomers. Two types of oligosaccharides are linked to the polyglutamine backbone (Figure 2). One oligosaccharide consists of an *N*-acetylglucosamine pentasaccharide at the reducing end and a galacturonic acid oligosaccharide at the nonreducing end. The second oligosaccharide has an *N*-acetylgalactosamine disaccharide at the reducing end and a maltose unit at the nonreducing end.

Usually, N-linked oligosaccharides of bacterial and archaeal cell-wall glycoconjugates are linked via galactosamine, glucose or rhamnose to the β-amide group of asparagine, while in the case of *N. occultus* the oligosaccharides are linked to the α-carboxylic groups of the polyglutamine backbone via an N-amide linkage. Therefore, the exopolymer of *N. occultus* represents a novel type of naturally occurring glycoconjugates.

3.2 Heteropolysaccharide

The Gram-positive cells of *Halococcus morrhuae* are surrounded by an electron-dense

Tab. 1 Microorganisms with glycosylated and nonglycosylated polyglutamate exopolymers

Domain/species	*Configuration*	*Linkages*	*Carbohydrates*
Bacteria			
Bacillus anthracis	D	γ	n.f.
Bacillus licheniformis	D	γ	n.f.
Bacillus megaterium	D	γ	n.f.
Bacillus subtilis	D	γ	n.f.
Flexithrix dorotheae	L	α	n.f.
Mycobacterium tuberculosis	L	α	n.f.
Planococcus halophilus	D	γ	n.f.
Sporosarcina halophila	D	γ	n.f.
Xanthobacter autotrophicus	L	α	n.f.
Xanthobacter flavus	L	α	n.f.
Archaea			
Natronococcus occultus	L	γ	D-GalNAc, D-GlcNAc, D-GalA, D-GlcA, D-Glc
Halococcus turkmenicus	L	γ	D-GlcNAc, D-Glc, D-Man

From Messner et al. (1997).n.f., not found.

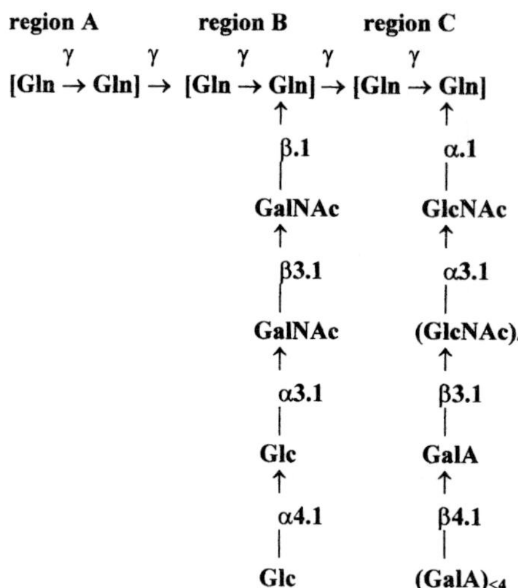

Fig. 2 Proposed structure of the repeating units (regions A–C) of the cell-wall polymer of *Natronococcus occultus*. The exact distribution of the repeating units was not determined. (Modified from Niemetz et al., 1997.)

cell wall with a thickness of about 50 nm. The cell-wall polymer is composed of a complex heterosaccharide consisting of a mixture of neutral and amino sugars, uronic acids, glycine and an aminuronic acid (Table 2) (Reistadt, 1972; Schleifer et al., 1982). The carbohydrate monomers are supposed to be arranged in three domains, which may be partly linked by *N*-glycyl-glucosaminyl bridges (Figure 3).

Tab. 2 Chemical composition of the cell walls of *Halococcus morrhuae* CCM 537^{T}

Component	*µmol per mg cell wall*
Glc	0.39
Man	0.21
Gal	0.18
GlcN	0.21
GalN	0.08
GulNA	0.09
Total UA	0.24
Gly	0.13
Sulfate	0.80

From Schleifer et al. (1982).

3.3 Methanochondroitin

Cells of *Methanosarcina* form often large, cuboid aggregates, the shape-maintaining component of which is a fibrillar nonsulfated polymer composed of a uronic acid and two *N*-acetylgalactosamine residues (Figure 4). A trimer which was identified as being the building block of the glycan (Kreisl and Kandler, 1986) forms a compact (or sometimes loose) matrix. Because of its resemblance to eukaryotic tissue – and especially to chondroitin sulfate – this was named “methanochondroitin”.

Modifications of the methanochondroitin are due to the additional occurrence of galacturonic acid in the two species *Methanosarcina mazei* and *Methanosarcina* sp. G1. Electron microscopic investigations indicate that, in some species, an S-layer may be located between the cytoplasmic membrane and the methanochondroitin matrix.

3.4 Pseudomurein

The orders Methanobacteriales and Methanopyrales contain different genera with anaerobic rod- or lancet-shaped bacteria or cocci. They live under conditions which extend from mesophilic to extremely thermophilic temperatures. The cells of these orders exhibit cell-wall sacculi with a thickness of about 15–20 nm and a density of 1.39–1.46 g cm^{-3} and a unit cell dimension of 4.5×10×21–22.5 Å. Conformational energy calculations have suggested a similar secondary structure, as for the eubacterial murein. Due to its chemical and three-dimensional structure. this cell-wall polymer must be classified as a peptidoglycan, which means that glycan and peptides each constitute about 50% of the polymeric material. The archaeal peptidoglycan was named

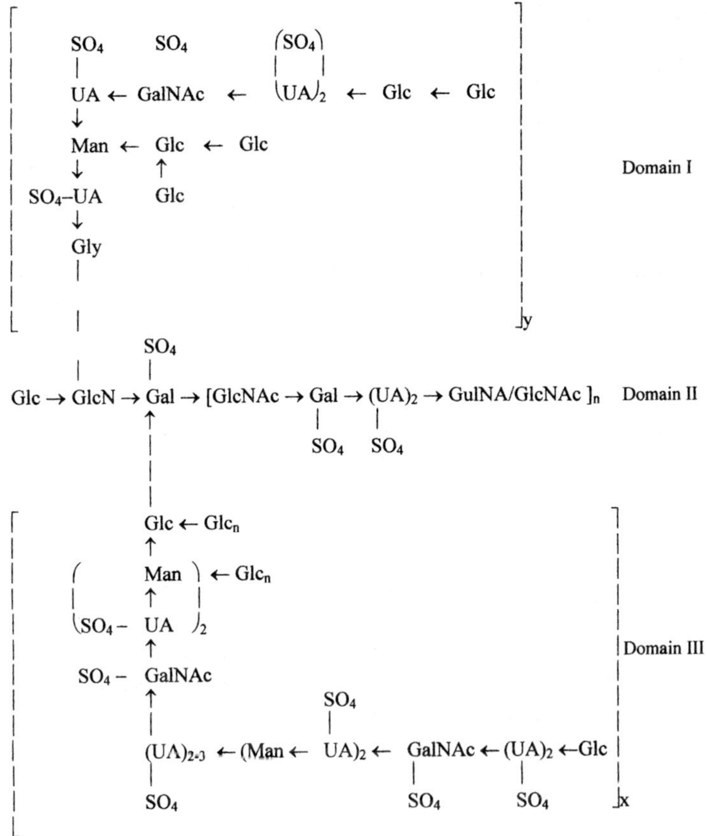

Fig. 3 Proposed domain structure of the cell-wall polymer of *Halococcus morrhuae* CCM 859. (Modified from Schleifer et al., 1982.)

pseudomurein; this possesses structural similarities to, but also differs significantly from, the eubacterial murein (Figure 5) (Kandler and König, 1985, 1993). The glycan consists of alternating β-(1 → 3)-linked *N*-acetyl-D-glucosamine and *N*-acetyl-L-talosaminuronic acid residues. The glycan strands are cross-linked by peptide subunits, which are often composed of a set of three L-amino acids: glutamic acid, alanine, and lysine.

The main differences between pseudomurein and murein are the occurrence of talosaminuronic acid instead of murein, the presence of a β(1 → 3)-linkage instead of a β(1 → 4)-linkage of the glycan components, the partial replacement of glucosamine by galactosamine, the lack of D-amino acids, and the accumulation of ε- and γ-peptide bonds. Chemical modifications are found in the glycan as well as in the peptide moiety (see Figure 5).

3.5 S-Layer

Many Archaea possess proteinaceous surface layers (S-layers), which form twodimensional regular arrays (Baumeister and Lembcke, 1992; Beveridge and Graham, 1991; Kandler and König, 1985, 1993; Mess-

a) [→)-β-D-GlcA-(1→3)-β-D-GalNAc-(1→4)-β-D-GalNAc-(1→]$_n$.

b) [D-GlcA-(1→3)-GalNAc-(1→4)-]$_{23\text{-}25}$

c) [β-D-GlcA-(1→3)-β-D-GalNAc-(1→4)-β-D-GlcA-(1→]$_n$
(4 or 6 sulfate)

Fig. 4 Proposed structure of the repeating units of (a) methanochondroitin, (b) teichuronic acid of *Bacillus licheniformis*, and (c) chondroitin sulfate of animal connective tissue. (From Kandler and König, 1993.)

```
(D-GalNAc)
-D-GlcNAc-(3←1)-β-L-NAcTalNA-
                          ↓
                          Glu (Asp)
                          ↓γ
                          Ala (Thr, Ser)
                          ↓ε
                          Lys←γ–Glu (— α,δ→ Orn)
                                 ↑
                                 Lys←γ– Glu
                                 ↑ε
                                 Ala (Thr, Ser)
                                 ↑γ
                                 Glu (Asp)
                                 ↑
                   -L-NAcTalNA-β-(1→3)-D-GlcNAc-
                                        (D-GalNAc)
```

Fig. 5 Proposed chemical structure of pseudomurein. (From Kandler and König, 1993; modifications in parentheses.)

α-D-3-O-MetMan*p*-(1→6)-α-D-3-O-MetMan*p*-[(1→2)- α-D-Man*p*]$_3$-(1→4)-D-GalNAc

Fig. 6 Proposed structure of the oligosaccharide of the S-layer glycoprotein of *Methanothermus fervidus*. (From Kärcher et al., 1993.)

ner and Sleytr, 1992; Sumper and Wieland, 1995). The chemical structure of S-layer glycoproteins has been determined from a few archaeal species, e.g., *Methanothermus fervidus* (Kärcher et al., 1993; cf. Kandler and König, 1993), *Halobacterium halobium*, and *Haloferax volcanii* (Lechner and Sumper, 1987; Sumper et al., 1990; cf. Sumper and Wieland, 1995).

The extreme thermophilic methanogen *M. fervidus* has a double-layered cell envelope, the pseudomurein being covered by an S-layer in hexagonal arrangement. The S-layer glycoprotein could be extracted with a solution of trichloroacetic acid followed by reverse-phase chromatography with aqueous formic acid as eluant. The mature protein consists of 593 amino acids. Compared with mesophilic S-layer proteins, a significantly greater amount of for example isoleucine, asparagine and cysteine, and also 14% more β-sheet structure, is present. The glycoprotein possesses also a typical leader peptide and 20 sequon structures as potential N-glycosylation sites. One type of oligosaccharide is present consisting of D-3-O-MetMan, D-Man, and D-GalNAc (Kärcher et al., 1993); this is linked via *N*-acetylgalactosamine to asparagine residues of the peptide moiety (Figure 6).

The surface glycoprotein of *Hb. halobium* (*Hb. salinarium*; cf. Sumper and Wieland, 1995) was the first glycosylated protein detected in prokaryotes. The polypeptide consists of a stretch of 12 hydrophobic amino acids at the C terminus, which functions as membrane anchor. Three different saccharide chains are linked to the peptide. A sulfated pentasaccharide repeating unit forms a glycosaminoglycan chain, which is linked to the asparagine residue at the second N-terminal position of the peptide chain via an *N*-acetylgalactosamine. Asparaginylglucose is the linkage unit of ten sulfated glucose, glucuronic acid and iduronic acid containing oligosaccharides. In addition to the two types of *N*-linked glycan chains, about 15 O-glycosidically linked glucosylgalactose disaccharides also occur (Figure 7). The disaccharides form a cluster close to the transmembrane domain. The glycoprotein of *Hb. halobium* is acidic because of the occurrence of more than 20% aspartic and glutamic acid residues and up to 50 mol of uronic acid and 50 mol sulfate residues per mol of peptide.

The mature polypeptide of the S-layer glycoprotein of *H. volcanii* is composed of 794 amino acids. Like the surface glycoprotein of *Hb. halobium*, a hydrophobic stretch of about 20 amino acids at the C terminus probably functions as a transmembrane domain. Glucosyl-(1 → 2)-galactose disaccharides are assumed to be linked to threonine residues clustering close to the membrane anchor. Sulfated and amino sugar-containing oligosaccharides are absent in the surface glycoprotein of *H. volcanii*.

3.6
Lipoglycan

Members of the thermoacidophilic genus *Thermoplasma*, like the mycoplasmas, lack a cell envelope, and stabilization of the cells is most likely maintained by a mannose-rich lipoglycan (Langworthy et al., 1982) and a glycoprotein (Yang and Haug, 1979). These two constituents are located in the cytoplasmic membrane, and their glycan chains are directed to the outside of the cell, forming a glycocalyx (Figure 8).

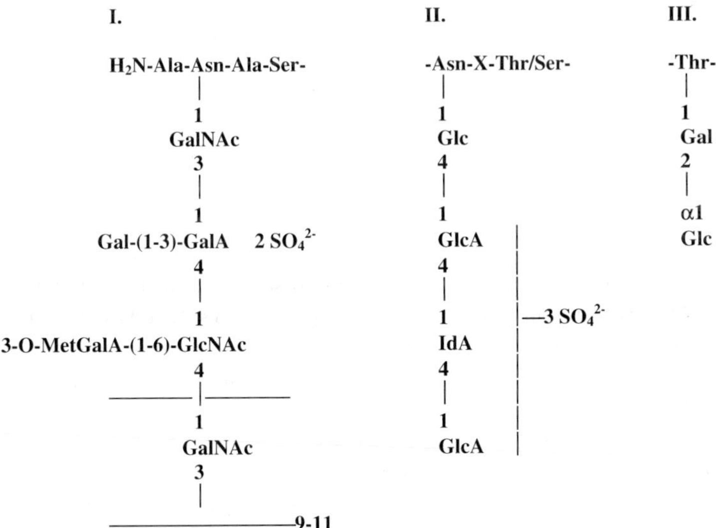

Fig. 7 Proposed structure of the three glycan moieties of the S-layer glycoprotein of *Halobacterium halobium*. The building block composed of five different sugars (glycan I) is linearly repeated 10–12 times. (Modified from Wieland, 1988.)

A.

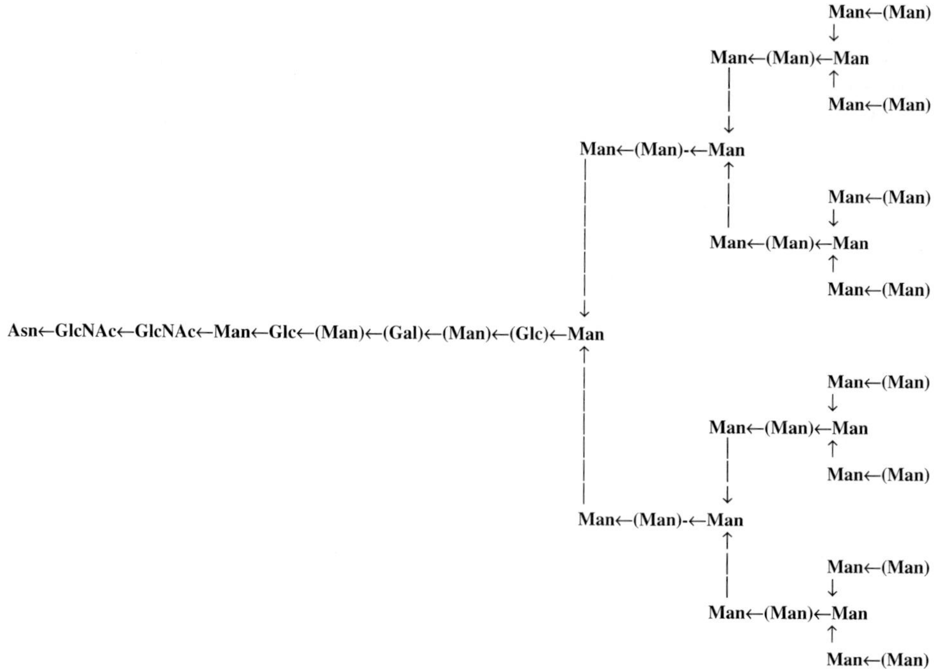

B.

$$[\text{Man-}(\alpha 1 \rightarrow 2)\text{-Man-}(\alpha 1 \rightarrow)\text{-Man-}(\alpha 1 \rightarrow 3)\text{-}]_8\text{Glc}(\alpha 1 \rightarrow 1)\text{-R}$$

Fig. 8 Glycan structures of the cytoplasmic membrane components of *Thermoplasma acidophilum*. (A) Glycan of the glycoprotein. (From Yang and Haug, 1979; parentheses indicate variable numbers of sugar residues.) (B) Glycan of the lipoglycan. (From Langworthy et al., 1982; modified; R = dibiphytanyldiglycerol tetraether.)

4 Occurrence

The Archaea possess carbohydrate-containing cell-surface polymers such as glutaminylglycan, heterosaccharide, methanochondroitin, pseudomurein, glycoproteins, or lipoglycan. The distribution of the different cell-wall polymers from selected archaeal genera is listed in Table 3.

5 Functions

The carbohydrate-containing cell-wall polymers of Archaea possess a shape-maintaining function. To maintain the rod shape, halobacteria require a high concentration of NaCl (>12%). Mg^{2+} ions may form salt bridges between sulfate and uronic acid residues of the oligosaccharides of the S-layer glycoprotein (Wieland, 1988) required for the integrity of the S-layer subunits. Lysis of the pseudomurein sacculi also leads to disruption of the cells (König et al., 1985).

Tab. 3 Occurrence of different cell surface polymers among selected archaeal genera

Archaea	*Polymer*
Halophiles	
Halobacterium	S-layer glycoprotein
Halococcus	Heterosaccharide
Haloferax	S-layer glycoprotein
Natronococcus	Glutaminylglycan
Methanogens	
Methanobacterium	Pseudomurein
Methanobrevibacter	Pseudomurein
Methanococcus	S-layer protein
Methanocorpusculum	S-layer glycoprotein
Methanoculleus	S-layer glycoprotein
Methanogenium	S-layer glycoprotein
Methanolacinia	S-layer glycoprotein
Methanolobus	S-layer glycoprotein
Methanomicrobium	S-layer protein
Methanoplanus	S-layer glycoprotein
Methanopyrus	Pseudomurein
Methanosaeta	(Glyco-)protein sheath
Methanothrix	(Glyco-)protein sheath
Methanosarcina	Methanochondroitin
Methanosphaera	Pseudomurein
Methanospirillum	(Glyco-)protein sheath
Methanothermus	Pseudomurein, S-layer glycoprotein
Sulfate reducers	
Archaeoglobus	S-layer glycoprotein
Extremely thermophilic sulfur metabolizers	
Desulfurococcus	S-layer glycoprotein
Pyrodictium	S-layer glycoprotein
Pyrobaculum	S-layer glycoprotein
Sulfolobus	S-layer glycoprotein
Thermoproteus	S-layer glycoprotein
Thermoplasmas	
Thermoplasma	Glycocalyx, lipoglycan

Methanosarcina mazei occurs in three morphological forms: single cells; laminae; and packets (Mayerhofer et al., 1992). Since the methanochondroitin matrix is degraded during the life cycle, leading to single disaggregated cells, the methanochondroitin matrix is responsible for the aggregate formation (Boone and Mah, 1987).

6 Biochemistry

Information concerning the biochemistry of archaeal cell-wall polymers is restricted to investigations on the biosynthesis of selected polymers; in addition, a few reports on the biological activity of the pseudomurein have also been produced.

6.1 Biosynthesis

The pathways of biosynthesis of archaeal cell-wall polymers show both similarities to, and striking differences from, the well-known biosynthetic pathways of structurally related bacterial and eukaryotic glycoconjugates. Some of the common and distinguishing features are outlined in this section.

6.1.1 Pseudomurein

A pathway for the biosynthesis of the pseudomurein has been proposed (Figure 9 (König et al., 1994). Biosynthesis of the

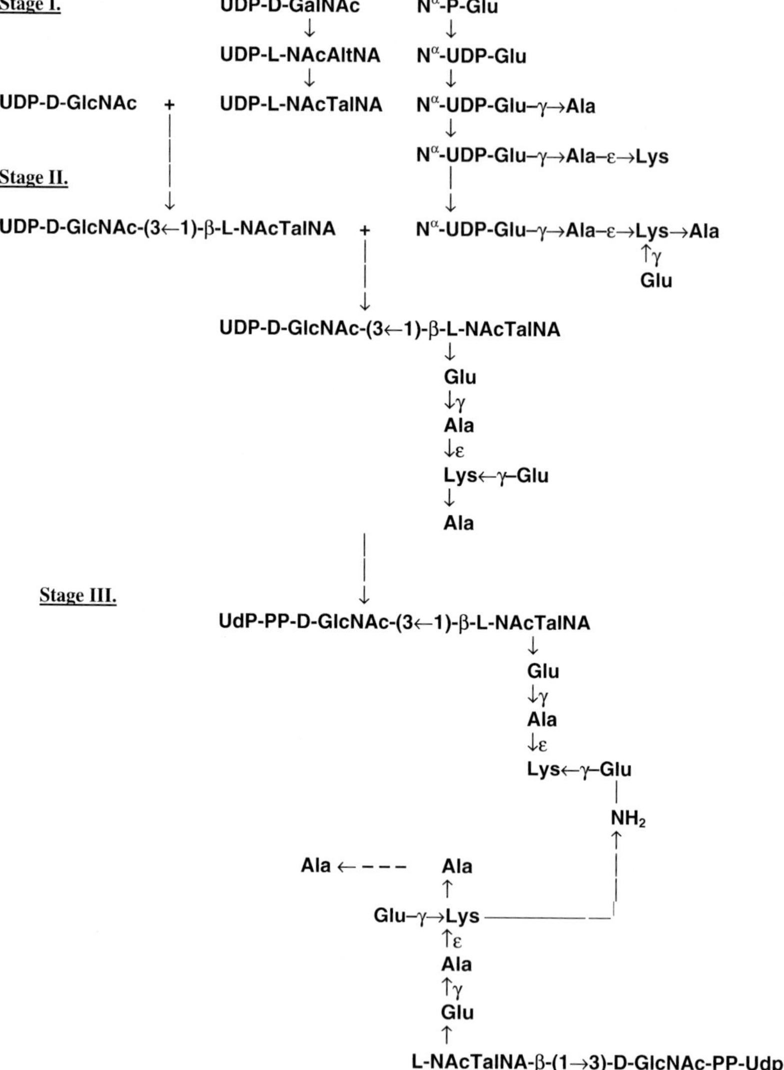

Fig. 9 Proposed biosynthetic pathway of pseudomurein (From König et al., 1994).

glycan intermediates starts with UDP-*N*-acetylglucosamine and UDP-*N*-acetyltalosaminuronic acid, which form the disaccharide UDP-GlcNAc(3 ← 1)β-NAcTalNA. The biosynthesis of UDP-*N*-acetyltalosaminuronic acid is achieved by epimerization and oxidation of UDP-*N*-acetylgalactosamine, where UDP-*N*-acetylaltrosaminuronic acid forms an intermediate. In parallel, a UDP-activated pentapeptide is formed. Biosynthesis of the pentapeptide starts with N^{α}-P-glutamic acid, which is transformed to N^{α}-UDP-glutamic acid, and to which the amino acid residues of alanine, lysine and glutamate are linked subsequently. The formation of a nucleotide-activated oligosaccharide and a nucleotide-activated pentapeptide are novel features of oligosaccharide and peptide biosynthesis, showing that a common origin of the archaeal pseudomurein and the eubacterial murein is unlikely.

6.1.2
Methanochondroitin

The biosynthesis of methanochondroitin starts with the formation of UDP-*N*-acetylchondrosine and UDP-glucuronic acid, followed by the transfer of *N*-acetylchondrosine to UDP-*N*-acetylgalactosamine, thereby yielding a UDP-activated trisaccharide which possesses the same structure as the repeating unit of methanochondroitin (Figure 10) (König et al., 1994). The trisaccharide is subsequently linked to the lipid carrier undecaprenol via pyrophosphate, transferred through the cytoplasmic membrane and polymerized.

6.1.3
S-layer Glycoproteins

The biosynthesis of the glycan of the glycoprotein from *Hb. halobium* has been studied in more detail, while in the case of *Methanothermus fervidus* only activated oligosaccharides have been isolated from cell extracts. In both species dolichol derivatives

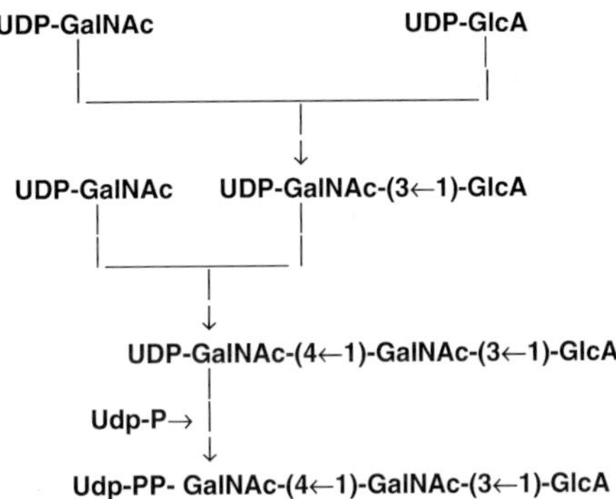

Fig. 10 Proposed scheme of the biosynthesis of methanochondroitin. (From König et al., 1994.)

serve as lipid carriers, while in the case of the pseudomurein undecaprenyl-pyrophosphate functions as a lipid carrier. It is clear that dolichyl-P-(P) is the universal lipid carrier in glycoprotein synthesis in prokaryotes and eukaryotes (König et al., 1994).

Halobacteria

Depending on the glycan chain, C_{60}-dolichyl-monophosphate and dolichylpyrophosphate serve as the lipid carriers for the glycosaminoglycan and the second sulfated oligosaccharide, respectively (see Figure 7). The complete glycosaminoglycan (including sulfation) is synthesized at the lipid carrier and then transferred to the nascent protein, the linkage to the protein taking place at the cell surface. The acceptor peptide is Asn(2)-Ala-Ser. Replacement of the serine residue of the consensus sequence by valine, leucine, and asparagine did not prevent N-glycosylation; N-glycosylation did not occur at Asn479, when Ser481 was removed (Zeitler et al., 1998), which indicates the presence of two different N-glycosyltransferases. In the case of the second sulfated oligosaccharide, again completely sulfated lipid-linked precursors are formed before transfer to the protein. Before transfer of this saccharide chain to the cell surface, some glucose residues are transiently methylated at C-3 (Wieland, 1988; Sumper and Wieland, 1995).

Methanothermus fervidus

Several nucleotide and lipid activated mono- and oligosaccharides have been isolated from cell extracts. Biosynthesis of the heterosaccharide is suggested to start with C-1-phosphate derivatives of the monosaccharides (König et al., 1994), these being converted to the corresponding nucleotide derivatives (UDP-, GDP-), which form UDP-activated oligosaccharides. At the lipid stage, the oligosaccharides form C_{55}-dolichyl-pyrophosphate-activated oligosaccharides. The lipid-activated oligosaccharides contain additional Glc residues, which are removed during further processing. The structure of the activated oligosaccharides and the exact pathway of the processing is still to be investigated, however.

6.2 Biological Activity

Prokaryotic cell-wall components such as murein or lipopolysaccharides exhibit a series of biological activities. In the case of Archaea, our knowledge is poor however, and is restricted mainly to pseudomurein

6.2.1 Pseudomurein

Pseudomurein exhibits several biological activities. In particular, pseudomurein – like murein – is antigenic in animals, the immunological characterization with monoclonal antibodies having revealed four distinctive determinants, namely the three glycan components GlcNAc, GalNAc, NAcTalNA and the C-terminal γ-glutamyl-alanine bond (Conway de Macario et al., 1983).

Pseudomurein also causes somnogenic and pyrogenic effects; intravenous injection of rabbits with suspensions of pseudomurein alter both sleep pattern and brain temperature (Johannsen et al., 1990), whilst acute inflammation was caused in a rat arthritis model (Stimpson et al., 1986).

In the case of *Bacillus anthracis,* the polyglutamate polymer plays an important role in pathogenesis. Whether the structurally related glutaminylglucan of the natronococci is a pathogenic factor has yet to be investigated, however.

7 Molecular Genetics

The genes of some S-layer proteins have been characterized (Lechner and Sumper, 1987; Sumper et al., 1990; Bröckl et al., 1991; Yao et al., 1994), though among the species referred to above, only in the case of *Methanobacterium thermoautotrophicum* strain ΔH has the complete genome sequence been published (Smith et al., 1997). Some enzymes with sequence similarities to those of eubacterial cell envelope-related enzymes have been summarized in the function category list (http://mbgd.genome.ad.jp/htbin/funccat.pl?species=mth). As yet, no unambiguous assignment to genes of enzymes involved in the biosynthesis or degradation of cell-wall carbohydrates of Archaea has been performed.

8 Biodegradation

An enzymatic degradation of cell-wall polymers has been observed in the genera *Methanosarcina* and *Methanobacterium*, and species of both genera have been shown to produce autolytic enzymes under certain circumstances.

8.1 Methanochondroitin

In substrate-depleted cultures of *Methanosarcina barkeri*, the cell-wall matrix is lost, yielding spontaneous protoplasts (Archer and King, 1984), whilst during the life cycle of *Methanosarcina mazei*, the matrix material is degraded, leading to disaggregation of the cells. For complete disaggregation, elevated concentrations of substrate (100 mM acetate or methanol) and divalent cations (Ca^{2+}) were required (Boone and Mah, 1987), thereby indicating that the matrix is responsible for aggregate formation.

8.2 Pseudomurein

As a consequence of its composition – which differs from that of murein – pseudomurein is resistant to cell wall-targeting antibiotics such as β-lactams, which interact with cytoplasmic intermediates. Pseudomurein is also insensitive against lysozymes and common proteases.

Methanobacterium bryantii undergoes spontaneous lysis when NH_4^+ or Ni^{2+} ions are exhausted in the culture medium (Jarrel et al., 1982). In the presence of D-sorbitol, *Methanobacterium thermoautotrophicum* forms protoplasts following the addition of lysozyme to growing cells (Sauer et al., 1984). An extracellular enzyme from a streptomycete was also found to lyse *Methanobacterium formicicum* (Bush, 1985).

An autolytic enzyme, which hydrolyzes the pseudomurein sacculi after depletion of the energy source, was found to be induced in *Methanobacterium wolfei*. This lytic enzyme was found to be a peptidase hydrolyzing the ε-bond between an alanine and a lysine residue of the peptide subunit (König et al., 1985; Kiener et al., 1987).

9 Production (Producers, World market, Applications, Patents)

To date, archaeal cell-wall components have not been used commercially.

10 Outlook and Perspectives

The cell envelopes of the Archaea are often directly exposed to extreme environmental conditions, and cannot be stabilized by cellular factors. Adaptation to extreme environments has not led to similar cell-surface structures. *Halobacterium* sp. and *Halococcus* sp. each thrive in saturated salt brines, but possess quite different cell-wall polymers such as glycoprotein, heterosaccharide, or glutaminylglycan. The same is true for extremely thermophilic archaebacteria, which may serve as models to elucidate survival strategies of natural compounds and may provide clues regarding the molecular mechanisms of resistance against high temperature, low and high pH, and high salt concentrations (Evrard et al., 1999). These investigations, in time, may lead to the development and application of these new biomaterials.

S-layers represent the most common cell surface layer of Archaea, and are the simplest biological membranes found in nature. A wide spectrum of applications for S-layers has begun to emerge. For example, isolated S-layer subunits assemble into monomolecular crystalline arrays in suspension, on surfaces and at interfaces. These lattices have functional groups on the surface in an identical position and orientation at the nanometer level, and such features have led to various applications including ultrafiltration membranes, immobilization matrices for functional molecules, affinity microcarriers and biosensors, conjugate vaccines, carriers for Langmuir-Blodgett films and reconstituted biological membranes, as well as pattering elements in molecular nanotechnology (Sleytr et al., 1994). In the past, such investigations have been performed almost exclusively with eubacterial S-layers, but as archaeal S-layer glycoproteins have been shown to be resistant often under extreme conditions, it is likely that an entirely new spectrum of future applications using archaeal S-layer glycoproteins will be developed.

Acknowledgements
The authors thank Heike Maria Kleinherne for assistance with the manuscript.

11 References

Archer, D. B., King, N. R. (1984) Isolation of gas vesicles from *Methanosarcina barkeri*, *J. Gen. Microbiol.* **130**, 167–172.

Baumeister, W., Lembcke, G. (1992) Structural features of archaebacterial cell envelopes, *J. Bioenerg. Biomembr.* **24**, 567–575.

Beveridge, T. J., Graham, L. L. (1991) Surface layers of bacteria, *Microbiol. Rev.* **55**, 684–705.

Boone, D. R., Mah, R. A. (1987) Effects of calcium, magnesium, pH, and extent of growth on the morphology of *Methanosarcina barkeri* S-6, *Appl. Environ. Microbiol.* **53**, 1699–1700.

Bröckl, G., Behr, M., Fabry, S., Hensel, R., Kaudewitz, H., Biendl, E., König, H. (1991) Analysis and nucleotide sequence of the genes encoding the surface-layer glycoproteins of the hyperthermophilic methanogens *Methanothermus fervidus* and *Methanothermus sociabilis*, *Eur. J. Biochem.* **199**, 147–152.

Brown, A. D., Cho, K. J. (1970) The walls of extremely halophilic cocci. Gram-positive bacteria lacking muramic acid, *J. Gen. Microbiol.* **62**, 267–270.

Bush, J. W. (1985) Enzymatic lysis of the pseudomurein containing methanogen *Methanobacterium formicicum*, *J. Bacteriol.* **163**, 27–36.

Conway de Macario, E., Macario, A. J. L., Magarinos, M. C., König, H., Kandler, O. (1983) Dissecting the antigenic mosaic of the archaebacterium *Methanobacterium thermoautotrophicum* by monoclonal antibodies of defined molecular specificity, *Proc. Natl. Acad. Sci. USA* **80**, 6346–6350.

Evrard, Ch., Declercq, J.-P., Debaerdemaeker, T., König, H. (1999) The first successful crystallization of a prokaryotic extremely thermophilic outer surface layer glycoprotein, *Z. Kristallogr.* **214**, 427–429.

Houwink, A. L. (1956) Flagella, gas vacuoles and cell-wall structure in *Halobacterium halobium*; an electron microscope study, *J. Gen. Microbiol.* **15**, 146–150.

Jarrel, K. F., Colvin, J. R., Sprott, G. D. (1982) Spontaneous protoplast formation in *Methanobacterium bryantii*, *J. Bacteriol.* **149**, 346–353.

Johannsen, L., Toth, L. A., Rosenthal, R. S., Opp, M. R., Obál Jr., F., Crady, A. B., Krueger, J. M. (1990) Somnogenic, pyrogenic, and hematologic effects of bacterial peptidoglycan, *Am. J. Physiol.* **258**, R182–R186.

Kärcher U, Schröder H, Haslinger E, Allmaier G, Schreiner R, Wieland F, Haselbeck A, König H (1993) Primary structure of the heterosaccharide of the surface glycoprotein of *Methanothermus fervidus*, *J. Biol. Chem.* **268**, 26821–26826.

Kandler, O., Hippe, H. (1977) Lack of peptidoglycan in the cell walls of *Methanosarcina barkeri*, *Arch. Microbiol.* **113**, 57–60.

Kandler, O., König, H. (1985) Cell envelopes of archaebacteria, in: *The Bacteria. A Treatise on Structure and Function. Archaebacteria.* (Woese, C. R., Wolfe, R. S., Eds.), New York: Academic Press, 413–457, Vol. VIII.

Kandler, O., König, H. (1993) Cell envelopes of Archaea: structure and chemistry, in: *The Biochemistry of Archaea (Archaebacteria)* (Kates, M., Kushner, D. J., Matheson, A. T., Eds), Amsterdam: Elsevier Science Publishers, 223–259.

Kiener, A., König, H., Winter, J., Leisinger, Th. (1987) Purification and use of *Methanobacterium wolfei* pseudomurein endopeptidase for lysis of *Methanobacterium thermoautotrophicum*, *J. Bacteriol.* **169**, 1010–1016.

König, H., Semmler, R., Lerp, C., Winter, J. (1985) Evidence for the occurrence of autolytic enzymes in *Methanobacterium wolfei*, *Arch. Microbiol.* **141**, 177–180.

König, H., Hartmann, E., Kärcher, U. (1994) Pathways and principles of the biosynthesis of methanobacterial cell wall polymers, *System. Appl. Microbiol.* **16**, 510–517.

Kreisl, P., Kandler, O, (1986) Chemical structure of the cell wall polymer of *Methanosarcina, Syst. Appl. Microbiol.* **7**, 293–299.

Langworthy, T. A., Tornabene, T. G., Holzer, G. (1982) Lipids of archaebacteria, *Zentralbl. Bakteriol., Mikrobiol. Hyg., Abt. 1, Orig.* **C3**, 228–244.

Lechner, J., Sumper, M. (1987) The primary structure of a procaryotic glycoprotein. Cloning and sequencing of the cell surface glycoprotein gene of halobacteria, *J. Biol. Chem.* **262**, 9724–9729.

Mayerhofer, L. E., Macario, A. L. J., Conway de Macario, E (1992) Lamina, a novel multicellular form of *Methanocarcina mazei* S-6, *J. Bacteriol.* **174**, 309–314.

Mescher, M. F., Strominger, J. L. (1976) Purification and characterisation of a prokaryotic glycoprotein from the cell envelope of *Halobacterium salinarium, J. Biol. Chem.* **251**, 2005–2014.

Messner, P., Sleytr, U. B. (1992) Crystalline bacterial cell-surface layers, *Adv. Microb. Physiol.* **33**, 213–274.

Messner, P., Allmaier, G., Schäffer, C., Wugeditsch, T., Lortal, S., König, H., Niemetz, R. Dorner, M. (1997) Biochemistry of S-layers, *FEMS Microbiol. Rev.* **20**, 25–46.

Niemetz, R., Kärcher, U., Kandler, O., Tindall, B. J., König, H. (1997) The cell wall polymer of the extremely halophilic archaeon *Natronococcus occultus, Eur. J. Biochem.* **249**, 905–911.

Reistadt, R. (1972) Cell wall of an extremely halophilic coccus. Investigation of ninhydrin-positive compounds, *Arch. Microbiol.* **82**, 24–30.

Sauer, F. D., Mahadevan, S., Erfle, J. D. (1984) Methane synthesis by membrane vesicles and a cytoplasmic cofactor isolated from *Methanobacterium thermoautotrophicum, Biochem. J.* **221**, 61–69.

Schleifer, K. H., Steber, J., Mayer, H. (1982) Chemical composition and structure of the cell wall of *Halococcus morrhuae, Zentralbl. Bakt., Mikrobiol. Hyg.*, Serie **C3**, 171– 178.

Sleytr, U. B., Sára, M., Messner, P., Pum, D. (1994) Two-dimensional protein crystals (S-layers): fundamentals and applications, *J. Cell. Biochem.* **56**, 171–176.

Smith, D. R. and 36 other authors (1997) Complete genome sequence of *Methanobacterium thermoautotrophicum* ΔH: functional and comparative genomics, *J. Bacteriol.* **179**, 7135–7155.

Stimpson, S. A., Brown, R. R., Anderle, S. K., Klapper, D. G., Clark, R. L., Cromartie, W. J., Schwab, J. H. (1986) Arthropathic properties of cell wall polymers from normal flora bacteria, *Infect. Immun.* **51**, 240–249.

Sumper, M., Berg, E., Mengele, R., Strobel, I. (1990) Primary structure and glycosylation of the S-layer protein of *Haloferax volcanii, J. Bacteriol.* **172**, 7111–7118.

Sumper, M., Wieland, F. T. (1995) Bacterial glycoproteins, in: *Glycoproteins* (Montreuil, J., Vliegenthart, J. F. G., Schachter, H., Eds.), Amsterdam: Elsevier, 455–473.

Wieland, F. T. (1988) The cell surfaces of halobacteria, in: *Halophilic bacteria* (Rodriguez-Valera, F., Ed.), Boca Raton: CRC Press Inc., 55–65, Vol. II.

Weiss, L. R. (1974) Subunit cell wall of *Sulfolobus acidocaldarius, J. Bacteriol.* **118**, 275–284.

Woese, C. R. (1987) Bacterial evolution, *Microbiol. Rev.* **51**, 221–271.

Yang, L. L., Haug, A. (1979) Purification and partial characterization of a procaryotic glycoprotein from the plasma membrane of *Thermoplasma acidophilum, Biochim. Biophys. Acta* **556**, 277–365.

Yao, R. Macario, A. J. L., Conway de Macario, E. (1994) An archaeal S-layer gene homolog with repetitive subunits, *Biochim. Biophys. Acta* **1219**, 697–700.

Zeitler, R., Hochmuth, E., Deutzmann, R., Sumper, M. (1998) Exchange of Ser-4 for Val, Leu and Asn in the sequon Asn-Ala-Ser does not prevent N-glycosylation of the cell surface glycoprotein from *Halobacterium halobium, Glycobiology* **8**, 1157–1164.

20 Index

a

h

i

j

k

m

r

u

v

w

x

y

z